D0074554

Eighth Edition

INQUIRY INTO
PHYSICS

VERN J. OSTDIEK | DONALD J. BORD

University of Michigan—Dearborn

CENGAGE
Learning®

Australia • Brazil • Mexico • Singapore • United Kingdom • United States

CENGAGE Learning

Inquiry Into Physics, Eighth Edition
Vern J. Ostdiek, Donald J. Bord

Product Director: Dawn Giovanniello

Product Manager: Rebecca Berardy Schwartz

Senior Content Developer: Susan Dust Pashos

Product Assistant: Caitlin N. Ghegan

Marketing Manager: Tom Ziolkowski

Senior Content Project Manager: Tanya Nigh

Digital Content Specialist: Justin Karr

Senior Art Director: Cate Barr

Manufacturing Planner: Doug Bertke

Production Service and Compositor:
MPS Limited

Intellectual Property Project Manager:
Erika Mugavin

Intellectual Property Analyst:
Christine Myaskovsky

Photo and Text Researcher: Lumina
Datamatics, Ltd.

Cover and Text Designer: Shawn Girsberger

Cover Image: © NASA. Morning aurora from the
International Space Station captured by
NASA astronaut, Scott Kelly.

© 2018, 2013, 2008 Cengage Learning

Unless otherwise noted, all art is © Cengage Learning.

ALL RIGHTS RESERVED. No part of this work covered by the copyright herein
may be reproduced or distributed in any form or by any means, except as
permitted by U.S. copyright law, without the prior written permission of the
copyright owner.

For product information and technology assistance, contact us at
Cengage Learning Customer & Sales Support, 1-800-354-9706.

For permission to use material from this text or product,
submit all requests online at **www.cengage.com/permissions**.
Further permissions questions can be e-mailed to
permissionrequest@cengage.com.

Library of Congress Control Number: 2016934469

Student Edition
ISBN: 978-1-305-95942-2

Loose-leaf Edition:
ISBN: 978-1-337-28964-1

Cengage Learning
20 Channel Center Street
Boston, MA 02210
USA

Cengage Learning is a leading provider of customized learning solutions
with employees residing in nearly 40 different countries and sales in more
than 125 countries around the world. Find your local representative at
www.cengage.com.

Cengage Learning products are represented in Canada by Nelson Education, Ltd.

To learn more about Cengage Learning Solutions, visit **www.cengage.com.**

Purchase any of our products at your local college store or at our preferred
online store **www.cengagebrain.com.**

Printed in the United States of America
Print Number: 02   Print Year: 2017

CONTENTS

CHAPTER 12

Relativity, Particle Physics, and Cosmology   456

Appendixes

Welcome to the eighth edition of *Inquiry Into Physics*. In this edition, the emphasis on the inquiry approach to learning physics is continued. Readers are encouraged to try things, to discover relationships between physical quantities on their own, and to look for answers in the world around them rather than seek them only in books or on the Internet. This book should not be treated as an answer key to the nature of physical phenomena (although it does offer many explanations of how things work), but as an extended invitation to examine things for yourself in a direct, hands-on fashion and to be daring and experiment on the various systems and structures that surround you. As you progress from topic to topic, give free rein to the innate curiosity that we all possess from birth, and don't be afraid to ask questions about the things you experience daily. Keep in mind that learning is a more exciting, complete, lasting, and enjoyable process when you are fully engaged in it and not merely a passive recipient of facts and statistics offered by your instructors, this text, and other print or electronic sources. Get in the game—of physics!

Our Intended Audience

This book is designed primarily for students who are taking an introductory physics course, perhaps for the first time, to satisfy collegiate requirements or who wish to satisfy their curiosity and thirst for understanding about the structure and range of interactions that characterize our physical universe. It offers a broad survey of the fundamental definitions, laws, and principles of the discipline of physics, as well as a large sampling of applications of these concepts in virtually every aspect of human experience. *Inquiry Into Physics* provides a first look into the things that comprise our universe, from the smallest subatomic particles to the largest galaxies, as well as the interactions that can occur between those things, from energy-releasing reactions among atomic nuclei to gut-wrenching collisions between automobiles, to gravitational interactions among stars and whole galaxies that result in catastrophic disruptions over time scales that dwarf modern humankind's 200,000-year history on Earth. In ways both large and small, physics principles and applications touch our lives literally from moment to moment: in the pumping action of the heart and the flow of blood in our arteries and veins, in the electrical impulses that enervate our muscles and organs, and in the processes that drive our senses of sight and sound and permit us to see and hear the goings-on in the world around us.

Physics also satisfies our intellectual curiosity by revealing Nature's secrets about things both mundane and exotic. Discoveries in physics, like the processes of nuclear fission and fusion, have shaped the political landscape of the world and are likely to continue to do so well into the future. A look back at the women and men who made many of these important discoveries connects us with our past and gives us deeper insight into the pathways that can lead to scientific understanding. These aspects and others are blended throughout the text in an effort to provide you with a sense of the unity both within the discipline of physics and between the subject and its human practitioners. The tools we use to present these facets of physics are basic ones: the written word; visualizations and models in the form of hundreds of photos, diagrams, and graphs; and simple mathematics to show how the conceptual side of physics is inextricably woven together with its quantitative side. By these means, we hope that you will complete your introductory study of physics, having achieved a greater appreciation not only for what we know about the physical universe, but also how we have come to know it.

What's New?

Each chapter has been thoroughly reviewed, and many sections have been rewritten to improve the clarity and accuracy of the prose. The text has been updated and in some cases expanded to reflect the latest scientific discoveries and achievements in the fields of physics and astronomy, and new material has been added to reflect current affairs and increasingly popular applications of physics principles in 21st-century life. Each major chapter section has been divided into subsections to better call out the content and organization of the main topic for students and to permit instructors to more finely tune reading assignments to their needs and preferences. The art program has been substantially improved throughout the book to offer more contemporary and relevant photos and figures to illustrate the connections between fundamental physical concepts and the modern world. In particular, in figures containing vectors, the arrows have been made bolder and the color scheme used to identify each vector expanded and made more internally consistent to aid in recognizing and relating such quantities. In addition, the format for the worked examples has been modified to better delineate the questions and the solutions, and the answers are now shaded to give them greater prominence.

The long-standing "Physics Potpourri" series has been reconceptualized and in some cases reorganized as a suite of "Applications" essays, focusing attention on ways that physics principles have been and continue to be applied to change—and hopefully enhance—our world. These text elements now also include one or two review questions to support student engagement. An expanded set of end-of-chapter (EOC) exercises, including more than 80 new questions and problems, has been provided, and the "explore-it-yourself" application boxes have been retitled "Physics to Go" to better emphasize the "take it with you" exploratory aspects of these features. The former "Physics Family Album" vignettes have been shortened in many cases and renamed "Profiles in Physics" to emphasize their focus on key personalities in the development of the discipline; like the "Applications" articles, the "Profiles" sections now contain review questions to better promote their use as important pedagogical elements of the book.

Some specific changes on a chapter-by-chapter basis include the following:

Prologue Updated *Application* highlighting the commercial aspects of the metric system and the new emphasis on units defined in terms of fundamental constants. Fifty percent of the art is new.

Chapter 1. More than a dozen new photos and line drawings are included. Three new EOC exercises and a new example have been added.

Chapter 2. Updated discussion of the *New Horizons* mission to Pluto, plus a new *Application* feature on *Chaos*. Five new EOC exercises and three new worked examples have been added.

Chapter 3. More than a dozen new figures have been introduced. An updated *Important Equations* section now appears to reflect new material on elastic energy and torque. Eight new EOC exercises and three new worked examples have been introduced.

Chapter 4. More than a dozen new figures have been included. The updated feature on element nomenclature includes the latest discoveries and naming conventions for the heaviest elements. New material on hydraulics and fluid flow has been introduced. Thirteen new EOC exercises and two new worked examples have been added.

Chapter 5. Expanded discussion of the ideal gas law. Expanded treatment of entropy and energy quality. Thirteen new EOC exercises and three new worked examples now appear.

Chapter 6. Twelve new figures and new material on the Hubble relation have been provided. Five new EOC exercises have been included.

Chapter 7. Revised and updated chapter opener on iProducts. More than a dozen new photos and refreshed line drawings. Updated information for Tables 7.1 and 7.2, plus additional material on superconducting power grid projects in the United States and Europe. Three new EOC exercises have been added.

Chapter 8. Nearly a dozen new or updated figures. Additional discussion of superconducting electromagnets and new information about the greenhouse effect and environmental evidence for climate change are presented. Two new EOC ranking exercises have been added.

Chapter 9. Explicit inclusion of mathematical treatment of double slit interference with a new worked example. Updated discussion of the Hubble Space Telescope focusing on instrumentation missions. Five new EOC exercises have been added.

Chapter 10. Nearly a dozen new or enhanced figures have been included.

Chapter 11. Updated discussion of clean-up work at the Fukushima Daiichi reactor site in Japan five years after the destructive earthquake and tsunami struck the plant. New material on radiation therapies for brain tumors using a gamma knife. Four new EOC exercises have been added.

Chapter 12. This chapter has been completely revised and merged with the former Epilogue on cosmology to produce a fresh and unified treatment of relativity, particle physics, and cosmology. A new section on general relativity with two new worked examples and a treatment of the LIGO discovery of gravitational waves now follows the opening discussion of special relativity. The subsequent three sections addressing particle physics and the Standard Model have been streamlined, while including material on the discovery of the Higgs boson and updates to the tabulated data for elementary particles and the Standard Model. Nearly a dozen new figures have been provided, and the *Summary* and *Important Equations* sections have been altered to reflect the new information. A new Mapping It Out! exercise dealing with the concepts and predictions of general relativity has been inserted, and 24 new EOC exercises have been added covering the new topics.

What's the Plan?

The traditional organization of topics followed in many introductory physics courses has been retained for this new edition. Once again, we have eschewed the use of shorter, more narrowly focused chapters and have continued with longer ones that offer the opportunity to present the overarching conceptual unity and continuity of the broad subdisciplines that make up our subject (kinematics and dynamics, energy, thermodynamics, optics, atomic and nuclear physics, etc.). Each of the 12 numbered chapters is built around the following common features:

Chapter Introductions. Opening each chapter, these examples of physics as it plays out in everyday life serve to motivate the reader to immediately engage with the material by showing how a common device or issue of current interest and importance connects to the concepts developed in the chapter.

Examples. Worked exercises that illustrate the roles of physics principles and simple mathematics in real-world situations are regularly presented as models for problem solving.

Physics to Go activities. These hands-on experiments and exercises give students the chance to see and do physics without the need for specialized equipment or highly sophisticated techniques. These investigations are generally strategically placed before the relevant text discussion, inviting the reader to directly experience the upcoming concepts and to begin to formulate their own understanding of them.

Learning Checks. Simple self-quizzes, designed to test the reader's basic comprehension of the material on a section-by-section basis, are included in each chapter.

Physics Applications essays. These self-contained features explore selected topics drawn from astronomy, the history of science, engineering, biophysics, environmental physics and other areas. These essays are provided to deepen and enrich the student's understanding of physics and especially its applications.

Concept Maps. Based on principles developed from educational research, these visual displays offer an alternative representation or organization of the relationships between important chapter concepts.

Profiles in Physics. The final content section of each chapter presents a look at the historical development and the human side of physics by describing aspects of the lives and work of some of the key men and women responsible for the discovery and expression of the laws and concepts presented in the chapter.

Summary. Each chapter summary is a brief, bulleted review of the key points in the chapter and a short, helpful list of the major concepts for students when preparing for tests and quizzes.

Important Equations. An annotated list of the equations included or developed in the chapter is presented for quick reference when problem solving.

Mapping It Out! These exercises, many intended for group collaboration, are included to help students use concept-mapping techniques to organize, unify, and improve their understanding of important concepts and relationships introduced in each chapter.

Questions. These end-of-chapter queries check students' basic understanding of the material and their ability to extend that understanding to new and different situations. A special icon (■) is used to distinguish the first type of question from the second for ease of identification and selection.

Problems. These end-of-chapter exercises offer students opportunities to hone their critical thinking skills using logic and simple mathematics to solve problems based on realistic applications of the physics developed in each chapter.

Challenges. More advanced questions and problems designed to test a reader's mastery of the material at a deeper level are included at the close of each chapter for the benefit of highly motivated students. Many of these exercises can be used as starting points for small group or whole class discussions.

Conveniently located in the appendices and front and back endsheets of the print version of this book are several additional resources that are helpful in problem solving or in gaining further appreciation of the scope of physics applications in the 21st century. These include:

- **Selected Applications** of physics concepts and principles ("Physics Connections to the Real World")
- **Tables of Conversion Factors** and other information (for example, tables of metric prefixes, physical constants, and other often-used data)
- **Periodic Table of the Elements**
- **Winners of the Nobel Prize in Physics**
- **Math Review**
- **Answers** to the odd-numbered Problems and to many of the Challenges and the Physics to Go activities (where appropriate).

There is more than enough material in the text than can usually be covered in a typical one-semester course. However, about 30 of the more specialized sections may be omitted with minimal impact on the later topics. Specifically, these include the **Applications** features and the **Profiles in Physics** segments and Sections 2.8, 3.8, 5.6, 5.7, 6.4, 6.5, 6.6, 8.4, 8.7, 9.5, 9.7, 10.7, 10.8, and 12.1–12.6.

What Else Is There?
What else is there? You've got it all.

MindTap® **MindTap® for Physics** is the digital learning solution that helps instructors engage and transform today's students into critical thinkers. Through paths of dynamic assignments and applications that you can personalize, real-time course analytics, and an accessible reader, MindTap® helps you turn cookie cutter into cutting edge, apathy into engagement, and memorizers into higher-level thinkers.

Developed and designed in response to years of research, MindTap® leverages modern technology and a powerful answer evaluation system to address the unmet needs of students and educators. The MindTap® assessments incorporate assorted just-in-time learning tools such as displayed solutions, solution videos for selected problems, and targeted readings and examples from the textbook. These just-in-time tools are embedded directly adjacent to each question to help students maintain focus while completing automatically-graded assignments.

The MindTap® Learning Path groups the most engaging digital learning assets and activities together by week and topic, including readings and automatically graded assignments, to help students master each learning objective. Interactive Video Vignettes encourage an active classroom where students can address their alternative conceptions outside the classroom. Interactive Video Vignettes include online video analysis and interactive individual tutorials to address learning difficulties identified by Physics Education Research (PER).

Easy to use, efficient, and informative, MindTap® provides instructors with the ability to personalize their course with dynamic online learning tools, videos, and assessments; a student diagnostic pre-test; and personalized improvement plans to help students improve their foundational arithmetic skills outside of class time, plus just-in-time resources available to students while doing homework.

Instructor Companion Site for Ostdiek & Bord Inquiry Into Physics, 8th edition

Everything you need for your course in one place! This collection of book-specific lecture and class tools is available online via www.cengage.com/login. Access and download PowerPoint presentations, images, the Instructor's Manual, videos, and more.

Each chapter of the **Instructor's Manual** by Thomas E. Sieland of Embry-Riddle Aeronautical University provides the following:

- Chapter outline
- Chapter overview
- Learning objectives
- Teaching suggestions and lecture hints
- Common misconceptions and how to counter them
- "Consider This" (additional tips and suggested resources)
- Answers to "Mapping It Out!"
- Answers to Questions
- Answers to even-numbered Problems
- Answers and brief solutions to Challenges

Cengage Learning Testing powered by Cognero is

a flexible, online system that allows you to author, edit, and manage test bank content from multiple Cengage Learning solutions; create multiple test versions in an instant; and deliver tests from your LMS, your classroom, or wherever you want.

CengageBrain.com CENGAGEbrain.com

To access your online learning solution or purchase materials for your course, visit **www.cengagebrain.com**.

Who's Responsible?

The production of this newest edition of *Inquiry Into Physics* involved the collaboration of many talented professionals at Cengage Learning. First and foremost, sincere thanks goes to Susan Dust Pashos, Senior Content Developer, for her steadfast support, keen eye, physics savvy, and deep understanding of what's possible and what's not in the publishing business. Without her wise advice, careful review, and effective advocacy, it is doubtful that this project would have come to fruition in as timely and successful manner as it has. Thanks are also due to Rebecca Berardy Schwartz, our Product Manager, whose confidence in this project and careful assembly of a team of consummate professionals within the Cengage Learning organization have produced what is arguably the best and most attractive edition of the book yet. Tanya Nigh, Content Project Manager, provided able and timely advisement and intervention on in-house production matters to resolve problems and maintain project momentum. The efforts of Michael Jacobs, Content Developer, to adapt content for the Learning Path in MindTap® for this edition of *Inquiry* are gratefully acknowledged, as are the contributions of Christine Myaskovsky, Intellectual Property Analyst, whose skill and perseverance in securing permissions for many critical images were unparalleled. The art program for this eighth edition has been updated and improved to an extent not seen since the first edition in 1985, and much of the credit for this transformation goes to Cate Barr, Senior Art Director, and Bruce Bond, Executive Director for Design, who, among other contributions, guided the thorough-going program of vector enhancements that are visible on practically every page of the book. Hats off to Cate as well for locating the exquisite cover image for this edition, which continues *Inquiry's* tradition of using striking images from Nature to call out the many ways in which physical principles are on display around us.

Kudos also go out to the vendors and contractors who have provided special services throughout the production of the book including Nitesh Sharma, project manager at our publishing and printing partner, Macmillan Publishing Solutions, in India, and Justin Karr and Susan Ordway who were instrumental in directing the XML coding for Mindtap®.

A deep debt of gratitude and appreciation is owed to Dr. Tom Sieland of Embry-Riddle Aeronautical University who once again, with skill and precision, developed the Instructor's Manual for the eighth edition of *Inquiry*. Tom's extensive experience as a physics teacher and his long-time classroom use of this book combine to infuse the Instructor's Manual with helpful pedagogical advice, accurate and authoritative solutions, and an extensive set of teaching suggestions and lecture hints specially matched to the book, thus making this ancillary a valuable resource for practitioners both new and more seasoned.

Thanks are also due to the many instructors and students who have used *Inquiry Into Physics* since its first appearance in 1985. Your advice, encouragement, and suggestions for improvement have been instrumental in keeping the book faithful to its original philosophy and purposes, up to date with regard to important developments in the field, and clear and accurate in its exposition. For this edition, a special expression of gratitude is owed to Dr. Vitaly Kresin at the University of Southern California and Dr. Brian Utter of Bucknell University for their evaluation of an early draft of the new Chapter 12; their suggestions for improvement have been instrumental in yielding a more coherently organized and vibrantly expressed presentation of relativity and high-energy physics than was initially presented.

I sincerely thank all of you and hope you enjoy this latest edition of *Inquiry Into Physics*.

Don Bord
September 2016

Dedication

This book is lovingly dedicated to my wife, Cheryl, who, through seven editions and nearly 30 years, has unstintingly and with great good humor offered me her support, encouragement, loyalty, advice, and foremost, her love, without which this text and so many of my other personal and professional achievements would never have been possible. Thank you, C.

PROLOGUE: GETTING STARTED

Pictorial Press Ltd/Alamy

Figure PO-0 "In a century that will be remembered foremost for its science and technology—in particular for our ability to understand and then harness the forces of the atom and universe—one person clearly stands out as both the greatest mind and paramount icon of our age: The kindly, absent-minded professor whose wild halo of hair, piercing eyes, engaging humanity, and extraordinary brilliance made his face a symbol and his name a synonym for genius, Albert Einstein." (*Time* magazine.)

P.1 Introduction

In 1999, *Time* magazine, known for naming an annual "person of the year," set out to choose its "person of the century." A daunting task it was, considering the events of the turbulent 20th century and the inevitable criticism that would come from those favoring someone not named. Would it be a world leader who shaped significant spans of the century—for good or bad—such as Franklin D. Roosevelt (a runner-up for person of the century) or Joseph Stalin (1939 and 1942 person of the year)? Perhaps a military figure such as Dwight D. Eisenhower (1944 person of the year) or a spiritual leader such as Pope John Paul II (1994 person of the year)? Would it be a champion of peace or justice such as Mahatma Gandhi (the other runner-up) or Martin Luther King, Jr. (1963 person of the year)? In the end, the selection was someone with enormous name recognition, but whose work most people would confess ignorance about: the physicist Albert Einstein (**Figure PO-0**).

Einstein was chosen because he symbolized the great strides made during the 1900s in deciphering and harnessing fundamental aspects of the material universe. But his style, his manner, and his allure had something to do with it as well. At times, he dominated physics with spectacular results, in a way that is reminiscent of certain "athletes of the 20th century" (Ali, Gretzky, Jordan, Montana, Navratilova, Nicklaus, Pele, Ruth . . .). In 1905, while working as a civil servant far removed from the great centers of physics research, Einstein had three scientific papers—on three different subjects—published in a German physics journal. They were so extraordinary that any one of them would likely have led to his receiving a Nobel Prize in physics—then, as now, the highest award in the field.

Had *Time* magazine been in business during previous centuries, the editors might well have honored other physicists in the same way, perhaps Galileo or Newton for the 17th century or Maxwell for the 19th. Such is the high regard that Western civilization has for the field of physics and those who excel in it. Partly, it is the impact their discoveries often have on our lives, by way of technological gadgets or civilization-threatening weaponry. But often it is the intellectual resonance we have with the revolutionary insights they give us about the universe (it is Earth that moves around the Sun, it is matter being converted into energy that makes the Sun shine . . .).

Welcome to the world of physics! You are embarking on an introduction to a field that continues to fascinate people in all walks of life. Tell a friend or family member that you are now studying physics. That will quite likely impress them in a way that most other subjects would not. Whether it should do so is an interesting question. We hope that when you finish this endeavor, you will answer yes.

P.2 Why Learn Physics?

The answer is easy for those majoring in physics, engineering, or other sciences: physics will provide them with important tools for their academic and professional lives.

Figure P.1 It took a lot of physics to get *Apollo* 17 to the Moon. Here mission commander Eugene A. Cernan makes a short checkout of the Lunar Roving Vehicle during the early part of the first *Apollo* 17 extravehicular activity at the Taurus-Littrow landing site in 1972.

The technology that our modern society relies on comes from applying the discoveries of physics and other sciences. From designing safe, efficient passenger jets to producing sophisticated, inexpensive tablets and cell phones, engineers apply physics every day.

Landing astronauts on the Moon and returning them safely to Earth, one of the greatest feats of the 20th century, is a good example of physics applied on many levels (**Figure P.1**). The machines involved—from powerful rockets to on-board computers—were designed, developed, and tested by people who knew a lot about physics. The planning of the orbits and the timing of the rocket firings to change orbits involved scientists with a keen understanding of basic physics such as gravity and the "laws of motion." Often, the payoff is not so tangible or immediate as a successful Moon landing. Behind great technological advances are years or even decades of basic research into the properties of matter.

For instance, take a portable Blu-ray DVD player (**Figure P.2**). A laser reads data off a spinning disc, integrated circuit chips inside "translate" the digital data into electrical signals, tiny magnets in the speakers help convert some of these signals into sound, and a liquid crystal display (LCD) provides information for the user. If you could send this device back in time to when your grandparents were children, it would astound physicists and electrical engineers of the day. But even at that time, scientists were studying the properties of semiconductors (the raw material for lasers and integrated circuit chips) and liquid crystals.

For you and others like you taking perhaps just one physics course in your life, the usefulness of physics is probably not a big reason for studying it. We will see that with even one course, you can use physics to determine, for example, how large a raft has to be to support you or whether using a toaster and a hair dryer at the same time will trip a circuit breaker. But you are not going to make a living with your understanding of physics, nor will you be using it (knowingly) every day. So why should *you* study physics? There are both aesthetic and practical reasons for learning physics. Seeing the order that exists in Nature and understanding that it follows from a relatively small number of "rules" can be fascinating—similar to learning the inspirations behind a musician's or an artist's work. Learning how common devices operate gives you a better understanding of how to use them and may reduce any frustration you have with them. An elementary knowledge of physics also helps you make more informed decisions regarding important issues facing you, your community, your nation, and the world. As you progress through this book, keep track of news events through the media of your choice. You may be surprised at just how often physics is in the news—directly or indirectly.

If you start this excursion into the world of physics with a sense of curiosity and a thirst for knowledge, you won't be disappointed. And you will have the two most important characteristics needed to make the endeavor both easy and successful. Learning how sunlight and raindrops combine to make a rainbow will deepen your appreciation of its beauty. Knowing about centripetal force will help you understand why ice or gravel on a curved road is dangerous. Learning the basics of nuclear physics will help you understand the danger of radon gas and the promise of nuclear fusion. Knowing the principles behind stereo speakers, iPhones, radar guns, MRIs, refrigerators, lasers, microwave ovens, acoustic guitars, and Polaroid sunglasses will give you a better appreciation of how these devices do what they do.

Throughout this book, we encourage the reader to be inquisitive. Just memorizing definitions and equations doesn't lead you to a real understanding of a subject, any more than memorizing a manual on playing soccer means you

Figure P.2 Portable Blu-ray DVD player: one product of decades of physics research.

can jump into a game and do well. You have to practice, try things, think of situations and how events would evolve, and so on. So it is with physics. Being able to recite Newton's third law of motion is good, but understanding what it means and how it works in the real world is what's really important.

Often, we pose questions or ask the reader to try something, so that when you realize what the answer or outcome is you will have truly learned something. You might think of it as "learning by inquiry." The *Physics to Go* activities are particularly designed for this purpose. Many of the questions at the ends of the chapters are inquiry-based. Once you get used to this method of learning, you will find that you will master the material faster and more deeply than before.

P.3 What Is Physics?

Because physics is one of the basic sciences, it is important to first have an idea of just what science is. Science is the *process* of seeking and applying knowledge about our universe. Science also refers to the *body of knowledge* about the universe that has been amassed by humankind. Pursuing knowledge for its own sake is pure or basic science; developing ways to use this knowledge is applied science. Astronomy is mostly a pure science, whereas engineering fields are applied science. The material we present in this book is a combination of fundamental concepts that we believe are important to know for their own sake and of examples of the many ways that these concepts are applied in the world around us.

There are other ways to classify the different areas of science besides pure and applied. There are physical sciences (physics and geology are just two examples), life sciences (biology and medicine), and behavioral sciences (psychology and sociology). As with most such schemes, there are overlaps: the subfield of biophysics is a good example.

Physics is not as easy to define as some areas of science such as biology, the study of living organisms. If you ask a dozen physicists to define the term, you are not likely to get two answers exactly alike. One suitable definition is that physics is the study of the fundamental structures and interactions in the physical universe. In this book, you will find much about the structures of things such as atoms and nuclei, along with close looks at how things interact by way of gravity, electricity, magnetism, and so on. Within physics, there is a wide range of divisions. Table P.1 lists some of the common areas based on one measure of research activity. There is a lot of overlap between the divisions, and some of them are clearly allied with other sciences like biology and chemistry.

Table P.1 Some commonly identified divisions of physics, ranked by number of doctorates earned each year. (Based on information from the American Institute of Physics.)

Area	Topics of Investigation
1. Condensed Matter	Structures and properties of solids and liquids
2. Particles and Fields	Fundamental particles and fields; high-energy accelerators
3. Astrophysics	Planets, stars, and galaxies; evolution of the universe
4. Nuclear Physics	Nuclei; nuclear matter and forces; quarks and gluons
5. Biological Physics	Physics of biological systems and phenomena
6. Atomic and Molecular Physics	Atoms and molecules; spectroscopy and quantum processes
7. Optics and Photonics	Study of light; laser technology
8. Applied Physics	Engineering applications; electronic devices; nanotechnology
9. Plasma and Fusion Physics	Laboratory and astrophysical plasmas; fusion research
10. Materials Science	Applications of condensed-matter physics

The field of physics is divided differently when the basics are being taught to beginners. The topics presented to students in their first exposure to physics are usually ordered according to their historical development (study of motion first, elementary particles and cosmology last). This ordering also approximates the ranking of areas by our everyday experience with them. We've all watched people in motion and things collide, but few people encounter the idea of quarks before taking a physics class—even though we and all of the objects we deal with are mainly composed of quarks.

The vast majority of students who take an introductory course in physics are not majoring in it. Most of those who do earn a degree in physics find employment in business, industry, government, or education. The latest data compiled by the American Institute of Physics indicate that the majority of individuals with bachelor's and master's degrees are employed in the first two areas, whereas those with doctorates were mainly found in the last two. In addition to the expected occupations like researcher and teacher, people with physics degrees also have job titles like engineer, manager, computer scientist, and technician. Often, physicists are hired not so much for their knowledge of physics as for their experience with problem solving and advanced technology.

P.4 How Is Physics Done?

So how does one "do" physics or science in general? How did humankind come by this mountain of scientific knowledge that has been amassed over the ages? A blueprint exists for scientific investigation that makes an interesting starting point for answering these questions. It is at best an oversimplification of how scientists operate. Perhaps we should regard it as a game plan that is frequently modified when the action starts. It is called the **scientific method.** One version of it goes something like this: careful *observation* of a phenomenon induces an investigator to question its cause. A *hypothesis* is formed that purports to explain the observation. The scientist devises an *experiment* that will test this hypothesis, hoping to show that it is correct—at least in one case—or that it is incorrect. The outcome of the experiment often raises more questions that lead to a *modification* of the hypothesis and further experimentation. Eventually, an accepted hypothesis that has been verified by different experiments can be elevated to a *theory* or a *law.* The term that is used—*theory* or *law*—is not particularly important in physics: physicists hold Newton's second *law* of motion and Einstein's special *theory* of relativity in roughly the same regard in terms of their validity and importance.

One nice thing about the scientific method is that it is a logical procedure that is practiced by nearly everyone from time to time (**Figure P.3**). Let's say that you get into your car and find that it won't start (observation). You speculate

Figure P.3 The basics of the scientific method, with an example.

that maybe the battery is dead (hypothesis). To see if this is true, you turn on the radio or the lights to see if they work (experiment). If they don't, you may look for someone to give you a jump start. If they do, you probably guess that something else, such as the starter, is causing the problem. Clearly, a good mechanic must be proficient at this way of investigating things. A health-care professional making a medical diagnosis uses similar procedures.

Does the outline of the scientific method appear in some "how-to" manual for scientific discovery? Are scientists required to take an oath to follow it faithfully every day at work? Of course not. But the individual elements of the method are essential tools of the scientist. They are useful to students as well. In the *Physics to Go* activities found throughout this book, we ask you to try things (experiment) and then draw conclusions based on the outcomes. Understanding how Nature works based on what you do and observe is a great way to learn.

One of the architects of the scientific method was Galileo Galilei (1564–1642). He thought a great deal about how science should be done and applied his ideas to his study of motion. Galileo believed that science had to have a strong logical basis that included precise definitions of terms and a mathematical structure with which to express relationships. He introduced the use of controlled experiments, which he applied with great success in his studies of how objects fall. By ingenious experimental design, he overcame the limitations of the crude timing devices that existed in his era and measured the acceleration of falling bodies. Galileo was a shrewd observer of natural phenomena from the swinging of a pendulum to the orbiting of the moons of Jupiter. By drawing logical conclusions from what he saw, he demonstrated that rules could be used to predict and explain natural phenomena that had long seemed mysterious or magical. We will take a closer look at Galileo's work and life in Chapter 1.

The scientific method is important, particularly in the day-to-day process of filling in the details about a phenomenon being studied. But it is not the whole story. Even a brief look at the history of physics reveals that there is no simple recipe that scientists have followed to lead them to breakthroughs. Some great discoveries were made by traditional physicists working in labs, proceeding in a "scientific method" kind of way. Occasionally, unplanned events come into play, such as accidents (Galvani discovered electric currents while performing biology experiments) or luck (Becquerel stumbled on nuclear radiation because of a string of cloudy days in Paris). Sometimes "thought experiments" were required because the technology of the time didn't allow "real" experiments to be performed. Newton predicted artificial satellites, and Einstein unlocked relativity in this way. Often, it was hobbyists, not professional scientists, who made significant discoveries; the statesman Benjamin Franklin and schoolteacher Georg Simon Ohm are examples. Sometimes, it is scientists correctly interpreting the results of others who failed to "connect the dots" themselves (Lise Meitner and her nephew Otto Frisch identified nuclear fission that way).

The point is that there are no "hard and fast" rules for making scientific discoveries, and this is no less true in physics than it is in chemistry, biology, or any other scientific discipline. One of the best ways to learn about the nature of scientific discovery is to study the past. Throughout this book, the *Profiles in Physics* sections, along with a few of the *Applications* features give you some idea of the variety of ways that discoveries have been made and glimpses of the personalities of the discoverers.

P.5 How Does One Learn Physics?

The goal of learning physics and any other science is to gain a better understanding of the universe and the things in it. We generally focus attention on only a small segment of the universe at one time, so that the structural complexities and interactions within it are manageable. We call this a *system*. Some examples of systems that we will talk about are the nucleus of an atom, the atom itself, a collection of atoms inside a laser, air circulating in a room, a rock moving near Earth's surface, and Earth with satellites in orbit around it (**Figure P.4**).

The kinds of things a person might want to know about a system include: (1) its structure or configuration, (2) what is going on in it and why, and (3) what will happen in/to it in the future. The first step is often relatively easy—identifying the objects in the system. Protons, electrons, chromium atoms, heated air, a rock, and the Moon are some of the things in the systems mentioned earlier. Often, the items in a physical system are already familiar to us. But a host of other intangible things in a system must also be identified and labeled before real physics can begin. We must define things like the *speed* of the rock, the *density* of the air, the *energy* of the atoms in the laser tube, and the *angular momentum* of a satellite to understand what is going on in a system and what its future evolution will be. We will call these things and others like them **physical quantities**. Most will be unfamiliar to you unless you have studied physics before. Together with the named objects, they form what can be called the *vocabulary of physics*. There are hundreds of physical quantities in regular use in the various fields of physics, but for our purposes in this book, we will need only a relatively small fraction of these.

Physics seeks to discover the basic ways in which things interact. Laws and principles express relationships that exist between physical quantities. For example, the law of fluid pressure expresses how the pressure at some location in a fluid depends on the weight of the fluid above. This law can be used to find the water pressure on a submerged submarine or the air pressure on a person's

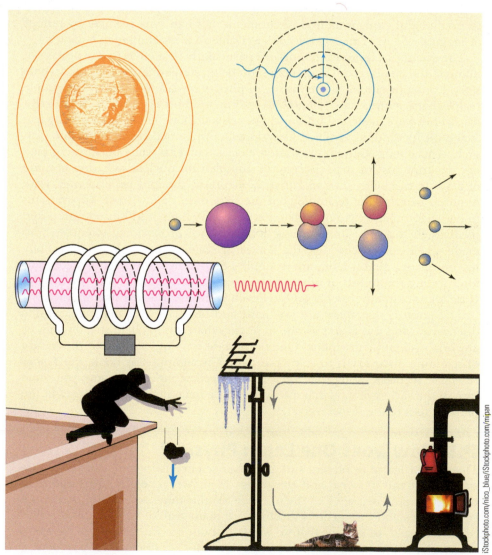

Figure P.4 Combination of six figures representing different systems we will be examining. The scale of the different parts varies greatly, from smaller than can be seen with a microscope to thousands of miles.

iStockphoto.com/nico_blue/iStockphoto.com/mipan

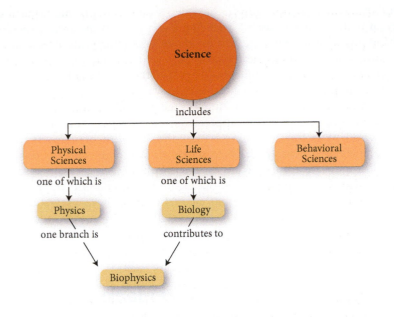

chest. These "rules" are used to understand the interactions in a system and to predict how the system will change with time in the future. The laws and principles themselves were formed after repeated, careful observations of countless systems by scientists throughout history. They withstood the test of time and repeated experimentation before being elevated to this status. You might regard physics as the continued search for, and the application of, basic rules that govern the interactions in the universe.

The process of learning physics has two main thrusts: the need to develop an understanding of the different physical quantities used in each area (establish a vocabulary), and the need to grasp the significance of the laws and principles that express the relationships among these physical quantities. Let us caution you again: memorizing the definitions and laws is only a first step—that alone won't do it. See, do, think, interact, visualize. Get involved in the physical world. That's how you learn physics. This book includes dozens of *Physics to Go* activities and worked-out examples based on real-world situations to help you in this process.

Another tool we use to help you visualize the relationships in physics is the *concept map*. Concept maps were developed in the 1960s and are used in a wide range of fields in a variety of ways. A concept map presents an overview of issues, examples, concepts, and skills in the form of a set of interconnected *propositions*. Two or more concepts joined by linking words or phrases make a proposition. The meaning of any particular concept is the sum of all the links that contain the concept. To "read" a concept map, start at the top with the most general concepts, and work your way down to the more specific items and examples at the bottom. Concept Map P.1 is one example. It is used to show some of the connections between the general concept, science, and one branch of physics, biophysics. This particular concept map could easily be expanded by, for example, showing all of the behavioral sciences or all of the branches of physics.

In this book, each chapter contains concept maps designed as summaries to help you organize the ideas, facts, and applications of physics. You should understand that many possible maps could be constructed from a given set of concepts. The maps drawn in this book represent one way of organizing and understanding a particular set of concepts.

You will have opportunities at the end of each chapter to develop lists of important concepts and to construct from them your own concept maps. Most people find it easier to understand relationships if they are displayed visually. You should find that the process of completing a concept map yourself gives you deeper insights into the ideas that are involved.

One of the main reasons physics has been so successful is that it harnesses the power of mathematics in useful ways. Many of the most important relationships involving physical quantities are best expressed mathematically. Predictions about the future conditions in a system usually involve math. The successes of Einstein, Newton, and others largely came about because they used mathematics to predict or explain things that no one could before (moving clocks appear to run slow, tides are caused by the Moon's gravity). An essential part of learning physics is developing an understanding of, and an appreciation for, this powerful side of physics.

The good news for beginners is that the simplest of mathematics—what most of us learned before age 16 or so—is all that is needed for this purpose. So hand in hand with the conceptual side of physics, we give you a taste of the mathematical side through worked-out examples and end-of-chapter problems. Over the years, we have found that even the most math-wary students often become very comfortable with this aspect of the material. An added benefit of your excursion into the world of physics is that you are likely to emerge with a better feel for the usefulness of simple mathematics.

P.6 Physical Quantities and Measurement

To be useful in physics, physical quantities must satisfy some conditions. A physical quantity must be *unambiguous*, its meaning clear and universally accepted. Understanding the meaning of a term involves more than just memorizing the words in its definition. To understand a concept, you must go beyond words. For example, the simple definition "energy is a measure of the capacity to do work" does not really convey our complete conceptual understanding of energy. Many physical quantities (speed, pressure, power, density, even energy) can be defined by an equation. Mathematical statements tend to be more precise than ones in words, making the meanings of these terms clearer.

Observation yields *qualitative* information about a system. Measurement yields *quantitative* information, which is central in any science that strives for exactness. Consequently, physical quantities must be measurable, directly or indirectly. One must be able to assign a numerical value that represents the amount of a quantity that is present. It is easy to visualize a measurement of distance, area, or even speed, but other quantities, like pressure, voltage, energy, or power, are a bit more abstract. Each of these can still be measured in prescribed ways, however. They would be useless if this were not so.

The basic act of measuring is one of comparison. To measure the height of a person, for instance, one would compare the distance from the floor to the top of the person's head against some chosen standard length such as a foot or a meter (**Figure P.5**). The height of the person is the number of units 1 foot or 1 meter long (including fractions) that have to be put together to equal that distance. The **unit of measure** is the standard used in the measurement—the foot or meter in this case. A complete measurement of a physical quantity, then, consists of a number and a unit of measure. For example, a person's height might be expressed as

$$\text{height} = 5.75 \text{ feet} \qquad \text{or} \qquad h = 5.75 \text{ ft}$$

Here h represents the quantity (height) and ft the unit of measure (feet). The same height in meters is

$$h = 1.75 \text{ m}$$

So when we introduce a physical quantity into our physics vocabulary—another "tool," so to speak—we must specify more than just a verbal definition. We should also give a mathematical

1 foot

1 meter

Thinkstock/Comstock/Thinkstock

Figure P.5 Measurement is an act of comparison. A person's height is measured by comparison with the length of a chosen standard. In this case, height equals five 1-foot lengths plus a segment 0.75 feet long. The same person's height is also equal to 1 meter plus 0.75 meter.

Table P.2 Common Metric Prefixes and Their Equivalents

1 *centi*meter = 0.01 meters	1 meter = 100 centimeters
1 *milli*meter = 0.001 meters	1 meter = 1,000 millimeters
1 *kilo*meter = 1,000 meters	1 meter = 0.001 kilometers
EXAMPLES	
189 centimeters = 1.89 meters	72.39 meters = 7,239 centimeters
25 millimeters = 0.025 meters	0.24 meters = 240 millimeters
7.68 kilometers = 7,680 meters	23.4 meters = 0.0234 kilometers

definition (if possible), relate it to other familiar physical quantities, and include the appropriate units of measure.

In the world today, there are two common systems of measure. The United States uses the **English system,** and the rest of the world, for the most part, uses the **metric system.** An attempt has been made in the United States to switch completely to the metric system, but so far it has not succeeded. The metric system has been used by scientists for quite some time, and we will use it a great deal in this book. It is a convenient system to use because the different units for each physical quantity are related by powers of 10. For example, a kilometer equals 1,000 meters, and a millimeter equals 0.001 meter. The prefix itself designates the power of 10. *Kilo-* means 1,000, *centi-* means 0.01 or $\frac{1}{100}$, and *milli-* means 0.001 or $\frac{1}{1000}$. A *kilo*meter, then, is 1,000 meters. Table P.2 illustrates the common metric prefixes. You may not know what an *ampere* is, but you should see immediately that a *milliampere* is one thousandth of an ampere.

The special *Applications* feature at the end of this section gives a brief look at the origins of the metric system. More than 20 of these special features appear throughout the book. They are intended to give you a deeper, richer view of selected topics in the history and applications of physics.

Having to use two systems of units is like living near the border between two countries and having to deal with two systems of currency. Most people who grew up in the United States have a better feel for the size of English-system units such as feet, miles per hour, and pounds than for metric-system units such as meters, kilometers per hour, and newtons. Often, the examples in this book will use units from both systems so that you can compare them and develop a sense of the sizes of the metric units. A *Table of Conversion Factors* relating the units in the two systems is included in the inside back cover of the print edition of this book. Fortunately, we won't have to deal with two systems of units after we reach electricity (Chapter 7).

A prologue is an introductory development. This prologue is an introduction to the field of physics, our approach to teaching it, and how to get started learning it. The groundwork has been laid, and we are now ready to proceed.

COMMERCIAL APPLICATIONS The Metric System: "For All Time, for All People."

The French Revolution, beginning with the storming of the Bastille on 14 July 1789, gave birth not only to a new republic but also to a new system of weights and measures. Eighteenth-century France's system of weights and measures had fallen into a chaotic state, with unit names that were confusing or superfluous and standards that differed from one region of the country to another. Seizing on the opportunity presented to them by the political and social turmoil accompanying the revolution, scientists and merchants, under the leadership of Charles-Maurice de Talleyrand, presented a plan to the French National Assembly in 1790 to unify the system. The plan proposed two changes: (1) the establishment of a decimal system of measurement and (2) the adoption of a "natural" scale of length. Neither of these two notions

was new to scholars of this period. The first had been discussed as early as 1585 by Simon Stevin, a hydraulic engineer in Holland, in a pamphlet called *De Thiende* (i.e., *The Tenth Part*). The second notion was introduced in 1670 by Abbé Gabriel Mouton, who proposed that a standard of length be defined in terms of the size of Earth—specifically a fraction of the length of the meridian arc extending from the North Pole to the equator.

After much give and take, the plan was finally adopted into law on 7 April 1795. The new legislation defined the meter as the measure of length equal to 1 ten-millionth of the meridian arc passing through Paris from the North Pole to the equator and the gram as the mass of pure water contained in a cube 1/100th of a meter (a

Juli/Shutterstock.com

*centi*meter) on a side at the temperature of melting ice. It also made this system obligatory in France.

The tasks of actually determining the sizes of these newly defined units were assigned to Jean-Baptiste Delambre and Pierre Méchain, who were to survey the length of the meridian arc through Paris, and to Louis Lefèvre-Gineau and Giovanni Fabbroni, who were to determine the absolute weight of water. As it turned out, these measurements were not made without difficulty and, sometimes, danger. For example, during the period between 1792 and 1798, Delambre and Méchain made measurements along the meridian between Dunkirk, France, and Barcelona, Spain, amid the riot and turmoil still present in many parts of Europe. They were frequently arrested as spies, often had their equipment confiscated, and were generally harassed at every turn. Finally in 1798, with the job done and the length of the centimeter accurately known, Lefèvre-Gineau and Fabbroni set about their work. They, too, encountered difficulties, largely having to do with reaching and maintaining the required measurement temperature, but they completed their task in only one year.

Sergey Kohl/Shutterstock.com

Figure P.7 Many countries have issued stamps to commemorate the development and/or adoption of the metric system. This one, from Romania, celebrates the centenary anniversary of this country's adoption of the metric system. It shows the symbols for the base units of measure in use at the time of its issuance (1966).

Beginning in 1798, an international committee with representatives from nine nations undertook to carry out the calculations required to produce the standards needed to define and extend the new system of weights and measures. It submitted its report to the French legislature for ratification on 27 June 1799, and the bill passed on 10 December of that year. That document is the first official text in which the metric system is mentioned. According to this law, the definitive standards of length and mass to be used in commercial and scientific interactions throughout France were "the meter and kilogram of platinum deposited with the legislative body."

National Institute of Standards and Technology

Figure P.6 The international 1-kilogram standard of mass, a platinum-iridium cylinder. This physical artifact will soon be replaced as the mass standard with a definition based on the fundamental unit of time, the second, and some agreed-upon fixed constants of Nature.

Since then, the definitions of length and mass have undergone several revisions, and other units of measure have been incorporated. For example, the numerical value of the meter is now established by the length of the second (see *Time Out!* in Section 1.1) and the fixed speed of light (299,792,458 meters per second), while the definition of international standard of mass, the kilogram, currently based on a physical prototype (**Figure P.6**), will soon (perhaps as early as 2018) be established in terms of the definitions of the second and the meter and another constant of Nature called *Planck's constant* (see Section 10.1). Despite these changes, however, the basic tenets of the metric system survive: simplicity and convenience stemming from its use of a decimal system of measure, and uniformity and reproducibility deriving from its reliance on an agreed upon set of standards, albeit increasingly ones involving natural constants instead of physical artifacts. Since its adoption in Europe, first in France, then in Holland (1816) and Greece (1836), the use of the metric system has proliferated so that no nation—even the United States, one of only a few countries that have not officially adopted the metric system for its manufacturing and commercial activities—is without knowledge of it (**Figure P.7**). The metric system has become, in the motto adopted by its founders, a system "for all time, for all people."

■ QUESTIONS

1. Describe the basic tenets or principles on which the metric system is based.
2. What were the first standards of mass and length specified under the metric system originally adopted into law in France in 1795?

THE STUDY OF MOTION

Marc Sanchez/Icon Sportswire 144/Marc Sanchez/Icon Sportswire/Newscom

Figure CO-1 Drag racer decelerating after completing a quarter-mile race.

CHAPTER INTRODUCTION:
Drag Racing

Acceleration, pure and simple. That's the point of drag racing. Increase the speed of the dragster as quickly as possible so it will travel the one-quarter of a mile (about 400 meters) faster than your opponent. These elite machines have engines with up to about 6,000 horsepower and, from a standing start, reach speeds of more than 300 mph in times shorter than it took you to read this sentence. In the process, they subject their drivers to average accelerations over four times that of gravity. At the end of their brief run, another critical feat must be accomplished: they have to be brought to a stop quickly and safely. How do you slow a vehicle that is going much faster than a passenger jet when it lands? One way is to deploy a parachute (**Figure CO-1**).

Imagine driving such a machine: five seconds of forward acceleration pressing you to the back of your seat, followed by many more seconds of acceleration in the opposite direction—backward—that strains the straps holding you in. And for much of that time you are going hundreds of miles per hour. The faint of heart need not apply for this job.

Drag racing is one of the most dramatic examples of human beings deliberately accelerating themselves. The large attendance at these events suggests that many people are thrilled to watch it happen. But, as we will see, accelerated motion is ubiquitous in everyday life, and after completing this chapter, you should have a good understanding of the concept of acceleration, as well as its relatives: speed and velocity.

This chapter begins our study of motion and its causes, a branch of physics called *mechanics*, which will be continued in Chapter 2. Here, we focus on the description of motion by introducing the concepts of speed, velocity, and acceleration, a field called *kinematics*.

1.1 Fundamental Physical Quantities

Our main task here is to take a closer look at motion—what it is, how one quantifies it, what the simplest kinds of motion are, and so on. Fortunately for beginners, most of the concepts and terms needed are already familiar because motion is such an important part of our everyday world. We have a sense of how fast 60 miles per hour (or 97 kilometers per hour) is and how to compute speed if we know the distance traveled and the time elapsed. The unit of measure pretty much tells us how to do that: "miles per hour" means divide the distance in miles by the time in hours. But there are some important subtleties that must be examined, and our basic ideas about velocity and acceleration need to be expanded. The ways that we will express relationships throughout the text, oftentimes involving things not so familiar as motion, will be presented. This is also a perfect time to take a closer look at those "physical quantities" introduced in the Prologue.

In physics, there are three basic aspects of the material universe that we must describe and quantify in various ways: *space*, *time*, and *matter*. All physical quantities used in this textbook involve measurements (or combinations of measurements) of space, time, and the properties of matter. The units of measure of all of these quantities can be traced back to the units of measure of distance, time, and two properties of matter called *mass* and *charge*. We will postpone our treatment of charge until Chapter 7.

19 mm

0.000019 km

iStockphoto.com/compucow

Figure 1.1 The two measurements represent the same distance, but the one in millimeters is more convenient to use and conceptualize.

1.1a Distance

Distance, time, and mass—known as **fundamental physical quantities**—are such basic concepts that it is difficult to define them, particularly time. **Distance** represents a measure of space in one dimension. Length, width, and height are examples of distance measurements. The following table lists the common distance units of measure, in both the metric system and the English system, and their abbreviations. (You might want to review the basics of the metric system presented in the Prologue.) This same format will be used for all physical quantities that have several common units.

Physical Quantity	Metric Units	English Units
Distance d (or l, w, h)	meter (m)	foot (ft)
	millimeter (mm)	inch (in.)
	centimeter (cm)	mile (mi)
	kilometer (km)	

Why are there so many different units in each system? Generally, it is easier to use a unit that fits the scale of the system being considered. The meter is good for measuring the size of a house, the millimeter for measuring the size of a coin, and the kilometer for measuring the distance between cities. Eighty kilometers is the same as 80,000 meters, but the former measurement is easier to conceptualize and to use in calculations. Similarly, it might be correct to say that a coin is 0.000019 kilometers in diameter, but 19 millimeters is a more convenient measure (**Figure 1.1**).

The sizes of all of the distance units, including the English units, are defined in relation to the meter. In this book, we will use meters most often in our examples. You should try to get used to distance measurements expressed in meters and have an idea, for example, of how long 25 and 0.2 meters are. **Table 1.1** shows some representative distances expressed in both metric and English units. (We will use scientific notation regularly throughout the book. See Appendix B for a review.)

It is a rather simple matter to convert a distance expressed in, say, meters to a distance expressed in another unit. For example, you might solve a problem and find that the answer is, "The sailboat travels 23 meters in 10 seconds." Just how far is 23 meters? Think of any numerical measurement as a number multiplied by a unit of measure. In this case, 23 meters equals 23 *times* 1 meter. Then the 1 meter can be replaced by the corresponding number of feet—a conversion factor found in the Table of Conversion Factors (on the inside back cover for users of print copies).

$$23 \text{ meters} = 23 \times 1 \text{ meter}$$

Table 1.1 Some Representative Sizes and Distances

Size/Distance	Metric	English
Size of a nucleus	1×10^{-14} m	4×10^{-13} in.
Size of an atom	1×10^{-10} m	4×10^{-9} in.
Size of a red blood cell	8×10^{-6} m	3×10^{-4} in.
Typical height of a person	1.75 m	5.75 ft
Tallest building	830 m	2,722 ft
Diameter of Earth	1.27×10^7 m	7,920 miles
Earth–Sun distance	1.5×10^{11} m	9.3×10^7 miles
Size of our galaxy	9×10^{20} m	6×10^{17} miles

But 1 meter = 3.28 feet; therefore:

$$23 \text{ meters} = 23 \times 1 \text{ meter} = 23 \times 3.28 \text{ feet}$$

$$23 \text{ meters} = 75.4 \text{ feet}$$

Two other physical quantities that are closely related to distance are *area* and *volume*. Area commonly refers to the size of a surface, such as the floor in a room or the outer skin of a basketball. The concept of area can apply to surfaces that are not flat and to "empty," two-dimensional spaces such as holes and open windows (**Figure 1.2**). Area is a much more general idea than "length times width," an equation you may have learned that applies only to rectangles. The area of something is just the number of squares 1 inch by 1 inch (or 1 meter by 1 meter, or 1 mile by 1 mile, etc.) that would have to be added or placed together to cover it.

In a similar manner, the volume of a solid is the number of cubes 1 inch or 1 centimeter on a side needed to fill the space it occupies. For common geometric shapes such as a rectangular box or a sphere, there are simple equations to compute the volume. We use volume measures nearly every time we buy food in a supermarket.

Figure 1.2 Every surface, whether it is flat or curved, has an area. The area of this rectangle and the area of the surface of the basketball are equal.

Physical Quantity	Metric Units	English Units
Area (A)	square meter (m²)	square foot (ft²)
	square centimeter (cm²)	square inch (in.²)
	square kilometer (km²)	square mile (mi²)
	hectare	acre

Physical Quantity	Metric Units	English Units
Volume (V)	cubic meter (m³)	cubic foot (ft³)
	cubic centimeter (cm³ or cc)	cubic inch (in.³)
	liter (L)	quart, pint, cup
	milliliter (mL)	teaspoon, tablespoon

Area and volume are examples of physical quantities that are based on other physical quantities—in this case, just distance. Their units of measure are called *derived units* because they are derived from more basic units (1 square meter = 1 meter × 1 meter = 1 meter²). Until we reach Chapter 7, all of the physical quantities will have units that are derived from units of distance, time, mass, or a combination of these.

1.1b Time

The measure of **time** is based on periodic phenomena—processes that repeat over and over at a regular rate. The rotation of Earth was originally used to establish the universal unit of time, the *second*. The time it takes for one rotation was set equal to 86,400 seconds (24 × 60 × 60). Both the metric system and the English system use the same units for time.

Physical Quantity	Metric Units	English Units
Time (t)	second (s)	second (s)
	minute (min)	minute (min)
	hour (h)	hour (h)

Figure 1.3 The regular swinging of the pendulum—by its repetitive motion—controls the speed of this clock.

University of Michigan-Dearborn

DEFINITION **Period** The time for one complete cycle of a process that repeats. It is abbreviated T, and the units are seconds, minutes, and so forth.

DEFINITION **Frequency** The number of cycles of a periodic process that occur per unit time. It is abbreviated f and has units s^{-1} or hertz (Hz).

Physics To Go 1.1

For this, you need a pendulum—something dangling on the end of a string about 1 to 2 feet (or 0.5 meter) long. A shoe hanging from its shoelace or keys tied to a thread work. You also need a timer—a digital watch that displays seconds is OK, or use the "stopwatch" feature on your cell phone.

1. Set the pendulum swinging back and forth, and time how long it takes for it to complete 10 full cycles (left to right then back to left is one cycle). Divide that time by 10 to get the *time for one cycle*. This should be around 1 to 2 seconds.
2. Now set it swinging again, and this time count how many full cycles happen in 10 seconds. (You can estimate fractions, like 7.5 cycles.) Divide this number by 10 to get the *number of cycles that occur per second*.
3. See if you can figure out the approximate mathematical relationship between the two numbers.
4. Make the string twice as long or half as long and repeat. What differences do you see in these new data relative to the previous set?

Clocks measure time by using some process that repeats. Many mechanical clocks use a swinging pendulum. The time it takes to swing back and forth is always the same. This is used to control the speed of a mechanism that turns the hands on the clock face (**Figure 1.3**). Mechanical wristwatches and stopwatches use an oscillating balance wheel for the same purpose. Quartz electric clocks and digital watches use regular vibrations of an electrically stimulated crystal made of quartz.

The first step in designing a clock is to determine exactly how much time it takes for one cycle of the oscillation. If it were 2 seconds for a pendulum, for example, the clock would then be designed so that the second hand would rotate once during 30 oscillations. The **period** of oscillation in that case is 2 seconds.

But there is another way to look at this. The clock designer must determine how many cycles must take place before 1 second (or 1 minute or 1 hour) elapses. In our example, one-half of a cycle takes place during each second. This is called the **frequency** of the oscillation.

The standard unit of frequency is the hertz (Hz), which equals 1 cycle per second.

$$1 \text{ Hz} = 1/s = 1 \text{ s}^{-1}$$

The frequency of AM radio stations is expressed in kilohertz (kHz) and those of FM stations in megahertz (MHz). *Mega-* is the metric prefix signifying 1 million. So, 91.5 MHz equals 91,500,000 Hz. Another metric prefix, *giga-*, is also commonly used with hertz. For example, the "speed" of a computer's processor—actually, the frequency of the electrical signal it uses—might be 2 gigahertz, which is 2 billion (2,000,000,000) hertz.

The relationship between the period of a cyclic phenomenon and its frequency is simple: the period equals 1 divided by the frequency, and vice versa.

$$\text{period} = \frac{1}{\text{frequency}}$$

$$T = \frac{1}{f}$$

Also,

$$f = \frac{1}{T}$$

EXAMPLE 1.1 A mechanical stopwatch uses a balance wheel that rotates back and forth 10 times in 2 seconds. What is the frequency of the balance wheel?

$$\text{frequency} = \text{number of cycles per time}$$

$$f = \frac{10 \text{ cycles}}{2 \text{ s}}$$

$$= 5 \text{ Hz}$$

What is the period of the balance wheel?

$$\text{period} = \text{time for one cycle}$$

$$= 1 \text{ divided by the frequency}$$

$$T = \frac{1}{5 \text{ Hz}}$$

$$= 0.2 \text{ s}$$

The balance wheel oscillates 300 times each minute.

1.1c Mass

The third basic physical quantity is mass. The **mass** of an object is basically a measure of how much matter it contains. (This statement illustrates the sort of circular definition that sometimes arises when one tries to define fundamental concepts.) We know intuitively that a large body such as a locomotive has a large mass because it is composed of a great deal of material. Mass is also a measure of what we sometimes refer to as **inertia**. The larger the mass of an object, the greater its inertia and the more difficult it is to speed up or slow down.

Physical Quantity	Metric Units	English Units
Mass (m)	kilogram (kg)	slug
	gram (g)	

Mass is not in common use in the English system; note the unfamiliar unit, the slug. Weight, a quantity that is related to but is *not* the same as mass, is used instead. We can contrast the two ideas: when you try to lift a heavy suitcase, you are experiencing its weight. When you pull it along and speed it up or slow it down, you are experiencing its mass (**Figure 1.4**). We will take another look at mass and weight in Chapter 2 to better clarify their relationship.

(a) (b)

Figure 1.4 (a) When lifting something, you must overcome its weight. (b) When changing its speed, you experience its mass.

■ Learning Check

Simple quizzes such as this one are included at the end of most sections. Though not designed to be a thorough examination of the prior material, if you have trouble answering these questions, it probably means you should go back and reread the section. Answers are given at the end of each Learning Check.

1. Three of the fundamental physical quantities in physics are
    (a) distance, time, and weight
    (b) distance, time, and mass
    (c) distance, time, and speed
    (d) distance, area, and volume
2. (True or False.) When the period of a pendulum is less than 1 second, the frequency is always greater than 1 hertz.
3. One kilogram is the same as _____ grams.
4. (True or False.) In mechanics, weight and mass measure the same property of matter.

ANSWERS: 1. (b) **2.** True **3.** 1,000 **4.** False

Although the notion of time may be hard to define, the process of timekeeping is straightforward, requiring only a system for accurately counting the cycles of regularly occurring events. The operative word here is *regularly*: to keep time or to measure a time interval, we need something that cycles or swings or oscillates at a constant rate. Up until 1956, the fundamental clock used to tell time was the Earth–Sun system, and the fundamental unit of time, the second, was defined as 1/86,400 of a *mean solar day*. A mean solar day is the average interval between successive crossings of your local meridian by the Sun. Unfortunately, the rate of rotation of Earth is not *strictly* constant, and a mean solar day varies in length over long time spans because of a gradual slowdown in Earth's rotation rate brought about by friction between its oceans and its crust. In addition, short-term variations in the rate of Earth's spin, some of which are seasonal, alter the length of a mean solar day.

For these reasons, in the late 1950s, a still more uniform cycle within the Earth–Sun system was sought, and the second was redefined as 1/31,556,925.9747 of the length of the year beginning in January 1900. This *ephemeris second* is based on the motion of Earth around the Sun, which is governed (as we shall see in Chapter 2) by Newton's laws of motion. As a result, its evaluation requires astronomical observation and calculation and is not easily obtained with high accuracy except after many years of careful work. Thus, although ephemeris time appears to be extremely uniform, it suffers from not being able to be found quickly and accurately.

In 1967, a new *atomic second* was defined that has become the basis for International Atomic Time (TAI). In this case, 1 second equals the interval of time containing 9,192,631,770 oscillations of light waves given off by isolated cesium atoms (see Chapter 10 for a discussion of atoms and their light-emitting properties). With modern cesium clocks, such as those in common use at the National Institute of Standards and Technology (NIST), errors in timing as small as 2×10^{-11} seconds per year can be achieved: such clocks neither gain nor lose a second in 150 million years. As reported in April 2015, a NIST-built strontium optical lattice clock set the current record for timing precision with variations equivalent to less than a second in over 15 *billion* years, a time comparable to the age of the universe.

Time measurement is so accurate that the length of the standard meter is set by how long it takes light to travel that distance—3.33564095 billionths of a second.

The introduction of atomic time leaves us with a problem. Our daily lives are generally governed by day–night cycles—that is, by the Earth–Sun clock, not by phenomena associated with cesium atoms. As Earth continues to spin down, the length of a mean solar day (and hence the mean solar second) will continue to grow larger relative to the atomic second. As time goes on, Earth–Sun clocks will gradually fall further and further behind atomic clocks. This is clearly not a desirable situation.

In the early 1970s, the French Bureau International de l'Heure, the world's official timekeepers, introduced Coordinated Universal Time (UTC). Under this system, the length or duration of the second is dictated by atomic time, but the time commonly reported by time services is required to remain within 0.9 seconds of mean solar time. In this system, a *leap second* is added or subtracted as needed to compensate for changes in the rate of Earth's rotation relative to atomic time. After the addition of a leap second on 30 June 2015, atomic time is 17 seconds ahead of UTC.

If this discussion seems a bit complicated, that's understandable. The topic of time and its determination is one that physicists and

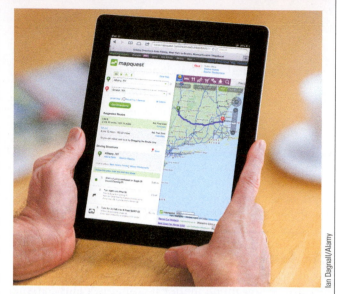

Figure 1.5 GPS navigation system display on a tablet.

philosophers have struggled with for centuries. The point to be emphasized is that as more sophisticated experiments have been devised to probe Nature's secrets, the need for ever more reliable and stable clocks has grown. The development of atomic clocks is the latest attempt to meet the requirements of scientists and engineers in this regard, and their use has now spilled over into our everyday lives. The adoption of UTC is an example. Another is the *global positioning system* (GPS). This system currently consists of an array of 31 active satellites, about 8 of which are always visible from any point on the globe. Each satellite circles Earth every 12 hours and carries as many as 4 cesium and rubidium atomic clocks to generate the accurate time signals used in making precise position measurements. Small handheld GPS receivers costing less than $100 are available that provide horizontal and vertical positional accuracies to better than 15 m and timing accuracies on the order of 15×10^{-9} s or less. Originally developed for military missions, the civilian use of GPS today extends to commercial ship and aircraft navigation; nationwide truck and freight-car tracking; "precision" farming, land surveying, and geological studies; and worldwide digital communication networks (including some that control banking functions at local ATMs). GPS receivers have even been integrated with databases of maps and street directories and installed in autos to assist drivers in reaching their destinations. GPS technology is now routinely available as part of personal cell phone and computer service (**Figure 1.5**). But regardless of how it is measured or used by humankind, one fact about time remains true: *tempus fugit*, "time flies."

■ QUESTIONS

1. The original "clock" used to define the length of the second was the daily rotation of Earth about its axis. Why has this "clock" been replaced by one based on the oscillation period of light waves emitted by atoms like cesium and rubidium?
2. What is a *leap second*, and why was it introduced into timekeeping practices by scientists in the 1970s?

Ian Dagnall/Alamy

1.2 Speed and Velocity

1.2a Speed

A key concept to use when quantifying motion is **speed**.

DEFINITION **Speed** Rate of movement. Time rate of change of distance from a reference point. The distance traveled divided by the time elapsed.

Physical Quantity	Metric Units	English Units
Speed (v)	meter per second (m/s)	foot per second (ft/s)
	kilometer per hour (km/h)	mile per hour (mph)

A couple of aspects of speed are worth highlighting. First, speed is *relative*. A person running on the deck of a ship cruising at 20 mph might have a speed of 8 mph relative to the ship, but the speed relative to the water or a nearby pier would be 28 mph (if headed toward the front of the ship). If we use the ship as the reference point, the speed is 8 mph. With the water or a pier as the reference point, the speed is 28 mph (**Figure 1.6**). If you are traveling at 55 mph on a highway and a car passes you going 60 mph, its speed is 5 mph relative to your car. Most of the time speed is measured relative to Earth's surface; this will be the case in this book, unless stated otherwise.

Second, it is important to distinguish between *average speed* and *instantaneous speed*. An object's average speed is the total distance it travels during some period of time divided by the time that elapses:

$$\text{average speed} = \frac{\text{total distance}}{\text{total elapsed time}}$$

If a 1,500-mile airline flight lasts 3 hours, the plane's average speed is 500 mph. Of course, the plane's speed changes frequently during the 3 hours, so its speed is not 500 mph at each instant in time. (You can average 75% on four exams without actually getting 75% on each or even any one test.) Similarly, a sprinter who runs the 100-meter dash in 10 seconds has an average speed of 10 m/s but is not traveling with that speed at each moment of the race. This gives rise to the concept of *instantaneous* speed—the speed that an object has at an instant in time. A car's speedometer actually gives instantaneous speed. When it shows 55 mph, it means that, if the car traveled with exactly that speed for 1 hour, it would go 55 miles. How can one determine the instantaneous speed of something that is not equipped with a speedometer? Instantaneous speed can't be measured exactly using the basic meaning of speed (distance traveled divided by the elapsed time) because an "instant" implies that *zero* time elapses. But we can get a good estimate of an object's instantaneous speed by timing how long it takes to travel a very short distance:

$$\text{instantaneous speed} \cong \frac{\text{very short distance}}{\text{very short time}}$$

Figure 1.6 Speed is relative. The speed of a person running on a ship is 8 mph relative to the deck. If the ship's speed is 20 mph, the person's speed relative to a stationary observer on the pier is either 28 mph (if the runner is headed toward the front of the ship) or 12 mph (if she is headed toward the rear).

For example, given sophisticated equipment, we might measure how long it takes a car to travel 1 meter. If that time is found to be 0.05 seconds (not an "instant" but a very short time), a good estimate of the car's instantaneous speed is

$$\text{instantaneous speed} = \frac{1 \text{ m}}{0.05 \text{ s}} = 20 \text{ m/s} = 44.8 \text{ mph}$$

In drag racing, the maximum instantaneous speed of a dragster is estimated by timing how long it takes to travel the final 60 feet (about 18 meters) of the quarter-mile race. This is typically less than 0.15 seconds, yielding an instantaneous speed of about 269 mph (120 m/s).

GPS receivers used by pilots, drivers, hikers, and others employ radio signals from satellites to determine location. They can compute the device's approximate instantaneous speed by computing how far it travels in a short time and then dividing that distance by that time. Some GPS receivers can update the position every second, so they can estimate instantaneous speed using 1 second as the "short time." A typical bicycle speedometer uses a magnetic sensor on the front wheel. The device can time how long it takes the wheel to make one rotation and estimate the instantaneous speed by using the circumference of the wheel (typically around 2.1 meters) as the "short distance."

The conversion from meters per second to miles per hour is done in the same way that was used with distance in Section 1.1a. From the Table of Conversion Factors 1 m/s = 2.24 mph; so

$$20 \text{ m/s} = 20 \times 1 \text{ m/s} = 20 \times 2.24 \text{ mph} = 44.8 \text{ mph}$$

Actually, this is the car's average speed during the short time span. In normal situations, a car's speed will increase or decrease by at most 1 mph during a twentieth of a second, so this is a pretty good estimate. Our answer is within ± 1 mph of the true value. During a collision, a car's speed can change a great deal during 0.05 seconds, so one would have to somehow use a much shorter time interval to calculate instantaneous speed in this case. An automaker might use high-speed digital cameras capable of recording 250,000 frames per second during a crash test to measure much shorter periods and corresponding distances. In general, "very short time" means an interval during which the object's speed won't change by an amount greater than the desired error limit of the estimate. The concept of instantaneous speed is what is important here, more so than the technical details of how it is measured in different situations.

▶ Physics To Go 1.2

Look up the record times for Olympic racing events in running, swimming, ice skating, or bicycling. (These can be found in a book of world records or online.) For several events, calculate the average speed and notice how it is lower for longer races. Why is this so? For comparison, compute the average speeds for the same distance in two different events such as 1,500 meters running and swimming. Why are they so different?

In some cases, an object may have been traveling for a while and already moved some distance before we start taking measurements to determine its speed (average or instantaneous). Perhaps we want to measure a sprinter's average speed during the last part of a race. Then the distance and time that we use would be the values at the end of the segment being timed (the *final* values) minus the values at the beginning of the segment (the *initial* values)—that is, the *changes* in distance and time. The general expression for speed is

$$\text{speed} = \frac{\text{change in distance}}{\text{change in time}} = \frac{d_{\text{final}} - d_{\text{initial}}}{t_{\text{final}} - t_{\text{initial}}}$$

$$v = \frac{\Delta d}{\Delta t}$$

(The symbol Δ is the capital Greek letter delta and is used to represent a "change in" a physical quantity.) This equation can represent both average and instantaneous speed. When Δt is the total elapsed time for a trip, v is the average speed. When Δt is a very short time, then v is the instantaneous speed.

EXAMPLE 1.2 An analysis of a videotape of Olympic gold-medal winner Usain Bolt (**Figure 1.7**) running the 100-meter dash in the 2009 World Championships in Athletics in Berlin yields the data in the accompanying table. Compute his average speed for the race, and estimate his peak instantaneous speed.

Segment (meters)	Time (seconds)
0–10	1.85
10–20	1.02
20–30	0.91
30–40	0.87
40–50	0.85
50–60	0.82
60–70	0.82
70–80	0.82
80–90	0.83
90–100	0.90
Total Distance: 100 m	Total Time: 9.69 s

SOLUTION For the entire race, this would be as follows:

$$\text{average speed} = v = \frac{\Delta d}{\Delta t} = \frac{d_{\text{final}} - d_{\text{initial}}}{t_{\text{final}} - t_{\text{initial}}}$$

$$v = \frac{100 \text{ m} - 0 \text{ m}}{9.69 \text{ s} - 0 \text{ s}} = \frac{100 \text{ m}}{9.69 \text{ s}}$$

$$= 10.32 \text{ m/s} (= 23.1 \text{ mph})$$

The nearly equally spaced times between 50 and 90 meters indicate that his speed was constant. Therefore, we can use any segment from this part of the race to compute his instantaneous speed. For the segment between 70 and 80 meters, this would be:

$$v = \frac{\Delta d}{\Delta t} = \frac{d_{\text{final}} - d_{\text{initial}}}{t_{\text{final}} - t_{\text{initial}}}$$

$$= \frac{80 \text{ m} - 70 \text{ m}}{0.82 \text{ s}} = \frac{10 \text{ m}}{0.82 \text{ s}}$$

$$= 12.2 \text{ m/s} (= 27.3 \text{ mph})$$

As the time for the final 10-m segment shows, Bolt slowed down as he approached the finish line after glancing back to confirm he had won the race.

Figure 1.7 Usain Bolt, who set world records in the 100-m and 200-m dashes.

Andy Lyons/Getty Images

The speed of a car being driven around in a city changes quite often. The instantaneous speed may vary from 0 mph (at stoplights) to 45 mph. The average speed for the trip is the total distance divided by the total time, maybe 20 mph.

When the speed of an object is constant, the average speed and the instantaneous speed are the same. In this case, we can express the relationship between the distance traveled and the time that has elapsed as follows:

$$d = vt \quad \text{(when speed is constant)}$$

This is an example of what is called a *direct proportionality:* we say that d is directly proportional to t (abbreviated $d \propto t$). If the time is doubled, then the distance is doubled. The constant speed v is called the *constant of proportionality.* We will encounter many examples in which one physical quantity is proportional to another.

> ## ▶ Physics To Go 1.3
>
> When lightning strikes (**Figure 1.8**), the flash of light reaches us in a fraction of a second, but the sound (thunder) is delayed. Why is that? The information in **Table 1.2** should help you answer that. The time delay between the flash and the sound can be used to estimate how far away a lightning strike is. The sound travels about 340 meters in 1 second or 1 mile in 5 seconds. So what is the simple rule that relates the distance of the lightning strike to the number of seconds between the light and the sound? (Note: Follow the safety guidelines whenever there is a potential for lightning: go into a building or a vehicle with a metal roof and stay away from windows, plumbing, telephones with cords, and so on.)

Most drivers are accustomed to speed measured in miles per hour or kilometers per hour. This is most convenient when talking about travel times for distances greater than a few miles. Often, it is more enlightening to use feet per second. For example, a car going 65 mph travels 130 miles in 2 hours (using the preceding equation). But for a potential accident, it is relevant to consider how far the car will travel in a few seconds. Because 65 mph equals 95.6 feet per second, during the 2 seconds it takes a driver to decide how to avoid an accident, the car will have traveled 191 feet—more than 10 car lengths.

Figure 1.8 A distant lightning strike provides a good demonstration of the difference between the speed of sound and the speed of light.

NOAA Photo Library

Table 1.2 Some Speeds of Interest

Description	Metric	English
Speed of light, c (in vacuum)	3×10^8 m/s	186,000 miles/second
Speed of sound (in air, room temperature)	344 m/s	771 mph
Highest instantaneous speeds:		
Running (cheetah)	28 m/s	75 mph
Swimming (sailfish)	30.6 m/s	68 mph
Flying—level (merganser)	36 m/s	80 mph
Flying—dive (peregrine falcon)	108 m/s	242 mph
Humans (approximate):		
Swimming	2.5 m/s	5.6 mph
Running	12 m/s	27 mph
Ice skating	14 m/s	31 mph

Try to develop a feel for the different speed units, particularly meters per second. You might keep in mind this comparison:

$$65 \text{ mph} = 95.6 \text{ ft/s} = 29.1 \text{ m/s} = 105 \text{ km/hr}$$

Here is one last example of doing conversions. From the Table of Conversion Factors, 1 mph = 0.447 m/s:

$$65 \text{ mph} = 65 \times 1 \text{ mph} = 65 \times 0.447 \text{ m/s} = 29.1 \text{ m/s}$$

It seems that the universe was created with an absolute speed limit—the speed of light in empty space. This speed is represented by the letter c. Nothing has ever been observed traveling faster than the speed of light. (For more on the constancy of the speed of light, see Section 12.1.) The value of c and some other speeds are included in Table 1.2.

1.2b Velocity

An important aspect of motion is *direction*. We will see that changing the direction of a body's motion can produce effects that are equivalent to changing the speed. **Velocity** is a physical quantity that incorporates both ideas.

The speed of a ship might be 10 m/s, whereas its velocity might be 10 m/s east. Whenever a moving body changes direction, such as a car going around a curve or someone walking around a corner, the velocity changes even if the speed does not. Being told that an airplane flies for 2 hours from a certain place and averages 100 mph does not give you enough information to determine where it is. Knowing that it travels in a straight line due north would allow you to pinpoint its location.

A speedometer alone gives the instantaneous speed of a vehicle. A speedometer used in conjunction with a compass would give velocity: speed and the direction of motion. In a car traveling along a winding road, the movement of the compass needle indicates that the velocity is changing even if the speed is not.

One of the standard displays on simple GPS receivers includes both the speed and the direction of motion—in other words, the velocity (**Figure 1.9**). Whenever either the speed or the heading (direction) is changing, the velocity is changing.

Velocity is an example of a physical quantity called a **vector**. Vectors have both a numerical size (magnitude) and a direction associated with them. Quantities that do not have a direction are called **scalars**. Speed by itself is a scalar. Only when the direction of motion is included do we have the vector velocity. Similarly, we can define the vector *displacement* as

DEFINITION **Velocity** Speed in a particular direction (same units as speed). Directed motion.

Figure 1.9 Hand-held GPS receiver capable of measuring speed and direction of motion, that is, velocity.

OvuOng/Shutterstock.com

v

v

Figure 1.10 Vector quantities can be represented by arrows. Arrow length indicates the vector's size, and arrow direction shows the vector's direction. In both figures, velocity is represented with an arrow. The car's velocity is greater than the pedestrian's, so the arrow representing the car's velocity is longer. Because both objects are moving in the same direction, their velocity vectors (arrows) are parallel.

distance in a specific direction. For the airplane referred to earlier, the *distance* it travels in 2 hours is 200 miles. Its actual location can be determined only from its *displacement*—200 miles due north, for example. The basic equation for speed, $v = \Delta d / \Delta t$, is also the equation for velocity (that's why v is used) with d representing a vector displacement.

We can classify most physical quantities as scalars or vectors. Time, mass, and volume are all scalars because there is no direction associated with them.

Vectors are represented by arrows in drawings, the length of the arrow being proportional to the size or magnitude of the vector (**Figure 1.10**). If a car is traveling twice as fast as a pedestrian, then the arrow representing the car's velocity is twice as long as that representing the pedestrian's velocity.

When an object can move forward or backward along a line, its velocity is positive when it is going in one direction and negative when it is going in the opposite direction. When going forward in a car, you might give the velocity as +10 m/s (positive). If the car stops and then goes backward, the velocity is negative, −5 m/s, for example. The velocity of a person on a swing is positive, then negative, then positive, and so on. Which direction is associated with positive velocity is somewhat arbitrary. When dribbling a basketball, you could say its velocity is positive when it is moving downward and negative when it is moving upward, or vice versa. Speed does not become negative when the velocity does. The negative sign is associated with the direction of motion.

When dealing with a system in which the direction of motion can change, even if forward and backward are the only possible directions, it is better to use velocity than speed because the + or − sign indicates the direction of motion. In a situation in which the direction of motion does not change, like a falling object, the terms *speed* and *velocity* are often used interchangeably. However, in the remainder of this text, *velocity* will be used in all situations in which the direction of motion can be important. In cases where an object's direction of motion doesn't change, we will use the direction of the object's initial motion as the positive direction (unless stated otherwise).

1.2c Vector Addition

Sometimes a moving body has two velocities at the same time. The runner on the deck of the ship in Figure 1.6 has a velocity relative to the ship and a velocity because the ship itself is moving. A bird flying on a windy day has a velocity relative to the air and a velocity because the air carrying the bird is moving relative to the ground. The velocity of the runner relative to the water or that of the bird relative to the ground is found by adding the two velocities together to give the *net*, or *resultant, velocity*. Let's consider how two velocities (or two vectors of any kind) are combined in *vector addition*.

When adding two velocities, you represent each as an arrow with its length proportional to the magnitude of the velocity—the speed. For the runner on the ship, the arrow representing the ship's velocity is $2\frac{1}{2}$ times as long as the arrow representing the runner's velocity because the two speeds are 20 mph and 8 mph, respectively. Each arrow can be moved around for convenience, provided its length and its direction are not altered. Any such change would make it a different vector. The procedure for adding two vectors is as follows.

Two vectors are added by representing them as arrows and then positioning one arrow so its tip is at the tail of the other. A new arrow drawn from the tail of the first arrow to the tip of the second is the arrow representing the resultant vector—the sum of the two vectors.

Figure 1.11 (a) The resultant velocity of a runner on the deck of a ship is found by adding the runner's velocity and the ship's velocity. The result is 28 mph forward. (b) Using the same procedure when the runner is headed toward the rear of the ship, the resultant velocity is 12 mph forward.

Figure 1.11 shows this for the runner on the deck of the ship. In Figure 1.11a, the runner is running forward in the direction of the ship's motion, so the two arrows are parallel. When the arrows are positioned "tip to tail," the resultant velocity vector is parallel to the others, and its magnitude—the speed—is 28 mph (8 mph + 20 mph). In Figure 1.11b, the runner is running toward the rear of the ship, so the arrows are in opposite directions. The resultant velocity is again parallel to the ship's velocity, but its magnitude is 12 mph (20 mph − 8 mph).

Vector addition is done the same way when the two vectors are not along the same line. **Figure 1.12a** shows a bird with velocity 8 m/s north in the air while the air itself has velocity 6 m/s east. The bird's velocity observed by someone on the ground, **Figure 1.12b**, is the sum of these two velocities. We determine this by placing the two arrows representing the velocities tip to tail as before and drawing an arrow from the tail of the first to the tip of the second (**Figure 1.12c** and **d**). The direction of the resultant velocity is toward the northeast. Watch for this when you see a bird flying on a windy day: often the direction the bird is moving is not the same as the direction its body is pointed.

What about the magnitude of the resultant velocity? It is not simply 8 + 6 or 8 − 6, because the two velocities are not parallel. With the numbers chosen for this example, the magnitude of the resultant velocity—the bird's speed—is

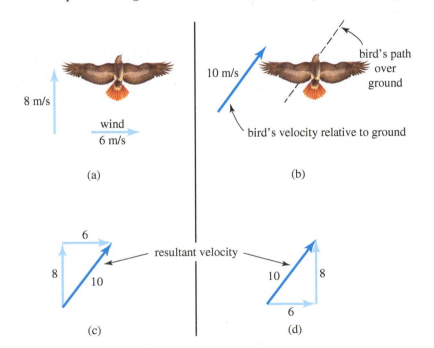

Figure 1.12 The velocity of a bird relative to the ground (b) is the vector sum of its velocity relative to the air and the velocity of the air (wind). Panels (c) and (d) show that the vectors can be added two different ways, but the resultant is the same vector.

Figure 1.13 Other examples of vector addition. The bird has the same speed and direction in the air, but the wind direction is different.

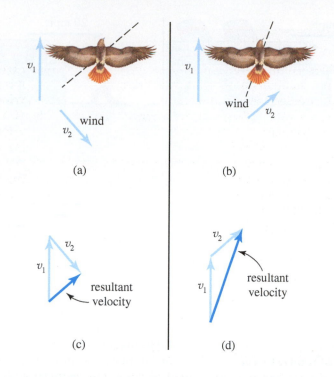

(a)　　　　(b)

(c)　　　　(d)

10 m/s. If you draw the two original arrows with correct relative lengths and then measure the length of the resultant arrow, it will be $\frac{5}{4}$ times the length of the arrow representing the 8 m/s vector. Then 8 m/s times $\frac{5}{4}$ equals 10 m/s.

Vector addition is performed in the same manner, no matter what the directions of the vectors. **Figure 1.13** shows two other examples of a bird flying with different wind directions. The magnitudes of the resultants in these cases are best determined by measuring the lengths of the arrows. There are many other situations in which a body's net velocity is the sum of two (or more) velocities (for example, a swimmer or boat crossing a river). Displacement vectors are added in the same fashion. If you walk 10 meters south, then 10 meters west, your net displacement is 14.1 meters southwest.

The process of vector addition can be "turned around." Any vector can be thought of as the sum of two other vectors, called *components* of the vector. When we observe the bird's single velocity in Figure 1.12b, we would likely realize that the bird has two velocities that have been added. Even when a moving body only has one "true" velocity, it may be convenient to think of it as two velocities that have been added together. For example, a soccer player running southeast across a field can be thought of as going south with one velocity and east with another velocity at the same time (**Figure 1.14**). A car going down a long hill has one velocity component that is horizontal and another that is vertical (downward).

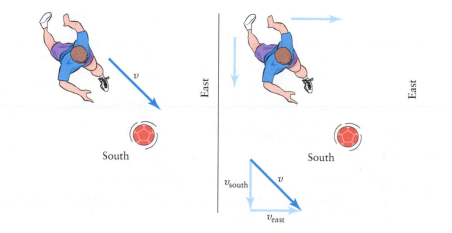

Figure 1.14 A soccer player running southeast can be thought of as having a velocity toward the east and a velocity toward the south at the same time. When these two velocities, called *components*, are added together, the resultant is the original velocity.

Learning Check

1. For objects moving with constant speed,
   (a) the instantaneous speed is always zero.
   (b) the average speed is always zero.
   (c) the instantaneous speed and average speed are always equal.
   (d) None of the above.
2. The highest speed that can be observed in Nature is
   (a) the speed of light.
   (b) the speed of sound.
   (c) 1,000,000 m/s.
   (d) There is no limit.
3. A physical quantity that has both a magnitude and a direction associated with it is called a
   _____.

4. (True or False.) As a car is traveling down a road, it is possible for its speed to be changing while its velocity is constant.
5. (Choose the *incorrect* statement.) When two velocities are added, the resultant velocity
   (a) can be parallel to one or both of the original velocities.
   (b) cannot have zero magnitude.
   (c) can have a greater magnitude than either of the original velocities.
   (d) can be perpendicular to one of the original velocities.

ANSWERS: 1. (c) **2.** (a) **3.** vector **4.** False **5.** (b)

1.3 Acceleration

The physical world around us is filled with motion. But think about this for a moment: cars, bicycles, pedestrians, airplanes, trains, and other vehicles all change their speed or direction often. They start, stop, speed up, slow down, and make turns. The velocity of the wind usually changes from moment to moment. Even Earth as it moves around the Sun is constantly changing its direction of motion and its speed, though not by much as reckoned on a daily basis. The main thrust of Chapter 2 is to show how the change in velocity of an object is related to the force acting on it. For these reasons, a very important concept in physics is **acceleration**.

DEFINITION **Acceleration** Rate of change of velocity. The change in velocity divided by the time elapsed.

$$a = \frac{\Delta v}{\Delta t}$$

Physical Quantity	Metric Units	English Units
Acceleration (a)	meter per second² (m/s²)	foot per second² (ft/s²)
		mph per second (mph/s)

Whenever something is speeding up or slowing down, it is undergoing acceleration. As you travel in a car, anytime the speedometer's reading is changing, the car is accelerating. Acceleration is a vector quantity, which means it has both magnitude and direction. Note that the relationship between acceleration and velocity is the same as the relationship between velocity and displacement. Acceleration indicates how rapidly velocity is changing, and velocity indicates how rapidly displacement is changing.

EXAMPLE 1.3 A car accelerates from 20 to 25 m/s in 4 seconds as it passes a truck (**Figure 1.15**). What is its acceleration?

20 m/s

25 m/s

SOLUTION Because the direction of motion is constant, the change in velocity is just the change in speed—the later speed minus the earlier speed.

$$a = \frac{\Delta v}{\Delta t} = \frac{\text{final speed} - \text{initial speed}}{\Delta t}$$

$$= \frac{25 \text{ m/s} - 20 \text{ m/s}}{4 \text{ s}}$$

$$= \frac{5 \text{ m/s}}{4\text{s}}$$

$$= 1.25 \text{ m/s}^2$$

This means that the car's speed increases 1.25 m/s during each second.

When something is slowing down (its speed is decreasing), it is undergoing acceleration. In everyday speech, the word *acceleration* is usually applied when speed is increasing, *deceleration* is used when speed is decreasing, and a change in direction of motion is not referred to as an acceleration. In physics, one word, *acceleration*, describes all three cases because each represents a change in velocity.

EXAMPLE 1.4 After a race, a runner, traveling in fixed positive direction, takes 5 seconds to come to a stop from a speed of 9 m/s. What is the runner's acceleration?

SOLUTION

$$a = \frac{\Delta v}{\Delta t} = \frac{0 \text{ m/s} - 9 \text{ m/s}}{5 \text{ s}}$$

$$= \frac{-9 \text{ m/s}}{5 \text{ s}}$$

$$= -1.8 \text{ m/s}^2$$

The minus sign means the acceleration and the velocity vectors are in opposite directions.

Perhaps the most important example of accelerated motion is that of an object falling freely near Earth's surface (**Figure 1.16**). By an object "falling freely," we mean that only the force of gravity is acting on it, and we ignore things like air resistance. A rock falling a few meters would satisfy this condition, but a feather would not.

Figure 1.16 Freely falling Sara has constant acceleration.

Table 1.3 Some Accelerations of Interest

Description	Acceleration	
	(m/s)²	(g)
Freely falling body (on the Moon)	1.6	0.16
Freely falling body (on Earth)	9.8	1
Space shuttle (maximum)	29	3
Drag-racing car (average for 0.25 mile)	32	4.2
Highest (sustained) survived by human without injury	245	46.2
Clothes—spin cycle in a typical washing machine	400	41
Tread of a typical car tire at 65 mph	2,800	285
Click beetle jumping	3,920	400
Bullet in a high-powered rifle	2,000,000	200,000
Mean acceleration of a proton in the Large Hadron Collider	1.86×10^9	1.9×10^8

Freely falling bodies move with a constant downward acceleration. The magnitude of this acceleration is represented by the letter g.

$$g = 9.8 \text{ m/s}^2 \qquad \text{(acceleration due to gravity; see Section 2.7)}$$

$$= 32 \text{ ft/s}^2 = 22 \text{ mph/s}$$

The downward velocity of a falling rock increases 22 mph each second that it falls. The letter g is often used as a unit of measure of acceleration. An acceleration of 19.6 m/s² equals 2 g. (Some representative accelerations are given in **Table 1.3**.)

> ### ▶ Physics To Go 1.4
>
> It's easy to compute the acceleration of a car (in g's) as it speeds up or slows down. Simply time how long (in seconds) it takes for the speed to change by 22 mph. Do this when you are a passenger and can see the speedometer. Try it when the car you're riding in is passing another vehicle on the highway. For example, it may take 4 seconds for a car to speed up from 20 mph to 42 mph. How do you find the acceleration in g's using these numbers? (Hint: If it took 1 second for the speed to increase by 22 mph, the acceleration would be 22 mph/s = 1 g.)

Another way to think of acceleration is in terms of the changes in the arrows drawn to represent velocity. When a car accelerates from one speed to a higher speed, the *length* of the arrow used to represent the velocity *increases* (**Figure 1.17**). If we place the two arrows side by side, the change in velocity can be represented by a third arrow, Δv, drawn from the tip of the first arrow to the tip of the second. The later velocity equals the initial velocity plus the change in velocity. The original arrow plus the arrow representing Δv equals the later arrow. The acceleration vector is this change in velocity divided by the time. Note that the acceleration vector points forward. Increasing the velocity requires a forward acceleration.

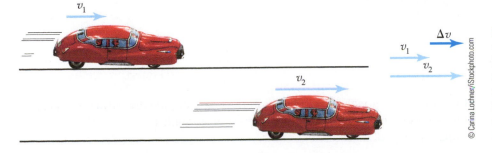

Figure 1.17 As a car accelerates while moving in a straight line, the length of the arrow representing its velocity increases. The arrow marked Δv represents the change in velocity from v_1 to v_2 and indicates the direction of the acceleration—forward.

© Carina Lochner/iStockphoto.com

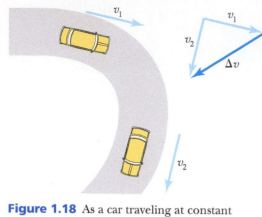

Figure 1.18 As a car traveling at constant speed rounds a curve, it undergoes centripetal acceleration. Here the arrow representing the car's velocity changes direction. Because the change in velocity, Δv, is directed toward the center of the curve, so is the acceleration.

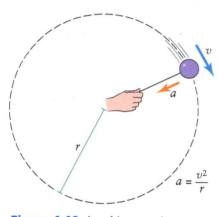

Figure 1.19 An object moving along a circular path is accelerated because its direction of motion, and therefore its velocity, are changing. Its centripetal acceleration equals the square of its speed, divided by the radius of its path. Doubling its speed would quadruple its acceleration. Doubling the radius of its path would halve the acceleration.

Figure 1.20 Car on a cloverleaf with a 20-meter radius.

1.3a Centripetal Acceleration

The concept of acceleration includes changes in *direction* of motion as well as changes in speed. A car going around a curve and a billiard ball bouncing off a cushion are accelerated, even if their speeds do not change.

We can once again use arrows to indicate the change in velocity. **Figure 1.18** shows a car at two different times as it goes around a curve. If we place the two arrows representing the car's velocity at each time tail to tail, we see that the direction of the arrow has changed. The change in velocity, Δv, is represented by the arrow drawn from the tip of v_1 to the tip of v_2. In other words, the original arrow plus the arrow representing Δv equals the later arrow, just like in straight-line acceleration (Figure 1.17).

Note the direction of the change in velocity Δv: it is pointed toward the center of the curve. Because the acceleration is in the same direction as Δv, it, too, is directed toward the center. For this reason, the acceleration of an object moving in a circular path is called **centripetal acceleration** (for "center-seeking"). The centripetal acceleration is always perpendicular to the object's instantaneous velocity.

So we know the direction of the acceleration of a body moving in a circular path, but what about its magnitude? The faster the body is moving, the more rapidly the direction of its velocity is changing. Consequently, the magnitude of the centripetal acceleration depends on the speed v. It also depends on the radius r of the curve (**Figure 1.19**). A larger radius means the path is not as sharply curved, so the velocity changes more slowly and the acceleration is smaller. The actual equation for the size of the acceleration is

$$a = \frac{v^2}{r} \quad \text{(centripetal acceleration)}$$

We can also describe the relationship between a, v, and r by saying that the *acceleration is proportional to the square of the speed*:

$$a \propto v^2$$

and the *acceleration is inversely proportional* to the *radius r*.

$$a \propto \frac{1}{r}$$

This means that when the speed is doubled, the acceleration becomes four times as large. If the radius is doubled, the acceleration becomes one-half as large.

EXAMPLE 1.5 Let's estimate the acceleration of a car as it goes around a curve. The radius of a segment of a typical cloverleaf highway interchange is 20 meters, and a car might take the curve with a constant speed of 10 m/s (about 22 mph; **Figure 1.20**).

SOLUTION Because the motion is circular,

$$a = \frac{v^2}{r} = \frac{(10 \text{ m/s})^2}{20 \text{ m}}$$

$$= \frac{100 \text{ m}^2/\text{s}^2}{20 \text{ m}} = 5 \text{ m/s}^2$$

If the car could go 20 m/s and stay on the road (highly doubtful), its acceleration would be four times as large: 20 m/s² or about 2 *g*.

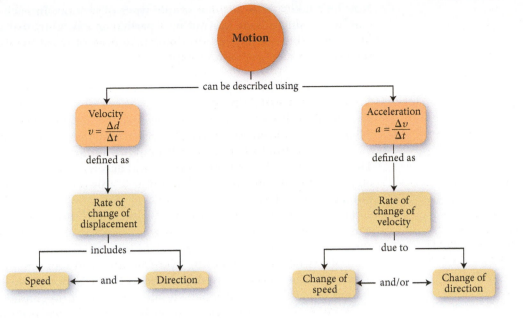

Many people have difficulty accepting the idea of centripetal acceleration, or they don't see how changing the direction of motion of a body is the same kind of thing as changing its speed. But some common experiences (like groceries in the back of your SUV spilling out of their bags when you make a sharp turn on the drive home from the supermarket) show that it is. Let's say that you are riding in a bus and that a book is resting on the slick seat next to you. The book will slide over the seat in response to the bus's acceleration. As the bus speeds up, the book slides backward. As the bus slows down, the book slides forward. In both cases, the reaction of the book is to move in the *direction opposite* the bus's acceleration. What happens when the bus goes around a curve? The book slides toward the *outside* of the curve, showing that the bus is accelerating toward the *inside* of the curve—a centripetal acceleration.

The cornering ability of a car is often measured by the maximum centripetal acceleration it can have when it rounds a curve. Automotive magazines often give the "cornering acceleration" or "lateral acceleration" of a car they are evaluating. A typical value for a sports car is 0.85 *g*, which is equal to 8.33 m/s².

Concept Map 1.1 summarizes the concepts of velocity and acceleration.

■ Learning Check

1. When an object's velocity is changing, we say that the object is _____.
2. A freely falling body experiences
   (a) constant velocity.
   (b) centripetal acceleration.
   (c) constant acceleration.
   (d) steadily increasing acceleration.
3. (True or False.) The acceleration of a slow-moving object can be greater than the acceleration of a fast-moving object.

4. When something moves in a circle with constant speed,
   (a) its acceleration is perpendicular to its velocity.
   (b) its acceleration is zero.
   (c) its velocity is constant.
   (d) its acceleration is parallel to its velocity.
5. An acceleration of 2.5 *g* is equivalent to _____ m/s².

ANSWERS: 1. accelerating **2.** (c) **3.** True **4.** (a) **5.** 24.5

1.4 Simple Types of Motion

Now let's take a look at some simple types of motion. In each example, we'll consider a single body moving in a particular way. Our goal is to show how distance and speed depend on time. Take note of the different ways that the relationships can be shown or expressed.

1.4a Constant Velocity

The simplest situation is one in which *no motion* occurs: a single body resting at a fixed position in space. We characterize this system by saying that the distance of the object, from whatever fixed reference point we choose, is unchanging. This means that the object's velocity (and acceleration) are zero. Because the object's velocity remains the same at 0 m/s, this is a trivial example of motion with constant velocity.

The next simplest case is *uniform motion*. Here the body moves with a constant non-zero velocity; that is, with a constant speed in a fixed direction. An automobile traveling on a straight, flat highway at a constant speed is a good example. A hockey puck sliding over smooth ice almost fits into this category, because it slows down only slightly because of friction. Note that the acceleration equals zero. That much should be obvious: no change in velocity, no acceleration. The interesting relationship here is between distance and time. We can express that relationship in four different ways: with words, mathematics, tables, and graphs.

Let's take the example of a runner traveling at a steady pace of 7 m/s. If you are standing on the side of the road, how does the distance from you to the runner change with time after the runner has gone past you? In words, the distance increases by 7 meters each second: two seconds after the runner passes by, the distance would be 14 meters (**Figure 1.21**). Stated mathematically, the distance in meters equals the time multiplied by the velocity, 7 m/s.

The time is measured in seconds, starting just as the runner passes you. The shorthand way of writing this is the mathematical equation

$$d = 7t \quad (d \text{ in meters, } t \text{ in seconds})$$

This is an example of the general equation that we saw earlier with $v = 7$ m/s:

$$d = vt$$

The same information can also be put in a table of values of time and distance. The accompanying table shows the values of distance (d) at certain times (t). These values all satisfy the equation $d = 7t$. You could make the table longer or shorter by using different time increments.

Time (s)	Distance (m)
0	0
1	7
2	14
3	21
4	28

$v = 7$ m/s

d

v

iStockphoto.com/A-Digit

Figure 1.21 The velocity of a runner is 7 m/s. This means that the runner's distance from a fixed point (you standing to the side of the runner) increases 7 meters each second.

The fourth way to show the relationship between distance and time is to graph the values in the table (**Figure 1.22a**).

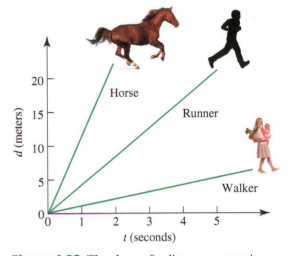

Figure 1.22 (a) Graph of distance versus time when velocity is a constant 7 m/s. (b) Same graph with the slope indicated.

In Figure 1.22b:

rise = 21 m − 7 m = 14 m

run = 3 s − 1 s = 2 s

$$\text{slope} = \frac{\text{rise}}{\text{run}} = \frac{14 \text{ m}}{2 \text{ s}}$$

$$\text{slope} = 7 \text{ m/s} = v$$

The usual practice is to graph distance versus time, which puts distance on the vertical axis. Note that the data points lie on a straight line. Remember this simple rule: when the speed is constant, the graph of distance versus time is a straight line. In general, when one quantity is proportional to another, the graph of the two quantities is a straight line.

An important feature of this graph is its *slope*. The slope of a graph is a measure of its steepness. In particular, the slope is equal to the *rise* between two points on the line divided by the *run* between the points. This is illustrated in **Figure 1.22b**. The rise is a distance, Δd, and the run is a time interval, Δt. So the slope equals Δd divided by Δt, which is also the object's velocity. The slope of a distance-versus-time graph equals the velocity.

The graph for a faster-moving body, a racehorse for instance, would be steeper—it would have a larger slope. The graph of d versus t for a slower object (a person walking) would have a smaller slope (**Figure 1.23**). When an object is standing still (when it has no motion), the graph of d versus t is a flat, horizontal line parallel to the time axis. The slope is zero because the velocity is zero.

Even when the velocity is not constant, the slope of a d versus t graph is still equal to the velocity. In this case, the graph is not a straight line because, as the slope changes (a result of the changing velocity), the graph curves or bends. The graph in **Figure 1.24** represents the motion of a car that starts from a stop sign, drives down a street, and then stops and backs into a parking place. When the car is stopped (the reference point), the graph is flat. The distance is not changing and the velocity is zero. When the car is backing up, the graph is slanted downward. The distance is decreasing, and the velocity is negative.

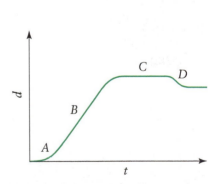

Figure 1.23 The slope of a distance-versus-time graph equals the velocity. The graph for a body with a higher velocity (the horse) is steeper—that is, it has a larger slope. When the velocity is very low, the graph is nearly a horizontal line.

1.4b Constant Acceleration

The next example of simple motion is *constant acceleration in a straight line*. This means that the object's velocity is changing at a fixed rate. A freely falling body is a good example. A ball rolling down a straight, inclined plane is another. Often, cars, runners, bicycles, trains, and aircraft have nearly constant acceleration when they are speeding up or slowing down.

Let's use free fall as our example. Assume that a heavy rock is dropped from the top of a building and that we can measure the instantaneous velocity of the rock and the distance that it has fallen at any time we choose (**Figure 1.25**). The rock falls with an acceleration equal to g. This is 9.8 m/s², which is equivalent to 22 mph per second. First, consider how the rock's velocity changes. The velocity increases 9.8 m/s, or 22 mph, each second. This means that the rock's velocity

Figure 1.24 Graph of distance versus time for a car with varying velocity. At point A on the graph, the slope starts to increase as the car accelerates. At B, its velocity is constant. The slope decreases to zero at point C when the car is stopped. At D, the car is backing up, so its velocity is negative.

Figure 1.25 A rock falls freely after it is dropped from the top of a building. The distance d is measured relative to the roof ledge. The rock's velocity increases at a steady rate—9.8 m/s each second.

equals the time (in seconds) multiplied by 9.8 (for m/s) or 22 (for mph). Mathematically,

$$v = 9.8t \quad (v \text{ in m/s, } t \text{ in seconds})$$

$$v = 22t \quad (v \text{ in mph, } t \text{ in seconds})$$

The general form of the equation that applies to any object that starts from rest and has a constant acceleration a is

$$v = at \quad (\text{when acceleration is constant})$$

So in constant acceleration, the velocity is proportional to the time. The proportionality constant is the acceleration a. The accompanying table gives the relevant values for this example.

Time	Velocity	
(s)	(m/s)	(mph)
0	0	0
1	9.8	22
2	19.6	44
3	29.4	66
4	39.2	88

The graph of velocity versus time is a straight line (**Figure 1.26**). The slope of a graph of velocity versus time equals the acceleration. This is because the rise is a change in velocity, Δv, and the run is a change in time Δt; so

$$\text{slope} = \frac{\Delta v}{\Delta t} = a$$

The corresponding graph for a body with a smaller acceleration—say, a ball rolling down a ramp—would have a smaller slope.

The similarity between the graphs in Figure 1.26 and the graphs in Figure 1.22 for uniform motion is obvious. But keep in mind that the graph of *distance* versus time is a straight line for uniform motion. Here it is the graph of *velocity* versus time that exhibits a linear behavior for uniformly accelerated motion.

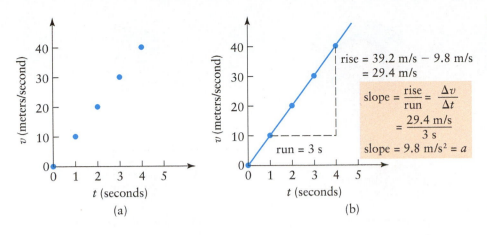

Figure 1.26 (a) Graph of velocity versus time for a freely falling body. (b) The same graph with the slope indicated.

rise = 39.2 m/s − 9.8 m/s = 29.4 m/s

run = 3 s

$$\text{slope} = \frac{\text{rise}}{\text{run}} = \frac{\Delta v}{\Delta t}$$

$$= \frac{29.4 \text{ m/s}}{3 \text{ s}}$$

$$\text{slope} = 9.8 \text{ m/s}^2 = a$$

If an object is already moving with velocity v_{initial} and then undergoes constant acceleration, its velocity after time t is $v = v_{\text{initial}} + at$. Clearly, in the case of an object starting from rest, v_{initial} is zero, and we recover the equation: $v = 0 + at = at$.

What is the relationship between distance and time when the acceleration is constant? It is a bit more complicated, as expected. **Figure 1.27** shows that the distance a falling body travels during each successive time interval grows larger as it falls. Because the velocity is continually changing, the distance equals the average velocity multiplied by the time. What is the average velocity? The object starts with velocity equal to zero; after accelerating for a time t, its velocity is at. Its average velocity is

$$\text{average velocity} = \frac{0 + at}{2} = \frac{1}{2} at$$

(If you take two quizzes and get 0 on the first and 8 on the second, your average grade is 4—one-half of 0 plus 8.)

The distance traveled is this average velocity times the time:

$$d = \text{average velocity} \times t = \frac{1}{2} at \times t$$

$$d = \frac{1}{2} at^2 \qquad \text{(when initial velocity is zero and acceleration is constant)}$$

In the case of a falling body, the acceleration is 9.8 m/s²; therefore,

$$d = \frac{1}{2} at^2 = \frac{1}{2} \times 9.8 \times t^2$$

$$= 4.9 t^2 \qquad (d \text{ in meters}, t \text{ in seconds})$$

So in the case of constant acceleration, the distance is proportional to the square of the time. The constant of proportionality is one-half the acceleration.

For any object that is already moving with velocity v_{initial} and then undergoes constant acceleration, the average velocity after time t is

$$v_{\text{average}} = \frac{v_{\text{initial}} + (v_{\text{initial}} + at)}{2} = v_{\text{initial}} + \frac{1}{2} at$$

Therefore,

$$d = v_{\text{average}} \, t = v_{\text{initial}} \, t + \frac{1}{2} at^2$$

A table of distance values for a falling body is shown in **Figure 1.28**. This distance increases rapidly. The graph of distance versus time curves upward (Figure 1.28b). This is because the velocity of the body is increasing with time, and the slope of this graph equals the velocity. **Table 1.4** and **Concept Map 1.2** summarize these three simple types of motion.

Richard Megna/Fundamental Photographs

Figure 1.27 A falling ball photographed with a flashing strobe light. Each image shows where the ball was at the instant when the light flashed. The images are close together near the top because the ball is moving more slowly at first. As it falls, it picks up speed and thus travels farther between flashes.

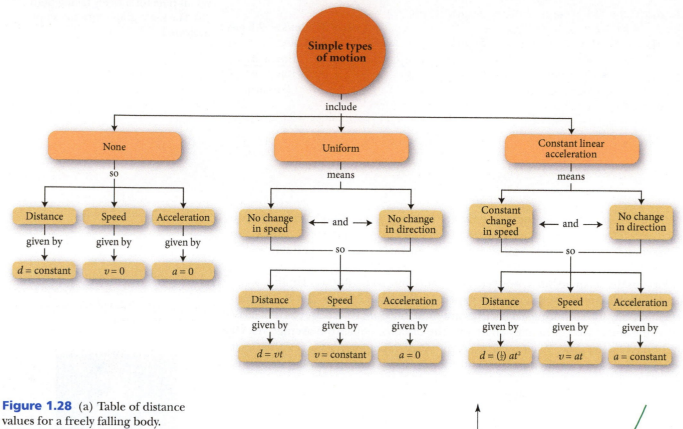

Figure 1.28 (a) Table of distance values for a freely falling body. (b) Graph of distance versus time for a freely falling body. The slope continually increases, indicating that the velocity increases throughout the motion.

Time (s)	Distance (m)
0	0
1	4.9
2	19.6
3	44.1
4	78.4

(a)

(b)

Table 1.4 Summary of Examples of Motion

Type of Motion	Behavior of Physical Quantities	Equations
Stationary object	Distance constant	$d = $ constant
	Velocity zero	$v = 0$
	Acceleration zero	$a = 0$
Uniform motion†	Distance proportional to time	$d = vt$
	Velocity constant	$v = $ constant
	Acceleration zero	$a = 0$
Uniform acceleration† (from rest)	Distance proportional to time squared	$d = \frac{1}{2}at^2$
	Velocity proportional to time	$v = at$
	Acceleration constant	$a = $ constant

†Distance measured from object's initial location.

1.4c More on Kinematics Graphs

Rarely does the acceleration of an object stay constant for long. As a falling body picks up speed, air resistance causes its acceleration to decrease (more on this in Section 2.5). When a car is accelerated from a stop, its acceleration usually decreases, particularly when the transmission is shifted into a higher gear. **Figure 1.29** shows the velocity of a car as it accelerates from 0 to 80 mph. Note that acceleration steadily decreases (the slope gets smaller). During the short time the transmission is shifted, the acceleration is zero.

Although graphs may be a good way to show the relationships between physical quantities, mathematics is the most precise way to express them, and this approach is not limited to two (or sometimes three) variables, as are graphs. However, math is inherently abstract and also somewhat like a language.

In this book, we will "translate" most of the mathematics we use into statements and also express many of the relationships graphically. But we will also retain some of the "original" language, mathematics, and show that even an understanding of high school math allows you to solve some interesting physics problems. The application of physics in our society by engineers and others requires mathematics. The Burj Khalifa, the world's tallest man-made structure, could not have been designed and built through the use of words only (**Figure 1.30**).

When you see a graph, the first thing you should do is take careful note of which quantities are plotted. You should notice that some of the graphs for the examples of motion are straight lines. When the graph shows *distance* versus time, a straight line means the *velocity* is constant. But when it shows *velocity* versus time, a straight line means that the *acceleration* is constant. Even though the shapes of the two graphs are similar, because the quantities plotted are different for each graph, they represent very different situations. To a business executive, a rising graph showing profits will inspire joy, whereas a rising graph showing expenses will cause concern.

The most important thing to look for in the shape of a graph is a trend. Is the slope always positive or always negative—is the graph continually going up or going down? If the slope changes, consider what that signifies. If it is a graph of velocity versus time, for example, choose a particular time and note whether the velocity is increasing, decreasing, or remaining the same in the vicinity of that point.

Let's look at an application of the kinds of graphs we've been using. During a karate demonstration, a concrete block is broken by a person's fist. **Figure 1.31a** shows the distance of the hand above the block versus time as measured using high-speed photography. The fist travels downward at nearly a constant speed until it contacts the block, at about 6 milliseconds. This causes a large acceleration as the fist is brought to a sudden stop. **Figure 1.31b**

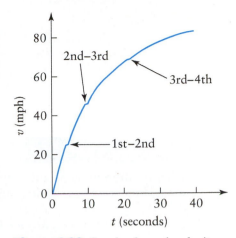

Figure 1.29 Graph of a car's velocity versus time as the car is accelerated. The notches in the curve occur at transmission shifts when the engine is momentarily disconnected from the drivetrain. Between the shifts, the acceleration (slope) decreases as the velocity increases.

Figure 1.30 Burj Khalifa is a megatall skyscraper in Dubai, United Arab Emirates. It is the tallest man-made structure in the world, standing at 830 m.

Figure 1.31 (a) Graph of the position of a fist versus time during a karate blow. Contact is made at about 6 milliseconds. (b) Graph of the velocity of the fist versus time. At contact, the fist accelerates rapidly. [From S. R. Wilke, R. E. McNair, M. S. Feld. "The Physics of Karate." *American Journal of Physics* vol. 51, pp. 783–790. Copyright © 1983 by American Association of Physics Teachers. Reprinted by permission.] (c) A karate demonstration showing how a concrete block can be broken by a person's fist.

shows the velocity of the fist, that is, the slope of the distance versus time graph at each time. Contact with the concrete is indicated by the steep part of the graph as the velocity goes to zero. If we take the slope of this segment of the velocity graph, we find that the magnitude of the acceleration of the fist at that moment is about 3,500 m/s², or 360 g (ouch!). What happened at about 25 milliseconds?

■ Learning Check

1. What is different about the graph of distance versus time for a fast-moving object compared to that of a slow-moving object?

2. (True or False.) A rock dropped from the top of a tall building travels the same distance during each 1-second interval that it falls.

3. If the graph of a car's velocity versus time curves upward, we can conclude its acceleration is _____.

4. Graphs of distance versus time and velocity versus time are useful for analyzing
   (a) the motion of a falling body.
   (b) the performance of a car.
   (c) laboratory measurements of an accelerating object.
   (d) All of the above.

ANSWERS: 1. larger slope **2.** False **3.** increasing **4.** (d)

Profiles in Physics Aristotle vs. Galileo

We finish this chapter with a brief look at how our method of analyzing motion evolved. The first great strides toward the development of exact sciences were made by the ancient Greeks. In the sixth century B.C.E., Pythagoras showed that numbers, often regarded merely as mental abstractions, are related to the natural world and even to human perception. His most famous discoveries were in geometry and acoustics. Other ancients, before the year 200 B.C.E., accomplished accurate measurements of the diameter of Earth, the Moon, and the Moon's orbit.

Aristotle

The philosopher Aristotle (384–322 B.C.E.; **Figure 1.32**) is regarded as the first person to attempt physics. In fact, he gave *physics* its name.

His ideas about physics were influenced a great deal by the methodology that he used in logic, biology, and other areas. Aristotle's analysis of motion, the first mechanics, rested on the distinction between *natural motion*, like that of a falling body, and *unnatural motion*, like that of a cart being pulled down a road.

According to Aristotle, heavy objects fall because they are seeking their natural place: Earth's center. In his model, the speed of a falling body is constant and depends on its weight and the medium through which it falls. Heavy objects fall faster than lighter ones. Also, a rock drops faster through air than it does through water. This analysis is only partly correct: it applies only after an object has been falling for a period of time and friction has become important. Speed first increases at a fixed rate—9.8 m/s each second, as we have seen. A marble and a rock dropped at the same time will build up speed together and hit the ground at the same time (**Figure 1.33**). Only after a rock has fallen a much greater distance will air resistance cause its speed to level off at some constant value. This final speed, called the *terminal speed*, depends as much on the shape and size of the object as it does on the weight. The same factors affect the terminal speeds of bodies falling through water, but such speeds are much slower (as

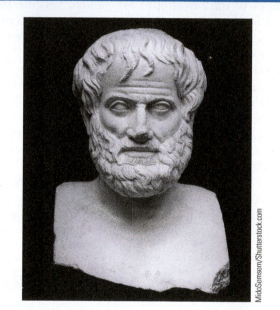

Figure 1.32 Bust of Aristotle.

Aristotle pointed out) and are reached much more quickly. We will take a closer look at this in Section 2.5.

Aristotle's description of the motion of falling bodies fits only after their motions are dominated by air resistance. In a similar way, his analysis of "unnatural motion" indirectly includes friction as the dominating influence. A rock's "natural" tendency is to remain at rest on the ground or to fall toward Earth's center if dropped. Making the rock move horizontally by pushing it is unnatural and requires some external agent or force. Aristotle's view that a force is always required to maintain horizontal motion fits well whenever there is a great deal of friction but not so well in cases such as those of a rock thrown horizontally or of a smooth, heavy ball rolling over a flat, hard surface. In these situations,

Figure 1.33 Strobe photograph of two falling bodies. Even though one is much heavier than the other, they have the same acceleration.

motion can continue for quite some time with no force acting to maintain it. Because of Aristotle's overwhelming reputation as a scholar, zealous supporters of his ideas allowed his theories to dominate physics, almost without question, for about 2,000 years.

Galileo

Galileo Galilei (1564–1642; **Figure 1.34**) lived during one of the most fruitful periods of human civilization—the Renaissance. Galileo made many important discoveries in mechanics and astronomy, some of which were disputed bitterly because they contradicted accepted views passed down from Aristotle's time. Galileo's strong support of the heliocentric (Sun-centered) model of the solar system and other factors led to his being placed on trial by leaders of the Inquisition. Though he was censured and forced to publicly recant his support of the heliocentric theory, Galileo's work gained wide acceptance, and he is considered one of the founders of modern physical science.

Galileo was one of the first to rely on observation and experimentation. In 1583, he noticed that a lamp hanging from the ceiling of the cathedral at Pisa would swing back and forth with a constant period, even though the length of the arc of its motion decreased. His discovery, that the frequency of a pendulum depends only on its length, is the basis for the pendulum clock.

Another important contribution by Galileo was his insistence that scientific terms, statements, and analyses be logically consistent. He sought to establish a "scientific method" and believed that, as a first step, the language of science should be unambiguous. To Galileo, mathematics was necessary to help accomplish this task.

Although his astronomical discoveries brought Galileo most of his fame (and controversy), his conclusions about motion are of interest

Figure 1.34 Galileo Galilei.

here. Galileo recognized that the simple rules that govern motion can be hidden by other phenomena. He realized that the two things that have the largest effect on objects moving on Earth are gravity and friction. More importantly, he reasoned that friction is often complicated and unpredictable and that it is best to first focus on systems in which friction doesn't dominate completely.

The physical system that Galileo used to analyze uniform motion and uniformly accelerated motion was a smooth, heavy ball rolling on a smooth, straight surface that can be tilted. If the surface is tilted (an inclined plane), the ball's speed will increase as it rolls down. The opposite occurs when the ball is initially rolling up the plane—its speed decreases. But what if the surface is level? Logically, the ball's speed should neither increase nor decrease but should stay the same (**Figure 1.35**). So Galileo reasoned correctly that, if there is no friction, an object moving on a level surface will proceed with constant speed.

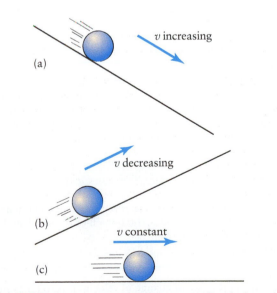

Figure 1.35 (a) As a ball rolls down an inclined plane, its speed increases. (b) As a ball rolls up an inclined plane, its speed decreases. (c) When a ball rolls on a level surface, it should therefore have a constant speed.

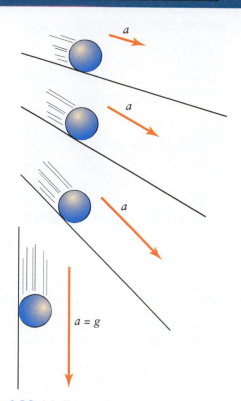

Figure 1.36 A ball is accelerated as it rolls down an inclined plane. Making the plane steeper increases the ball's acceleration. When the plane is vertical, the acceleration equals *g*.

Galileo discovered the law of falling bodies by also using a ball rolling on an inclined plane. He understood that a ball rolling down an inclined plane is also accelerated because of gravity, and he reasoned that tilting the surface more and more steeply would make the ball's motion become closer to that of free fall (**Figure 1.36**). Using the inclined plane and clocks he invented himself, Galileo discovered the correct relationship between distance and time for uniformly accelerated motion.

Legend has it that Galileo once dropped two different-sized objects from the top of the Leaning Tower of Pisa to show that they would hit the ground at nearly the same time. Whether he actually

Figure 1.37 Astronaut David Scott proving that all objects fall at the same rate under the influence of gravity alone.

did this or not is unknown. But, his conclusion about the motion of falling bodies is correct whenever air resistance is negligible. This was illustrated dramatically by astronaut David R. Scott while performing experiments on the Moon. He simultaneously dropped a hammer and a feather. Both fell at the same rate and hit the lunar surface at the same time because there is no air on the Moon to slow the feather (**Figure 1.37**).

Galileo used an approach that is now standard procedure in physics: to understand a real physical process, first consider an idealized system in which complicating factors (like friction) are absent. He promoted the idea of imagining how bodies would move in a perfect vacuum, devoid of air resistance and other forms of friction. Only by understanding a simplified model of reality can one hope to comprehend its complexities and subtleties.

Aristotle and other ancient Greeks initiated the science of physics. Galileo corrected Aristotle's mechanics and established the importance of mathematics and experimentation in physics. But the greatest name in the development of mechanics is that of Isaac Newton, who was born on Christmas Day in the year that Galileo died.

■ **QUESTIONS**
1. To what extent was Aristotle's model of falling bodies correct? How was it wrong?
2. Describe two major discoveries or contributions that Galileo made to science and the study of mechanics.

SUMMARY

» **Distance, time,** and **mass** are the three **fundamental physical quantities** used in mechanics, the branch of physics dealing with the motion of objects.

» **Speed, velocity,** and **acceleration** are derived quantities (based on distance and time) that are used to specify how an object moves.

» **Acceleration,** the rate of change of velocity, is zero for bodies that are moving with a constant speed in a straight line.

» On Earth, freely falling objects have a constant downward acceleration of 9.8 m/s² or 32 ft/s².

» Velocity and acceleration are **vectors,** meaning that they have direction as well as magnitude.

» A body moving along a circular path undergoes **centripetal acceleration** because the direction of its velocity is changing.

» Aristotle was the first to attempt a scientific analysis of motion. His conclusions about "natural motion" and "unnatural motion," although flawed, were still on the right track.

» Galileo made several important discoveries in physics and correctly described the motion of freely falling bodies. He was one of the first to use logic, observation, experimentation, and mathematical analysis in his work, all of which are important components of modern scientific methodology.

Equation	Comments	Equation	Comments
Fundamental Equations		*Special-Case Equations*	
$T = \dfrac{1}{f}$	Relates period and frequency	$d = vt$	Distance from starting point when velocity is constant
$f = \dfrac{1}{T}$	Relates frequency and period	$d = \dfrac{1}{2}at^2$	Distance when acceleration is constant and object starts from rest
$v = \dfrac{\Delta d}{\Delta t}$	Definition of velocity or speed	$v = at$	Velocity when acceleration is constant and object starts from rest
$a = \dfrac{\Delta v}{\Delta t}$	Definition of acceleration	$a = g$	Acceleration of freely falling body
		$a = \dfrac{v^2}{r}$	Centripetal acceleration (object moving in a circular path)

MAPPING IT OUT!

1. Consider Concept Map 1.1, which provides an overview of motion. Review Sections 1.2 and 1.3 and make a list of at least five additional concepts you could add to the map. Now consider where to place these concepts on the map. Keep in mind that general, more inclusive concepts go toward the top, whereas more specific ideas and examples go toward the bottom. Place your concepts on the map. Draw the possible connections between your five new concepts and the rest of the map. Remember to use "linking words" to create propositions and to express your ideas clearly. When you've finished updating your map, compare your revised concept map with that of a classmate's. Are they identical? Would you expect them to be?

2. Review the material introduced in Section 1.1 on physical quantities. Based on your understanding of the meaning of the concepts and examples given in this section, complete Concept Map 1.3. Fill in the missing concepts or linking phrases so that the propositions and connections make good physical sense.

CONCEPT MAP 1.3 Mapping It Out! Exercise 2.

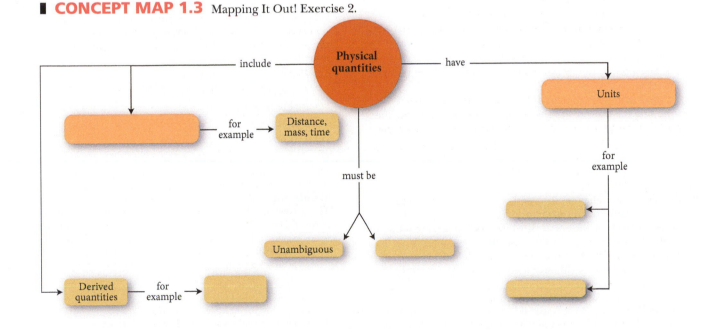

(■ Indicates a review question, which means it requires only a basic understanding of the material to answer. Questions without this designation typically require integrating or extending the concepts presented thus far.)

1. Two rectangular rugs are on display in a showroom. If one rug is twice as long as the other, does this necessarily mean that its area is also twice as large as that of the second? Explain.

2. ■ Explain what a "derived unit" of measure is.

3. A pendulum clock is taken to a repair shop. Its pendulum is replaced by a shorter one that oscillates with a smaller period than the original. What effect, if any, does this have on how the clock runs?

4. ■ What are the "basic" or "fundamental" physical quantities? Why are they called that?

5. A wind is blowing from the north (the air is moving toward the south). When a person is walking toward the north, is the relative speed of the wind that the person senses greater than, the same as, or less than the speed the person senses when not walking? How about when the person is walking toward the south?

6. Scenes in films or television programs sometimes show people jumping off moving trains and having unpleasant encounters with the ground. If someone is on a moving flatbed train car and wishes to jump off, how could the person use the concept of relative speed to make a safer dismount?

7. ■ List the physical quantities identified in this chapter. From which of the fundamental physical quantities is each derived? Which of them are vectors, and which are scalars?

8. ■ What is the distinction between speed and velocity? Describe a situation in which an object's speed is constant but its velocity is not.

9. A ball thrown at a brick wall bounces directly back with the same speed it had when it struck the wall. Has the velocity of the ball changed? Explain.

10. ■ What is "vector addition" and how is it done?

11. Can the resultant of two velocities have zero magnitude? If so, give an example.

12. A swimmer heads for the opposite bank of a river. Make a sketch showing the swimmer's two velocities and the resultant velocity.

13. A basketball player shoots a free throw. Make a sketch showing the basketball's velocity just after the ball leaves the player's hands. Draw in two components of this velocity, one horizontal and one vertical. Repeat the sketch for the instant just before the ball reaches the basket. What is different?

14. ■ What is the relationship between velocity and acceleration?

15. As a stop light changes from red to green, a car starts to cross through the intersection. An instant before it begins to move, its velocity is zero. Must its acceleration at that time also be zero. Why or why not? Explain.

16. ■ How does the velocity of a freely falling body change with time? How does the distance it has fallen change? How about the acceleration?

17. Using concepts and physical quantities discussed in this chapter, explain why it is usually safe for a person standing on the seat of a chair to jump horizontally and land on the floor, but not for a person standing on the roof of a tall building to jump horizontally and land on the ground.

18. ■ What is centripetal acceleration? What is the direction of the centripetal acceleration of a car going around a curve?

19. During 200-meter and 400-meter races, runners must stay in lanes as they go around a curved part of the track. If runners in two different lanes have exactly the same speed, will they also have exactly the same centripetal acceleration as they go around a curve? Explain.

20. As a car goes around a curve, the driver increases its speed. This means the car has two accelerations. What are the directions of these two accelerations?

21. The following are speeds and headings displayed on a GPS receiver. (Heading gives the direction of motion based on north = 0°, east = 90°, south = 180°, etc.) In each case, indicate whether the receiver was accelerating during the time between the displays and, if it was, describe in what way the receiver was accelerating.

   (a) Initially: 60 mph, 70°; 5 seconds later: 50 mph, 70°.

   (b) Initially: 50 mph, 70°; 5 seconds later: 70 mph, 70°.

   (c) Initially: 60 mph, 70°; 5 seconds later: 60 mph, 90°.

22. In Figure 1.19, arrows show the directions of the velocity and the acceleration of a ball moving in a circle. Make a similar sketch showing these directions for a car (a) speeding up from a stop sign and (b) slowing down as it approaches a stop sign.

23. ■ If a ball is thrown straight up into the air, what is its acceleration as it moves upward? What is its acceleration when it reaches its highest point and is stopped at an instant?

24. ■ What does the slope of a distance-versus-time graph represent physically?

25. Sketch a graph of velocity versus time for the motion illustrated in Figure 1.24. Indicate what the car's acceleration is at different times.

26. The following data describe different situations where a person is walking or running through a train car while the train is in motion. The speed and direction of motion of the person (P) and the train (T) are given in each case, with east being reckoned as positive, west as negative. An observer is standing beside the track watching the train and its occupant pass by. Rank these situations on the basis of the velocity of the walkers/runners with respect to the stationary observer outside the train from largest to smallest. If any of the walkers/runners have the same velocity, give them the same ranking.

   (a) P: 2 m/s, west; T: 20 m/s, east

   (b) P: 6 m/s, east; T: 30 m/s, west

   (c) P: 4 m/s, west; T: 30 m/s, west

   (d) P: 10 m/s, west; T: 10 m/s, east

   (e) P: 2 m/s, east; T: 20 m/s, west

   (f) P: 4 m/s, east; T: 25 m/s, east

27. Eight arrows are successively shot straight up into the air. All the arrows have the same size and shape, but are made of different materials and so have different masses (M). The arrows also have different upward speeds (V) as they leave the bow. The data for the eight arrows is shown here. Ignoring any effects associated with air resistance, rank the arrows according to the maximum height that each achieves from greatest to smallest.

   Arrow A: M = 0.075 kg; V = 16 m/s
   Arrow B: M = 0.180 kg; V = 12 m/s
   Arrow C: M = 0.100 kg; V = 18 m/s
   Arrow D: M = 0.075 kg; V = 12 m/s
   Arrow E: M = 0.120 kg; V = 10 m/s
   Arrow F: M = 0.090 kg; V = 16 m/s
   Arrow G: M = 0.180 kg; V = 10 m/s
   Arrow H: M = 0.050 kg; V = 8 m/s

PROBLEMS

1. A yacht is 20 m long. Express this length in feet.
2. Express your height in (a) meters and (b) centimeters.
3. A convenient time unit for short time intervals is the *millisecond*. Express 0.0452 s in milliseconds.
4. One mile is equal to 1,609 m. Express this distance in kilometers and in centimeters.
5. A hypnotist's watch hanging from a chain swings back and forth every 0.8 s. What is the frequency of its oscillation?
6. The quartz crystal used in an electric watch vibrates with a frequency of 32,768 Hz. What is the period of the crystal's motion?
7. A passenger jet flies from one airport to another 1,200 miles away in 2.5 h. Find its average speed.
8. At the 2006 Winter Olympics in Torino, Italy, U.S. speed skater Apolo Ohno took the gold medal for the 500-m sprint by completing the course in 41.935 s. What was his average speed for this event?
9. A runner in a marathon passes the 5-mile mark at 1 o'clock and the 20-mile mark at 3 o'clock. What is the runner's average speed during this time period?
10. The Moon is about 3.8×10^8 m from Earth. Traveling at the speed of light, 3.0×10^8 m/s, how long does it take a laser beam to go from Earth to the Moon and back again? The same physics was responsible for the noticeable delay in communications signals between lunar astronauts and controllers at the Houston Space Flight Center.
11. In Figure 1.13, assume that $v_1 = 8$ m/s and $v_2 = 6$ m/s. Use a ruler to estimate the magnitudes of the resultant velocities in (c) and (d).
12. On a day when the wind is blowing toward the south at 3 m/s, a runner jogs west at 4 m/s. What is the velocity (speed and direction) of the air relative to the runner?
13. How far does a car going 25 m/s travel in 5 s? How far would a jet going 250 m/s travel in 5 s?
14. A long-distance runner has an average speed of 4 m/s during a race. How far does the runner travel in 20 min?
15. Draw an accurate graph showing distance versus time for the car in Problem 13. What is the slope of the graph?
16. The graph in **Figure 1.38** shows the distance versus time for an elevator as it moves up and down in a building. Compute the elevator's velocity at the times marked a, b, and c.

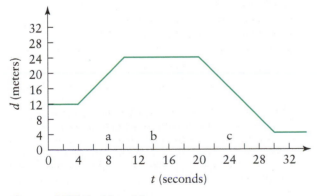

Figure 1.38 Problem 16.

17. A high-performance sports car can go from 0 to 100 mph (44.7 m/s) in 7.9 s.
    (a) What is the car's average acceleration?
    (b) The same car can come to a complete stop from 30 m/s in 3.2 s. What is its average acceleration?

18. As a baseball is being thrown, it goes from 0 to 40 m/s in 0.15 s.
    (a) What is the acceleration of the baseball?
    (b) What is the acceleration in g's?
19. A child attaches a rubber ball to a string and whirls it around in a circle overhead. If the string is 0.5 m long and the ball's speed is 10 m/s, what is the ball's centripetal acceleration?
20. A child sits on the edge of a spinning merry-go-round that has a radius of 1.5 m. The child's speed is 2 m/s. What is the child's acceleration?
21. A runner is going 10 m/s around a curved section of track that has a radius of 35 m. What is the runner's acceleration?
22. During a NASCAR race, a car goes 50 m/s around a curved section of track that has a radius of 250 m. What is the car's acceleration?
23. A rocket accelerates from rest at a rate of 60 m/s².
    (a) What is its speed after it accelerates for 40 s?
    (b) How long does it take for the rocket to reach a speed of 7,500 m/s?
24. Initially stationary, a train has a constant acceleration of 0.5 m/s².
    (a) What is its speed after 15 s?
    (b) What is the total time required for the train to reach a speed of 25 m/s?
25. (a) Draw an accurate graph of the speed versus time for the train in Problem 24.
    (b) Draw an accurate graph of the distance versus time for the train in Problem 24.
26. Draw an accurate graph of the velocity versus time for the elevator in Problem 16.
27. A skydiver jumps out of a helicopter and falls freely for 3 s before opening the parachute.
    (a) What is the skydiver's downward velocity when the parachute opens?
    (b) How far below the helicopter is the skydiver when the parachute opens?
28. A rock is dropped off the side of a bridge and hits the water below 2 s later.
    (a) What was the rock's velocity when it hit the water?
    (b) What was the rock's average velocity as it fell?
    (c) What is the height of the bridge above the water?
29. The roller coaster in **Figure 1.39** starts at the top of a straight track that is inclined 30° with the horizontal. This causes it to accelerate at a rate of 4.9 m/s² (1/2 g).
    (a) What is the roller coaster's speed after 3 s?
    (b) How far does it travel during that time?

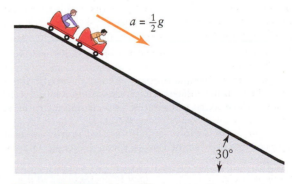

Figure 1.39 Problem 29.

30. During takeoff, an airplane goes from 0 to 50 m/s in 8 s.
    (a) What is its acceleration?
    (b) How fast is it going after 5 s?
    (c) How far has it traveled by the time it reaches 50 m/s?

31. The graph in **Figure 1.40** shows the velocity versus time for a bullet as it is fired from a gun, travels a short distance, and enters a block of wood. Compute the acceleration at the times marked a, b, and c.

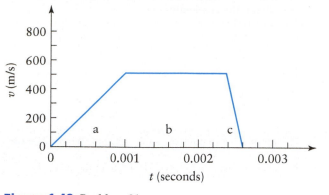

Figure 1.40 Problem 31.

32. A bungee jumper falls for 1.3 s before the bungee cord begins to stretch. Until the jumper has bounced back up to this height, the elastic cord causes the jumper to have an average acceleration upward of 4 m/s².
    (a) How fast is the jumper going when the bungee cord begins to stretch?
    (b) How far below the diving platform is the jumper at that moment?
    (c) How long after the bungee cord begins to stretch does the jumper reach the low point of the drop?
    (d) How far below the diving platform is the jumper at the instant the speed is zero?

33. A drag-racing car goes from 0 to 300 mph in 5 s. What is its average acceleration in *g*'s?

CHALLENGES

1. Most cars can decrease their speed much more rapidly than they can increase it (see Problem 17). What factors contribute to cause this?

2. The Moon's mass is 7.35×10^{22} kg, and it moves in a nearly circular orbit with radius 3.84×10^8 m. The period of its motion is 27.3 days. Use this information to determine the Moon's (a) orbital speed and (b) its acceleration.

3. A car is stopped at a red light. When the light turns green, the car begins to accelerate with a constant acceleration of 4.0 m/s². Just as the car starts moving, it is passed by a small truck traveling uniformly with a speed of 12 m/s.
    (a) If each vehicle continues to move as described, find the distance traveled by each after 1 s, 3 s, 5 s, and 7 s. Assume $t = 0$ just as the truck passes the car.
    (b) Plot the data for each vehicle on the same distance-versus-time graph and determine the approximate time at which the car overtakes the truck. Can you see a way to solve this problem algebraically to find the precise time using only the kinematic equations?

4. A sports car is advertised to have a maximum cornering acceleration of 0.85 *g*.
    (a) What is the maximum speed that the car can go around a curve with a 100-m radius?
    (b) What is its maximum speed for a 50-m radius curve?
    (c) If wet pavement reduces its maximum cornering acceleration to 0.6 *g*, what do the answers to (a) and (b) become?

5. A spacecraft lands on a newly discovered planet orbiting the star 51 Pegasi. To measure the acceleration from gravity on the planet, an astronaut drops a rock from a height of 2 m. A precision timer indicates that it takes the rock 0.71 s to fall to the surface. What is the acceleration from gravity on this planet?

6. When an object is thrown straight *upward*, gravity causes it to decelerate at a rate of 1 *g*—its speed decreases 9.8 m/s or 22 mph each second.
    (a) Explain how we can use the equations, tables, and graphs for a freely falling body to analyze this motion.
    (b) A baseball is thrown vertically upward at a speed of 39.2 m/s (88 mph). How much time elapses before it reaches its highest point, and how high above the ground does it get?

7. For an object starting at rest with a constant acceleration, derive the equation that relates its speed to the distance it has traveled. In other words, eliminate time in the two equations relating d to t and v to t. Test the equation by showing that an object reaches a final speed of 9.8 m/s if it is dropped from a height of 4.9 m.

8. A race car starts from rest on a circular track with radius 100 m and begins to increase its speed by 5 m/s each second. At what point in time is the car's vector acceleration directed 45° away from straight ahead? What is the magnitude of the resultant acceleration at that moment?

9. Air Force Col. John P. Stapp was a pioneer in the study of the effects of large, rapid accelerations and decelerations on humans, as well as one of the early designers and advocates for pilot/passenger safety harnesses in planes and automobiles in the late 1940s and early 1950s. In many of the rocket sled experiments that he conducted, Stapp himself volunteered to ride the sled. On 10 December 1954, Stapp rode a jet-powered sled that achieved a maximum speed of 282.4 m/s (632 mph!), setting a land speed record for this achievement. After reaching peak velocity, the sled was rapidly decelerated, coming to rest in a mere 1.4 s.
    (a) Calculate the average (negative) acceleration that Col. Stapp experienced as the sled came to a stop. How many *g*'s is this?
    (b) Find the distance Stapp traveled during the time it took to brake the sled from its maximum speed to rest.
    Because the acceleration of the sled as it was brought to a stop was not constant, the average acceleration computed in part (a) is smaller than the maximum (negative) acceleration that Col. Stapp suffered. On-board measurements showed that Stapp survived up to 46.2 *g*'s of acceleration during the final moments of his run (see Table 1.3).

2

NEWTON'S LAWS

JHUAPL/SwRI/Nasa Images

Styx   Nix   Kerberos   Hydra

10 miles
10 km

Charon

JHUAPL/SwRI/Nasa Images

Figure CO-2 (a) *New Horizons* image of Pluto taken on 14 July 2015 near time of closest approach. (b) Comparative image of the five moons of Pluto as recorded by the *New Horizons* spacecraft during its six-month reconnaissance of this dwarf planet and its environs in 2015.

CHAPTER INTRODUCTION: New Horizons—Old Physics

Since the 1960s, robotic space probes have been sent throughout the solar system to examine the Sun, our Moon, other planets and their moons, asteroids, and comets. Among the most recent and dramatic example of such scientific exploration has been the rendezvous of the *New Horizons* spacecraft with the distant dwarf planet Pluto in July 2015. Launched in January 2006, and accelerated on its way by a "gravity assist" from the planet Jupiter in February 2007, *New Horizons* conducted a six-month-long reconnaissance of Pluto and its moons, producing images of unprecedented clarity and revealing an incredible diversity of geological landforms unlike anything seen elsewhere in the solar system (Figure CO-2). Equally stunning were the images of Pluto's five moons, dominated by Charon, nearly 50 percent the size of Pluto itself and exhibiting an unexpectedly rich landscape covered with mountains, canyons, landslides, surface-color variations, and more.

The scientific instruments on *New Horizons* and the rockets that sent it on its journey are products of modern technology. But the principles behind interplanetary travel are rooted in laws established by Sir Isaac Newton more than 300 years ago. His third law of motion explains how a vehicle can accelerate in empty space where there is nothing to "push" against. His second law makes it possible to figure out how long to fire a rocket engine to change a spacecraft's velocity by a desired amount. His law of universal gravitation is the tool for taking into account the effect of Earth, the Sun, and other planets on a spacecraft's path. As if that weren't enough, Newton also invented calculus, the branch of mathematics essential for performing the latter two tasks.

The *New Horizons* high-resolution imagery has dazzled and puzzled both scientists and the public with its revelations about Pluto and its environment. Even more exciting and startling information about the Kuiper Belt and its vast collection of ancient, icy mini-worlds will likely be in store as the spacecraft heads farther out into space to examine that system at least a billion miles beyond Neptune's orbit. It remains the case, however, that without the basic physics first articulated in the 17th century the current and continuing achievements of this mission would not have been possible.

2.1 Force

Sir Isaac Newton (1642–1727) was an English scholar who made many fundamental discoveries in both physics and mathematics. Newton is often regarded as the father of modern physical science; for this reason, it is difficult to overestimate his importance in the development of today's civilization. The dominant scientific, industrial, and technological advancements of the last two centuries were triggered in part by Newton's work.

In formulating his mechanics, Newton began with Galileo's ideas about motion and then sought systematic rules that govern motion and, more important, *changes in motion*. The key concept in Newtonian mechanics is **force**.

> **DEFINITION** **Force** A push or a pull acting on a body. Force usually causes some distortion of the body, a change in its velocity, or both. Force is a vector.

Physical Quantity	Metric Units	English Units
Force (F)	newton (N)	pound (lb)
	dyne	ounce (oz)
	metric ton	ton

The conversion factors between the primary units of force in the two measurement systems are

$$1 \text{ N} = 0.225 \text{ lb} \qquad 1 \text{ lb} = 4.45 \text{ N}$$

This means that a force of 150 pounds is equal to one of 668 newtons.

The distortion caused by a force is often obvious, such as the compression of a sofa cushion when you sit on it. Sometimes it cannot be easily observed without help. For example, high-speed photography reveals that one side of a tennis ball is flattened as a racquet hits it (**Figure 2.1**). Spring scales, like the produce scales in supermarkets, use distortion to measure forces. The greater the force acting to stretch or compress a spring, the greater the distortion (**Figure 2.2**).

Force is a bit difficult to define and is sometimes regarded as a fundamental quantity like time and distance. The English-system units should indicate to you that the idea of force is quite common. The words *push, shove, lift, pull,* and *drag* are everyday synonyms that we use to describe force. Force and energy (see Chapter 3) are probably the two most ubiquitous and useful concepts in all of physics. Newton's laws of motion are simple, direct statements about forces in general and their relationship to motion.

Let's consider some examples of force to show how versatile the concept is. We exert forces on objects in dozens of everyday situations, such as in pushing or pulling a door open, lifting a book, pulling out a drawer, and throwing a ball (**Figure 2.3**). Many machines we use are designed to exert forces. Cranes, hoists, jacks, and vises are all examples of such machines. Automobiles and other propelled vehicles function by causing forces to act on them. If this last statement seems a bit odd to you, come back to it after you have read about Newton's third law of motion in Section 2.6.

Visuals Unlimited/Encyclopedia/Corbis

Figure 2.1 Force on the tennis ball distorts it and accelerates it.

(a)   (b)

Vern Ostdiek/University of Michigan-Dearborn

Vern Ostdiek/University of Michigan-Dearborn

Figure 2.2 (a) A force of 5 newtons stretches the spring 9 centimeters. (b) Doubling the force on the spring doubles the distance it is stretched. The scale itself has a spring inside and makes use of this principle.

Figure 2.3 We exert forces every day.

2.1a Weight

The most common force in our lives is **weight.** Most of us measure the weight of our bodies regularly. Many things that we buy, such as coffee, dog food, and nails, are sold by weight.

The direction of the force of gravity is what determines our conceptions of "up" and "down." We are so accustomed to living with this force that pulls everything toward Earth's surface that we often forget that it is a force (**Figure 2.4**). A child's observation that things naturally

DEFINITION | **Weight** The force of gravity acting on a body. Symbolized by *W*.

Figure 2.4 Weight (*W*) is the downward force of gravity. This force acts on objects whether they are stationary, moving horizontally, or moving vertically. (Arrows are not to scale.)

"fall down" leads one to Aristotle's idea of motion toward a natural place. But Newton made the important observation that objects are pulled toward Earth by a force in much the same way that a sled is pulled by a person. The *law of universal gravitation*, the topic of Section 2.7, generalizes the concept of gravitational force.

The weight of an object depends on two things: the amount of matter comprising the object (its mass) and the distribution of external, mass-bearing agents with which the object is interacting gravitationally. The second point means that a body's weight depends on where it is. The weight of an object on the Moon is about one-sixth the weight it would have on Earth because the gravitational pull of the Moon is less than Earth's. Even on Earth, the weight of an object varies slightly with location. A 190-pound person would weigh about 1 pound less at the equator than at the North or South Pole.

2.1b Friction

Another important common force is **friction.**

DEFINITION	**Friction** A force of resistance to relative motion between two bodies or substances in physical contact.

Friction is at work when a chair slides across a floor, when brakes keep a car from rolling down a hill, when air resistance slows down a baseball, and when a boat glides over the water. Frictional forces arise at the surfaces or boundaries of the materials involved. We can distinguish two types of friction: static and kinetic. When there is no relative motion between two objects, the friction that acts is **static friction.** In order to push a refrigerator across a floor, you must first overcome the force of static friction between it and the floor to initiate the motion. A person who is walking or running relies on the static friction between his or her shoes and the ground to maintain motion. Even though the person's body is moving relative to the ground, there is no relative motion between the shoes and the ground when they are in contact. It is difficult to walk on ice because the force of static friction is reduced, often leading to slipping (and falling). Even a moving car relies on the static friction between its tire surfaces and the pavement. As long as the tires are not skidding or spinning, there is no relative motion between them and the pavement where they are in contact with one another.

> ## ▶ Physics To Go 2.1

1. Place a coin, an eraser, a small block, or some other object on a large hardcover book or notebook and gradually tilt the book until the object slides off. Note how steep the book has to be before the object starts sliding. Do this with several different objects. Is the steepness of the book when the different objects begin to slide the same? If not, can you correlate the maximum tilt angle with the roughness of the surface of the sliding object?
2. Now repeat this with something that can roll, such as a small ball or a pencil. What is different about your observations this time? What can you conclude about the difference between the resistive force of sliding friction and that of rolling friction?

A block of wood resting on an inclined plane is a simple system that relies on static friction (**Figure 2.5**). If the plane is horizontal and there is no force acting to slide the block along the surface, there is no friction. When

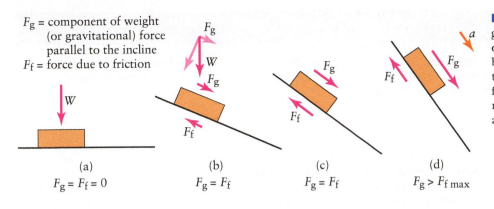

F_g = component of weight (or gravitational) force parallel to the incline
F_f = force due to friction

(a) $F_g = F_f = 0$

(b) $F_g = F_f$

(c) $F_g = F_f$

(d) $F_g > F_{f\,max}$

Figure 2.5 Static friction opposes gravity, which acts to pull the block down the incline. As the plane is tilted, both forces (F_g and F_f) increase until the maximum possible force of static friction is reached (c). Further tilting makes F_g larger than F_f, and the block accelerates down the incline (d).

the plane is tilted a small amount, friction acts to keep the block from sliding down. Here the force of friction opposes the component of the force of gravity—weight—which acts parallel to the plane. The steeper the plane is, the greater the force of static friction needed to keep the block from sliding down. When the angle of the plane reaches a certain critical value, the block will begin to move. The component of the weight parallel to the plane will have exceeded the maximum force of static friction. Notice that the force of static friction between two surfaces can have any value between zero and some characteristic maximum that depends critically on the nature of the two surfaces in contact.

Kinetic friction acts when there is relative motion between two substances in contact, such as an aircraft when moving through the air, a fish swimming underwater, or a tire skidding on pavement. In Figure 2.5d, kinetic friction acts on the block of wood when it is sliding down the inclined plane. The force of kinetic friction that acts between two solids is usually less than the maximum static friction that can act. This is why a car can be stopped more quickly when its tires are not skidding.

The effects of kinetic friction are often undesirable. A car that is traveling on a flat road at a constant speed consumes fuel mainly because it must act against the forces of kinetic friction—air resistance acting on the car's exterior and friction between various moving parts in the axles, transmission, and engine. This also applies to aircraft and ships. Brakes represent a useful application of kinetic friction. On most bicycles and cars, the brakes consist of pads that rub against the wheel rims or disks attached to the axles (**Figure 2.6**). When the brakes are applied, the force of kinetic friction between the pads and the rim or disk slows the vehicle.

It is often difficult to include the effects of friction in mathematical models of physical systems. Consequently, we will find it helpful to consider situations in which frictional forces are small enough to be ignored. It is easier to analyze the motion of a falling rock than that of a falling feather or snowflake. We need to understand these simple systems before we can incorporate the complexities of friction into real-world situations.

Figure 2.6 Bicycle brakes use kinetic friction as the pads rub against the rim of the wheel.

Vern Ostdiek/University of Michigan-Dearborn

■ **Learning Check**

1. (True or False.) Every force that acts on anything has a specific direction associated with it.
2. Which of the following is *not* a unit of force?
   (a) newton   (b) kilogram   (c) ounce
   (d) ton   (e) All are units of force.
3. The force of gravity acting on an object is called _____.

4. The air exerts a force on a person standing in a strong wind. This type of force is one example of _____ friction.

ANSWERS: 1. True **2.** (b) **3.** weight **4.** kinetic

"Can't live with it and can't live without it." This old adage can easily be used to describe friction. On the one hand, friction can be a nuisance. Friction and the general wear it produces on machine parts sliding past one another have enormous economic impact and affect our national security and quality of life. It has been estimated that the equivalent of between 2 and 6 percent of the U.S. gross domestic product could be saved each year by reducing friction and its associated wear and energy losses. Adopting even the lowest estimates, this amounts to a staggering $120 billion or more annually. On the other hand, without friction it would be impossible to walk or drive a car, play a violin, or skydive. In some cases, it's beneficial to maximize friction instead of minimize it, as in the friction between the tires of a car and the road during braking.

Tribology (from the Greek *tribein*, to slide or rub) is a branch of science dedicated to the study of friction and wear and how to control them by lubrication. The use of friction by human beings dates to neolithic times when fires were started by rubbing sticks together or by generating sparks by striking flint. The value of lubricants was recognized more than 4,000 years ago by Sumerian and Egyptian engineers who used oil and mud to ease the transportation of large stones on sledges at construction sites (**Figure 2.7**). The first systematic study of friction is found in the work of Leonardo da Vinci more than 500 years ago. (Among other things, Leonardo appears to have been the first to experiment with objects sliding on inclined planes as shown in Figure 2.5.) However, many aspects of friction are still not well understood today, including its origins in the electrical forces that dominate at the submicroscopic level.

Because Leonardo's experiments lay undiscovered in his notebooks until the 1960s, first credit for describing the nature of the frictional interaction between two solid surfaces sliding against one another is usually given to French physicist Guillaume Amontons. In 1699, Amontons noted that the force of friction between two surfaces in relative motion is directly proportional to the force pressing the surfaces together. The constant of proportionality is called the *coefficient of friction*. Amontons also reported that the friction force was independent of the apparent area of contact: a small block experiences the same frictional force as a large one made of the same material as long as the force pressing each of them against a third surface is the same.

Additional aspects of the frictional interaction were discovered more than 50 years later by Charles Augustin Coulomb, better known for his work on electrostatics (see Section 7.2), in a comprehensive experimental study based on Newtonian mechanics. While confirming the earlier results of Amontons, Coulomb found in addition that the friction force is generally independent of the speed with which the two objects slide past one another, at least as long as the relative velocity is not too high. Coulomb also studied the differences between static and kinetic friction. He was the first to begin to model the frictional interaction between solids by imagining that the surfaces were covered with elastic fibers (think Velcro) that became intermeshed, thus impeding the smooth slippage of one object past the other.

Although these "laws" of sliding friction have been known for centuries, attempts to explain them have only relatively recently begun to show promise. The beginnings of a proper explanation of why the friction force is independent of the apparent area of contact was not presented until around 1940. At this time, Frank Bowden and David Tabor suggested that when two surfaces touch one another, the actual microscopic area of contact, determined when *asperities*—atomic pits or mountains—on one surface meet those of the other, is far smaller than the apparent macroscopic area (**Figure 2.8**). At each of these microscopic contact points, a local bond is formed that welds the two surfaces together. According to this so-called *adhesion model* of friction, to initiate sliding motion between two objects, a sufficiently large force must be applied to sever the existing microscopic bonds (overcome the static frictional force) between them. Once in relative motion, a force (albeit of lesser magnitude) must still be applied to keep the surfaces sliding past one another because new contact regions are continuously being formed as others are being broken.

What now seems clear is that Amontons's observations that the friction force is proportional to the force pressing the surfaces together can be understood from an atomic perspective: the harder you squeeze two surfaces together, the greater the true area of microscopic contact (asperity to asperity) and the greater the frictional force. Even with the

Figure 2.7 Portion of a relief from the tomb of Ti, a royal hairdresser and later steward in Egypt's 5th dynasty. Dating from around 2400 B.C.E., the image shows one of the first documented uses of lubrication to reduce friction. In particular, a worker is seen pouring an unknown liquid ahead of and beneath the sledge on which Ti's statue is being dragged to the tomb.

Werner Forman/Art Resource, NY

Figure 2.8 A highly simplified schematic of our modern conception of interacting surfaces. Here the contact points (or *asperities*) are the locations where friction occurs between two rough surfaces sliding past one another (upper). If the force that squeezes the two surfaces together increases, then so too does the total area of microscopic contact. That increase, and not merely the degree of surface roughness, governs the size of the resulting frictional force.

deeper insight into friction that modern surface studies have afforded, however, we still do not have a comprehensive physical model for friction that allows us to connect directly our knowledge of the microscopic contact point forces and the macroscopic "laws" of friction. All told, it still seems fair to say that friction remains a sticky business.

You can get some idea of the frictional forces that exist between different surfaces under various circumstances by comparing the coefficients of friction measured by scientists and engineers over the years. For example, the static friction coefficient for rubber on dry concrete is about 1.0, whereas the value for rubber on wet concrete is 30% lower, only 0.7. Little wonder, then, why it's harder to start and stop your car on wet pavement than on dry. Similarly, the static friction coefficient for shoes on wood is 0.9, and that for shoes on ice is 0.1—results not too surprising to anyone who's taken a fall on an icy sidewalk in winter. By contrast, the coefficient of friction for Teflon on most substances is only 0.04 or less, making this material an ideal coating for cookware, gears, and the "feet" of computer gaming mice. Go online and search out some other values of the coefficients of friction for other systems. Do the tabulated values confirm your personal experience? Try to secure some measurements for biological systems, if possible. Joint wear and the ameliorating effects of fluid-film lubrication are currently among the hottest topics in the field of biomechanics. Good luck!

■ **QUESTIONS**
1. Give three important "laws" that characterize frictional interactions between two solids sliding past one another.
2. Describe the *adhesion model* of friction.

2.2 Newton's First Law of Motion

As we have just seen, many forces act in everyday situations. Newton's laws make no reference to specific types of forces. They are general statements that apply whether a force is caused by gravity, friction, or a direct push or pull.

> **LAWS** **Newton's First Law of Motion** An object will remain at rest or in uniform motion with constant velocity unless acted on by a net external force.

The last phrase in this statement needs some explanation. An *external* force is one that is caused by some agent outside the object or system in question. Weight is an external force because it is caused by something outside an object (Earth, for example). If your car stalls, you cannot move it by sitting in the driver's seat and pushing on the windshield. This would be an example of an internal force. You must exit the car and push from the outside. The *net* force is the vector sum of all the external forces acting on the body. If one person pushes on the front of your car while another pushes on the back with equal effort, the net horizontal force on the vehicle is zero. If two forces act in the same direction, the net force is the sum of the two (**Figure 2.9**). If two forces act in

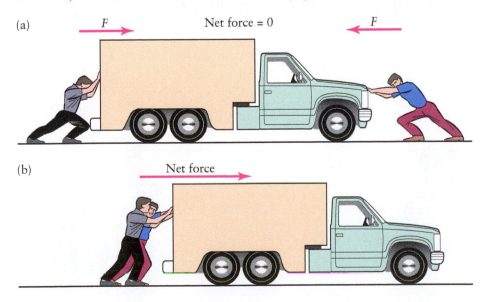

Figure 2.9 (a) Equal forces in opposite directions produce a net force equal to zero. (b) Forces in the same direction add together.

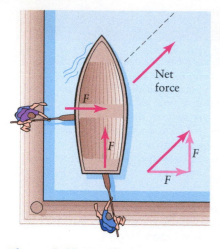

Figure 2.10 Two sailors push on a boat in different directions. The net force is the vector sum of the two applied forces.

different directions, the net force is found by adding the two vectors together as discussed in Section 1.2 (**Figure 2.10**).

Upon first reading Newton's first law does not seem to be terribly profound. Obviously, an object will remain stationary unless a net force causes it to move. But the law also states that anything that is already moving with a certain velocity will not speed up, slow down, or change direction unless a net force acts on it. This means that the states of no motion and of uniform motion are equivalent as far as forces are concerned. Aristotle's (flawed) concept of motion (see Chapter 1, Profiles in Physics) implies that a force is required to maintain an object's motion. Newton's first law implies that a force is required only to *change* the state of motion. This may seem to run counter to your intuition, but that is because you rarely see a moving object with *no* net force acting on it. A car traveling at a constant velocity has zero net force acting on it because the various forces (including air resistance and gravity) cancel each other.

As you throw a ball, your hand exerts a net force that is in the same direction as the ball's velocity. The ball's speed increases because the force acts in the same direction as its motion. When you catch the ball, the force is opposite to the direction of the ball's velocity. Here the force slows down the ball.

Another important point implied by the first law is that a force is required to change the direction of motion of an object. Velocity (a vector) includes direction. Constant velocity implies constant direction as well as constant speed. To make a moving object change its direction of motion, a net external force must act on it. To deflect a moving soccer ball, a player must exert a sideways force on it with the foot or head (**Figure 2.11**). Without such a force, the ball will not change direction.

> ### ◗ Physics To Go 2.2
>
> For this you need a sock, a piece of thread about an arm's length or more long, a reasonably sharp knife (be careful!), and a place where a flying, rolled-up sock isn't going to damage anything. Roll up the sock and tie it to one end of the thread. Grasp the other end and whirl the sock in a horizontal circle above your head. With your free hand, or with the aid of an assistant, quickly (but carefully!) move the knife into the path of the thread so that the knife cuts the thread near its middle at the moment the sock is passing directly in front of you. Stow the knife, and then find the sock. In which direction did it go after the force on it was removed (i.e., after the string was cut)? Not sure? Tie the two pieces of thread together and repeat the experiment.

2.2a Centripetal Force

The implication of Newton's first law is that a net external force must act on an object to speed it up, slow it down, or change its direction of motion. Because each of these is an acceleration of the object, the first law tells us that

Figure 2.11 Force is needed to change the direction of motion.

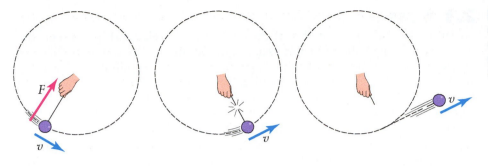

Figure 2.12 Centripetal force must act on any object to keep it moving on a circular course. If the force is removed, the object will move at the same speed in a straight line.

a force is required to produce any acceleration. The centripetal acceleration inherent in circular motion requires a force; it is called the **centripetal force.** Like centripetal acceleration (see Section 1.3a), the centripetal force on an object is directed toward the center of the object's path. As a car goes around a curve, the centripetal force that acts on it is a sideways frictional force between the tires and the road. Without this friction, as is nearly the case on ice, there is no centripetal force, and the car will go in a straight line—off the curve.

Imagine a rubber ball tied to a string and whirled around in a horizontal circle overhead. The ball "wants" to move in a straight line (by the first law) but is prevented from doing so by the centripetal force acting through the string. If the string breaks, the path of the ball will be a straight line in the direction the ball was moving at the instant the string broke (**Figure 2.12**). You may be tempted to think that the ball would move in a direct line away from the center of its previous circular motion or perhaps continue to curve around along its original path for a time. But that is not the case. If you ever see a hammer thrower, a discus thrower, or a catapult in action, notice the path of the projectile after it is released. It moves along a line tangent to its original circular path.

There are other examples of this effect. Children (or even physics professors) riding on a spinning merry-go-round must hold on because their bodies "want" to travel in a straight line, not in a circle. Similarly, clothes are partially dried during the spin cycle in an automatic washer because the water droplets tend to travel in straight lines and move through the holes in the side of the tub. The clothes are "pulled away" from the water as they spin. The same effect could be used to create an "artificial gravity" on a space station by making it spin. (For movie buffs, the classic film *2001: A Space Odyssey* showed this effect particularly well.) You may have tried out a "spinning room" at an amusement park; it uses the same principle (**Figure 2.13**).

Kyle Grantham/Corbis Wire/ZUMA PressCorbis

Figure 2.13 Children at an amusement park enjoying the benefits of centripetally-induced forces that hold them against the wall of a "spinning room" as the floor drops away beneath them.

Learning Check

1. For an object to move with constant velocity,
   (a) the net force on it must be zero.
   (b) a constant force must act on it.
   (c) it must move in a circle.
   (d) there must be no friction acting on it.

2. (True or False.) It is possible for the net force on an object to be zero even if several forces are acting on it at the same time.

3. The force that must act on a body moving along a circular path is called the _____.

ANSWERS: **1.** (a) **2.** True **3.** centripetal force

2.3 Mass

Newton's second law of motion expresses the exact relationship between the net force acting on an object and its acceleration. But before stating the second law, we will take another look at mass (cf. Section 1.1c). Imagine the effect of a small net force acting on a car (**Figure 2.14**). The resulting acceleration would be quite small. The same net force would cause a much larger acceleration if it acted on a shopping cart.

The property of matter that makes it resist acceleration is often referred to as *inertia*. A car is more difficult to accelerate than a shopping cart because it has more inertia. The concept of inertia is embodied in the physical quantity **mass**.

> **DEFINITION** **Mass** A measure of an object's resistance to acceleration. A measure of the quantity of matter in an object.

Physical Quantity	Metric Units	English Units
Mass (*m*)	kilogram (kg)	slug
	gram (g)	

The concept of mass is not in everyday use in the English system of units. For example, flour or sugar is purchased by weight (pounds), not by mass (slugs). In the metric system, the situation is the reverse. Mass is commonly used instead of weight. Flour is purchased by the kilogram, not by the newton. It is both useful and important to "get a feel for" the size of the kilogram, and so we offer the following "mixed" conversion: 1 kilogram *weighs* 2.2 pounds on Earth. One kilogram *does not equal* 2.2 pounds, but on Earth, anything with that mass has a *weight* of 9.8 newtons, which happens to equal 2.2 pounds.

The actual mass of a given object depends on its size (volume) and its composition. A rock has more mass than a pebble because of its size; it has more mass than an otherwise identical piece of Styrofoam because of its composition. The rock is harder to accelerate in the sense that a larger force is required to achieve a given acceleration. Fundamentally, the mass of a body is determined mainly by the total numbers of subatomic particles that comprise it (protons, neutrons, and electrons).

Mass is *not* the same as weight. Mass and weight are related to each other in that the weight of an object is proportional to its mass. But weight is a force that arises from a gravitational interaction. Mass is an intrinsic property of matter that does not depend on any external phenomenon. One major cause of confusion is that users of both the English system and the metric system often incorrectly lump together the two concepts in everyday use. Weight is often incorrectly used in place of mass in countries using the English system, and mass is often incorrectly used in place of weight in countries using the metric system. It usually doesn't matter because they are proportional to each other, but it will if sometime in the future people routinely travel to the Moon or to other places where the acceleration from gravity is different from that on Earth.

If a hammer has a mass of 1 kilogram, that will not change if it is taken into space aboard a spacecraft and then to the Moon's surface. Its mass is 1 kilogram wherever it is. The hammer's weight, however, varies with the location, because weight depends on gravity (**Figure 2.15**). Its weight on Earth is 9.8 newtons, its weight would appear to be zero in a spacecraft orbiting Earth, and its weight would be only 1.6 newtons on the Moon's surface. Though "weightless" in the spacecraft, the hammer is not massless and would still resist acceleration as it does on Earth.

Figure 2.14 The effect of a net force on an object depends on the object's mass. If the mass is very large, the acceleration of the object will be small. Conversely, if the mass is small, the acceleration will be large.

Another way of approaching mass and weight is to think of them as two different characteristics of matter. We might call these aspects *inertial* and *gravitational*. Mass is a measure of the inertial property of matter—how difficult it is to change its velocity. Weight illustrates the gravitational aspect of matter: any object experiences a pull by Earth, the Moon, or any other body near it.

Learning Check

1. If object A has a larger mass than object B,
   (a) A will be easier to accelerate than B.
   (b) A will weigh less than B.
   (c) A will be harder to keep moving in a circle.
   (d) All of the above.

2. (True or False.) The mass of an object depends on where the object is.

ANSWERS: 1. (c) 2. False

2.4 Newton's Second Law of Motion

2.4a Force and Acceleration

We are now ready to state Newton's second law of motion. This law is our most important tool for applying mechanics in the real world.

> **LAWS** **Newton's Second Law of Motion** An object is accelerated whenever a net external force acts on it. The net force equals the object's mass times its acceleration.
>
> $$F = ma$$

This law expresses the exact relationship between force and acceleration. For a given body, a larger force will cause a proportionally larger acceleration (**Figure 2.16**). Force and acceleration both are vectors. The direction of the acceleration of an object is the same as the direction of the net force acting on

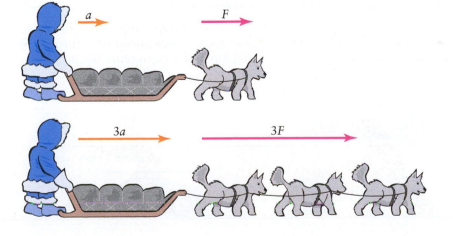

Figure 2.16 The acceleration of an object is proportional to the net force. Tripling the force on an object triples its acceleration.

it. The unit of measure of force, the newton, a derived unit, is established by this law. One newton is the force required to give a 1-kilogram mass an acceleration of 1 m/s². In other words:

$$1 \text{ newton} = 1 \text{ kilogram-meter/second}^2$$

EXAMPLE 2.1 An airplane with a mass of 2,000 kilograms is observed to be accelerating at a rate of 4 m/s². What is the net force acting on it?

SOLUTION

$$F = ma = 2,000 \text{ kg} \times 4 \text{ m/s}^2$$
$$= 8,000 \text{ N}$$

Another way to state Newton's second law is that an object's acceleration is equal to the net force acting on the object divided by its mass:

$$a = \frac{F}{m}$$

This means that a large force acting on a large mass can result in the same acceleration as a small force acting on a small mass. A force of 8,000 newtons acting on a 2,000-kilogram airplane causes the same acceleration as a force of 4 newtons acting on a 1-kilogram toy.

The following example illustrates what can be done with the mechanics that we have learned so far.

EXAMPLE 2.2 An automobile manufacturer decides to build a car that can accelerate uniformly from 0 to 60 mph in 10 s (**Figure 2.17**). In metric units, this is from 0 to 27 m/s. The car's mass is to be about 1,000 kilograms. What is the force required?

SOLUTION First, we must determine the acceleration and then use Newton's second law to find the force. As we did in Chapter 1:

$$a = \frac{\Delta v}{\Delta t} = \frac{27 \text{ m/s} - 0 \text{ m/s}}{10 \text{ s}}$$
$$= 2.7 \text{ m/s}^2$$

So the force needed to cause this acceleration is

$$F = ma = 1,000 \text{ kg} \times 2.7 \text{ m/s}^2$$
$$= 2,700 \text{ N}$$

(This is equal to 2,700 × 0.225 lb = 607.5 lb.) In the next chapter, we will use this information to determine the size (power output) of the car's engine.

Newton's second law establishes the relationship between mass and weight. Recall that any freely falling body has an acceleration equal to g (9.8 m/s²). By the second law, the size of the gravitational force needed to cause this acceleration is

$$F = ma = mg$$

Figure 2.17 Car accelerating from rest to 60 mph or 27 m/s.

We call this force the object's *weight*. So in this case the expression "force is equal to mass times acceleration" translates to "weight is equal to mass times acceleration due to gravity":

$$F = ma \rightarrow W = mg$$

On Earth, where the acceleration from gravity is 9.8 m/s², the weight of an object is

$$W = m \times 9.8 \qquad \text{(on Earth, } m \text{ in kg, } W \text{ in N)}$$

The weight of a 2-kilogram brick is

$$W = 2 \text{ kg} \times 9.8 \text{ m/s}^2 = 19.6 \text{ N}$$

On the Moon, the acceleration from gravity is 1.6 m/s². Therefore,

$$W = m \times 1.6 \qquad \text{(on the Moon, } m \text{ in kg, } W \text{ in N)}$$

▶ Physics To Go 2.3

For this you need the same "sock on a string" used in Physics to Go 2.2, but no knife. Whirl the sock in circles overhead at a low speed (less than one complete circle per second). Then do it at a higher speed. (If the thread is weak, it may break.) What do you feel in the string when the sock's speed is high compared to when its speed is low? What does this tell you about the relationship between the speed of an object moving in a circle and the size of the centripetal force that acts on it?

Forces that cause deceleration or centripetal acceleration also follow the second law. To slow down a moving object, a force must act in the direction opposite the object's velocity (**Figure 2.18**). A force acting sideways on a moving body causes a centripetal acceleration (see Section 1.3a). The object moves along a circular path as long as the net force remains perpendicular to the velocity. The size of the centripetal force required is

$$F = ma$$

and because

$$a = \frac{v^2}{r}$$

$$F = \frac{mv^2}{r} \qquad \text{(centripetal force)}$$

EXAMPLE 2.3 In Example 1.5, we computed the centripetal acceleration of a car going 10 m/s around a curve with a radius of 20 meters. If the car's mass is 1,000 kilograms, what is the centripetal force that acts on it?

SOLUTION

$$F = \frac{mv^2}{r} = \frac{1,000 \text{ kg} \times (10 \text{ m/s})^2}{20 \text{ m}}$$

$$= 5,000 \text{ N} = 1,120 \text{ lb}$$

Figure 2.19 The centripetal force necessary to keep a car on a curved road is supplied by static friction between the tires and the road. The size of the force is proportional to the square of the speed. One car going twice as fast as the other would need four times the centripetal force.

We had already computed the acceleration ($a = 5.0$ m/s^2), so we could have used $F = ma$ directly:

$$F = ma = 1,000 \text{ kg} \times 5.0 \text{ m/s}^2$$

$$= 5,000 \text{ N}$$

The centripetal force acting on a car going around a flat curve is supplied by the friction between the tires and the road. Note that the faster the car goes, the greater the force required to keep it moving in a circle. If two identical cars go around the same curve, and one is going two times as fast as the other, then the faster car needs four times the centripetal force (**Figure 2.19**). The required centripetal force is inversely proportional to the radius of the curve. A tighter curve (one with a smaller radius) requires a larger force or a smaller speed. If a car goes around a curve too fast, the force of friction will be too small to maintain the needed centripetal force, and the car will go off the outside of the curve.

The form of the second law introduced here, $F = ma$, is not the version originally formulated by Newton, but it is the most useful one for our purposes. It applies only when the mass of the object doesn't change. This is almost always the case, but there are important exceptions. For example, the mass of a rocket decreases rapidly as it consumes its fuel. Consequently, its acceleration will increase even if the net force on the rocket doesn't change. More advanced mathematics than we will use here is needed to deal with such cases.

2.4b The International System of Units (SI)

The metric unit of force—the newton—is the force required to cause a 1-kilogram mass to accelerate 1 m/s^2. So when a mass in kilograms is multiplied by an acceleration in meters per second squared, the result is a force in newtons. If grams or centimeters per second squared (cm/s^2) were used instead, the force unit would not be newtons.

At this point, you may be experiencing some confusion over the different units of measure, particularly when different physical quantities are combined in an equation. To help alleviate this problem, a separate system of units within the metric system was established. It is called the **International System,** or **SI,** after the French title *Système International d'Unités*. This system associates only one unit of measure with each physical quantity (**Table 2.1**). Each unit is chosen so that the system is internally consistent—that is, when two or more physical quantities are combined in an equation, the result will be in SI units if the original physical quantities are in SI units. For example, the SI units of distance and time are the meter and the second, respectively. Consequently, the SI unit of speed is the meter per second. Table 2.1 gives the SI units of the physical quantities that we have used so far.

Table 2.1 SI Units (Partial List)

Physical Quantity	SI Unit
Distance (d)	meter (m)
Area (A)	square meter (m^2)
Volume (V)	cubic meter (m^3)
Time (t)	second (s)
Frequency (f)	hertz (Hz)
Speed and velocity (v)	meter per second (m/s)
Acceleration (a)	meter per second squared (m/s^2)
Force (F) and weight (W)	newton (N)
Mass (m)	kilogram (kg)

You may notice that, for the most part, we have been using SI units in our examples. From now on, we will use SI units exclusively when working in the metric system.

Learning Check

1. When a given net force acts on a body, the body's acceleration depends on its _____.
2. The weight of an object equals its mass times _____.
3. Identical cars A and B go around the same curve, with A traveling two times as fast as B. The centripetal force on A
   (a) is two times as large as the centripetal force on B.
   (b) is four times as large as the centripetal force on B.
   (c) is one-half as large as the centripetal force on B.
   (d) is the same as that on B, because they are on the same curve and have the same mass.

ANSWERS: 1. mass **2.** acceleration of gravity **3.** (b)

2.5 Examples: Different Forces, Different Motions

In Chapter 1, we considered three simple states of motion, zero velocity, constant velocity, and constant acceleration. Now, let's see how force is involved in these cases and then look at some other types of motion. Here the particular cause of the force is not important. A constant net force acting on a moving object has the same effect whether it is the result of gravity, friction, or someone's pushing on the object. What interests us now is the relationship between the force—both its magnitude and direction—and the motion. Our concerns are the direction of the force compared to the object's velocity, whether the force is constant, and the way in which the force varies if it is not constant.

The acceleration of a stationary object or of one moving with uniform motion (constant velocity) is zero. By the second law, this means that the net force is also zero (**Figure 2.20**). It is that simple. When the net force on a body is zero, we say that it is in *equilibrium*, whether it is stationary or moving with constant velocity.

In uniform acceleration, the second law implies that a constant force acts to produce a constant acceleration. A steady net force acting in a fixed direction will make an object move with a constant acceleration. A freely falling body is a good example: the force (weight) is constant and always acts in a downward direction.

A net force that acts opposite to the direction of motion of an object will cause it to slow down. If the force continues to act, the object will come to a stop at an instant and then accelerate in the direction of the force—that is, opposite its original direction of motion. This is what happens when you throw a ball straight up into the air (**Figure 2.21**). It is accelerating downward the whole time. While on the way up, its speed decreases until it reaches its highest point. There it has zero speed at an instant and then begins to fall with increasing speed. Its acceleration is always *g*, even at the instant that it is stopped.

2.5a Projectile Motion Revisited

What happens when an object is thrown upward at an angle to Earth's surface? This is one example of a classic mechanics problem—projectile motion. The motion is a composite of horizontal and vertical motions. The path that a projectile takes has a characteristic arc shape that is nicely illustrated by the stream of water from a water fountain (**Figure 2.22**). This shape, an important one in mathematics, is called a *parabola*.

Figure 2.20 Whether a boat is moving with constant velocity or sitting at rest, the net force remains zero.

Figure 2.21 When a ball is thrown straight up from Earth's surface, its acceleration is 1 g downward. This causes it to slow down on its way upward, stop at an instant, and then speed up on its way downward.

The key to understanding projectile motion is to realize that the vertical force of gravity has no effect on the horizontal motion. An object initially moving horizontally has the same downward acceleration g as an object that is simply dropped. So, ignoring air resistance, a projectile moves horizontally with constant speed and vertically with constant downward acceleration (**Figure 2.23**). As an object moves along its path, the vertical component of its velocity decreases to zero (at the highest point of the arc) and then increases downward. The horizontal component of its velocity stays constant. Over level ground, a projectile will travel farthest if it starts at an angle of 45° to the ground. When the force of air resistance is large enough to affect the motion, such as when a softball is thrown very far, the maximum range occurs at smaller angles.

Vern Ostdiek/University of Michigan-Dearborn

Figure 2.22 The stream of water from a water fountain shows the shape of the path of projectiles—a parabola.

▶ Physics To Go 2.4

Investigate projectile motion with a garden hose over level ground on a day with no wind. Shoot the water up at an angle to the ground, and note how the distance to where the stream hits the ground varies with the angle of the nozzle. What angle gives the maximum range? For any range shorter than this there are two different angles that will give the same range. Demonstrate this.

Figure 2.23 (a) The path of a projectile is shown with its velocity vector at selected moments. At the very top its velocity is horizontal. (b) The same projectile with the vertical and horizontal components of the velocity shown separately. The horizontal velocity component stays constant, while the vertical component decreases, and then increases downward, because of the constant downward acceleration due to gravity. (c) A strobe view of three "projectiles" being juggled.

Richard Megna/Fundamental Photographs

2.5b Simple Harmonic Motion

Here is a simple situation with a force that is not constant. Imagine an object—say, a block—attached to the end of a spring (Figure 2.24). When it is not moving, the block is in equilibrium, and the net force on it is zero. If you lift the block up a bit and then release it, it will experience a net force downward, toward its original (rest) position. The reverse occurs if you pull it down a bit below its rest position and release it. Then the net force on it will be up, again toward its original position. Such a force is called a *restoring force* because it acts to restore the system to the original configuration. In this case, the net force is proportional to the displacement from the rest position. The farther the block is displaced, the greater the force acting to move it back. (Figure 2.2 also shows this.)

The equation describing this force may be written as follows:

$$F = -kd,$$

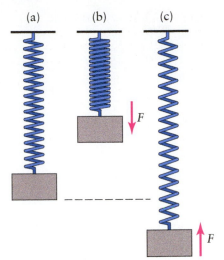

Figure 2.24 The net force that acts on a block hanging from a spring depends on the displacement of the block from its rest (equilibrium) position (a). If raised and then released, the block experiences a net force downward (b). If the block is pulled downward and then released (c), the net force is upward.

where the minus sign serves to remind us that the direction of the force (a vector) is opposite to that of the displacement (or stretch). Here, k is the *spring constant*, which measures the stiffness of the spring; from the force equation, it can readily be seen that the units of k are newtons per meter (N/m). Strong springs like those found on passenger cars and heavy-duty trucks have large values of k, whereas weak springs, like those used in kitchen scales, have small k's. For a given displacement, a larger value of k yields a greater restoring force. Similarly, for a given value of k, a larger displacement produces a bigger force. This is another example of a direct proportionality in physics; a plot of force versus stretch gives a straight line whose slope is the spring constant.

This relationship is often referred to as *Hooke's law*, after Robert Hooke, a contemporary of Newton and a noted scientist in his own right. Unlike Newton's second law, which is a general statement about the organization of the physical world, Hooke's law is special case that applies only when elastic media (springs, rubber bands, bungee cords, muscle fibers, etc.) are not stretched to their breaking points or beyond their *elastic limits* where they become permanently deformed and lose their ability to return to their original shape. Within these limits, the spring force is linear and the spring constant is truly "constant."

So, what kind of motion does this force cause? If an object is pulled down and then released, the upward force will cause it to accelerate. It will pick up speed as it moves upward, but the force and acceleration will decrease as it nears the rest position. When it reaches that point, the force is zero, and it stops accelerating but continues its upward motion (Newton's first law). Once it has moved past the rest position, the force is downward. The object will slow down, stop at an instant, and then gain speed downward. This process is repeated over and over: the object oscillates up and down.

This type of motion, which is very important in physics, is called **simple harmonic motion.** It occurs in many other systems—a pendulum swinging through a small angle, a cork bobbing up and down in water, a car with very bad shock absorbers, and air molecules vibrating with the sound from a tuning fork. In fact, simple harmonic motion is involved in all kinds of waves, as we shall see.

The graphs of distance versus time and velocity versus time show the characteristic oscillation (Figure 2.25). Both graphs have what is known as a *sinusoidal* shape. This shape arises often in physics, and we will see it again when we talk about waves (Chapter 6).

Simple harmonic motion is cyclical motion with a constant frequency. In our example, the frequency of oscillation depends only on the mass of the object and the strength of the spring, specifically.

$$f = \frac{1}{2\pi} \times \sqrt{\frac{k}{m}} = 0.159 \times \sqrt{\frac{k}{m}}$$

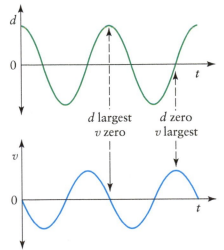

Figure 2.25 Graphs of d versus t and v versus t for simple harmonic motion. Notice that the velocity is zero when the distance is largest, and vice versa. At the high and low points of the motion, the mass is not moving at an instant. As it passes through the rest position, it has its highest speed.

The same mass on a stronger spring will have a higher frequency. For a given spring, a larger mass will have a smaller (lower) frequency. This principle is employed in an inertial balance, a device that uses simple harmonic motion to measure the mass of an object. To use it, one measures the frequency of oscillation of the object and then compares this frequency to those of known masses listed in a table or plotted on a graph. In an orbiting spacecraft, this sort of scale must be used because of the weightless environment (**Figure 2.26**).

EXAMPLE 2.4 A mass of 0.25 kg is connected to a spring and displaced 0.30 m from equilibrium. (a) If the spring constant for the system is 2.4 N/m, find the size of the restoring force acting on the mass. (b) The mass is released and executes simple harmonic motion in response to the applied force. What are the frequency and period of the resulting oscillation?

SOLUTION (a) Applying Hooke's law, we find the magnitude of the applied force to be

$$F = kd = 2.4 \text{ N/m} \times 0.30 \text{ m} = 0.72 \text{ N}$$

(b) The frequency of the oscillation of the mass is

$$f = 0.159 \times \sqrt{k/m} = 0.159 \times \sqrt{(2.4 \text{ N/m})/(0.25 \text{ kg})}$$

$$f = 0.159 \times \sqrt{9.6 \text{ s}^{-2}} = 0.159 \times 3.10 \text{ s}^{-1}$$

$$f = 0.493 \text{ Hz}$$

The period is then given by

$$T = 1/f = 1/0.493 \text{ Hz} = 2.03 \text{ s}$$

2.5c Falling Body with Air Resistance

The force of air resistance that acts on things moving through the atmosphere, like a thrown baseball or a falling skydiver, is one example of kinetic friction. This force is in the opposite direction of the object's velocity and will cause the object to slow down if no other force opposes it. The faster an object goes, the larger the force of air resistance. There is no simple equation for the size of the force of air resistance. For things like baseballs, bicyclists, cars, and aircraft, the force of air resistance is approximately proportional to the square of the object's speed relative to the air. For example, on a day with no wind the force of air resistance on a car going 60 mph is about four times as large as when it is going 30 mph.

Figure 2.26 European Space Agency astronaut Andre Kuipers uses an inertial balance to measure his body mass in the Zvezda Service Module of the International Space Station.

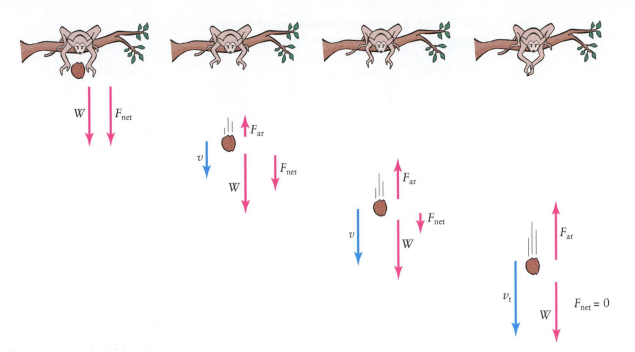

Figure 2.27 Successive views of a falling body affected by air resistance (ar). The upward force of air resistance increases as the object's speed increases. When this force is large enough to offset the downward weight (W), the net force (F_{net}) is zero (far right). The object's speed is now constant and is called the terminal speed (v_t).

▶ Physics To Go 2.5

You'll need a shoe or a small book or other similar object and two coffee filters to conduct this experiment. Wad up one coffee filter tightly to the approximate size of a golf ball. Leave the other filter in the shape it typically has when placed in the basket of your coffee maker.

1. Hold the two filters above your head, one in each hand with the flat bottom of the open filter facing down, and drop them simultaneously. Do they reach the floor at the same time? Does the weight of a body alone determine the effect of air resistance on it as it falls?
2. Similarly, drop the shoe or book and the wadded coffee filter. Do they reach the floor at the same time? What does this demonstrate about the effect of air resistance on falling bodies?

Without air resistance, the constant force on a falling body (its weight) gives it a constant acceleration g. Its speed increases steadily until it hits the ground. But as a body falls through the air, the force of air resistance grows as the speed increases and eventually affects the motion (**Figure 2.27**). This increasing force acts opposite to the downward force of gravity, so the net force decreases. This continues as the body gains speed until the force of air resistance acting upward equals the weight force acting downward. At this point, the net force is zero, so the acceleration vanishes and the speed remains constant from then on. This speed is called the *terminal speed* of the body. **Figure 2.28** shows a graph of speed versus time for this motion.

Rocks and other dense objects have large terminal speeds and may fall for many seconds before air resistance affects their motion appreciably. Feathers, dandelion seeds, and balloons take less than a second to reach their terminal speeds, which are quite low. If a skydiver jumps out of a hovering helicopter or hot-air balloon, he or she will fall for about 2 or 3 s before the force of air resistance starts to have a major effect. The terminal speed depends on the skydiver's size and orientation when falling, but it is typically around 120 mph (about 54 m/s). Then, when the parachute opens, the increased air resistance slows the skydiver to a much lower terminal speed—maybe 10 mph. (The terminal speed depends on the density of the air. In October 2014, Alan Eustace jumped from a balloon at 135,890 feet and achieved at record speed of 822 mph during his fall. As he descended into denser air, he slowed down even before he deployed his parachute.)

Figure 2.28 Graph of speed versus time for a falling body with air resistance. At first the speed increases rapidly as in free fall (compare to Figure 1.26b). But the increasing force of air resistance gradually reduces the acceleration to zero, at which point the speed is constant—the terminal speed v_t.

Table 2.2 Summary of Examples of Forces

Nature of Net Force	Description of Motion
Zero net force	Constant velocity: stationary or motion in straight line with constant speed

Constant net force
  Force parallel to velocity
  Force opposite to velocity
  Force perpendicular to velocity
	Constant acceleration
	Motion in a straight line with increasing speed
	Motion in a straight line with decreasing speed
	Motion in a circle: radius depends on speed and force

Restoring force proportional to displacement	Simple harmonic motion (oscillation)
Net force decreases as speed increases	Acceleration decreases: velocity reaches a constant value

In the spring-and-block example in Figure 2.24, the net force depends on the object's distance from its equilibrium position. When air resistance affects the motion of a falling body, the net force depends on the body's speed. These examples illustrate how interconnected the different physical quantities are in real physical systems. Things can get quite complicated: imagine a spring and mass suspended underwater. Here the net force would depend on both the block's position and its speed. See if you can describe how the block would move. Table 2.2 summarizes the types of forces that we have considered thus far.

In all of the examples in this section, the position of the object can be predicted for any instant of time in the future through the use of Newton's laws of motion and appropriate mathematical techniques—provided we have accurate information about the initial position and velocity of the object and the forces that affect its motion. This ability to predict the position of the object is the main reason Newton's laws are so important. It allows us to send spacecraft to planets billions of miles away and to predict what a roller coaster will do before it is built. But discoveries in the last century show that we can't always predict the future configurations of systems. In Chapter 10, we describe how the science of quantum mechanics, the essential tool for dealing with systems on the scale of atoms or smaller, tells us that such arbitrarily accurate predictions cannot be made because it is impossible to know simultaneously both the precise position and velocity of particles such as electrons. The study of chaos has revealed that even in some relatively simple systems there is inherent randomness: the future configuration at some instant in time cannot be predicted no matter how accurately we know the forces, initial positions, and initial velocities. However, the mechanics based on Newton's laws remain one of the most valuable tools for applying physics in the world around us.

Learning Check

1. (True or False.) The speed of a projectile remains constant as it travels even though its velocity is changing.
2. The motion of a child on a swing moving back and forth is an example of _____.
3. A napkin is dropped and falls to the floor. As it falls,
   (a) the force of air resistance acting on it increases.
   (b) its speed increases.
   (c) the net force on it decreases.
   (d) All of the above.
4. In which of the following situations is the net force constant?
   (a) simple harmonic motion
   (b) freely falling body
   (c) a dropped object affected by air resistance
   (d) All of the above.

ANSWERS: 1. False **2.** simple harmonic motion **3.** (d) **4.** (b)

Chaos. The word conjures up images of extreme disorder, complete unpredictability, utter randomness. But *chaos science*? Science seems to be the antithesis of chaos. A well-ordered, logical, deterministic process that, when applied properly, yields precise, rationale outcomes or conclusions. How are we to reconcile these two very different meanings and integrate them into the subject first brought to major international attention by James Gleick in a 1987 book entitled *Chaos: The Making of a New Science*?

Consider for a moment the kinematic equation giving the position of an object of mass m at time t when a force F is impressed upon it:

$$x = x_0 + v_0 t + \frac{1}{2}\left(\frac{F}{m}\right)t^2$$

Here x_0 is the initial position of the object at time $t = 0$, and v_0 is its initial velocity. We have used Newton's second law to write the acceleration a as (F/m). For any time t, this expression makes an unambiguous prediction about where the object will be for a specified set of initial conditions, viz., x_0 and v_0. For a fixed applied force, the location of the object will always be the same given the same set of values (x_0, v_0). Such a situation is deterministic in the sense that inputting the same information at the start produces the same outcome. Period.

Imagine now altering the initial conditions by some small but arbitrary amount. What might you expect the new prediction of this equation to be relative to the old one? Your intuition would suggest that a small change in x_0 and v_0 might induce an equally small change in the final position x. Moreover, knowing how the changed initial values compared to the original ones, you might also be able to anticipate the direction (larger or smaller) of the change in x. And for truly deterministic circumstances, you'd be right.

Imagine your surprise, however, if you discovered that a tiny alteration in the initial position or velocity of the object produced a large and unexpected change in its predicted location. Even more surprising would be the fact that, in general, the system exhibited a profound sensitivity to the given initial conditions to the extent that small deviations in the starting values gave rise to time evolutions of the system to a seemingly random array of ending points. This behavior—an extreme sensitivity to initial conditions—is called the *butterfly effect* and is one of the principal hallmarks of a chaotic system. The name for this property was given by Edward Lorenz, one of the pioneers in the development of chaos theory and a weather and climate expert, who presented a paper in 1972 with the title: *Does the flap of a butterfly's wings in Brazil set off a tornado in Texas?* The point Lorenz sought to emphasize was that small perturbations of local meteorological conditions, the flapping of a butterfly's wings in South America, can, in principle, produce catastrophic weather events in places thousands of miles away—a tornado in Texas, for example—because climate system dynamics are inherently chaotic by their very nature.

Applications of chaos science (or chaos theory) as elaborated by Lorenz, Benoit Mandelbrot, James Yorke (who first coined the word "chaos" as used in mathematics), Mitchell Feigenbaum, and others over the last 40 years now span the gamut from astrophysics and celestial mechanics (chaotic orbits of planets and asteroids) to biology and environmental science

Figure 2.29 A common design for a compound pendulum. The motion of the system is confined to a vertical plane, and the lower rod is free to pivot about its attachment point through 360° (i.e., to flip over). The angles θ_1 and θ_2 set the initial conditions for the release and subsequent motion of the system.

(population evolution with chaotic characteristics) to computer science (cybersecurity and cryptology) to engineering and robotics and on to mathematics, meteorology, psychology, and, of course, physics. In physics, examples of chaotic systems include apparently simple ones like the double pendulum and far more complex ones, for example, associated with turbulent flow in fluids. Let us briefly examine the chaotic properties of the compound pendulum.

Figure 2.29 shows a common design of such a system. It consists of two uniform rods attached to one another with a universal joint; the opposite end of one of the rods is connected by another universal joint to a fixed point. Depending on the nature of the joints, the motion can be in two dimensions or three; we will consider the case of two-dimensional motion in the plane of the page only. The lower rod is free to rotate through a full 360° (that is, to flip over) during the course of the motion. The angles θ_1 and θ_2 that each rod makes with respect to the vertical set the initial conditions for the subsequent motion. **Figure 2.30** shows the path of the free end of the lower rod after the pendulum has been

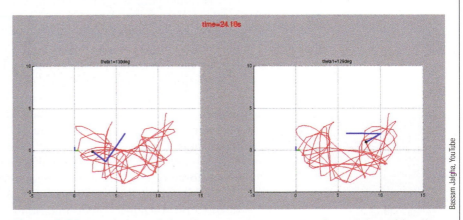

Figure 2.30 Simulations of chaotic motion in a compound pendulum (blue line segments). *Left*: Time sequence of positions of the tip of the lower, articulated pendulum rod (defined by the blue dot) after release for a run time of about 24 s *Right*: Time sequence for an identical pendulum for the same run time, but with a release angle that differs from the first by 1°. The paths taken in the two cases are demonstrably different, showing the sensitivity of the motion to small changes in the initial conditions.

released from rest. The left image shows the path of the tip of the lower rod as traced out for an elapsed time of about 24 s; the right image shows a similar plot over the same time period for an identical pendulum for which the angle θ_2 differs by 1°, yielding an angle between the rods of 129° instead of 130° as in the left image. As may be seen, the motion of the two pendulums is strikingly different, demonstrating the sensitivity of the behavior of the system to small changes in the initial conditions (the starting angles) and thus its chaotic behavior.

Newtonian mechanics is a deterministic theory in so far as the equations of motion that emerge from it offer universal, unambiguous, and definite functional relationships between the system variables (position, velocity, acceleration, force, etc.). However, the quality of the predictions (output) yielded by these relationships turns out to be only as good as the precision of the specification of the initial conditions (input). The existence of chaotic behavior in dynamical systems in no way diminishes the power, range of applicability, or validity of Newton's laws of motion, but it does reinforce the well-known adage that what you get out of a process or activity often (always?) depends on what you put in.

■ QUESTION

1. Name a key property of physical systems that exhibit dynamical chaos. Give an example of such a system and how it demonstrates chaotic behavior.

2.6 Newton's Third Law of Motion

Newton's third law of motion is a statement about the nature of forces in general. It is a simple law that adds an important perspective to the understanding of forces.

> **LAWS** **Newton's Third Law of Motion** Forces always come in pairs: when one object exerts a force on a second object, the second exerts an equal and oppositely directed force on the first.

F on wall ← → *F* on person

If object A causes a force on object B, then object B exerts an equal force in the opposite direction on A.

$$F_{\text{B on A}} = -F_{\text{A on B}}$$

If you push against a wall with your hand, the wall exerts an equal force back on your hand (**Figure 2.31**). A book resting on a table exerts a downward force on the table equal to the weight of the book. The table exerts an upward force on the book also equal to the book's weight. Earth pulls down on you with a force that is called your weight. Consequently, you exert an equal but upward force on Earth. When you dive off a diving board, Earth's gravitational force accelerates you downward. At the same time, your equal and opposite force upward on Earth accelerates it toward you. From a practical standpoint, however, because Earth's mass is about 100 thousand billion billion (10^{23}) times your mass, its acceleration is negligible and goes unnoticed.

The third law gives a new insight into what actually happens in many physical systems. If you are on roller skates and push hard against a wall, you accelerate backward (**Figure 2.32**). Think about this for a moment: a *forward* force by your hands makes you accelerate in the *opposite direction*. On the surface, this seems like a violation of Newton's second law of motion. In reality, the equal and opposite force of the wall on your hands causes the acceleration. If you stand in the middle of the floor, you can't use your hands to move yourself because you need to push against something. Similarly, when a car speeds up, the engine causes the tires to push backward on the road. It is the road's equal and opposite force on the tires that causes the car to accelerate forward. The same thing happens when the brakes slow the vehicle. The recoil of a gun is caused by the third law: the large force accelerating the bullet forward produces an equal and opposite force backward on the gun—the "kick."

F on person ← → *F* on drawer

Figure 2.31 If one body exerts a force on a second body, the second exerts an equal and opposite force on the first.

Figure 2.33 illustrates Newton's third law. A small cart is fitted with a spring-loaded plunger that can be pushed into the cart and retained. When the spring

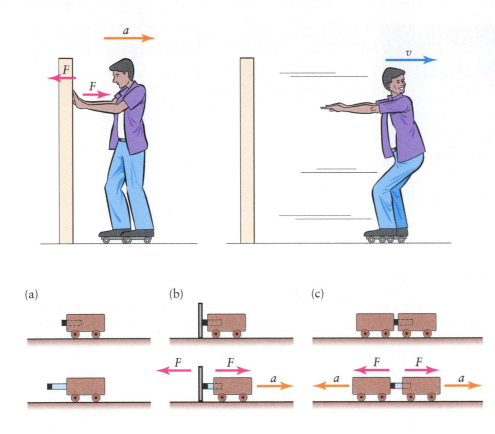

(a) (b) (c)

Figure 2.33 A small cart is fitted with a spring-loaded plunger and trigger. The cart is accelerated only if the plunger can exert a force on something else (b and c); then the equal and opposite force on the cart causes an acceleration.

is released, the plunger moves out. If there is nothing for the plunger to push against (Figure 2.33a), the cart does not move afterward. The plunger can exert no force, so there is no equal and opposite force to accelerate the cart. If the cart is next to a wall when the spring is released (Figure 2.33b), the force on the wall results in an opposite force on the cart, and the cart accelerates away from the wall. If a second cart is next to the plunger (Figure 2.33c), both carts feel the same force in opposite directions, and they accelerate away from each other. In Chapter 3, we will show that the ratio of the speeds of the carts depends on the ratio of their masses. If one has twice the mass of the other, it will have half the speed.

Rockets and jet aircraft are propelled by ejecting combustion gases at a high speed (**Figure 2.34**). The engine exerts a force that ejects the gases, and the gases exert an equal and opposite force on the rocket or jet. Unlike cars, they do not need to push against anything to be propelled. The act of expelling the propellant gases yields the force needed to drive the rocket in the opposite direction.

Birds, airplanes, and gliders can fly because of the upward force exerted on their wings as they move through the air (**Figure 2.35a**). As air flows under and over a wing, it is deflected (forced) downward. This results in an equal and opposite upward force on the wing called *lift* (**Figure 2.35b**). Propellers on airplanes, helicopters, boats, and ships employ the same basic idea. They pull or push the air or water in one direction, resulting in a force on the propeller in the opposite direction. Propellers are useless in a vacuum.

▶ Physics To Go 2.6

When riding in a car on a highway, put your hand a short distance out of the window, palm down, and use it as a wing to illustrate lift. The vertical force is the lift; the force acting backward is known as *drag*. Vary the angle of your hand and notice how the lift varies. At a certain large angle, the lift actually goes to zero: this is known as a *stall*.

NASA

Figure 2.34 Rockets use Newton's third law.

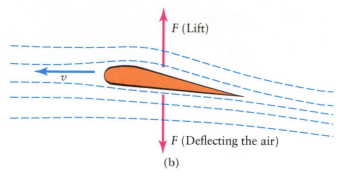

F (Lift)

v

F (Deflecting the air)

(a)

(b)

Figure 2.35 (a) A hang glider having fun with Newton's third law. (b) A wing on a flying aircraft deflects the air downward. This downward force on the air causes an equal and opposite upward force on the wing.

Whenever an object is accelerating, it exerts an equal and opposite force on whatever is accelerating it. You have probably noticed this when riding in a car, bus, or airplane as it accelerates: the seatback exerts a forward force on you that causes you to accelerate. Your body pushes back on the seat with an equal and opposite force. It seems that there is some force "pulling" you back against the seat. This is not a real force, just a reaction of your mass to acceleration. The same effect is observed when an object is undergoing a centripetal acceleration. As a car or bus goes around a curve, you seem to be "pulled" to the side. Again, it is just a reaction to a net force causing an acceleration.

Like Newton's first law, the third law is mainly conceptual rather than mathematical. It emphasizes that forces arise only during interactions between two or more things. By thinking in terms of pairs of forces, we can more easily distinguish the causes and effects in such interactions.

EXAMPLE 2.5 During an EVA ("extravehicular activity") to repair a camera on board the Hubble Space Telescope, an astronaut exerts a force of 125 N on the spacecraft. (a) How much force, if any, does the HST exert on the astronaut during this interaction? (b) If the astronaut (plus life-support equipment) has a mass of 85 kg and the mass of the spacecraft is 11,600 kg, find the resultant acceleration of each body just after the interaction.

SOLUTION (a) Applying Newton's third law to this interaction, if the astronaut applies a force of 125 N on the spacecraft, the spacecraft exerts an equal, but oppositely directed, force on the astronaut. Thus, the astronaut experiences a 125-N force from the HST.

(b) Now, using Newton's second law, we may calculate the individual accelerations of the astronaut and the spacecraft. For the HST:

$$F_{HST} = m_{HST} \times a_{HST}$$

$$125 \text{ N} = 11{,}600 \text{ kg} \times a_{HST}$$

$$a_{HST} = (125 \text{ N}) \div 11{,}600 \text{ kg} = 0.0108 \text{ m/s}^2$$

For the astronaut:

$$F_{astro} = m_{astro} \times a_{astro}$$

$$125 \text{ N} = 85 \text{ kg} \times a_{astro}$$

$$a_{astro} = (125 \text{ N}) \div 85 \text{ kg} = 1.52 \text{ m/s}^2$$

Concept Map 2.1 summarizes the relationship between force and Newton's laws of motion.

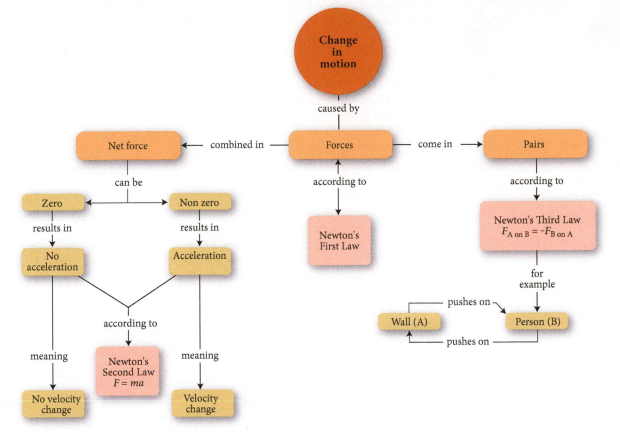

■ Learning Check

1. (True or False.) At this moment you are exerting a force upward on Earth.
2. In the process of throwing a ball forward
   (a) the ball exerts a rearward force on your hand.
   (b) the ball exerts a forward force on your hand.
   (c) your hand exerts a rearward force on the ball.
   (d) the only force present is your force on the ball.
3. At one moment in a football game, player A exerts a force of 200 N to the east on player B. The size of the force that B exerts on A is _____, and it is directed to the _____.
4. (True or False.) When you jump upward, the floor exerts a force on you that makes you accelerate.

ANSWERS: 1. True 2. (a) 3. 200 N, west 4. True

2.7 The Law of Universal Gravitation

Newton's fourth major contribution to the study of mechanics is not a law of motion but a law relating to gravity. Newton made an important intellectual leap: he realized that the force that pulls objects toward Earth's surface also holds the Moon in its orbit. Moreover, he claimed that every object exerts an attractive force on every other object. This concept is called *universal gravitation:* gravity acts everywhere and on all things. The force that Earth exerts on objects near its surface—weight—is just one example of universal gravitation.

What determines the size of the gravitational force that acts between two bodies? Newton used his deep understanding of mathematics and mechanics along with information about the orbits of the Moon and the planets to deduce what the force law must be. The third law of motion states that when two bodies exert forces on one another, the forces are equal in size. If the gravitational force

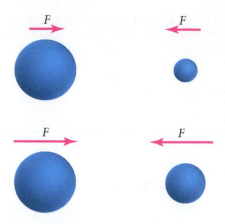

Figure 2.36 The equal and opposite gravitational forces that act on any pair of objects depend on the masses of both objects. If either object is replaced by another with a larger mass, the two forces are proportionally larger.

depends on an object's mass, Newton reasoned, then the force on each object in a gravitationally interacting pair is proportional to each object's mass (**Figure 2.36**). If the mass of either object is doubled, the sizes of the forces on both are doubled.

The size of the gravitational force between two bodies must also depend on the distance between them. For example, the Sun is much more massive than Earth, but its force on you is much less than Earth's, because you are much closer to Earth. For geometric reasons, Newton felt that the size of the gravitational force is inversely proportional to the square of the distance between the bodies. But he needed proof—proof that he got from examining the Moon's orbit.

Long before Newton's time, astronomers had measured the radius of the Moon's orbit and, knowing this and the period of its motion, had determined its orbital speed. So Newton could calculate the Moon's (centripetal) acceleration, v^2/r, which turned out to be about $g/3{,}600$. In other words, the Moon's acceleration is about 3,600 times smaller than that of an object falling freely near Earth. Newton also realized that the Moon is about 60 times farther from Earth's center than an object on Earth's surface. The acceleration of the Moon is 60 squared, or 3,600, times smaller because it is 60 times as far away (**Figure 2.37**). Because acceleration is proportional to force, Newton had his proof that the gravitational force is inversely proportional to the square of the distance between the centers of the two objects.

LAWS | **Newton's Law of Universal Gravitation** Every object exerts a gravitational pull on every other object. The force is proportional to the masses of both objects and inversely proportional to the square of the distance between their centers:

$$F \propto \frac{m_1 m_2}{d^2}$$

where m_1 and m_2 are the masses of the two objects, and d is the distance between their centers.

This force acts between Earth and the Moon, Earth and you, two rocks in a desert—any pair of objects. Your weight thus depends on the distance between you and Earth's center. If you were twice as far from Earth's center (8,000 miles instead of 4,000 miles), you would weigh one-fourth as much (**Figure 2.38**). If you were three times as far away, you would weigh one-ninth as much.

In 1798, English physicist Henry Cavendish performed precise measurements of the actual gravitational forces acting between masses. Cavendish used a delicate torsion balance (**Figure 2.39**). Two masses were balanced on a beam that was suspended from a thin wire attached to the beam's midpoint. Two larger masses were placed next to the original ones on the beam. The gravitational forces on the smaller masses were strong enough to rotate the beam slightly and twist the wire. Cavendish used the amount of twist to measure the size of the gravitational force. His results showed that the force between two 1-kilogram masses 1 meter apart would be 6.67×10^{-11} newtons. With this result, Newton's law of universal gravitation can be expressed as an equation:

$$F = \frac{6.67 \times 10^{-11} \, m_1 m_2}{d^2} \quad \text{(SI units)}$$

The constant of proportionality in the equation is called the gravitational constant G:

$$G = 6.67 \times 10^{-11} \text{ N-m}^2/\text{kg}^2$$

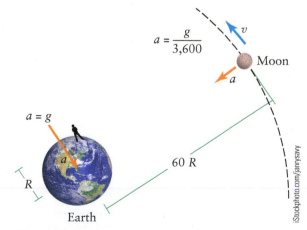

Figure 2.37 The gravitational force that Earth exerts on other objects is inversely proportional to the square of the distance from Earth's center to the object's center. That is why the Moon's centripetal acceleration is much smaller than g. (Figure not drawn to scale.)

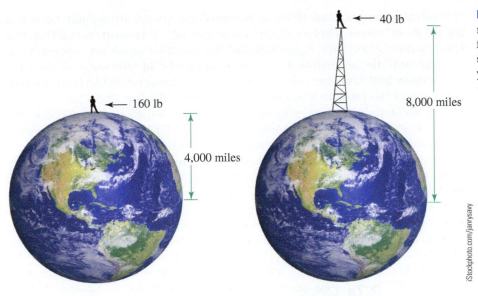

iStockphoto.com/jannysavy

Figure 2.38 On a tower 4,000 miles high, you would be twice as far from Earth's center as you are when standing on its surface. On the tower, you would weigh one-fourth as much. (Figure not drawn to scale.)

Because *G* is such a small number, the gravitational force between objects is usually quite small. For example, the force between two persons, each with mass equal to 70 kilograms, when they are 1 m apart is only 0.00000033 newtons, or 0.0000012 ounces. Earth's gravitational force on us (and our force on Earth) is so large because Earth's mass is huge.

In fact, we can use the law of universal gravitation to compute the mass of the entire Earth. Begin by computing the gravitational force exerted on a body of mass *m* by Earth—the body's weight—using the equation $W = mg$. This force can also be calculated using

$$F = \frac{GmM}{R^2}$$

Here *M* represents Earth's mass, and *R* is Earth's radius, the distance the body is from Earth's center. The value of *R*, first measured by the Greeks more than 2,000 years ago, is about 6.4×10^6 meters. Because these two forces are equal to each other,

$$W = F$$

$$mg = \frac{GmM}{R^2}$$

Canceling the *m:*

$$g = \frac{GM}{R^2}$$

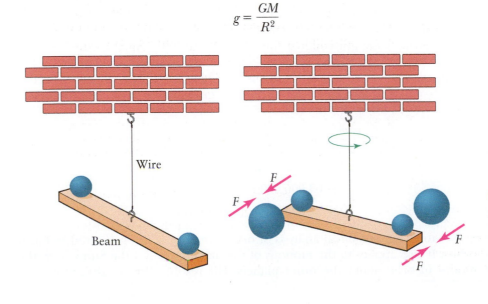

Wire

Beam

F

F

F

F

Figure 2.39 Simplified sketch of the torsion balance used by Cavendish to measure the gravitational force between masses. The two large objects exert forces on the two small ones, causing the suspending wire to twist.

Table 2.3 Surface Gravity in Our Solar System

Location	Acceleration due to Gravity (g)
Earth	1.0
Sun	27.9
Moon	0.16
Mercury	0.38
Venus	0.88
Mars	0.39
Jupiter	2.65
Saturn	1.05

The values of g, G, and R can be inserted and the resulting equation solved for M, Earth's mass. The result: $M = 6 \times 10^{24}$ kg. (It is worth remarking that the m's on the left and right sides of the previous equation represent the inertial and the gravitational masses, respectively, of the body in question [cf. Section 2.3]. The fact that they are equal and can be divided out is a consequence of the *principle of equivalence* discussed in Section 12.2b.)

The acceleration from gravity on the Moon is not the same as it is on Earth. Likewise, each of the other planets has its own "*g*." This variation exists simply because the masses and the radii are all different. The values of the acceleration from gravity have been computed for the Moon, the Sun, and the planets, using the preceding equation and each one's mass M and radius R (Table 2.3 for some of these).

Newton used his law of universal gravitation to explain a variety of phenomena that had only been mysteries before. These included the cause of tides, the motion of comets, and the theoretical basis for orbital motion.

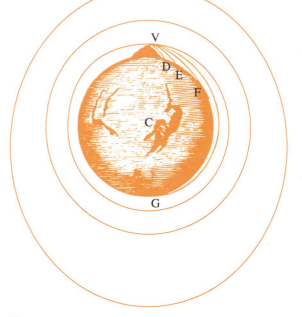

Figure 2.40 A re-creation of the original drawing of Newton's cannon "thought experiment." The cannon is placed on a very high mountaintop (V). The path of the cannonball depends on its initial speed.

2.7a Orbits

Newton used an elegant "thought experiment" to illustrate that orbital motion around Earth is actually an extension of projectile motion. Imagine that a cannon is placed at the top of a very high mountain and that it can shoot a cannonball horizontally at any desired speed (Figure 2.40). If the cannonball just rolls out of the barrel, it will fall vertically in a straight line to Earth. Given some small initial speed, its trajectory to Earth will be a parabola (path D in Figure 2.40). But if its speed is continuously increased, the ball travels farther and farther around Earth before it hits the ground (paths E, F, and G). If it were possible to shoot a cannonball with a high enough speed, it would travel in a full circle around Earth and hit the cannon in the rear. It would be in orbit around Earth, just as the Moon is. An object in orbit is continually "falling" toward Earth. On an even higher mountain, one could place the cannonball in an orbit with a larger radius.

Realistically, this could not be done because of the force of air resistance. To orbit Earth, an object must be above (outside) Earth's atmosphere. But the idea is a handy way of showing the connection between a terrestrial phenomenon—projectile motion—and the motion of the Moon around Earth. It also shows what Newton predicted—that we can put satellites into orbit.

It is quite easy to estimate how fast something has to move to stay in orbit. When an object is moving in a circle around Earth, the centripetal force acting on it is Earth's gravitational pull. For a satellite orbiting just above Earth's surface, the gravitational force is approximately equal to the satellite's weight when it is on Earth's surface—mg. Also, the radius of the satellite's orbit is approximately equal to Earth's radius, R (Figure 2.41). The required speed is found by equating the centripetal force to the gravitational force. The centripetal force is mv^2/R, where R is the radius of Earth. So

$$\frac{mv^2}{R} = mg$$

$$v^2 = gR = 9.8 \text{ m/s}^2 \times (6.4 \times 10^6 \text{ m})$$

$$= 63{,}000{,}000 \text{ m}^2/\text{s}^2$$

$$v = 7{,}900 \text{ m/s} = 17{,}700 \text{ mph}$$

Figure 2.41 Satellite in a low, circular orbit around Earth. The gravitational (centripetal) force on the satellite is about equal to the satellite's weight when on Earth, mg, and the radius of the orbit is about equal to Earth's radius, R.

Newton's mathematical analysis of orbital motion is not restricted to Earth orbits. It also applies to the motions of the planets around the Sun and to the moons in orbit about the other planets. His results allowed astronomers to

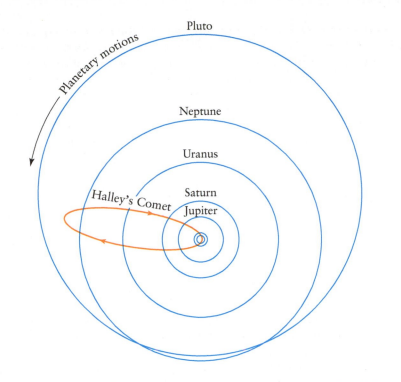

calculate the orbits of celestial objects with higher accuracy and with fewer observational data than before. He showed that comets, such as Halley's Comet, are moving in orbits around the Sun. The orbits of most of them are flattened ellipses with the Sun near one end, at a point called the *focus* of the ellipse (**Figure 2.42**). This is why Halley's Comet is near enough to Earth to be seen only every 76 years. It spends most of its time far from the Sun and Earth.

The orbits of all eight planets and Pluto are actually ellipses, although they are much closer to being circular than the orbit of Halley's Comet. The Sun is at one focus of these orbits as well. Note in Figure 2.42 that the elliptical nature of Pluto's orbit causes it to spend part of the time inside the orbit of Neptune.

▶ Physics To Go 2.7

You can draw the correct shape of an ellipse using two tacks or pins, some string or thread, a ruler, a piece of paper, and a surface into which you can push the tacks (e.g., a bulletin board).

1. Place the tacks into the paper 10 centimeters apart. Wrap the string around the tacks and tie it to form a tight loop around them. Move the tacks a few centimeters closer together. Insert a pen into the loop, move it outward until the string is again taut and draw the complete path around the tacks while holding the pen against the string (**Figure 2.43**). The resulting figure is an ellipse with each tack at one focus of the ellipse.

2. Place the tacks 4 centimeters apart. The ellipse you draw has the shape of Pluto's orbit. (The Sun would be at one of the pins.) Note that it is hard to distinguish this shape from a circle. How should you place the pins to draw a true circle?

3. Place the tacks 8.6 centimeters apart to draw the shape of the orbit of Nereid, a moon of Neptune that has a highly elliptical orbit.

2.7b Gravitational Field

Gravitation is an example of action at a distance. Objects exert forces on each other, even though they may be far apart and there is no matter between them to transmit the forces. The other forces we have talked about involve direct contact between things. Gravitation does not.

How is force possible without contact? One way to get better insight into this situation is to use the concept of a *field*. Imagine an object situated in space. The matter in the object causes an effect or a disturbance in the space around it. We call this a *gravitational field*. This field extends out in all directions but

Figure 2.43 Drawing an ellipse.

Vern Ostdiek/University of Michigan-Dearborn

Figure 2.44 Two ways of showing the gravitational field around an object: (a) with arrows and (b) with field lines.

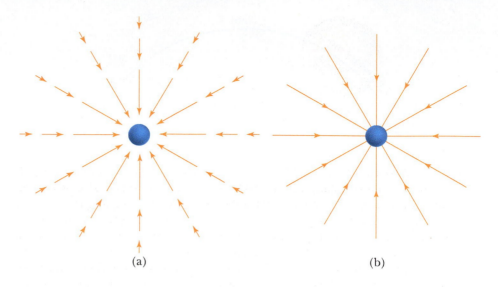

(a)   (b)

becomes weaker at greater distances from the object. In this model, the field itself causes a force to act on any other object. It plays the role of an invisible agent for the gravitational force. Whenever a second body is in the field of the first, it experiences a gravitational force. But the field is present even when there is no other object around to experience its effect.

We might call this a "force field" because it causes forces on other bodies. But do not imagine it to be the kind of "invisible barrier" one finds in science-fiction movies. One way to represent the shape of the gravitational field around an object is by drawing arrows (vectors) at different points in space. They show the magnitude and the direction of the force that *would act* on anything placed at each point (**Figure 2.44a**). These arrows are long near the object and short farther away because the gravitational force decreases as the distance increases. The arrows all point inward toward the object because the gravitational force is attractive and will tend to draw other bodies toward the first. Another way to represent the gravitational field is to connect the arrows, making so-called field lines. The direction of the field line at any point in space again shows the direction of the force that would act on an object placed there. The strength of the gravitational field is represented by the spacing of the lines: the lines are farther apart where the field is weaker (**Figure 2.44b**).

All of the forces in nature can be traced to four fundamental forces (see Chapter 12). The gravitational force is one of them. The others are the electromagnetic force (which is responsible for electric and magnetic effects), the strong nuclear force, and the weak nuclear force. These will be discussed in later chapters. But, you might ask, what about friction and other forces involving direct contact? Most of these can be traced back to the electromagnetic force. This force determines the sizes and shapes of atoms and molecules. For example, a stretched spring pulls back because of the electrical forces between the atoms in it.

Learning Check

1. (True or False.) If the distance between a spacecraft and Earth is doubled, the gravitational force on the spacecraft will be one-half as large.
2. (Choose the *incorrect* statement.) The gravitational force exerted on a satellite by Earth
   (a) is directed toward Earth's center.
   (b) depends on the satellite's speed.
   (c) depends on the satellite's mass.
   (d) depends on Earth's mass.

3. (True or False.) Gravity supplies the centripetal force needed to keep a planet in a circular orbit.
4. Every object creates a _____ in the space around it.

ANSWERS: 1. False **2.** (b) **3.** True **4.** gravitational field

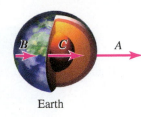

Earth

Moon

Figure 2.45 Gravitational forces exerted by the Moon on material at three different locations on Earth. (Figure not drawn to scale.)

2.8 Tides

For the most part, tides are the result of gravitational forces exerted on Earth by the Moon. To understand this, consider **Figure 2.45**. Points *A*, *B*, and *C* all lie on a line through Earth's center and the Moon's center. According to Newton's law of universal gravitation, Earth material at *A* experiences a stronger attractive force toward the Moon than does material at *C* because it is closer to the Moon. Similarly, material at *C* experiences a greater attractive force toward the Moon than does material at *B*. Thus, the material at *A* is pulled away from material at *C*, whereas the material at *C* is in turn pulled away from the material at *B*. The net effect is to separate these three points. Thus, Earth's shape is elongated or stretched by the gravitational pull of the Moon along a line connecting the two bodies.

This view is one that might be seen by a distant observer looking in at the Earth–Moon system. One might ask what an observer at point *C* sees. Such an observer would notice that point *A* is pulled toward the Moon and away from *C*. The observer would also see that point *B*, relative to *C*, seems to be pushed away from *C* in the direction away from the Moon.

We can redraw our earlier figure now indicating how an Earthbound observer at *C* views the forces exerted by the Moon at points *A* and *B* (**Figure 2.46**). Notice that the same stretching or elongation of Earth along the direction of the Moon occurs as before, but now the symmetry of the situation as seen from *C* is manifested.

But what has this to do with tides? Suppose we were to perform this same kind of analysis for points *D*, *E*, *F*, *G*, *H*, and *J*. We would find that, relative to *C*, forces along the directions of the arrows shown in Figure 2.46 would exist because of the gravitational presence of the Moon. Now imagine Earth is covered with an initially uniform depth of water. How would this fluid move in response to these forces? Again, applying the laws of mechanics, we are led to the following, perhaps somewhat startling, results. (1) Water at points *D* and *E* would weigh slightly more than water elsewhere because, in addition to Earth's own gravitational pull toward *C*, there is a small component of force toward *C* produced by the Moon. (2) Conversely, water at points *A* and *B* would weigh slightly less than water elsewhere because the Moon exerts small forces in directions opposite to Earth's own inward gravitational attraction toward *C*. These forces work to counteract Earth's gravity and, hence, to reduce the weight of the water. (3) Water at points *F*, *G*, *H*, and *J* would experience a force parallel to Earth's surface and would begin to flow in the directions of the arrows at each

Earth

Moon

iStockphoto.com/jannysavy

Figure 2.46 Gravitational forces exerted by the Moon on material at different locations on Earth, relative to point *C*, Earth's center. (Figure not drawn to scale.)

Figure 2.47 (a) Tidal bulges in the oceans caused by the gravitational pull of the Moon. (b) As Earth rotates, places on its surface move from low tide to high tide, back to low tide, and so on. (Figures not drawn to scale.)

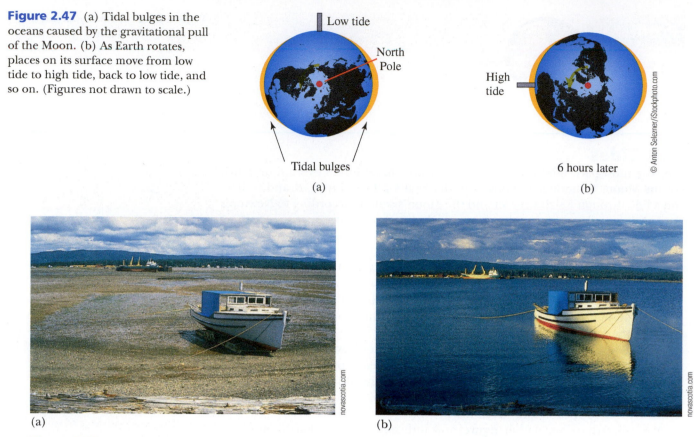

(a)

(b)

© Anton Selezner/iStockphoto.com

(a)

(b)

novascotia.com

Figure 2.48 Low tide (a) and high tide (b) off the coast of Nova Scotia.

point, much as water flows down any inclined surface in response to gravity. This flow causes the water to pile up at opposite sides of Earth along the line joining Earth to the Moon. These "piles" of water are called *tidal bulges,* and as Earth rotates, points on its surface move into and then out of the bulges, resulting in the high and low tides observed at intervals of roughly 6 hours (**Figure 2.47** and **Figure 2.48**).

Many complications to this simple picture cause the behavior of real tides to deviate somewhat from that predicted by this model. In particular, we have ignored such things as tidal effects caused by the Sun, the influence of Earth's rotation on the motion of the tidal bulges, and the effects on the local heights of tides caused by Earth's rugged, uneven surface. The first of these complications leads to such phenomena as the lower-than-average *neap tides,* which occur when the Moon is in either the first- or third-quarter phase (and hence 90 degrees removed from the Sun in the sky), and the higher-than-average *spring tides,* which happen when the Moon is in either the new or the full phase (and hence in line with the Sun in the sky). Despite having ignored these problems, we are still able to understand the basic characteristics of tides in terms of this simple Newtonian model.

Learning Check

1. (True or False.) Tides are caused by the fact that Earth's center and the different parts of its oceans are not all the same distance from the Moon.
2. Before going to the beach, you try to estimate the tide level (high, low, or in between) by noting where in the sky the Moon is. It is low tide

(a) when the Moon is near its highest point in the sky.
(b) when the Moon is near the horizon.
(c) whenever the Moon is not visible.
(d) None of the above.

ANSWERS: 1. True **2.** (b)

saac Newton was born in rural Woolsthorpe, England, on Christmas Day in 1642 (**Figure 2.49**). Newton's father died shortly before Isaac's birth, and when his mother remarried three years later, he was entrusted to the care of his maternal grandparents. His interest in science was kindled while he attended a boarding school, and he entered Trinity College at Cambridge in 1661. There he studied mathematics and physics until the university was closed in the summer of 1665 because of the Great Plague. Newton returned to Woolsthorpe and during the next two years made his great intellectual discoveries in mechanics—the laws of motion and gravitation. As if this were not enough for someone in his early 20s, he also made important contributions to mathematical analysis, independently invented calculus, and explained how prisms produce the spectrum of colors from sunlight. Any of these accomplishments alone would have ensured him a place in the history of science and mathematics. The fact that they were performed in only two years by an "amateur" in isolation from the great intellectual centers of the world makes Newton's feat one of the greatest in the history of human thought.

After Cambridge reopened in 1667, Newton returned to pursue his studies, but he did not publish his discoveries. He recognized their importance but was apparently reluctant to reveal them before they were completely developed. One of the details that he eventually proved, using his calculus, was that the gravitational force between spherical bodies depends on the distance between their centers. Depending on how you look at it, this may or may not seem logical, but Newton wanted proof. If this were not true, Newton's law of universal gravitation would not tie together Earth's force on the Moon and Earth's force on objects near its surface. Newton's instructors were aware of his accomplishments, and one of his mathematics professors relinquished his position to Newton in 1669.

During his 20-year tenure as a professor of mathematics at Cambridge, Newton concentrated more on the study of light than on mechanics. He invented the reflecting telescope, giving astronomers one of their most valuable tools (**Figure 2.50**). Nearly all of the great telescopes in use today, including the Hubble Space Telescope, are reflectors. Newton's work with lenses and the prismatic analysis of light were important contributions to optics. He did publish some of these discoveries, but they were met with contentious criticism. As a result, Newton resolved to withhold his other works. His major critic argued that he had made some interesting findings about how light

Universal Images Group/Getty Images

Figure 2.50 Newton's reflecting telescope, which he presented to the Royal Society in London.

behaved but hadn't dealt with the fundamental question of what precisely light is. Newton realized that one must first know as much as possible about the properties and behavior of light before one can hope to explain what it is.

The great astronomer Edmond Halley, after whom Halley's Comet is named, recognized the importance of Newton's mechanics to physics in general and to astronomy in particular. Halley succeeded in convincing Newton to publish a treatise on mechanics. After 18 months of prodigious concentration and labor, Newton completed his greatest work, the *Philosphiae Naturalis Principia Mathematica* (*Mathematical Principles of Natural Philosophy;* **Figure 2.51**). Published in 1687, the *Principia* was soon recognized as one of the greatest books ever written. In it, Newton presented his basic principles of mechanics and gravitation and used them to explain a number of physical phenomena that had puzzled scientists for centuries. Most of the material in this chapter is based on the *Principia*.

In 1696, Newton left Cambridge and the academic world and became Warden of the Mint in London. Three

SSPL/Getty Images

Figure 2.49 Isaac Newton (1642–1727). His achievements have been recognized the world over and commemorated in sculpture and art, on stamps and coins, and here on the British one-pound note, an especially fitting tribute given his service as Master of the Mint.

PHILOSOPHIÆ
NATURALIS
PRINCIPIA
MATHEMATICA

AS400 DB/Bettmann/Corbis

Figure 2.51 Title page of Newton's *Principia Mathematica.*

years later, he was appointed Master of the Mint and served in this capacity until his death. Newton put England on the gold standard and oversaw the introduction of milling on the edges of coins to discourage people from shaving off the precious metals. His duties included sending convicted counterfeiters to the gallows. In 1704, he published another important book, *Opticks*, which summarized his work with light and color that had spanned several decades. In 1705, he was knighted by Queen Anne. Sir Isaac Newton was buried with full honors at Westminster Abbey in 1727.

It is a bit difficult to fully appreciate Newton's work at this point in history because we are on the inside looking out, so to speak. The very manner in which we "do" science is based, to a large extent, on Newton's discoveries. He showed that Nature has special built-in rules (laws) that govern diverse phenomena. Individual types of motion need not be considered in isolation; they are all subject to three basic laws. He fully united the terrestrial world and the cosmos by showing that these rules apply in both realms. Newton revealed the immense usefulness and power to be gained by incorporating advanced mathematics into physics. He broke from the abstract, speculative approach to science and opened a new way with a combination of ingenious experimentation and theoretical analysis. One can argue that Newton's intellectual accomplishments are unsurpassed in their influence on history.

■ QUESTIONS

1. Name a key invention that Newton contributed to optics that astronomers continue to exploit today.
2. What important mathematical "tool" did Newton develop that made possible many of his discoveries in mechanics and gravitation?

SUMMARY

» **Force** is the most important physical quantity in Newtonian mechanics. Forces are all around us; **weight** and **friction** play dominant roles in our physical environment.

» The three **laws of motion** are direct, universal statements about forces in general and how they affect motion.

» **Newton's first law of motion** states that an object's velocity is constant unless a net force acts on it.

» Newton's **second law of motion** states that the net force on an object equals its mass times its acceleration. This law is the key to the application of mechanics. By knowing the size of the force acting on a body, one can determine its acceleration. Then its future velocity and position can be predicted using the concepts in Chapter 1.

» Although Newton's second law holds for all forces, the resultant motion it predicts depends importantly on the specific mathematical expression of the force. For an elastic medium like a spring or bungee cord, which exerts a restoring force proportional to the amount of compression or extension it experiences, the resulting motion of an attached mass is oscillatory with a frequency proportional to the square root of the ratio of the spring constant to the mass. For an object falling through a fluid medium like air or water in which the resistive force is proportional to the square of the object's relative velocity, the net force can reach zero yielding motion at a constant speed called the terminal speed.

» **Newton's third law of motion** states that whenever one body exerts a force on a second one, the second exerts an equal and opposite force on the first. This deceptively simple statement gives us new insight into the nature of forces.

» **The law of universal gravitation** states that equal and opposite forces of attraction act between every pair of objects. The size of each force is proportional to the masses of the objects and inversely proportional to the square of the distance between their centers.

» The nature of orbits and the causes of tides are just two of the many phenomena that can be explained using the law of universal gravitation.

» **Isaac Newton** is the founder of modern physical science. His laws of motion, law of gravitation, and calculus form a nearly complete system for solving problems in mechanics. He also made important discoveries in optics. Newton's combination of logical experimentation and mathematical analysis has shaped the way science has been done ever since.

IMPORTANT EQUATIONS

Equation	Comments	Equation	Comments
Fundamental Equations		*Special-Case Equations*	
$F = ma$	Newton's second law of motion	$F = \dfrac{mv^2}{r}$	Centripetal force (object moving in a circular path)
$W = mg$	Relates weight and mass	$F = -kd$	Hooke's law
$F = \dfrac{Gm_1 m_2}{d^2}$	Law of universal gravitation		

MAPPING IT OUT!

1. Reread Section 2.7 on the law of universal gravitation and make a list of concepts and examples that might serve as a basis for developing a concept map summarizing the material in this section. After creating your list, reorder the items, ranking them from most general to least general (that is, most specific).

2. In this chapter, you've encountered a large number of concepts related to forces and motion. Organizing a concept map might help clarify the meanings of many of these concepts for you. As a start, you examine **Concept Map 2.2** pertaining to the concept of "net force" created by a student who had previously taken an introductory physics course. Unfortunately, this student had some misconceptions about this topic, so there are some blatant errors in the concept map. Locate and correct as many of these errors as you can. (*Hint*: Carefully inspect the linking words used to connect the concepts and consider the meanings of the "propositions" they make.)

■ **CONCEPT MAP 2.2** Mapping It Out! Exercise 2.

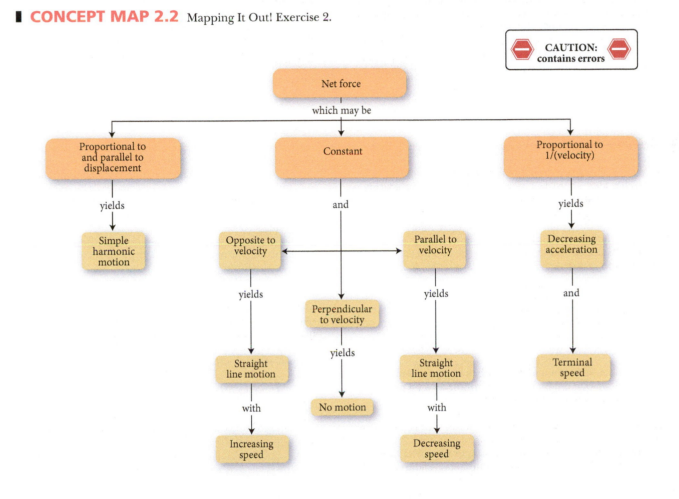

QUESTIONS

(■ **Indicates a review question, which means it requires only a basic understanding of the material to answer. Questions without this designation typically require integrating or extending the concepts presented thus far.**)

1. ■ What is force? Identify several of the forces that are acting on or around you right now.
2. ■ What is weight? Under what circumstances might something be weightless?
3. A person places a book on the roof of a car and drives off without remembering to remove it. As the book and the car move down a street at a steady speed, two horizontal forces are acting on together the book. What are they?

4. ■ Name the two types of friction. Can both types act on the same object at the same time? If so, give an example of such a circumstance.
5. A person places a hand on a closed book resting on a table and then presses downward while pushing forward. Either the book slides across the table or the hand slides across the book. What determines which event happens? Which type(s) of friction is (are) involved?
6. ■ What does a physicist mean by "external" force? In light of Newton's third law of motion, why can't an internal force alone produce an acceleration of a system?
7. At one moment in a football game, player A exerts a force to the east on player B. At the same time, a teammate of A

exerts the same-sized force to the south on player B. In what direction is B likely to go because of these forces? Draw a diagram to support your answer.

8. ■ How does an object move when it is subject to a steady centripetal force? How does it move if that force suddenly disappears?

9. A woman is riding on a train while watching the display on her GPS unit. She notices that both the "speed" and the "direction" readings are not changing. What can the she conclude about the net force acting on the train car?

10. ■ Discuss the distinction between mass and weight.

11. ■ Two astronauts in an orbiting space station "play catch" (throw a ball back and forth to each other). Compared to playing catch on Earth, what effect, if any, does the "weightless" environment have on the process of accelerating (throwing and catching) the ball?

12. An "extreme" roller coaster is moving along its track. During a brief period the track exerts a downward force on the cars. Describe what is happening. (What is the shape of the track at this point?)

13. A single-engine airplane usually has its propeller in the front. Ocean freighters usually have their propeller(s) in the rear. From the perspective of Newton's second law of motion, is this significant?

14. As a rocket ascends, its acceleration increases even if the net force on it stays constant. Why?

15. ■ What is the international system of units (SI)?

16. An archer aims an arrow precisely horizontally over a large, flat field and lets it fly. At the same instant, the archer's watchband breaks and the watch falls to the ground. Does the watch hit the ground before, at the same time as, or after the arrow hits the ground? Defend your answer.

17. ■ Describe the variation of the net force on and the acceleration of a mass on a spring as it executes simple harmonic motion.

18. If a 0.5-kg object hanging from a spring stretches it by 0.30 m, then by how much will the spring be stretched if a 1-kg object is suspended from it?

19. ■ Explain how the change in the force of air resistance on a falling body causes it to eventually reach a terminal speed.

20. The terminal speed of a ping-pong ball is about 20 mph. From the top of a tall building on a windless day, a ping-pong ball is thrown *downward* with an initial speed of 50 mph. Describe what happens to the ball's speed as it moves downward from the moment it is thrown to the moment when it hits the ground.

21. At least two forces are acting on you right now. What are these forces? Describe the relationship between the size and direction of these two forces.

22. As any car travels with constant velocity on a straight, flat section of highway, the road still exerts both a vertical (upward or downward) force and a horizontal (forward or backward) force on the car. Identify the specific direction of each of these forces.

23. How is Newton's third law of motion involved when you jump straight upward?

24. Jane and John are both on roller skates and facing each other. First Jane pushes John with her hands and they move apart. They get together again, and John pushes Jane equally hard with his hands and they move apart. Is their subsequent motion any different in the two cases? Why or why not?

25. ■ Describe how the magnitude of the gravitational force between objects was first measured.

26. Imagine that two identical containers, one filled with marbles and the other filled with Styrofoam beads, are released simultaneously from rest from the same height—say, 3 m— above Earth's surface.
    (a) Which container, if either, experiences the greater gravitational force? Justify your answer.
    (b) Which container, if either, experiences the greater acceleration from gravity? Again, justify your answer.

27. If suddenly the value of *G*, the gravitational constant, increased to a billion times its actual value, give several examples of the kinds of things that would happen. (Think locally as well as globally.)

28. The first "Lunar Olympics" is to be held on the Moon inside a huge dome. Of the usual Olympic events—track and field, swimming, gymnastics, and so on—which would be drastically affected by the Moon's lower gravity? In which events do you think Earth-based records would be broken? In which events would the performances be no better—or perhaps worse— than on Earth?

29. ■ In the broadest terms, what causes tides?

30. The Sun's mass is very much larger than the Moon's, yet the tides on Earth produced by the Sun are much lower than those caused by the Moon. Why?

31. We have studied four different laws authored by Sir Isaac Newton. For each of the following, indicate which law is best for the task described.
    (a) Calculating the net force on a car as it slows down.
    (b) Calculating the force exerted on a satellite by Earth.
    (c) Showing the mathematical relationship between mass and weight.
    (d) Explaining the direction that a rubber stopper takes after the string that was keeping it moving in a circle overhead is cut.
    (e) Explaining why a gun recoils when it is fired.
    (f) Explaining why a wing on an airplane is lifted upward as it moves through the air.

32. The graphs in **Figure 2.52** show plots of force versus acceleration for several objects. Rank these displays using the identifying numbers according to the mass of the affected object from smallest to largest. If any objects share the same mass, give them the same ranking. For reference, the graphs all have the same scale for their respective *F* and *a* axes. Explain your reasons for your rankings.

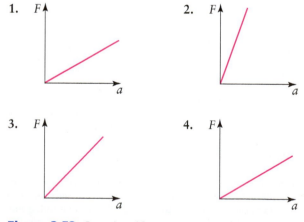

Figure 2.52 Question 32.

33. Six identical cars are towing identical trailers at constant, albeit different, speeds along level roads in the desert. The trailers carry different loads, so their masses vary. The circumstances of each vehicle are shown here. Using the letter identifiers, rank the **difference** between the magnitude of the force exerted by the car on the trailer and the

magnitude of the force exerted by the trailer on the car for each situation from greatest to least. If any situations have the same difference, give them the same ranking. Explain your reasoning in forming your rankings.
(a) Trailer load = 1000 kg; vehicle speed = 20 m/s
(b) Trailer load = 2000 kg; vehicle speed = 20 m/s
(c) Trailer load = 4000 kg; vehicle speed = 40 m/s
(d) Trailer load = 4000 kg; vehicle speed = 10 m/s
(e) Trailer load = 2000 kg; vehicle speed = 10 m/s
(f) Trailer load = 1000 kg; vehicle speed = 10 m/s

34. A mass m is attached to a spring with spring constant k, as shown in **Figure 2.53**. The mass is pulled to the right a distance of 0.2 m and released. Rank the following spring–mass combinations according to their oscillation periods from shortest to longest. If any combinations have the same period, give them the same rank. You should assume that there is no friction between the mass and the horizontal surface.

Figure 2.53 Question 34.

(a) $k = 0.5$ N/m; $m = 0.25$ kg
(b) $k = 0.5$ N/m; $m = 0.50$ kg
(c) $k = 0.5$ N/m; $m = 1.00$ kg
(d) $k = 1.0$ N/m; $m = 0.25$ kg
(e) $k = 1.0$ N/m; $m = 0.50$ kg

PROBLEMS

1. Express your weight in newtons. From this determine your mass in kilograms.
2. A child weighs 300 N. What is the child's mass in kilograms? In slugs?
3. Suppose an airline allows a maximum of 30 kg for each suitcase a passenger brings along.
   (a) What is the weight in newtons of a 30-kg suitcase?
   (b) What is the weight in pounds?
4. The mass of a certain elephant is 1,130 kg.
   (a) Find the elephant's weight in newtons.
   (b) Find its weight in pounds.
5. The mass of a subway car and passengers is 40,000 kg. If its acceleration as it leaves a station is 0.9 m/s², what is the net force acting on it?
6. A motorcycle and rider have a total mass equal to 300 kg. The rider applies the brakes, causing the motorcycle to accelerate at a rate of −5 m/s². What is the net force on the motorcycle?
7. As a 2-kg ball rolls down a ramp, the net force on it is 10 N. What is the acceleration?
8. In an experiment performed in a space station, a force of 60 N causes an object to have an acceleration equal to 4 m/s². What is the object's mass?
9. The engines in a supertanker carrying crude oil produce a net force of 20,000,000 N on the ship. If the resulting acceleration is 0.1 m/s², what is the ship's mass?
10. The Kingda Ka roller coaster in New Jersey is the world's tallest ride of its kind. As the passenger cars are launched from rest at the start, they are accelerated uniformly to a speed of 57 m/s (128 mph) in just 3.5 s.
    (a) What is the acceleration experienced by passengers on this ride in m/s²? In g's?
    (b) If a certain passenger has a mass of 65 kg, what is the force in newtons that acts on him during the launch phase of this ride? What is the force in pounds?
11. A person stands on a scale inside an elevator at rest (**Figure 2.54**). The scale reads 800 N.
    (a) What is the person's mass?
    (b) The elevator accelerates upward momentarily at the rate of 2 m/s². What does the scale read then?
    (c) The elevator then moves with a steady speed of 5 m/s. What does the scale read now?

800 N

$a = 2$ m/s²

Figure 2.54 Problem 11.

12. A jet aircraft with a mass of 4,500 kg has an engine that exerts a force (thrust) equal to 60,000 N.
    (a) What is the jet's acceleration when it takes off?
    (b) What is the jet's speed after it accelerates for 8 s?
    (c) How far does the jet travel during the 8 s?
13. At the end of Section 1.4, we mentioned that the maximum acceleration of a fist during a particular karate blow was measured to be about 3,500 m/s² just before impact with the concrete block. If the mass of the fist was approximately 0.7 kg, what was the maximum force delivered to the concrete block?
14. A sprinter with a mass of 80 kg accelerates uniformly from 0 m/s to 9 m/s in 3 s.
    (a) What is the runner's acceleration?
    (b) What is the net force on the runner?
    (c) How far does the sprinter travel during the 3 s?
15. As a baseball is being caught, its speed goes from 30 to 0 m/s in about 0.005 s. Its mass is 0.145 kg.
    (a) What is the baseball's acceleration in m/s² and in g's?
    (b) What is the size of the force acting on it?

16. On aircraft carriers, catapults are used to accelerate jet aircraft to flight speeds in a short distance. One such catapult takes a 18,000-kg jet from 0 to 70 m/s in 2.5 s.
    (a) What is the acceleration of the jet (in m/s² and g's)?
    (b) How far does the jet travel while it is accelerating?
    (c) How large is the force that the catapult must exert on the jet?

17. At the end of an amusement park ride, it is desirable to bring a gondola to a stop without having the acceleration exceed 2 g. If the total mass of the gondola and its occupants is 2,000 kg, what is the maximum allowed braking force?

18. An airplane is built to withstand a maximum acceleration of 6 g. If its mass is 1,200 kg, what size force would cause this acceleration?

19. Under certain conditions, the human body can safely withstand an acceleration of 10 g.
    (a) What net force would have to act on someone with mass of 50 kg to cause this acceleration?
    (b) Find the weight of such a person in pounds, then convert the answer to (a) to pounds.

20. A race car rounds a curve at 60 m/s. The radius of the curve is 400 m, and the car's mass is 600 kg.
    (a) What is the car's (centripetal) acceleration? What is it in g's?
    (b) What is the centripetal force acting on the car?

21. A hang glider and its pilot have a total mass equal to 120 kg. While executing a 360° turn, the glider moves in a circle with an 8-m radius. The glider's speed is 10 m/s.
    (a) What is the net force on the hang glider?
    (b) What is the acceleration?

22. A 0.1-kg ball is attached to a string and whirled around in a circle overhead. The string breaks if the force on it exceeds 60 N. What is the maximum speed the ball can have when the radius of the circle is 1 m?

23. On a highway curve with radius 50 m, the maximum force of static friction (centripetal force) that can act on a 1,000-kg car going around the curve is 8,000 N. What speed limit should be posted for the curve so that cars can negotiate it safely?

24. A centripetal force of 200 N acts on a 1,000-kg satellite moving with a speed of 5,000 m/s in a circular orbit around a planet. What is the radius of its orbit?

25. As a spacecraft approaches a planet, the rocket engines on it are fired (turned on) to slow it down so it will go into orbit around the planet. The spacecraft's mass is 2,000 kg, and the thrust (force) of the rocket engines is 400 N. If its speed must be decreased by 1,000 m/s, how long must the engines be fired? (Ignore the change in the mass as the fuel is burned.)

26. A space probe is launched from Earth headed for deep space. At a distance of 10,000 miles from Earth's center, the gravitational force on it is 600 lb. What is the size of the force when it is at each of the following distances from Earth's center?
    (a) 20,000 miles
    (b) 30,000 miles
    (c) 100,000 miles

27. A hand exerciser utilizes a coiled spring. A force of 89.0 N is required to compress the spring by 0.0191 m. Find the force needed to compress the spring by 0.0508 m. What is the value of the spring constant for this unit?

28. A mass of 0.75 kg is attached to a relaxed spring with $k = 2.5$ N/m. The mass rests on a horizontal, frictionless surface. If the mass is displaced by 0.33 m, what is the magnitude of the force exerted on the mass by the spring? If the mass is then released to execute simple harmonic motion along the surface, with what frequency will it oscillate?

CHALLENGES

1. The force on a baseball as it is being hit with a bat can be more than 8,000 lb. No human can push on a bat with that much force. What is happening in this instance?

2. Two forces, one equal to 15 N and another equal to 40 N, act on a 50-kg crate resting on a horizontal surface as shown in **Figure 2.55**.
    (a) What is the net horizontal force on the crate?
    (b) What is its horizontal acceleration?
    (c) If the crate starts from rest, what is its horizontal speed after 5 s?
    (d) How far has the crate traveled along the surface in this time?

Figure 2.55 Challenge 2.

3. Why does banking a curve on a highway allow a vehicle to successfully negotiate the turn at a higher speed?

4. As a horse and wagon are accelerating from rest, the horse exerts a force of 400 N on the wagon (**Figure 2.56**). Illustrating Newton's third law, the wagon exerts an equal and opposite force of 400 N. Because the two forces are in opposite directions, why don't they cancel each other and produce zero acceleration (i.e., no motion)?

Figure 2.56 Challenge 4.

5. Perform the calculation of the force acting between two 70-kg people standing 1 m apart to verify the result given in Section 2.7 (3.3×10^{-7} N). Show your work.

6. Perhaps you've noticed that the rockets used to put satellites and spacecraft into orbit are usually launched from pads near the equator. Why is this so? Is the fact that rockets are usually launched to the east also important? Why?

7. The acceleration of a freely falling body is not exactly the same everywhere on Earth. For example, in the Galapagos Islands at the equator, the acceleration of a freely falling body is 9.780 m/s², whereas at the latitude of Oslo, Norway, it is 9.831 m/s². Why aren't these accelerations the same?

8. A 200-kg communications satellite is placed into a circular orbit around Earth with a radius of 4.23×10^7 m (26,300 miles) (**Figure 2.57**).

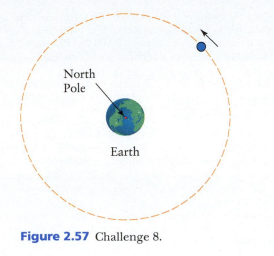

North
Pole

Earth

Figure 2.57 Challenge 8.

(a) Find the gravitational force on the satellite. (There is some useful information in Section 2.7.)

(b) Use the equation for centripetal force to compute the speed of the satellite.

(c) Show that the period of the satellite—the time it takes to complete one orbit—is 1 day. (The distance it travels during one orbit is 2π, or 6.28, times the radius.) This is a *geosynchronous orbit:* the satellite stays above a fixed point on Earth's equator.

9. Complete the calculation of the mass of Earth as outlined in Section 2.7.

10. During our discussion of the motion of a falling body near Earth's surface, we said that the gravitational force acting on it—its weight—is constant. But the law of universal gravitation tells us that the gravitational force on a body increases as it gets closer to Earth. Is there a contradiction here? Explain. (*Hint:* Calculate the gravitational force an a 1-kg mass located on Earth's surface and then repeat the calculation for the same mass located 20 m above Earth's surface. Compare your answers.)

3

ENERGY AND CONSERVATION LAWS

Figure CO-3 Collisional physics.

CHAPTER INTRODUCTION:
Forensic Physics

The squealing tires. The screeching brakes. The crunching of metal and popping of glass. These are the sounds of a car crash (**Figure CO-3**). Is anyone hurt? How badly? Has someone called the EMTs? These questions and more are quickly asked in the aftermath of a vehicle collision as immediate efforts are made to save lives and property. But eventually other questions surface: How did this happen? Was someone speeding? Who is at fault? Who is responsible for the damages? The answers to these questions often require the intervention of law enforcement officials, legal counselors and officers of the court, and perhaps surprisingly, physicists. In particular, it requires *forensic physicists* who specialize in applying the laws of physics to unravel the why's and how's associated with criminal acts and traffic accidents.

One of the principal services that forensic physicists provide is accident reconstruction: determining how fast and in what directions two vehicles were moving just before a crash. Using data and measurements acquired at the accident scene (like the lengths and densities of skid marks, the nature and distribution of scattered debris, even the condition of brake light filaments that can indicate whether the light was on or off at the time of the crash), forensic physicists apply the laws of conservation of momentum and conservation of energy to analyze the collision and to

assess what was going on just before the crash, based on information about how things are immediately afterward. In so doing, it becomes possible to understand what the pre-crash circumstances were like and to establish, for example, whether one car was exceeding the speed limit, whether one or both drivers were applying their brakes or attempting to take evasive action, or if one car had drifted across the median into the lane of the other, oncoming car. Such conclusions can significantly influence the outcomes of legal proceedings following a serious accident in which lives or property are lost.

In this chapter, we introduce the use of conservation laws of linear momentum, energy, and angular momentum in the study of how objects move. These laws are simple but powerful tools for analyzing complex interactions of all types—not just collisions—that we could not handle using only the concepts from Chapter 2. Work and power, two important physical quantities related to energy, are introduced as well. The concept of energy is one of the most important in physics, and its usefulness extends well beyond the area of mechanics.

3.1 Conservation Laws

Newton's laws of motion, in particular the second law, govern the instantaneous behavior of a system. They relate the forces that are acting at any instant in time to the resulting changes in motion. Conservation laws involve a different approach to mechanics, more of a "before-and-after" look at systems. A conservation law states that the total amount of a certain physical quantity present in a system stays constant (is conserved). For example, we might state the following conservation law.

PRINCIPLES The total mass in an isolated system is constant.

In this case, an "isolated system" is one for which no matter enters or leaves the system. The law states that the total mass of all the objects in a system doesn't change regardless of the kinds of interactions that go on within it. (As we shall see in Chapter 11, Einstein showed that energy must be included in this law because in some interactions—particularly those involving subatomic particles—energy can be converted into matter and vice versa. We will not need this refinement until then, however.)

A simple example of this seemingly obvious conservation law is given by aerial refueling. One aircraft, called a tanker, pumps fuel into a second aircraft while both are in flight (**Figure 3.1**). If we ignore the small amount of fuel that both aircraft consume during the refueling, this can be regarded as an isolated system. The law of conservation of mass tells us that the total mass of both aircraft remains constant. So if the receiving aircraft gains 2,000 kilograms of fuel, we automatically know that the tanker loses 2,000 kilograms of fuel. If one tanker refuels several aircraft in a formation, we include all of the aircraft in the system. The total mass of fuel dispensed by the tanker equals the total mass of fuel gained by the other aircraft. This fact may be useful. Let's say that the fuel gauge on one of the receiving aircraft is faulty. To determine how much fuel that aircraft was given, the crews use the conservation of mass: the total mass of fuel unloaded from the tanker minus the total mass of fuel given to the other aircraft in the formation equals the mass of fuel given to the aircraft in question.

The preceding example illustrates how we can use a conservation law. Without knowing the details about an interaction (for example, the actual rate at which fuel is transferred between the aircraft), we can still extract quantitative information by simply comparing the total amount of mass before and after. Three more conservation laws are presented in this chapter. Although they are a bit less intuitive and a bit more complicated to use than the law of conservation of mass, they are applied in the same way.

Figure 3.1 Aerial refueling represents a simple example of the conservation of mass.

3.2 Linear Momentum

The conservation law for **linear momentum** follows directly from Newton's laws of motion, and we will consider it first.

> **DEFINITION**   **Linear Momentum**   The mass of an object times its velocity. Linear momentum is a vector.
>
> $$\text{linear momentum} = mv$$

Linear momentum is often referred to simply as *momentum.* It is a vector quantity (because velocity is a vector), and its SI unit of measure is the kilogram-meter/second (kg-m/s).

Linear momentum incorporates both mass and motion. Anything that is stationary has zero momentum. The faster a body moves, the larger its momentum. A heavy object moving with a certain velocity has more momentum than a light object moving with the same velocity (**Figure 3.2**). For example, the momentum of a bicycle and rider with a total mass of 80 kilograms and a speed of 10 m/s is

$$mv = 80 \text{ kg} \times 10 \text{ m/s} = 800 \text{ kg-m/s}$$

The linear momentum of a 1,200-kilogram car with the same speed is 12,000 kg-m/s. The bicycle and rider would have to be going 150 m/s (not likely) to have this same momentum.

Figure 3.2 Linear momentum
depends on mass and velocity. If a car
and a bicycle have the same velocity,
the car has the larger momentum
because it has the larger mass.

momentum

momentum

3.2a Newton's Second Law, Revisited

Newton originally stated his second law of motion using linear momentum. In particular, we can restate this law as follows.

> **LAWS** **Newton's Second Law of Motion (alternate form)** The net external force acting on an object equals the rate of change of its linear momentum.
>
> $$\text{force} = \frac{\text{change in momentum}}{\text{change in time}}$$
>
> $$F = \frac{\Delta(mv)}{\Delta t}$$

To change an object's linear momentum, a net force must act on it. The larger the force, the faster the momentum will change. If the mass of the object stays constant, which is true in many cases, this equation is equivalent to the first form of Newton's second law (Section 2.4) because

$$\frac{\Delta(mv)}{\Delta t} = m\frac{\Delta v}{\Delta t} = ma \qquad \text{(if mass is constant)}$$

Either form of the second law could be used for cars, airplanes, baseballs, and so on. For rockets and similar things with changing mass, only the alternate form should be used.

We can get yet another useful form of the second law by multiplying both sides by Δt to obtain

$$\Delta(mv) = F\Delta t$$

The quantity on the right side is called the *impulse*. The impulse equals the change in momentum. The same change in momentum can result from a small force acting for a long time or a large force acting for a short time. This equation is useful for analyzing what goes on during impacts that use balls or clubs.

When you throw a tennis ball, a small force acts on it for a relatively long period of time (as your hand moves through the air). When the ball is served with a racket at the same speed, a large force acts for a short period of time. In either case, the result is a change in momentum of the ball, $\Delta(mv)$. One reason to have good follow-through on a shot is to prolong the time of contact, Δt, between the ball and the racket. This leads to a greater change in momentum, so the ball will leave the racket with a higher speed.

EXAMPLE 3.1 Let's estimate the average force on a golf ball as it is driven off the fairway (Figure 3.3). The ball's mass is 0.045 kg, and it leaves the club head with a horizontal speed of, say, 50 m/s (110 mph). High-speed photographs indicate that the contact time is about 5 milliseconds (0.005 s). (We will ignore any spin imparted to the ball upon impact.)

SOLUTION Because the ball starts with zero horizontal speed (and hence zero momentum), its change in momentum is

$$\Delta(mv) = \text{momentum afterwards}$$
$$= 0.045 \text{ kg} \times 50 \text{ m/s}$$
$$= 2.25 \text{ kg-m/s}$$

Because the impulse equals the change in momentum, the average force is

$$F = \frac{\Delta(mv)}{\Delta t} = \frac{2.25 \text{ kg-m/s}}{0.005 \text{ s}}$$
$$= 450 \text{ N} = 101 \text{ lbs}$$

Figure 3.3 A golf club exerts a large force on the golf ball for a short time.

3.2b Conservation of Linear Momentum: Collisions

Our main application of the idea of linear momentum is based on the following conservation law.

> **LAWS** **Law of Conservation of Linear Momentum** The total linear momentum of an isolated system is constant.

For this law, an isolated system means that there are *no outside forces* causing changes in the linear momenta of the objects inside the system. The momentum of an object can change only because of interaction with other objects *in the system*. For example, once the cue ball on a pool table has been shot (given some momentum), the pool table and the balls form an isolated system. If the cue ball collides with another ball initially at rest, its momentum is changed (decreased) (Figure 3.4). This change occurs because of an interaction with another object in the system. The momentum of the ball with which it collides is increased, but the total linear momentum of the system remains the same. If someone put a hand on the table and stopped the cue ball, the system would no longer be isolated, and we couldn't assume that its total linear momentum was constant.

The most important use of the law of conservation of linear momentum is in the analysis of collisions. Two billiard balls colliding, a traffic accident, and two skaters running into each other are familiar examples of collisions. We will

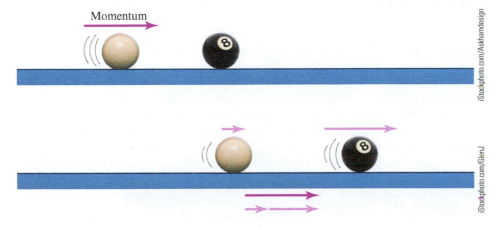

Figure 3.4 Before (above) and after views of a collision. The cue ball loses momentum, and the eight ball gains momentum. The total momentum of the two balls is the same before and after. The momentum vector before the collision equals the sum of the two momentum vectors after as shown by the arrows at the bottom of the figure.

limit our examples to collisions involving only two objects that are moving in one dimension (along a line).

During any collision, the colliding objects exert equal and opposite forces on each other that cause them to accelerate (in opposite directions). These forces are usually quite large and are often the result of direct contact between the bodies, as is the case with billiard balls and automobiles. "Action-at-a-distance" forces between objects that don't actually come into contact, such as the gravitational pull between a spacecraft and a planet, can also be involved in collisions. The following statement is the key to applying the law of conservation of linear momentum to a collision:

LAWS The total linear momentum of the objects in a system before the collision is the same as the total linear momentum after the collision.

$$\text{total } mv \text{ before} = \text{total } mv \text{ after}$$

EXAMPLE 3.2 We can use the law of conservation of linear momentum to analyze a simple automobile collision. A 1,000-kilogram automobile (car 1) runs into the rear of a stopped car (car 2) that has a mass of 1,500 kilograms. Immediately after the collision, the cars are hooked together, and their subsequent speed is estimated to be 4 m/s (**Figure 3.5**). What was the speed of car 1 just before the collision?

SOLUTION Our conservation law tells us that the total linear momentum of the system will be constant:

$$\text{total } (mv)_{\text{before}} = \text{total } (mv)_{\text{after}}$$

Before the collision, only car 1 is moving. The total linear momentum before the collision is

$$(mv)_{\text{before}} = m_1 \times v_{\text{before}} = (1{,}000 \text{ kg}) \times v_{\text{before}}$$

After the collision, both cars are moving together as one body. The total momentum after the collision is

$$(mv)_{\text{after}} = (m_1 + m_2) \times v_{\text{after}} = (1{,}000 \text{ kg} + 1{,}500 \text{ kg}) \times 4 \text{ m/s}$$

$$= 2{,}500 \text{ kg} \times 4 \text{ m/s} = 10{,}000 \text{ kg-m/s}$$

These two linear momenta are equal. Therefore,

$$1{,}000 \text{ kg} \times v_{\text{before}} = 10{,}000 \text{ kg-m/s}$$

$$v_{\text{before}} = \frac{10{,}000 \text{ kg-m/s}}{1{,}000 \text{ kg}}$$

$$= 10 \text{ m/s}$$

Figure 3.5 A 1,000-kilogram car collides with a 1,500-kilogram car that is stationary. Afterward, the two cars are hooked together and move with a speed of 4 m/s. Conservation of linear momentum allows us to determine that the speed of the first car was 10 m/s.

v_b

$v_a = 4 \text{ m/s}$

iStockphoto.com/Galinovma79

This type of analysis is routinely used to reconstruct traffic accidents (remember the chapter-opening discussion). For example, it can be used to determine whether a vehicle was exceeding the speed limit just before a collision. In the previous example, the first car was going 10 m/s, or about 22 mph. If the accident happened in a 15-mph speed zone, the driver of car 1 would have been speeding.

The speed of a bullet or a thrown object can be measured similarly. A bullet is fired into, and becomes embedded in, a block of wood hanging from a string (**Figure 3.6**). If one measures the masses of the bullet and the wood block and the speed of the block immediately afterward, the initial speed of the bullet can be found by using the conservation of linear momentum. The key is measuring the speed of the block of wood after the collision. One can do this most easily by using the law of conservation of energy. We will show how in Section 3.5.

In Section 2.6, we described a simple experiment in which we used two carts (Figure 2.33c). One cart has a spring-loaded plunger that pushes on the other cart, causing both carts to be accelerated. By the law of conservation of linear momentum,

$$(mv)_{\text{before}} = (mv)_{\text{after}}$$

Before the spring is released, neither cart is moving, so the momentum is zero:

$$(mv)_{\text{before}} = 0$$

Therefore, the total linear momentum afterward must be zero as well. But because both carts are moving, this momentum equals:

$$(mv)_{\text{after}} = 0 = (mv)_1 + (mv)_2$$

So,

$$(mv)_1 = -(mv)_2$$

The momentum of one of the carts is negative. This has to be the case, because they are moving in opposite directions. (Remember: linear momentum is a vector.) Because $(mv)_1 = m_1 \times v_1$:

$$m_1 \times v_1 = -m_2 \times v_2$$

$$v_1 = -\frac{m_2}{m_1} \times v_2$$

$$\frac{v_1}{v_2} = -\frac{m_2}{m_1}$$

So we have derived the statement made in Section 2.6: the ratio of the speeds of the two carts is the inverse of the ratio of their masses. If the mass of cart 2 is 3 kilograms and the mass of cart 1 is 1 kilogram, the ratio is 3 to 1. Cart 1 moves away with three times the speed of cart 2 (**Figure 3.7**). Note that this gives us only the ratio of the speeds, not the actual speed of each cart, which

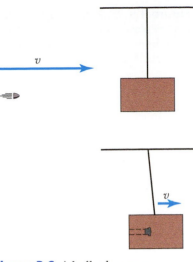

Figure 3.6 A bullet becomes embedded in a block of wood. If the speed of the block and the masses of the block and the bullet are measured, the initial speed of the bullet can be computed using conservation of linear momentum.

Figure 3.7 The mass of cart 2 on the right is three times the mass of cart 1. After the spring is released, the speed of the lighter cart is three times the speed of the more massive cart.

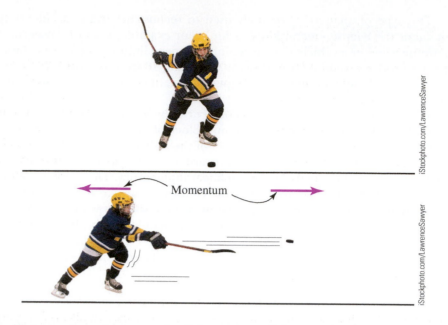

Momentum

would depend on the strength of the spring. With a weak spring, the speeds might be 1 m/s and 3 m/s. With a stronger spring, the speeds might be 2.5 m/s and 7.5 m/s.

We can use linear momentum conservation to get a different view of some of the situations described in Section 2.6. There we used forces and Newton's third law of motion. When a gun is fired, the bullet acquires momentum in one direction, and the gun gains equal momentum in the opposite direction—the "kick" of the gun against the shoulder or hand. If a hockey player shoots a puck forward, he or she will move backward with equal momentum (Figure 3.8). Rockets and jets give momenta to the ejected exhaust gases. They in turn gain momentum in the forward direction. For a rocket with no external forces on it, the increase in momentum during each second will depend on how much gas is ejected—which equals the mass of fuel burned—and on the speed of the gas (Figure 3.9).

Why is linear momentum conserved? We can use Newton's second and third laws of motion to answer this question. When two objects exert forces on each other by colliding or via a spring plunger, the forces are equal and opposite. They push on each other with the same size force but in opposite directions. By the second law (alternate form), these equal forces cause the momenta of both objects to change at the same rate. As long as the objects are interacting, *they change one another's momentum by the same amount but in opposite directions*. The momentum gained (or lost) by one object is exactly offset by the momentum lost (or gained) by the other. The total linear momentum is not changed. In Example 3.2, the 1,000-kilogram car is slowed from 10 m/s before the collision to 4 m/s after the collision (Figure 3.5). Its momentum is decreased by 6,000 kg-m/s (1,000 kilograms × 10 m/s − 1,000 kilograms × 4 m/s). The 1,500-kilogram car goes from 0 m/s before to 4 m/s after. So its momentum is increased by 6,000 kg-m/s (1,500 kilograms × 4 m/s).

These examples illustrate the usefulness of conservation laws. The approach is different from that used with Newton's second law in Chapter 2, where it was necessary to know the size of the force that acts on the object at each moment to determine its velocity. With the law of conservation of linear momentum, we do not have to know the details of the interactions—how large the forces are and how long they act. All we need is some information about the system before and after the interaction. Note that in Example 3.2 we used information from after the collision to determine the speed of the car *before* the collision. In the example with the carts, we determined the ratio of the speeds *after* the interaction.

Figure 3.9 As rocket engines give momentum to exhaust gases, they gain momentum in return in the opposite direction.

1. (True or False.) A conservation law can be used without knowing all of the details about what is occurring inside a system.
2. The linear momentum of a truck will always be greater than that of a bus if
   (a) the truck's mass is larger than the bus's but its speed is the same.
   (b) the truck's speed is larger than the bus's but its mass is the same.
   (c) both its mass and its speed are larger than the bus's.
   (d) Any of the above.
3. In a collision, the total _____ is the same before and after.
4. (True or False.) During a collision between two football players, the total linear momentum of the system cannot be zero.

ANSWERS: 1. True **2.** (d) **3.** linear momentum **4.** False

3.3 Work: The Key to Energy

The law of conservation of energy is arguably the most important of the conservation laws. It is not only useful for solving problems, but also is a powerful theoretical statement that can be used to understand widely diverse phenomena and to show what hypothetical processes are or are not possible. As we mentioned earlier, the concept of energy is one of the most important in physics. This is because energy takes many forms and is involved in all physical processes. One could say that every interaction in our universe involves a transfer of energy or a transformation of energy from one form to another.

We can compare the concept of energy to that of financial assets, which can take the form of cash, real estate, material goods, or investments, among other things. The study of economics is partly a study of these forms of financial assets and how they are transferred and transformed. Much of physics deals with the forms of energy and the energy transformations that occur during interactions.

3.3a Work

When first encountered, the concept of energy is a bit difficult to understand because there is no simple way to define it. As an aid, we will first introduce work, a physical quantity that is quite basic and that gives us a good foundation for understanding energy.

The idea of work in physics arises naturally when one considers *simple machines* such as the lever and the inclined plane when used in situations with negligible friction. Let's say that you use a lever to raise a heavy rock (**Figure 3.10**). If you

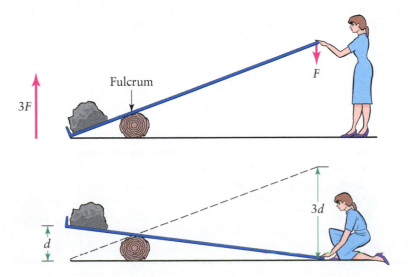

Figure 3.10 When a lever is used to raise a rock, a small force on the right end results in a larger force on the left end. But the right end moves a greater distance than the left end. The force multiplied by the distance moved *is the same* for both ends.

Figure 3.11 In using a crowbar to remove a nail from a piece of wood, a person does the same amount of work to complete the task as would be done in pulling it out using, say, a pair of pliers, but with a much smaller applied force.

place the fulcrum close to the rock, you find that a small, downward force on your end results in a larger, upward force on the rock. However, the distance your end of the lever moves as you push it down is correspondingly larger than the distance the rock is raised. By measuring the forces and distances, we find that dividing the larger force by the smaller force gives the same number (or ratio) as dividing the larger distance by the smaller distance. In particular,

$$\frac{F \text{ on left end}}{F \text{ on right end}} = \frac{d \text{ right end moves}}{d \text{ left end moves}}$$

We can rearrange this equation with a bit of algebra to get the following result:

$$(F \text{ on left}) \times (d \text{ left moves}) = (F \text{ on right}) \times (d \text{ right moves})$$

$$F_{\text{left}} \, d_{\text{left}} = F_{\text{right}} \, d_{\text{right}}$$

In other words, even though the two forces and the two distances are different, the quantity *force times distance* has the same value for both ends of the lever. We might say that raising the rock is a fixed task. One can perform the task by lifting the rock directly or by using a lever. In the former case, the force is large, equal to the rock's weight, but the distance moved is small. When using the lever, the applied force is smaller, but the distance through which the force acts is larger. The quantity *Fd* is the same, regardless of which way the task is accomplished. A similar circumstance would hold when using a crowbar to remove a nail embedded in a two-by-four (**Figure 3.11**).

We reach the same conclusion when considering an inclined plane. Let's say that a barrel must be placed on a loading dock (**Figure 3.12**). Lifting the barrel directly requires a large force acting through a small distance, the height of the dock. If the barrel is rolled up a ramp, a smaller force is needed, but the barrel must be moved a greater distance. Again, the product of the force and the distance moved is the same for the two methods.

$$(F \text{ lifting}) \times \text{height} = (F \text{ rolling}) \times (\text{ramp length})$$

$$F_{\text{lifting}} \, d_{\text{lifting}} = F_{\text{rolling}} \, d_{\text{rolling}}$$

The quantity *force times distance* is obviously a useful way of measuring the "size" of a task. It is called **work**.

DEFINITION	**Work** The force that acts times the distance moved in the direction of the force:

$$\text{work} = Fd$$

Physical Quantity	Metric Units	English Units
Work	joule (J) [SI unit]	foot-pound (ft-lb)
	erg	British thermal unit (Btu)
	calorie (cal)	
	kilowatt-hour (kWh)	

Figure 3.12 Two identical barrels need to be placed on the loading dock. One is lifted directly, requiring a large force. The other is rolled up the ramp, an inclined plane. A smaller force is needed, but it must act on the barrel over a longer distance.

EXAMPLE 3.3 Because of friction, a constant force of 100 newtons is needed to slide a box at a steady speed across a room (**Figure 3.13**). If the box moves 3 meters, how much work is done?

SOLUTION

$$\text{work} = Fd$$
$$= 100 \text{ N} \times 3 \text{ m}$$
$$= 300 \text{ N-m}$$

The unit in the answer to Example 3.3 is the newton-meter (N-m). This combination is called the *joule*.

$$1 \text{ joule} = 1 \text{ newton-meter} = 1 \text{ newton} \times 1 \text{ meter}$$
$$1 \text{ J} = 1 \text{ N-m}$$

The joule is a derived unit of measure, and it is the SI unit of work and energy. Just remember that if you use only SI units for force, distance, and so on, your unit for work and energy will always be the joule (J).

EXAMPLE 3.4 Let's say that the barrel in Figure 3.12 has a mass of 30 kilograms and that the height of the dock is 1.2 meters. How much work would you do when lifting the barrel?

SOLUTION For a constant lift speed, the force is just the weight of the barrel, *mg*.

$$F = W = mg = 30 \text{ kg} \times 9.8 \text{ m/s}^2 = 294 \text{ N}$$

Hence,

$$\text{work} = Fd = Wd$$
$$= 294 \text{ N} \times 1.2 \text{ m}$$
$$= 353 \text{ J}$$

The work you do when rolling the barrel up the ramp would be the same. The force would be smaller, but the distance would be larger.

Figure 3.14 When you carry a box across a room, the force on the box is perpendicular to the direction the box moves. No work (as a physicist defines the term) is done on the box by the force you exert when holding it up.

Physics To Go 3.1

You will have to be in a multistory building with stairs to complete this exercise. Walk around on a completely level surface (perhaps back and forth in a hallway) over a distance of 8 meters or so (about 25 ft). Wait a minute or two, and then climb the stairs upward for about 8 meters. Do you feel any different after completing the two tasks? In which "exercise" do you think you did the most work? In which did you exert the greater force in the direction of motion? How/where were these forces applied? (Be careful.)

3.3b Different Forces, Different Work

Because force is a vector, work equals the distance moved times the *component* of the force parallel to the motion. Work itself, however, is *not* a vector. There is no direction associated with work, although work may be positive or negative. When the force and the motion are in the same direction, the work is positive, as in Examples 3.3 and 3.4. When they are in the opposite direction, the work is negative.

But what about when there is *no* component of force along the direction of motion? For example, suppose the applied force is perpendicular to the displacement of the object. Then in this case, *no* work is done on the object by the force. When you carry a box across a room, your force on the box is vertical, whereas the displacement of the box is horizontal. There is no component of force along the direction of motion; hence, no work is done on the box (**Figure 3.14**). (You may wonder then why you feel so much more tired when carrying a box across a room than when simply walking the same distance. Revisit Physics to Go 3.1 if you need to.)

Uniform circular motion is another situation in which a force acts on a moving body but no work is done. Recall that a centripetal force must act on anything to keep it moving uniformly along a circular path (**Figure 3.15**). This force is always toward the center of the circle and perpendicular to the object's velocity at each instant. Therefore, the force does not do work on the object.

Work can be done in a circular motion by a force that is not a centripetal force. When you turn the crank on a pencil sharpener, a fishing reel, or an old-fashioned ice cream maker, for instance, you exert a force on the handle that is in the same direction as the handle's motion. Hence, you do work on the handle.

Work *is done* on an object when it is accelerated in a straight line. The following example shows how the amount of work can be computed. (In the next section, we will see that there is an easier way to do this.)

Figure 3.15 The string exerts a force on the ball that is perpendicular to its motion. No work is done by this force.

EXAMPLE 3.5 In Example 2.2, we used Newton's second law to compute the force needed to accelerate a 1,000-kilogram car from 0 to 27 m/s in 10 seconds. Our answer was $F = 2,700$ newtons. Using the definition of work, compute how much work is done in this process.

SOLUTION To find the distance that the car travels, we use the fact that the car accelerates steadily at 2.7 m/s² for 10 seconds. Using the equation from Section 1.4:

$$d = \frac{1}{2}\, at^2$$
$$= \frac{1}{2} \times 2.7 \text{ m/s}^2 \times (10 \text{ s})^2$$
$$= 1.35 \text{ m/s}^2 \times 100 \text{ s}^2 = 135 \text{ m}$$

So the work done is

$$\text{work} = Fd$$
$$= 2,700 \text{ N} \times 135 \text{ m}$$
$$= 364,500 \text{ J}$$

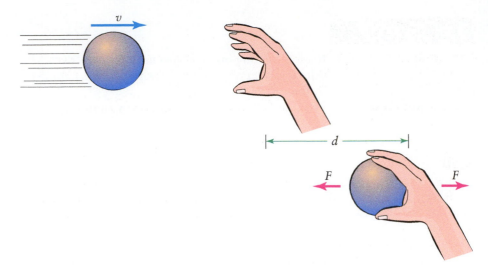

We have seen that work is done (a) when a force acts to move something against the force of gravity (Figures 3.10 and 3.12), (b) when a force acts to move something against friction (Figure 3.13), and (c) when a force accelerates an object. There are many other possibilities. When a force distorts something, work is done. For example, to compress or stretch a spring, a force must act on it. This force produces a displacement in the same direction as the force, so positive work is done.

Work is also done when a force causes something to slow down. When you catch a ball, your hand exerts a force on the ball. As the ball slows down, it pushes your hand back with an equal and opposite force (**Figure 3.16**). In this case, the ball does positive work on your hand. Your hand does negative work on the ball. The amount of work that the ball does on your hand is equal to the amount of work that was originally done on the ball to accelerate it (ignoring the effect of air resistance). If you allow your hand to move back as you catch the ball, the force of the ball on you will be less than if you try to keep your hand stationary. The work that the ball will do on your hand is the same either way. Since work = force × distance, the force on your hand will be smaller if the distance that the ball moves while you catch it is larger.

One last example: when an object falls freely, the force of gravity does work on it. As in Example 3.5 with the car, this work goes to accelerate the object. In particular, if a body falls a distance d, the work done on it by the force of gravity is

$$\text{work} = Fd$$

But

$$F = W = mg$$

So:

$$\text{work} = Wd = mgd$$

The work that the force of gravity does on an object as it falls is equal to the work that was done to lift the object the same distance. When something is lifted, we say that work is done *against* the force of gravity. The movement is in the opposite direction of this force. When something falls, work is done *by* the force of gravity. The movement is in the same direction as the force.

In summary, work is done by a force whenever the point of application moves in the direction of the force. Work is done against the force whenever the point of application moves opposite the direction of the force. Forces always come in pairs that are equal and opposite (Newton's third law of motion). Consequently, when work is done by one force, this work is being done against the other force in the pair.

Learning Check

1. (True or False.) The parking brake in most automobiles makes use of a lever so that the force the driver must exert when setting it is reduced. Consequently, the amount of work the driver must do is also reduced.

2. Work is done on an object when
   (a) it moves in a circle with constant speed.
   (b) it is accelerated in a straight line.
   (c) it is carried horizontally.
   (d) All of the above.

3. If a force acting on an object and the object's velocity are in opposite directions, the work done by the force is _____.

4. (True or False.) The force of gravity cannot do work.

ANSWERS: 1. False **2.** (b) **3.** negative **4.** False

DEFINITION **Energy** The measure of a system's capacity to do work. That which is transferred when work is done. Abbreviated E.

Figure 3.17 A crane can do work because it possesses stored chemical energy in its fuel.

3.4 Energy

In Section 3.3, we saw that work can often be "recovered." The work that is done when a ball is thrown is equal to the work done by the ball when it is caught. The work done when lifting an object against the force of gravity is equal to the work done by the force of gravity when the object falls. (This is the principle behind the operation of pile drivers at construction sites.) When work is done on something, it gains **energy**. This energy can then be used to do work. A thrown ball is given energy. This energy is given up to do work when the ball is caught.

The units of energy are the same as the units of work. Like work, energy is a scalar; it has no direction associated with it, although it can be positive or negative.

The more work done on something, the more energy it gains, and the more work it can do in return. We might say that energy is "stored work." To be able to do work, an object must have energy. When you lift a heavy tool such as a soil or gravel compactor/tamper, you are doing work on the tool and transferring some energy to it. The tamper can then do work in return when it is released and falls to the ground.

There are many forms of energy corresponding to the many ways in which work can be done. Some of the more common forms of energy are chemical, electrical, nuclear, and gravitational, as well as the energy associated with heat, sound, light, and other forms of radiation. Anything possessing any of these forms of energy is capable of doing work (**Figure 3.17**).

3.4a Mechanical Energy: Kinetic and Potential

In mechanics, there are two main forms of energy, which we can classify under the single heading of *mechanical energy*. Anything that has energy because of its motion or because of its position or configuration has mechanical energy. We refer to the former as **kinetic energy** and to the latter as **potential energy**.

Anything that is moving has kinetic energy. The simplest example is an object moving in a straight line. The amount of kinetic energy an object has depends on its mass and speed. In particular,

DEFINITION **Kinetic Energy** Energy resulting from motion. Energy that an object has because it is moving. Abbreviated KE.

$$KE = \frac{1}{2} mv^2 \qquad \text{(kinetic energy)}$$

The kinetic energy that an object has is equal to the work done when accelerating the object from rest. So another way to determine the amount of work done when accelerating an object is to compute its kinetic energy. This also shows that the work done depends only on the object's final speed and not on how rapidly or slowly it was accelerated.

EXAMPLE 3.6 In Example 3.5, we computed the work that is done on a 1,000-kilogram car as it accelerates from 0 to 27 m/s. Let's compute the total kinetic energy gained by the car during its acceleration.

SOLUTION The car's kinetic energy when it is traveling 27 m/s is

$$KE = \frac{1}{2} \, mv^2 = \frac{1}{2} \times 1,000 \text{ kg} \times (27 \text{ m/s})^2$$
$$= 500 \text{ kg} \times 729 \text{ m}^2/\text{s}^2$$
$$= 364,500 \text{ J}$$

This is the same as the work done as the car accelerates. The car can do 364,500 joules of work because of its motion.

The kinetic energy of a moving body is proportional to the square of its speed. If one car is going twice as fast as a second, identical car, the faster one has four times the kinetic energy. It takes four times as much work to stop the faster car. Note that the kinetic energy of an object can never be negative. (Why? Because m is always positive and v^2 is positive even when v is negative.)

Because speed is relative, kinetic energy is also relative. A runner on a moving ship has KE relative to the ship and a different KE relative to a person standing on the dock (see Figure 1.6).

Another way that an object can have kinetic energy is by rotating. To make something spin, work must be done on it. A dancer or skater performing a pirouette has kinetic energy (**Figure 3.18**). A spinning top has kinetic energy, as do Earth, the Moon, the Sun, and other astronomical objects as they spin about their axes. The amount of kinetic energy that a spinning object has depends on its mass, its rotation rate, and the way its mass is distributed. Many common objects, from simple children's toys to modern hybrid vehicles, employ rotating elements, like disks and gears, as a means to "store" rotational kinetic energy and/or to smooth out the delivery of energy in systems where the source is discontinuous (e.g., reciprocating engines). **Figure 3.19** shows a toy car and an exercise machine that operate this way.

Figure 3.18 A spinning dancer has kinetic energy, even though she stays in one place.

(a)

Flywheel

(b)

Figure 3.19 This transparent toy car (a) and exercise machine (b) use rotational kinetic energy stored in spinning flywheels. The car's flywheel is the disc with the spiral yellow, orange, and red pattern.

DEFINITION Potential Energy
Energy resulting from an object's position or orientation. Energy that a system has because of its configuration. Abbreviated *PE*.

The amount of **potential energy** that a system acquires is equal to the work done to put it into a given orientation, position, or configuration relative to other objects. When an object is lifted, it is given potential energy. The energy thus acquired can then be used to do work. For example, when the weights on a cuckoo clock or any other gravity-powered clock are raised, they are given potential energy (**Figure 3.20**). As they slowly fall, they do work in operating the clock.

In Section 3.3, we computed the work done when an object is lifted. This is equal to the potential energy that it is given. Because the work is done against the force of gravity, it is called **gravitational potential energy**.

$$PE = \text{work done} = \text{weight} \times \text{vertical displacement} = Wd = mgd$$

$$PE = mgd \qquad \text{(gravitational potential energy)}$$

Gravitational potential energy is a common type of potential energy, and it is often referred to simply as potential energy. Potential energy is a relative quantity because height can be measured with respect to different starting levels.

Figure 3.20 The source of energy to run this clock is the potential energy of the weights.

EXAMPLE 3.7 A 3-kilogram brick is lifted to a height of 0.5 meters above a table (**Figure 3.21**). What is its potential energy relative to the table top? to the floor?

SOLUTION

$$PE = mgd$$

$$PE = 3 \text{ kg} \times 9.8 \text{ m/s}^2 \times 0.5 \text{ m}$$

$$= 14.7 \text{ J} \quad \text{(relative to the table)}$$

But the tabletop itself may be 1 meter above the floor. The brick's height above the floor is 1.5 meters, and so its potential energy relative to the floor is

$$PE = mgd$$

$$= 3 \text{ kg} \times 9.8 \text{ m/s}^2 \times 1.5 \text{ m}$$

$$= 44.1 \text{ J} \quad \text{(relative to the floor)}$$

A person sitting in a chair has gravitational potential energy relative to the floor, to the basement of the building, and to the level of the oceans. Usually, some convenient reference level is chosen for determining potential energies. In a room it is handy to use the floor as the reference level for measuring heights and, consequently, potential energies.

An object's potential energy will be negative when it is *below* the chosen reference level. Often the reference level is chosen so that negative potential energies signify that an object is "trapped" or "bound" in the system. For example, it is convenient to measure potential energy relative to the ground level on a flat area outdoors. Anything on the ground has zero potential energy, and anything above the ground has positive potential energy. If there is a hole in the ground, any object in the hole will have negative potential energy (**Figure 3.22**). Any object that has zero or positive potential energy can move about horizontally if it has any kinetic energy. Objects with negative potential energy are confined to the hole. They must be given enough energy to get out of the hole before they can move horizontally.

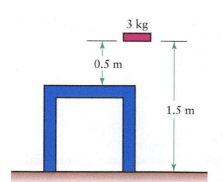

Figure 3.21 The potential energy (*PE*) of an object depends on its height with respect to some specified reference level. The brick has 14.7 J of *PE* relative to the table, but 44.1 J of *PE* relative to the floor.

3.4b Other Forms of Potential Energy

Springs, rubber bands, and other flexible media can possess another type of potential energy, **elastic potential energy**. Work must be done on a spring to stretch or compress it. This gives the spring potential energy. This "stored energy" can then be used to do work. The actual amount of potential energy a spring has depends on two things: how much it is stretched or compressed and how strong it is. In Figure 3.7, the combined kinetic energies of the carts after the spring is released equal the original elastic potential energy of the spring. A stronger spring would possess more potential energy and would give the carts more *KE*, so they would go faster.

Using Hooke's law (Section 2.5b), it may be demonstrated that the equation for the potential energy stored in a spring stretched or compressed by a distance *d* from its relaxed state is

$$PE_{\text{spring}} = \frac{1}{2}\,kd^2$$

where *k* is the spring constant. Thus, if the stretch of a spring is doubled, its potential energy will be quadrupled. This energy dependence is similar in form to that of the kinetic energy, which varies as the square of the velocity.

Many devices use elastic potential energy. Toy dart guns have a spring inside that is compressed by the shooter. When the trigger is pulled, the spring is released and does work on the dart to accelerate it. The potential energy of the spring is converted to kinetic energy of the dart. The bow and arrow operate this same way, with the bow acting as a spring (Figure 3.23). Windup devices such as clocks, toys, and music boxes use energy stored in springs to operate the mechanism. Usually, the spring is in a spiral shape, but the principle is the same. Rubber bands provide lightweight energy storage in some toy airplanes.

Michael Steele/Getty Images Sport/Getty Images

EXAMPLE 3.8 In Example 2.4, a spring with spring constant of 2.4 N/m was stretched from equilibrium by a distance of 0.30 m. How much energy was stored in the spring under these conditions?

SOLUTION

$$PE_{\text{spring}} = \frac{1}{2}\,kd^2 = \frac{1}{2} \times 2.4\ \text{N/m} \times (0.30\ \text{m})^2$$

$$= \frac{1}{2} \times 2.4\ \text{N/m} \times 0.09\ \text{m}^2$$

$$= 0.108\ \text{N-m} = 0.108\ \text{J}$$

Figure 3.23 Yun Ok-Hee of South Korea takes aim during the women's archery individual bronze medal match at the 2008 Olympic Games on 14 August 2008 in Beijing, China. While pulling the bowstring back, she does work on the bow. This gives the bow elastic potential energy.

Figure 3.24 The Leonid meteor shower. A meteoroid leaves a glowing trail in the night sky. Air resistance causes its kinetic energy to be converted into internal energy (heat) and radiant energy (light).

Another form of energy important in mechanical systems is **internal energy**. Internal energy, heat, and temperature are discussed formally in Chapter 5. Basically, the internal energy of a substance is the total energy of all the atoms and molecules in the substance. To raise something's temperature, to melt a solid, or to boil a liquid all require increasing the internal energy of the substance. Internal energy decreases when a substance's temperature decreases, when a liquid freezes, or when a gas condenses.

Internal energy is involved whenever there is kinetic friction. In Figure 3.13, the work done on the box as it is pushed across the floor is converted into internal energy because of the friction between the box and the floor. The result is that the temperatures of the floor and the box are raised (although not by much). As a car or a bicycle brakes to a stop, its kinetic energy is converted into internal energy in the brakes. Automobile disc brakes can become red hot under extreme braking. Meteors ("shooting stars") are a spectacular example of kinetic energy being converted into internal energy (Figure 3.24). When they enter the atmosphere at very high speed, air resistance heats them enough to glow and melt. Gravitational potential energy can also be converted into internal energy. A box slowly sliding down a ramp has its gravitational potential energy converted into internal energy by friction. If you climb a rope and then slide down it, you can burn your hands severely as some of your potential energy is converted into internal energy because of the friction between your hands and the rope.

Internal energy can also be produced by internal friction when something is distorted. The work you do when stretching a rubber band, pulling taffy, or crushing an aluminum can generates internal energy. When you drop something like a book and it doesn't bounce, most of the energy the object had is converted into internal energy on impact.

Internal energy arising from frictional interactions or physical deformations, unlike kinetic energy and potential energy, usually cannot be recovered. Work done to lift a box and give it potential energy can be recovered as work or some other form of energy. Work done to slide a box across a floor becomes internal energy that is "lost" (made unavailable). In every mechanical process, some energy is converted into internal energy. It has been estimated that in the United States the annual financial losses associated with overcoming friction are several hundred billion dollars. (See "Friction: A Sticky Subject" at the end of Section 2.1.)

In summary, work always results in a transfer of energy from one thing to another, in a transformation of energy from one form to another, or both. In our earlier analogy in which we compared energy to financial assets, work plays the role of a transaction such as buying, selling, earning, or trading. These transactions can be used to increase or decrease the net worth of an individual or to convert one form of asset into another. Work done *on* a system increases its energy. Work done *by* the system decreases the energy of the system. Work done *within* the system results in one form of energy being changed into another.

Learning Check

1. To be able to do work, a system must have _____.

2. Identical cars A and B are traveling down a highway, with A going two times as fast as B. The kinetic energy of A is
   (a) 2 times as large as the kinetic energy of B.
   (b) 4 times as large as the kinetic energy of B.
   (c) one-half as large as the kinetic energy of B.
   (d) the same as that of B, because they have the same mass.

3. (True or False.) When you lift an object, its gravitational potential energy decreases.

4. A hiker loads camping gear into her SUV. In the process, the rear springs of her vehicle are compressed by a factor of two over their usual length. The potential energy stored in each of the springs under this condition relative to its normal value is
   (a) half as large.
   (b) twice as large.
   (c) one-fourth as large.
   (d) four times as large.

5. As a baseball player slides into home, the player's kinetic energy is converted into _____ energy.

ANSWERS: 1. energy, **2.** (b) **3.** False **4.** (d) **5.** internal

3.5 The Conservation of Energy

In the preceding section, we described several situations in which one form of energy was converted into another. These included a dart gun (potential energy in a spring converted to kinetic energy of the dart), a car braking (kinetic energy converted into internal energy), and a box sliding down an inclined plane (gravitational potential energy becoming internal energy). There are countless situations involving many other forms of energy.

Many devices in common use are simply energy converters. Some examples are listed in Table 3.1. Some of these involve more than one conversion. In a hydroelectric dam (see Figure 3.25), the potential energy of the water behind the dam is converted into kinetic energy of the water. The moving water then hits a turbine (propeller), which gives it rotational kinetic energy. The rotating turbine turns a generator, which converts the kinetic energy into electrical energy.

Internal energy is an intermediate form of energy in both the car engine and the nuclear power plant. In a car engine, the chemical energy in fuel is first converted into internal energy as the fuel burns (explodes). The heated gases expand and push against pistons (or rotors in rotary engines). These in turn make a crankshaft and flywheel rotate to ultimately turn the drive shaft and wheels. In a nuclear power plant, nuclear energy is converted into internal energy in the reactor core. This internal energy is used to boil water into steam. The steam is used to turn a turbine, which turns a generator that produces electrical energy.

Internal energy is also a "wasted" by-product in all of the devices in Table 3.1. Any device that has moving parts has some kinetic friction. Some of the input

Table 3.1 Some Energy Converters

Device	Energy Conversion
Lightbulb	Electrical energy to radiant energy
Car engine	Chemical energy (in fuel) to kinetic energy
Battery	Chemical energy to electrical energy
Elevator	Electrical energy to gravitational potential energy
Generator	Kinetic energy to electrical energy
Electric motor	Electrical energy to kinetic energy
Solar cell	Radiant energy (light) to electrical energy
Flute	Kinetic energy (of air) to acoustic energy (of sound wave)
Nuclear power plant	Nuclear energy to electrical energy
Hydroelectric dam	Gravitational potential energy (of water) to electrical energy

Figure 3.25 A hydroelectric power station uses several energy conversions.

energy is converted into internal energy by this friction. The generator, electric motor, car engine, and the electrical power plants all suffer unavoidable frictional losses (Figure 3.26). In some of the devices, internal energy is produced because of the nature of the process. More than 95% of the electrical energy used by an incandescent lightbulb is converted to internal energy (heat), not usable light. More than 60% of the available energy in coal-fired and nuclear power plants goes to unused internal energy. We will investigate this issue further in later chapters.

Even though there are many different forms of energy and countless devices that involve energy conversions, the following law always holds.

> **LAWS** **Law of Conservation of Energy** Energy cannot be created or destroyed, only converted from one form to another. The total energy in an isolated system is constant.

For a system to be isolated, energy cannot enter or leave it. For a mechanical system, work cannot be done on the system by an outside force, nor can the system do work on anything outside of it.

This law means that energy is a commodity that cannot be produced from nothing or disappear into nothing. If work is being done or a form of energy is "appearing," then energy is being used or converted somewhere. Unlike money, which you can counterfeit or burn, you cannot manufacture or eliminate energy.

The law of conservation of energy is both a practical and a theoretical tool. It can be used to solve engineering problems, notably in mechanics, and it is a necessary condition that proposed theoretical models must satisfy. As an example of the latter, a theoretical astrophysicist may develop a model that explains how stars convert nuclear energy into heat and radiation. A first test of the validity of the model is whether it conserves energy or not when applied.

3.5a Applications of Energy Conservation

We now illustrate the practical usefulness of the law by considering some mechanical systems. In each case, we assume that friction is negligible so we do not have to take into account any conversion of potential energy or kinetic energy into internal energy.

The basic approach in using the law of conservation of energy is the same as that used with the corresponding law for linear momentum. If there is a conversion of energy in the system from one form to another, then the total energy before the conversion equals the total energy after the conversion.

$$\text{total energy before} = \text{total energy after}$$

Figure 3.26 This wind generator converts kinetic energy of the wind into electrical energy plus some internal energy because of friction.

PE = 18 J
KE = 0

3 m

PE = 12 J
KE = 6 J

PE = 0
KE = 18 J

iStockphoto.com/BenGoode

iStockphoto.com/Gmcoop

Figure 3.27 A basketball rolls off the rim and falls to the floor. Initially, it has potential energy only. As it falls, its potential energy decreases as its kinetic energy increases. Just before it hits the floor, it has kinetic energy only. At each point as it falls, its total energy, kinetic plus potential, is the same, 18 joules.

A good example that we have encountered before (in Sections 1.4 and 2.5) is the motion of a freely falling body. If an object is raised to a height d, it has gravitational potential energy. If it is released, the object will fall and convert its potential energy into kinetic energy. It is a continuous process. As the object falls, its potential energy decreases because its height decreases; at the same time, its kinetic energy increases because its speed increases. If it is falling freely, there is no air resistance, and the only two forms of energy are kinetic and gravitational potential. Energy conservation then means that the sum of the kinetic energy and the potential energy of the object is always the same (**Figure 3.27**).

$$E = KE + PE = \text{constant}$$

We can use this fact to show how the object's speed just before it hits the floor depends on the height d from which it is released. We do this by looking at the object's energy just as it is dropped and then just before it hits. At the instant it is released, its kinetic energy is zero because its speed is zero and its potential energy is mgd. So its initial energy is

$$E_i = KE + PE = 0 + PE$$
$$= PE = mgd \quad \text{(just released)}$$

When it reaches the floor, at the instant before impact, its potential energy is zero. So, its final energy is

$$E_f = KE + PE = KE + 0$$
$$= KE = \frac{1}{2}mv^2 \quad \text{(just before impact)}$$

These two quantities are equal because the object's total energy has not changed, only its form. The potential energy the object had when it was released equals the kinetic energy it has just before impact.

$$\frac{1}{2}mv^2 = mgd$$

Dividing both sides by m and multiplying by 2, we get

$$v^2 = 2gd$$
$$v = \sqrt{2gd} \quad \text{(speed after falling distance } d\text{)}$$

EXAMPLE 3.9 In 2003, a man went over Horseshoe Falls, part of Niagara Falls, and survived. He is thought to be the first person to do so without the aid of any safety devices. The height of the falls is about 50 meters. Estimate the speed of the man when he hit the water at the bottom of the falls.

SOLUTION Assuming that the air resistance is too small to affect the motion appreciably, we can use the preceding result:

$$v = \sqrt{2gd} = \sqrt{2 \times 9.8 \text{ m/s}^2 \times 50 \text{ m}}$$

$$= \sqrt{980 \text{ m}^2/\text{s}^2}$$

$$= 31.3 \text{ m/s} \qquad \text{(about 70 mph)}$$

The speed of a freely falling body does not depend on its mass. It only depends on how far it has fallen (d) and the acceleration resulting from gravity. Table 3.2 shows the speed of an object after it has fallen various distances. In Section 1.4 we illustrated the relationships between speed and time and between distance and time. This completes the picture.

We have the reverse situation when an object is thrown or projected straight up. It starts with kinetic energy, rises until all of that has been converted into potential energy, and then falls (Figure 2.21). The conservation of energy tells us that the kinetic energy that the object begins with equals its potential energy at the highest point. This results in the same equation relating the initial speed v to the maximum height reached, d:

$$v^2 = 2gd$$

We can rewrite this as

$$d = \frac{v^2}{2g}$$

To compute how high something will go when thrown straight upward, insert its initial speed for v in this equation. We can also use Table 3.2 "backward": a ball thrown upward at 12 mph will reach a height of 5 feet.

Table 3.2 Speed Versus Distance for a Freely Falling Body

A. SI Units		B. English Units	
Distance d (m)	Speed v (m/s)	Distance d (ft)	Speed v (mph)
0	0	0	0
1	4.4	1	5.5
2	6.3	2	7.7
3	7.7	3	9.5
4	8.9	4	11
5	9.9	5	12
10	14	10	17
20	20	20	24
100	44	100	55

Note. Parts A and B are independent. The distances and speeds in A and B are not equivalent.

Figure 3.28 Setup for Physics to Go 3.2.

Battery

Backstop

l

l

▶ Physics To Go 3.2

Battery race. You will need two cylindrical batteries (AAs or AAAs work well), a ramp for the batteries to roll on (a piece of cardboard or a thin hardcover book), and a book or some other kind of backstop.

1. Place the backstop in front of the ramp a distance from the ramp's lower end equal to its length (**Figure 3.28**). Release a battery from the top of the ramp and then from places lower down the ramp. Is it traveling the same speed at the bottom each time? Why?

2. Hold one battery at the top of the ramp and the other one one-fourth of the way up from the bottom of the ramp. Release them at the same time. Which one wins the race to the backstop? Repeat with the lower battery starting at different places on the ramp. What happens in these races?

Now let's consider a similar problem. A roller coaster starts from rest at a height d above the ground. It rolls without friction or air resistance down a hill (**Figure 3.29**). What is its speed when it reaches the bottom?

Again, the only forms of energy are gravitational potential energy and kinetic energy, because we assume there is no friction. The total energy, which is kinetic energy plus potential energy, is constant. The kinetic energy of the roller coaster at the bottom of the hill must equal its potential energy at the top:

$$KE \text{ (at bottom)} = PE \text{ (at top)}$$

$$\frac{1}{2} mv^2 = mgd$$

This is the same result obtained for a freely falling body. The speed at the bottom is consequently given by the same equation:

$$v = \sqrt{2gd}$$

It would have been impossible to solve this problem using only the tools from Chapter 2. To use Newton's second law, $F = ma$, one needs to know the net force that acts at every instant. The net force driving the roller coaster along its path varies as the slope of the hill changes, so it is a very complicated

d

v

v

Figure 3.29 As a roller coaster travels down a hill, its potential energy is converted into kinetic energy. If there is no friction, its kinetic energy at the bottom equals its potential energy at the top. Its speed at the bottom is the same as that of an object dropped from the same height.

Stone/Getty Images

problem. The principle of energy conservation allows us to easily solve a problem that we could not have solved before. In the process, we also come up with the following general result: the law of conservation of energy tells us that for an object affected by gravity but not friction, the speed that it has at a distance (*d*) below its starting point is given by the preceding equation *regardless of the path it takes.* The speed of a roller coaster that rolls down a hill is the same as that of an object that falls vertically the same height. The roller coaster does take more time to build up that speed (its acceleration is smaller), and the falling body therefore reaches the ground sooner.

The motion of a pendulum involves the continuous conversion of gravitational potential energy into kinetic energy and back again. Let's say that a child on a swing is pulled back (and up; **Figure 3.30**). The child then has gravitational potential energy because of his or her height above the rest position of the swing. When released, the child swings downward, and the potential energy is converted into kinetic energy. At the lowest point in the arc, the child has only kinetic energy, which equals the original potential energy. The child then swings upward and converts the kinetic energy back into potential energy. This continues until the swing stops at a point nearly level with the starting point. This process is repeated over and over as the child swings. Air resistance takes away some of the kinetic energy. If the child is pushed each time, the work done puts energy into the system and counteracts the effect of the air resistance. Without air resistance or any other friction, the child would not have to be pushed each time and would continue swinging indefinitely.

The maximum height that a pendulum reaches (at the turning points) depends on its total energy. The more energy a pendulum has, the higher the turning points. In Section 3.2 we described a way to measure the speed of a bullet or a thrown object (Figure 3.6). The law of conservation of linear momentum is used to relate the speed of the bullet before the collision to the speed of the block (and bullet) afterward. If the wood block is hanging from a string, the kinetic energy it gets from the impact causes it to swing up like a pendulum. The more energy it gets, the higher it will swing. We can determine the speed of the block after impact by measuring how high the block swings.

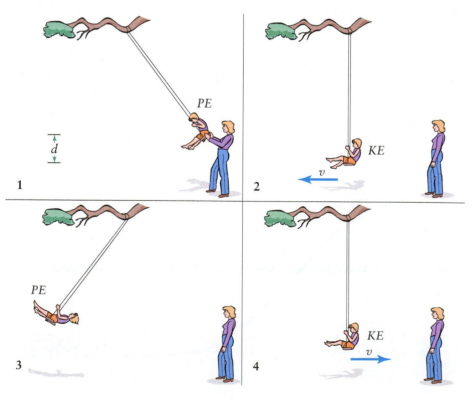

Figure 3.30 As a child swings back and forth, gravitational potential energy is continually converted into kinetic energy and back again. The potential energy at the highest (turning) points equals the kinetic energy at the lowest point.

The potential energy of the block (and bullet) at the high point of the swing equals the kinetic energy of the block (and bullet) right after impact. This results in the same equation relating the speed at the low point to the height reached:

$$v = \sqrt{2gd}$$

In Section 2.5b, we discussed the motion of an object hanging from a spring (Figure 2.24). The motion also consists of a continual conversion of potential energy into kinetic energy and back again.

Figure 3.31 shows a ball being putted at a miniature golf course. Because the ball rests in a small valley below ground level, its potential energy is negative relative to the level ground. When the ball is not moving, its total energy is negative because its kinetic energy is zero and its potential energy is negative. The golfer gives the ball kinetic energy by hitting it with a club. A weak putt gives it enough energy to roll back and forth but not to "escape" from the hole (Figure 3.31a). The ball's total energy is larger but still negative. In Figure 3.31b, the golfer gives the ball more energy by hitting it harder. Because its total energy is still negative, the ball again oscillates back and forth, although reaching a higher point on each side before turning around. In Figure 3.31c, the golfer hits the ball just hard enough for the ball to roll out of the valley and stop once it is out. The ball is given just enough kinetic energy to make its total energy equal to zero, and it "escapes" from the little valley. If the ball were hit even harder, it would escape and have excess kinetic energy—it would continue to roll on the level ground.

This is the principle behind rocking a car when it is stuck. If a tire is in a hole, it is best to make the car oscillate back and forth. By giving it some energy during each cycle, by pushing or by using the engine, one can often provide the car enough energy to leave the hole.

There are many analogous systems in physics in which an object is bound unless its total energy is equal to or greater than some value. A satellite in orbit around Earth is an important example. The satellite's motion from one side of Earth to the other and back is similar to the motion of the golf ball in the valley. If it is given enough energy, the satellite will escape from Earth and move away, much like the golf ball. The minimum speed that will give a satellite enough energy to leave Earth is called the *escape velocity*. Its value is approximately 11,200 m/s or 25,000 mph.

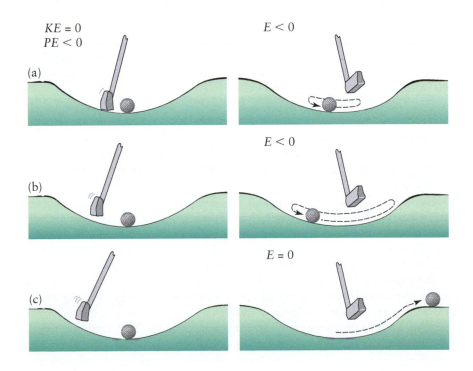

$KE = 0$
$PE < 0$

(a)

$E < 0$

(b)

$E < 0$

(c)

$E = 0$

Figure 3.31 A golf ball at rest in the small valley has negative potential energy. Hitting the golf ball gives it kinetic energy, but it oscillates inside the valley if its total energy is negative, (a) and (b). If the golf ball is given enough kinetic energy to make its total energy zero, it rolls out of the valley and stops (c).

When water boils, the individual water molecules are given sufficient energy to break free from the liquid (Chapter 5). Sparks and lightning occur only after electrons are given enough energy to break free from their atoms (Chapter 7). The transition of a system from a bound state to a free state is quite common in physics and is not limited to mechanical systems.

Learning Check

1. As a skier gains speed while gliding down a slope, _____ energy is being converted into _____ energy.
2. (True or False.) Work done inside an isolated system can increase the total energy in the system.
3. (Choose the *incorrect* statement.) The total kinetic energy plus potential energy of a body
   (a) can be negative.
   (b) always remains constant if the body is falling freely.

(c) always remains constant if friction is acting.
(d) can remain constant even if the body's speed is decreasing.

4. Diver A jumps off a platform and is going 5 m/s when entering the water. To be going 10 m/s when entering the water, diver B would have to jump off a platform that is _____ times as high as the platform A used.

ANSWERS: 1. potential, kinetic **2.** False **3.** (c) **4.** four.

3.6 Collisions: An Energy Perspective

Earlier in this chapter, we pointed out that the main "tool" for studying all collisions is the law of conservation of linear momentum (Section 3.2). In this section, we look at collisions from an energy standpoint. In some collisions, the only form of energy involved, before and after, is kinetic energy. In other collisions, forms of energy like potential energy and internal energy play a role.

3.6a Types of Collisions

Collisions may be classified as follows.

> **DEFINITION** **An Elastic Collision** is one in which the total kinetic energy of the colliding bodies after the collision *equals* the total kinetic energy before the collision.
>
> **An Inelastic Collision** is one in which the total kinetic energy of the colliding bodies after the collision *is not equal* to the total kinetic energy before. The total kinetic energy after can be greater than, or less than, the total kinetic energy before.

In an elastic collision, kinetic energy is conserved. The total energy is *always* conserved in both types of collisions, but in elastic collisions no energy conversions take place that make the total kinetic energy after different from the total kinetic energy before.

Figure 3.32 illustrates examples of these two types of collisions. Two equal-mass carts traveling with the same speed but in opposite directions collide. In both collisions, the total linear momentum before the collision equals the total after. (This total is equal to zero. Why?) In Figure 3.32a, the carts bounce apart because of a spring attached to one of them. After the collision, each cart has the same speed it had before, but it is going in the opposite direction. Consequently, the total kinetic energy of the two carts is the same after the collision as it was before. This is an elastic collision.

Figure 3.32b is an example of an inelastic collision. This time the two carts stick together (because of putty on one of them) and stop. The total kinetic

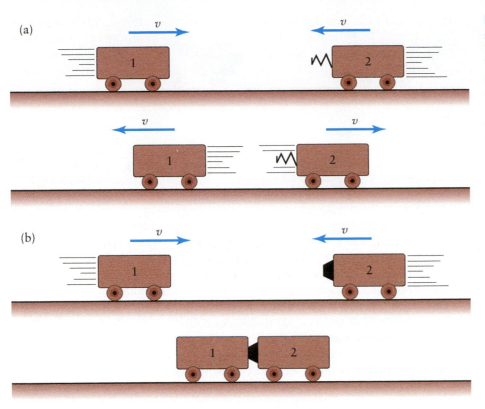

v

1

v

2

v

1

v

2

(b)

v

1

2

1

2

Figure 3.32 (a) Two carts with the same mass m and speed v collide head on and bounce apart. The total kinetic energy of the carts is the same before and after the collision. (b) This time the two carts stick together after the collision. In this case, the kinetic energy after the collision is zero.

energy after the collision is zero in this case. The automobile collision analyzed in Example 3.2 (Figure 3.5) is also an inelastic collision. To see this, use the information to calculate the total kinetic energy before and after the collision.

EXAMPLE 3.10   Recall the automobile collision analyzed in Example 3.2 (Figure 3.5). Compare the amounts of kinetic energy in the system before and after the collision.

The kinetic energy before the collision was

SOLUTION

The kinetic energy before the collision was

$$KE_{\text{before}} = \frac{1}{2} \times 1{,}000 \text{ kg} \times (10 \text{ m/s})^2$$

$$= 50{,}000 \text{ J}$$

The kinetic energy after the collision was

$$KE_{\text{after}} = \frac{1}{2} \times 2{,}500 \text{ kg} \times (4 \text{ m/s})^2$$

$$= 20{,}000 \text{ J}$$

So 30,000 joules (60%) of the kinetic energy before the collision was converted into other forms of energy.

In these two examples of inelastic collisions, part or all of the original kinetic energy of the colliding bodies is converted into other forms of energy, mostly internal energy, and also some sound (the "crash" that we would hear). In Figure 3.32b, *all* of the kinetic energy is converted into other forms of energy.

In some collisions, the total kinetic energy after the collision is greater than the total kinetic energy before the collision. If our old friend the cart, with its plunger pushed in ("loaded"), is struck by a second cart, the plunger will be

Figure 3.33 Cart 2 has energy stored in its spring-loaded plunger. When this cart is struck by cart 1, this potential energy is converted into kinetic energy, which is then shared by both carts. The total kinetic energy after the collision is greater than the total kinetic energy before the collision.

released by the shock (**Figure 3.33**). The plunger's potential energy is transferred to both carts as kinetic energy. Therefore, the total kinetic energy after the collision is greater than the total kinetic energy before the collision. It is an *inelastic* collision. The stored energy is released by the collision.

Collisions are also responsible for other phenomena such as gas pressure and the conduction of heat. Air molecules colliding with the inner surface of a balloon keep the balloon inflated. When you touch a piece of ice, the molecules in your finger collide with, and lose energy to, the molecules of the ice. This lowers the temperature of your finger.

3.6b Noncontact Collisions

The use of collisions is an invaluable tool in studying the structure and properties of atoms and nuclei in cases where the forces "act at a distance" and there is no direct contact between the interacting bodies. Much of the information in Chapters 10, 11, and 12 was gleaned from the careful analysis of countless collisions of this type. Linear accelerators, cyclotrons, tevatrons, and other devices produce high-speed collisions between atoms, nuclei, and subatomic particles. The collisions, which involve electrical and nuclear forces, are recorded and analyzed using the law of conservation of linear momentum and other principles. If a collision is inelastic, the amount of kinetic energy "lost" or "gained" in the collision is useful for determining the properties of the colliding particles.

Elastic collisions involving the force of gravity (and no physical contact) are used in space exploration. Called the *slingshot effect* or *gravity assist*, the technique involves having a spacecraft overtake a planet and pass it on its side away from Earth. The spacecraft gains kinetic energy from the planet, the way the eight ball gains kinetic energy in the collision depicted in Figure 3.4. The planet loses kinetic energy, but the planet is so huge that its decrease in speed is imperceptible—like the effect of a beach ball bouncing off the front end of a moving semitruck. Space probes sent to the outer part of the solar system (beyond Jupiter) such as the *New Horizons* craft discussed in the introduction to Chapter 2 have relied on gravity assists. The *Voyager 2* spacecraft used gravity assists from Jupiter, Saturn, and Uranus on its journey to Neptune and beyond. (Its speed was increased by 10 miles per second by the Jupiter gravity assist.)

Recent missions have exploited gravity assists from the inner planets, thereby allowing heavy spacecraft to be started on their journeys using relatively small rockets. On the way to its 1995 arrival at Jupiter, the *Galileo* spacecraft used two gravity assists from Earth and one from Venus. The 6-ton *Cassini* spacecraft used two gravity assists from Venus, one from Earth, and one from Jupiter to reach Saturn in 2004. Gravity assists can also be used to slow the motion of spacecraft. The *Messenger* probe launched in 2004 to study the chemistry and geological history of Mercury's surface and the nature of its magnetic field, among other things, used such maneuvers in flybys of Venus in 2006 and 2007 to put the spacecraft on trajectory for its first flyby of Mercury in January 2008.

Figure 3.34 The path of the *Messenger* probe on its journey to conduct an exploration of the surface of Mercury. Along the way, the spacecraft carried out several complex gravity assist maneuvers to reduce its speed, the most significant occurring during its 5 June 2007 flyby of Venus.

Two additional flybys of Mercury, the last in September 2009, paved the way for the probe's insertion into orbit around Mercury in March 2011 (**Figure 3.34**).

During its four-year survey of the topography, atmosphere, and magnetosphere of Mercury, *Messenger* made numerous important discoveries about this planet. Two of the primary new findings include: (1) the detection of water ice near the polar regions in craters that are permanently shadowed from the Sun, and (2) the determination that Mercury's composition includes a larger-than-predicted abundance of volatile elements with low boiling points, placing this planet in line with the other terrestrial planets in our solar system (Venus, Earth, and Mars) and suggesting a common formation scenario for all four bodies. *Messenger's* mission ended on 3 June 2015 when the spacecraft was guided to impact on the surface of Mercury.

■ Learning Check

1. In a collision that is inelastic, the total _____ after the collision is not the same as before the collision.

2. In an elastic collision between two bodies, which of the following is *not* true:
   (a) the total linear momentum of the bodies is the same before and after the collision.
   (b) the total kinetic energy of the bodies is the same before and after the collision.
   (c) the total energy (all forms) of the bodies is the same before and after the collision.
   (d) the two bodies must have equal masses.

3. (True or False.) A spacecraft can gain or lose energy by passing near a planet.

ANSWERS: 1. kinetic energy **2.** (d) **3.** True

3.7 Power

We have seen many examples of work being done and energy being transformed into other forms. The amount of time involved in these processes has not entered into the discussion until now. Let's say that a ton of bricks needs to be loaded from the ground onto a truck (**Figure 3.35**). This might be done in one of two ways. First, a person could lift the bricks one at a time and place them on the truck. This might take the person an hour. Second, a forklift could be used to load the bricks all at once. This might take only 10 seconds. In both

Figure 3.35 A ton of bricks is loaded onto a truck in two different ways. The work done is the same, but because the forklift does the job much faster, the power is much greater.

cases, the same amount of work is done. The force on each brick (its weight) times the vertical distance it is moved (the height of the truck bed) is the same whether the bricks are loaded one at a time or all at once. The work done, *Fd*, is the same in both cases, but the **power** is different.

> **DEFINITION** **Power** The rate of doing work. The rate at which energy is transferred or transformed. Work done divided by the time. Energy transferred divided by the time.
>
> $$P = \frac{\text{work}}{t} \qquad P = \frac{E}{t}$$

Physical Quantity	Metric Units	English Units
Power (*P*)	watt (W)	foot-pound/second (ft-lb/s)
		horsepower (hp)

In SI units, the weight of a ton of bricks is 8,900 newtons. If the height of the truck bed is 1.2 meters, then the work done is

$$\text{work} = Fd = 8,900 \text{ N} \times 1.2 \text{ m} = 10,680 \text{ J}$$

If the forklift does this work in 10 seconds, the power is

$$P = \frac{\text{work}}{t} = \frac{10,680 \text{ J}}{10 \text{ s}}$$
$$= 1,068 \text{ J/s} = 1,068 \text{ W}$$

The unit joule per second (J/s) is defined to be the watt (W), the SI unit of power.

$$1 \text{ watt} = \frac{1 \text{ joule}}{1 \text{ second}}$$
$$1 \text{ W} = 1 \text{ J/s}$$

The watt will likely be familiar to you because it is commonly used to measure the power consumption of electrical devices. A 60-watt lightbulb uses electrical energy at the rate of 60 joules each second. A 1,600-watt hair dryer uses 1,600 joules of energy each second (**Figure 3.36**).

Horsepower is the most commonly used unit of power in the English system. Engines in automobiles, lawn mowers, water pleasure craft, and many other motorized devices are rated in horsepower. The basic power unit, foot-pound per second, is the unit of work, foot-pound, divided by the unit of time, second. The conversion factors are

$$1 \text{ hp} = 550 \text{ ft-lb/s} = 746 \text{ W}$$

Figure 3.36 The number, 1600, indicates the maximum electric power this hair dryer may safely draw from the circuit.

A device that could raise 550 pounds a distance of 1 foot in 1 second would produce 1 horsepower. Raising 110 pounds a distance of 5 feet in 1 second would also require 1 horsepower.

The mathematical relationship between power and work (or energy) is the same as that between speed and distance. In each case, the former is the rate of change of the latter.

$$P = \frac{\text{work}}{\text{time}} \qquad \longleftrightarrow \qquad v = \frac{\text{distance}}{\text{time}}$$

A runner and a bicyclist can both travel a distance of 10 miles, but the latter can do it much faster because a bicycle is capable of much higher speeds. A person and a forklift can both do 10,680 joules of work raising the bricks, but the forklift can do it much faster because it has more power.

EXAMPLE 3.11 In Examples 2.2 and 3.5, we computed the acceleration, force, and work for a 1,000-kilogram car that goes from 0 to 27 m/s in 10 seconds. Let us now determine the required power output of the engine.

SOLUTION The work, 364,500 joules, is done in 10 seconds. Hence, the power is

$$P = \frac{\text{work}}{t} = \frac{364,500 \text{ J}}{10 \text{ s}}$$

$$= 36,450 \text{ W} = 48.9 \text{ hp}$$

The car's kinetic energy when going 27 m/s is also 364,500 joules (see Example 3.6). It takes the engine 10 seconds to give the car this much kinetic energy, so we get the same result using energy divided by time.

Figure 3.37 The *Light Eagle* in flight in October 2013. Designed as part of the Massachusetts Institute of Technology's Daedalus Project, this aircraft is similar to another built by MIT that was used in April 1988 to set both distance and duration records for human-powered flight of 115.11 km (71.53 mi) and 3.9 h, respectively, in a trip from Crete to the island Santorini.

Given enough food or fuel, there is usually no limit to how much work a device or a person can do. But there is a limit on how fast the work can be done. The power is limited. In other words, only so much work can be done each second. The power output can be anything from zero (no work) to some maximum. For example, a 100-horsepower automobile engine can put out from 0 to 100 horsepower. When accelerating as fast as possible, the engine is putting out its maximum power. While cruising down a flat highway, the engine may be putting out only 10–20 horsepower, enough to counteract the effects of air resistance and other frictional forces.

The human body has a maximum power output that varies greatly from person to person. In the act of jumping, a well-conditioned athlete can develop more than 8,000 watts, but only for a fraction of a second. The same athlete would have a maximum of less than 800 watts if the power level had to be maintained for an hour. The average person can produce 800 watts or more for a few seconds and perhaps 100–200 watts for an hour or more (**Figure 3.37**). When running, the body uses energy to overcome friction and air resistance. In short races, the best runners can maintain a speed of 10 m/s for about 20 seconds (cf. Usain Bolt's speed and time in Example 1.2). In longer races, the speeds are lower because the power level has to be maintained longer. For a race lasting about 30 minutes, the best average speed is about 6 m/s.

EXAMPLE 3.12 On 12 June 1979, Bryan Allen flew the *Gossamer Albatross* across the English Channel, turning the propeller of the aircraft using a bicycle-like mechanism. The flight lasted 2.82 hours. If Allen produced a steady 250 watts of power during the flight, how much work did he do to complete the trip?

SOLUTION

$$\text{work} = Pt = 250 \text{ W} \times 2.82 \text{ h} = 250 \text{ W} \times 10,140 \text{ s}$$

$$= 2,535,000 \text{ J} = 606 \text{ Calories}$$

(1,000 calories equals 1 Calorie, sometimes called a "food calorie.")

> ▶ **Physics To Go 3.3**
>
> It is relatively easy to compute your own power output when walking or running up a flight of stairs (**Figure 3.38**). First you need to compute the work you do by measuring the vertical height of the stairs and then multiplying this number by your weight. (You may want to use SI units—meters and newtons.) The power is this work divided by the time it takes to climb the stairs.
>
> If you go hiking or biking up a long steep hill, you can estimate your steady power output by performing the same calculation. The vertical distance can be found using a topographic map or a GPS app on your cell phone. (Do not forget to include the weight of your backpack and/or bicycle.)

Figure 3.38 To measure your power output when going up a flight of stairs, multiply the height of the stairs by your weight, and then divide by the time it takes you to climb the stairs.

1. (True or False.) When more people ride upward in an elevator, it requires more power.

2. Identical cars A and B are being driven up the same steep hill. If the power output of A is larger than that of B, what must be happening?

ANSWERS: **1.** True **2.** A is going faster, accomplishing the climb in less time.

ENVIRONMENTAL APPLICATIONS Exponential Growth and the Energy "Crisis"

We hear a lot about the need to conserve resources, particularly energy, in the news these days. In the light of our discussion in this chapter, one might wonder why people are concerned about conservation of energy. After all, energy *must* be conserved according to fundamental physical principles, mustn't it? What's all the fuss about something that happens naturally? The paradox here derives from the ways physicists and the general public use and interpret the term *energy conservation*. It is an example of the kind of confusion that can arise when the same words are used with different meanings by different groups.

As we have seen in Section 3.5, conservation of energy as a *physical law* refers to the fact that energy can neither be created nor destroyed in any interaction but merely changed in form. Put another way, in any system the total amount of energy remains constant. However, conservation of energy as an *economic, environmental, or social principle* involves reducing our reliance on certain types of energy sources such as coal, oil, and natural gas and becoming more efficient in their use in cases where switching to other sources is deemed unacceptable or unfeasible. For the general public, then, conservation of energy means husbanding precious "nonrenewable" natural resources and, overall, using less energy to accomplish the myriad tasks we carry out daily in our society.

Among the elements that have prompted society's attempts to use less energy is the way energy consumption, particularly in the United States, has increased over the last 150 years. **Figure 3.39** displays the annual amount of energy consumed in the United States since 1850 from sources such as wood, coal, oil, and natural gas, as well as from hydroelectric, geothermal, and nuclear power generating plants. Note that the graph is *not* a straight line. From one year to the next, the amount of energy consumed does not increase by a fixed amount as would be the case if the relationship between energy use and time were linear. Instead, energy consumption grows increasingly rapidly with each passing year.

If one analyzes this behavior carefully, it may be demonstrated that the *change* in energy consumption during any given interval of time depends directly on the amount of energy being used at the start of the time interval. Symbolically, using the notation of Chapter 1, we may write

$$\Delta E/\Delta t \propto E$$

where E is the energy consumption, Δ means "change in," and $\propto$ is the mathematical symbol for "proportional to." A quantity that increases (or decreases) in this manner is said to exhibit *exponential* growth (or decay). Evidently, energy consumption in the United States has grown exponentially since 1850, and in the judgment of many experts the origins of the energy crisis are to be found in the exponential growth of energy consumption.

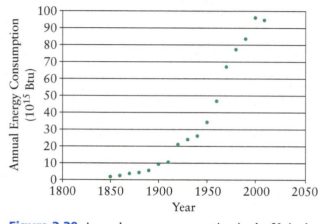

Figure 3.39 Annual energy consumption in the United States since 1850. The total consumption from all sources rose exponentially until about 1980, with an annual growth rate of 4.3%. Since then, the growth rate has slowed, and, for the last several years, has actually fallen below its high of 101.5×10^{15} Btu in 2007.

The problem, as these investigators see it, lies in the so-called doubling time for energy use. It is possible to show that the doubling time is related to the percent increase of the quantity in a given interval of time (a year, a week, a day, a second, and so on). Specifically, if we let t_D represent the doubling time, then

$$t_D \cong 70/PC$$

where PC is the percent change over some unit of time. The units of time for t_D will be those used to express PC. (The "$\cong$" in the formula means "approximately equal to.") For example, if a certain quantity grew steadily at a rate of 10 percent per year, it would double in size in 70/10 or 7 years.

The data shown in Figure 3.39 yield a growth rate in energy consumption in the United States of a little more than 4 percent a year, at least for many years preceding 1980. At this rate, the doubling time for energy use in this country would be about 16 years. This means that if such growth were sustained, every 16 years the amount of energy consumed in the United States would double in size, requiring a similar doubling of the production capacity of all the energy industries in the nation. Similarly, at this rate, in only 80 years (five doubling times), the U.S. energy consumption (and production) would increase by a factor of $2 \times 2 \times 2 \times 2 \times 2 = 2^5$ or 32. Exponential quantities rise (or fall) extremely quickly, even for what might be considered modest growth (or decay) rates.

Pushing this example just a bit further, we can begin to see why scientists, engineers, economists, government officials, and others have called for profound changes in the way our society uses energy to avoid a severe energy crisis: at 4 percent annual growth, in just 320 years the amount of energy required by our citizenry would be more than 1 million times what it is today! This is far beyond even the most optimistic estimates of our capacity to find and perfect new energy sources.

Clearly, for exponentially increasing quantities, as time goes on, their values quickly become enormous, eventually approaching infinity. For this reason, it is impossible for any real quantity to continue to grow exponentially for a long period of time, much less forever. The ability of any system to sustain such growth is rapidly outstripped, and the growth is halted. Since the 1970s, the public's appreciation of the severity of the energy problem has deepened considerably, as has its commitment to adopting measures to reduce the growth in energy consumption. In part because of energy conservation measures implemented during the past 20 years, particularly the production of more fuel-efficient cars, total energy consumption in the nation has slowed considerably. In fact, for the 5-year period from 2004 through 2008, energy consumption in this country plateaued at about 100×10^{15} Btu before dropping nearly 6 percent in 2009 to 94.2×10^{15} Btu. The 5-year average from 2010 through 2014 has remained at close to this level at 96.9×10^{15} Btu an encouraging sign that we as a nation are beginning to appreciate the need to curb our appetite for energy to avoid the dire consequences of unrestrained exponential growth.

■ **QUESTIONS**
1. Distinguish between what a physicist and a politician might each mean when using the term "energy conservation."
2. If the population in a certain country was discovered to be increasing at a steady rate of 6 percent every decade, how many years would it take for the population of that nation to double in size?

3.8 Rotation and Angular Momentum

Our final conservation law applies to rotational motion. A spinning ice skater and a satellite moving in a circular path around Earth are examples. You might say that this law is the rotational analogue or counterpart of the law of conservation of linear momentum.

> **LAWS** **Law of Conservation of Angular Momentum** The total angular momentum of an isolated system is constant.

For a system to be isolated so that the law of conservation of angular momentum applies, the only net external force that can act on the object must be directed toward or away from the center of the object's motion. The centripetal force required to keep an object moving in a circle fits this condition. A force that acts in any direction other than toward or away from the center of motion produces what is called **torque** (from the Latin word for "twist"). When a spacecraft fires its rocket engines to reenter the atmosphere, the force on it acts opposite to its direction of motion and therefore produces a torque that decreases its angular momentum. Torque is the rotational analogue of force: a net external force changes an object's linear momentum, and a net external torque changes an object's angular momentum.

> **DEFINITION** **Torque** The net external torque acting on a system equals the rate of change of its angular momentum.
>
> $$\text{torque} = \frac{\text{change in angular momentum}}{\text{change in time}}$$

Like force, torque is a vector. The direction of the torque is perpendicular to the plane of the object's orbit and dependent on the direction of rotation produced by the torque or twist. Positive torques are associated with counter-clockwise rotations as seen from above the axis of spin; negative torques correspond to clockwise rotations as seen from the same perspective.

But what exactly is **angular momentum**? We first introduce angular momentum for the simple case of a body moving in a circle and then extend it to motion along noncircular paths. Imagine a small object moving along a circular path, like a satellite in orbit around Earth. In this case, the angular momentum equals the product of the object's mass, its orbital speed, and the radius of its path:

$$\text{angular momentum} = mvr \qquad \text{(circular path)}$$

Notice that the angular momentum of the object is also equal to its linear momentum (mv) multiplied by the radius of its circular path.

Again, like linear momentum, angular momentum is a vector. Its direction is reckoned in the same manner as the torque that produces it.

To illustrate the law of conservation of angular momentum, imagine a ball circling overhead on the end of a string that passes through a tube (**Figure 3.40a**). (We assume there is no friction or air resistance.) The faster the ball goes, the greater its orbital angular momentum. Using a longer string for the same speed would also make its angular momentum larger. Imagine suddenly shortening the string by pulling downward on the end with your free hand, letting the string slide through the tube. This makes the ball move in a circle with a smaller radius but with a higher speed (**Figure 3.40b**). Because the force exerted on the ball is directed toward the center of its motion, angular momentum is conserved—it has the same value before and after the change. Because the radius of the ball's path is now smaller, its speed must be higher. If you let the string out so the radius is larger, the ball will slow down, keeping the angular momentum constant.

In the process of pulling the string downward, you do work. This work goes to increase the kinetic energy of the ball.

With caution, we can use this definition of angular momentum—mvr—for an object moving in a path other than a circle. **Figure 3.41** shows the elliptical orbit of a satellite moving around Earth. At points A and B, the satellite's velocity will be perpendicular to a line joining the satellite to Earth's center. At these points the satellite's path is like a short segment of a circle. Consequently, its angular momentum is mvr. At point B, r is smaller than at point A. Because the angular momentum is the same at A as at B, the satellite's speed is greater at B. For example, if the satellite is 13,000 kilometers (about 8,000 miles, twice Earth's radius) from Earth's center at point B and 26,000 kilometers from Earth's center at point A, its speed at B will be two

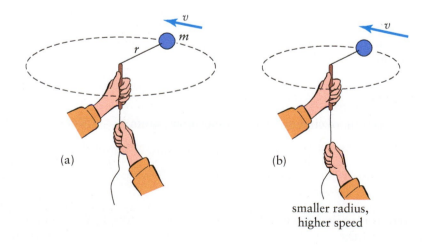

(a)   (b)

smaller radius,
higher speed

Figure 3.40 (a) The angular momentum of an object moving in a circle equals mvr. (b) If the radius is decreased, the object speeds up so that the angular momentum stays the same.

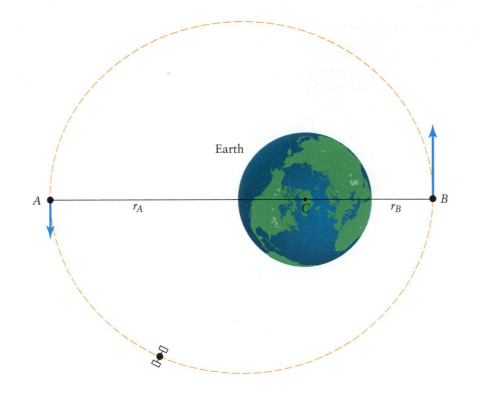

Figure 3.41 A satellite in orbit around Earth is twice as far from Earth's center at point *A* as it is at point *B*. Conservation of angular momentum then tells us that its speed at *A* is one-half its speed at *B*. (Figure not to scale.)

times its speed at A. The actual values for the speeds are about 6,400 m/s (about 14,000 mph) when it is closest to Earth and about 3,200 m/s when it is farthest away.

> ### ▶ Physics To Go 3.4
>
> For this, you need a chair that can spin in circles easily. (*Caution*: Don't try this if you are prone to dizziness.) Move it to a place at least 5 feet from walls, furniture, and other objects. Sit in it with both of your arms straight out from your sides; then use your feet to make the chair spin. Lift your feet and then pull your arms tight against your body. What happens? Extend your arms out to the side again. What happens? (You can enhance the effect by holding a weight [e.g., a can of beans] in each hand.)

An object spinning about an axis, like a toy top or an ice skater doing a pirouette, has angular momentum. We can think of each part of the object as moving in a circle with a certain radius and having orbital angular momentum. For example, a spinning skater's hands, arms, shoulders, and other body parts are all moving in circles. The combined angular momentum of the parts of a spinning body remains constant if no torque acts on it. The rate of spinning can be increased or decreased by repositioning parts of the object. For example, the spinning ice skater can start with arms extended outward. When the arms are pulled in closer to the body, the skater spins faster. Each part of the skater's arms has a certain amount of angular momentum as it moves in a circle with some radius. Pulling the arms in decreases the radius, which makes the angular momentum of that part of the arm decrease. Consequently the rest of the skater's body spins faster, so that the total angular momentum remains the same. The reverse happens if the arms are moved out—the skater slows down.

Concept Map 3.1 illustrates the three conservation laws for linear momentum, energy, and angular momentum.

■ CONCEPT MAP 3.1

■ Learning Check

1. The centripetal force acting on a body moving in a circle changes its linear momentum, but it does not change its _____ momentum.
2. (True or False.) If both the linear momentum and the kinetic energy of a satellite in orbit around Earth increase, then its angular momentum must also increase.

3. Complete this comparison: Force is to _____ as linear momentum is to angular momentum.

ANSWERS: 1. angular **2.** False **3.** torque

ASTRONOMICAL APPLICATION Starquakes: A *Glitch* in Time

Pulsars are astronomical objects that emit a regular sequence of pulses of radio energy. The pulse period (the interval of time between one pulse and the next) is usually quite short (a second or less for most pulsars), and it is generally accepted that pulsars are rapidly rotating *neutron stars*.

A neutron star is a remnant of a star that may have ended its life in a brilliant flash called a "supernova explosion." Though small (about 20 kilometers across), neutron stars are very dense: a teaspoon of neutron star stuff could easily weigh a billion tons.

The most probable mechanism by which pulsars produce the radio energy we receive involves the very strong magnetic fields that these objects are also known to possess. Charged particles (mostly electrons) near the surface of the neutron star experience a force from the magnetic field (see Section 8.2) and are accelerated toward the magnetic poles. As they do so, they begin to radiate energy. Specifically, radio waves are beamed out into space from what might be termed "radio hot spots" near the magnetic poles of the neutron star. As these hot spots are rotated across our line of sight by the spinning neutron star, we receive regularly spaced, short bursts of radio radiation. What we observe then is an astronomical "lighthouse effect" (**Figure 3.42**).

When pulsars were first discovered in 1967, their most prominent feature was the precision with which the pulses were spaced.

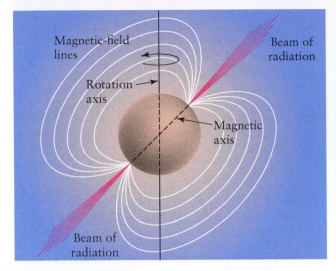

Figure 3.42 A simplified "lighthouse" model of a pulsar. Radio "hot spots" associated with the strong magnetic field of the neutron star emit copious quantities of radio radiation. As these hot spots are carried across our line of sight by the spin of the neutron star, we receive sharp pulses of radio energy.

Figure 3.43 The Crab Nebula—debris of a supernova explosion that was observed and recorded by the Chinese in the year 1054.

However, after continued observation, scientists found that these "pulsar clocks" were slowing down. Their periods were gradually increasing. We now think that the neutron star associated with a pulsar is losing energy through interactions with its environment and rotating more and more slowly with time. Astronomers expected a continual decay in the rotation rates of these compact stars and a corresponding lengthening of the periods of pulsars. The analogy with a toy top slowly spinning down because of frictional losses to the table on which it rotates is perhaps not a bad one to have in mind here.

But Nature is full of surprises. In 1969, a pulsar in the constellation of Vela suddenly sped up—that is, its period abruptly decreased! Afterward, it once again resumed its steady decline in spin at the same rate as before. This was not an isolated event. At least 16 similar speed ups in the Vela pulsar have been observed. They have also been seen in more than 60 other pulsars, including the one embedded in the Crab Nebula supernova remnant (**Figure 3.43**). These brief events in which a pulsar increases its spin by a small amount are called *glitches*. Their explanation has its root in the law of conservation of angular momentum.

Current models for the structure of neutron stars suggest that they have solid, crystalline crusts up to a few kilometers thick. Beneath the crust, and to some extent permeating it, is a sea of superfluid neutrons (see Section 4.5 for more on superfluidity). During the steady slowdown in the spin of the neutron star crust, there is a lag in the response of the interior fluid so that it is always spinning somewhat faster than the crust. (The dynamics are similar to that of an uncooked egg that, when set spinning, slows down much more slowly than a hardcooked one because of differential rotation of the liquid white and yolk inside. Try it! This is a neat way to distinguish cooked eggs from uncooked ones.)

The first proposed cause of pulsar spin-ups, and one that may play a role in explaining small glitches like those observed in the Crab pulsar, was "starquakes" (or sometimes "crustquakes"). Most spinning objects, including Earth and the Sun, are oblate in shape—that is, they have a larger equatorial radius than polar radius (cf. our discussion of tidal bulges in Section 2.8). Because of their rotation, they are flattened at the poles and bulged out along the equator. As a spinning body slows down, the oblateness decreases, and the object becomes more nearly spherical in shape. In a rapidly rotating neutron star, the crust is sufficiently rigid that it cannot respond to the necessary reduction in oblateness in a smooth and continuous manner, so internal stresses build up until the crust cracks and the "bulges" abruptly move slightly inward. To the extent that the neutron star may be considered an isolated body, conservation of angular momentum must apply to it, yielding an increase in the star's rotation rate as a result. This effect is analogous to an ice skater pulling in her arms during a pirouette and spinning faster.

Major glitch events like those seen in the Vela pulsar probably require an additional, more complicated explanation in which the inner neutron superfluid becomes coupled to the crust and transfers angular momentum to it, causing it to speed up and spin faster. But again, all this occurs within the framework of conservation of angular momentum.

■ **QUESTIONS**

1. Describe the basic features of the "lighthouse" model of a pulsar, and indicate how they combine to give rise to the observed sharply pulsed radio emission from these sources.

2. What is a "glitch" as related to the rotation of a pulsar? Give a physical explanation or model for glitches like those associated with the Crab Nebula pulsar.

Most of the principles discussed in this chapter were developed piecemeal by a host of physicists and mathematicians in the 17th, 18th, and 19th centuries. They are extensions of Newton's mechanics, not fundamentally new formulations. As mentioned near the end of Section 3.2, the law of conservation of linear momentum has its roots in Newton's laws of motion. The ideas of work and potential energy arise naturally from Newton's second law of motion when the force on an object depends only on its position.

Before Newton's *Principia*, the importance of momentum and kinetic energy had been discovered experimentally. Galileo used the product of weight and velocity as momentum, a quantity that is nearly the same as our linear momentum, because weight and mass are proportional. In 1665, Dutch physicist Christian Huygens reported that the quantity mv^2 was conserved during the collision of two hard balls. (Huygens later acquired fame for his work in optics. His explanation of the fundamental nature of light was contrary to Newton's but eventually proven to be correct.) The actual use of the word energy, the identification of kinetic energy as being equal to one-half the mass times the speed squared, and the statement of the law of conservation of energy came about around the middle of the 19th century.

The simple machines discussed in Section 3.3 were first analyzed by the famous Greek mathematician and scientist Archimedes

(287?–212 B.C.). A citizen of Syracuse, Sicily, Archimedes (**Figure 3.44**) explained how a force could be amplified using a lever, a combination of pulleys, or other simple machines. The historian Plutarch described how Archimedes demonstrated the usefulness of pulleys by moving a large, dry-docked ship by

Figure 3.44 Archimedes.

himself. This so impressed King Hiero of Syracuse that he persuaded Archimedes to construct machines to defend the city. These proved to be quite useful when the Romans attacked Syracuse by land and sea. Plutarch includes this account in his *Life of Marcellus*:

> Archimedes began to ply his engines, and shot against the land forces of the assailants all sorts of missiles and immense masses of stones, which came down with incredible din and speed: nothing whatever could ward off their weight, but they knocked down in heaps those who stood in their way, and threw their ranks into confusion. At the same time huge beams were suddenly projected over the ships from the walls, which sank some of them with great weights plunging down from on high: others were seized at the prow by iron claws, or beaks like the beaks of cranes, drawn straight up into the air, and then plunged stern foremost into the depths, or were turned round and round by means of enginery within the city, and dashed upon the steep cliffs that jutted out beneath the wall of the city, with great destruction of the fighting men on board, who perished in the wrecks.

Archimedes' machines forced Marcellus, the Roman general, to abandon the direct assault in favor of a long siege. Two years later Syracuse fell, and Archimedes was killed by the conquerors.

Archimedes' most famous discovery was the principle of buoyancy that is named after him (Section 4.5). We should also note that he was one of the first to engage in two pursuits that have been central to physics since that time. First, as an engineer, he used his discoveries to construct useful devices. Second, he was a scientific consultant for the military. To this day, the military is a primary source of funding for research in physics and other sciences. Many great discoveries in physics and engineering, some quite terrifying, have come from efforts to devise offensive and defensive weapons of war.

■ QUESTION

1. Who was Archimedes? Give two significant contributions to the field of physics that Archimedes made that justify his inclusion as an important "profile in physics."

SUMMARY

» **Conservation laws** are powerful tools for analyzing physical systems, particularly those in mechanics. Their main advantage is that it is not necessary to know the details of what is going on in the system at each instant in time to apply them effectively.

» The use of conservation laws is based on a "before-and-after" approach: the total amount of the conserved physical quantity *before* an interaction is equal to the total amount *after* the interaction.

» **Linear momentum, energy,** and **angular momentum** are physical quantities that are defined and used mainly because they are conserved in isolated systems.

» The main application of the **law of conservation of linear momentum** is to collisions. The total linear momentum before a collision equals the total linear momentum after the collision,

if the system is isolated. This applies to all collisions, both elastic and inelastic.

» **Work** is done whenever a force acts through a distance in the same direction as the force.

» To be able to do work, a device or a person must have **energy**. The act of doing work involves the transfer of energy from one thing to another, the transformation of energy from one form to another, or both.

» In mechanics, the main forms of energy are **kinetic energy, potential energy,** and **internal energy**. There are many other forms of energy corresponding to different sources of work. Any form of energy can be used to do work if a suitable conversion device is available.

» The **law of conservation of energy** tells us that in all such conversions, the total amount of energy remains constant.

» **Power** is the rate of doing work or using energy. It is the measure of how fast energy is transferred or transformed.

» **Angular momentum** is a conserved quantity in rotational motion. Just as an applied force is needed to change the linear momentum of a system, an applied torque is required to change the angular momentum of a system. Torque is the analog of force for angular motion.

IMPORTANT EQUATIONS

Equation	Comments	Equation	Comments
Fundamental Equations		*Special-Case Equations*	
linear momentum $= mv$	Definition of linear momentum	$v = \sqrt{2gd}$	Speed after falling freely a distance d
$F = \dfrac{\Delta(mv)}{\Delta t}$	Alternate form of Newton's second law	$d = \dfrac{v^2}{2g}$	Height reached given initial speed v
$\Delta(mv) = F\Delta t$	Impulse-momentum relation	angular momentum $= mvr$	Angular momentum (circular motion)
work $= Fd$	Definition of work	torque $= \dfrac{\Delta(mvr)}{\Delta t}$	Newton's second law for rotation (circular path)
$KE = \dfrac{1}{2}mv^2$	Kinetic energy		
$PE = Wd = mgd$	Gravitational potential energy		
$PE = \dfrac{1}{2}kd^2$	Elastic potential energy		
$P = \dfrac{\text{work}}{t}$; $P = \dfrac{E}{t}$	Definition of power		

MAPPING IT OUT!

1. Re-examine Section 3.3 on work. Make a list of at least *10* key concepts and applications relating to work; write each of the concepts or examples on small Post-it® notes. Rank the items on your list from most inclusive to least inclusive, and organize your notes on a large piece of paper or poster board in a manner that reflects your ranking. Leave enough space between concepts to permit the insertion of appropriate linking words and phrases. Now, on separate Post-its,® perhaps of a different color, write down the required linking words to form meaningful propositions with your concepts; place these notes at the correct locations on the chart paper or poster board. You have just constructed a concept map for work! Examine the map carefully. Do all the propositions make sense? Are the most general concepts located near the top of the map? Are major concepts or connections missing? After you have finished refining your concept map, compare it with one that a classmate has produced. What similarities do they share? What differences are there between them? How might the maps be connected to ones designed to promote understanding of the concept of energy?

2. Repeat Exercise 1 for Section 3.2 on linear momentum and relate your concept map directly to Concept Map 3.1 at the end of Section 3.8.

QUESTIONS

(■ **Indicates a review question, which means it requires only a basic understanding of the material to answer. Questions without this designation typically require integrating or extending the concepts presented thus far.**)

1. ■ What is a conservation law? What is the basic approach taken when using a conservation law?
2. ■ Why is the alternate form of Newton's second law of motion given in this chapter the more general form?
3. Could the linear momentum of a turtle be greater than the linear momentum of a horse? Explain why or why not.
4. An astronaut working with many tools some distance away from a spacecraft is stranded when the "maneuvering unit" malfunctions. How can the astronaut return to the spacecraft by sacrificing some of the tools?
5. ■ For what type of interaction between bodies is the law of conservation of linear momentum most useful?
6. Describe several things you have done today that involved doing work. Are you doing work right now?
7. If we know that a force of 5 N acts on an object while it moves 2 meters, can we calculate how much work was done with no other information? Explain.
8. During a head-on collision between two automobiles, the occupants are decelerated rapidly. Use the idea of impulse to explain why an air bag that quickly inflates in front of an occupant reduces the likelihood of injury.

9. When climbing a flight of stairs, do you do work on the stairs? Do the stairs do work on you?
10. People and machines around us do work all the time. But is it possible for things such as magnets and Earth to do work? Explain.
11. Identify as many different ways as you can for giving energy to a basketball.
12. ■ Identify as many different forms of energy as you can that are around you at this moment.
13. When you throw a ball, the work you do to accelerate it equals the kinetic energy the ball gains. If you do *twice* as much work when throwing the ball, does it go *twice as fast*? Explain.
14. Describe the motion of an object that possesses kinetic energy yet undergoes no net displacement.
15. ■ How can the gravitational potential energy of something be negative?
16. ■ What is elastic potential energy? Identify some of the things that currently surround you that possess elastic potential energy.
17. ■ If a spring is compressed to half its length, by how much does the amount of energy stored in the spring change?
18. Identify the energy conversions taking place in each of the following situations. Name all of the relevant forms of energy that are involved.
    (a) A camper rubbing two sticks together to start a fire.
    (b) An arrow shot straight upward, from the moment the bowstring is released by the archer to the moment when the arrow reaches its highest point.
    (c) A nail being pounded into a board, from the moment a carpenter starts to swing a hammer to the moment when the nail has been driven some distance into the wood by the blow.
    (d) A meteoroid entering Earth's atmosphere.
19. Solar-powered spotlights have batteries that are charged by solar cells during the day and then operate lights at night. Describe the energy conversions in this entire process, starting with the Sun's nuclear energy and ending with the light from the spotlight being absorbed by the surroundings. Name all of the forms of energy that are involved.
20. Truck drivers approaching a steep hill that they must climb often increase their speed. What good does this do, if any?
21. If you hold a rubber ball at eye level and drop it, it will bounce back, but not to its original height. Identify the energy conversions that take place during the process, and explain why the ball does not reach its original release level.
22. A ball is thrown straight upward from the surface of the Moon. Is the maximum height it reaches less than, equal to, or greater than the maximum height reached by a ball thrown upward on Earth with the same initial speed? (Ignore air resistance in both cases.) Explain.
23. ■ Describe the distinction between elastic and inelastic collisions. Give an example of each.
24. Many sports involve collisions between things—such as balls and rackets—and between people—as in football or hockey. Characterize the various sports collisions as elastic or inelastic.
25. Carts A and B stick together whenever they collide. The mass of A is twice the mass of B. How could you roll the carts toward each other in such a way that they would be stopped after the collision? (Assume there is no friction and that the carts move on level ground.)
26. Is it possible for one object to gain mechanical energy from another without touching it? Explain.
27. ■ How are the physical concepts power and speed similar?

28. Two cranes are lifting identical steel beams at the same time. One crane is putting out twice as much power as the other. Assuming friction is negligible, what can you conclude is happening to explain this difference?
29. A person runs up several flights of stairs and is exhausted at the top. Later, the same person walks up the same stairs and does not feel as tired. Why is this? Ignoring air resistance, does it take more work or energy to run up the stairs than to walk up?
30. ■ How can a satellite's speed decrease without its angular momentum changing?
31. ■ Why do divers executing midair somersaults pull their legs in against their bodies?
32. It is possible for a body to be both spinning and moving in a circle in such a way that its total angular momentum is zero. Describe how this can be.
33. Five identical boxes with the same speeds slide along a frictionless horizontal surface. The mass of each box is 10 kg. The same magnitude force, F, is applied to each box, but along different directions. Rank the five situations described here from greatest to smallest according to the work done on the box by the force while the box moves through the distance d indicated each description. For this analysis, take motion/distance directed to the right as positive and force directed up as positive. If any of the situations result in the same work being done, give them the same ranking.
    (a) F to the right, and $d = 5$ m to the right
    (b) F to the right, and $d = 10$ m to the right
    (c) F up, and $d = 10$ m to the right
    (d) F to the left, and $d = 5$ m to the right
    (e) F down, and $d = 5$ m to the right
34. Eight cars move along smooth horizontal roadways in the same direction at specified speeds, v, toward identical barriers. All the cars have the same size and shape, but carry different loads and, hence, have different masses, m. The cars collide with the barriers and come to a stop after having traveled some distance. Assume the force exerted on each of the cars by their respective barrier is the same. Rank the situations described here from greatest to smallest according to the stopping distance of each vehicle. If any situations yield the same stopping distance, give them the same ranking.
    (a) $m = 1000$ kg; $v = 10$ m/s
    (b) $m = 1000$ kg; $v = 20$ m/s
    (c) $m = 1000$ kg; $v = 30$ m/s
    (d) $m = 3000$ kg; $v = 10$ m/s
    (e) $m = 2000$ kg; $v = 10$ m/s
    (f) $m = 2000$ kg; $v = 20$ m/s
    (g) $m = 2000$ kg; $v = 30$ m/s
    (h) $m = 3000$ kg; $v = 20$ m/s
35. Six blocks with different masses, m, each start from rest at the top of smooth, frictionless inclines having length d and vertical height h and slide down. Rank the order, from greatest to smallest, of the final kinetic energies of the masses when they reach the bottom of the inclines after having traveled their full lengths. If any of the situations yield the same kinetic energies, give them the same ranking.
    (a) $m = 10$ kg; $h = 1$ m; and $d = 10$ m
    (b) $m = 10$ kg; $h = 1$ m; and $d = 5$ m
    (c) $m = 5$ kg; $h = 0.5$ m; and $d = 10$ m
    (d) $m = 1$ kg; $h = 2$ m; and $d = 5$ m
    (e) $m = 1$ kg; $h = 0.5$ m; and $d = 5$ m
    (f) $m = 15$ kg; $h = 0.75$ m; and $d = 7.5$ m

PROBLEMS

1. A sprinter with a mass of 65 kg reaches a speed of 10 m/s during a race. Find the sprinter's linear momentum.
2. Which has the larger linear momentum: a 2,000-kg houseboat going 5 m/s or a 600-kg speedboat going 20 m/s?
3. In Section 2.4, we computed the force needed to accelerate a 1,000-kg car from 0 to 27 m/s in 10 s. Compute the force using the alternate form of Newton's second law. The change in momentum is the car's momentum when traveling 27 m/s minus its momentum when going 0 m/s.
4. A runner with a mass of 80 kg accelerates from 0 to 9 m/s in 3 s. Find the net force on the runner using the alternate form of Newton's second law.
5. In Section 1.4, we considered the collision of a karate expert's hand with a concrete block. Based on the graphs in Figure 1.31, the initial downward speed of the fist with mass 0.75 kg is about −13 m/s and the collision time is approximately 25 ms. Find the impulse and the average force exerted on the block by the fist during the collision.
6. A basketball with a mass of 0.62 kg falls vertically to the floor where it hits with a speed of −6 m/s. (We take the positive direction to be upward here.) The ball rebounds, leaving the floor with a speed of 4.5 m/s. (a) What impulse acts on the ball during its collision with the floor? (b) If the ball is in contact with the floor for 0.04 s, what is the average force of the ball on the floor?
7. A pitcher throws a 0.5-kg ball of clay at a 6-kg block of wood. The clay sticks to the wood on impact, and their joint velocity afterward is 3 m/s. What was the original speed of the clay?
8. A 3,000-kg truck runs into the rear of a 1,000-kg car that was stationary. The truck and car are locked together after the collision and move with speed 9 m/s. What was the speed of the truck before the collision?
9. A 50-kg boy on roller skates moves with a speed of 5 m/s. He runs into a 40-kg girl on skates. Assuming they cling together after the collision, what is their speed?
10. Two persons on ice skates stand face to face and then push each other away (**Figure 3.45**). Their masses are 60 and 90 kg. Find the ratio of their speeds immediately afterward. Which person has the higher speed?

Figure 3.45 Problem 10.

11. A loaded gun is dropped on a frozen lake. The gun fires, with the bullet going horizontally in one direction and the gun sliding on the ice in the other direction. The bullet's mass is 0.02 kg, and its speed is 300 m/s. If the gun's mass is 1.2 kg, what is its speed?
12. A running back with a mass of 80 kg and a speed of 8 m/s collides with, and is held by, a 120-kg defensive tackle going in the opposite direction. How fast must the tackle be going before the collision for their speed afterward to be zero?
13. A motorist runs out of gas on a level road 200 m from a gas station. The driver pushes the 1,200-kg car to the gas station. If a 150-N force is required to keep the car moving, how much work does the driver do?
14. In Figure 3.10, the rock weighs 100 lb and is lifted 1 ft by the lever.
    (a) How much work is done?
    (b) The other end of the lever is pushed down 3 ft while lifting the rock. What force had to act on that end?
15. A weight lifter raises a 100-kg barbell to a height of 2.2 m. What is the barbell's potential energy?
16. A microwave antenna with a mass of 80 kg sits atop a tower that is 50 m tall. What is the antenna's potential energy?
17. A personal watercraft and rider have a combined mass of 400 kg. What is their kinetic energy when they are going 15 m/s?
18. As it orbits Earth, the 11,000-kg Hubble Space Telescope travels at a speed of 7,900 m/s and is 560,000 m above Earth's surface.
    (a) What is its kinetic energy?
    (b) What is its potential energy?
19. The kinetic energy of a motorcycle and rider is 60,000 J. If their total mass is 300 kg, what is their speed?
20. In compressing the spring in a toy dart gun, 0.5 J of work is done. When the gun is fired, the spring gives its potential energy to a dart with a mass of 0.02 kg.
    (a) What is the dart's kinetic energy as it leaves the gun?
    (b) What is the dart's speed?
21. An archer using a simple bow exerts a force of 180 N to draw back the bow string 0.50 m.
    (a) What is the average work done by the archer in preparing to launch her arrow? (*Hint:* Compute the average work as you would any average quantity: average work $= \frac{1}{2}$ [final work − initial work].)
    (b) If all the work is converted into the kinetic energy of the arrow upon its release, what is the arrow's speed as it leaves the bow? Assume the mass of the arrow is 0.021 kg and ignore any kinetic energy in the bow as it relaxes to its original shape.
    (c) If the arrow is shot straight up, what is the maximum height achieved by the arrow? Ignore any effects due to air resistance in making your assessment.
22. A worker at the top of a 629-m-tall television transmitting tower in North Dakota accidentally drops a heavy tool. If air resistance is negligible, how fast is the tool going just before it hits the ground?
23. A 25-kg child uses a pogo stick to bounce up and down. The spring constant, k, of the toy equals 8750 N/m.
    (a) By how much would the spring be compressed by the child if she simply balanced herself vertically on the pedals of the stick?
    (b) How much energy is stored in the spring under this circumstance?
24. A student drops a water balloon out of a dorm window 12 m above the ground. What is its speed when it hits the ground?

25. A child on a swing has a speed of 7.7 m/s at the low point of the arc (**Figure 3.46**). How high will the swing be at the high point?

Figure 3.46 Problem 25.

26. The cliff divers at Acapulco, Mexico, jump off a cliff 26.7 m above the ocean. Ignoring air resistance, how fast are the divers going when they hit the water?
27. At NASA's Zero Gravity Research Facility in Cleveland, Ohio, experimental payloads fall freely from rest in an evacuated vertical shaft through a distance of 132 m.
    (a) If a particular payload has a mass of 45 kg, what is its potential energy relative to the bottom of the shaft?
    (b) How fast will the payload be traveling when it reaches the bottom of the shaft? Convert your answer to mph for a comparison to highway speeds.
28. The fastest that a human has run is about 12 m/s.
    (a) If a pole vaulter could run this fast and convert all of her kinetic energy into gravitational potential energy, how high would she go?
    (b) Compare this height with the world record in the pole vault.
29. A bicycle and rider going 10 m/s approach a hill. Their total mass is 80 kg.
    (a) What is their kinetic energy?
    (b) If the rider coasts up the hill without pedaling, how high above its starting level will the bicycle be when it finally rolls to a stop?
30. In January 2003, an 18-year-old student gained a bit of fame for surviving—with only minor injuries—a remarkable traffic accident. The vehicle he was driving was "clipped" by another one, left the road, and rolled several times. He was thrown upward from the vehicle (he wasn't wearing a seat belt) and ended up dangling from an overhead telephone cable and a ground wire about 8 meters above the ground. Rescuers got him down after 20 minutes. It is estimated that he reached a maximum height of about 10 meters.
    (a) Estimate the driver's vertical speed when he was thrown from the vehicle.
    (b) If he had not landed in the wires, how fast would he have been going when he hit the ground?
31. The ceiling of an arena is 20 m above the floor. What is the minimum speed that a thrown ball would need to just reach the ceiling?
32. Compute how much kinetic energy was "lost" in the inelastic collision in Problem 8.
33. Compute how much kinetic energy was "lost" in the inelastic collision in Problem 9.
34. A 1,000-W motor powers a hoist used to lift cars at a service station.
    (a) How much time would it take to raise a 1,500-kg car 2 m?
    (b) If the original motor is replaced with one rated at 2,000 W, how long would it take to complete this task?
35. How long does it take a worker producing 200 W of power to do 10,000 J of work?
36. An elevator is able to raise 1,000 kg to a height of 40 m in 15 s.
    (a) How much work does the elevator do?
    (b) What is the elevator's power output?
37. A particular hydraulic pile driver uses a ram with a mass of 1040 kg. If the maximum pile energy is 11,780 J, how high must the ram be raised to achieve this value? Assuming it takes 0.62 s for the pile driver's winch motor to raise the ram at a constant speed to this height, what is the power output by the motor in completing this task? Express your answer in both watts and horsepower.
38. A compact car can climb a hill in 10 s. The top of the hill is 30 m higher than the bottom, and the car's mass is 1,000 kg. What is the power output of the car?
39. In the annual Empire State Building race, contestants run up 1,575 steps to a height of 1,050 ft. In 2003, Australian Paul Crake completed the race in a record time of 9 min and 33 s. Mr. Crake weighed 143 lb (65 kg).
    (a) How much work did Mr. Crake do in reaching the top of the building?
    (b) What was his average power output (in ft-lb/s and in hp)?
40. It takes 100 minutes for a middle-aged physics professor to ride his bicycle up the road to Alpe d'Huez in France. The vertical height of the climb is 1,120 m, and the combined mass of the rider and bicycle is 85 kg. What is the bicyclist's average power output?
41. Two small 0.25-kg masses are attached to opposite ends of a very lightweight rigid rod 0.5 m long. The system is spinning in a horizontal plane around a vertical axis perpendicular to the rod located halfway between the masses. Each mass is moving in a circle of radius 0.25 m at a speed of 0.75 m/s. What is the total angular momentum of this system?

CHALLENGES

1. Rank the following three collisions in terms of the extent of damage that the car would experience. Explain your reasons for ranking the collisions as you did.
   (a) A car going 10 m/s striking an identical car that was stationary on level ground.
   (b) A car going 10 m/s running into an immovable concrete wall.
   (c) A head-on collision between identical cars, both going 10 m/s.
2. A bullet with a mass of 0.01 kg is fired horizontally into a block of wood hanging on a string. The bullet sticks in the wood and causes it to swing upward to a height of 0.1 m. If the mass of the wood block is 2 kg, what was the initial speed of the bullet?
3. In a head-on, inelastic collision, a 4,000-kg truck going 10 m/s east strikes a 1,000-kg car going 20 m/s west.
   (a) What is the speed and direction of the wreckage?
   (b) How much kinetic energy was lost in the collision?
4. A person on a swing moves so that the support rods are horizontal at the turning points (**Figure 3.47**). Show that the centripetal acceleration of the person at the low point of the arc is exactly 2 g, regardless of the length of the rods. This means that the force on the rods at the low point is equal to *three times* the person's weight. (*Hint:* The vertical distance

between the turning point and the low point equals the length of the rods.)

Figure 3.47 Challenge 4.

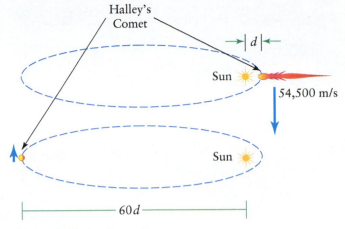

Figure 3.48 Challenge 7.

5. Assume that as a car brakes to a stop it undergoes a constant acceleration (deceleration). Explain why the stopping distance becomes four *times* as large if the initial speed is *doubled*.

6. The "shot" used in the shot-put event is a metal ball with a mass of 7.3 kg. When thrown in Olympic competition, it is accelerated to a speed of about 14 m/s. As an approximation, let's say that the athlete exerts a constant force on the shot while throwing it and that it moves a distance of 3 m while accelerating.
   (a) What is the shot's kinetic energy?
   (b) Compute the force that acts on the shot.
   (c) It takes about 0.5 s to accelerate the shot. Compute the power required. Convert your answer to horsepower.

7. At the point in its orbit when it is closest to the Sun, Halley's Comet moves with a speed of 54,500 m/s (**Figure 3.48**). When it is at its most distant point, the separation between it and the Sun is about 60 times larger than when it is at its closest point. What is the speed of the comet at the distant point?

8. The *work-energy theorem:* Show explicitly that the work, Fd, done by a constant force to accelerate an object of mass m from rest to a final velocity v is equal to the kinetic energy, $(1/2)mv^2$, of the object. (*Hint:* Use Newton's second law, combined with the kinematic equations for the distance traveled by and the final velocity of a uniformly accelerated object in time t.)

9. A series of five 0.1-kg spheres are arrayed along a thin, lightweight rigid rod with length 0.5 m at intervals of 0.1 m from one end of the rod. The system spins about an axis perpendicular to and passing through the unoccupied end of the rod with a period of 0.3 s. In this way, each mass moves in the same plane along a circular path whose radius is its distance from the rotation axis.
   (a) How far does each mass move during one revolution?
   (b) What is the speed of each mass as it orbits the axis?
   (c) What is the total angular momentum of the system?

Karl R. Martin/Shutterstock.com

Figure CO-4 The Goodyear blimp after launch from Akron, Ohio.

CHAPTER INTRODUCTION: Airships

You've probably seen them so often on TV you hardly notice them anymore. Blimps, floating lazily above major sporting events (**Figure CO-4**), functioning as a combination camera platform and giant billboard in the sky. But such lighter-than-air craft have a long and colorful history. Humans first took to the air in controlled flight aboard balloons more than a century before the Wright brothers launched their "aeroplane" from the sands of Kitty Hawk, North Carolina in 1903. The era of powered human flight began before the Civil War when an airship was equipped with a steam engine. The first regular airline service across the Atlantic Ocean featured giant rigid airships called *zeppelins*, each carrying dozens of passengers in luxurious accommodations. More than 100 zeppelins were built, some being used to bomb London and other European cities during World War I (that's right, I, not II). In the 1930s the U.S. Navy operated two flying aircraft carriers—huge, 200-ton airships that could launch and retrieve small airplanes. (Both were destroyed in storms.) And one of the first—and still one of the most spectacular—disasters caught on film was the flaming destruction of the *Hindenburg* zeppelin in 1937.

So how can a metal aircraft weighing so many tons float in the air like a party balloon? Why are modern airships spared the fate of the *Hindenburg*? Answers to these and many more related questions are found in this

chapter, as we look into the physics of extended matter. The ideas presented in this chapter—particularly the law of fluid pressure, Archimedes' principle, and the concept of density—make it easy to understand how things such as airships, altimeters, and atomizers function.

4.1 Matter: Phases, Forms, and Forces

4.1a Phases of Matter

The subject of this section is matter—anything that has mass and occupies space. Earth, the water we drink, the air we breathe, our bodies—everything we touch is composed of matter. Obviously, matter exists in different forms that have different physical properties. We can classify matter into four categories: **solid, liquid, gas,** and **plasma.** These are called the four **phases,** or *states,* of matter. (In physics, *gas* isn't a colloquialism for gasoline, and *plasma* does *not* refer to the liquid part of blood.) Briefly, we can distinguish the four phases as follows.

DEFINITION **Solids** Rigid; retain their shape unless distorted by a force. Examples: rock, wood, plastic, iron.

Liquids Flow readily; conform to the shape of a container; have a well-defined boundary (surface); are not easily compressed. Examples: water, alcohol, gasoline, blood.

Gases Flow readily; conform to the shape of a container; do not have a well-defined surface; can be readily compressed (squeezed into a smaller volume). Examples: air, carbon dioxide, helium, radon.

Plasmas Have the properties of gases and also conduct electricity; interact strongly with magnetic fields; commonly exist at higher temperatures. Examples: gases in operating fluorescent, neon, and vapor lights (**Figure 4.1**); matter in the Sun and stars.

Nearly all of the matter in our everyday experience appears as solid, liquid, or gas. Traditionally, these have been referred to as the three states of matter, with plasmas being a special "fourth" state of matter. Although plasmas are rare on Earth, most of the visible matter in the universe is in the form of plasmas in stars. In the last 50 years, the study of plasmas has grown to be one of the major subfields of physics because of the interest in nuclear fusion (the topic of Section 11.7). Nuclear fusion is the source of energy for stars, our Sun included. One of the main goals of plasma physics is to produce laboratory-scale versions of starlike plasmas in which fusion can occur.

Admittedly, many substances do not fit easily into one of these four categories. Granulated sugar and salt flow readily and take the shape of a container. But they are considered to be solids because a single granule of each substance does fit the description of a solid, and because both can be crystallized into larger chunks. Tar and molasses do not flow readily, particularly when they are cold, and so they act somewhat like solids. But given time they do flow and take the shape of their container, so they are considered liquids.

Many substances are composites of matter in two different phases. Styrofoam behaves like a solid but is composed mostly of gas trapped in millions of tiny, rigidly connected bubbles. The water in many rivers carries along tiny, solid particles that will settle if the water is allowed to stand. The mists that fill a shower stall and comprise fog and low clouds consist of millions of small droplets of water mixed in with the air. Apples and potatoes are solid but contain a great deal of liquid.

Another factor that complicates our neat classification of matter is that the phase of a given substance can change with temperature and pressure. Water is a good example. Normally a liquid, water becomes a solid (ice) when cooled below 0°C (32°F; Figure 4.2). Under normal pressure, water becomes a gas (steam) when its temperature is raised above 100°C (212°F). Even at room temperature, water can be made to boil if the air pressure is reduced, however. Propane, carbon dioxide, and many other gases can be forced into the liquid phase at room temperature by increasing the pressure. Most refrigerators and air conditioners depend on the pressure-induced liquification of a gaseous refrigerant for their proper functioning. (More on this in Section 5.7.)

To simplify our discussion in this chapter, we will consider only matter in one "pure" phase: solid like rock, liquid like pure water, or gaseous like carbon dioxide. When we say that a substance exists in a particular phase, we mean its phase at normal room temperature and pressure unless stated otherwise.

The phases of matter refer to the macroscopic (external) form and properties of matter. These in turn are determined by the microscopic (internal) composition of matter. Around 2,500 years ago, some Greek philosophers theorized that all matter as we see and experience it is composed of tiny, indivisible pieces. It turns out that this is true up to a point: diamond, water, and oxygen are all composed of extremely tiny "building blocks" or units that are the smallest entities that retain the identity of the substance. (But, as we shall see, the building blocks themselves are composed of still smaller particles.) Water is a liquid, and diamond is a solid because of the properties of the small units that comprise each. Thus, we can reclassify all substances by the nature of their intrinsic composition. We will begin with the simplest class of matter and proceed in order of increasing complexity.

4.1b Forms of Matter

The chemical **elements** represent the simplest and purest forms of everyday matter. At this time, scientists have identified 118 different elements and have agreed on names for 114 of them. Some of the common substances used in our society

Figure 4.1 Compact fluorescent lights employ glowing plasmas.

Figure 4.2 The phase of any substance, like water shown here, depends upon its temperature.

Figure 4.3 Simplified diagram of an atom. An atom consists of electrons in orbit around a compact nucleus, which in turn contains protons and neutrons. All atoms of a particular element have the same number of protons. This number is the element's atomic number. (Figure not drawn to scale.)

are elements. These include hydrogen, helium, carbon, nitrogen, oxygen, neon, gold, iron, mercury, and aluminum. Each element is composed of incredibly small particles called **atoms**. There are 118 different atoms, one for each known element. Only about 90 of the elements exist naturally on Earth. The others are artificially produced in laboratories. The majority of the elements are quite rare and have names familiar only to chemists and other scientists (see the special feature "What's in a Name" at the end of this section).

The atom is not an indivisible particle: it has its own internal structure. Every atom has a very dense, compact core called the **nucleus** that is surrounded by one or more particles called **electrons**. The nucleus itself is composed of two kinds of particles, **protons** and **neutrons** (**Figure 4.3**). (Protons and neutrons are themselves composed of smaller particles called quarks, but more on this in Chapter 12.) The protons and electrons have equal but opposite electric charges and attract each other. The electrons are much lighter than protons and neutrons, and they move in orbits with the attraction of the protons supplying the required centripetal force.

Every atom associated with a particular element has the same unique number of protons. For example, atoms that have 2 protons are atoms of helium, those with 8 protons are atoms of oxygen, those with 79 protons are atoms of gold, and so on. The *atomic number* of an element is the number of protons that are in each atom of the element. The atomic number of helium is 2 because every atom of helium has two protons in it. For oxygen it is 8, for gold it is 79, and so on. Each element is also given a brief identifier called its *chemical symbol*. **Table 4.1** contains the chemical symbols and atomic numbers of some familiar elements. It also includes the phase of each element at room temperature and normal pressure. The Periodic Table of the Elements on the back inside cover

Table 4.1 Some Common Chemical Elements

Element	Symbol	Atomic Number	Phase
Hydrogen	H	1	Gas
Helium	He	2	Gas
Carbon	C	6	Solid
Nitrogen	N	7	Gas
Oxygen	O	8	Gas
Neon	Ne	10	Gas
Sodium	Na	11	Solid
Aluminum	Al	13	Solid
Silicon	Si	14	Solid
Chlorine	Cl	17	Gas
Calcium	Ca	20	Solid
Iron	Fe	26	Solid
Nickel	Ni	28	Solid
Copper	Cu	29	Solid
Zinc	Zn	30	Solid
Silver	Ag	47	Solid
Gold	Au	79	Solid
Mercury	Hg	80	Liquid
Lead	Pb	82	Solid
Uranium	U	92	Solid

of the print edition lists all of the known elements and some of their properties. We will take a closer look at the structure of atoms in Chapters 7 and 10.

Chemical **compounds** are the next simplest form of everyday matter. There are millions of different compounds, including many common substances such as water, salt, sugar, and alcohol. Compounds are similar to elements in that each compound also has a unique building block called a **molecule**. Every molecule of a particular compound consists of the same unique combination of two or more atoms held together by electrical forces. For example, each molecule of water consists of two atoms of hydrogen attached to one atom of oxygen (**Figure 4.4**). Similarly, one atom of carbon attached to one atom of oxygen forms a molecule of carbon monoxide. Each compound can be represented by a "formula"—a shorthand notation showing both the kinds and the numbers of atoms in each of the compound's molecules. You are probably familiar with the formula for water, H_2O. Some others are NaCl (table salt), CO_2 (carbon dioxide), CO (carbon monoxide), $C_{12}H_{22}O_{11}$ (sugar), and C_2H_5OH (ethyl alcohol). The study of how atoms combine to form molecules is a major part of chemistry.

Some elements in the gas phase are also composed of molecules. The oxygen in the air we breathe is an element, but most of the oxygen atoms are paired up to form O_2 molecules. Ozone (O_3) is a rarer form of oxygen that is present in air pollution and also in the ozone layer, which is found about 15 miles above Earth's surface. Each ozone molecule is composed of three oxygen atoms. Nitrogen gas also exists in the form of molecules, N_2. Helium and neon are both gaseous elements whose atoms do not form molecules, but remain separate.

Many substances, such as air, stone, and seawater, are composed of two or more different compounds or elements that are physically mixed together. These are classified as **mixtures** and **solutions**. Air consists of dozens of different gases that are mixed together. **Table 4.2** shows the composition of clean, dry air. The air we breathe also contains water vapor and pollutants (such as carbon monoxide). The amount of these gases present in the air varies considerably from place to place and from day to day. The air in the Sahara Desert does not have quite the same composition as the air in Los Angeles.

A mixture of two elements is not the same as a compound. If you mix hydrogen and oxygen, you simply have a gaseous mixture. The individual hydrogen atoms and oxygen molecules remain separate. If you ignite the mixture, the hydrogen and oxygen atoms will combine to form water molecules. Energy is released explosively in the form of heat (fire) because the atoms have less total energy when they are bound to each other.

The example of hydrogen, oxygen, and water also illustrates that the properties of a compound (water) are usually quite different from the properties of the constituent elements (hydrogen and oxygen). Sodium (Na) is a solid that reacts violently with water. Chlorine (Cl) is a gas that is used to kill bacteria in drinking water. Either element alone could be fatal. But when sodium and chlorine are combined chemically, the result is table salt (NaCl), a necessary part of our diet (**Figure 4.5**).

Figure 4.4 Water molecules consist of two hydrogen atoms attached to an oxygen atom. This structure is common to all molecules of water, ice, and steam.

Table 4.2 Composition of Clean, Dry Air

Gas	Percent Composition (by volume)
Nitrogen (N_2)	78.1
Oxygen (O_2)	20.9
Argon (Ar)	0.93
Carbon dioxide (CO_2)	0.03
Other gases (Ne, He, and so on)	0.04

Sodium   Chlorine   Table salt

+

MARTYN F. CHILLMAID/SCIENCE PHOTO LIBRARY; ANDREW LAMBERT PHOTOGRAPHY/ SCIENCE PHOTO LIBRARY; istockphoto.com/ eyewave

Figure 4.5 Separately, sodium and chlorine are poisonous. When chemically combined, they form table salt.

Figure 4.6 Think about how many golf balls it would take to fill up a hollow ball the size of the entire Earth. That's about how many atoms there are in each golf ball.

iStockphoto.com/fambros; iStockphoto.com/pederk; iStockphoto.com/jannysavy

The basic unit of life is the cell, and each cell is composed of many different compounds. Many of these contain billions of atoms in each molecule, the DNA molecule being an important example. The main constituents of such "organic" molecules are hydrogen, carbon, oxygen, and nitrogen. Many other elements are present in smaller amounts. For example, your bones and teeth contain calcium.

So atoms are basic to compounds as well as to elements. It is difficult to imagine how small atoms are and just how many there are in common objects. Atoms are about 1 ten-millionth of a millimeter in diameter. A ball 1 inch in diameter is about midway between the size of an atom and the size of Earth (**Figure 4.6**). In other words, if a 1-inch ball were expanded to the size of Earth, each atom in the ball would expand to about1 inch in diameter.

Because atoms are so small, huge numbers of them are present in anything large enough to be seen. There are about 100 thousand billion billion (one followed by 23 zeros) atoms in each of your fingernails. Even the smallest particles that can be seen with the naked eye contain far more atoms than there are people on Earth.

Atoms are nearly indestructible—only nuclear reactions affect them—and they are continually recycled. Earth and everything on it are believed to be composed of the debris of stars that exploded billions of years ago. Chemical processes such as fire, decay, and growth result in atoms being combined with or dissociated from other atoms. A single carbon atom in your earlobe may once have been part of a dinosaur, a redwood tree, a rose, or Leonardo da Vinci—or all four.

4.1c Behavior of Atoms and Molecules

The constituent particles of matter, atoms and molecules, exert electrical forces on each other. The nature of these forces determines the properties of the substance. The forces between atoms in an element depend on the configuration of the electrons in each atom. In a compound, the size and shape of the molecules, as well as the forces between the molecules, affect its observed form and properties. We can relate the three common phases of matter to the interparticle forces as follows.

DEFINITION **Solids** Attractive forces between particles are very strong; the atoms or molecules are rigidly bound to their neighbors and can only vibrate.

Liquids The particles are bound together, though not rigidly; each atom or molecule can move about relative to the others but is always in contact with other atoms or molecules.

Gases Attractive forces between particles are too weak to bind them together; atoms or molecules move about freely with high speed and are widely separated; particles are in contact only briefly when they collide.

PHOTOTAKE Inc./Alamy Stock Photo

(a) (b) (c)

Figure 4.7 (a) The atoms or molecules in a crystalline solid are arranged in a regular three-dimensional pattern, similar to the way rooms are arranged in a large apartment building. The atoms or molecules exert forces on each other. (b) A crystal behaves much like an array of particles that are connected to each other by springs. (c) Image produced with a scanning tunneling microscope (STM) at the IBM Thomas J. Watson Research Center showing the geometric arrangement of silicon atoms.

A standard model for representing the interparticle forces in a solid consists of each atom connected to its neighbors by a spring. The atoms are free to oscillate like a mass hanging from a spring. (The vibration of the atoms or molecules is related to the temperature, as we shall see in Chapter 5.) Often atoms or molecules in a solid form a regular geometric pattern called a *crystal* (**Figure 4.7**). Table salt is a crystalline compound in which the sodium and chlorine atoms alternate with each other. In solids that do not have a regular crystal structure, called *amorphous solids*, the atoms or molecules are "piled together" in a random fashion. Glass is a good example of such a solid.

Carbon is a very interesting element because it has two common crystalline forms (graphite and diamond) that possess very different properties, plus a large number of recently discovered molecular forms. In diamond, each carbon atom is strongly bonded to each of its four nearest neighbors resulting in a crystalline solid that is the hardest known natural material (**Figure 4.8a**). In graphite, the primary ingredient in the misnamed "lead" pencils, the carbon atoms form sheets, with each atom strongly bonded to its three nearest neighbors in the same layer to form a mesh of hexagons (**Figure 4.8b**). These sheets can easily be forced to slide relative to each other, making graphite an excellent "dry" lubricant. Carbon atoms can also bond together to form large molecules, the most famous being C_{60}, named *buckminsterfullerene* ("buckyball" for short) after the famous engineer and philosopher R. Buckminster Fuller. The molecule reminded its discoverers of geodesic domes invented by Fuller. The 60 atoms in each buckminsterfullerene molecule arrange themselves into a soccer-ball shape consisting of 12 pentagons and 20 hexagons. The C_{60} molecules can in turn bond together to form a crystal (**Figure 4.8c**). Dozens of larger and smaller hollow carbon molecules can also form, as well as "carbon nanotubes" consisting of graphitelike sheets of carbon atoms rolled up into tubes. Fullerenes have been used extensively in several biomedical applications, including the design of high-performance

Diamond
(a)

magnetix/Shutterstock.com

Graphite
(b)

Fullerene
(c)

Figure 4.8 Three forms of solid carbon. Each small sphere represents a carbon atom, and the connecting rods represent the force that holds pairs of carbon atoms together: (a) diamond; (b) graphite; and (c) C_{60} crystal.

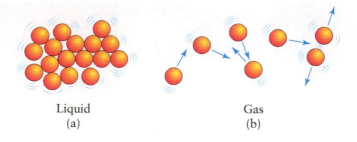

Figure 4.9 (a) The atoms or molecules in a liquid remain in contact with each other, but they are free to move about. (b) In a gas, the atoms or molecules are not bound to each other. They move with high speed and interact only when they collide.

Liquid (a)

Gas (b)

Jochen Tack/Alamy Stock Photo

Figure 4.10 A collection of spherical magnets share some properties of atoms or molecules in a liquid, although the forces that bind them have different origins.

MRI contrast agents, x-ray imaging contrast agents, and photodynamic therapy, where photosensitive C_{60} molecules are absorbed by cancer cells and then illuminated to produce in situ reactive oxygen molecules that cause cell death.

In a liquid, the forces between the particles are not strong enough to bind them together rigidly. The atoms or molecules are free to move around as well as vibrate (**Figure 4.9a**). This is similar to a collection of ball-shaped magnets clinging to each other. The forces between the particles are responsible for surface tension and the spherical shape of small drops (**Figure 4.10**).

Many compounds have an interesting intermediate phase between solid and liquid called the *liquid crystal* phase. The molecules have some mobility, as in a liquid, but they are arranged regularly, as in a solid. Liquid crystal displays (LCDs) in calculators, flat-panel video displays, cell phones, and digital watches use electrically induced movement of the molecules to change the optical properties of the liquid crystal. (More on this in Section 9.1.)

In a gas, the atoms and molecules are widely separated and move around independently, except when they collide (**Figure 4.9b**). Their speeds are surprisingly high: the oxygen molecules you are now breathing have an average speed of more than 1,000 mph. As each molecule collides randomly with other molecules, its direction of travel is altered and its speed is sometimes increased and other times decreased.

Often, the surface of a solid or a liquid forms a boundary for a gas. The surface of water in a glass, a lake, or an ocean forms a lower boundary for the air above. The inside of the walls of a tire are a boundary for the air inside (**Figure 4.11**). The high-speed atoms or molecules in the gas exert a force on the surface as their random motions cause them to collide with it. This is the basis for gas pressure, as mentioned in Section 3.6. In other words, the weight of a car is supported by the collisions of air molecules with the inner walls of the tires.

At normal temperatures and pressures, the average distance between the centers of the atoms or molecules in a gas is about 10 times that in a solid or liquid (**Figure 4.12**). There is a great deal of empty space in a gas, and that is why gases can be compressed easily.

Figure 4.11 The air molecules inside a tire collide with the walls and exert forces on it. These forces are responsible for the pressure that keeps the tire from collapsing.

Figure 4.12 Gases can be compressed because the atoms or molecules are widely separated. Increasing the force (pressure) on the boundaries of a gas squeezes the particles closer together.

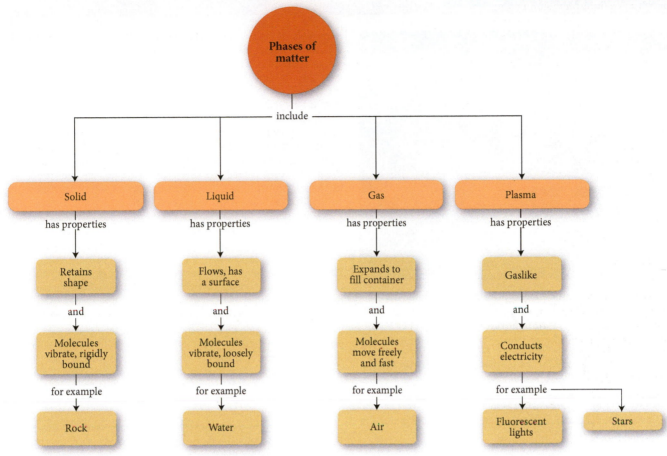

Let's summarize the information in this section. The 118 different types of atoms (the elements) form the basis of matter as we experience it. Molecules are composed of two or more atoms "stuck" together to form the basic unit of compounds. The nature of the forces between the building blocks (atoms or molecules) determines whether the substance is a solid, a liquid, or a gas. Mixtures and solutions consist of different elements or compounds mixed together. Concept Map 4.1 summarizes this section in another way.

■ Learning Check

1. The three common phases of matter on Earth are _____ , _____ , and _____ .
2. (True or False.) The number of protons in the nucleus of an atom determines which element it is.
3. The molecule is the basic unit of all
   (a) elements.
   (b) compounds.
   (c) solutions.
   (d) mixtures.

4. Graphite, diamond, and buckminsterfullerene are different forms of the element _____ .
5. (True or False.) Air can be compressed easily because there is a great deal of empty space between air molecules.

ANSWERS: 1. solid, liquid, gas 2. True 3. (b) 4. carbon 5. True

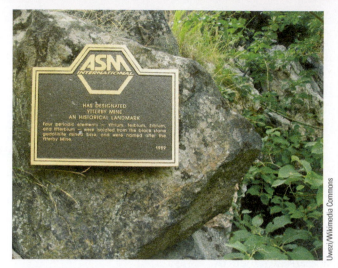

Figure 4.13 A plaque placed outside a mine near the village of Ytterby, Sweden, by the American Society for Metals International commemorates the discovery of some of the most exotic and rare elements in the periodic table.

What's in a name? Well, for the elements, names contain clues to their chemical and physical properties, to their places and means of discovery, to the origins of their chemical symbols, and to the scientists who figured prominently in their isolation. In short, a great deal can be learned about the chemical elements from a study of their names.

Consider the element chlorine (Cl). The word *chlorine* derives from the Greek word *chloros* meaning "pale green"—an apt description for this greenish-yellow gas employed as a lethal weapon during World War I (the infamous "mustard gas") and now used as a disinfectant and water-purifying agent to help save lives. The element iridium (Ir) takes its name from the Greek word for rainbow, *iris*. This is also a rather poetic description of an element whose crystalline salts are brilliantly colored, spanning the entire rainbow of hues from deep blue to dark red. Or take zirconium (Zr). Its name comes from the Persian word *zargun* meaning "goldlike." This, too, seems an appropriate way to characterize the element responsible for the luster and sheen of the mineral and gemstone that we now call *zircon* but that was well known to the ancients and is described in biblical writings.

Many elements obtain their names from their places of discovery. Probably the best examples are the elements erbium (Er), terbium (Tb), ytterbium (Yb), and yttrium (Y), all of which take their names from the village of Ytterby in Sweden near Stockholm (**Figure 4.13**). Ytterby is the site of a quarry that yielded many unusual minerals containing these and other uncommon elements. Further examples of elements named after places first associated with their discovery are californium (Cf), named for the state and university in which it was first produced by a team of Berkeley scientists in 1950, and strontium (Sr), whose name comes from the mineral stronianite found outside the village of Strontian in Scotland, where the element was first discovered in 1790. A final example of this type is helium (He), meaning "the Sun"; it serves to remind us that the element was first detected in 1868 in the outer atmosphere of the Sun during a solar eclipse.

The symbols of most of the chemical elements bear a direct correspondence to their names, usually consisting of the first letter and, at most, one additional letter from the name. However, for many of the common elements that have been known for centuries, this is not the case. Classic examples include iron (Fe), gold (Au), mercury (Hg), and copper (Cu). For such elements, the symbols often originate from their Latin equivalents and date from a time hundreds of years ago when Latin was the language of scholarship the world over. Iron is an Anglo-Saxon word whose Latin equivalent is *ferrum*, hence the symbol Fe. Similarly, gold derives from an Anglo-Saxon word of the same spelling, but its symbol, Au, is taken from the Latin word *aurum* meaning "shining dawn." Mercury, named after the fleet-footed Roman god, takes its symbol, Hg, from the Latin word *hydrargyrum*, which translates to "liquid silver"—a perfect description of this shiny, liquid metal. Lastly, the symbol for copper, Cu, comes from the Latin *cuprum*; this word is a shortened corruption of the phrase *aes Cyprium*, meaning "metal from Cyprus." The Romans obtained virtually all of their supply of copper from this Mediterranean island. Other instances of this type of symbol derivation include silver (Ag from the Latin *argentum*) and lead (Pb from the Latin *plumbum*—compare our word *plumber*).

A number of the element names derive from the names of scientists who have contributed to the discovery or the isolation of the elements. Specific cases include mendelevium (Md), named after Dmitri Mendeleev, a Russian chemist who developed the periodic table of the elements; curium (Cm), named for Pierre and Marie Curie, pioneers in the discovery and characterization of radioactive elements; and gadolinium (Gd) after Finnish chemist Johan Gadolin, the discoverer of yttrium.

As mentioned in the text, 118 elements have been identified. Traditionally, the individual or group responsible for first isolating a new element earns the right to propose its name. At the time of this writing, names have been established for 114 elements, those of elements 104–114 having only been recently agreed on by the Committee on Inorganic Chemistry Nomenclature of the International Union of Pure and Applied Chemistry. The new names follow the same general patterns observed in elements with atomic numbers less than 103. For example, element 104 is named rutherfordium (Rf), after Ernest Rutherford, a New Zealand physicist who won the Nobel Prize in chemistry in 1907 for establishing that radioactivity causes a transmutation of an element (see Chapter 11 for more on this topic). Element 105, dubnium (Db), is named after the city of Dubna, where Russian investigators first identified this species. Hassium (Hs), element 108, comes from the Latin name for the state of Hesse in Germany, the location of the Laboratory for Heavy Ion Research (GSI), which first produced elements 107–112 (**Figure 4.14**). Recently, element 110 was named darmstadtium (Ds), honoring the site of the GSI. Element 112, copernicium (Cn),

Figure 4.14 GSI facility in Darmstadt, Germany.

is named after 16th-century Polish astronomer Nicolaus Copernicus, who presented the first comprehensive scientific treatise arguing for a heliocentric model of the solar system. Names for elements 113 to 118 have yet to be agreed upon (see the Periodic Table of Elements located on the inside back cover of the book).

In summary, there can be a lot in a name, especially if it is the name of a chemical element. Perhaps this short excursion into the lexicology of the elements will make you want to know more about this subject or will at least make it a little easier for you to remember the symbols of the more common elements that make up our environment.

■ **QUESTIONS**

1. List at least three sources of names and symbols for the elements in the periodic table and give an example of each.

2. Fill in the blanks of this little story with the names of the appropriate elements. BeSTe OF LuCK In YOURe SeArCH!
   "The poor fellow crashed his (Ag) _____
   (Ne) _____ into a wall. Luckily, a good
   (Sm) _____ came along and tried to (He)
   _____, but, alas, it was too late. All the
   hapless rescuer could do was (Ba) _____ in a
   (Kr) _____ the hill nearby."

4.2 Pressure

Now that we have established some of the basic properties of solids, liquids, and gases, it is time to consider how the mechanics we've presented in the preceding chapters relates to extended matter. We have already seen in Chapters 1, 2, and 3 that extended objects can often be treated as particles (for example, a rock as it falls, satellites and even planets as they move in orbits, trucks during collisions, and so on). Furthermore, the law of conservation of angular momentum (Section 3.8) allows us to examine interesting types of rotational motions of solid bodies. But what about gases and liquids? The fluids in and around us—blood in our veins and arteries, the atmosphere, streams and oceans—are in constant motion. Can we extend our study of mechanics to fluid motion? The answer is a qualified yes.

Newton's laws and the conservation laws can be applied to fluids using appropriate extensions of physical quantities such as mass and force, but the mathematics is much more complicated and well beyond the level of this text. (Some of the most powerful computers in the world are used exclusively to solve problems involving fluids, such as the motions of the atmosphere.) So we will limit our study of fluid motion to one (restricted) conservation law (Section 4.7). However, there are several phenomena related to fluids at rest that are important in our daily lives and that can be dealt with using simple mathematics. These phenomena are the principal topics of most of the rest of this chapter. We first introduce the physical quantities **pressure** and **density**, which are extensions of the concepts of force and mass, respectively. These are two essential quantities for studying fluids both at rest and in motion.

4.2a Defining Pressure

Forces that are exerted by gases, liquids, and solids normally are spread over a surface. The force that the floor exerts on you when you are standing is distributed over the bottoms of your feet. A boat floating on water has an upward force spread over its submerged surface (**Figure 4.15**). The air around a floating balloon exerts forces on the balloon's surface. In situations like these, the physical quantity **pressure** is quite useful.

Figure 4.15 The water exerts an upward force that is spread over the submerged surface of the boat.

| DEFINITION | **Pressure** The force per unit area when the force acts perpendicular to a surface. The perpendicular component of a force acting on a surface divided by the area of the surface. |

$$p = \frac{F}{A}$$

Jack Hollingsworth/Corbis

Table 4.3 Some Pressures of Interest

Description	p (Pa)	p (psi)	p (atm)
Lowest laboratory pressure	7×10^{-12}	1×10^{-14}	7×10^{-16}
Atmospheric pressure at an altitude of 100 km	0.06	9×10^{-6}	6×10^{-7}
Lowest recorded sea level atmospheric pressure	0.87×10^5	12.6	0.86
Average sea level atmospheric pressure	1.01×10^5	14.7	1
Highest recorded sea level atmospheric pressure	1.08×10^5	15.7	1.07
Typical pressure inside a tire	3.1×10^5	45	3.1
Pressure at 11 km underwater (Mariana Trench)	1.1×10^8	1.6×10^4	1.1×10^3
Highest laboratory water jet pressure	6.8×10^{10}	9.9×10^6	6.8×10^5
Earth's center	1.7×10^{11}	2.5×10^7	1.7×10^6
Center of the Sun	2.5×10^{16}	3.6×10^{12}	2.5×10^{11}

Pressure, like work, is a scalar; there is no direction associated with it. We will use p as the symbol for pressure to distinguish it from power denoted by P.

Physical Quantity	Metric Units	English Units
Pressure (p)	pascal (Pa) (1 Pa = 1 N/m²)	pound per square foot (lb/ft²)
		pound per square inch (psi)
	millimeters of mercury (mm Hg)	inches of mercury (in. Hg)

The standard pressure units consist of a force unit divided by an area unit. The SI unit of pressure is the pascal (Pa), which equals 1 newton per square meter. The English unit for pressure (psi) may be the most familiar to you. Air pressure in tires is commonly measured in psi. For comparison,

$$1 \text{ psi} = 6{,}890 \text{ Pa}$$

This number is so large because a force of 1 pound on each square inch would produce a huge force on 1 square meter—a much larger area. The two pressure units involving mercury will be explained in Section 4.4. Another common unit of pressure is the atmosphere (atm). It equals the average air pressure at sea level. (Later, we will see that the air pressure is lower at higher altitudes.) It is related to the other units as follows:

$$1 \text{ atm} = 1.01 \times 10^5 \text{ Pa} = 14.7 \text{ psi}$$

This unit is like the g used as a unit of acceleration. We are all subject to the pressure of the air around us and to the acceleration from gravity, so it is natural to compare other pressures and accelerations to them. The old practice of using the length of a person's foot as a unit of distance is another example of taking a unit of measure provided by Nature. Table 4.3 lists some representative pressures.

EXAMPLE 4.1 A 160-pound person stands on the floor. The area of each shoe that is in contact with the floor is 20 square inches. What is the pressure on the floor?

SOLUTION Assuming the person's weight is shared equally between the two shoes, the force of one shoe is 80 pounds. So,

$$p = \frac{F}{A} = \frac{80 \text{ lb}}{20 \text{ in.}^2}$$

$$= 4 \text{ psi} \qquad \text{(standing on both feet)}$$

By Newton's third law of motion, the floor exerts an equal and opposite force on the shoes. Thus the pressure of the floor on the shoes is also 4 psi.

If the person stands on one foot instead, that shoe has all of the weight and the pressure is

$$p = \frac{160 \text{ lb}}{20 \text{ in.}^2}$$

$$= 8 \text{ psi} \qquad \text{(standing on one foot)}$$

What if the person put on high-heeled shoes and balanced on the heel of one shoe (Figure 4.16)? The bottom of the heel might measure 0.5 inches by 0.5 inches. The area is then 0.25 square inches. So the pressure in this case would be

$$p = \frac{160 \text{ lb}}{0.25 \text{ in.}^2}$$

$$= 640 \text{ psi} \qquad \text{(balanced on a narrow heel)}$$

Figure 4.16 The pressure on the floor is much higher with narrow heels.

> ### Physics To Go 4.1
>
> Lightly squeeze a push pin or thumbtack between two fingers, as shown in **Figure 4.17**. The force is the same on each finger by Newton's third law. Do you feel the same sensation in each finger? How is pressure involved?

Example 4.1 and Physics to Go 4.1 show that the same force causes a much higher pressure when it acts over a smaller area. We can think of pressure as a measure of how "concentrated" a force is.

There are many situations in which a liquid or a gas is under pressure and exerts a force on the walls of its container. The relationship between force and pressure can be used to determine the force on a particular area of the walls. Because pressure equals force divided by area, the pressure times the area equals the force. In other words, the total force on a surface equals the force on each square inch (the pressure) times the total number of square inches (the area).

$$F = pA$$

EXAMPLE 4.2 Over the past 25 years, there have been several spectacular (and tragic) aircraft mishaps involving rapid loss of air pressure in passenger cabins (Figure 4.18). (The causes included failure of a cargo door, outer skin rupture due to corrosion or cracks or both, and small bombs.) Because the cabins are pressurized, there are large outward forces acting on windows, doors, and the aircraft skin. Let's estimate the sizes of these forces.

SOLUTION The pressure inside a passenger jet cruising at high altitude (about 7,500 meters or 25,000 feet) is about 6 psi (0.41 atmospheres) greater than the pressure outside. What is the outward force on a window measuring 1 foot by 1 foot and on a door measuring 1 meter by 2 meters? The area of the window is

$$A = 1 \text{ ft} \times 1 \text{ ft} = 12 \text{ in.} \times 12 \text{ in.}$$

$$= 144 \text{ in.}^2$$

(The area has to be in square inches because the pressure is in pounds per square *inch.*)

Figure 4.17 When a tack is squeezed between two fingers, the force on each finger is the same but the pressure is not.

The net outward force on the window is

$$F = pA = 6 \text{ psi} \times 144 \text{ in.}^2$$

$$= 864 \text{ lb}$$

A rather small pressure causes a large force on the window. For the door, we use SI units:

$$p = 6 \text{ psi} = 6 \times 1 \text{ psi} = 6 \times 6{,}890 \text{ Pa}$$

$$= 41{,}340 \text{ Pa}$$

$$A = 1 \text{ m} \times 2 \text{ m} = 2 \text{ m}^2$$

Consequently, the force on the door is

$$F = pA = 41{,}340 \text{ Pa} \times 2 \text{m}^2$$

$$= 82{,}680 \text{ N}$$

The force on the door is greater than the weight of 10 small automobiles, or 100 people.

$p_{out} = 14$ psi

30 psi

$p_{in} = 44$ psi

Figure 4.19 An absolute pressure inside of 44 psi produces a gauge pressure of 30 psi.

4.2b Gauge Pressure

Pressure is a relative quantity. When you test the air pressure in a tire, you are comparing the pressure of the air inside the tire with the pressure outside the tire. When a tire is flat, there is still air inside it, but the pressure inside is the same as the atmospheric pressure outside.

For example, let us say that the atmospheric pressure is 14 psi. A tire is tested, and the air pressure gauge shows 30 psi. This means that the air pressure inside the tire is 30 psi higher than the air pressure outside the tire. So the actual pressure on the inner walls of the tire is 30 + 14 = 44 psi. This is sometimes referred to as the *absolute pressure*. The pressure relative to the outside air (30 psi in this case) is then called the *gauge pressure* (**Figure 4.19**).

We can look at this another way. The pressure inside the tire, 44 psi, causes an outward force of 44 pounds on each square inch of the tire wall. The air pressure outside the tire causes an inward force of 14 pounds on each square inch. The net force from air pressure on each square inch of the tire is then 30 pounds.

The gauge pressure of the air in a tire changes if the outside air pressure changes. What happens if the car is driven into a large chamber and the air pressure in the chamber is increased from 14 psi to 44 psi? The gauge pressure in the tires will then be zero, and the tires will go flat even though no air has been removed from them. When the air pressure is reduced to 14 psi, the tires will expand again to their normal shape.

In the previous example of the pressurized aircraft, the 6 psi is the gauge pressure. The absolute pressure inside the cabin might be 11 psi, and the air pressure outside might be 5 psi. The atmospheric pressure at 26,000 feet is about 5 psi.

Vern Ostdiek/University of Michigan

> ### Physics To Go 4.2
>
> Put an ounce or so of water into an empty aluminum can. Heat the can until the water is boiling vigorously and you can clearly see mist coming out of the opening. Let the water boil for about a minute. With a gloved hand, quickly but carefully turn the (hot!) can upside down over a pan or sink with cold water in it and plunge the top of it an inch or so into the water. What happens? (**Figure 4.20**.) How is the atmosphere involved in this?

Figure 4.20 The outcome of *Physics to Go 4.2.*

The standard pen-shaped tire-pressure tester nicely illustrates some of the physics that we have considered so far. It consists of a hollow tube (cylinder) fitted with a piston that can slide back and forth in the cylinder (**Figure 4.21**). Air from the tire enters the cylinder at the left end and pushes on the piston. The right end allows air from the outside to push on the other side of the piston. If the pressure inside the tire is greater than the outside air pressure, there is a net force to the right on the piston. A spring placed behind the piston is compressed by this net force. The greater the net force on the piston, the greater the compression of the spring. A calibrated shaft extends from the right side of the piston and out of the right end of the cylinder. When the piston is pushed to the right, the shaft protrudes from the right end the same distance that the spring is compressed. The length of shaft showing indicates the gauge pressure in the tire, because that is what causes the force on the piston.

We conclude this section with one last important note about pressure. Because gases are compressible, the volume of a gas can be changed (Figure 4.12). Whenever the volume of a fixed amount of gas is changed, the pressure in the gas changes also. Increasing the volume of a gas reduces the pressure. Decreasing the volume increases the pressure. To understand why this is the case, recall from the previous section that the collisions of the atoms and molecules in a gas with a surface cause gas pressure. If the volume is decreased, the particles are squeezed together so there are more of them near each square inch of the boundaries. More collisions occur each second, which means more force on each square inch and higher pressure. The opposite happens when the volume is increased.

The temperature of a gas influences the speeds of the atoms or molecules. Consequently, the pressure is also affected by the gas temperature. When the

Piston

F

Figure 4.21 In a common tire-pressure tester, the higher pressure of the air in the tire pushes the piston to the right and compresses the spring. The higher the pressure, the greater the force and the greater the compression of the spring.

temperature of a given quantity of gas is kept constant, however, the pressure p is related to the volume V as follows:

$$pV = \text{a constant} \qquad \text{(gas at fixed temperature)}$$

This means that the *volume of a gas is inversely proportional to the pressure.* If the pressure is doubled, the volume is halved. This relationship was discovered by Robert Boyle, a mentor and colleague of Isaac Newton, and is now referred to as *Boyle's law.* In Chapter 5, we will see precisely what effect temperature has on pressure and volume.

■ Learning Check

1. If the same-sized force is made to act over a smaller area,
   (a) the pressure is decreased.
   (b) the pressure is not changed.
   (c) the pressure is increased.
   (d) the result depends on the shape of the area.
2. (True or False.) A sealed box contains air under pressure. The outward forces on the walls will be largest on those walls with the smallest area.

3. The pressure difference between the inside and the outside of a tire is the _____ pressure.
4. (True or False.) If the mass and temperature of the gas in a container stay constant while the volume is decreased, the pressure will also decrease.

ANSWERS: 1. (c) **2.** False **3.** gauge **4.** False

4.3 Density

4.3a Mass Density

Pressure is an extension of the idea of force. Similarly, **mass density** is an extension of the concept of mass. Just as pressure is a measure of the concentration of force, density is a measure of the concentration of mass.

> **DEFINITION** **Mass Density** The mass per unit volume of a substance. The mass of a quantity of a substance divided by the volume it occupies.
>
> $$D = \frac{m}{V}$$

Physical Quantity	Metric Units	English Units
Mass density (D)	kilogram per cubic meter (kg/m³)	slug per cubic foot
	gram per cubic centimeter (g/cm³)	

We use D as the symbol for density, reserving d for distance. Density is a scalar quantity.

To find the mass density of a substance, one measures the mass of a representative part or "sample" of it and divides by the volume of that part or sample. It doesn't matter how much is used: the greater the volume, the greater the mass.

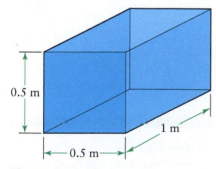

0.5 m

1 m

0.5 m

Figure 4.22 The mass of the water in this aquarium is 250 kilograms. Using this and the volume of the aquarium, we can compute the mass density of water.

EXAMPLE 4.3 The dimensions of a rectangular aquarium are 0.5 meters by 1 meter by 0.5 meters. The mass of the aquarium is 250 kilograms larger when it is full of water than when it is empty (**Figure 4.22**). What is the density of the water?

SOLUTION First, the volume of water is

$$V = l \times w \times h$$

$$= 1\ \text{m} \times 0.5\ \text{m} \times 0.5\ \text{m}$$

$$= 0.25\ \text{m}^3$$

So:

$$D = \frac{m}{V} = \frac{250 \text{ kg}}{0.25 \text{ m}^3}$$

$$= 1{,}000 \text{ kg/m}^3 \qquad \text{(mass density of water)}$$

If a tank with twice the volume were used, the mass of the water in it would be twice as great, and the density would be the same. The mass density of any amount of pure water is $1{,}000 \text{ kg/m}^3$.

When the same tank is filled with gasoline, the mass of the gasoline is found to be 170 kilograms. Therefore, the density of gasoline is

$$D = \frac{170 \text{ kg}}{0.25 \text{ m}^3}$$

$$= 680 \text{ kg/m}^3$$

Except for small changes caused by variations in temperature or pressure, the mass density of any pure solid or liquid is constant. It is an identifying trait of that substance. The mass densities of pure gases, on the other hand, vary greatly with changes in temperature or pressure. We have seen that doubling the pressure on a gas halves the volume. That would double the mass density as well. In Chapter 5, we will describe how changing the temperature of a gas also alters its volume and, therefore, its density. By convention, the tabulated densities of gases are given at the standard temperature and pressure (STP): 0°C temperature and 1 atmosphere pressure.

The mass density of each element or compound is fixed. Water, lead, mercury, salt, oxygen, gold, and so on, all have unique mass densities that have been measured and cataloged. The density of a mixture containing two or more substances depends on the density and percentage of each component. Most metals in common use are alloys, consisting of two or more metallic elements and other elements such as carbon. For example, 14-karat gold is only about 58% gold. However, the mass density of a mixture with a particular composition is constant.

Table 4.4 lists the mass densities (column 3) of several common substances. The values for mixtures can vary and so are merely representative. The mass densities of the gases are for standard conditions.

Having a list of the densities of common substances is quite useful for three reasons. First, one can use mass density to help *identify* a substance. For example, one can determine whether or not a gold ring is solid gold by measuring its density and comparing it to the known density of pure gold. Second, density measurement is used routinely to determine how much of a particular substance is present in a mixture. The coolant in an automobile radiator is usually a mixture of water and antifreeze. These two liquids have different densities (Table 4.4), so the density of a mixture depends on the ratio of the amount of water to the amount of antifreeze. The higher the density, the greater the antifreeze content and the lower the freezing temperature of the coolant (Figure 4.23). By simply measuring the coolant density, one can determine the coolant's freezing temperature. If you have donated blood to a blood bank, part of the screening included checking to see whether the hemoglobin content of your blood was high enough. This is done by determining whether the blood's density is greater than an accepted minimum value.

Third, one can *calculate the mass* of something if one knows what its volume is. The mass of a substance equals the volume that it occupies times its mass density.

$$m = V \times D$$

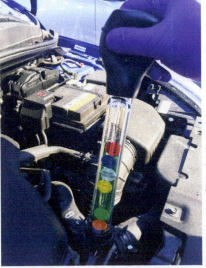

Figure 4.23 The freezing point of a car's coolant can be determined by measuring the coolant's density. This indicates the relative amount of antifreeze in the mixture. (See Section 4.5 for more.)

© Christine Myaskovsky/Cengage Learning

Table 4.4 Densities of Some Common Substances

Substance	Type*	Mass Density, D (kg/m³)	Weight Density, D_W (lb/ft³)	Specific Gravity
Solids				
Styrofoam	m	37	2.3	0.037
Juniper wood	m	560	35	0.56
Ice	c	917	57.2	0.917
Ebony wood	m	1,200	75	1.2
Concrete	m	2,500	156	2.5
Aluminum	e	2,700	168	2.7
Granite	m	2,700	168	2.7
Diamond	e	3,400	210	3.4
Iron	e	7,860	490	7.86
Brass	m	8,500	530	8.5
Nickel	e	8,900	555	8.9
Copper	e	8,930	557	8.93
Silver	e	10,500	655	10.5
Lead	e	11,340	708	11.34
Uranium	e	19,000	1,190	19
Gold	e	19,300	1,200	19.3
Liquids				
Gasoline	m	680	42	0.68
Ethyl alcohol	c	791	49	0.791
Water (pure)	c	1,000	62.4	1.00
Seawater	m	1,030	64.3	1.03
Antifreeze	m	1,100	67	1.1
Sulfuric acid	c	1,830	114	1.83
Mercury	e	13,600	849	13.6
Gases (at 0°C and 1 atm)				
Hydrogen	e	0.09	0.0056	0.00009
Helium	e	0.18	0.011	0.00018
Nitrogen (N_2)	c	1.25	0.078	0.00125
Air	m	1.29	0.08	0.00129
Oxygen (O_2)	c	1.43	0.089	0.00143
Carbon dioxide	c	1.98	0.12	0.00198
Radon	e	10	0.627	0.010

*Note: "e" stands for element, "c" for compound, and "m" for mixture.

EXAMPLE 4.4 The mass of water needed to fill a swimming pool can be computed by measuring the volume of the pool. Let's say a pool is going to be built that will be 10 meters wide, 20 meters long, and 3 meters deep. How much water will it hold?

SOLUTION The volume of the pool will be

$$V = l \times w \times h = 20 \text{ m} \times 10 \text{ m} \times 3 \text{ m} = 600 \text{ m}^3$$

$$m = V \times D = 600 \text{ m}^3 \times 1{,}000 \text{ kg/m}^3$$

$$= 600{,}000 \text{ kg}$$

That is a lot of water (about 3,000 bathtubs full).

4.3b Weight Density and Specific Gravity

In some cases, it is practical to use another type of density called **weight density**. It is commonly used in the English system of units because weight is in more common use than mass.

> **DEFINITION** **Weight Density** The weight per unit volume of a substance. The weight of a quantity of a substance divided by the volume it occupies.
>
> $$D_W = \frac{W}{V}$$

Physical Quantity	Metric Units	English Units
Weight density D_W	newton per cubic meter (N/m³)	pound per cubic foot (lb/ft³)
		pound per cubic inch (lb/in.³)

The weight density of a substance is equal to the mass density times the acceleration from gravity. This is because the weight of a substance is just g times its mass.

$$W = m \times g \rightarrow D_W = D \times g$$

Column 4 in Table 4.4 shows representative weight densities in English units. These can be used in the same ways that mass densities are used. One interesting exercise is to compute the weight of the air in a room. (Often people do not realize that air and other gases *do* have weight.)

EXAMPLE 4.5 A college dormitory room measures 12 feet wide by 16 feet long by 8 feet high. What is the weight of air in it under normal conditions?

SOLUTION The weight is related to the volume by the equation:

$$W = D_W \times V$$

But

$$V = l \times w \times h = 16 \text{ ft} \times 12 \text{ ft} \times 8 \text{ ft}$$

$$= 1,536 \text{ ft}^3$$

So the weight is

$$W = D_W \times V = 0.08 \text{ lb/ft}^3 \times 1,536 \text{ ft}^3$$

$$= 123 \text{ lb}$$

This is for sea-level pressure and 0°C temperature. At normal temperature, 20°C, the weight would be 115 pounds.

> ▶ **Physics To Go 4.3**
>
> The next time you go grocery shopping, rank several different packaged foods or ingredients by density: different breakfast cereals, flour, beans, and marshmallows are possibilities (**Figure 4.24**). Before you go, devise a method that you can use. Is the type of container important? Can you always determine if one substance has a higher density than another without using measuring devices?

Figure 4.24 Can you rank these items according to their densities? (*Hint:* Each package weighs one pound.)

Traditionally, the term *density* has referred to mass density for those using the metric system and to weight density for those using the English system. When comparing the densities of different substances, yet another quantity is often used: **specific gravity**. The specific gravity of a substance is the ratio of its density to the density of water. Column 5 of Table 4.4 shows the specific gravities of the substances. Diamond is 3.4 times as dense as water, so its specific gravity is 3.4. This means that a certain volume of diamond will have 3.4 times the mass and 3.4 times the weight of an equal volume of water.

Water is used as the reference for specific gravities simply because it is such an important and yet common substance in our world. The density of water is another example of a "natural" unit of measure like the atmosphere (atm) and the acceleration of gravity (g). As a matter of fact, the original definition of the unit of mass in the metric system, the gram, was based on the density of water. (See the special feature on "The Metric System" in the Prologue.) The gram was defined to be the mass of 1 cubic centimeter of water at 0°C. This is why the density of water is exactly 1,000 kg/m³.

One final note on the concept of density in general: we have considered only *volume* density—the mass or weight per unit volume. For matter that is primarily two-dimensional, such as flooring material or wrapping paper, it is often convenient to use *surface* density—the mass or weight per unit area. Similarly, for strings, ropes, and cables, one can use a *linear* density—the mass or weight per unit length. The basic concept of density is that it is a measure of the concentration of matter. It relates mass or weight to physical size.

Demetrius Pashos/Cengage Learning

Learning Check

1. The mass density of an object equals its _____ divided by its _____.
2. (True or False.) An alcoholic beverage is a mixture of water and ethyl alcohol. A "stronger" beverage (one with more alcohol in it) is likely to have a higher mass density than a "weaker" one.
3. Solid object A weighs more than solid object B. Consequently, we can conclude that
   (a) A's weight density must be greater than B's.
   (b) A's volume must be greater than B's.
   (c) A's weight density and volume must both be greater than B's.
   (d) Measurements would be required in order to determine the cause of the weight difference.
4. The specify gravity of a substance is computed by dividing its density by the density of _____.

ANSWERS: 1. mass, volume **2.** False **3.** (d) **4.** water.

4.4 Fluid Pressure and Gravity

A fluid is any substance that flows readily. All gases and liquids are fluids, as are plasmas. Granulated solids such as salt or grain can be considered fluids in some situations because they can be poured and made to flow.

Fluids are very important to us: without air and water, life as we know it would be impossible. In the remainder of this chapter, we will discuss some of the properties of fluids in general. Usually, the statements will apply to both liquids and gases, but the fact that gases are compressible and liquids are generally not is sometimes important.

4.4a The Law of Fluid Pressure

We live in a sea of air—the atmosphere. Though we are usually not aware of it, the air exerts pressure on everything in it. This pressure varies with altitude and can cause one's ears to "pop" when riding in a fast elevator. When swimming underwater, the same phenomenon can occur if you go deeper under the surface. In both the atmosphere and underwater, the pressure is caused by the force of gravity. Without gravity, air and water would not be pulled to Earth. They would simply float off into space, just as we would. The pressure in any fluid increases as you go deeper in it and decreases when you rise through it.

Before considering the exact relationship between depth and pressure, we state two general properties of pressures in fluids. First, fluid pressures act in all directions. When you put a hand under water, the pressure acts not only on the top of your hand but also on the sides and bottom. Second, the force of gravity causes the pressure in a fluid to vary with depth only, not with horizontal position. In liquids, the pressure depends on the vertical distance from the surface and is independent of the shape of the container.

One can illustrate both of these principles by filling a rubber boot with water. When holes are punched in the boot, the pressure causes water to run out (**Figure 4.25**). Water will run out of holes in the top, sides, and bottom of the toe because the pressure acts in all directions on the inner surface. Water comes out faster from holes that are farther below the surface because the pressure is greater. The speed is the same for all holes that are at the same level, regardless of their orientation (up, down, or sideways) and their location (toe, heel, etc.). This is because the pressure in a particular fluid depends only on the depth, not on the lateral position.

The following law explains how the pressure in a fluid is related to gravity.

Figure 4.25 A rubber boot is filled with water. The pressure of the water acts on all parts of the inner surface of the boot and forces water out of any hole punched in the boot. The speed of the water coming out of a hole depends on how far the hole is below the surface of the water.

Vern Ostdiek/University of Michigan

LAWS **Law of Fluid Pressure** The (gauge) pressure at any depth in a fluid at rest equals the weight of the fluid in a column extending from that depth to the "top" of the fluid divided by the cross-sectional area of the column.

▶ **Physics To Go 4.4**

Take an empty water or soft-drink container (plastic or aluminum) and punch three round holes in the side, one near the bottom, another halfway up the side, and the third near the top. Hold it under a faucet over a sink and run water into it just fast enough to keep it full as water runs out through the holes. What is different about the streams of water coming from the three holes? What does that tell you about the pressure inside the container at the levels of these holes?

This law is as much a prescription for determining the pressure in a fluid as it is a description of what causes pressure in a fluid. For liquids, we can use it to derive the simple relationship between pressure and depth. Let's say a tank is filled with a liquid so that the bottom is some distance h below the liquid's surface. On the bottom, we look at a rectangular area that has length l and width w (**Figure 4.26**). All of the liquid directly above the rectangle is in a column with dimensions l by w by h. The weight of this liquid pushes down on the rectangle. This causes a pressure on the bottom that is equal to the weight of the liquid in the column divided by the area of the rectangle.

$$p = \frac{F}{A} = \frac{W}{A} = \frac{\text{weight of liquid}}{\text{area of rectangle}}$$

It does not matter what the actual area is: a larger rectangle will have a proportionally larger amount of liquid in the column above. The actual height of the column of liquid is what determines the pressure. We compute the pressure using the fact that the weight of the liquid equals the weight density D_W of the liquid times the volume V of the column.

$$F = W = D_W \times V = D_W \times l \times w \times h$$

$$A = \text{area of rectangle} = l \times w$$

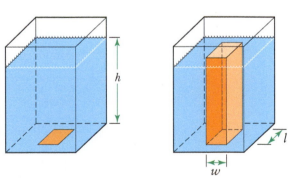

Figure 4.26 A tank is filled with a liquid to a depth h. Any rectangular area on the bottom supports the weight of all of the fluid directly above it. So the pressure on the bottom equals the weight of the liquid in the column divided by the area of that rectangle.

Figure 4.27 The pressure at the bottom of each of these tubes is the same if the vertical heights of the fluid surfaces are equal, regardless of the shapes of the containers.

So,

$$p = \frac{W}{A} = \frac{D_W \times l \times w \times h}{l \times w}$$

$$p = D_W h \qquad \text{(gauge pressure in a liquid)}$$

Or, because $D_W = Dg$,

$$p = Dgh \qquad \text{(gauge pressure in a liquid)}$$

This gives the pressure from the liquid above. If there is also pressure on the liquid's surface (like atmospheric pressure), this will be passed on to the bottom as well. Also, there is nothing special about the bottom: at any level above the bottom, the liquid above exerts a force and, therefore, pressure on the liquid below that level. We can summarize our result as follows.

| **PRINCIPLES** | In a liquid, the absolute pressure at a depth h is greater than the pressure at the surface by an amount equal to the weight density of the liquid times the depth. |

$$p = D_W h = Dgh \qquad \text{(gauge pressure in a liquid)}$$

It is important to note that the pressure at a given point in a fluid on or near Earth's surface only depends upon the depth and the density of the fluid. For a given fluid, the pressure at the bottom of a container—regardless of its shape—is determined by the height of the free-standing surface of the fluid (**Figure 4.27**). Thus the pressure at the bottom of an inverted cone filled with water is the same as the pressure at the bottom of a narrow tube filled to the same level.

EXAMPLE 4.6   Let's calculate the gauge pressure at the bottom of a typical swimming pool—one that is 10 feet (3.05 meters) deep.

SOLUTION   The gauge pressure at that depth, using Table 4.4, is

$$p = D_W h = 62.4 \text{ lb/ft}^3 \times 10 \text{ ft}$$

$$= 624 \text{ lb/ft}^2$$

To convert this to psi, we use the fact that 1 square foot equals 144 square inches.

$$p = 624 \frac{\text{lb}}{\text{ft}^2} = 624 \frac{\text{lb}}{144 \text{ in.}^2}$$

$$= 4.33 \text{ lb/in.}^2 = 4.33 \text{ psi (gauge pressure)}$$

At a depth of 20 feet, the gauge pressure would be twice as large: 8.66 psi. The absolute pressure is the gauge pressure plus the atmospheric pressure, 14.7 psi for the 10 ft swimming pool at sea level. (Absolute pressure = 4.33 psi + 14.7 psi = 19.03 psi.)

If we do this calculation in SI units:

$$p = Dgh = 1,000 \text{ kg/m}^3 \times 9.8 \text{ m/s}^2 \times 3.05 \text{ m}$$

$$= 29,900 \text{ Pa}$$

The general result for the increase in pressure with depth in water is

$$p = 0.433 \text{ psi/ft} \times h \qquad \text{(for water, } h \text{ in ft, } p \text{ in psi)}$$

For every 10 feet of depth in water, the pressure increases 4.33 psi. **Figure 4.28** shows a graph of the pressure underwater versus the depth. It is a straight line, because the pressure is directly proportional to the depth. In seawater, the density is slightly higher, and the pressure increases 4.47 psi for every 10 feet. Submarines and other devices that operate underwater must be designed to withstand high pressures. For example, at a depth of 300 feet in seawater, the pressure is 134 psi ($p = 0.447 \times 300$). The force on each square foot of a surface is more than 9 tons.

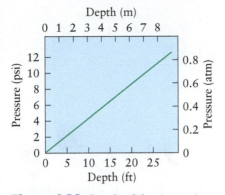

Figure 4.28 Graph of the (gauge) pressure underwater versus the depth. The pressure increases 0.433 psi with every foot of depth. All liquids have graphs with the same shape, but the slope is greater if the density is higher.

EXAMPLE 4.7 At what depth in pure water is the gauge pressure 1 atmosphere?

SOLUTION

$$p = 0.433 \text{ psi/ft} \times h$$

$$14.7 \text{ psi} = 0.433 \text{ psi/ft} \times h$$

$$\frac{14.7 \text{ psi}}{0.433 \text{ psi/ft}} = h$$

$$h = 33.9 \text{ ft} = 10.3 \text{ m}$$

Because mercury is 13.6 times as dense as water, the gauge pressure is 1 atmosphere in this liquid at a depth of:

$$h = \frac{33.9 \text{ ft}}{13.6} = 2.49 \text{ ft} = 29.9 \text{ in.} = 0.76 \text{ m} \qquad \textbf{(Figure 4.29)}$$

4.4b Fluid Pressure in the Atmosphere

The law of fluid pressure has a simple form in liquids. For gases, things are a bit more complicated. We know that the density of a gas depends on the pressure. At greater depths in a gas, the increased pressure causes increased density. The total weight of a vertical column of gas can be computed only with the aid of calculus.

Earth's atmosphere is a relatively thin layer of a mixture of gases. The decrease in air pressure with altitude is further complicated by variations in the temperature and composition of the gas. The heating of the air by the Sun,

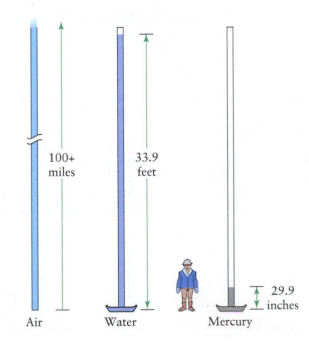

100+ miles — Air

33.9 feet — Water

29.9 inches — Mercury

Figure 4.29 The pressure at the bottom of each column is 1 atmosphere.

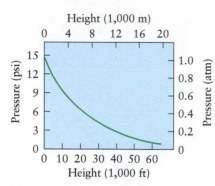

Height (1,000 m)

Height (1,000 ft)

Figure 4.30 Graph of the absolute air pressure versus height above sea level. The graph is not a straight line because the density decreases with altitude. The pressure never quite reaches zero.

Earth's rotation, and other factors cause the air pressure at a given place to vary slightly from hour to hour. Also, the pressure is not always the same at points with the same altitude.

In spite of the complexity of the situation, we can make some statements about the general variation of air pressure with altitude. Because the atmosphere is a gas, there is no upper surface or boundary. The air keeps getting progressively thinner—there are fewer molecules per unit volume—as you go higher in the atmosphere. At 9,000 meters (30,000 feet) above sea level, the density of the air is only about 35 percent of that at sea level. At this altitude, the average person cannot remain conscious because there is not enough oxygen in each breath. At about 160 kilometers (100 miles) above sea level, the density is down to one-billionth of the sea-level density. This is often regarded as the effective upper limit of the atmosphere. Spacecraft can remain in orbit at this altitude because the thin air causes very little air resistance.

Figure 4.30 is a graph of the air pressure versus altitude for the lower atmosphere. (When comparing this graph with Figure 4.28, remember that the depth in a liquid is measured downward from the surface, but the height in the atmosphere is measured upward from sea level.) The pressure rapidly decreases with increasing height at low altitudes, where the density is still fairly high. Remember that the pressure at any elevation depends on the weight of all of the air above. At 9,000 meters, the pressure is about 0.3 atmospheres. This means that only 30 percent of the air is above 9,000 meters. Even though the atmosphere extends more than 100 miles up, most of the air (70 percent) is less than 6 miles up.

Air pressure is measured with a *barometer*. The simplest type is the mercury barometer. It consists of a vertical glass tube with its lower end immersed in a bowl of mercury. All of the air is removed from the tube, so there is no air pressure acting on the surface of the mercury in the tube (**Figure 4.31**). The air pressure on the mercury in the bowl is transmitted to the mercury in the tube. This forces the mercury to rise up in the tube. (When drinking through a straw, you reduce the pressure in your mouth, and the atmospheric pressure forces the drink up the straw.) The mercury will rise in the tube until the pressure at the base of the tube equals the air pressure. Hence, the air pressure is determined by measuring the height of the column of mercury.

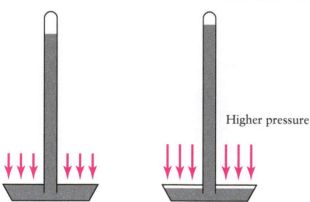

Higher pressure

Figure 4.31 A mercury barometer. If there is no air in the tube, the air pressure on the mercury in the bowl forces the mercury to rise up the tube. The pressure at the bottom of the tube equals the air pressure. The higher the air pressure, the higher the column of mercury.

When the air pressure is 1 atmosphere (14.7 psi), the column of mercury is 760 millimeters (29.9 inches) long. At lower pressures, the column is shorter, so we can use the length of the column of mercury as a measure of the pressure: 760 millimeters of mercury equals 1 atmosphere. Other liquids will work, but the tube has to be much longer if the liquid's density is small. For example, it would take a column of water 10.3 meters (33.9 feet) long to produce, and therefore measure, a pressure of 1 atmosphere (Figure 4.29).

This unit of measure is not limited to measuring air pressure. Blood pressure is also given in millimeters of mercury. If your blood pressure is 100 over 60, it means that the pressure drops from 100 millimeters of mercury during each heartbeat to 60 millimeters of mercury between beats.

A more portable type of barometer consists, more or less, of a very short metal can with the air removed from the inside (**Figure 4.32**). The air pressure

Figure 4.32 Schematic of an aneroid barometer. With higher atmospheric pressure, the top of the can is pushed in more, causing the pointer to indicate higher pressure.

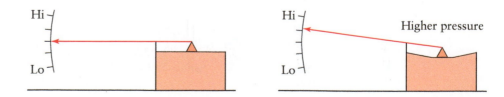

Hi

Lo

Hi

Lo

Higher pressure

Figure 4.33 An altimeter (left) measures altitude by measuring the air pressure. A vertical airspeed indicator (right) measures vertical speed by measuring how rapidly the air pressure is increasing or decreasing.

causes the ends to be squeezed inward. The higher the pressure, the greater the distortion of the ends. A pointer is attached to one end and moves along a scale as the distortion of the end changes. This is called an *aneroid barometer*.

The variation of air pressure with height is used to measure the altitude of aircraft. An *altimeter* is an aneroid barometer with a scale that registers altitude instead of pressure (left of **Figure 4.33**). For example, if an aneroid barometer in an airplane indicated the pressure was 0.67 atmospheres, one could infer from Figure 4.30 that the altitude was 10,000 feet. Each pressure reading on the barometer would correspond to a different altitude.

A more sophisticated instrument is used to measure the vertical speed of an aircraft—how fast it is going up or down. This device, called a *vertical airspeed indicator* in airplanes (right of Figure 4.33) and a *variometer* in gliders, senses changes in the air pressure. When the aircraft is going up, the measured air pressure decreases. The instrument converts the rate of change of the air pressure into a vertical speed.

The law of fluid pressure applies to granulated solids in much the same way that it does to liquids. The walls of storage bins and grain silos are reinforced near the bottom because the pressure is higher there. Again, it is the force of gravity pulling on the material above that causes the pressure.

▶ Physics To Go 4.5

Fill a glass with water, hold a piece of cardboard or stiff paper on the top with your hand, and invert it. You might want to do this over a sink. Make sure there is a good seal all around the top of the glass before you turn it over. When you remove your hand, why doesn't the water fall out?

■ Learning Check

1. (True or False.) Gravity causes the pressure in the ocean to vary with depth.
2. Two aquariums are completely full of water, but the pressure at the bottom of aquarium A is greater than at the bottom of aquarium B. The cause of this could be that
   (a) the water in A is deeper than it is in B.
   (b) A contains freshwater and B contains seawater.
   (c) A is shaped like a box and B is shaped like a cylinder.
   (d) Any of the above.

3. (True or False.) The atmospheric pressure at 5,000 m above the ground is exactly twice as large as it is at 10,000 m above the ground.
4. A barometer is a device that measures _____ .

4. atmospheric pressure

ANSWERS: 1. True **2.** (a) **3.** False

4.5 Archimedes' Principle

4.5a Buoyancy

The force of gravity causes the pressure in a fluid to increase with depth. This in turn causes an interesting effect on substances partly or totally immersed in a fluid. In some cases, the substance floats—wood or oil on water, blimps or hot-air balloons in air. In other cases, the substance seems lighter—a 100-pound rock can be lifted with a force of about 60 pounds when it is underwater. Obviously, some force acts on the foreign substance to oppose the downward force of gravity (weight). This force is called the **buoyant force**.

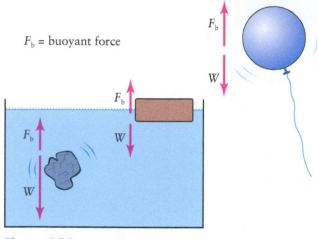

F_b = buoyant force

DEFINITION **Buoyant Force** The upward force exerted by a fluid on a substance partly or completely immersed in it.

An upward buoyant force acts on anything immersed in any gas or liquid. As long as no other forces are acting on the substance, there are three possibilities (**Figure 4.34**). If the buoyant force is *less than* the weight of the substance, it will *sink*. If the buoyant force is *equal* to the weight, the substance will *float*. If the buoyant force is *greater than* the weight of the substance, it will *rise* upward. Examples of each case (in the same order) are a rock in water, a piece of wood floating on water, and a helium-filled balloon rising in air. Both the rock and the balloon experience a net force because the weight and the buoyant force do not cancel each other. This net force causes each to accelerate momentarily until the force of friction offsets the buoyant force and a terminal speed is reached.

Figure 4.34 The rock sinks because the buoyant force on it is smaller than its weight. The wood floats because the buoyant force on it is equal to its weight. The helium balloon rises because the buoyant force on it is greater than its weight.

▶ Physics To Go 4.6

For this you need a glass or ceramic coffee mug, a rubber band, and a sink with several inches of water standing in it. Loop the rubber band around the handle so that the cup hangs sideways (**Figure 4.35**). The rubber band should be strong enough to support the mug while also stretching noticeably.

1. Measure how long the stretched rubber band is.
2. Lower the mug into water, making sure no air is trapped in the cup and that it doesn't come in contact with the bottom or walls of the sink. When the mug is fully submerged, measure how long the rubber band is. What do you find? How can you interpret your results?

At this point, we might pose two questions. First, what causes the buoyant force? Second, what determines the magnitude of the buoyant force? The answer to the first question can be arrived at rather simply in light of the previous section. Consider an object that is completely immersed (**Figure 4.36**). Because its bottom surface is deeper in the fluid than its top surface, the pressure on the bottom surface will be greater. Hence the upward force on its bottom surface caused by this pressure is greater than the downward force on its top surface. In short, the difference in fluid pressure acting on the surfaces of the object causes a net upward force. (The forces on the sides of the object are equal and opposite, so they offset one another.)

When something floats on the surface of a liquid, only its lower surface experiences the fluid pressure of the liquid. This causes an upward force.

4.5b Archimedes' Principle

As for the second question, the size of the buoyant force is determined by a law formulated in the third century BCE by Greek scientist Archimedes (see *Profiles in Physics*, Chapters 3 and 4).

Figure 4.35 Setup for *Physics to Go 4.6*.

Vern Ostdiek/University of Michigan

Figure 4.36 The fluid pressure causes a force on each surface of an immersed object. The pressure on the lower surface is greater than the pressure on the upper surface. Consequently, the upward force on the bottom is greater than the downward force on the top. The net upward force is the buoyant force.

F_{down}

$F_{\text{net}} = F_{\text{up}} - F_{\text{down}}$
$= F_{\text{b}}$

F_{up}

PRINCIPLES　**Archimedes' Principle**　The buoyant force acting on a substance in a fluid at rest is equal to the weight of the fluid displaced by the substance.

$$F_{\text{b}} = \text{weight of displaced fluid}$$

When a piece of wood is placed in water, it displaces some of the water; that is, part of the wood occupies the same space or volume formerly occupied by water. The weight of this displaced water equals the buoyant force acting on the wood. Obviously, any object that is completely submerged in a fluid will displace a volume of fluid equal to its own volume.

If the buoyant force on an object is less than its weight, it sinks, but the *net* downward force is reduced. **Figure 4.37** shows an object hanging from a scale. Its weight is 10 newtons. As the object is lowered into a beaker of water, it displaces some of the water. The scale reading is reduced by an amount equal

Figure 4.37 The weight of an object is 10 newtons. When the object is immersed in water, the scale reads a smaller force because of the upward buoyant force. The buoyant force is equal to the weight of the water that the object displaces.

to the buoyant force. When the scale shows 6 newtons, the buoyant force is 4 newtons.

$$\text{scale reading} = \text{weight} - \text{buoyant force}$$

$$6\,\text{N} = 10\,\text{N} - 4\,\text{N}$$

The weight of the water that spills over the side of the beaker is also 4 newtons. If the object is lowered farther into the water, the buoyant force will increase, and the scale reading will decrease.

Notice that the buoyant force acting on a substance doesn't depend on what the substance is, only on how much fluid it displaces when immersed. Identical balloons filled to the same size with helium, air, and water all have the same buoyant force acting on them. Only the helium-filled balloon floats in air because its weight is less than the buoyant force. Also, the weight of a substance alone does not determine whether or not it will float. A tiny pebble will sink in water, but a 2-ton log will float.

The key to understanding these circumstances is the concept of density. Let's take the simple case of a solid object submerged in a liquid. The volume of the liquid displaced by the object is, of course, equal to the object's volume. So,

$$\text{weight of object} = \text{weight density (of object)} \times \text{volume}$$

$$W = D_W(\text{object}) \times V$$

The buoyant force is

$$\text{buoyant force} = \text{weight of displaced fluid} \qquad (\text{Archimedes' principle})$$

$$F_b = \text{weight density (of fluid)} \times \text{volume}$$

$$= D_W(\text{fluid}) \times V$$

From this, we conclude that if the weight density of the object is greater than the weight density of the fluid, the object's weight is greater than the buoyant force. The object sinks. If its density is less than the fluid's density, it floats. By simply comparing the densities (or specific gravities) of the substance and the fluid, one can determine whether or not the substance will float. Any substance with a smaller density than that of a given fluid will float in (or on) that fluid.

In Table 4.4 we see that juniper, ice, gasoline, ethyl alcohol, and all of the gases will float on water because their densities are less than that of water. (Actually, the alcohol would simply mix in with the water. In the case of a fluid being immersed in another fluid, it is best to imagine the first being enclosed in a balloon or some other container that has negligible weight but keeps the fluids from mixing.)

The density of mercury is so high that everything in the table except gold and uranium will float on it (Figure 4.38). Hydrogen and helium both float in air, whereas radon and carbon dioxide sink.

Ships and blimps float, even though they are constructed of materials with higher densities than the fluids in which they float. But they are composites of substances: most of the volume of a ship is occupied by air, and most of the volume of a blimp is occupied by helium. In both cases, the *average* density of the object is less than the density of the fluid.

As a ship is loaded with cargo, it sinks lower into the water. This makes it displace more water, thereby increasing the buoyant force and countering the added weight (Figure 4.39).

4.5c Applications of Archimedes' Principle

Archimedes' principle is routinely used to measure the densities or specific gravities of solids that sink. The object is hung from a scale, and the reading is recorded. (It does not matter whether the scale gives mass or weight. Here

Figure 4.38 A steel ball floats in mercury because the density of mercury is greater than that of steel.

Charles D. Winters/Science Source

(a) Loaded cargo ship

(b) Unloaded cargo ship

we assume it reads weight.) Then the object is completely immersed in water, and the new scale reading is recorded (Figure 4.37). The *difference* in these two readings equals the weight of the displaced water according to Archimedes' principle.

W of displaced water = scale reading out of water − scale reading in water

W of displaced water = scale (out) − scale (in)

So the weight of the object is known, and the weight of an *equal* volume of water is also known. Density is weight divided by the volume. Since the volumes are the same, the density of the object divided by the density of the water *equals* the weight of the object divided by the weight of the water.

$$\frac{\text{density of object}}{\text{density of water}} = \frac{\text{scale reading out of water}}{\text{scale (out)} - \text{scale (in)}}$$

This ratio is just the specific gravity of the object. So,

$$\text{specific gravity} = \frac{\text{scale (out)}}{\text{scale (out)} - \text{scale (in)}}$$

Figure 4.40 The number of balls that float in this antifreeze tester depends on the density of the coolant. Higher-density coolant (right) has a lower freezing point.

Vern Ostdiek/University of Michigan

Vern Ostdiek/University of Michigan

In Section 4.3, we described how the density of the coolant in an automobile engine indicates the antifreeze content. The density is measured with a simple device that employs Archimedes' principle. **Figure 4.40** shows an antifreeze tester that consists of a narrow glass tube containing five balls. The balls are specially made so that each one has a slightly higher density than the one above it. The coolant is drawn up into the glass tube with a suction bulb. Each ball will float if its density is less than the density of the coolant. The number of balls that float depends on the density of the coolant. If the antifreeze content is high, the coolant's density is high, and all of the balls float. A similar strategy is used in so-called "state-of-charge" indicators in automotive and marine batteries.

To check the hemoglobin content of blood, a drop of it is placed in a liquid that has the correct minimum density. If the blood sinks, its density is greater than that of the liquid, and the hemoglobin content is high enough.

The following examples use Archimedes' principle.

EXAMPLE 4.8 A contemporary Huckleberry Finn wants to construct a raft by attaching empty plastic 1-gallon milk jugs to the bottom of a sheet of plywood. The raft and passengers will have a total weight of 300 pounds. How many jugs are required to keep the raft afloat on water?

SOLUTION The buoyant force on the raft must be at least 300 pounds. Consequently, the raft must displace a volume of water that weighs 300 pounds.

$$F_b = \text{weight of water displaced}$$

$$= D_W (\text{water}) \times \text{volume of water displaced}$$

$$300 \text{ lb} = 62.4 \text{ lb/ft}^3 \times V$$

$$V = \frac{300 \text{ lb}}{62.4 \text{ lb/ft}^3}$$

$$= 4.8 \text{ ft}^3$$

To keep 300 pounds afloat, 4.8 cubic feet of water must be displaced. One cubic foot equals 7.48 gallons. So about 36 (= 4.8 × 7.48) one-gallon jugs will be needed to keep Huck and his party dry.

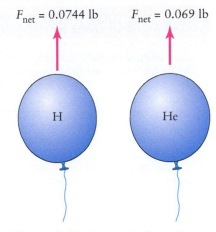

$F_{net} = 0.0744$ lb $F_{net} = 0.069$ lb

H   He

Figure 4.41 One cubic foot of hydrogen can lift 0.0744 pounds. One cubic foot of helium can lift 0.069 pounds.

EXAMPLE 4.9 Before the spectacular and tragic destruction of the German airship *Hindenburg* on 6 May 1937, blimps, zeppelins, and balloons were filled with hydrogen. Now helium is used. Let's compare the two gases in terms of their buoyancy in air. Each cubic foot of hydrogen gas weighs 0.0056 pounds at 0°C and 1 atm (Table 4.4).

SOLUTION In an airship, a cubic foot of hydrogen will displace a cubic foot of air. So each cubic foot of hydrogen sustains a buoyant force of 0.08 pounds, the weight of 1 cubic foot of air. Therefore, the net force on each cubic foot of hydrogen gas is

$$\text{net force} = 0.08 \text{ lb} - 0.0056 \text{ lb}$$

$$F = 0.0744 \text{ lb} \qquad \text{(hydrogen)}$$

Each cubic foot of hydrogen can lift 0.0744 pounds (**Figure 4.41**). If helium is used instead, the buoyant force is still the same, but the weight of each cubic foot of helium is 0.011 pounds. So,

$$\text{net force} = 0.08 \text{ lb} - 0.011 \text{ lb}$$

$$F = 0.069 \text{ lb} \qquad \text{(helium)}$$

Each cubic foot of hydrogen can lift 8 percent more than a cubic foot of helium. A balloon filled with hydrogen can lift 8 percent more than an identical balloon filled with helium. However, the big factor that tilts the scale in favor of helium is that it is not combustible like hydrogen.

The air exerts a buoyant force on everything in it, not just balloons. This force, 0.08 pounds for each cubic foot, is so small that it can be ignored except when dealing with other gases. For example, the volume of a person's body is typically around 2 or 3 cubic feet. (The density of the human body is about the same as that of water. That is why we just barely float in water. So the approximate volume of your body is your weight divided by the weight density of water.) This means that the buoyant force on a person because of the air is between 0.16 and 0.24 pounds.

> ## Physics To Go 4.7

(You have to do this one outside—and no smoking!) Fill a glass about two-thirds full with equal amounts of water and gasoline. These two liquids do not mix, and you can clearly see that the gasoline does float on the water. Now put an ice cube into the glass. What happens to the ice cube? Does this agree with what you might have predicted using information in Table 4.4?

Learning Check

1. The _____ acting on a boat in water at rest is equal to the weight of the water displaced by the boat.
2. One balloon is filled with pure water and a second is filled with antifreeze. What does each balloon do after being released by a scuba diver beneath the surface of the ocean?
3. As a person climbs into a rowboat, it must sink lower into the water so that
    (a) the net force on the boat remains zero.
    (b) it displaces more water, consequently increasing the buoyant force on it.
    (c) its bottom is deeper in the water where the higher pressure causes a larger upward force.
    (d) All of the above.
4. (True or False.) The buoyant force on a hydrogen-filled balloon is larger than that on a helium-filled balloon with the same size.

ANSWERS: 1. buoyant force **2.** pure water balloon rises, the other sinks **3.** (d) **4.** False

Supermarkets. Superbowls. Superglue. Superhighways. Superstars. Superfluids? The prefix *super* has been attached to many nouns to indicate qualities that go beyond those ordinarily attributed to them. Its use draws attention to the fact that these things possess a *super*abundance of the usual kinds of properties that define them or that the defining characteristics are somehow *superior* to those of common, garden-variety examples of these entities. Indeed, the word super comes from the Latin, meaning "over" or "above."

Taken at face value, then, we might expect that a *super*fluid would be one whose "fluidity" is more pronounced than in ordinary cases; it might, for example, be able to flow more freely than common fluids. And this is basically true. In technical terms, we speak of a superfluid as having no friction or resistance to flow—that is, no *viscosity*. Unlike ordinary fluids, a superfluid can flow through narrow channels with no pressure difference. It also apparently defies gravity by flowing as a film over the edges of containers (the so-called *creep effect*; **Figure 4.42**) and by spouting dramatically when heated (the *fountain effect*). As we shall discuss later, superfluids display some very interesting rotational phenomena as well.

Discovered in 1938 by P. Kapitza (who received a 1978 Nobel Prize for his work) and J. Allen, superfluids are a subclass of what are called *quantum fluids*. (Superconductors, discussed in Chapter 7, are also members of this broad class.) They exhibit striking macroscopic effects because of special microscopic ordering. To see these effects requires that the substance under investigation remain in a liquid state at temperatures that are low enough to permit the ordering to occur. For most elements and compounds, the needed temperatures are so low that the substances condense into the solid phase of matter before reaching these critical temperatures. An exception to this behavior is the element helium, which condenses into a normal liquid at around −269°C (and 1 atmosphere pressure) and then undergoes another "phase" transition to the superfluid state below about −271°C.

The approach of a system of particles to superfluidity is associated with the onset of what is termed *Bose–Einstein condensation* in the liquid: a macroscopic quantity of liquid particles all come to occupy the same microscopic (quantum) energy state. (For more on quantum physics, see Chapter 10.) An analogy given by University of Illinois physicist Anthony Leggett, who shared the 2003 Nobel Prize in physics, may be helpful here in understanding what is going on.

Imagine you are atop a high mountain overlooking a distant city square on market day. As you examine the shopping crowd, you see people milling about in all directions seemingly at random: each individual is doing his or her own thing. Suppose you return to the mountain and inspect the town square on the day of a parade when the milling crowd is replaced by a marching band or a precision drill team. Now every member of the group is doing the same thing at the same time, and it's much easier to see and understand what is going on. The comparable situation in fluid physics is one in which a normal liquid plays the role of the market-day crowd—every atom doing something different. A superfluid is like the drill team—all the atoms forced into the same state, every atom doing exactly the same thing at the same time. A superfluid, then, acts like one large collective atom instead of billions of independent individual atoms. This collective action makes the effects of this microscopic phase change noticeable on macroscopic levels.

How does this collective condensation account for flow without friction? Viscosity is absent in a superfluid because it costs the fluid too much energy to have all of its atoms respond together to the frictional forces. Imagine the following experiment. Put some liquid helium at a temperature above the superfluid transition temperature in a cup and then place the container on the axis of a phonograph turntable. Start the turntable spinning. The normal helium liquid will begin to rotate with the container after a few minutes, dragged around by the frictional interactions between the fluid and the walls of the container; the motion is subsequently communicated throughout the fluid volume by viscous friction between the helium atoms. Next, cool the liquid below the transition temperature to a superfluid state. The fluid will continue to rotate, with all the helium atoms occupying the same energy state, equal to the rotational energy of the cup. Now stop the rotation of the container. The superfluid helium will continue to spin, unlike a normal fluid for which friction with the walls of the cup would gradually bring it to a halt. This is a graphic illustration of the ineffectiveness of friction to act on a superfluid. The Bose–Einstein condensed helium atoms remain rotating because to collectively pass to a nonrotating state would require an enormous investment of additional energy, which the system doesn't have. The key to this behavior is the cooperative interaction of all the helium atoms. If only a few individual atoms were involved, the energy required to respond to the frictional forces could be found, but for 10^{23} atoms it cannot!

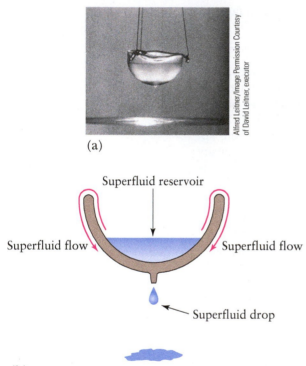

Alfred Leitner/Image Permission Courtesy of David Leitner, executor

(a)

Superfluid reservoir

Superfluid flow ← → Superfluid flow

Superfluid drop

(b)

Figure 4.42 (a) The "creep effect" in superfluid helium. (b) Simplified diagram of the "creep effect." Although superfluid helium has no viscosity, the liquid surface still possesses surface tension which allows it rise up the sides of its container without any normal frictional resistance to flow. This permits the liquid to travel up and over the top edge of the container and then down the outside surface, finally to drip from the lowest point on the container to the surface below. This process will continue until the vessel is eventually emptied of the superfluid liquid.

Superfluidity is an important topic of pure research, and increasingly for practical application. For example, in the former category, the study of superfluids in the laboratory will have significance for our understanding of the interior structure of neutron stars, which are thought to be in superfluid states (see the application on "Starquakes" at the end of Section 3.8). In addition, because superfluids represent systems where gravitational and electromagnetic effects are minimized, investigating them may elucidate the characteristics of the weak nuclear force (see Chapter 12) in macroscopic domains. The discovery and characterization of superfluidity, in a form of helium called *helium-3,* won David Lee, Robert Richardson, and Douglas Osherhoff the Nobel Prize in physics in 1996.

On the practical side, the excellent heat-transfer properties of superfluids have already been exploited in cooling other systems to temperatures below −270°C. Moreover, the friction-free flow of superfluid helium have been used to measure Earth's rotation and may soon be used in gyroscopes as part of spacecraft navigation systems. Recent studies at the University of California at Berkeley have also opened up the possibility of using oscillations in superfluid helium to define an international standard of pressure based on precisely known quantum-mechanical constants.

■ **QUESTIONS**
1. Superfluids exist in a state called a Bose-Einstein condensate. Describe what this means and give an analogy to help clarify its meaning.
2. Give two applications of superfluids in science and technology.

4.6 Pascal's Principle

When a force acts on a surface of a solid, the resulting pressure is "transmitted" through the solid only in the original direction of the force. When you sit on a stool, your weight causes pressure in the stool legs that is transmitted to the floor. This pressure doesn't act to the side of the legs, only downward.

In a fluid, any pressure caused by a force is transmitted everywhere throughout the fluid and acts in all directions. When you squeeze a tube of toothpaste, the pressure is passed on to all points in the toothpaste. An inward force on the sides of the tube can cause the toothpaste to come out of the end. This property of fluids should be familiar to you and might even be used to distinguish fluids from solids. Pascal's principle is a formal statement of this phenomenon.

PRINCIPLES **Pascal's Principle** Pressure applied to an enclosed fluid is transmitted undiminished to all parts of the fluid and to the walls of the container.

This property of fluids is exploited in widely used hydraulic systems. Hydraulic jacks, dental and barber chairs, and the brake systems in automobiles are common examples. The basic components of such systems are piston and cylinder combinations. **Figure 4.43** shows a small piston and cylinder on the left connected via a tube to a larger piston and cylinder on the right. (The effect of gravity can be ignored.) A liquid is in both cylinders and the connecting tube. When a force acts on the left piston, the resulting pressure is passed on throughout the liquid. This pressure causes a force on the right piston. Because this piston is larger than the one on the left, the force ($F = pA$) will also be larger. For example, if the area of the right piston is five times that of the left piston, the force will be five times as large. This system behaves like a lever (Figure 3.10). A small force acting at one place causes a larger force at another place. As with the lever, the smaller piston will move a correspondingly

Figure 4.43 The pressure caused by the force on the small piston is transmitted throughout the fluid and acts on the larger piston. The resulting force on this piston is larger than the force on the smaller piston.

Master cylinder

Tire

Brake pedal

Wheel cylinder

Brake pad

Disc (rotor)

(a)

F

Brakes applied

(b)

Figure 4.44 (a) Simplified diagram of an automotive hydraulic brake system. Only one wheel, equipped with disc brakes, is shown. (b) When the brake pedal is pushed, the pressure increase in the master cylinder is passed on through the fluid to the wheel cylinder. The piston is pushed outward, and the brake pads squeeze the disc.

greater distance than the larger piston. This ensures that the work done is the same for both pistons.

In mathematical terms, the mechanical advantage of a hydraulic system can be quantified simply as follows. Because Pascal's principle guarantees that the pressure in a closed system like that shown in Figure 4.43 is the same throughout, the ratio of the force on each piston to its respective cross-sectional area must be the same:

$$\frac{F_1}{A_1} = p = \frac{F_2}{A_2}$$

or

$$F_1\left(\frac{A_2}{A_1}\right) = F_2$$

When A_2 is greater than A_1 as in the picture, F_2 will be larger than F_1 in the same proportion as noted earlier. This relationship holds when the confined fluid is incompressible (e.g., is not a gas) and the pistons are at the same elevation so that no work is done by gravity during the application of the forces.

In automotive brake systems, the brake pedal is connected to a piston that slides in the "master" cylinder. This cylinder is connected via tubing to "wheel" cylinders on the brakes on the wheels (**Figure 4.44**). (Actually, there are two cylinders in tandem in the master cylinder. Each is connected to two wheels: if one subsystem fails, the other two wheels will still have braking power.)

Brake fluid fills the cylinders and tubing. When the brake pedal is pushed, the piston produces pressure in the master cylinder that is transmitted to the wheel cylinders. The piston in each wheel cylinder is attached to a mechanism that applies the brakes. In disc brakes the piston squeezes disc pads against the sides of a rotating disc attached to the wheel. This action is very similar to that of rim brakes on bicycles. The mechanism in drum brakes is a bit more complicated.

A hydraulic brake system offers two advantages. Using wheel cylinders with larger diameters than those of the master cylinder gives a mechanical advantage: the forces on the wheel-cylinder pistons are larger than the force applied to the master-cylinder piston. Also, this is an efficient way to transmit a force from one location in the car to four other locations.

F_1

F_2

Figure 4.45 Force multiplication in a hydraulic lift system.

EXAMPLE 4.10 In a hydraulic car lift (**Figure 4.45**), the input piston has cross-sectional area $A_1 = 0.0025$ m² and negligible weight. The output plunger has area $A_2 = 0.0625$ m², and the weight of the car and the output plunger together is 17,500 N. If the input piston and output plunger are at the same level as shown in the figure, what is the size of the input force F_1 required to produce an output force F_2 capable of supporting the car and the output plunger?

SOLUTION Pascal's law provides that the pressure throughout the closed hydraulic system must be the same. Thus the pressure at the input piston p_1 must be equal to the pressure at the output plunger p_2. Using the definition of pressure

$$p_1 = p_2$$

$$\frac{F_1}{A_1} = \frac{F_2}{A_2}$$

Solving for F_1,

$$F_1 = F_2 \left(\frac{A_1}{A_2} \right)$$

$$F_1 = (17,500 \text{ N}) \frac{(0.0025 \text{ m}^2)}{(0.0625 \text{ m}^2)}$$

$$= \boxed{7000 \text{ N}}$$

■ Learning Check

1. Squeezing the sides of a flexible water bottle to make water come out of the top is an example of exploiting _____ principle.

2. (True or False.) When the pressure in the fluid in two cylinders in a hydraulic system is the same, the forces on the two pistons are not necessarily equal.

ANSWERS: 1. Pascal's **2.** True

4.7 Bernoulli's Principle

In this section, we present a simple principle that applies to moving fluids. Water moving in a stream or through pipes to a water faucet, air traveling through heating ducts or as a cool breeze in the summer, blood being pumped through your arteries and veins are all examples of fluids flowing. It is common for each of these fluids to speed up and slow down as it flows. Accompanying any change in the speed of the fluid is a change in pressure of the fluid. This is stated in the following principle, named after Swiss physicist and mathematician Daniel Bernoulli (1700–1782).

PRINCIPLES | **Bernoulli's Principle** For a fluid undergoing steady flow, the pressure is lower where the fluid is flowing faster.

In the context of this principle, *steady flow* means that there is no random swirling of the fluid and that no outside forces increase or decrease the rate of flow. As in the application of other fluid principles in this chapter, we again assume that the fluid has constant density (is incompressible) and has no viscosity.

This principle is based on the conservation of energy. A fluid under pressure has what can be called *pressure potential energy*. The higher the pressure, the greater the potential energy of any given volume of fluid. (When you open a faucet, water rushes out as the potential energy that results from pressure is converted into the kinetic energy of running water. Low water pressure makes the water come out slowly, with low kinetic energy, because it starts with low potential energy.) Moving fluids have both kinetic and potential energy. When a fluid speeds up, its kinetic energy increases. Its total energy remains constant under the circumstances described, so its potential energy and, therefore, its pressure decrease.

One of the best examples of Bernoulli's principle is that depicted in **Figure 4.46**. Water flowing through a pipe passes through a smaller section spliced into the pipe. The water speeds up when it enters the narrow region and then slows down when it reenters the wide region. This occurs because the volume of fluid passing through each part of the pipe each second is the same. Where the cross-sectional area is smaller, the fluid must flow faster if the same number of cubic inches of fluid is to get through each second. You've seen this already if you've ever used your thumb to partially block water coming out of the end of a garden hose.

To quantify this relationship between flow velocity and cross-sectional area, we note that the *volume flow rate*, that is, the volume of fluid passing through a given cross-sectional area, A, of a horizontal pipe, is the product Av, where v is the fluid speed at that point along the pipe. This may be verified by examining the units of the quantity: $(meter)^2 \times (meters/second) = (meters)^3/second =$ volume/time. (We continue to assume here that the fluid density is constant; that is, the fluid is incompressible.)

Under these circumstances, the volume flow rate is a constant of the fluid motion, and for any two points along the pipeline:

$$A_1 v_1 = A_2 v_2$$

This equation is sometimes referred to as the *equation of continuity*. Rearranging, we have

$$\frac{v_1}{v_2} = \frac{A_2}{A_1}$$

This relation predicts that if area A_2 is three times larger than area A_1, then the speed of the fluid at point 1 will be three times greater than that at point 2. Example 4.11 gives an application of this proportionality.

EXAMPLE 4.11 A common garden hose has an opening with a cross-sectional area of 5.1×10^{-4} m². When the spigot is opened, the water emerges from the hose with a speed of 0.85 m/s. If the gardener places her finger over the opening and reduces the area to 2.0×10^{-4} m², how fast will the water now exit the hose?

SOLUTION Since water may be considered an incompressible fluid under these circumstances, we may apply the equation of continuity

$$A_1 v_1 = A_2 v_2$$

Taking v_2 to be the speed after the opening area is reduced and solving,

$$v_2 = v_1 \left(\frac{A_1}{A_2} \right)$$

$$v_2 = (0.85 \, \text{m/s}) \frac{(5.1 \times 10^{-4} \, \text{m}^2)}{(2.0 \times 10^{-4} \, \text{m}^2)}$$

$$= 2.17 \, \text{m/s}$$

Figure 4.47 Perfume atomizers use Bernoulli's principle. Air is forced to move through the horizontal tube when the bulb is squeezed. In this tube, the air has to move fast, so the pressure is low. Normal air pressure on the perfume in the bottle forces it to rise upward into the stream.

Bernoulli's principle tells us that the pressure in the moving water is *smaller in the narrow section* than in the wide sections upstream and downstream. Pressure gauges placed in the pipe show this to be so. This is one of those rare situations in which physical fact runs counter to one's intuition. On first thought, most people would predict that the pressure should be higher in the narrow part of the pipe because the fluid is "squeezed into a smaller stream." But that is not the case. The pressure is actually lower in the narrow part.

Atomizers on perfume bottles utilize Bernoulli's principle. When a bulb is squeezed, air moves through a horizontal tube (**Figure 4.47**). The air moves fast, so the air pressure is low. A second small tube runs from the horizontal tube down into the perfume. Because the air pressure is reduced in the horizontal tube, the normal air pressure acting on the surface of the perfume forces the liquid to rise upward in the vertical tube and to enter the moving air. Carburetors used on lawn mowers and the automatic shutoff mechanism on gas pump nozzles also make use of Bernoulli's principle.

Concept Map 4.2 summarizes this chapter's concepts related to pressure.

▶ Physics To Go 4.8

Hold the top of a piece of paper horizontally just below your lips so that the paper hangs limp (**Figure 4.48**). Blow hard over the top of the paper. What does the paper do? What causes this?

Vern Ostdiek/University of Michigan

Figure 4.48 Angie is set to illustrate Bernoulli's principle.

Learning Check

1. Bernoulli's principle is exploited by devices such as _____.

2. (True or False.) At a place in a particular ventilation duct where the air pressure is lowest, the air is likely to be moving fastest.

ANSWERS: **1.** perfume atomizer or carburetor. **2.** True

Profiles in Physics The Rise of the Modern Atomic Theory and the Development of Fluid Physics

Greek philosopher Democritus, who lived about a century before Aristotle, developed the theory that all matter is composed of particles too small to be seen. He called these particles *atoms*, from the Greek word for *indivisible*, because he thought them to be the final result of any repeated subdivision of matter. The atomic theory as expounded by Democritus and other Greek philosophers was not accepted by Aristotle. For this and other reasons, the atomic theory was abandoned for more than 2,000 years.

The successes of Galileo and Newton opened a new era of experimentation and advancement in all of the sciences. The development of chemistry in the 17th and 18th centuries revived the atomic theory because it could explain many of the new discoveries. Robert Boyle (1626–1691), an Irish-born physical scientist, identified "elements" as those substances that could not be further decomposed by chemical analysis. A century later, French chemist Antoine Lavoisier (1743–1794; **Figure 4.49**) refined the concept of the element and showed that mass is conserved during chemical reactions. When paper burns, the total mass of the combustion products, ashes, and smoke is the same as the total mass of the paper and oxygen that were combined during the burning.

During the 18th century, several of the elements were correctly classified: iron, gold, mercury, carbon, and oxygen to name a few.

Salt and several other compounds were mistakenly identified as elements because they could not be decomposed into their constituent elements with the techniques then available.

In 1808, John Dalton (1766–1844), an English schoolteacher (**Figure 4.50**), presented what was essentially the modern atomic theory. He correctly stated that the properties of each element are determined by the properties of its atoms, the atoms of different elements have different masses, and two or more different atoms in combination form the basic units of chemical compounds. The actual structure of atoms and the fact that they aren't indestructible were only determined during the last 125 years.

Fluids

The scientific study of fluids probably began with Archimedes' discoveries about buoyancy. In his book *On Floating Bodies*, he proves the famous principle that has immortalized his name. Apparently, Archimedes' interest in the topic began when King Hiero of Syracuse presented him with a problem. The king had given a certain weight of gold to a craftsman to be made into a crown. The finished product was presented to the king, but he suspected that some of the gold had been replaced by an equal weight of silver. King Hiero asked Archimedes to find a way to determine the composition of the crown without damaging it. The story goes that Archimedes thought of the solution when he noticed how water

Figure 4.49 Lavoisier with his wife Marie, who was also his assistant.

Figure 4.50 English scientist John Dalton developed the concepts of atoms and molecules and their relationship to elements and compounds.

was displaced as he sank into a bathtub. He supposedly rushed home naked, shouting, "Eureka!" ("I have found it!").

The concept of specific gravity was developed in Arabia approximately 1,000 years ago. Arabian scientists also described how to use Archimedes' principle to measure the specific gravities of solids and estimated how high the atmosphere extends.

French mathematician Blaise Pascal (1623–1662) made several discoveries about pressure in liquids, the most famous being the principle named after him (Section 4.6). He described how this principle implies that a force can be "amplified" using combinations of different-sized pistons and cylinders (Figure 4.43). Pascal also showed experimentally that the pressure in a liquid depends only on the depth and not on the shape of the container.

For 2,000 years after Aristotle's time, the philosophical principle *horror vacui*—that "Nature abhors a vacuum"—obscured the understanding of air pressure. It was used to explain why a liquid could be drawn up into a tube by removing some of the air. When a straw is being used, the drink is supposedly pulled into the straw because Nature does not allow the creation of a vacuum. Galileo was puzzled by reports that a suction pump could not raise water in a pipe higher than about 33 feet. This indicated to him that there seemed to be a limit to *horror vacui*.

The invention of the mercury barometer by Italian mathematician Evangelista Torricelli (1608–1647), a student of Galileo's, proved to be the key to discovering the nature of atmospheric pressure. Torricelli correctly suggested that the pressure of the "sea of air" (the atmosphere) forces liquids to rise in a tube when air is removed from it. Pascal learned of Torricelli's work and performed many experiments with his own barometers (**Figure 4.51**). In one of them, he used red wine in a glass tube 46 feet long. Pascal reasoned that the column of mercury in a barometer should be shorter at higher elevations. Yet he could not detect a difference in the length when he moved his barometer to the top of a church tower. Pascal's brother-in-law took a barometer to the Puy de Dôme, a mountain in south-central France, and did observe the effect, however, vindicating Pascal's prediction.

German physicist Otto von Guericke (1602–1686) performed several experiments that dramatically illustrated that atmospheric pressure can cause large forces. In one such experiment, he filled a copper sphere with water, sealed it, and then pumped out the water. The unbalanced air pressure on the outside caused the sphere to implode. Von Guericke invented air pumps and used them to produce near vacuums. In 1654, he demonstrated his *Magdeburg hemispheres* to Emperor Ferdinand III. He fitted two hemispheres together and pumped the air out of the spherical

Figure 4.51 A large barometer built by Pascal in Rouen, France, to measure air pressure.

cavity (**Figure 4.52**). The force of the air pressure acting on the two hemispheres was so great that eight horses pulling on each one were unable to separate them.

■ QUESTIONS

1. Describe two additional contributions to fluid physics developed by Blaise Pascal besides the principle that bears his name.
2. What important contribution did John Dalton make to our understanding of the structure and composition of matter?

Figure 4.52 Otto von Guericke demonstrating that air is stronger than horses.

SUMMARY

» The matter that makes up the material universe can be classified in different ways. In terms of the external, observable properties of matter, we can identify four **phases: solid, liquid, gas,** and **plasma**.

» In terms of the submicroscopic composition of ordinary matter, there are **elements** (consisting of **atoms**), **compounds** (consisting of **molecules**), and combinations of these.

» The nature of the constituent particles—atoms and molecules—and the forces that act between them determine the physical properties of a given element or compound.

» **Pressure, mass density,** and **weight density** are extensions of the concepts of force, mass, and weight.

» Pressure is a measure of the "concentration" of any force spread over an area.

» Mass density is the mass of a substance per unit volume. Weight density is the weight per unit volume.

» Liquids and gases are **fluids:** they flow readily and take the shapes of their containers.

» The force of gravity causes the pressure in a fluid to increase with depth. The **law of fluid pressure** states how the pressure at any point in a fluid is determined by the weight of the fluid above.

» In liquids, the pressure is proportional to the distance below the liquid's surface. Gases are compressible, and the pressure increases with depth in a different way because the density is not constant.

» Anything partly or completely immersed in a fluid experiences an upward **buoyant force.** By **Archimedes' principle,** this force is equal to the weight of the fluid that is displaced. Ships, submarines, blimps, and hot-air balloons all rely on this principle. Archimedes' principle is also routinely used to measure the densities of solids and liquids.

» **Pascal's principle** states that any additional pressure in a fluid is passed on uniformly throughout the fluid. This principle is the basis for hydraulic brake systems.

» **Bernoulli's principle** is exploited by carburetors, atomizers, and other devices. It states that the pressure in a moving fluid decreases when the speed increases and vice versa.

IMPORTANT EQUATIONS

Equation	Comments	Equation	Comments
Fundamental Equations		*Special-Case Equations*	
$p = \dfrac{F}{A}$	Definition of pressure	$p = D_W h = Dgh$	Pressure (gauge) at a depth h in a liquid
$D = \dfrac{m}{V}$	Definition of mass density	$p = 0.433 \text{ psi/ft} \times h$	Pressure (gauge) in psi underwater at a depth h in feet
$D_W = \dfrac{W}{V}$	Definition of weight density	$A_1 v_1 = A_2 v_2$	Equation of continuity (constant density)
$F_b = W_{\text{fluid displaced}}$	Archimedes' principle	$\dfrac{F_1}{A_1} = \dfrac{F_2}{A_2}$	Hydraulic equation (constant density)

MAPPING IT OUT!

1. In Section 4.1 in the description of matter, the following terms were introduced: *elements, atoms, electrons, protons, neutrons, nucleus, molecules, compounds, mixtures,* and *solutions.* Create a concept map explaining the composition of matter by appropriately organizing and linking these concepts to form meaningful propositions. After completing your concept map, compare your map with that of a classmate or the instructor. Are they the same? Should they be? Discuss the similarities and differences that you find between the maps.

2. Review Section 4.3 carefully. Based on your understanding of this material, develop a concept map to distinguish and relate the concepts of *mass density, weight density,* and *specific gravity.* Your map should address such issues as the basic definitions of these concepts and their applications in physics and in everyday life.

QUESTIONS

(■ **Indicates a review question, which means it requires only a basic understanding of the material to answer. Questions without this designation typically require integrating or extending the concepts presented thus far.**)

1. ■ Describe the four phases of matter. Compare their external, observable properties. Compare the nature of the forces between atoms or molecules (or both) in the solid, liquid, and gas phases.

2. Identify some of the elements that exist in pure form (not in compounds) around you.

3. ■ What is the difference between a mixture of two elements and a compound formed from the two elements?

4. If you classify everything around you as an element, a compound, or a mixture, which category do you think would have the largest number of entries? Why?

5. ■ Why can gases be compressed so much more readily than solids or liquids?

6. Suppose you are in the International Space Station in orbit around Earth and a fellow astronaut gives you what appears to be an inflated balloon. Describe how you could determine whether the balloon contains a gas, a liquid, or a solid.

7. ■ Use the concept of pressure to explain why snowshoes are better than regular shoes for walking in deep snow.
8. Why is it that a person can lie still on a "bed" of nails (**Figure 4.53**) without suffering any serious injuries but would incur severe puncture wounds to his feet if he tried to stand barefoot on the same "bed"?

Figure 4.53 Question 8.

9. Bicycle tires are often inflated to pressures as high as 75 psi, whereas automobile tires, which must support much heavier loads, are pumped to only 32 psi. Why is that? Explain.
10. The same bicycle tire pump is used to inflate a mountain bike tire to 40 psi and then a road bike tire to 100 psi. What difference would the user notice when using the pump on the two tires?
11. ■ Explain the difference between gauge pressure and absolute pressure.
12. ■ How can you use the volume of some quantity of a pure substance to calculate its mass?
13. The mass density of a mixture of ethyl alcohol and water is 950 kg/m³. Is the mixture mostly water, mostly alcohol, or about half and half? What is your reasoning?
14. Believe it or not, canoes have been made out of concrete (and they actually float). But even though concrete has a lower density than aluminum, a concrete canoe weighs a lot more than an aluminum one of the same size. Why is that?
15. Would the weight density of water be different on the Moon than it is on Earth? What about the mass density? Explain.
16. The way pressure increases with depth in a gas is different from the way it does in a liquid. Why?
17. Workers are to install a hatch (door) near the bottom of an empty storage tank. In choosing how strong to make the hatch, does it matter how tall the tank is? How wide it is? Whether it is going to hold water or mercury? Explain.
18. If the acceleration due to gravity on Earth suddenly increased, would this affect the atmospheric pressure? Would it affect the pressure at the bottom of a swimming pool? Explain.
19. If Earth's atmosphere warmed up and expanded to a larger total volume but its total mass did not change, would this affect the atmospheric pressure at sea level? Would this affect the pressure at the top of Mount Everest? Explain.
20. Is there a pressure variation (increase with depth) in a fuel tank on a spacecraft in orbit? Why or why not?
21. ■ Explain how a barometer can be used to measure altitude.
22. An oak barrel is filled with water and then tightly sealed. A hole is drilled in the barrel's top, and a narrow, open tube is fitted securely into the hole. Water is then slowly added to the tube. When the height of the water column reaches a certain point, the barrel bursts apart. Explain this occurrence using the principles of fluid physics. (This experiment is called *Pascal's barrel* and was allegedly performed by Pascal

in 1646, although there is no mention of this demonstration anywhere in his archived papers.)
23. The "suction cup" (**Figure 4.54**) is a common device used to suspend pictures, plants, and other objects from walls, ceilings, windows, and other smooth surfaces. Why is the name "suction cup" a bit of a misnomer? Is anything really "sucking"? How exactly does a "suction cup" work? Explain.

Figure 4.54 A "suction cup" at work; see Question 23.

24. Why does the buoyant force always act upward?
25. ■ What substances would sink in gasoline but float in water?
26. It is easier for a person to float in the ocean than in an ordinary swimming pool. Why?
27. ■ A ship on a large river approaches a bridge, and the captain notices that the ship is about a foot too tall to fit under the bridge. A crew member suggests pumping water from the river into an empty tank on the ship. Would this help? Why or why not?
28. In "The Unparalleled Adventure of One Hans Pfaall" by Edgar Allen Poe, the hero discovers a gas whose density is "37.4 times" less than that of hydrogen. How much better at lifting would a balloon filled with the new gas be compared to one filled with hydrogen?
29. A brick is tied to a balloon filled with air and then tossed into the ocean. As the balloon is pulled downward by the brick, the buoyant force on it decreases. Why?
30. Venus's atmosphere is much more dense than Earth's whereas that of Mars is far less dense. Suppose it is decided to send a probe to each planet that, once it arrived, would be carried around in the planet's atmosphere by a helium-filled balloon. How would the size of each balloon compare to the size that would be needed on Earth?
31. ■ What is Pascal's principle?
32. ■ How does a car's brake system make use of Pascal's principle?
33. ■ What important thing happens when the speed of a moving fluid increases?
34. The pressure in the air along the upper surface of an aircraft's wing (in flight) is lower than the pressure along the lower surface. Compare the speed of the air flowing over the wing to that of the air flowing under the wing.
35. ■ How does a perfume atomizer make use of Bernoulli's principle?
36. When two trains, going in opposite directions, are passing on tracks that are laid out close together, the train cars can often be seen to be leaning in toward one another where they are in proximity. How might the air passing through the narrow gap separating the two trains contribute to the observed attraction between their cars?
37. Smoking was a common habit in the United States in times past, and nonsmokers riding in cars where another passenger had lit up quickly learned to open their window a crack

while the car was moving to exhaust the smoke from the compartment. Why does this action help to draw the smoke out of the car? Explain using principles of fluid physics.

38. Corrugated plastic pipes are commonly used to carry water away from the foundation areas of houses. These accordion-like pipes typically vary in diameter, alternating between smaller and larger widths, at regular intervals of a few centimeters or so, making them much more flexible than straight plastic pipes. Assuming the water in such pipes undergoes steady flow, describe the variations in the speed and pressure of the water as it moves along.

39. Explain why a smoothly flowing stream of water from a faucet often gets progressively narrower in cross-section as the water falls (**Figure 4.55**).

Cengage Learning

Figure 4.55 Question 39.

40. Six hollow cylinders having diameters D and heights h are filled with water. The cylinders all have a hole drilled in their sides; the holes are all the same size and are all located at the same height above the bases of the cylinders. The holes are plugged with rubber stoppers. The values of D and h for the six cylinders are provided. Rank the combinations according to the pressure exerted by the water on the stopper from largest to smallest. If two (or more) combinations yield the same pressure, give them the same ranking. Explain the reasoning you used in arriving at your rankings.
   (a) $D = 0.25$ m; $h = 0.40$ m.
   (b) $D = 0.20$ m; $h = 0.50$ m.
   (c) $D = 0.50$ m; $h = 0.35$ m.
   (d) $D = 0.25$ m; $h = 0.35$ m.
   (e) $D = 0.40$ m; $h = 0.50$ m.
   (f) $D = 0.25$ m; $h = 0.60$ m.

41. Five containers each hold the same volume of various liquids. Five blocks composed of various solids are fully submerged, one in each of the five containers. The blocks all have the same size and are all held at the same depth in the containers. The volume of each block is one-fourth the volume of the liquids in the containers. For the given combinations of block mass, M_b, and liquid mass, M_l, rank the buoyant forces on the blocks from largest to smallest. If two or more circumstances yield the same buoyant forces, give them the same ranking. Explain your reasoning in setting your ordering.
   (a) $M_b = 0.04$ kg; $M_l = 0.20$ kg.
   (b) $M_b = 0.03$ kg; $M_l = 0.15$ kg.
   (c) $M_b = 0.04$ kg; $M_l = 0.12$ kg.
   (d) $M_b = 0.02$ kg; $M_l = 0.08$ kg.
   (e) $M_b = 0.03$ kg; $M_l = 0.12$ kg.

42. Pistons are fitted to two cylindrical chambers connected through a horizontal tube to form a hydraulic system. The piston chambers and the connecting tube are filled with an incompressible fluid. The cross-sectional areas of piston 1 and piston 2 are A_1 and A_2, respectively. A force F_1 is exerted on piston 1. Rank the resultant force F_2 on piston 2 that results from the combinations of F_1, A_1, and A_2 given from greatest to smallest. If any of the combinations yield the same force, give them the same ranking.
   (a) $F_1 = 2.0$ N; $A_1 = 0.5$ m²; and $A_2 = 1.0$ m².
   (b) $F_1 = 1.0$ N; $A_1 = 0.5$ m²; and $A_2 = 0.25$ m².
   (c) $F_1 = 1.0$ N; $A_1 = 1.0$ m²; and $A_2 = 2.0$ m².
   (d) $F_1 = 2.0$ N; $A_1 = 0.25$ m²; and $A_2 = 1.0$ m².
   (e) $F_1 = 2.0$ N; $A_1 = 0.25$ m²; and $A_2 = 0.5$ m².
   (f) $F_1 = 1.0$ N; $A_1 = 1.0$ m²; and $A_2 = 0.5$ m².

PROBLEMS

1. A grain silo is filled with 2 million pounds of wheat. The area of the silo's floor is 400 ft². Find the pressure on the floor in pounds per square foot and in psi.

2. A bicycle tire pump has a piston with area 0.44 in.². If a person exerts a force of 30 lb on the piston while inflating a tire, what pressure does this produce on the air in the pump?

3. A large truck tire is inflated to a gauge pressure of 80 psi. The total area of one sidewall of the tire is 1,200 in.². What is the outward force on the sidewall because of the air pressure?

4. The water in the plumbing in a house is at a gauge pressure of 300,000 Pa. What force does this cause on the top of the tank inside a water heater if the area of the top is 0.2 m²?

5. A box-shaped metal can has dimensions 8 in. by 4 in. by 10 in. high. All of the air inside the can is removed with a vacuum pump. Assuming normal atmospheric pressure outside the can, find the total force on one of the 8-by-10-in. sides.

6. A viewing window on the side of a large tank at a public aquarium measures 50 in. by 60 in. (**Figure 4.56**). The average gauge pressure from the water is 8 psi. What is the total outward force on the window?

iStockphoto.com/andipantz

Figure 4.56 Problem 6.

7. A large chunk of metal has a mass of 393 kg, and its volume is measured to be 0.05 m³.
   (a) Find the metal's mass density and weight density in SI units.
   (b) What kind of metal is it?

8. A small statue is recovered in an archaeological dig. Its weight is measured to be 96 lb, and its volume 0.08 ft³.
   (a) What is the statue's weight density?
   (b) What substance is it?

9. A large tanker truck can carry 20 tons (40,000 lb) of liquid.
   (a) What volume of water can it carry?
   (b) What volume of gasoline can it carry?

10. The total mass of the hydrogen gas in the *Hindenburg* zeppelin was 18,000 kg. What volume did the hydrogen occupy?

11. A large balloon used to sample the upper atmosphere is filled with 900 m³ of helium. What is the mass of the helium?

12. A certain part of an aircraft engine has a volume of 1.3 ft³.
   (a) Find the weight of the piece when it is made of iron.
   (b) If the same piece is made of aluminum, what is its weight? Determine how much weight is saved by using aluminum instead of iron.

13. The volume of the Drop Tower "Bremen" (a 100-meter-tall tube used to study processes during free fall) is 1,700 m³.
   (a) What is the mass of the air that must be removed from it to reduce the pressure inside to nearly zero (1 Pa compared to 100,000 Pa)?
   (b) What is the weight of the air in pounds?

14. It is determined by immersing a crown in water that its volume is 26 in.³ = 0.015 ft³.
   (a) What would its weight be if it were made of pure gold?
   (b) What would its weight be if its volume were half gold and half lead?

15. Find the gauge pressure at the bottom of a swimming pool that is 12 ft deep.

16. The depth of the Pacific Ocean in the Mariana Trench is 36,198 ft. What is the gauge pressure at this depth?

17. Calculate the gauge pressure at a depth of 300 m in seawater.

18. A storage tank 30 m high is filled with gasoline.
   (a) Find the gauge pressure at the bottom of the tank.
   (b) Calculate the force that acts on a square access hatch at the bottom of the tank that measures 0.5 m by 0.5 m.

19. The highest point in North America is the top of Mount McKinley in Alaska, 20,320 ft above sea level. Using Figure 4.30, find the approximate air pressure there.

20. The highest altitude ever reached by a glider (as of this writing) is 15,460 m. What is the approximate air pressure at that altitude?

21. An ebony log with volume 12 ft³ is submerged in water. What is the buoyant force on it?

22. An empty storage tank has a volume of 1,500 ft³. What is the buoyant force exerted on it by the air?

23. A blimp used for aerial camera views of sporting events holds 200,000 ft³ of helium.
   (a) How much does the helium weigh?
   (b) What is the buoyant force on the blimp at sea level?
   (c) How much can the blimp lift (in addition to the helium)?

24. A modern-day zeppelin holds 8,000 m³ of helium. Compute its maximum payload at sea level.

25. A box-shaped piece of concrete measures 3 ft by 2 ft by 0.5 ft.
   (a) What is its weight?
   (b) Find the buoyant force that acts on it when it is submerged in water.
   (c) What is the net force on the concrete piece when it is underwater?

26. A juniper-wood plank measuring 0.25 ft by 1 ft by 16 ft is totally submerged in water.

(a) What is its weight?
(b) What is the buoyant force acting on it?
(c) What is the size and the direction of the net force on it?

27. The volume of an iceberg is 100,000 ft³ (**Figure 4.57**).
   (a) What is its weight, assuming it is pure ice?
   (b) What is the volume of seawater it displaces when floating? (*Hint:* You know what the weight of the seawater is.)
   (c) What is the volume of the part of the iceberg out of the water?

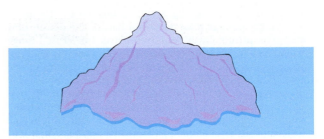

Figure 4.57 Problem 27.

28. A boat (with a flat bottom) and its cargo weigh 5,000 N. The area of the boat's bottom is 4 m². How far below the surface of the water is the boat's bottom when it is floating in water?

29. A scale reads 100 N when a piece of aluminum is hanging from it. What does it read when it is lowered so that the aluminum is submerged in water?

30. A rectangular block of ice with dimensions 2 m by 3 m by 0.2 m floats on water. A person weighing 600 N wants to stand on the ice. Would the ice sink below the surface of the water?

31. A dentist's chair with a person in it weighs 1900 N. The output plunger of a hydraulic system starts to lift the chair when the dental assistant's foot exerts a force of 45 N on the input piston. Neglecting any difference in the heights of the piston and the plunger, what is the ratio of the area of the plunger to the area of the piston?

32. A booster pump on a brake system designed to be used horizontally consists of two cylinders capped by pistons connected by a hose. The cylinders and hose are filled with an incompressible fluid. The system produces a mechanical advantage of five times; that is, the output force is five times as great as the input force. If the cross-sectional area of the input piston is 4.0 in.², how large must the area of the output piston be?

33. The wing of an airplane has an average cross-sectional area of 12 m² and experiences a lift force of 75,000 N. What is the average difference in the air pressure between the top and bottom of the wing?

34. The volume flow rate in an artery that supplies blood to the brain is 3.5×10^{-6} m³/s. If the cross-sectional area of the artery is 8.5×10^{-5} m², what is the average speed of the blood through this vessel? Find the average blood speed through a narrowed section of the artery where the cross-sectional area is reduced by a factor of 3.

35. Air flows through a heating duct with a square cross-section with 8-inch sides at a speed of 5.0 ft/s. Just before reaching an outlet in the floor of a room, the duct widens to assume a square cross-section with sides equal to 12 inches. Compute the speed of the air flowing into the room, assuming that we can treat the air as an incompressible fluid.

1. When exactly 1 cup of sugar is dissolved in exactly 1 cup of water, *less* than 2 cups of solution result. Why?

2. Near the end of Section 4.3, we stated that, because the gram was defined to be the mass of 1 cm^3 of water, the density of water is exactly 1,000 kg/m^3. Verify this.

3. What would be the mass density, weight density, and specific gravity of aluminum on the Moon? The acceleration of gravity there is 1.6 m/s^2.

4. Earth's mass is 6×10^{24} kg, and its radius is 6.4×10^6 m. What is the average mass density of Earth? The density of the rocks comprising Earth's outermost layer (its "crust") ranges from 2,000 to 3,500 kg/m^3. Based on your answer, what can you conclude about the material deep inside Earth's interior?

5. In 1993, a football coach in the United States accused an opposing team's punter of using a football inflated with helium instead of air. Estimate how much lighter a football would be if inflated to a gauge pressure of 1 atm with helium instead of air. The volume of a football is approximately 0.1 ft^3.

6. Two swimming pools are 8 ft deep, but one measures 20 ft by 30 ft, and the other measures 40 ft by 60 ft. Identical drain valves at the bottom of each pool are 10 in.^{2} in area. Compare the force on each valve.

7. Scuba divers take their own supply of air with them when they go underwater. Why couldn't they just take a long hose with them from the surface and breathe through it (**Figure 4.58**)?

Figure 4.58 Challenge 7.

8. A motorist driving through Colorado checks the tire pressure in Denver (elevation 5,000 ft) and then again at the Eisenhower Tunnel (elevation 11,000 ft). Would the pressures be the same? What two main factors affect the tire pressure as the car climbs?

9. A glass contains pure water with ice floating in it. After the ice melts, will the water level be higher, lower, or the same? (Ignore evaporation.)

10. At one point in the novel *Slapstick* by Kurt Vonnegut, the force of gravity on Earth suddenly "increased tremendously." The result:

    . . . elevator cables were snapping, airplanes were crashing, ships were sinking, motor vehicles were breaking their axles, bridges were collapsing, and on and on.

    Would a ship really sink if the force of gravity were increased?

11. A brick rests on a large piece of wood floating in a bucket of water. The brick slides off and sinks. Does the water level in the bucket go up, go down, or stay the same?

12. As a helium balloon rises up in the air, work is done on it against the force of gravity. What is doing the work? What energy transfer or transformation is taking place?

13. Prove Archimedes' principle. That is, show that the buoyant force is equal to the weight of the displaced fluid. Assume the object submerged in the fluid is a rigid rectangular box with dimensions l, w, and h with its top and bottom faces oriented horizontally (Figure 4.36).

14. Water flows straight down from an open faucet (Figure 4.55). The cross-sectional area of the faucet is 1.8×10^{-4} m^2, and the speed of the water as it leaves the faucet is 0.85 m/s. Ignoring air resistance, find the cross-sectional area of the water stream at a point 0.10 m below the faucet opening.

15. Use conservation of energy to show that the speed with which water flows out of a tap near the bottom of an open container equals $\sqrt{2gh}$ where h is the height of the free-standing surface of the water above the tap. Assume that the water at the surface is moving with negligible velocity. Refer to Section 3.5 and Figure 3.29 for guidance.

5

TEMPERATURE AND HEAT

Figure CO-5 Hurricane Katrina off the Gulf Coast in August 2005.

CHAPTER INTRODUCTION: Hurricanes

Viewed from space, a hurricane—one type of tropical cyclone—has an element of beauty to it. But for millions of people who live near vulnerable coastal areas around the world, such storms pose a deadly threat. Hurricane Katrina, shown in **Figure CO-5**, devastated New Orleans and nearby communities in Louisiana and Mississippi in August 2005, killing an estimated 1,800 people. The death tolls from several tropical cyclones that struck South Asia over the last century totaled more than 100,000 each. High winds that destroy buildings and cause coastal flooding take many lives, but often it is the inland flooding caused by torrential rains that is the most deadly.

How do these monster storms develop and what is the source of their energy? Warm ocean water evaporates into the air and is carried high into towering clouds. There the water vapor condenses into drops, releasing energy that causes the surrounding air to move upward. Air near Earth's surface rushes in to replace the rising air, leading to the high winds characteristic of these storms. The total energy involved is huge—roughly the equivalent of 30 World-War-II–era atomic bombs exploding each minute.

To more fully understand the physical processes that drive natural phenomena such as hurricanes, as well as a host of devices in common everyday use such as refrigerators and thermometers, we must examine the concepts of temperature and heat in a systematic way. This is the goal of this chapter.

5.1 Temperature

Next to time, **temperature** may be the most commonly used physical quantity in our daily lives. We might loosely define temperature as a measure of hotness or coldness. But the concepts of hot and cold are themselves rather vague, subjective, and relative. In the summer, an air temperature of 70°F feels cool, whereas the same temperature in the winter feels warm.

The idea of temperature is distinct from that of heat or internal energy. For example, when one drop is removed from a cup of hot water, both the drop and the cup have the *same* temperature, but the amount of internal energy associated with each is very different (**Figure 5.1**). Placing the drop in the palm of your hand would not have nearly the same effect (pain and injury) as pouring the cup of water into it.

Thermometers are devices that measure the temperature of a substance. To function, they all depend on some physical property that changes with temperature. A common thermometer design exploits the fact that a liquid, frequently red-colored alcohol, will expand or contract when heated or cooled, thereby rising or falling in a glass tube as the temperature varies. Some thermometers are based on other temperature-dependent physical properties, including the volume of solids, the pressure or volume of a gas, the electrical properties of metals, the amount and frequency of radiated energy, and the speed of sound in a gas.

5.1a Temperature Scales

There are three different temperature scales in common use for calibrating thermometers: Fahrenheit, Celsius, and Kelvin. The normal freezing and boiling temperatures of water—called the *phase transition* temperatures—may be

Figure 5.1 Water in the cup and water in the dropper have the same temperature, but the water in the cup can transfer much more energy to its surroundings.

used to compare the three scales. In the Fahrenheit scale, the boiling point of water under a pressure of 1 atmosphere is 212°—designated 212°F. The freezing point of water under this pressure is 32°F. So there are 180 units, called *degrees*, that separate the two temperatures. The Celsius scale, formerly called the *centigrade* scale, is metric based; it uses 100 degrees between the freezing and boiling points of water. Zero degrees Celsius—designated 0°C—is the freezing temperature, and 100°C is the boiling temperature.

▶ Physics To Go 5.1

Go to the thermometer display in a department store, hardware store, or other retail outlet. Look carefully at the temperatures displayed by the different thermometers. Do they all show exactly the same temperature? Should they be the same? What does this tell you about the accuracy of these instruments?

Most of us spend our lives subjected to temperatures within the range of −60 to 120°F (−51 to 49°C). Much higher temperatures exist in common places: the interiors of stoves and automobile engines, the filaments of lightbulbs, the flame of a candle, and so on. The Sun's surface temperature is about 10,000°F (5,700°C), and its interior is at about 27,000,000°F (15,000,000°C). Temperatures this high have been produced on Earth in experiments with plasmas and in nuclear explosions. There is no upper limit on temperature.

At the other extreme, there *is* a limit on cold temperatures. The coldest temperature, called **absolute zero,** is −459.67°F (−273.15°C). For reasons to be discussed later, because it is impossible to go below this temperature, the Kelvin scale is a convenient one because it uses absolute zero as its starting point (zero). (This scale is also referred to as the "absolute temperature scale.") The size of the unit in the Kelvin scale is the same as that in the Celsius scale, except it is called a kelvin (K) instead of a degree.

Any temperature in the Kelvin scale equals the corresponding Celsius value plus 273.15:

$$T(K) = T(°C) + 273.15$$

The normal boiling and freezing temperatures of water are 373.15 K and 273.15 K, respectively. (Note that the degree symbol is not used with the Kelvin scale.)

Figure 5.2 shows a comparison of the three temperature scales. Note that the Fahrenheit and Celsius scales agree at −40°. **Table 5.1** lists some representative temperatures. As a physical quantity, temperature is represented by the letter *T*. (*T* is used to represent both temperature and period [Section 1.1] in this book. Hopefully, the contexts in which these symbols are used will be sufficiently clear so as to avoid any confusion about which quantity is being referenced.)

5.1b Temperature and Energy

What determines the temperature of matter? In other words, what is the difference between a cup of coffee when it is hot (200°F) and the same cup of coffee when it is cold (70°F)? The atoms and molecules that compose matter have kinetic energy. In gases, they move about randomly with high speed. In liquids and solids, they vibrate much like a mass on a spring or an object oscillating in a hole (Section 3.5). At higher temperatures, the atoms and molecules in matter move faster and have higher kinetic energies.

PRINCIPLES The Kelvin temperature of matter is proportional to the average kinetic energy of the constituent particles.

Kelvin scale temperature ∝ average *KE* of atoms and molecules

Figure 5.2 Comparison of the three temperature scales. The Celsius and Kelvin scales have the same sized unit. The Fahrenheit degree is five-ninths the Celsius degree.

K	°C	°F	
373	100	212	Boiling point of water at sea level
363	90	194	
353	80	176	
343	70	158	
333	60	140	134°F (57°C) Highest (weather-related) temperature
323	50	122	recorded in the world
313	40	104	A hot day
303	30	86	Average body temperature 98.6°F (37°C)
293	20	68	Average room temperature
283	10	50	
273	0	32	Freezing (melting) point of water (ice) at sea level
263	−10	14	
253	−20	−4	
243	−30	−22	A bitterly cold day
233	−40	−40	
223	−50	−58	
213	−60	−76	
203	−70	−94	
193	−80	−112	
183	−90	−130	−129°F (−89°C) Lowest (weather-related) temperature
173	−100	−148	recorded in the world

Because of collisions between the particles, during which energy is exchanged, the particles do not all have exactly the same kinetic energy at each instant, nor does the energy of a given particle stay exactly the same from one moment to the next. But the *average* kinetic energy of all of the particles is constant as long as the temperature stays constant.

Table 5.1 Representative Temperatures in the Three Temperature Scales

Description	°F	°C	K
Absolute zero	−459.67	−273.15	0
Helium boiling point*	−452	−268.9	4.25
Nitrogen boiling point	−320.4	−195.8	77.35
Oxygen boiling point	−297.35	−182.97	90.18
Alcohol freezing point	−175	−115	158
Mercury freezing point	−37.1	−38.4	234.75
Water freezing point	32	0	273.15
Normal body temperature	98.6	37	310.15
Water boiling point	212	100	373.15
"Red hot" (approx.)	800	430	700
Aluminum melting point	1,220	660	933
Iron melting point	2,797	1,536	1,809
Sun's surface (approx.)	10,000	5,700	6,000
Sun's interior (approx.)	27×10^6	15×10^6	15×10^6
Highest laboratory temperature	7.2×10^9	4×10^9	4×10^9

* All boiling points are for 1 atm pressure.

Figure 5.3 At lower temperatures, atoms and molecules have lower average kinetic energy (*KE*). At absolute zero, their kinetic energies would be zero: they would be stationary.

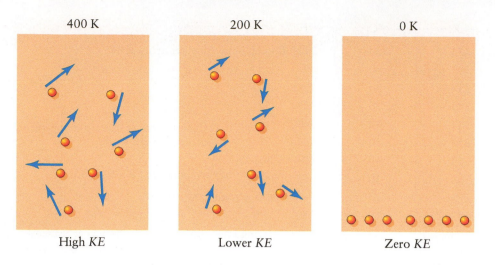

400 K	200 K	0 K
High *KE*	Lower *KE*	Zero *KE*

So when a cup of coffee is hot, the molecules in it have higher average kinetic energy than when it is cold. If you put your finger into hot coffee, the atoms and molecules in the coffee pass on their higher kinetic energy to the atoms and molecules in your finger by way of collisions: your finger is warmed.

This fact—that temperature depends on the average kinetic energy of atoms and molecules—is very important. It should help you understand many of the phenomena we will discuss in this chapter. In gases, higher kinetic energy means that the atoms and molecules move about with higher speeds. For liquids and solids, the molecules vibrate through a greater distance like a pendulum swinging through a larger arc. You may recall that when particles oscillate like this, they also have potential energy. This is the case here, too, but the potential energy of "bound" atoms and molecules in liquids and solids is not directly related to the temperature. This potential energy is important when a substance undergoes a change of phase—such as freezing or boiling. More on this later.

This principle also accounts for the existence of an absolute zero. At colder temperatures, the average kinetic energy of the particles is smaller. If they stopped moving altogether, the average kinetic energy would be zero. This would be the lowest possible temperature—absolute zero (**Figure 5.3**). The word *would* is used because, as it turns out, absolute zero can never be reached. The atoms and molecules in a substance cannot be completely stopped. Researchers do get very close to absolute zero—within a billionth of a degree or less—but they cannot reach it exactly.

At very low temperatures, many substances acquire unusual properties. Plastic and rubber become as brittle as glass. Below about 2 K, helium is a "superfluid" liquid; it flows without friction (see the application at the end of Section 4.5). Some materials become "superconductors." They conduct electricity without resistance (more on this in Chapter 7).

The nature of matter is also different at very high temperatures. Above about 6,000 K, the kinetic energies of atoms are so high that they cannot bind together, so there can be no solids or liquids—or even molecules. Above about 20,000 K, the electrons begin to break free from atoms, and only plasmas can exist.

Learning Check

1. The average kinetic energy of the atoms in a gold ring determines its _____ .
2. (Choose the *incorrect* statement.) Absolute zero is
   (a) the lowest possible temperature.
   (b) 0° on the Celsius scale.
   (c) the temperature at which atoms and molecules would not be moving.
   (d) more than 400 degrees below zero on the Fahrenheit scale.
3. (True or False.) If the temperature of something was raised 50°C, that means its temperature was raised 90°F.

ANSWERS: 1. temperature **2.** (b) **3.** True

As mentioned in Chapter 4, we live at the bottom of a "sea" of air that provides us with one of the basic ingredients of sustained life: oxygen. Indeed, had Earth been unable to retain an atmosphere for more than several billion years, the evolution of life as we know it could not have taken place at all. But what physical conditions or quantities determine whether or not a planet is capable of holding an appreciable atmosphere? More specifically, why does Earth possess a substantial atmosphere, whereas Mercury does not (**Figure 5.4**)? Is the composition of such an atmosphere related to these physical conditions? In other words, is it possible to understand, in terms of our physics, why the atmosphere of Jupiter is primarily hydrogen while that of Earth is made up principally of heavier gases such as nitrogen and oxygen (see Table 4.2)?

Figure 5.4 The planet Earth (top) has an appreciable atmosphere, but the planet Mercury (bottom) does not.

Every planetary body in our solar system receives some radiation from the Sun. Those closer to the Sun receive more radiation and hence have higher surface temperatures than those farther away. Given the definition of temperature, it is clear that atoms and molecules in the atmosphere of a planet near the Sun will generally have higher average kinetic energies (and higher average speeds) than those in the atmosphere of a planet far from the Sun.

The ability of a planet to retain these atmospheric atoms and molecules depends primarily on the strength of its gravitational field. If the average speeds of the atmospheric particles are high and the planet's gravity is low, the atmosphere will gradually escape. Conversely, if the average speeds of the atmospheric particles are low and the gravity is high, the atmosphere will be retained. This is why Mercury and the Moon do not possess primordial atmospheres. Mercury is a small planet very close to the Sun: it is hot (having a surface temperature of more than 400 K) and has a low surface gravity (1 *g* on Mercury equals 3.7 m/s^2 as opposed to 9.8 m/s^2 on Earth). Any atmosphere it may once have had escaped long ago because the particles were far too energetic to have been held in by the planet's weak gravity. Similarly, any original atmosphere the Moon may have had must have dissipated billions of years ago despite its cooler temperature (about 300 K, like that of Earth) because its gravitational field is only about one-sixth as strong as that of Earth.

For planets with moderate temperatures and gravities, some gases can be retained and others cannot. It is possible to predict which gases will be retained in a given circumstance by comparing the average molecular speed of a gas with the escape velocity of the planet. The *escape velocity* (as discussed in Section 3.5) is the minimum speed needed to overcome the gravitational pull of the planet. The higher the gravity of a planet, the greater its escape velocity. In general, a planet will retain a particular gas in its atmosphere over millions of years only if its escape velocity is four to six times larger than the average molecular speed of that gas. Only under these circumstances will the planet's gravity be high enough to prevent the gas from eventually leaking off into space.

Figure 5.5, a plot of characteristic velocity versus temperature, displays this rule. The labeled points show the average surface temperature and escape velocity (divided by six) for the eight planets and three satellites: Earth's moon; Titan, a moon of Saturn; and Triton, a moon of Neptune. Superimposed on this plot are lines for different gases tracing the average particle speeds of these gases as a function of temperature.

Several things can be noted: (1) for a given gas, as the temperature increases (far left), the average molecular speed increases. This is to be expected, of course, on the basis of what is meant by the Kelvin temperature of a gas. (2) At a given temperature, the speeds of the lighter species, such as hydrogen and helium, are greater than those of heavier ones, such as nitrogen and carbon dioxide. This, too, is easily understood in terms of what we know about the physics of gases. For a fixed temperature, the average kinetic energy of, say, a helium atom will be equal to that of a carbon dioxide molecule. But the kinetic energy of any particle is proportional to the product of its mass and its velocity squared. For a particular value of kinetic energy, higher masses correspond to smaller velocities. Hence, at any temperature, the more massive carbon dioxide molecule will be moving more slowly than the lighter He atom. (3) For a planet to retain a particular gas, its position on the graph must lie above the line for that gas. Notice that the points corresponding to the planets Jupiter and Saturn lie well above all the lines. These planets are so massive and so cold (being far from the Sun) that they can retain all the gases considered here in their atmospheres—even hydrogen, which has the lightest atoms. Earth, on the other hand, can hold heavier gases such

NASA

Johns Hopkins University Applied Physics Laboratory/Carnegie Institution of Washington/NASA

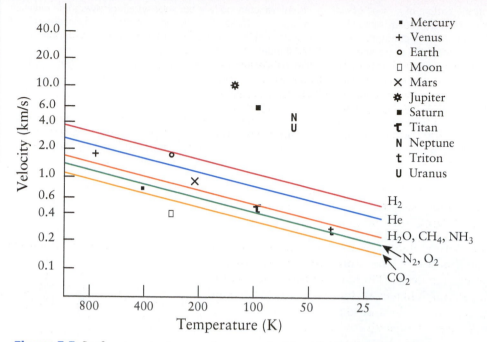

Figure 5.5 Surface temperatures and escape velocities (divided by 6) for the eight planets and three satellites. The solid lines show the variation of molecular speed as a function of temperature for several common atmospheric gases.

to keep in mind is that, armed only with the physics we have developed so far, we are in a position to make important predictions about the likely existence of atmospheres on other planetary bodies within the solar system. Such predictions figure prominently in attempts to identify locations beyond Earth where life may have evolved.

The search for planets orbiting Sun-like stars that may harbor conditions appropriate for the development of carbon-based life forms is one of the most active areas in astronomy today. During the past 15 years, more than 2,000 *extrasolar planets* have been discovered using a variety of astronomical techniques. Most of these systems have orbital characteristics quite different from those of the planets revolving around the Sun, but a few exhibit properties similar to our own solar system. These are clearly the most exciting because they offer the best prospects for locales where other forms of life may exist.

as nitrogen and oxygen but not hydrogen. And, as discussed above, the Moon and the planet Mercury are incapable of holding onto any of the common atmospheric constituents, whereas Titan and Triton possess atmospheres dominated by nitrogen.

The analysis presented here has ignored many factors that complicate the question of planetary atmosphere retention (such as interactions between planetary atmospheres and the *solar wind*, a continuous stream of electrons and other charged particles that leave the Sun and propagate outward into the interplanetary medium, which are particularly important for a nearby planet like Mercury), but it is correct in its broad outline. The important point

■ QUESTIONS

1. Explain why the Moon and Mercury possess only very weak, transient atmospheres consisting of constituents temporarily captured from the solar wind or released by collisions with interplanetary debris.

2. The dwarf planet Pluto has an average surface temperature of 44 K and an escape speed (divided by six) of 0.2 km/s. Based on Figure 5.5, which gases, if any, would this body be likely to have retained as an atmospheric constituent over the past 3 billion years?

5.2 Thermal Expansion

Thermal expansion is an important phenomenon that is exploited by the common types of thermometers and by a variety of other useful devices. In almost all cases, substances that are not constrained expand when their temperatures increase. (Exceptions include water below 4°C and some compounds of tungsten.) The air in a balloon expands when heated, the mercury in a thermometer expands upward in the glass tube when placed in a hot liquid, and sections of bridges become longer in the summer. If the substance is constrained sufficiently, it will not expand, but forces and pressures will be created in response to the constraint. For example, if an empty pressure cooker is sealed and then heated, the air inside is prevented from expanding. But the pressure inside increases and causes larger and larger outward forces on the inner surfaces of the pressure cooker.

We can see qualitatively why this expansion occurs. At higher temperatures, the atoms and molecules in a solid or a liquid vibrate through a larger distance and so push each other apart slightly. In gases, they move with higher and higher speeds as the temperature rises. A balloon will expand when heated because the higher-speed air molecules will cause higher pressure and push the balloon's surface outward as they collide with it. In these cases, the expansion occurs in all three dimensions. For example, as a brick is heated, its length, width, and thickness all increase proportionally.

Figure 5.6 A metal rod has length l when the temperature is T. When the temperature is increased by an amount ΔT, the length of the rod increases by a proportional amount Δl.

5.2a Solids: Linear Expansion

We can use logic and basic mathematics to predict the amount of expansion that occurs. Let us first consider the simplest case—the thermal expansion of a long, thin solid such as a metal rod. The main expansion will be an increase in its length l (**Figure 5.6**). This increase, designated Δl, depends on three factors:

1. The *original length l*. The longer the rod is to begin with, the greater the change in length will be.

2. The *change in temperature*, designated ΔT. The larger the increase in temperature, the greater the increase in length.

3. The *substance*. For example, the increase in length of an aluminum rod will be more than twice that of an identical iron rod under the same conditions.

Point 3 can be tested through experimentation. The expansions of different solids are measured under similar conditions. The results are used to assign a **coefficient of linear expansion** to each material. The value of this coefficient is a fixed parameter of each substance, much like mass density or specific gravity. It is represented by the Greek letter alpha, α. Because aluminum expands more than iron under the same circumstances, the coefficient of linear expansion of aluminum is larger than that of iron (**Figure 5.7**).

Based on this analysis, the equation that gives the change in length in terms of the change in temperature and the coefficient of linear expansion may be written as

$$\Delta l = \alpha\, l\, \Delta T$$

The change in length is proportional to the change in temperature and to the original length. The coefficient of linear expansion, α, is the constant of proportionality. **Table 5.2** lists α for several different solids. In the equation, the units of l and Δl must be the same. Therefore, the units of temperature, usually °C, must cancel those of α; that is to say, the units of α must be inverse temperature—for example, 1/°C.

Figure 5.7 Because aluminum expands more than iron, given the same increase in temperature, its coefficient of linear expansion is larger.

EXAMPLE 5.1 The center span of a steel bridge is 1,200 meters long on a winter day when the temperature is −5°C. How much longer is the span on a summer day when the temperature is 35°C?

SOLUTION First, the change in temperature is the final temperature minus the initial temperature:

$$\Delta T = 35 - (-5) = 35 + 5 = 40°C$$

From Table 5.2, the coefficient of linear expansion for steel is

$$\alpha = 12 \times 10^{-6}/°C$$

So,

$$\Delta l = \alpha\, l\, \Delta T$$
$$= (12 \times 10^{-6}/°C) \times 1{,}200 \text{ m} \times 40°C$$
$$= (12 \times 10^{-6}/°C) \times 48{,}000 \text{ m-}°C$$
$$= 576{,}000 \times 10^{-6} \text{ m}$$
$$= 0.576 \text{ m}$$

Table 5.2 Some Coefficients of Linear Expansion

Solid	α ($\times 10^{-6}/°C$)
Aluminum	25
Brass or bronze	19
Brick	9
Copper	17
Glass (plate)	9
Glass (Pyrex)	3
Ice	51
Iron or steel	12
Lead	29
Quartz (fused)	0.4
Silver	19

Figure 5.8 Expansion joints allow for the thermal expansion of bridges and other elevated roadways. Each end of a section of the roadway is connected to a metal "comb." The teeth of the two combs fit together and can move back and forth as the lengths of the sections change. For comparison, the thermometer is 30 centimeters (1 foot) wide.

The change in length in Example 5.1 is considerable and must be allowed for in the bridge design. Expansion joints, which act somewhat like loosely interlocking fingers, are placed in bridges, elevated roadways, and other such structures to allow thermal expansion to safely occur (**Figure 5.8**).

The equation for thermal expansion also works when the temperature decreases. When this occurs, the change in temperature is negative, so the change in length is also negative: the solid becomes shorter. Another way of saying this is that thermal expansion is a reversible process. If something becomes longer when heated, it will get shorter when cooled. Most of the phenomena discussed in this chapter are reversible. If something happens when the temperature increases, the reverse will happen when the temperature decreases.

> ▶ **Physics To Go 5.2**
>
> If you live near a bridge or elevated roadway on which you can walk safely, find an expansion joint. (Figure 5.8 shows one type.) Photograph it or sketch what it looks like on a hot afternoon and then sometime later when it is much cooler. What differences do you find?

The *bimetallic strip* is an ingenious and widely used application of thermal expansion. As its name implies, a bimetallic strip consists of two strips of different metals bonded to one another (**Figure 5.9**). The two metals have different coefficients of linear expansion, so they expand by different amounts when heated. The result is that the bimetallic strip bends—one way when heated and the other way when cooled. The greater the change in temperature, the greater the bending. For example, if brass and iron are used, the brass will expand and contract more than the iron will. The brass will be on the outside of the curve when the strip is hot and on the inside of the curve when it is cold.

Thermostats, thermometers, circuit breakers, lamp flashers, and choke-control mechanisms on small engines and older automobiles often contain a bimetallic strip that is curled into a spiral. The coil will either partly unwind or wind up more tightly when the temperature changes. To make a thermometer, a pointer is attached to one end of the spiral, and the other end is held fixed. As the temperature varies, the pointer moves over a scale that indicates the temperature (**Figure 5.10**). In thermostats, the movement of the end of the spiral is used to turn a switch on or off. The switch might activate a heater, an air conditioner, a fire alarm, or the cooling unit in a refrigerator.

As mentioned, thermal expansion occurs in all three dimensions. A bridge not only becomes longer but also wider and thicker as the temperature increases. The area of any surface increases, as does the volume of the solid. If a

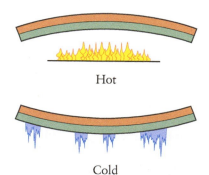

Figure 5.9 In this bimetallic strip, the metal composing the upper layer has a larger coefficient of thermal expansion than the metal composing the lower layer. When heated, the strip curves downward because the upper layer undergoes a greater change in length. When cooled, the strip curves upward.

Vern Ostdiek/University of Michigan

Figure 5.10 This thermometer uses a bimetallic strip in the shape of a spiral. Even small temperature changes cause the pointer attached to the outer end of the spiral to rotate noticeably.

solid has a hole in it, thermal expansion will make the hole bigger, contrary to most people's intuition (**Figure 5.11**). This is because thermal expansion causes every point in a solid to move *away* from every other point. (It is similar to what happens when a picture is enlarged.) A point on one side of a hole moves away from any point on the other side of the hole.

5.2b Liquids

The behavior of liquids is quite similar to that of solids. Because liquids do not hold a certain shape, it is best to consider the change in volume caused by thermal expansion. In general, liquids expand considerably more than solids. This means that when a container holding a liquid is heated, the level of the liquid will usually rise because the increase in volume of the liquid typically exceeds the increase in volume of the (solid) container. When a mercury thermometer is heated, the mercury in the tube and in the glass bulb at the bottom expands more than does the confining glass, so the level of mercury rises. If the glass expanded more than the mercury, the column would go down at higher temperatures instead of up.

At the beginning of this section, we implied that there are exceptions to the general rule that matter expands when heated. The most important example of this is water when it's near its freezing temperature. Above 4°C (39°F), water expands when heated like ordinary liquids. But between 0°C and 4°C, water actually contracts when heated and expands when cooled. The volume of a given amount of water at 3°C is *less* than the volume of the same water at 1°C. This anomaly accounts for the fact that lakes, ponds, and other bodies of water freeze on top first. As long as the average water temperature is above 4°C, the warmer (less dense) water is buoyed to the surface, while the cooler water

Figure 5.11 The hole becomes larger when the object is heated.

Figure 5.12 Above 4°C, the warmer water rises to the surface. Below 4°C, the cooler water rises to the surface.

Figure 5.13 If the pressure in a gas is kept constant, the volume that the gas occupies is proportional to the Kelvin temperature. A 10-percent increase in temperature will cause the volume of a balloon to increase by 10 percent.

settles to the bottom (**Figure 5.12**). As the air gets cooler in autumn, the water at the surface is cooled. When the average temperature of the water is below 4°, the cooler water (closer to freezing) is now less dense and rises to the surface. Consequently, the surface water freezes first because it is cooler than the water below, and it is in contact with the cold air.

Water is also unusual in that its density when in the solid phase (ice) is less than its density when in the liquid phase. Because of this, ice floats in water, whereas most solids (for example, candle wax) sink in their own liquid.

5.2c Gases: The Ideal Gas Law

The volume expansion of gases is larger than that of solids and liquids. Also, the amount of expansion does not vary with different gases (except at very low temperatures or very high pressures). Instead of relating the expansion to a change in temperature, it is simpler to state the relationship between the volume occupied by the gas and the temperature. In particular, as long as the pressure remains constant, the volume occupied by a given amount of gas is *proportional* to its temperature (in kelvins).

$$V \propto T \quad \text{(gas at fixed pressure)}$$

This relationship is often referred to as *Charles' law*, after Jacques Charles (1746–1823) who discovered it in around 1787. Note that the temperature **must** be in kelvins. When the temperature of a gas is increased by some percentage, the volume increases by the same percentage. The volume of a balloon at 303 K (86°F) is about 10 percent larger than the volume of the same balloon at 273 K (32°F) (**Figure 5.13**). By comparison, the volume change of typical solids is less than 1 percent over the same temperature range. In liquids, it is a maximum of 5 percent.

> ### ▶ Physics To Go 5.3
>
> For this you need a balloon and a pitcher or some other container. It's easiest if the container is transparent enough to see the level of water in it from the outside (**Figure 5.14**).
>
> 1. Inflate the balloon to a point where it will fit completely inside the pitcher but with some space around it and below it; tie it so it won't deflate. The balloon should not be fully inflated: the air inside must be able to expand.
> 2. Slowly fill the pitcher with cold tap water as you hold the balloon down, its top level with the pitcher's top, until the pitcher overflows. Hold the balloon completely submerged for a minute or so to allow the air inside the balloon to reach the temperature of the water, adding more water if necessary to keep the pitcher full.
> 3. Remove the balloon and mark the level of the water.
> 4. Repeat procedures 2 and 3 using warm or tolerably hot water.
> 5. Is the water level the same both times? If not, can you explain the difference? What happened to the air in the balloon when it was in warm water?

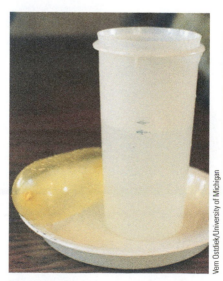

Figure 5.14 Illustrating thermal expansion. (See Physics to Go 5.3).

The *ideal gas law* expresses the interdependence of the pressure, volume, and temperature of a gas.

LAWS **Ideal Gas Law** In a gas with a density that is low enough that interactions between its constituent particles can be ignored, the pressure, volume, and temperature of the gas are related by the following equation:

$$pV = (\text{constant}) \; T$$

The constant depends on the quantity of gas present but not its specific type.

The utility and importance of this general relationship comes from the fact that it is applicable to most real gases under most conditions. Because the constant appearing in this equation only depends on the mass of gas present and not on what kind of gas it is, the equation can be used equally well for hydrogen, helium, oxygen, carbon dioxide, and other gases. Moreover, except under extreme circumstances, the ranges over which the parameters p, V, and T vary are limited enough in practice to ensure that the gas particles generally remain far enough apart that their interactions are negligible. Together, these facts make the ideal gas law useful for describing the thermodynamic properties of gases in the atmospheres of the Sun and stars, the air in a classroom, and the helium in a floating blimp.

EXAMPLE 5.2 Upon launch, the Goodyear blimp typically contains 5400 m³ of helium at an absolute pressure of 1.1×10^5 Pa (1.0 atm). The temperature of the gas is 280 K (7°C). If the blimp is flying at an altitude of 3,000 m (10,000 ft) where the pressure is 0.66×10^5 Pa (0.6 atm) and the temperature is 270 K (−3°C), what is the volume of helium while the blimp is airborne, assuming the mass of the gas remains constant?

SOLUTION Assuming the helium behaves as an ideal gas, and given that the amount of gas does not change, then the ideal gas law may be rewritten as

$$\frac{pV}{T} = \text{constant}$$

Applying this relationship to the initial (i) and final (f) conditions as prescribed

$$\left(\frac{p_i}{T_i}\right) V_i = \left(\frac{p_f}{T_f}\right) V_f$$

Solving for the unknown final volume and inserting the values yields

$$V_f = \left(\frac{p_i}{p_f}\right)\left(\frac{T_f}{T_i}\right) V_i$$

$$= \left(\frac{1.0 \text{ atm}}{0.6 \text{ atm}}\right)\left(\frac{270 \text{ K}}{280 \text{ K}}\right)(5400 \text{ m}^3)$$

$$= (1.67)(0.96)(5400 \text{ m}^3)$$

$$= 8660 \text{ m}^3$$

The gas expands in volume by nearly 60 percent due primarily to the lower pressure at higher altitudes. Blimp manufacturers must factor this expansion into their designs to prevent structural damage to the airship during flight.

A given amount of gas then can have any combination of pressure, volume, and temperature so long as the three values satisfy the ideal gas law. For example, if the volume of the gas is fixed, the pressure will increase whenever the temperature increases. In other words, the pressure is proportional to the temperature (in kelvins) as long as the volume stays constant. In the earlier example of heating air in a pressure cooker, the volume of the air inside would

Figure 5.15 A hot-air balloon makes use of thermal expansion.

Elsa Hoffmann/Shutterstock.com

be nearly constant. (This is because the volume expansion of the metal is quite small compared to that of a gas.) So the pressure inside would increase proportionally with the temperature.

Regardless of which phase of matter is involved, changing the temperature of a substance will not change its mass or its weight. Because thermal expansion causes the volume to increase while the mass and weight stay the same however, the mass density and weight density decrease. The mass density of a hot piece of iron is slightly less than the mass density of the same piece when it is cold. The mass density of the balloon referred to earlier is about 10 percent less when its temperature is 303 K than when its temperature is 273 K.

The reduction in density of a gas at constant pressure because of heating is exploited in hot-air balloons. The air in the balloon, which is basically a large bag with an opening at the bottom, is heated with a burner (**Figure 5.15**). The pressure inside remains equal to the atmospheric pressure because of the opening. Consequently, the air inside the balloon expands and its density decreases. The balloon can float in the air because it is filled with a gas (hot air) that has a smaller density than the surrounding fluid (cooler air). As the air inside cools, the balloon will sink toward Earth until the burner heats the air again.

Learning Check

1. A key is placed in a hot oven and warmed to 100°C. As a result
   (a) the hole in the key becomes smaller.
   (b) the key is bent into an arc.
   (c) the key becomes longer.
   (d) All of the above.
2. (True or False.) A bimetallic strip is made out of two metals that have different coefficients of linear expansion.
3. What odd thing happens to water when its temperature is increased from 1°C to 2°C?

4. Heating the air in a hot-air balloon reduces the air's _____ .
5. (True or False) Under constant pressure conditions, the ideal gas law predicts that the volume occupied by a given amount of gas is inversely proportional to its Kelvin temperature.

ANSWERS: 1. (c) **2.** True **3.** It contracts (volume decreases) **4.** density **5.** False

5.3 The First Law of Thermodynamics

We have discussed what temperature is and how changes in temperature can affect the physical properties of matter—density in particular. The next item to consider is how the temperature of matter is changed. This will lead us back to the concept of energy.

5.3a Heat and Internal Energy

There are two general ways to increase the temperature of a substance:

1. By exposing it to something that has a higher temperature.
2. By doing work on it in certain ways.

The first way is very familiar to you. When you heat something on a stove, warm your hands over a heater, or feel the Sun warm your face, the temperature increase is caused by exposure to something that has a higher temperature: the stove, the heater, or the Sun. (More on this in Section 5.4.) As mentioned before, the atoms and molecules in the substance being warmed gain kinetic energy from those in the hotter substance, albeit sometimes through the mediating action of other particles. The temperature of a substance will rise only if its atoms and molecules gain kinetic energy.

Friction is an effective means of raising the temperature of a substance by the second way (doing work on it). When something is heated by kinetic friction, work is done on it. This was discussed in more detail at the end of Section 3.4.

> ### Physics To Go 5.4
>
> For this, you need a bicycle tire pump and a flat bicycle tire or other deflated object such as a large beach ball or air mattress that you can inflate. First touch the lower end of the pump (or the metal fitting on the hose if it has one) and notice how warm or cold it feels. Pump up the tire or other object to the recommended pressure and then feel the lower end (or metal fitting) again. What has changed?

Another example of raising the temperature of a substance by doing work on it is the compression of a gas. When a gas is squeezed (quickly) into a smaller volume, its temperature increases (**Figure 5.16**). In diesel engines, air is compressed so much that its temperature is raised above the combustion temperature of diesel fuel. When the fuel is injected into the compressed hot air, it ignites.

Computation of the temperature rise of a gas when it is compressed a certain amount is beyond the scope of this text. But we can illustrate the type of heating that occurs by stating the results of one example: if air at 27°C is

Higher temperature

Figure 5.16 Work is done on a gas when it is compressed in a cylinder. This work causes the temperature of the gas to increase.

Figure 5.17 The air in a diesel engine has an initial temperature of 27°C. It is compressed by the piston until it occupies one-twentieth of its original volume. This raises the temperature to 721°C. (Compression not to scale.)

compressed to one-twentieth of its initial volume in a diesel engine, its temperature will increase to more than 700°C (**Figure 5.17**).

Both processes can be reversed to cause the temperature of a substance to decrease. The cooler air in a refrigerator lowers the temperature of a pitcher of tea. Air escaping from a tire is cooled as it expands.

Temperature depends on the average kinetic energy of atoms and molecules. For matter to undergo a change in temperature, its atoms and molecules must gain energy (increased temperature) or lose energy (decreased temperature). In gases, the constituent particles have kinetic energy only. All of the energy given to the atoms and molecules acts to increase the temperature of the gas. Things are different in solids and liquids. The atoms and molecules have kinetic energy *and* potential energy because they are bound to each other and oscillate. Energy given to the particles goes to increase both their *PE* and their *KE*. The concept of **internal energy,** which we introduced in Section 3.4, incorporates both forms of energy.

> **DEFINITION** **Internal Energy** The sum of the kinetic energies and potential energies of all the atoms and molecules in a substance.

Internal energy is represented by *U.* Its units of measure, the joule, calorie or ft-lb, are the same as those of energy and work.

In gases, the internal energy is the total of the kinetic energies only: the atoms and molecules do not have potential energy. (Gravitational potential energy is not included in internal energy.) In solids and liquids, both the kinetic energy and the potential energy of the particles contribute to the internal energy.

> **DEFINITION** **Heat** The form of energy that is transferred between two substances because they have different temperatures.

As the temperature of a substance rises, its internal energy increases. If this is accomplished by exposure to a hotter substance, we say that **heat** has flowed from the hotter substance into the cooler substance.

Heat is symbolized by *Q.* Its units are the same as those of work and energy. Traditionally, the calorie, kilocalorie (also written Calorie), and British thermal unit (Btu) were used exclusively as units of heat. The joule and the foot-pound were used for work and the other forms of energy. Now the joule is becoming the standard unit for heat as well.

The internal energy of something can increase when heat flows into it and decrease when heat flows out of it (**Figure 5.18**). Heat flow is the spontaneous transfer of energy from a hotter (higher temperature) substance to a cooler (lower temperature) substance. In this respect, heat is much like work in mechanics. As work is being done,

iStockphoto.com/KateLeigh; iStockphoto.com/KateLeigh; iStockphoto.com/ Givaga; iStockphoto.com/PLAINVIEW; iStockphoto.com/khorzhevska

Figure 5.18 Heat flows into the water from the burner, increasing the water's internal energy. Heat flows out of the water into the ice, lowering the water's internal energy.

energy is transferred from one thing to another or is transformed from one form to another. A hot pizza does not contain heat, just as a battery and a wound-up spring do not contain work. The battery and the spring contain potential energy, which means they can be used to do work (given the appropriate motor or other mechanism). In the same manner, the hot pizza contains internal energy, and it can therefore transfer heat to something that is cooler. *Heat and work are energy in transition whereas internal energy and potential energy are stored energy.*

5.3b The First Law of Thermodynamics

Two substances with the same temperature are said to be in *thermal equilibrium.* No heat transfers between them. An abandoned cup of hot coffee cools off as heat is transferred from it to the surrounding air. Once the coffee and cup reach the same temperature as the air, the transfer stops and they are in thermal equilibrium.

The first law of thermodynamics is a formal summary of the preceding discussion.

> **LAWS** **First Law of Thermodynamics** The change in internal energy of a substance equals the work done on it plus the heat transferred to it:
>
> $$\Delta U = \text{work} + Q$$

The work referred to in this law must be the type that transfers energy directly to atoms and molecules, such as compressing a gas. Work is done on an object when it is lifted, but this does not affect its internal energy—just its gravitational potential energy.

The work is positive if it is done *on* the substance and negative if the substance does work on something else. When gas in a cylinder is compressed by a piston, the work that is done is positive. If the piston is then released, the gas will expand and push the piston out. In this case, the gas does work on the piston, and the work is negative (**Figure 5.19**). Similarly, Q is positive when heat flows into the substance and negative when heat flows out of it. If you place a brick in a hot oven, heat flows into it, and Q is positive. Placing the brick in a freezer would result in a flow of heat out of it and a negative Q.

The first law of thermodynamics is nothing more than a restatement of the law of conservation of energy as it applies to thermodynamic systems. Work done on, or heat transferred to, a substance is "stored" in it as internal energy. In addition to its theoretical significance, the first law of thermodynamics is an important tool used in the analysis of things such as internal combustion engines and air conditioners.

Work done on the gas   Work done by the gas

Figure 5.19 When the piston compresses the gas, the work done on the gas is positive. When the gas pushes the piston back, it does work on the piston. In this case, the work done on the gas is negative.

EXAMPLE 5.3 During a tournament match, a racquetball player loses 6.2×10^5 J of internal energy. In the process, the player transfers 3.6×10^5 J of heat to his surroundings. How much work has the player done during the competition?

SOLUTION We apply the first law of thermodynamics to this situation:

$$\Delta U = \text{work} + Q$$

The system here is the athlete, so ΔU and Q are both negative: the player *loses* internal energy and heat flows *from* the athlete to his surroundings. Thus,

$$-6.2 \times 10^5\,\text{J} = \text{work} - 3.2 \times 10^5\,\text{J}$$
$$\text{work} = -6.2 \times 10^5\,\text{J} + 3.2 \times 10^5\,\text{J}$$
$$= -2.6 \times 10^5\,\text{J}$$

The work is negative because the athlete has done work *on* objects in the environment, for example, by exerting forces to accelerate the racquetball.

This section began with a consideration of how the temperature of matter is changed. How does internal energy fit in with this? Temperature depends on the kinetic energies of the atoms and molecules in a substance. Because these kinetic energies are part of the internal energy of the substance, raising the temperature of something increases its internal energy because it raises the kinetic energies of the atoms and molecules.

Perhaps you are wondering why the concept of internal energy is used at all, because temperature is determined only by the kinetic energies of particles. Internal energy is useful when considering phase transitions. When water boils on a stove, for example, heat is transferred to it, but the temperature stays the same. This means that the average kinetic energy of the water molecules also stays the same. The heat transferred during a change in phase increases the internal energy by increasing only the potential energies of the molecules. The energy is used to break the bonds between the water molecules and to free them. We will take a closer look at this in Section 5.6.

Concept Map 5.1 summarizes the concepts presented in this section.

▌CONCEPT MAP 5.1

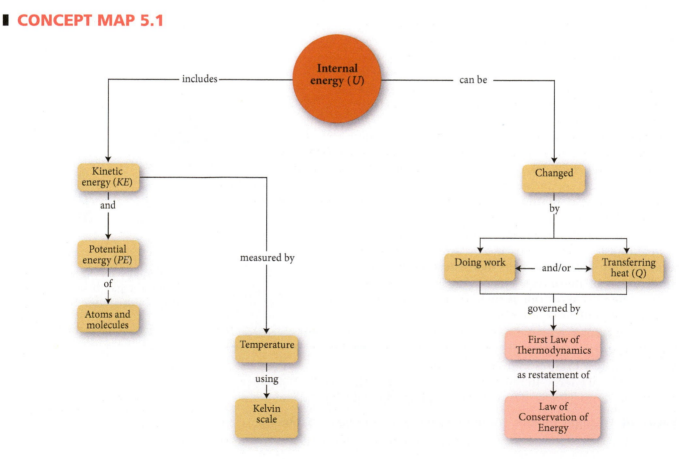

Learning Check

1. (True or False.) The internal energy of a solid body equals the sum of the kinetic energies of all of its constituent atoms and molecules.
2. _____ is transferred from a substance if it is exposed to something that has a lower temperature.
3. The internal energy of air can be increased by
   (a) compressing it.
   (b) allowing heat to flow out of it.
   (c) lifting it.
   (d) allowing it to expand.
4. The first law of thermodynamics is a special version of the law of conservation of _____ .

ANSWERS: 1. False **2.** Heat **3.** (a) **4.** energy

5.4 Heat Transfer

Transferring heat is the more common of the two ways to change the temperature of something. Heat transfer occurs whenever there is a temperature difference between two substances or between parts of the same substance. In this section, we discuss the three different mechanisms for heat transfer: **conduction, convection,** and **radiation.**

> **DEFINITION** **Conduction** The transfer of heat between atoms and molecules in direct contact.
>
> **Convection** The transfer of heat by buoyant mixing in a fluid.
>
> **Radiation** The transfer of heat by way of electromagnetic waves.

5.4a Conduction

Conduction occurs when a pan is placed on a hot stove, when you put your hands into cold water, and when an ice cube comes into contact with warm air. The atoms and molecules in the warmer substance transfer some of their energy directly to the particles in the cooler substance. In these examples, the conduction takes place across the boundary between the two substances, where the atoms and molecules collide with each other. Conduction also is responsible for the transfer of heat from one part of a solid to another part. (Conduction also happens in fluids but is not as important as convection.) Even though only the bottom of a pan is in contact with a hot burner, the heat flows through the metal and soon raises the temperature of all parts of the pan (**Figure 5.20**). As the atoms and molecules on the bottom are heated, their constant jostling passes some of their increased kinetic energy to neighboring atoms and molecules.

The ease with which heat flows within matter varies greatly. Materials through which heat moves slowly are called *thermal insulators.* Wool, Styrofoam, and bundles of fiberglass strands are all good insulators because they contain large amounts of trapped air or other gases. Conduction is poor within a gas because the atoms and molecules are not in constant contact with each other. Diamond and metals such as iron and copper are *thermal conductors:* heat flows readily through them. Concrete, stone, wood, and glass are between the two extremes. A vacuum completely prevents conduction because no atoms or molecules are present to pass along the energy.

Metals are good conductors of heat for the same reason that they are good conductors of electricity. Some of the electrons in the atoms in metals are free to move about from one atom to the next. The motion of these "conduction electrons" constitutes an electric current (Chapter 7). These electrons can also carry internal energy from the warmer part of a metal object to the cooler parts.

Figure 5.20 Heat is conducted to the pan from the flame in contact with it. Conduction also takes place within the pan: heat flows from the hot bottom up the sides and into the handle.

Figure 5.21 A rug feels warmer than a bare floor because it reduces the conduction of heat from your feet.

The conduction of heat within matter is similar to the flow of a fluid, except that nothing material moves from one place to another. The rate at which heat flows from a hot part of an object to a cold part depends on several things—the difference in temperature between the two places, the distance between them, the cross-sectional area through which the heat flows, and how good a thermal conductor the substance is.

Heat conduction is important in our daily lives. In cold weather, we wear clothes that slow the conduction of heat from our warm bodies to the cold air. Handles on metal pans are often made of wood or plastic to reduce the conduction of heat from the burner to your hand. One reason carpets and rugs are used is that they feel warm when you step on them with bare feet. A rug is no warmer than the bare floor next to it, but it is a poorer thermal conductor. When you step on the rug, very little heat is conducted away from your feet, so they stay warm. When you step on bare wood or tile, materials that are better thermal conductors, your feet are cooled more because heat is removed from them more rapidly (**Figure 5.21**).

5.4b Convection

Convection is the dominant mode of heat transfer within fluids. Whenever part of a fluid is heated, its thermal expansion causes its density to decrease, so it rises. (Water below 4°C is an exception.) The result is a natural mixing of the fluid. Conduction can then occur between the warmer fluid and the cooler fluid around it.

A room with a wood stove or heater is warmed by convection. The air that is heated is less dense and rises to the ceiling; cooler, denser air near the floor moves toward the heat source to replace the rising air. The result is a natural circulation of air along the ceiling, floor, and walls of the room (**Figure 5.22**). The warmed, rising air cools when it contacts the ceiling and walls and then sinks to the floor. The same type of circulation can occur in heated aquariums and swimming pools.

You may have heard the statement "heat rises." This is *not* physically correct. Heat is not something material that can rise or fall. "Hot air rises" or "heated fluids rise" are better statements for conveying the idea. Remember also that it is the denser surrounding fluid that pushes the warmer, less-dense fluid upward. (Remember: Gases exert pressure [a *push*] on their surroundings; they cannot "pull" on objects in contact with them.)

Mechanical mixing of a fluid that causes heat transfer is an example of what is called *forced* convection. Stirring cool cream into hot coffee with a spoon is forced convection—the mixing is not caused by thermal buoyancy. Another example is

Figure 5.22 Convection causes a natural circulation of air in a room heated with a wood stove. Air heated by contact with the stove rises to the ceiling. Air cooled by contact with an outside wall sinks to the floor. This results in lateral movement of air along the ceiling and the floor.

hot or cold air being blown around the interior of a building or vehicle, causing heat transfer by the mixing of the warmer air with the cooler air.

Convection in Earth's atmosphere is a major cause of clouds, wind, thunderstorms, and other meteorological phenomena. White, puffy, cumulus clouds are formed when warm air rises into cooler air above and the water vapor in it condenses into droplets.

Sea breezes—steady winds blowing into shore along coasts—are caused by convection. Sunshine raises the temperature of the land more than the sea, so the air over the ground is heated and rises upward. This reduces the air pressure over the land, so the higher pressure over the sea forces air to move inland (**Figure 5.23**, top). At night, the land cools more than the sea, so the process is reversed, and a *land breeze* is produced. Air over the cooler ground sinks and forces air to move out to sea (Figure 5.23, bottom). (Note: Only the surface of the ground is heated by the Sun because the soil is a poor thermal conductor. Solar energy transferred to the sea is quickly spread deeper below the surface by currents and convective mixing, reducing the temperature increase at the surface. We will see in Section 5.5 that water also requires a great deal of heat transfer to raise its temperature.)

Large-scale convection takes place within the ocean itself. Water near the equator is heated by the Sun, rises to near the ocean's surface, and flows towards the poles. There, the water cools and sinks deeper and then flows back towards the equator. (This simple flow pattern is complicated by a number of factors, including the rotation of Earth, but is correct in its broad outline.)

5.4c Radiation

Radiation is the transfer of heat via electromagnetic waves. We've all felt the warmth of the Sun on our faces on a calm day and the heat radiating from a hot fire. This is heat radiation—a type of wave related to radio waves and x-rays. (We will discuss electromagnetic waves in more detail in Chapters 8 and 10.) This is the only one of the three types of heat transfer that can operate through a vacuum. The Sun's radiation passes through 150 million kilometers (93 million miles) of nearly empty space and heats Earth and everything on it. Without the Sun's radiation, Earth would be a cold, lifeless rock. (More on this in Section 8.7.)

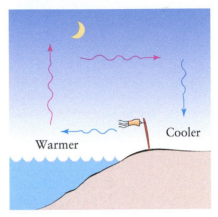

Figure 5.23 (top) Sea breeze. Warmed air rises from the heated land, causing cooler air to be driven in from the sea. (bottom) The land cools at night, so the flow is reversed.

> ## ▶ Physics To Go 5.5
>
> 1. Hold your hand a few inches from the side of a bare lightbulb.
> 2. Remove your hand, let it cool off a bit, then place it the same distance directly above the bulb (**Figure 5.24**). What is different about the two circumstances? Why?

Infrared heat lamps warm things by emitting radiation. From a distance, you can feel the heat from a camp fire or other hot source because the radiation from it warms your hands and face.

You might think of the radiation as a "vehicle" or "carrier" of internal energy. Internal energy of atoms and molecules is converted into electromagnetic energy—radiation. The radiation then carries the energy through space until it is absorbed by something. When absorbed, the energy in the radiation is converted into internal energy of the atoms and molecules of the absorbing substance.

Everything emits electromagnetic radiation: the Sun, Earth, your body, a textbook, your computer keyboard, etc. The amount and type of radiation depend on the temperature of the emitter. The hotter something is, the more electromagnetic radiation it emits. Things below about 800°F (430°C) emit mostly infrared light, which we cannot see. Hotter substances emit *more* infrared light and also visible light. That is why we can see things in the dark that are "red hot" or "white hot." Things that are even hotter, like the Sun at 10,000°F, not only emit more infrared and visible light but also emit ultraviolet light. We will take a closer look at heat radiation in Chapter 8.

Figure 5.24 A glowing lightbulb transfers heat to your hand by both radiation and convection.

Vern Ostdiek/University of Michigan

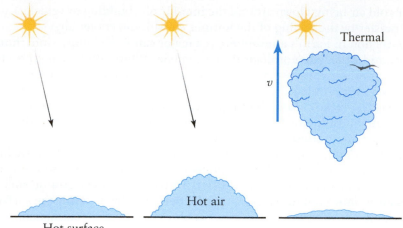

Figure 5.25 Thermals are rising "bubbles" of air formed on surfaces heated by the Sun. They give a free upward ride to soaring birds and aircraft. Finding a thermal and then staying in it are a bit difficult because thermals are invisible.

Everything around you is constantly and currently emitting radiation and absorbing the radiation emitted by other things. How does this lead to a net transfer of heat? Emission of radiation cools an object, and absorption of radiation warms it. If something absorbs radiation faster than it emits radiation, it is heated. Your face is warmed by the Sun because it absorbs more radiant energy than it emits.

5.4d Combinations

All three mechanisms of heat transfer are involved in soaring. The Sun warms Earth via radiation. Air in contact with hot ground is heated by conduction and expands. If conditions are favorable, the hot air will form an invisible "bubble" that breaks free from the surface and rises upward into the air, causing convection. (This is similar to the formation of steam bubbles on the bottom of a pan of boiling water.) These bubbles of rising heated air are called *thermals* (**Figure 5.25**). Hang-glider and sailplane pilots and soaring birds such as eagles, hawks, and vultures seek out thermals and circle around in them. The upward speed of a typical thermal is around 5 m/s (11 mph), so they provide an easy, free ride upward.

The different mechanisms of heat transfer become important in reducing heating and cooling costs in buildings. The more energy that flows out of a building in cold weather or into a building in warm weather, the more it costs to heat or cool the building. Reducing this flow of heat saves money (**Figure 5.26**). Putting insulation in the walls

Figure 5.26 Heating and cooling costs for a home can be lowered by keeping in mind the mechanisms of heat transfer.

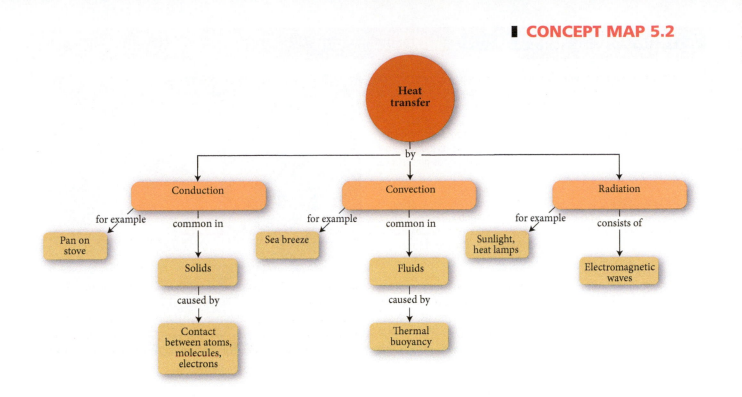

and ceilings reduces conduction. Using thicker insulation in the ceiling counteracts one effect of convection: the warmest air in the room is at the ceiling, so that is where heat would be transferred most rapidly out of the room during the winter. Using window shades, blinds, and canopies to keep direct sunlight out during the summer reduces heat transfer by radiation.

Thermos bottles are also designed to limit the flow of heat into or out of their contents. They use a near-vacuum between their inner and outer walls that almost eliminates heat flow resulting from conduction. The inner glass chamber is coated with silver or aluminum to reflect radiation.

The basic concepts of heat transfer are summarized in Concept Map 5.2.

■ Learning Check

1. A material that conducts heat slowly is called a
   _____.
2. (True or False.) Convection is the only mechanism of heat transfer than can take place in a vacuum.
3. For each of the following situations, indicate which mechanism of heat transfer is the dominant one.
   (a) Clothes being dried by hanging them well above a campfire.
   (b) A piece of chocolate melting in your mouth.

(c) The surface of the Moon cooling off after the Sun has set.
(d) After a fire alarm sounds, touching a door knob to see if there is fire on the other side of the door.
4. (True or False.) A liquid cannot be heated simultaneously by conduction, convection, and radiation.

ANSWERS: 1. thermal insulator **2.** False
3. (a) convection (b) conduction (c) radiation
(d) conduction **4.** False

As discussed in the text, energy can be transported from one location to another by three mechanisms, conduction, convection, and radiation. Usually, the bulk of the energy in any system is transferred by the mechanism that is the most efficient. In a star like our own Sun, energy is moved primarily by radiation throughout most of its interior but by convection during the last 30 percent or so of its journey from the Sun's core. The specific mechanism at work depends on the local physical conditions in the star.

The Sun is an average star, one of more than 10 billion in the Milky Way galaxy. It is in what is called the *main sequence* phase of its lifetime. We might call this the Sun's "adulthood." In this period, energy is produced at the center, or core, of the star by nuclear fusion reactions involving hydrogen nuclei (protons) (see Chapter 11 for more on fusion). This energy is transported from the core out to regions just beneath the solar surface—a distance of some 70 percent of the Sun's radius—by radiation, that is, in the form of electromagnetic waves, working their way outward until they reach what is known as the *convection zone*. This process is exceedingly slow. The electromagnetic waves travel less than 1 centimeter before they undergo collisions with protons, helium nuclei, and free electrons in the solar interior that randomly scatter them, altering their directions of motion. It requires about a million years for energy produced in the Sun's core to reach its outer parts.

When the energy arrives at the outer 30 percent of the Sun's interior, radiation becomes less efficient as a transport mechanism than convection. The reason for this has to do with what is called the *opacity* of the gas. The opacity is a measure of how effectively the gas absorbs electromagnetic waves. Although the gas in the solar interior is fairly opaque, radiation is still more efficient than convection at moving the energy outward. However, in the very outermost parts of the Sun, the gas temperatures are low enough that hydrogen atoms can exist. The formation of these atoms suddenly makes the gas extremely opaque.

The energy is effectively "dammed up," creating a locally overheated region beneath the Sun's surface. This condition starts the upward movement of the heated, low-density gas and its subsequent mixing with the cooler, overlying fluid. A convection zone develops. From here, the energy finally bursts forth into space at the top of the solar atmosphere.

Photographs of the solar surface taken from high-altitude balloons, rockets, and orbiting satellites permit us to see the top of this convection zone (**Figure 5.27**). The rolling, bubbling gases can be clearly seen in the form of light and dark convection cells. Such mottling has been variously described as "granulation" or "salt-and-pepper" patterns or as "oatmeal-like" in texture. The brighter areas are warmer, updrafting material. The entire pattern continually changes as the cells rise and fall on time scales of about 10 minutes. Each bubble has an irregular shape and varies in size from 800 to 1,200 kilometers across—nearly one-quarter the size of the United States! These are no small cauldrons of boiling brew!

Conduction does not play a significant role in the transport of energy in stars such as the Sun because the Sun's material is gaseous and not extremely dense. Conduction does become an important mechanism in the terminal stages of most stars' lives. In this final phase of their evolution, all but the most massive stars end up as small (Earth-sized or less), dense (more than a billion times as dense as water) objects called *white dwarfs*. Because of the compactness of these common end points of stellar evolution, their properties are in some ways similar to those of ordinary solid matter. It may not be too surprising then, given our discussion of heat flow in solids, to find that conduction is an important process

(a)

NASA

(b)

Alan Title, Lockheed Solar Observatory/NASA

Figure 5.27 (a) High-resolution image of the surface of the Sun showing a pattern of convection cells called *granulation*. (b) Time variations in the pattern of solar granulation. These images were taken at two-minute intervals and show how the granulation pattern on the Sun changes with time. Note particularly the explosive behavior of the central granule. For scale, the upper left panel shows the size of the state of Texas.

controlling their structure and the time it takes for these stars to cool off and finally go dark. In fact, conduction in these terminal stars is so efficient at transporting energy that their interiors have nearly uniform temperatures throughout. Were it not for the thin insulating blanket of normal atmospheric material covering them, they would quickly lose all their residual heat and become blackened stellar cinders.

At this stage, you might ask how we know that stars such as the Sun transport energy in the manner we've described. Up until recently, *direct* evidence to support the accuracy of this picture was hard to come by. However, with the *Global Oscillations Network Group* (GONG) project and the *Solar and Heliospheric Observatory* (SOHO) spacecraft in the late 1990s, the science of *helioseismology*, the study of surface vibrations on the Sun to understand the nature of the solar interior, reached maturity and provided just the data needed to check our models. It turns out that the roiling motion of the granulation cells near the Sun's surface generates low-amplitude sound waves that propagate through our star. These vibrations can be detected as tiny shifts in the wavelengths of the light emitted by the Sun. (For more on sound waves and wavelength shifts—that is, the Doppler effect—see Chapter 6.) Comparisons between the observed oscillations and the predictions of our best models show extraordinarily good agreement (within ~0.2%!) and place the boundary between the radiative and convection zones at 71.3 percent of the Sun's radius.

■ **QUESTIONS**
1. In the Sun, which energy transport mechanism is most efficient in the deep inner regions? In the outer near-surface regions?
2. What is helioseismology? How have helioseismic data helped confirm our understanding of how energy is moved from the center to the surface of solar-type stars?

5.5 Specific Heat Capacity

Transferring heat to a substance or doing work on it increases its internal energy. In this section, we describe how the temperature of the substance is changed as a result. To simplify matters, we assume that no phase transitions take place.

We might state the topic now under consideration in the form of a question: to increase the temperature of a substance by some amount ΔT, what quantity of heat Q must be transferred to it? (We could just as well ask how much work must be done on it.)

The amount of heat needed is proportional to the *temperature increase*. It takes twice as much heat to raise the temperature 20°C as it does to raise it 10°C. So,

$$Q \propto \Delta T$$

The amount of heat needed also depends on the *quantity* (mass) of the substance to which the heat is transferred. For a given increase in temperature, 2 kilograms of water will require twice as much heat transfer as 1 kilogram. So,

$$Q \propto m$$

The required heat transfer also depends on the *substance*. It takes more heat to raise the temperature of water 1°C than it does to raise the temperature of an equal mass of iron 1°C. As with thermal expansion (Section 5.2), a characteristic value can be assigned to each substance indicating the relative amount of heat needed to raise its temperature. This number, called the *specific heat capacity C*, is determined experimentally for each substance. The larger the specific heat capacity of a substance, the greater the amount of heat transfer needed to raise its temperature by a given amount. Thus,

$$Q \propto C$$

We can combine the three proportionalities into the following equation:

$$Q = C\,m\,\Delta T$$

The amount of heat required equals the specific heat capacity of the substance times the mass of the substance times the temperature increase. The SI unit of specific heat capacity is the joule per kilogram-degree Celsius (J/kg-°C). If the specific heat capacity of a substance is 1,000 J/kg-°C, then it takes 1,000 joules of energy to raise the temperature of 1 kilogram of that substance 1°C. (Because the kelvin is the same size temperature unit as the °C, it also can be used to specify the heat capacity of a substance.) **Table 5.3** lists the specific heat capacities of some common substances.

Table 5.3 Some Specific Heat Capacities

Substance	C (J/kg-°C)
Solids	
Aluminum	890
Concrete	670
Copper	390
Ice	2,000
Iron and steel	460
Lead	130
Silver	230
Liquids	
Gasoline	2,100
Mercury	140
Seawater	3,900
Water (pure)	4,180

EXAMPLE 5.4 Let's compute how much energy it takes to make a cup of coffee or tea. Eight ounces of water has a mass of about 0.22 kilograms. How much heat must be transferred to the water to raise its temperature from 20°C to the boiling point, 100°C?

SOLUTION The change in temperature is

$$\Delta T = 100 - 20 = 80°C$$

The specific heat capacity C of water is 4,180 J/kg-°C (from Table 5.3). Therefore,

$$Q = C\, m\, \Delta T$$
$$= 4{,}180 \text{ J/kg-°C} \times 0.22 \text{ kg} \times 80°C$$
$$= \boxed{73{,}600 \text{ J}}$$

Figure 5.28 The energy needed to heat five cups of water to boiling from room temperature is about the same as the energy needed to accelerate a small car to 60 mph.

It takes an enormous amount of energy to heat water. You may recall that in Example 3.6 we computed the kinetic energy of a small car traveling at highway speed. The answer was 364,500 joules. This much energy is only enough to bring about five cups of water to the boiling point from 20°C (**Figure 5.28**). Usually, a relatively large amount of mechanical energy does not produce a large temperature change when it is converted into thermal energy. Example 5.5 shows this another way.

EXAMPLE 5.5 A 5-kilogram concrete block falls to the ground from a height of 10 meters. If all of its original potential energy goes to heat the block when it hits the ground, what is its change in temperature?

SOLUTION There are two energy conversions. The block's gravitational potential energy, $PE = mgd$, is converted into kinetic energy as it falls. When it hits the ground, the kinetic energy is converted into internal energy in the inelastic collision (**Figure 5.29**). Actually, this internal energy would be shared between the block and the ground, but we assume that it all goes to the block. So the equivalent amount of heat transferred to the block equals the original potential energy.

$$Q = PE = mgd = 5 \text{ kg} \times 9.8 \text{ m/s}^2 \times 10 \text{ m}$$
$$= \boxed{490 \text{ J}}$$

Figure 5.29 A concrete block falls 10 meters to the ground. If all of the block's energy is converted into heat that is absorbed by the block only, its temperature is raised 0.15°C.

The increase in temperature, ΔT, of the block is

$$Q = C\,m\,\Delta T$$

$$490\,\text{J} = 670\,\text{J/kg-°C} \times 5\,\text{kg} \times \Delta T$$

$$490\,\text{J} = 3{,}350\,\text{J/°C} \times \Delta T$$

$$\frac{490\,\text{J}}{3{,}350\,\text{J/°C}} = \Delta T$$

$$\Delta T = \boxed{0.15°\text{C}}$$

At the end of Section 3.4, we discussed how mechanical energy is often converted into internal energy. At that time you may have wondered why you usually don't notice a temperature increase when you slide something across a floor or drop a book on a table. The quantities of energy or work that we typically deal with do not go far in changing the temperatures of the objects involved.

Let's consider one last example in which the amount of mechanical energy is so large that considerable heating does take place.

EXAMPLE 5.6 A satellite in low Earth orbit experiences a slight but continuous air resistance and eventually reenters Earth's atmosphere. As it moves downward through the increasingly dense air, the frictional force of air resistance converts the satellite's kinetic energy into internal energy. If the satellite is mostly aluminum and all of its kinetic energy is converted into internal energy, what would be its temperature increase?

SOLUTION Again, some of the heat is transferred to the air and some to the satellite. Just to get some idea of the potential heating, we assume that all of the heat from the friction flows into the satellite. We do not need to know the mass of the satellite, because m divides out, as we will see.

The heat transferred to the satellite equals its original kinetic energy. In Section 2.7 we computed the speed of a satellite in low Earth orbit— 7,900 m/s. So,

$$Q = KE = \frac{1}{2}\,mv^2 = \frac{1}{2} \times m \times (7{,}900\,\text{m/s})^2$$

$$= (31{,}200{,}000\,\text{J/kg}) \times m$$

The increase in temperature caused by this much heat is

$$Q = (31{,}200{,}000\,\text{J/kg}) \times m = C\,m\,\Delta T$$

$$31{,}200{,}000\,\text{J/kg} = 890\,\text{J/kg-°C} \times \Delta T$$

$$\frac{31{,}200{,}000\,\text{J/kg}}{890\,\text{J/kg-°C}} = \Delta T$$

$$\Delta T = \boxed{35{,}000°\text{C}}$$

Of course, its temperature could not actually increase this much: the satellite would start to melt. Even if 90 percent of the heat went to the air, the remaining 10 percent would still be enough to melt the satellite. (This would make $\Delta T = 3{,}500°\text{C}$. The melting point of aluminum from Table 5.1 is 660 degrees Celsius.)

The purpose of Example 5.6 was to show you why unprotected satellites and meteoroids usually disintegrate when they enter Earth's atmosphere. The friction from the air transforms their kinetic energy into enough internal energy to raise their surface temperature by several thousand degrees. The objects simply start melting on the outside. Meteors can be seen at night because of their extreme temperatures: they leave behind a trail of hot, glowing air and meteoroid fragments (Figure 3.24).

Before the middle of the 19th century, the concepts of heat and mechanical energy were not connected to each other. It was thought that heat dealt with changes in temperature only and had nothing directly to do with mechanical energy. (The explanations of heating caused by friction were a bit strange as a result.) Water once again was used as a basis for defining a unit of measure: the calorie was defined to be the amount of heat needed to raise the temperature of 1 gram of water 1°C. (The "food calorie," denoted with a capital C, that dieters count is actually the kilocalorie. It is used to measure the energy content of foods: 1 Calorie = 1000 calories). Similarly, the British thermal unit (Btu) was defined to be the amount of heat needed to raise the temperature of 1 pound of water 1°F. This made the specific heat capacity of water equal to 1 cal/g-°C = 1 Btu/lb-°F.

In 1843, James Joule announced the results of experiments in which a measurable amount of mechanical energy was used to raise the temperature of water by stirring it. The amount of internal energy given to the water equaled the mechanical energy expended. The result allowed Joule to calculate what was called the "mechanical equivalent of heat"—the relationship between the unit of energy and the unit of heat:

$$1 \text{ cal} = 4.184 \text{ J}$$

It was only later that the metric unit of energy was renamed in honor of Joule.

You may have noticed that the specific heat capacity of water is quite high, nearly twice as high as that of anything else in Table 5.3. This ability to absorb (or release) large amounts of internal energy is another property of water that adds to its uniqueness—and its usefulness. Water is used as a coolant in automobile engines, power plants, and countless industrial processes partly because it is plentiful and partly because its specific heat capacity is so high. Engine parts near where the fuel burns are exposed to very high temperatures. The metal would be damaged, or even melt, if there were no way of cooling the parts. Water is circulated to these areas of the engine, where it absorbs heat from the hotter metal, thereby cooling the parts. The heated water flows to the radiator, where it transfers the heat to the air via cooling fins and fans.

▪ Learning Check

1. A 10-g piece of copper heated to 50°C would feel hotter in your hand than a 10-g piece of silver also heated to 50°C, even though copper and silver conduct heat equally well. This is because copper's _____ is greater than silver's.

2. The amount of internal energy that a solid object would have to gain in order for its temperature to rise a certain amount depends on
(a) how large the temperature change is.
(b) what the object's mass is.
(c) what material it is made of.
(d) All of the above.

3. (True or False.) The temperature of a moving baseball increases as it is caught.

4. (True or False.) One reason water is useful for cooling things is that its specific heat capacity is relatively low.

ANSWERS: 1. specific heat capacity **2.** (d) **3.** True **4.** False

5.6 Phase Transitions

A phase transition or "change of state" occurs when a substance changes from one phase of matter to another. Table 5.4 lists the common phase transitions, the phases involved, and the effect of the transition on the internal energy of the substance.

Let's say that a pan of water is placed on a stove and that heat is transferred to the water at some rate. This causes the temperature to rise steadily until the water starts to boil. Then the temperature stays the same ($100°C = 212°F$ at sea level) even though heat is continuing to be transferred to the water. The added energy is no longer increasing the kinetic energy of the water molecules: it is breaking the "bonds" that hold the molecules in the liquid state. The molecules are given enough energy to escape from the water's surface and become free molecules of steam (Figure 5.30a).

Below the boiling temperature, the water molecules are bound to each other and so have negative potential energies. During boiling, each molecule in turn is given enough energy to break free of the bonds. The average kinetic energy of the molecules stays the same during boiling, and the temperature of the water therefore stays constant.

A similar process occurs when a solid melts. The atoms or molecules go from being rigidly bound to each other in the solid state to being rather loosely bound in the liquid state. As in boiling, the increase in internal energy that occurs during melting goes to increase *only the potential energies* of the atoms or molecules. So the temperature of ice remains at $0°C$ ($32°F$) while it is melting.

Condensation and freezing are simply the reverse processes of boiling and melting, respectively. Here the potential energies of the atoms and molecules decrease, but their kinetic energies (and hence temperature) remain the same.

The temperature at which a particular phase transition occurs depends on the properties of the atoms or molecules in the substance—particularly the masses of the particles and the forces acting between them. When these forces are very strong, such as in table salt, the melting and boiling temperatures are quite high. When the forces are very weak, such as in helium, the phase-transition temperatures are very low—near absolute zero.

The boiling temperature of each liquid varies with the pressure of the air (or other gas) that acts on its surface. When the pressure is 1 atmosphere, water boils at $100°C$. At an elevation of 3,000 meters (10,000 feet) above sea level, where the pressure is about 0.67 atmospheres, the boiling point of water is reduced to $90°C$. That is why it takes longer to cook food by boiling at higher elevations. The temperature of the boiling water is lower, so conduction of heat into the food is slower. If the pressure is *increased* to 2 atmospheres such as in a pressure cooker, the boiling point of water is $120°C$ and food cooks faster (Figure 5.30b). One type of nuclear power plant, called a *pressurized water reactor*, uses high pressure to keep water from boiling even at several hundred degrees Celsius.

Table 5.4 Common Phase Transitions[a]

Name	Phases Involved	Effect
Boiling	Liquid to gas	Increases U
Melting	Solid to liquid	Increases U
Condensation	Gas to liquid	Decreases U
Freezing	Liquid to solid	Decreases U

[a] Sublimation is a fairly rare phase transition in which a solid goes directly to a gas, bypassing the liquid phase. The atoms or molecules break free from the rigid forces binding them in a crystal and move off in the gas phase. Dry ice (solid carbon dioxide) undergoes sublimation at temperatures above $-78.5°C$ under 1 atmosphere. This means that carbon dioxide cannot exist in the liquid phase under normal pressure. Mothballs, formerly containing naphthalene but now commonly made from dichlorobenzene crystals, also undergo sublimation at room temperature.

Figure 5.30 (a) The temperature of boiling water stays constant at 100°C even as heat flows into it. (b) Higher pressure inside the pressure cooker raises the boiling point of water to about 120°C.

(a) (b)

This dependence of the boiling temperature on the pressure makes it possible to induce boiling or condensation simply by changing the pressure. For example, water exists as a liquid at 110°C when under 2 atmospheres of pressure. If the pressure is reduced to 1 atmosphere, the boiling temperature is lowered to 100°C, and the water begins to boil. In the same manner, steam at 110°C and 1 atmosphere pressure will start to condense if the pressure is increased to 2 atmospheres. As we will see in Section 5.7, such pressure-induced phase transitions are the basis for the operation of refrigerators and other "heat movers."

5.6a Latent Heats

A specific amount of internal energy must be added or removed from a particular substance to complete a phase transition. For example, 334,000 joules of heat must be transferred to each kilogram of ice at 0°C to melt it. This quantity is called the **latent heat of fusion** of water. A much larger amount, 2,260,000 joules, must be transferred to each kilogram of water at 100°C to convert it completely into steam. This is the **latent heat of vaporization** of water. During the reverse processes, freezing and condensation, the same amounts of internal energy must be extracted from the water and steam, respectively.

EXAMPLE 5.7   Ice at 0°C is used to cool water from room temperature (20°C) to 0°C. How much water can be cooled by 1 kilogram of ice?

SOLUTION   Heat flows into the ice as it melts, cooling the water in the process. The maximum amount of water it can cool to 0°C corresponds to all of the ice melting. So the question is, how much water will be cooled by 20°C when 334,000 joules of heat are transferred from it?

$$Q = C\,m\,\Delta T$$

$$-334{,}000\text{ J} = 4{,}180\text{ J/kg-°C} \times m \times -20°\text{C}$$

$$-334{,}000\text{ J} = -83{,}600\text{ J/kg} \times m$$

$$\frac{-334{,}000\text{ J}}{-83{,}600\text{ J/kg}} = m$$

$$m = 4.0\text{ kg}$$

Ice at 0°C can cool about four times its own mass of water from 20°C to 0°C—provided this takes place in a well-insulated container to minimize heat conduction from the outside.

These large latent heats of ice and water have common applications. The reason that ice is so good at keeping drinks cold is that it absorbs a large amount of internal energy while melting. As long as there is ice in the drink, the temperature remains near 0°C. One reason water is ideal for extinguishing fires is that it absorbs a huge amount of internal energy when it vaporizes. When poured on a hot, burning substance, the water absorbs heat as it boils, thereby cooling the substance (**Figure 5.31**).

> ### ▶ Physics To Go 5.6
>
> You need an electric kitchen range or hot plate, an empty aluminum soft drink can, an oven mitt or glove, and some kind of clock for timing.
>
> 1. Put a tablespoon or so of water into the can.
> 2. Turn the surface heating unit to high and wait a minute for it to heat up.
> 3. Place the can on the unit and time how long it takes before the water starts boiling (you should be able to hear it and see mist coming out). This might be about 1 minute.
> 4. "Restart" the clock and time how long it takes for all of the water to be boiled away (when you can no longer hear water boiling). Turn the unit off and **immediately** remove the can from the heating unit using the mitt or glove. If you don't, the can will quickly get hot enough for the paint to start smoking.
> 5. How do the times compare? What does this tell you about the amount of internal energy it takes to heat water to boiling compared to the amount it takes to convert water at the boiling point to steam?

Here's an example that summarizes the relationship between the temperature of water, its phase, and its internal energy. A block of ice at a temperature of −25°C is placed in a special chamber. The pressure in the chamber is kept at 1 atmosphere while heat is transferred to the ice at a fixed rate. **Figure 5.32** shows a graph of the temperature of the water versus the amount of heat transferred to it.

Between points a and b on the graph, the heat transferred to the ice simply raises its temperature. At point b, the ice starts to melt, and the temperature stays fixed at 0°C until all of the ice is melted, point c. From c to d, the heat that is transferred to the water goes to increase its temperature. Between d and e, the water boils while the temperature remains at 100°C. As soon as all of the water is converted into steam, at e, the temperature of the steam starts to rise.

We could reverse the process by placing steam in the chamber and transferring heat from it. The result would be like moving along the graph from right to left.

Figure 5.31 Firefighters routinely exploit water's vaporization to save lives and protect property.

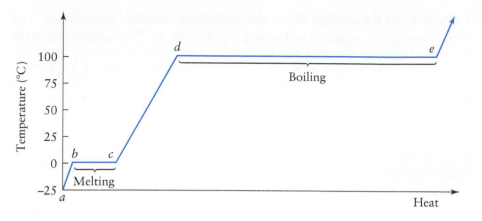

Figure 5.32 Graph of the temperature of water versus the heat transferred to it. The phase transitions correspond to the two places where the graph is flat: the internal energy increases, but the temperature stays constant.

5.6b Humidity

At temperatures below their boiling points, liquids can gradually go into the gas phase through a process known as **evaporation.** Water left standing will eventually "disappear" because of this. How can this phase transition occur at temperatures below the boiling point? Individual atoms or molecules in a liquid can go into the gas phase if they have enough energy. At temperatures below the boiling point, some of the atoms or molecules do have enough energy to do this. Even though the *average* energy of the particles as measured by the liquid's temperature is too low for boiling to occur, some of them have more energy than the average, and some have less. Atoms or molecules with higher-than-average energy can break free from the liquid if they are near the surface. Once in the air, the atoms or molecules can remain in the gas phase even though the temperature is below the boiling point.

Because of evaporation, water vapor is always present in the air. The amount varies with geographic location (proximity to large bodies of liquid water), climate, and weather. **Humidity** is a measure of the amount of water vapor in the air.

> **DEFINITION** **Humidity** The mass of water vapor in the air per unit volume. The density of water vapor in the air.

The unit of humidity is the same as that of mass density. The humidity generally ranges from about 0.001 kg/m³ (cold day in a dry climate) to about 0.03 kg/m³ (hot, humid day). Note that these densities are much less than the normal density of the air, 1.29 kg/m³. Even in humid conditions, water vapor is only a small component of the air—less than 5 percent.

At any given temperature, there is a maximum possible humidity, called the *saturation density.* This upper limit exists because the water molecules in the air "want" to be in the liquid phase. If there are too many molecules in the air, the probability is high that several will get close enough to each other for the attractive force between them to take over. They begin to form droplets and condense onto surfaces. This is what happens during the formation of fog and dew, or the mist in a shower (**Figure 5.33**). The saturation density

Figure 5.33 Fog in the cooler air in a valley. Water droplets form when the humidity exceeds the saturation density and there is too much water vapor in the air.

NOAA Photo Library, NOAA Central Library; OAR/ERL/National Severe Storms Laboratory (NSSL)

Table 5.5 Saturation Density of Water Vapor in the Air

Temperature		Saturation Density	Temperature		Saturation Density
(°C)	(°F)	(kg/m³)	(°C)	(°F)	(kg/m³)
−15	5	0.0016	15	59	0.0128
−10	14	0.0022	20	68	0.0173
−5	23	0.0034	25	77	0.0228
0	32	0.0049	30	86	0.0304
5	41	0.0068	35	95	0.0396
10	50	0.0094	40	104	0.0511

is *higher* at higher temperatures because the water molecules are moving faster and are less likely to "stick" together when they collide. Hence, more molecules can be present in any volume of air without droplet formation. Table 5.5 lists the saturation density of water vapor in the air at several different temperatures.

The reverse of evaporation also takes place: water molecules in the air near the surface of water can be deflected into the water and "captured." When the humidity is well below the saturation density at that temperature, evaporation occurs faster than reabsorption of water molecules. If the humidity increases to the saturation density, water molecules are reabsorbed into the water at the same rate that they evaporate. The water does not disappear. This is why damp towels dry slowly in humid environments.

The upshot of this is that the humidity alone doesn't determine how rapidly water evaporates. What matters is how close the humidity is to the saturation density. The **relative humidity** is a good indicator of this relation.

DEFINITION **Relative Humidity** The humidity expressed as a percentage of the saturation density.

$$\text{relative humidity} = \frac{\text{humidity}}{\text{saturation density}} \times 100\%$$

When the relative humidity is 40 percent, this means that 40 percent of the maximum amount of water vapor is present.

EXAMPLE 5.8 What is the relative humidity when the humidity is 0.009 kg/m³ and the temperature is 20°C?

SOLUTION From Table 5.5, the saturation density at 20°C is 0.0173 kg/m³. Therefore,

$$\text{relative humidity} = \frac{0.009 \text{ kg/m}^3}{0.0173 \text{ kg/m}^3} \times 100\%$$

$$= 52\%$$

The same humidity in air at 15°C would make the relative humidity 70 percent. If the air were cooled to 9°C, the relative humidity would be 100 percent.

When air is cooled and the water vapor content stays constant, the relative humidity *increases*. When air is heated and the humidity stays constant, the relative humidity *decreases*. This is why heated buildings often feel dry in the winter.

Cold air from the outside enters the building and is heated. Unless water vapor is artificially added to this air with a humidifier, the relative humidity will be very low.

If air is cooled while the humidity stays constant, eventually condensation begins to occur. The temperature at which this happens is called the *dew-point* temperature. Often on clear nights, the air temperature drops until the dew point is reached and condensation begins. The result is dew on plants and other surfaces—hence the name "dew" point. The dew point is easily predicted if the humidity is known: it is the temperature at which that humidity reaches the saturation density. Table 5.5 or **Figure 5.34** can be used for this purpose. For example, when the humidity is 0.0128 kg/m³, the dew point is 15°C (59°F).

Figure 5.34b shows how the dew point can be estimated graphically. The point ○ represents the air described in Example 5.8. After the air is cooled from 20°C to 15°C, its state is marked by ●. The point moved horizontally to the lower temperature. Continued cooling would cause the air to reach the saturation density curve at a temperature of about 9°C, so that is the dew point.

Water droplets often form on the sides of cans and other containers holding cold drinks. This happens when the temperature of the container's sides is below the dew-point temperature. Air in contact with the surface is cooled until the dew point is reached and condensation begins. (This process also occurs when car windows "fog over" in cold weather.) If the air is very dry (low relative humidity), condensation doesn't occur because the dew point is below the temperature of the cold drink.

The process of evaporation cools a liquid. Atoms or molecules in the gas phase have more energy than when in the liquid phase. When water molecules evaporate, they take away some internal energy from the water, thereby cooling it. Here is another way of looking at it: because the molecules with the higher kinetic energies are the ones that evaporate, the average kinetic energy of the water molecules that remain is lowered.

Our bodies are cooled by the evaporation of perspiration. On hot days, we feel hotter if the humidity is high because evaporation is inhibited. A breeze feels cool because it removes air near our skin that has a higher humidity because of perspiration. The drier air that replaces it allows for more rapid evaporation—and cooling.

Concept Map 5.3 summarizes the basic phase transitions and their effect on the potential energy of atoms and molecules.

Figure 5.34 (a) Graph of the saturation density versus temperature (data in Table 5.5). (b) Same graph showing the situation described in Example 5.8.

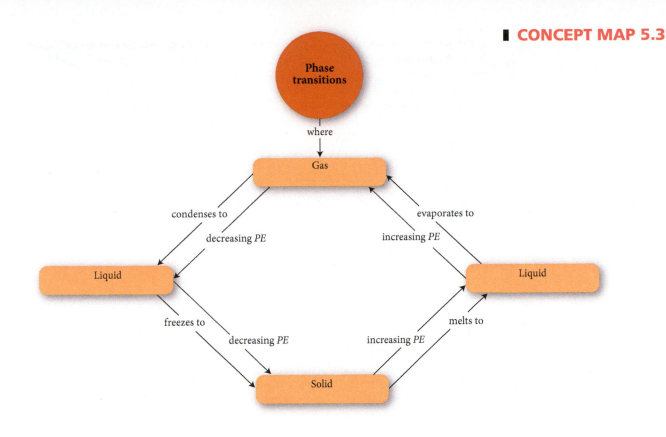

Learning Check

1. If heat is flowing into water but its temperature is not changing, what can you conclude is happening?
2. (True or False.) Water's very high latent heat of vaporization is a major reason it is very effective at extinguishing fires.
3. The relative humidity of the outside air usually increases during the night. Why?
4. When droplets form on the outside of a glass containing ice water, this means that in the air right next to the glass

(a) the humidity is approximately equal to the saturation density.
(b) the relative humidity is about 100 percent.
(c) the temperature is approximately equal to the dew point.
(d) All of the above.

ANSWERS: 1. It is boiling. **2.** True **3.** Air temperature decreases. **4.** (d)

5.7 Heat Engines and the Second Law of Thermodynamics

Along with the efforts made in the recent past to conserve energy used to heat and cool buildings, much has been done to improve the efficiency of other ways that we use energy. In this section, we will consider some of the basic theoretical principles that are involved in energy-conversion devices that use heat and mechanical energy.

5.7a Heat Engines

Most of the energy used in our society comes from fossil fuels—coal, oil, and natural gas. A fraction of these fuels is burned directly for heating, such as in gas stoves and oil furnaces. But most of these fuels are used as the energy input for devices that are classified together as **heat engines.**

DEFINITION **Heat Engine** A device that transforms heat into mechanical energy or work. It absorbs heat from a hot source such as burning fuel, converts some of this energy into usable mechanical energy or work, and outputs the remaining energy as heat to some lower-temperature reservoir.

Figure 5.35 Simplified sketch of an electric power plant showing the energy (heat) inputs and outputs.

Gasoline engines, diesel engines, jet engines, and steam-electric power plants are all heat engines. In gasoline and diesel engines, some of the heat from burning fuel is converted into mechanical energy. The remainder of the heat is ejected to the air from the exhaust pipe, the radiator, and the hot surfaces of the engine. Coal, nuclear, and some types of solar power plants produce electricity by using steam to turn a generator (**Figure 5.35**). Heat from burning coal, fissioning nuclear fuel, or the Sun is used to boil water. The steam is piped to a turbine (basically a propeller) that is given rotational energy by the steam. A generator connected to the turbine converts rotational energy into electrical energy. After the steam leaves the turbine, it is condensed back into the liquid phase via cooling with water or the air. Cooling towers, which function somewhat like automobile radiators, are used in the latter case. Most of the heat transferred to the water to boil it is ejected from the plant as waste heat when the steam is condensed.

Even though the actual inner workings of heat engines are quite complicated, from an energy point of view we can represent them with a simple schematic diagram (**Figure 5.36**). Energy in the form of heat from some source is input into a mechanism. The mechanism converts some of the heat into

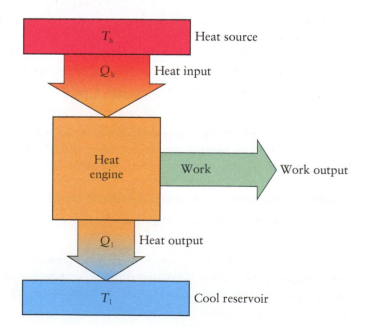

Figure 5.36 Diagram of a heat engine. The mechanism absorbs energy from the heat source, uses some of it to do work, and releases the remainder to some lower-temperature reservoir.

mechanical energy and rejects the remainder. In this simplified representation of a heat engine, the heat source has some fixed (high) temperature T_h and the reservoir that absorbs the rejected heat has some fixed (lower) temperature T_l.

During a given period while the engine is operating, some quantity of heat Q_h is absorbed from the hot source. Some of this energy is converted into usable work, and the remainder of the original energy input is ejected as waste heat Q_l. This wasted heat is unavoidable. The following law is a formal statement of this.

> **LAWS** **Second Law of Thermodynamics** No device can be built that will repeatedly extract heat from some high-temperature source and deliver mechanical work or energy without ejecting some heat to a lower-temperature reservoir.

The energy efficiency of any device or process is the usable output divided by the total input, times 100 percent:

$$\text{efficiency} = \frac{\text{energy or work output}}{\text{energy or work input}} \times 100\%$$

If the efficiency of a device is 25 percent, then one-fourth of the input energy is converted into usable forms. The remaining three-fourths are "lost"—released as waste heat.

For heat engines, the input energy is Q_h, and the output is the work. Therefore,

$$\text{efficiency} = \frac{\text{work}}{Q_h} \times 100\% \qquad \text{(heat engine)}$$

As we have seen, there are many different types of heat engines. Some of them use processes that are inherently more efficient than others. However, there is a theoretical upper limit on the efficiency of a heat engine. This maximum efficiency is called the *Carnot efficiency* after French engineer Sadi Carnot. Carnot discovered that the efficiency of a perfect heat engine is limited by the temperatures of the heat source and of the lower-temperature reservoir. In particular,

$$\text{Carnot efficiency} = \frac{T_h - T_l}{T_h} \times 100\%$$

(T_h and T_l must be in kelvins.)

Real heat engines have friction, imperfect insulation, and other factors that reduce their efficiencies. But even if these were completely eliminated, a heat engine could not have an efficiency of 100 percent because this would imply that the temperature of the reservoir was zero or that the temperature of the heat source was infinitely high, neither of which is achievable in practice.

EXAMPLE 5.9 A typical coal-fired power plant uses steam at a temperature of 1,000°F (810 K). The steam leaves the turbine at a temperature of about 212°F (373 K). What is the theoretical maximum efficiency of the power plant?

SOLUTION Here, $T_h = 810$ K and $T_l = 373$ K. So,

$$\text{Carnot efficiency} = \frac{810 - 373}{810} \times 100\%$$

$$= 54\%$$

This is the ideal efficiency. The actual efficiency of a coal-fired power plant is usually around 40 percent. Modern diesel power plants are up to 49 percent efficient.

Figure 5.37 For every three railroad cars of coal burned at an electric power plant, the energy contained in about two of them goes to waste heat. This is a consequence of the second law of thermodynamics, as well as an imperfect heat engine.

Nearly two-thirds of the energy input to a power plant is lost as waste heat (**Figure 5.37**). The unused energy causes thermal pollution by artificially heating the environment. The actual top efficiencies of the other heat engines in common use are similar: diesel engine, 35 percent; jet engine, 23 percent; gasoline (piston) engine, 25 percent.

There are two general ways to improve the efficiency of a heat engine. One is to improve the process and reduce energy losses so that the efficiency gets closer to the Carnot efficiency. The other way is to increase the Carnot efficiency by raising T_h or lowering T_l, or both. The efficiencies of steam-based heat engines have been improved a great deal through the use of higher-temperature steam.

5.7b Heat Movers

Refrigerators, air conditioners, and heat pumps are devices that act much like heat engines in reverse. They use an input of energy to cause heat to flow from a cooler substance to a warmer substance—opposite the natural flow from hotter to colder objects. The purpose of refrigerators and air conditioners is to extract heat from an area and thereby cool it. Heat pumps are designed to heat an interior space in the winter as well as cool it in the summer. In these devices, as heat is transferred from a substance being cooled, a larger amount of heat flows into a different substance that is then warmed. A refrigerator removes heat from its interior and ejects heat into the room. We will call these devices **heat movers,** because the term describes what they do.

The mechanisms in the three devices just mentioned are much the same. A gas called the *refrigerant* is forced to undergo phase changes in a cyclic process. The gas must have a fairly low boiling point and be condensed easily when the pressure on it is increased. Freon (CCl_2F_2) was the most commonly used refrigerant, although it has been largely replaced by other compounds that do not destroy ozone after being released into the atmosphere. The refrigerant is compressed into the liquid phase by a pump and forced to flow through a small opening called an *expansion valve* (**Figure 5.38**). The pressure on the other side of the valve is kept low so that the refrigerant quickly goes into the gas phase (boils). This phase transition cools the refrigerant; in the process, it absorbs

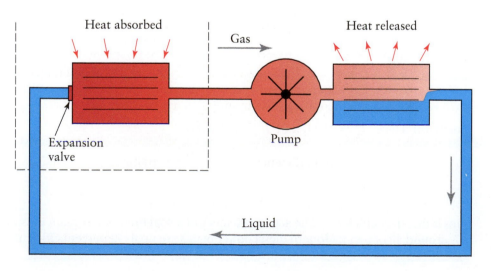

Figure 5.38 Refrigerators remove heat from their interiors by exploiting the phase transition of the refrigerant. As the refrigerant vaporizes, it absorbs heat from the refrigerator's interior. As it condenses, it releases heat to the air in the room.

heat from the surroundings. The gas flows back to the pump, where it is recompressed into the liquid phase. This raises the temperature of the refrigerant, causing heat to flow from it to the surroundings. After the refrigerant is cooled and liquefied, it flows through the expansion valve, and the cycle repeats.

Although the net result is a flow of heat from something cooler to something warmer, the actual heat flow into, and then out of, the refrigerant is always a flow from something warmer to something cooler. Basically, the refrigerant is first cooled to a temperature below that of the substance being cooled and then heated to a temperature above that of the substance being warmed. Also note that energy must be supplied to the pump to make the process work.

Heat movers can be represented with a diagram similar to that for heat engines (**Figure 5.39**). Unlike heat engines, the mechanism in heat movers has an *input* of energy or work. During a given period of time, a quantity of heat Q_l is absorbed from a cool reservoir with temperature T_l. An amount of work is done on the refrigerant by the pump, and an amount of heat Q_h is released to a warm reservoir with temperature T_h. In refrigerators, the cool reservoir is the air in their interiors. The warm reservoir is the air in the room. Air conditioners absorb heat from the air that they circulate inside a building, a vehicle, and so on, and transfer heat to the warmer outside air. When in the "heating mode," heat pumps remove heat from the outside air, ground, or groundwater and release heat inside a building. In the "cooling mode," heat pumps reverse the heat flow: heat is transferred from the inside of the building and ejected to the outside.

The law of conservation of energy tells us that the amount of heat released, Q_h, is equal to the amount of heat absorbed, Q_l, plus the energy input to the mechanism. The relative values of Q_h, Q_l, and the energy input depend on the efficiency of the pumping mechanism, the difference between the two temperatures, T_h and T_l, and other factors. In the average refrigerator, the amount of heat transferred from the interior is about *three times* the amount of energy input. In the heating mode, the amount of heat delivered by a good heat pump is about *three times* the amount of energy consumed by the mechanism. Energy is not being "created": the heat pump is simply extracting heat from a cooler substance (making it cooler still), and then transferring a larger amount of heat to a warmer substance. Remember, any system or substance with a temperature greater than absolute zero stores some internal energy, implying that it can serve as a source of heat transfer if a suitable heat pump can be devised.

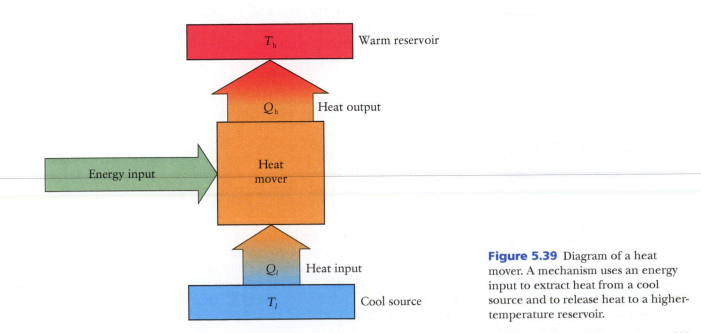

Figure 5.39 Diagram of a heat mover. A mechanism uses an energy input to extract heat from a cool source and to release heat to a higher-temperature reservoir.

5.7c Usable Energy

The outputs of a heat engine are heat flowing into a low-temperature reservoir, thereby increasing its internal energy, and useful work or energy. The latter eventually becomes internal energy as well because of friction and other processes. So the overall effect of a heat engine is to convert internal energy in a higher-temperature source into internal energy in a lower-temperature reservoir. In turn, this internal energy is no longer available to do as much useful work: Carnot tells us that lower temperatures would give us lower efficiencies. Even though energy is not truly "lost" or "destroyed" in this process, it is made less available for similar use again.

This is a general result of energy transformations: the energy that remains is less usable than the original energy. Electrical energy is a highly convenient, easily used, and versatile form of energy. Lightbulbs, heaters, and fans convert electrical energy into other desirable forms (radiant energy by which to read a book, heat energy to warm a room, and the kinetic energy of moving air to provide cooling on a hot day) plus the inevitable (and less beneficial) internal energy associated with friction (as, for example, in the rotating parts of a fan). Certainly in the case of the internal energy, but also for the other "useful" forms of energy, the ease with which the remaining energy can be harnessed to drive other, desired processes is nowhere near that of the original electrical energy. For example, the efficiency with which the kinetic energy of the moving air from a fan can be collected and transformed into mechanical energy using a high-efficiency wind-turbine system (<60 percent) is much smaller than the efficiency with which the original electrical energy can be converted into rotational energy of the fan motor (>90 percent). In this sense, we might say that the *quality* of electrical energy is higher than that associated with wind energy.

The physical quantity **entropy,** symbolized as S, can be used as a means of assessing the amount of "hard-to-use" or low-quality or *dispersed* energy in a system at a specific temperature; that is, the energy that cannot be readily harnessed to do external work. As previously noted, natural processes tend to be accompanied by a dispersal of energy and hence an increase in the entropy of a system. Commonly, entropy is taken to be a measure of a system's disorder (or "mixedupedness" to use the terminology of 19th-century American physicist J. Willard Gibbs): increasing entropy is associated with increasing disorder (**Figure 5.40**). In this view, the more disordered the system, the more its energy is spread out over different possible internal states and the greater its entropy. Considerations of this type have lead to an alternative formulation of the second law of thermodynamics:

> **LAWS** **Second Law of Thermodynamics (alternate form)** In any thermodynamic process, the total entropy of the system and its environment either increases or remains constant: $\Delta S \geq 0$.

Entropy changes are zero only for reversible processes, but no natural, spontaneous process is truly reversible. Thus, we expect the entropy of the universe as a whole to become greater with time.

Although it may not be obvious that this statement is equivalent to our earlier one for the second law in Section 5.7a, it can be shown to be true. Indeed, the association of the concept of entropy with the amount of "useless" energy (energy unavailable to do "useful" work) leads to the conclusion that for any process to go forward, a quantity of energy equal to $\Delta S \times T_R$, where ΔS is the entropy change in the system and T_R is the Kelvin temperature of the external environment (the "reservoir," R), **must** be transferred to the system's

Figure 5.40 A student's study area tends toward increasing disorder as time goes by. At the beginning of the semester (top), the desk is well organized, but a few days or weeks later (bottom), it is messy and disorganized. The entropy of the area has increased with time.

Cengage Learning/Charles D Winters

surroundings. If this condition is not fulfilled, then the process (or the device designed to carry out the process) cannot be realized.

The SI unit for entropy is the joule per kelvin (J/K). Thus, the product of an entropy change and a temperature measured in kelvins yields a quantity whose units are those of energy.

The entropy of part of a system can be decreased temporarily at the expense of the rest of the system. As living things grow, they organize the nutrients and life-sustaining chemicals they take in to become more highly ordered; their entropy decreases. But in doing so, they disperse the energy in the food they consume and thereby increase the entropy of these substances. After the organisms die, the entropy of their remains once again increases as they decay. As a car battery is charged, its entropy decreases, but the charging system is itself consuming energy to drive the process with the result that entropy is increasing somewhere else in the environment.

EXAMPLE 5.10 As a result of heat flowing spontaneously through a metal rod from a hot reservoir at 550 K to a cold reservoir with temperature of 350 K, 130 J of energy is transferred to the environment (and hence unavailable to do useful work). Determine the amount by which this irreversible process changes the entropy of the universe. Assume that the ambient temperature of surroundings where the transfer occurs is 290 K and that no other thermodynamic changes accompany this process.

SOLUTION Because the hot-to-cold flow is irreversible, the quantity of energy transferred to the universe is

$$\Delta E = \Delta S \times T_R$$

where T_R is the temperature of the surroundings. Rearranging to solve for the entropy change,

$$\Delta S = \left(\frac{\Delta E}{T_R}\right)$$

$$= \frac{130 \text{ J}}{290 \text{ K}}$$

$$= 0.45 \text{ J/K}$$

This irreversible process causes the entropy of the universe to increase as expected: ΔS is positive.

The term *energy shortage* that frequently appears in connection with statements about the rate at which our civilization is consuming its energy resources and the risks that behavior poses for the sustainability of future generations is something of a misnomer. There is, in fact, as much energy around now as ever before. (After all, the total energy of the universe is constant; see the application at the end of Section 3.5.) But our society is using up high-quality, conveniently stored energy in fossil fuels, uranium, and forests, for example, and producing lower-quality, more dispersed internal energy. We are irreversibly increasing the entropy of the world, leaving less *usable* energy for our progeny. However, by tapping "renewable" energy sources such as solar, wind, geothermal, and tidal power, we can exploit continuously available (at least for the next several billion years) supplies of energy that would otherwise go to waste—and we reduce greenhouse gas emissions in the process. Solar energy eventually becomes high-entropy internal energy, whether it is absorbed by the ground or by solar cells. If absorbed by high-efficiency solar cells, though, we get as an intermediate step some low-entropy, high-quality electrical energy to meet our needs.

1. (True or False.) The efficiency of a heat engine could be increased by lowering the temperature of its cool reservoir.
2. Which of the following devices is *not* a heat mover? (a) air conditioner (b) heat engine (c) heat pump (d) refrigerator
3. (True or False.) For each joule of electrical energy that a refrigerator uses, more than 1 joule of heat

can be transferred from the food and beverages stored inside it.
4. As a heat engine operates, it does not change the amount of energy there is in the universe, but it does increase the _____ of the universe.

ANSWERS: 1. True **2.** (b) **3.** True **4.** entropy

Profiles in Physics Thermometry and Calorimetry

The first primitive thermometer was invented by none other than Galileo in 1592. It made use of the expansion and contraction of air with temperature. A glass bulb equipped with a long, thin neck was partially filled with water and then inverted into a container of water (**Figure 5.41**). The water level would move up and down in the neck as changes in temperature caused the air in the bulb to expand and contract. The device was also affected by changes in atmospheric pressure, so it was a combination thermometer and barometer. Galileo did not equip his device with a temperature scale.

In 1657, a number of disciples of Galileo formed an "academy of experiment" in Florence and developed functional thermometers. These were based on the thermal expansion of alcohol in sealed tubes and were not affected by changes in atmospheric pressure. The academicians established temperature scales for their thermometers by choosing two fixed points and dividing the interval into a number of degrees.

The Florentine thermometers became quite popular in Europe and stimulated many scientists to attempt improvements. An astronomer in Paris named Ismaël Boulliau was the first to use mercury in a thermometer. Perhaps the most famous person in the history of the development of the modern thermometer is

Gabriel Daniel Fahrenheit (1686–1736), a scientist and manufacturer of meteorological instruments. Fahrenheit experimented extensively with both alcohol and mercury thermometers, as well as with fixed points and temperature scales. Fahrenheit made the important discovery that the boiling temperature of liquids varies with atmospheric pressure. This accounted for the confusion that other experimenters encountered when trying to use the boiling point of a liquid as a fixed point on a temperature scale. The scale that he developed used as its fixed points the temperature of an ice–saltwater mixture and the temperature of the human body. Fahrenheit divided the interval into 96 steps and made the first fixed point 0 degrees. The scale was later adjusted after measurement errors were discovered. The result is our modern Fahrenheit scale, with the temperature of melting ice at 32° and that of boiling water (at normal atmospheric pressure) at 212°.

Several experimenters sought to establish a scale based on 100° between fixed points—a "centigrade" scale. Swedish astronomer Andreas Celsius (1701–1744) introduced a scale that used 100 degrees between the freezing and boiling points of water, but it read backward: 100° was the freezing temperature and 0° was the boiling temperature. This scale was later inverted and is our modern Celsius scale. But no one temperature scale has ever quite gained universal acceptance. During the 18th century, more than a dozen different scales were in use.

The existence of an "absolute zero" was first indicated in experiments with air thermometers such as Galileo's. The volume of a fixed amount of a gas was known to decrease with temperature. One

Figure 5.41 Galileo's "thermometer."

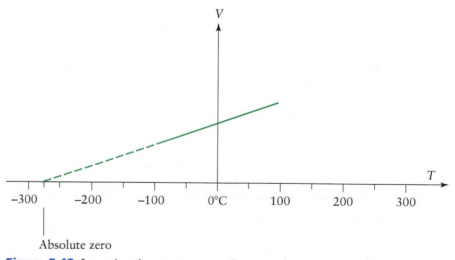

Figure 5.42 Lowering the temperature of a gas under constant pressure reduces the volume it occupies. The dotted line indicates that the volume would go to zero at some very low temperature—absolute zero.

could simply graph the volume of the gas versus the temperature and extrapolate the straight line to the point where the volume would be zero (**Figure 5.42**). The corresponding temperature would be the coldest possible temperature—absolute zero. It was known to exist even before it was possible to get within 200 degrees of it. This temperature is the zero point of the Kelvin scale, named after British physicist Lord Kelvin, who first established an absolute temperature scale.

Heat

The development of the understanding of the basic nature of heat presents an interesting counterexample to the general progress of science. Originally, Democritus and other Greeks thought that heat was some kind of material substance that flowed into objects when they were heated. This view was challenged by many leading scientists of the 17th century, including such familiar ones as Isaac Newton and Robert Boyle, who considered heat to involve some kind of internal motion. Although this was the correct approach, it was replaced in the 18th century by a reversion to the material model of heat. Joseph Black (1728–1799), a Scottish scientist, and Antoine Lavoisier were two of the strongest proponents of the *caloric theory* of heat. Heat was supposedly an invisible, massless fluid called *caloric*. This model, though incorrect, was adequate for explaining many thermodynamic phenomena. Lavoisier measured the specific heat capacities of many substances, and Black successfully measured the quantity of heat needed to melt a given amount of ice. The science of thermodynamics progressed quite nicely during the next century, even though the basic concept of heat was wrong.

The Achilles's heel of the caloric theory was friction. How was heat, and therefore caloric, produced by friction? Among the first to dispute the caloric theory was Benjamin Thompson, later titled Count Rumford (**Figure 5.43**). Rumford (1753–1814), one of the first

Figure 5.44 James Prescott Joule.

Americans to leave his mark in science, was born in Massachusetts and was employed there until he fled to Europe during the Revolutionary War. Rumford spent the rest of his life in England, Austria, Bavaria (where he was given his title), and finally France. He worked as a scientist, an industrialist, and in various capacities in the military. At one point, Rumford supervised the construction of cannons at a factory in Munich. The cannon barrels were made by boring out the centers of solid iron cylinders. During this process, friction between the drill bit and the cannon barrel produced heat. The caloric theory implied that the friction released caloric from the metal. Rumford performed many experiments and found that the amount of heat that could be produced when a dull bit was used was unlimited. How could the iron contain an infinite quantity of a material substance (caloric)? Rumford reasoned that the heat was produced by the action of the drill bit and that caloric did not exist.

Several experiments by English brewer and amateur scientist James Prescott Joule (1818–1889; **Figure 5.44**) indicated that heat is a form of energy. His most famous experiment consisted of a paddle wheel that was made to churn water, thereby heating it, by the action of a falling weight. Joule measured the temperature change of the water and used its specific heat capacity to relate the amount of work done by the falling weight to the heat transferred to the water. The result is the famous "mechanical equivalent of heat" (Section 5.5). In other experiments, Joule measured the heat produced by the compression of air and by electrical currents. This, along with the work of many other scientists, led to the first law of thermodynamics.

A great impetus to the science of thermodynamics was provided by attempts to improve the performance of steam engines. The SI unit of power is named after James Watt, a Scottish engineer who was responsible for several design changes. Nicolas Léonard Sadi Carnot (1796–1832; **Figure 5.45**) was a French engineer who introduced the theoretical analysis of heat engines. Initially a follower of the caloric theory, Carnot was nonetheless able to determine what factors affect the efficiency of a heat engine. His analysis

Figure 5.43 Benjamin Thompson, later Count Rumford, at a cannon factory.

Figure 5.45 Sadi Carnot.

was later modified to include the fact that heat is associated with internal energy, not caloric.

Research in thermodynamics has received new impetus in recent decades as concerns about energy supplies and costs have made energy efficiency ever more important. Success in this area is manifested by the fact that most of today's cars, buildings, and devices use less energy to perform the same function than those of 50 years ago.

■ **QUESTIONS**

1. Discuss some of the early developments in the field of thermometry that have led to the Fahrenheit and Celsius temperature scales currently in use today.
2. Describe the caloric theory of heat. How did Rumford's experiments bring about the demise of this model as a valid representation of heat flow?

SUMMARY

» **Temperature** is the basis of our sense of hot and cold. Physically, the temperature of a substance is proportional to the average kinetic energy of its atoms and molecules.

» Three different temperature scales are in common use today—the **Fahrenheit, Celsius,** and **Kelvin** scales. The Kelvin temperature scale uses the lowest possible temperature, **absolute zero,** as its zero point.

» In most cases, matter expands when its temperature is raised. The amount of expansion depends on the substance and the temperature change. This property is exploited in mercury and alcohol thermometers and in bimetallic strips.

» The **internal energy** of a substance is the total potential and kinetic energies of its atoms and molecules. It can be increased by doing work on the substance and by transferring **heat** to it.

» The **first law of thermodynamics** states that the change in internal energy of a substance equals the work done on it plus the heat transferred to it.

» Heat can be transferred from one place to another in three ways. **Conduction** is the transfer of heat via contact between atoms or molecules. **Convection** is the transfer of heat via buoyant mixing in a fluid. **Radiation** is the transfer of heat via electromagnetic radiation. In some situations, all three processes occur at the same time.

» Except during phase transitions, the temperature of matter increases whenever its internal energy increases. The specific heat capacity is a characteristic of each substance that relates the mass and temperature increase to the heat transferred.

» During phase transitions, the potential energies of the atoms and molecules change while their average kinetic energy remains constant. So the internal energy increases while the temperature stays the same.

» **Evaporation** is a liquid-to-gas phase transition that occurs below the boiling temperature. It is responsible for the water vapor present in the air. **Humidity** and **relative humidity** are two measures of the water-vapor content in air.

» **Heat engines** are devices that use heat from a hot source to do work. They release heat at a cooler temperature in the process. The maximum efficiency of a theoretically "perfect" heat engine is determined by the temperatures of the hot and cold reservoirs.

» **Heat movers** use an energy input to remove heat from a cool substance and transfer it to a warmer substance.

» **Entropy** is a measure of the disorder in a system. Heat engines do not "consume" energy, but they do increase entropy, thus reducing the quality or ease of the use of the energy. Increases in the entropy (disorder) of the universe result from all natural processes.

IMPORTANT EQUATIONS

Equation	Comments
$\Delta l = \alpha \, l \, \Delta T$	Thermal expansion of a rod
$pV = (\text{constant}) \, T$	Pressure, volume, and temperature of a fixed amount of a gas (ideal gas law)
$\Delta U = \text{work} + Q$	First law of thermodynamics
$Q = C \, m \, \Delta T$	Heat needed to raise the temperature of a substance by ΔT
$\text{relative humidity} = \dfrac{\text{humidity}}{\text{saturation density}} \times 100\%$	Definition of relative humidity
$\text{efficiency} = \dfrac{\text{energy or work output}}{\text{energy or work input}} \times 100\%$	Definition of efficiency of a heat engine
$\text{efficiency} = \dfrac{T_h - T_1}{T_h} \times 100\%$	Carnot efficiency
$\Delta E = \Delta S \times T_R$	Energy transfer accompanying entropy change

1. In Section 5.2, we discussed the phenomenon of thermal expansion. **Concept Map 5.4** was to have been used to help you organize and understand this material better. When the map was printed, however, we realized that several key concepts and linking words and phrases had been left out. Because it was too late to make any changes in the book, it now becomes your task to complete the map, entering the missing information in the appropriate places. If you have any trouble doing so, review the reading on thermal expansion and make a rank-ordered list of the important concepts you find there. Use the list in addressing the omissions in the map provided.

2. Figures 5.36 and 5.39 schematically show the operation or function of *heat engines* and *heat movers,* respectively. An alternative way to capture the information and meaning contained in these flowcharts is through the use of concept maps. Pick one of these figures, and, based on your understanding of it, produce a proper concept map that represents the concepts involved in either heat engines or heat movers.

CONCEPT MAP 5.4 Mapping It Out! Exercise 1.

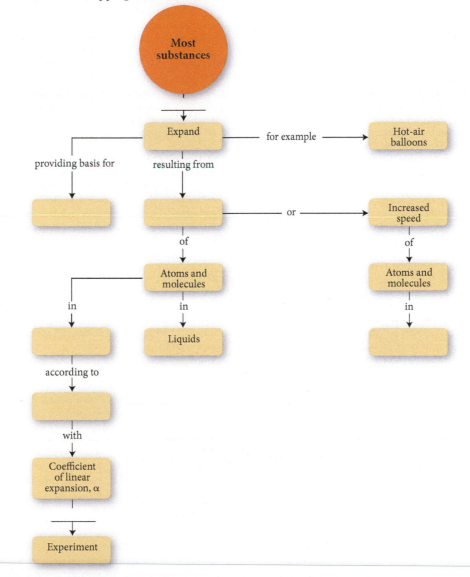

QUESTIONS

(■ **Indicates a review question, which means it requires only a basic understanding of the material to answer. Questions without this designation typically require integrating or extending the concepts presented thus far.**)

1. ■ What are the three common temperature scales? What are the normal boiling and freezing points of water in each scale?

2. The Fahrenheit and Celsius temperature scales agree at $-40°$ ($-40°C = -40°F$). Do the Fahrenheit and Kelvin temperature scales ever agree? How about the Celsius and Kelvin scales?

3. ■ What is the significance of absolute zero?

4. ■ What happens to the atoms and molecules in a substance as its temperature increases?

5. Air molecules in a warm room (27°C = 300 K) typically have speeds of about 500 m/s (1,100 mph). Why is it that we are unaware of these fast-moving particles continuously colliding with our bodies?

6. A special glass thermometer is manufactured using a liquid that expands less than glass when the temperature increases. Assuming the thermometer does indicate the correct temperature, what is different about the scale on it?

7. ■ Explain what a bimetallic strip is and how it functions.

8. Refer to the thermometer in Figure 5.10. Which metal in the bimetallic strip has the larger coefficient of linear expansion: the metal on the outer side of the spiral or the metal on the inner side? How can you determine this?

9. ■ A certain engine part made of iron expands 1 mm in length as the engine warms up. What would be the approximate change in length if the part were made of aluminum instead of iron?

10. ■ What is unusual about the behavior of water below the temperature of 4°C?

11. A company decides to make a novelty glass thermometer that uses water instead of mercury or alcohol.
    (a) The thermometer would include a warning informing the user that it should not be exposed to temperatures below 0°C. Why?
    (b) Suppose the thermometer is taken outside where the temperature is 1°C. Describe how the level of the water would change as it adjusts to the new temperature and how at some point it would behave very differently than a mercury- or alcohol-filled thermometer.

12. You may have noticed warning labels in aerosol cans of spray paint or air freshener that caution the user: "Do not expose or store container to/in high temperatures." Explain why the manufacturers include this advisory on their products.

13. After sitting outside on a cold night, the pressure in the tires of your car is 26 psi. If you drive your vehicle to work, a distance of 25 miles, at highway speeds, would you expect the tire pressure to increase, decrease, or remain the same? Explain the rationale for your choice. Assume that the outside air temperature remains constant during your trip.

14. ■ What are the two general ways to increase the internal energy of a substance? Describe an example of each.

15. Air is allowed to escape from an inflated tire. Is the temperature of the escaping air higher than, lower than, or equal to the temperature of the air inside the tire? Why?

16. Is it possible to compress air without causing its internal energy to increase? If so, how?

17. According to the first law of thermodynamics, under what conditions would it be possible for a system to absorb heat from its surroundings yet suffer no change in its internal energy?

18. ■ Describe the three methods of heat transfer. Which of these are occurring around you at this moment?

19. A potato will cook faster in a conventional oven if a large nail is inserted into it. Why?

20. Double-paned thermal insulating windows consist of two glass panels separated by a narrow gap from which most of the air has been removed to create a near vacuum in the space between the panes. Explain how this design helps to better insulate a home than a window that is made from a single sheet of glass.

21. A coin and a piece of glass are both heated to 60°C. Which will feel warmer when you touch it?

22. A submerged heater is used in an aquarium to keep the water above room temperature. Should it be placed near the surface of the water or near the bottom to be most effective? Explain.

23. On a cool night with no wind, people facing a campfire feel a breeze on their backs. Why?

24. Suppose you hang a bag of ice inside a room in which the air is at normal room temperature. If you position the palm of your hand a few inches to the side of the ice, what would you feel? Why? Would you feel anything different if you placed your palm the same distance away but below the ice?

25. When heating water on a stove, a full pan of water takes longer to reach the boiling point than a pan that is half full. Why?

26. A 1-kg piece of iron is heated to 100°C, and then submerged in 1 kg of water initially at 0°C. The iron cools and the water warms until they are at the same temperature (in thermal equilibrium). Assuming there is no other transfer of heat involved, is the final temperature closer to 0°C, 50°C, or 100°C? Why?

27. In Example 5.5, is it really necessary to know the mass of the concrete block to solve this problem? Put another way, would the answer be different if it was a 10-kg block that was dropped instead of a 5-kg one?

28. A piece of aluminum and a piece of iron fall without air resistance from the top of a building and stick into the ground on impact. Will their temperatures change by the same amount? Explain.

29. The specific heat capacity of water is extremely high. If it were much lower, say, one-fifth as large, what effect would this have on processes such as fighting fires and cooling automobile engines?

30. ■ Why does the temperature of water not change while it is boiling?

31. ■ Describe how changing the air pressure affects the temperature at which water boils.

32. One way to desalinate seawater—remove the dissolved salts so that the water is drinkable—is to distill it: boil the seawater and condense the steam. The salts stay behind. This technique has one major disadvantage. It consumes a large amount of energy. Why is that?

33. ■ What is saturation density? How does it change when the temperature increases?

34. ■ What effect does heating the air in a room have on the relative humidity?

35. *Wood's metal* is an alloy of the elements bismuth, lead, tin, and cadmium that has a melting point of 70°C. Describe how it might be used in an automatic sprinkler system for fire suppression.

36. When trying to predict the lowest temperature that will be reached overnight, forecasters pay close attention to the dew-point temperature. Why is the air temperature unlikely to drop much below the dew point? (The high latent heat of vaporization of water is important.)

37. ■ Explain what a heat engine does and what a heat mover does.

38. ■ In winter, the amount of internal energy a heat pump delivers to a house is greater than the electrical energy it uses. Does this violate the law of conservation of energy? Explain.

39. ■ What is entropy? In general, how does the entropy of a system change with time?

40. When a parcel of water freezes to become ice, it becomes more ordered and its entropy decreases. Does this violate the alternate form of second law of thermodynamics? Explain.

41. In a galaxy far, far away, a process occurs that results in a decrease in the entropy of a system by 8 J/K; that is, $\Delta S = -8$ J/K. According the entropy form of the second law of thermodynamics, which of the following (there may be more than one) is a possible value for the entropy change, $\Delta S_{universe}$, for the rest of the universe? Explain your choice(s).
    (a) -8 J/K
    (b) 0 J/K
    (c) $+8$ J/K
    (d) $+16$ J/K.

42. A homeowner plans to design and build a new patio using 8 in. by 4 in. by 2 in. paving blocks. The blocks are delivered in a large pile by the quarry operator and then laid by the stone mason in the geometrical pattern specified by the owner.

After 25 years, the home is sold to new owners who wish to put in a deck. They have the patio completely demolished, breaking up the blocks into gravel, which is used as fill at another construction site. In this scenario, over 25 years, the blocks existed in three different states: (1) as a pile of as-yet-unused building material; (2) as a decorative patio; and (3) as broken-up gravel. Which of the following represents the proper ranking of these three states in order of *increasing* entropy? Explain your choice.
(a) 1,2,3
(b) 1,3,2
(c) 2,1,3
(d) 2,3,1
(e) 3,2,1
(f) 3,1,2

43. Is our society truly facing an "energy crisis," assuming by this term we mean that we are running out of energy? What *is* happening to our energy resources as a result of the increasing industrialization of the world?

44. Six insulated containers hold 1500 g of water at 24°C. A small copper cylinder is placed in each cup; the masses and initial temperatures of the cylinders vary as given below. Rank the containers according to the maximum temperature of the water in each container after the cylinder is added, from largest to smallest. You may assume that the cylinder is completely submerged in the water.
Container A: $m = 100$ g; $T = 30°C$
Container B: $m = 200$ g; $T = 60°C$
Container C: $m = 300$ g; $T = 90°C$
Container D: $m = 200$ g; $T = 15°C$
Container E: $m = 300$ g; $T = 30°C$
Container F: $m = 100$ g; $T = 60°C$

45. Six cylinders each contain the same mass of helium gas but have different volumes and temperatures as given below. All of the containers are made of the same material and stored under the same environmental conditions. Rank the containers according to their pressures from highest to lowest. If any of the containers have the same pressure, give them the same ranking. Explain how you formed your rankings.
Container A: $T = 200$ K; $V = 2000$ cm³
Container B: $T = 300$ K; $V = 15,000$ cm³
Container C: $T = 400$ K; $V = 5000$ cm³
Container D: $T = 300$ K; $V = 10,000$ cm³
Container E: $T = 200$ K; $V = 4000$ cm³
Container F: $T = 500$ K; $V = 10,000$ cm³

46. Five Carnot engines operate between reservoirs with temperatures T_h and T_l as given below. Rank these engines according to their efficiency from most efficient to least efficient. If any of the engines have the same efficiency, give them the same ranking. Explain how you arrived at your rankings.
Engine A: $T_h = 100°C$; $T_l = 0°C$
Engine B: $T_h = 250°C$; $T_l = 150°C$
Engine C: $T_h = 75°C$; $T_l = -25°C$
Engine D: $T_h = 300°C$; $T_l = 100°C$
Engine E: $T_h = 300°C$; $T_l = 0°C$

PROBLEMS

1. Your jet is arriving in London, and the pilot informs you that the temperature is 30°C. Should you put on your jacket? Use Figure 5.2 to determine the temperature in degrees Fahrenheit.

2. On a nice winter day at the South Pole, the temperature rises to −60°F. What is the approximate temperature in degrees Celsius?

3. An iron railroad rail is 700 ft long when the temperature is 30°C. What is its length when the temperature is −10°C?

4. A copper vat is 10 m long at room temperature (20°C). How much longer is it when it contains boiling water at 1 atm pressure?

5. A machinist wishes to insert a steel rod with a diameter of 5 mm into a hole with a diameter of 4.997 mm. By how much would the machinist have to lower the temperature of the rod to make it fit the hole?

6. An aluminum wing on a passenger jet is 30 m long when its temperature is 20°C. At what temperature would the wing be 5 cm (0.05 m) shorter?

7. A fixed amount of a particular ideal gas at 16°C and a pressure of 1.75×10^5 Pa occupies a volume of 2.75 m³. If the volume is increased to 4.20 m³ and the temperature is raised to 26.4°C, what will be the new pressure of the gas?

8. The volume of an ideal gas enclosed in a thin, elastic membrane in a room at sea level where the air temperature is 18°C is 8×10^{-3} m³. If the temperature of the room is increased by 10°C, what is the new volume of the gas?

9. A gas is compressed inside a cylinder (Figure 5.16). An average force of 50 N acts to move the piston 0.1 m. During the compression, 2 J of heat are conducted away from the gas. What is the change in internal energy of the gas?

10. Air in a balloon does 50 J of work while absorbing 70 J of heat. What is its change in internal energy?

11. How much heat is needed to raise the temperature of 5 kg of silver from 20°C to 960°C?

12. A bottle containing 3 kg of water at a temperature of 20°C is placed in a refrigerator where the temperature is kept at 3°C. How much heat is transferred from the water to cool it to 3°C?

13. (a) Compute the amount of heat needed to raise the temperature of 1 kg of water from its freezing point to its normal boiling point.
(b) How does your answer to (a) compare to the amount of heat needed to convert 1 kg of water at 100°C to steam at 100°C?

14. Aluminum is melted during the recycling process.
(a) How much heat must be transferred to each kilogram of aluminum to bring it to its melting point, 660°C, from room temperature, 20°C?
(b) About how many cups of coffee could you make with this much energy (see Example 5.4)?

15. A 1,200-kg car going 25 m/s is brought to a stop using its brakes. Let's assume that a total of approximately 20 kg of iron in the brakes and wheels absorbs the heat produced by the friction.
(a) What was the car's original kinetic energy?
(b) After the car has stopped, what is the change in temperature of the brakes and wheels?

16. A 0.02-kg lead bullet traveling 200 m/s strikes an armor plate and comes to a stop (**Figure 5.46**). If all of the bullet's

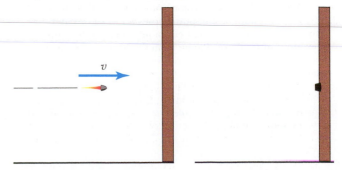

Figure 5.46 Problem 16.

energy is converted to heat that it alone absorbs, what is its temperature change?

17. A 10-kg lead brick is dropped from the top of a 629-m-tall television transmitting tower in North Dakota and falls to the ground. Assuming all of its energy goes to heat it, what is its temperature increase?

18. Water flowing over the Lower Falls in Yellowstone National Park drops 94 m. If all of the water's energy goes to heat it, what is its temperature increase?

19. On a winter day, the air temperature is −15°C, and the humidity is 0.001 kg/m³.
    (a) What is the relative humidity?
    (b) When this air is brought inside a building, it is heated to 20°C. If the humidity isn't changed, what is the relative humidity inside the building?

20. On a summer day in Houston, the temperature is 35°C and the relative humidity is 77 percent.
    (a) What is the humidity?
    (b) To what temperature could the air be cooled before condensation would start to take place? (That is, what is the dew point?)

21. Inside a building, the temperature is 20°C, and the relative humidity is 40 percent. How much water vapor is there in each cubic meter of air?

22. On a hot summer day in Washington, D.C., the temperature is 86°F, and the relative humidity is 70 percent. How much water vapor does each cubic meter of air contain?

23. An apartment has the dimensions 10 m by 5 m by 3 m. The temperature is 25°C, and the relative humidity is 60 percent. What is the total mass of water vapor in the air in the apartment?

24. The total volume of a new house is 800 m³. Before the heat is turned on, the air temperature inside is 10°C, and the relative humidity is 50 percent. After the air is warmed to 20°C, how much water vapor must be added to the air to make the relative humidity 50 percent?

25. The temperature of the air in thermals decreases about 10°C for each 1,000 m they rise. If a thermal leaves the ground with a temperature of 30°C and a relative humidity of 31 percent, at what altitude will the air become saturated and the water vapor begin to condense to form a cloud? (In other words, at what altitude does the temperature equal the dew point?)

26. In cold weather, you can sometimes "see" your breath. What you are seeing is a mist of small water droplets, the same as in clouds and fog. Suppose air leaves your mouth with temperature 35°C and humidity 0.035 kg/m³ and mixes with an equal amount of air at 5°C and humidity 0.005 kg/m³.
    (a) What is the relative humidity of the mixed air if its temperature and humidity equal the averages of those of the two original air masses?
    (b) Represent what happens by plotting three points in a graph like Figure 5.34.

27. What is the Carnot efficiency of a heat engine operating between the temperatures of 300°C (573 K) and 100°C (373 K)?

28. What is the maximum efficiency that a heat engine could have when operating between the normal boiling and freezing temperatures of water?

29. As a gasoline engine is running, an amount of gasoline containing 15,000 J of chemical potential energy is burned in 1 s. During that second, the engine does 3,000 J of work.
    (a) What is the engine's efficiency?
    (b) The burning gasoline has a temperature of about 4,000°F (2,500 K). The waste heat from the engine flows into air at about 80°F (300 K). What is the Carnot efficiency of a heat engine operating between these two temperatures?

30. A proposed ocean thermal-energy conversion (OTEC) system is a heat engine that would operate between warm water (25°C) at the ocean's surface and cooler water (5°C) 1,000 m below the surface. What is the maximum possible efficiency of the system?

31. An irreversible process takes place by which the entropy of the universe increases by 1.67 J/K. If the temperature of the environment in which the process occurred is 220 K, how much energy was made unavailable for useful work as a result?

32. The temperature in the deep interiors of some giant molecular clouds in the Milky Way galaxy is 50 K. Compare the amount of energy that would have to be transferred to this environment to the amount that would have to transferred to a room temperature environment to bring about a 1.0 J/K increase in the entropy of the universe in each case.

CHALLENGES

1. A solid cube is completely submerged in a particular liquid and floats at a constant level. When the temperature of the liquid and the solid is raised, the liquid expands more than the solid. Will this make the solid rise upward, remain floating, or sink? Explain the rationale for your choice.

2. Pyrex glassware is noted for its ability to withstand sudden temperature changes without breaking. Explain how its coefficient of linear expansion contributes to this ability.

3. During a workout, a shot-putter loses 0.175 kg of water through evaporation. The heat required to produce the evaporation (a phase change from liquid water to water vapor) comes from the athlete's body. During his exercise, the shot-putter does 1.25×10^5 J of work.
   (a) Assuming it requires 2.42×10^6 J of energy to vaporize one kilogram of perspiration, find the amount of heat transferred from the athlete to evaporate the sweat and to help cool him.
   (b) Find the change in the internal energy of the shot-putter during his workout.
   (c) Determine the minimum number of food Calories that must be consumed to replace this loss of internal energy.

4. As air rises in the atmosphere, its temperature drops, even if no heat flows out of it.
   (a) Based on what you learned in Sections 4.4 and 5.3, explain why this is so.
   (b) Cumulus clouds form when rising air is cooled to the point where water droplets form because of condensation. Why are these clouds usually much higher above the ground in dry climates than in wet ones?

5. If air at 35°C and 77 percent relative humidity is cooled to 25°C, what mass of water would condense out of the air in a room that measures 5 m by 4 m by 3 m? (See Problem 20.)

6. The door of a refrigerator is left open. Assuming the refrigerator is in a closed room, will the air in the room eventually be cooled?

6

WAVES AND SOUND

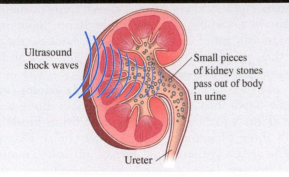

Figure CO-6 ESWL being used to fragment kidney stones too large to pass out of the body naturally. The ultrasound shock waves employed in this procedure are produced and focused outside the body by several physical techniques.

CHAPTER INTRODUCTION: Sound Medicine

The diagnosis: kidney stones, a disease that sometimes afflicts people in their 20s and 30s. The condition is very painful, and, if left untreated, can lead to death. One treatment option is open surgery, which, because of the possibility of complications and a four-to-six-week recovery time, is now used mainly in cases of very large stones that block the flow of urine. More commonly, since the 1980s, there has been an alternative that works in most cases: you can have the stones pulverized with sound. No scalpels involved.

The process is called *extracorporeal shock-wave lithotripsy* (ESWL). A device called a lithotripter focuses intense sound waves on the stones, which are broken into tiny fragments that can then pass easily out of the patient's body (**Figure CO-6**). The sound is produced and focused outside of the body—hence the term *extracorporeal*. Some ESWL systems make use of a reflector based on the shape of an ellipse (Figure 2.43). An intense sound pulse (shock wave) produced at one focus bounces off the reflector and converges on the other focus. The reflector is positioned so that the stone is at that second focus. Other lithotripters focus the sound with an "acoustic lens" in much the same way a magnifying glass can be used to focus sunlight to start a fire. In

both systems, the sound is produced and focused inside a water-filled "cushion," similar to a balloon, that is pressed against the patient's body. The sound never travels in air, and for good reason, as we shall see later in this chapter.

The effective use of sound in this healing procedure has much to do with its wave-like nature. Waves are an integral part of our everyday lives. Whether playing a guitar, listening to a radio, clocking the speed of a thrown baseball, or having a kidney stone shattered, we are using a wave of some kind. Our two most often used senses—sight and hearing—are highly developed wave-detection mechanisms. In the first part of this chapter, we look at simple waves and examine some of their general properties. The remainder of the chapter is about sound—how it is produced, how it travels in matter, and how it is perceived by humans.

6.1 Waves—Types and Properties

Ripples moving over the surface of a still pond, sound traveling through the air from a radio speaker, a pulse "bouncing" back and forth on a piano string, light from the Sun illuminating and warming Earth—these are all waves (**Figure 6.1**). We can feel the effects of some waves, such as earthquake tremors (called *seismic waves*), as they pass. Others, such as sound and light, we sense directly with our ears and eyes. Technology has given us numerous devices that produce or detect waves that we cannot sense (microwaves, ultrasound, x-rays).

What are waves? Though many and diverse, they share some basic features. They all *involve vibration or oscillation* of some kind. Floating leaves show the vibration of the water's surface as ripples move by. Our ears respond to the oscillation of air molecules and give us the perception of sound. Also, waves move and *carry energy* yet do not have mass. The sound from a loudspeaker can break a wineglass even though no matter moves from the speaker to the glass. We can define a wave as follows.

Figure 6.1 Like all waves, these water ripples involve oscillation.

Mario7/Shutterstock.com

<table>
</table>

| DEFINITION | **Wave** A traveling disturbance consisting of coordinated vibrations that transmit energy with no net movement of matter. |

Sound, water ripples, and similar waves consist of vibrations of matter—air molecules or the water's surface, for example. The substance through which such waves travel is called the *medium* of the wave. Particles of the medium vibrate in a coordinated fashion to form the wave.

A rope stretched between two people is a handy medium for demonstrating a simple wave (**Figure 6.2**). A flick of the wrist sends a wave pulse down the rope. Each short segment of the rope is pulled upward in turn by its neighboring segment. The forces between the parts of the medium are responsible for "passing along" the wave. This kind of wave is not unlike a row of dominoes knocking each other over, except that the medium of a wave does not have to be "reset" after a wave goes by.

Many waves—sound, water ripples, waves on a rope—require a material medium. They cannot exist in a vacuum. On the other hand, light, radio waves, microwaves, and x-rays can travel through a vacuum because they do not require a medium for their propagation. We will take a close look at these special waves—called *electromagnetic waves*—in Chapter 8.

Waves occur in a great variety of substances: in gases (sound), liquids (water ripples), and solids (seismic waves through rock). Some travel along a line (a wave on a rope), some across a surface (vibrations on a drum head) and some throughout space in three dimensions (light). Many more examples could be listed. Clearly, waves are everywhere, and they are diverse in nature.

A wave can be short and fleeting, called a *wave pulse,* or steady and repeating, called a *continuous wave.* The sound of a bursting balloon, a tsunami (large, solitary ocean wave generated by an earthquake), and the light from a camera flash are examples of wave pulses. The sound from a tuning fork and the light from the Sun are continuous waves. **Figure 6.3** shows a wave pulse and a continuous wave on a long rope. You can see that a continuous wave is like a series or "train" of wave pulses, one after another. (For now, we can treat light as a simple wave. In Chapter 10, we will present the modern view of light in terms of a stream of particles called *photons* as developed by Albert Einstein and others.)

6.1a Wave Types and Speed

If we take a close look at many different types of waves, we find that they can be classified according to the orientation of the wave oscillations. There are two main wave types: **transverse** and **longitudinal**.

| DEFINITION | **Transverse Wave** A wave in which the oscillations are perpendicular (transverse) to the direction the wave travels. Examples: waves on a rope, electromagnetic waves, some seismic waves. |

Longitudinal Wave A wave in which the oscillations are along the direction the wave travels. Examples: sound in the air, some seismic waves.

Vern Ostdiek/University of Michigan

Figure 6.2 A wave pulse traveling on a ropelike spring.

Both types of waves can be produced on a Slinky—a short, fat spring that was a popular toy a generation or more ago. If a Slinky is stretched out on a flat, smooth tabletop, a transverse wave can be produced by moving one end from side to side, perpendicular to the Slinky's length (**Figure 6.4a**). A longitudinal wave is produced by pushing and pulling one end back and forth, first toward the other end, then back (**Figure 6.4b**). For each type of wave, one can produce either a wave pulse or a continuous wave.

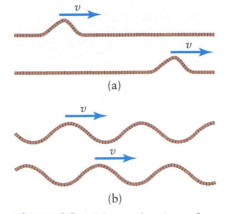

Figure 6.3 (a) Successive views of a wave pulse as it travels on a rope. (b) A continuous wave traveling on a rope.

▶ Physics To Go 6.1

For this, you'll need to borrow or buy a Slinky, preferably the metal kind. On a smooth table, large desk, or bare floor, stretch the Slinky out about 5 feet, with the other end held by a partner or fastened rigidly to the surface.

1. Send a transverse pulse down the Slinky by quickly moving your hand to the side and back. Send a longitudinal pulse by quickly moving your hand toward the other end and back. Do the two pulses seem to travel at the same speed? (Do they take about the same amount of time to get to the other end?)
2. Again send a transverse pulse down the Slinky. Watch what happens when it reaches the other end. Does it reflect? If so, is the reflected pulse identical to the original pulse?

A Slinky is not the only medium that can carry both transverse and longitudinal waves. Both kinds of waves can travel in any solid. Earthquakes and underground explosions produce both longitudinal and transverse seismic waves that travel through Earth. Simple waves that involve oscillation of atoms and molecules *must* be longitudinal to travel in liquids and gases because of the absence of rigid bonds between the particles.

Many waves are neither purely longitudinal nor purely transverse. Although a water ripple appears to be a simple transverse wave, individual parcels of water actually move in circles or ellipses—they oscillate forward and backward as well as up and down. Waves in plasmas and in the atmosphere are even more complicated. But the two simple types of waves described here are common and well suited for illustrating wave phenomena.

The speed of a wave is the rate of movement of the disturbance. (Do not confuse this with the speed of individual particles as they oscillate.) For a given type of wave, the speed is determined by the properties of the medium. In the waves that we have been discussing, the masses of the particles that oscillate

(a)

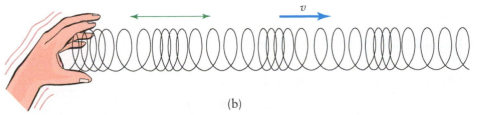

(b)

Figure 6.4 (a) A continuous transverse wave on a Slinky. Each coil oscillates up and down as the wave travels to the right. (b) A continuous longitudinal wave on a Slinky. Each coil oscillates left and right as the wave travels to the right.

and the forces that act between them affect the wave speed. As a longitudinal wave, for example, travels on a Slinky, each coil is accelerated back and forth by its neighbors. Basic mechanics tells us that the mass of each coil and the size of the force acting on it will determine how quickly it—and therefore the wave—moves. In general, weak forces or massive particles in a medium cause the wave speed to be low.

Often, the speed of waves in a medium can be predicted by measuring some other properties of the medium. After all, the factors that affect wave speed—particle masses and interparticle forces—also affect other properties of a substance. For example, the speed of waves on a stretched rope or a Slinky or on a taut wire can be computed by using the force F that must be exerted to keep it stretched and its **linear mass density ρ**, which equals its mass m divided by its length l. (The symbol ρ represents the Greek letter rho, pronounced like *row*.) In particular,

$$v = \sqrt{\frac{F}{\rho}} \qquad \left(\text{wave on a rope or spring; } \rho = \frac{m}{l}\right)$$

Increasing this force, also called the *tension*, will cause the waves to move faster. This is how stringed instruments such as guitars and pianos are tuned. (More on this in Section 6.4.)

EXAMPLE 6.1 A student stretches a Slinky out on the floor to a length of 2 meters. The force needed to keep the Slinky stretched is measured and found to be 1.2 newtons. The Slinky's mass is 0.3 kilograms. What is the speed of any wave sent down the Slinky by the student?

SOLUTION First, we compute the Slinky's linear mass density:

$$\rho = \frac{m}{l} = \frac{0.3 \text{ kg}}{2 \text{ m}}$$

$$= 0.15 \text{ kg/m}$$

The speed of waves on the Slinky is then

$$v = \sqrt{\frac{F}{\rho}} = \sqrt{\frac{1.2 \text{ N}}{0.15 \text{ kg/m}}}$$

$$= \sqrt{8 \text{ m}^2/\text{s}^2} = 2.8 \text{ m/s}$$

The speed of sound in air or any other gas depends on the ratio of the pressure of the gas to the density of the gas. But for each type of gas, this ratio depends only on the temperature. In particular, the speed of sound in a gas is proportional to the square root of the Kelvin temperature. For air,

$$v = 20.1 \times \sqrt{T} \qquad \text{(speed of sound in air; } v \text{ in m/s, } T \text{ in kelvins)}$$

Although the air is thinner at higher altitudes, the speed of sound there is actually lower because the air is colder at these elevations.

EXAMPLE 6.2 What is the speed of sound in air at room temperature (20°C = 68°F)?

SOLUTION The temperature in kelvins is

$$T = 273 + 20 = 293 \text{ K}$$

Therefore,

$$v = 20.1 \times \sqrt{T}$$
$$= 20.1 \times \sqrt{293} = 20.1 \times 17.1$$
$$= 344 \text{ m/s } (770 \text{ mph})$$

The numerical factor (20.1) in the equation in Example 6.2 is determined by the properties of the molecules that comprise air and therefore applies to air only. The speed of sound in any other gas will be different, and the corresponding equation for v will have a different numerical factor. Two examples:

Speed of sound in helium: $v = 58.8 \times \sqrt{T}$ (SI units)

Speed of sound in carbon dioxide (CO_2): $v = 15.7 \times \sqrt{T}$ (SI units)

> ### ▶ Physics To Go 6.2
>
> (*Note: The idea for this activity came from the* Constructing Physics Understanding *website at San Diego State University.*)
>   For this, you need two identical narrow-necked glass bottles, the ability to produce a tone in a bottle by blowing across its opening, and a seltzer tablet.
>
> 1. Put water into both bottles until they are about half full. "Tune" them so they produce the same tone or note when you blow across their openings by adding or removing water.
> 2. Add a seltzer tablet to one of the bottles and wait a minute for the fizzing to die down. Carbon dioxide (CO_2) gas produced during the process replaces most of the air in the bottle.
> 3. Now blow across each bottle as before. Do they still produce the same tone? If not, how do they differ? What physical property of the sound in each bottle has changed? How might this cause the perceived tonal differences?

6.1b Amplitude, Wavelength, and Frequency

For the remainder of this section, we will take a look at some of the properties of a continuous wave. A convenient example is a transverse wave on a Slinky produced by moving one end smoothly side to side. **Figure 6.5** shows a "snapshot" of such a wave. It shows the shape of the Slinky at some instant in time. Note that the wave has the same sinusoidal shape you've seen before (Figure 2.25).

The high points of the wave are called *peaks* or *crests,* and the low points are called *valleys* or *troughs.* The dashed line through the middle represents the equilibrium configuration of the medium—its shape when there is no wave.

In addition to wave speed, there are three other important parameters of a continuous wave that can be measured: **amplitude**, **wavelength**, and **frequency**. At any moment, the different particles of the medium are generally displaced from their equilibrium positions by different amounts. The maximum displacement is called the *amplitude* of the wave.

The amplitude is just a distance equal to the height of a peak or the depth of a valley, which are the same for a pure wave. The amplitude of a particular type of wave can vary greatly. For water waves, it can be a few millimeters for ripples to tens of meters for ocean waves (**Figure 6.6**). When we hear a sound, its loudness depends on the amplitude of the sound wave: louder sounds have larger amplitudes.

There is also a large variation in the wavelengths of particular types of waves. The wavelengths of sound (in air) that can be heard by humans range from about 2 centimeters (very high pitch) to about 17 meters (very low pitch).

> **DEFINITION** **Amplitude** The maximum displacement of points on a wave measured from the equilibrium position.

> **DEFINITION** **Wavelength** The distance between two successive "like" points on a wave. For example, the distance between two adjacent peaks or two adjacent valleys. Wavelength is represented by the lowercase Greek letter lambda (λ).

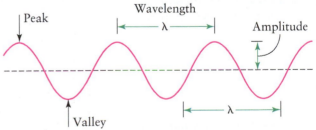

Figure 6.5 "Snapshot" of a transverse wave. The dashed line shows the equilibrium configuration—the position of the medium when no wave is present.

DEFINITION

Frequency The number of cycles of a wave passing a point per unit time. The number of oscillations per second in the wave.

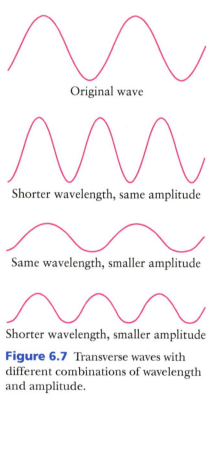

Figure 6.6 A wave with a large amplitude.

Original wave

Shorter wavelength, same amplitude

Same wavelength, smaller amplitude

Shorter wavelength, smaller amplitude

Figure 6.7 Transverse waves with different combinations of wavelength and amplitude.

Figure 6.8 The amplitude of a longitudinal wave on a Slinky equals the greatest lateral displacement of the coils.

Typical wavelengths for radio waves are 3 meters for FM stations and 300 meters for AM stations.

Any segment of a wave that is one wavelength long is called one *cycle* of the wave. As each cycle of a wave passes by a given point in the medium, that point makes one complete oscillation—up, down, and back to the starting position. Figure 6.5 shows three complete cycles of a wave.

Amplitude and wavelength are independent features of a wave: a short-wavelength wave can have a small or a large amplitude (**Figure 6.7**).

To understand what the frequency of a wave is, we must "unfreeze" the wave and imagine it as it moves along. The rate at which the wave cycles pass a point is the frequency of the wave. Recall from Section 1.1 that the unit of measure of frequency is the hertz (Hz).

If you move the end of a Slinky back and forth three times each second, you will produce a wave with a frequency of 3 hertz. The note A above middle C on a modern piano has a frequency of 440 hertz. This means that 440 cycles of the sound wave reach your ear each second. The piano wires producing the sound and the air molecules in the room all vibrate with the same frequency: 440 hertz.

Under ideal conditions, a person with good hearing can hear sounds with frequencies as low as 20 hertz or as high as 20,000 hertz. Frequency is important in other kinds of waves as well. Each radio station broadcasts a radio wave with a specific frequency—for example, 1,100 kilohertz = 1,100,000 hertz, or 92.5 megahertz = 92,500,000 hertz.

Amplitude, wavelength, and frequency can be identified for both transverse waves and longitudinal waves, although the amplitude of a longitudinal wave is a bit difficult to visualize. It is still the maximum displacement from the equilibrium position, but in this case the displacement is along the direction the wave is traveling. **Figure 6.8** shows a close-up of a Slinky with no wave and then one with a longitudinal wave traveling on it. The amplitude is the farthest distance that any coil is displaced to the right or left of its equilibrium position. The regions where the coils are squeezed together are called **compressions**, and the regions where they are spread apart are called **expansions** or rarefactions. The wavelength is the distance between two adjacent compressions or two adjacent expansions.

The speed of a wave, its wavelength, and its frequency are related to each other in a simple way. Imagine a continuous wave passing by a point, perhaps ripples moving by a plant stem. The speed of the wave equals the number of cycles that pass by each second multiplied by the length of each cycle. For example, if five cycles pass the stem each second and the peaks of the ripples are 0.03 meters apart, the wave speed is 0.15 m/s (**Figure 6.9**). In general,

wave speed = number of cycles per second × length of each cycle

The two quantities on the right of the equal sign are the frequency of the wave and the wavelength, respectively. Therefore,

$$v = f\lambda$$

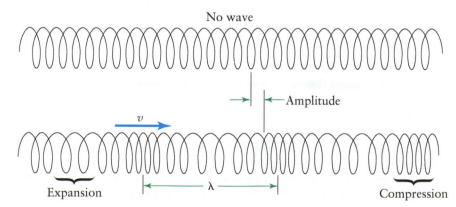

No wave

Amplitude

v

Expansion

λ

Compression

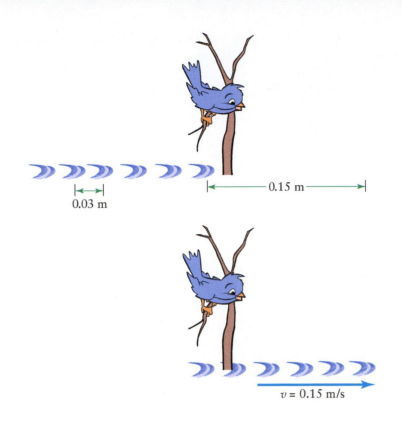

0.15 m

0.03 m

$v = 0.15$ m/s

The velocity of a continuous wave is equal to the frequency of the wave times the wavelength.

In many cases, all waves that travel in a particular medium have the same speed. Sound is an important example of this; sound pulses, low-frequency sounds, and high-frequency sounds travel through the air with the same speed: 344 m/s at room temperature. Similarly, light, radio waves, and microwaves travel with the same speed in a vacuum: 3×10^8 m/s. According to the equation $v = f\lambda$, when the wave speed is the same for all waves, *higher* frequency waves must have proportionally *shorter* wavelengths. A 20-hertz sound wave has a wavelength of about 17 meters, whereas a 20,000-hertz sound wave has a wavelength of about 1.7 centimeters.

EXAMPLE 6.3 Before a concert, musicians in an orchestra tune their instruments to the note A, which has a frequency of 440 hertz. What is the wavelength of this sound in air at room temperature?

SOLUTION The speed of sound at this temperature is 344 m/s, so

$$v = f\lambda$$
$$344 \text{ m/s} = 440 \text{ Hz} \times \lambda$$
$$\frac{344 \text{ m/s}}{440 \text{ Hz}} = \lambda$$
$$\lambda = 0.78 \text{ m} = 2.6 \text{ ft}$$

The wavelength of sound with a frequency of 220 hertz is twice as large—1.56 meters.

In some media, the speed of wave propagation varies with frequency so that the relationship between frequency and wavelength is not as simple as that discussed here. In such cases, the wave undergoes *dispersion*, an effect that can change the shape of a wave pulse as it moves through or on the medium. We

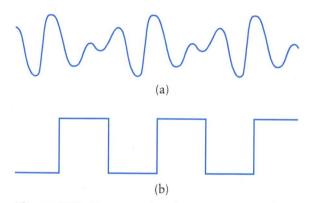

(a)

(b)

Figure 6.10 Two examples of complex waves. The lower wave, a "square wave," is used in many electronic devices.

shall encounter additional dispersive phenomena in Chapter 9 in connection with our discussion of light. For now, we will ignore such complications in our treatment of sound waves.

Not all continuous waves have the simple sinusoidal shape shown in Figure 6.5. In fact, waves with precisely that shape are relatively rare. Any continuous wave that does not have a sinusoidal shape is called a *complex wave*. **Figure 6.10** shows two examples. Note that there are three different-sized peaks in each cycle of the upper wave. The shape of a wave is called its *waveform*. The two complex waves in the figure have about the same wavelength and amplitude, but they have very different waveforms. The waveform is another feature that is needed when comparing complex waves. We will take a closer look at this in Section 6.6.

Concept Map 6.1 summarizes the characteristics of waves.

■ **Learning Check**

1. (True or False.) The flash from a camera is an example of a continuous wave.

2. A _____ wave is one with the oscillations perpendicular to the direction of wave propagation.

3. The only way to make sound travel faster through air is to increase the air's _____.

4. (True or False.) Regions in a longitudinal wave where the particles of the medium are squeezed together are called compressions.

5. Two sound waves traveling through air have different frequencies. Consequently, they must also have different
   (a) amplitudes.
   (b) speeds.
   (c) wavelengths.
   (d) All of the above.

ANSWERS: 1. False **2.** transverse **3.** temperature **4.** True **5.** (c)

6.2 Aspects of Wave Propagation

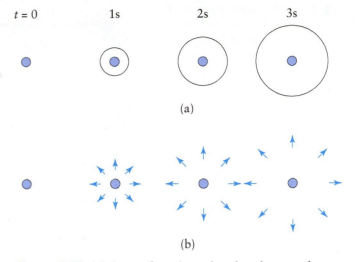

(a)

(b)

Figure 6.11 (a) A wave front is used to show how a pulse spreads over water. (b) The same wave pulse, at the same times, represented with rays. The rays point in the direction the wave travels and are perpendicular to the wave front.

6.2a Wave Fronts and Rays

In this section, we consider what waves do as they travel. For waves traveling along a surface or throughout space in three dimensions, it is convenient to use two different ways to represent the wave. We will call these the **wave-front model** and the **ray model**. **Figure 6.11** shows how each is used to illustrate a wave pulse on water as it travels from the point where it was produced. The wave front is a circle that shows the location of the peak of the wave pulse. A ray is a straight arrow that shows the direction a given segment of the wave is traveling. A laser beam and sunlight passing through a small hole in a window shade both approximate individual rays of light that we can see if there is dust in the air. On the other hand, the rays of water ripples are not visible, but we do see the wave fronts (Figure 6.1).

For a continuous water wave, the wave fronts are concentric circles around the point of origin (the "source" of the wave) that represent individual peaks of the wave (**Figure 6.12**). The largest circle shows the position of the first peak that was produced. Each successive wave front is smaller because it came later and has not traveled as far. The distance between adjacent wave fronts is equal to the wavelength of the wave. Again, a continuous wave is like a series of wave pulses produced one after another. The rays used to represent a continuous wave are lines radiating from the source of the wave (the blue arrows in Figure 6.12). The wave fronts arriving at a point far from the source are nearly straight lines (far right in Figure 6.12). The corresponding rays are nearly parallel.

For a wave moving in three-dimensional space, like the sound traveling outward from you in all directions as you shout or whistle, the wave fronts are spherical shells surrounding the source of the wave. The wave front of a wave pulse, such as the sound from a hand clap, expands like a balloon that is being inflated very rapidly. For continuous three-dimensional waves such as a steady whistle, the wave fronts form a series of concentric spherical shells that expand like the circular wave fronts of a wave on a surface. A 440-hertz tuning fork produces 440 of these wave fronts each second. The surface of each wave front expands outward with a speed of 344 m/s (at room temperature). As with waves on a surface, the rays used to represent a continuous wave in three dimensions are lines radiating outward from the wave source in all directions.

One inherent aspect of the propagation of waves on a surface or in three dimensions is that the amplitude of the wave necessarily decreases as the wave gets farther from the source. A certain amount of energy is expended to create a wave pulse or each cycle of a continuous wave. This energy is distributed over the wave front and determines the amplitude of the wave: the greater the amount of energy given to a wave front, the larger the amplitude. As the wave front moves out, it gets larger, so this energy is spread out more and becomes less concentrated. This attenuation accounts for the decrease in loudness of sound as a noisy car moves away from you and for the decrease in brightness of a lightbulb as you move away from it.

Waves that require a medium for their support and propagation often transfer some of their energy to their surroundings, which also reduces the wave amplitude. Such energy losses further *attenuate* the amplitude beyond what would be predicted from a simple consideration of the

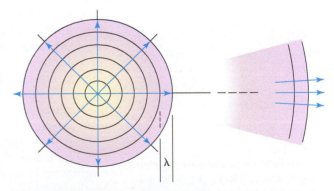

Figure 6.12 Wave fronts and rays for a continuous wave on a surface. Far away from the wave source, the wave fronts are nearly planar, and the rays are nearly parallel.

Figure 6.13 A wave pulse traveling on a rope is reflected at a fixed end. In this case, the pulse is inverted.

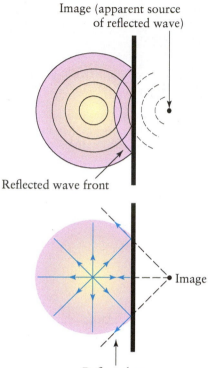

Image (apparent source of reflected wave)

Reflected wave front

Image

Reflected ray

Figure 6.14 The reflection of water ripples off the side of a pool. Both models of the wave show that the reflected wave appears to diverge from a point behind the wall called the *image*.

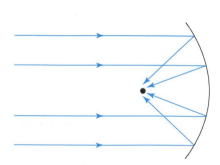

Figure 6.15 A sound wave reflecting off a concave surface. The reflected rays converge toward a point, indicating that the wave's amplitude increases.

effects of distance alone. For the most part, we shall ignore such complications in the discussion that follows.

One can infer when the amplitude of a wave is changing by noting changes in the wave front or the rays. If the wave fronts are growing larger, then the amplitude is getting smaller. The same thing is indicated when the rays are diverging (slanting away from each other).

At great distances from the source of a three-dimensional wave, the wave fronts become nearly flat and are called *plane waves*. The corresponding rays are parallel, and the wave's amplitude stays constant. The light and other radiation we receive from the Sun come as plane waves because of the great distance between Earth and the Sun.

With this background, we will look at several phenomena associated with wave propagation.

6.2b Reflection

Think about how many times you looked in a mirror today. That's a very common use we make of the reflection of waves, but it's not the only one. As we will see, the sound we hear inside rooms is affected by reflection, musical instruments like guitars make use of it when producing sound, radar and sonar systems use it for a variety of purposes (such as checking how fast we are driving), and so on.

A wave is reflected whenever it reaches a boundary of its medium or encounters an abrupt change in the properties (density, temperature, and so on) of its medium. A wave pulse traveling on a rope is reflected when it reaches a fixed end (**Figure 6.13**). It "bounces" off the end and travels back along the rope. Notice that the reflected pulse is inverted. When the end of the rope is attached to a very light (but strong) string instead, the reflected pulse is not inverted. (The incoming pulse causes two pulses to leave the junction, a reflected pulse and a pulse that continues into the light string.) This reflection occurs because of an abrupt change in the density of the medium from high density (for the heavy rope) to low density (for the light string).

Similarly, a wave on a surface or a wave in three dimensions is reflected when it encounters a boundary. The wave that "bounces back" is called the *reflected wave*. Rays are more commonly used to illustrate reflection because they nicely show how the direction of each part of the wave is changed. When a wave is reflected from a straight boundary (for surface waves) or a flat boundary (in three dimensions), the reflected wave appears to be expanding out from a point behind the boundary (**Figure 6.14**). This point is called the *image* of the original wave source. An echo is a good example: sound that encounters a large flat surface, such as the face of a cliff, is reflected and sounds like it is coming from a point behind the cliff.

Our most common experience with reflection is that of light from a mirror. The image that you see in a mirror is a collection of reflected light rays originating from the different points on the object you see. (More on this in Section 9.2.)

Reflection from surfaces that are not flat (or straight) can cause interesting things to happen to waves. **Figure 6.15** shows a wave being reflected by a curved surface. Note that the rays representing the reflected part of the wave are converging toward each other. This means that the amplitude of the wave is increasing—the wave is being "focused." Parabolic microphones seen on the sidelines of televised football games use this principle to reinforce the sounds made on the playing field. Satellite receiving dishes do the same with radio waves (**Figure 6.16**).

A reflector in the shape of an ellipse has a useful property. (We saw in Section 2.8 that the orbits of satellites, comets, and planets can be ellipses.) An ellipse has two points in its interior called *foci* (the plural of *focus*). If a wave is produced at one focus, it will converge on the other focus after reflecting off the elliptical surface. All rays originating from one focus reflect off the ellipse and pass through the other focus (**Figure 6.17**). A room shaped like an ellipse is called a *whispering chamber* because a person standing at one focus can hear faint sounds—even whispering—produced at the other focus. This property of the ellipse is also used in the medical treatment of kidney stones (recall the Chapter Introduction).

Figure 6.16 Parabolic dish antennas focus radio waves by reflection.

> ### Physics To Go 6.3
>
> Locate an isolated building with a large, flat side. Stand more than 15 meters (50 feet) away and clap your hands once. You should hear an echo shortly after the clap. Slowly walk toward the building, clapping your hands occasionally. What is different about the time between a clap and when you hear its echo? At some point you should notice that you no longer hear the echo because it comes too quickly after the direct sound. About how far are you from the reflecting surface when this happens?

6.2c Doppler Effect

Can you recall the last time a fast-moving emergency vehicle with its siren blaring passed near you? If so, you may remember that the pitch or tone of its sound dropped suddenly as it went by—although you may be so used to this phenomenon that you didn't notice it. This is a manifestation of the Doppler effect: the apparent change in the frequency of wave fronts emitted by a moving source, perhaps a tugboat floating down a river or a train traveling along a track, each blowing its horn. Each wave front expands outward from the point where the source was when it emitted that wave front. In contrast to what is shown in Figure 6.12, where the source is stationary, ahead of the moving source, the wave fronts are bunched together (**Figure 6.18**). This means that the wavelength is *shorter* than when the source is at rest, and therefore the frequency of the wave is *higher*. Behind the moving source, the wave fronts are spread apart: the wavelength is *longer*, and the frequency is *lower* than when the source is motionless. In both places, the higher the speed of the wave source, the greater the change in frequency. (*Note*: The speed of a wave in a medium is constant and is not affected by any motion associated with the wave source. Thus, if the wavelength goes up, the frequency must go down, and vice versa, to yield a constant wave speed: $v = \lambda f$.)

The frequency of sound that reaches a person in front of a moving train is higher than that perceived when the train is not moving. A person behind

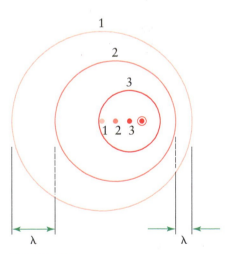

Figure 6.18 The Doppler effect with a moving source. A wave source moves to the right with constant speed. Each dot indicates the source's location when a wave front was emitted. Wave front 1 was emitted when the source was at position 1, and similarly for 2 and 3. Ahead of the source (to the right), the wavelength of the wave is decreased. Behind the source (to the left), the wavelength is increased.

Figure 6.17 The focusing property of an elliptical reflector. Each ray of a wave produced at one focus reflects off the ellipse and passes through the other focus.

Figure 6.19 The Doppler effect with moving observers. The person in the car on the left hears a higher frequency than the pedestrian. The person in the car on the right hears a lower frequency.

the moving train hears a lower frequency. As a train or a fast car moves by, you hear the sound shift from a higher frequency (pitch) to a lower frequency. The change in the loudness of the sound, which you also hear, is *not* part of the Doppler effect: it involves a separate process.

A similar shift in frequency of sound occurs if *you* are moving toward a stationary sound source (**Figure 6.19**). This Doppler shift happens because the speed of the wave relative to you is higher than that when you are not moving. The wave fronts approach you with a speed equal to the wave speed plus your speed. Because the wavelength is not affected, the equation $v = f\lambda$ tells us that the frequency of the wave is increased in proportion to the speed of the wave relative to you. By the same reasoning, when one is moving away from the sound source, the frequency is reduced.

The Doppler effect occurs for both sound and light and is routinely taken into account by astronomers. The frequencies of light emitted by stars that are moving toward or away from Earth are shifted. If the speed of the star is known, the original frequencies of the light can be computed. If the frequencies are known instead, the speed of the star can be computed from the amount of the Doppler shift. Such information is essential for determining the motions of stars in our galaxy or of entire galaxies throughout the universe (see the Astronomy Application at the end of this section).

Echolocation is the process of using the waves reflected from an object to determine its location. *Radar* and *sonar* are two examples. Basic echolocation uses reflection only: a wave is emitted from a point, reflected by an object of some kind, and detected on its return to the original point. The time between the emission of the wave and the detection of the reflected wave (the round-trip time) depends on the speed of the wave and the distance to the reflecting object. For example, if you shout at a cliff and hear the echo 1 second later, you know that the cliff is approximately 172 meters away. This is because the sound travels a total of 344 meters (172 meters each way) in 1 second (at room temperature). If it takes 2 seconds, the cliff is approximately 344 meters away, and so on (**Figure 6.20**).

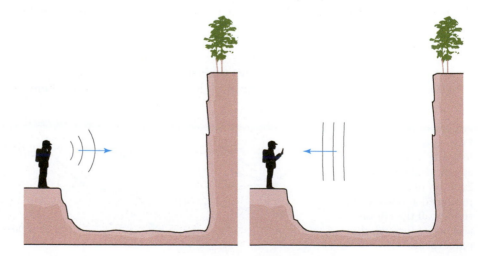

Figure 6.20 Simple echolocation. You can determine the distance to the cliff by timing the echo.

With sonar, a sound pulse is emitted from an underwater speaker, and any reflected sound is detected by an underwater microphone. The time between the transmission of the pulse and the reception of the reflected pulse is used to determine the distance to the reflecting object. Basic radar uses a similar process with microwaves that reflect off aircraft, raindrops, and other things.

Incorporating the Doppler effect in echolocation makes it possible to immediately determine the speed of an approaching or departing object. A moving object causes the reflected wave to be Doppler shifted. If the frequency of the reflected wave is *higher* than that of the original wave, the object is moving *toward* the source. If the frequency is *lower,* then the object is moving *away.*

Doppler radar uses this combination of echolocation and the Doppler effect. The time between transmission and reception gives the distance to the object, whereas the amount of frequency shift is used to determine the speed. Law-enforcement officers use Doppler radar to check the speeds of vehicles (**Figure 6.21**), and Doppler radar is also used in baseball, tennis, and other sports to clock the speed of a ball. Dust, raindrops, and other particles in air reflect microwaves, making it possible to detect the rapidly swirling air in a tornado with Doppler radar. Another potentially life-saving application is the detection of wind shear—drastic changes in wind speed near storms that have caused low-flying aircraft to crash. (There are a few more examples of the application of this physics at the end of Section 6.3.)

Figure 6.21 Radar unit in use.

6.2d Bow Waves and Shock Waves

In the previous discussion, we have implicitly assumed that the speed of the wave source is much less than the wave speed itself. However, if you've ever heard a sonic boom or been jostled by the wake of a passing watercraft while floating in the water, you've had experience with circumstances where the reverse is true. **Figure 6.22a** shows another series of wave fronts produced by a moving wave source. This time the speed of the wave source is *greater than* the wave speed. The wave fronts "pile up" in the forward direction and form a large-amplitude wave pulse called a **shock wave**. This is what causes the V-shaped bow waves produced by swimming ducks and moving boats (**Figure 6.22b**).

Aircraft flying faster than the speed of sound produce a similar shock wave. In this case, the three-dimensional wave fronts form a conical shock wave, with the aircraft at the cone's apex. This conical wave front moves with the aircraft and is heard as a sonic boom (a sound pulse) by persons on the ground.

6.2e Diffraction

Think about walking down a street and passing by an open door or window with sound coming from inside. You can hear the sound even before you get to

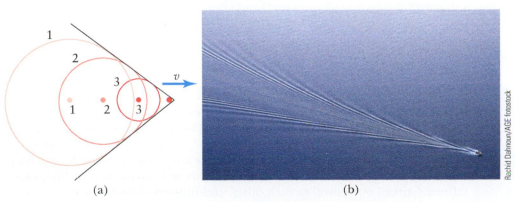

(a)　　　(b)

Figure 6.22 (a) A shock wave is produced when the speed of a wave source exceeds the wave speed. Parts of the wave fronts combine along the two black lines to form a V-shaped wave front. (b) A bow wave formed by a boat (wave source) moving faster than the speed of the waves along the water's surface.

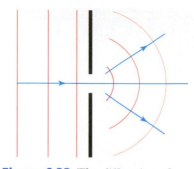

Figure 6.23 The diffraction of a wave as it passes through an opening in a barrier. The wave fronts spread out to the sides after passing through.

the opening, as well as after you've passed it. The sound doesn't just go straight out of the opening like a beam; it spreads out to the sides. This is **diffraction**. **Figure 6.23** shows wave fronts as they reach a gap in a barrier. These might be sound waves passing through a door or ocean waves encountering a breakwater. The part of the wave that passes through the gap actually sends out wave fronts to the sides as well as ahead. The rays that represent this process show that the wave "bends" around the edges of the opening.

The extent to which the diffracted wave spreads out depends on the ratio of the size of the opening to the wavelength of the wave. When the opening is much larger than the wavelength, there is little diffraction: the wave fronts remain straight and do not spread out to the sides appreciably. This is what happens when light comes in through a window. The wavelength of light is less than a millionth of a meter, and consequently, there is little diffraction. When the wavelength is roughly the same size as the opening, the diffracted wave spreads out much more (**Figure 6.24**). The sizes of windows and doors are well within the range of the wavelengths of sound waves, so sound diffracts a great deal after passing through them. Higher frequencies (shorter wavelengths) are not diffracted as much as the lower frequencies.

Figure 6.24 Diffraction of water waves through a gap in a barrier. When the size of the gap is about equal to the wavelength of the waves, the waves are appreciably diffracted (spread out) beyond the barrier (left). When the size of the opening is much larger than the wavelength, there is little diffraction (right). The waves continue to move straight ahead with little spreading to the right or left beyond the barrier.

Andrew Lambert Photography/Science Source

> ## Physics To Go 6.4
>
> Stand near an open window or door of a building in which a continuous sound, such as recorded music, is being produced (**Figure 6.25**).
>
> 1. Move back and forth past the opening and notice that you can hear sound when you are well off to one side. Where is the sound the loudest?
> 2. Move back and forth past the opening again and this time pay close attention to how much treble (high frequency) and bass (low frequency) there are in the sound. Is there a difference in the balance depending on where you are relative to the middle of the opening?

6.2f Interference

Interference arises when two continuous waves, usually with the same amplitude and frequency, arrive at the same place. The sound from a stereo with the same steady tone coming from each speaker is an example of this situation. Another way to cause interference is to direct a continuous wave at a barrier with two openings in it. The two waves that emerge from the two openings will diffract (spread out), overlap each other, and undergo interference.

Consider the case of identical, continuous water waves produced by two small objects made to oscillate up and down in unison on the surface of the water. As these two waves travel outward, each point in the surrounding water moves up and down under the influence of both waves. If we move around in an arc about the wave sources, we find that at some places the water is moving up and down with a large amplitude. At other places, the water is actually still—it is not oscillating at all (**Figure 6.26**).

To see why this characteristic pattern of large-amplitude and zero-amplitude motion arises, consider **Figure 6.27a**—a sketch showing two waves at one moment in time. The thicker lines represent peaks of the waves, and the thinner lines represent the valleys. In **Figure 6.27b**, the straight lines labeled C indicate the places where the two waves are "in phase"—the peak of one wave matches the peak of the other, and valley matches valley. The two waves reinforce each other, and the amplitude is large. This is called **constructive interference**. On the straight lines labeled D, the waves are "out of phase"—the peak of one wave matches the valley of the other. The two waves cancel each other. (Whenever one wave has upward displacement, the other has downward displacement, and vice versa. Therefore, the net displacement is always zero.) This is called **destructive interference**. **Figure 6.27c** shows the same waves a short time later after the waves have traveled one-half of a wavelength. The pattern of constructive and destructive interference is not altered as the waves travel outward. If the photograph shown in Figure 6.26 had been taken earlier or later, it would look the same.

Whether the two waves are in phase or out of phase depends on the relative distances they travel. To reach any point on line C_1 in Figure 6.27b, the two waves travel the same distance and consequently arrive with peak matching peak and valley matching valley. Along the line C_2, the wave from the source on the left must travel a distance equal to one wavelength farther than the wave from the source on the right. The reverse is true along the line on the left labeled C. In general, there is *constructive interference* at all points where one wave travels one, or two, or three . . . wavelengths farther than the other wave.

Figure 6.25 Experiencing sound diffraction.

Richard Megna/Fundamental Photographs

Figure 6.26 Interference pattern of water waves from two nearby sources (labeled as X). The thin lines of calm water indicate destructive interference. Between these lines are regions of large-amplitude waves caused by constructive interference.

Figure 6.27 Interference of two waves.

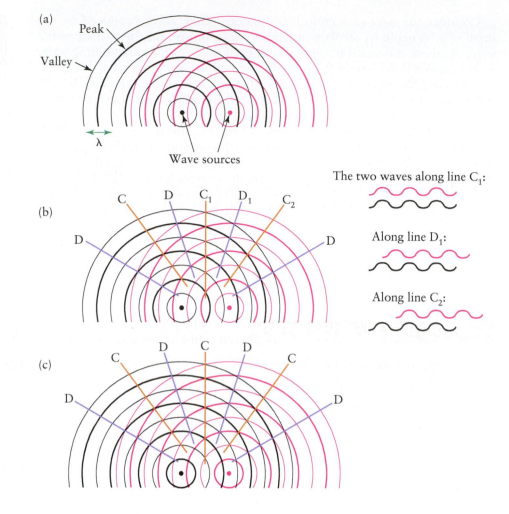

(a) Peak Valley λ Wave sources

(b)

The two waves along line C_1:

Along line D_1:

Along line C_2:

(c)

On the other hand, along the line of destructive interference labeled D_1, the wave from the source on the left has to travel one-half wavelength farther than the wave from the source on the right. They arrive with peak matching valley and cancel each other. Along the far-right line labeled D, the wave from the source on the left has to travel $1\frac{1}{2}$ wavelengths farther, so the two waves again arrive out of phase. The reverse is true for the lines showing destructive interference on the left. In general, there is *destructive interference* at all points where one wave travels $\frac{1}{2}$, or $1\frac{1}{2}$, or $2\frac{1}{2}$, . . . wavelengths farther than the other wave. At places in between constructive and destructive interference, the waves are not completely in phase or out of phase, so they partially reinforce or cancel each other.

Sound and other longitudinal waves can undergo interference in the same way. We can imagine Figure 6.27 representing sound waves with the peaks corresponding to compressions and the valleys corresponding to expansions. Along the lines of constructive interference, one would hear a loud, steady sound. Along the lines of destructive interference, one would hear no sound at all. In Chapter 9, we will apply a similar analysis to understand the interference of light waves.

These are some of the more important phenomena associated with waves as they propagate. Later in this chapter and in Chapter 9, we will take a closer look at some of these and introduce others that are particularly important for light.

1. As a sound wave or water ripple travels out from its source, its _____ decreases.
2. Which of the following can change the frequency of a wave?
   (a) interference
   (b) Doppler effect
   (c) diffraction
   (d) All of the above.
3. (True or False.) The amplitude of a sound wave can be increased by making it reflect off a curved surface.

4. A sophisticated echolocation system can determine the following about an object by reflecting a wave off of it:
   (a) in what direction it is located
   (b) how far away it is
   (c) how fast it is approaching or moving away
   (d) All of the above.
5. When two identical waves undergo _____ interference, the net amplitude is zero.

ANSWERS: 1. amplitude 2. (b) 3. True 4. (d) 5. destructive

ASTRONOMICAL APPLICATION   The Hubble Relation—Expanding Our Horizons

We mentioned in this section that the Doppler effect can be used to determine the line-of-sight speeds for individual stars bright enough to allow accurate measurement of the shifts in frequency of the light emitted by them. The question of whether or not the speeds of approach or recession of large aggregates of stars (galaxies) could be measured from similar shifts in the combined light emitted by all the members of the galaxy was first investigated by Vesto Slipher beginning in about 1912. By 1925, Slipher had carefully studied the radiation emitted by 40 galaxies and found that he could indeed determine the speeds with which they were moving. What he discovered was quite surprising: a number of the galaxies had very high speeds, as much as 5,700 km/s (13 million mph), and 38 out of the 40 showed frequency shifts that indicated they were moving away from us.

During the last 90 years, much has been accomplished in extragalactic astronomy, but the essentials of Slipher's results remain true. The vast majority of galaxies—and certainly all of those farther from us than about 3 million light-years (Mly)—are moving away from us and are doing so with velocities that are, in many cases, significant fractions of the speed of light. One of the more important additional facts to be discovered about galaxies during this period concerns the relationship between the recessional speed of a galaxy and its distance from us. Shortly after Slipher reported his results, Edwin Hubble noticed that galaxies with low recessional speeds were all relatively close to us, whereas those with high recessional speeds were more remote. By 1929, Hubble, in whose honor the Hubble Space Telescope was named, had amassed enough data to show a direct proportionality between the speeds with which galaxies are receding from us and their distances. This connection between speed and distance for all but the nearest galaxies has been verified by modern observations and is now referred to as the *Hubble relation*.

In mathematical form, this relation is usually written as

$$v = H_0 d$$

where v is the speed of recession of the galaxy, H_0 is the Hubble parameter, and d is the distance to the galaxy. This is the equation of a straight line (cf. Section 1.4a) with slope equal to H_0 (**Figure 6.28**). Because the velocities and distances to even the nearest galaxies are large, their speeds are usually measured in kilometers per second (km/s) and their distances in millions of light-years (Mly). Observations of thousands of galaxies have yielded a current value for the Hubble parameter of between 20 and 21 km/s/Mly. If either the velocity or the distance to a candidate galaxy can be determined by astronomical means, the other quantity can be computed by simple arithmetic.

What is the significance of the Hubble relation? Only that the universe is expanding! To understand this, we must recognize that galaxies (or more properly, clusters of galaxies) are the "atoms" of the universe; just as one speaks of an "expanding gas" when the atoms of the gas are moving away from one another, so one can talk of an "expanding universe" in which the atoms of the universe, the galaxies, are observed to be systematically separating from each other. And this is precisely what the Hubble relation is telling us. No matter in what direction we look, we find galaxies rushing away from us with speeds proportional to their distances: the farther away they are, the faster they are receding from us. You may find the "raisin bread" analogy depicted in **Figure 6.29** helpful in visualizing what is happening here.

The Hubble relation (and its interpretation) has had a profound influence on *cosmology*, the study of the structure and evolution of the universe. Indeed, by studying the motion of the most remote galaxies observable and checking for departures from a strictly linear trend in the Hubble relation, it is now possible to determine whether the universe will continue to expand forever or gradually slow to a stop and then begin to collapse in on itself. Recent data (c. 2010) show that at large distances the Hubble relation curves in such a way as to indicate that the universe will expand forever. For the latest on cosmology and the expansion of the universe, see Chapter 12.

■ QUESTIONS
1. An astronomer measures the speed of recession of a remote galaxy to be 365 km/s using the Doppler principle. According to the Hubble relation, about how far away is the galaxy?
2. Describe the significance of the Hubble relationship for our understanding of the structure and evolution of the universe.

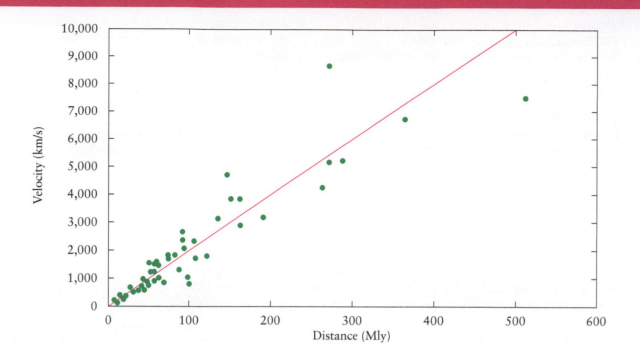

Figure 6.28 The Hubble relation. Plotted here are the recessional velocities of galaxies versus their distances. (1 light-year = 9.46 × 10^{12} km.) The (red) line represents a best fit to the data and yields a slope of about 20 km/s/Mly. More recent measurements including many more galaxies out to distances of more than 20,000 Mly give a linear relation with a slightly larger slope of ~21 km/s/Mly.

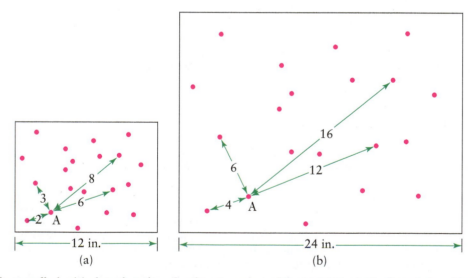

Figure 6.29 The so-called raisin bread analogy for the expansion of the universe. As the bread rises (through the action of the yeast in the dough), it doubles in size in a time equal to, say, 1 hour. Consequently, all the raisins move farther apart. Some characteristic distances from one of the raisins, say, A, to several others are shown. Because each distance doubles during an hour, each raisin must move away from the one selected at a speed proportional to its distance. This remains true no matter which raisin is chosen as the origin. The point is that a uniform expansion leads naturally to a Hubble-like relationship between velocity and distance. (*Note:* The raisins themselves, like the galaxies they represent, are not expanding. Only the space between them is growing and stretching with time.)

6.3 Sound

Our most common experience with sound is in air, but it can travel in any solid, liquid, or gas. For example, when you speak, much of what you hear is sound that travels to your ears through the bones and other tissues in your head. That is why a recording of your voice does not sound the same to you as what you hear when you are talking. Table 6.1 lists the speed of sound in some common substances.

> ### ▶ Physics To Go 6.5
>
> You might want to do this when no one is around to see you or hear you.
>
> 1. Speak a sentence in a normal way and pay attention to how your voice sounds.
> 2. Now block the sound into both ears. (One way is to cover both of them with the palms of your hands.) Speak the sentence again. What is different?
> 3. Record your voice electronically and then listen to the recording. Now what is different?

The speed of sound in any substance depends on the masses of its constituent atoms or molecules and on the forces between them. The speed of sound is generally higher in solids than in liquids and gases because the forces between the atoms and molecules in solids are very strong. Sound in gases and liquids is a longitudinal wave, whereas in solids it can be either longitudinal or transverse. In the rest of this chapter, we will concentrate mainly on sound in air.

6.3a Pressure Waves

Sound is produced by anything that is vibrating and causing the air molecules next to it to vibrate. Figure 6.30 shows a representation of a sound wave that was emitted by a vibrating tuning fork. The shading represents the air molecules that we, of course, cannot see. The wave looks very much like a longitudinal wave on a Slinky (Figure 6.8). These compressions and expansions travel at 344 m/s (at room temperature).

The air pressure in each compression is higher than normal atmospheric pressure because the air molecules are squeezed closer together. Similarly, the pressure in each expansion is below atmospheric pressure. Beneath the sketch is a graph of the air pressure along the direction the wave is traveling. Note that

Table 6.1 Speed of Sound in Some Common Substances

Substance	Speed* (m/s)	(mph)
Air		
At −20°C	320	715
At 20°C	344	770
At 40°C	356	795
Carbon dioxide	269	600
Helium	1,006	2,250
Water	1,440	3,220
Human tissue	1,540	3,450
Aluminum	5,100	11,400
Granite	4,000	9,000
Iron and steel	5,200	11,600
Lead	1,200	2,700

*At room temperature (20°C) except as indicated.

Figure 6.30 A representation of part of the sound wave emitted by a tuning fork (left). The air pressure is increased in each compression and reduced in each expansion as shown in the graph and the oscilloscope display.

Leonard Lessin/Science Source

it has the characteristic sinusoidal shape. So a sound wave can be represented by a series of pressure peaks and valleys—a pressure wave. It is more convenient to think of sound as regular fluctuations of air pressure than as vibrations of molecules, although it is both. The amplitude of a sound wave is the maximum pressure change. For a very loud sound, this is only about 0.00002 atmospheres.

It is these pressure variations that our ears detect and convert into the sensation of sound. The eardrum is a flexible membrane that responds to pressure changes. The oscillating pressure of a sound wave forces the eardrum to vibrate in and out. A remarkable set of physiological structures within the ear converts this oscillation of the eardrum into an electrical signal to the brain that is perceived as sound.

The **waveform** of a sound wave is the graph of the air-pressure fluctuations caused by the sound wave. The easiest way to display the waveform of sound is to connect a microphone to an oscilloscope, an electronic device often seen displaying heartbeats in television hospital shows. (Most personal computers have this capability as well; **Figure 6.31a**.) The oscilloscope shows a graph of the pressure variations detected by the microphone.

We can classify sounds by their waveforms (**Figure 6.31b, c, d**). A **pure tone** is a sound with a sinusoidal waveform. A tuning fork produces a pure tone, as

(a)

Vern Ostdiek/University of Michigan-Dearborn

Figure 6.31 Examples of the three types of waveforms. (a) Computer used to display waveforms of sounds. (b) Pure tone: sound from a tuning fork. (c) Complex tone: a spoken "ooo" sound. (d) Noise: sound of air rushing over microphone.

does a person carefully whistling a steady note. A **complex tone** is a complex sound wave. The waveform of a complex tone repeats itself but is not sinusoidal. Therefore, any complex tone also has a definite wavelength and a definite frequency. Most steady musical notes are complex tones.

The third type of sound is called *noise*. Noise has a random waveform that does not repeat over and over. For this reason, noise does not have a definite wavelength or frequency. The sound of rushing air is a good example of noise. (In everyday speech, "noise" is often used to describe any unwanted sound, even if it is a pure tone or a complex tone.)

Sound with frequencies outside the range of 20 to 20,000 hertz cannot be heard by people. Inaudible sound with frequency less than 20 hertz is called **infrasound**. High-amplitude infrasound can be felt, rather than heard, as periodic pressure pulses. The hearing ranges of elephants and whales extend into the infrasound region. Sound with frequencies higher than 20,000 hertz (20 kilohertz) is called ultrasound. The audible ranges of dogs, cats, moths, mice, and bats extend into ultrasound frequencies; they can hear very high frequency sounds that humans cannot.

6.3b Sound Applications

Before going into greater detail about sounds mainly meant for human hearing, we consider some of the many other uses of sound.

A variety of animals use sound for echolocation—to "see" their surroundings and to find prey. Dolphins and some other marine animals emit clicking sounds that reflect off fish and other objects. By paying attention to how long it takes for reflected sound to return, to the direction from which it comes, and to how strong the reflected sound is, a dolphin can get a very good idea of the sizes and locations of nearby objects. Bats use very high frequency sound, usually ultrasound, in a highly sophisticated echolocation system that employs the Doppler effect (**Figure 6.32**). Each species uses a characteristic range of frequencies, anywhere from 16 to 150 kilohertz. The bat emits a short burst of sound that reflects off surrounding objects such as the ground, trees, and flying insects. The bat detects these echoes and uses the time it takes the sound to make the round-trip to determine the distances to the objects. The shift in the frequency of sound reflected off moving objects is used by the bat to track down its dinner—flying insects. The bat even compensates for the Doppler shift in the frequency of the emitted sound caused by its own motion.

Although most applications of sound in science and technology use ultrasound, a few interesting devices have been developed that use lower-frequency sound. Special-use refrigerators that utilize sound waves in a gas instead of a pump circulating a refrigerant (refer to Figure 5.38) are now on the market. A large-amplitude sound wave inside a chamber produces huge pressure oscillations in a gas such as helium. The system is tuned in such a way that during the part of the cycle when the pressure is decreasing, the gas expands and absorbs heat from the substance to be cooled. This heat is transferred to a different part of the chamber where it is released. There are no moving parts, no lubricants, and no environmentally harmful refrigerants. The best of such thermoacoustic refrigerators have efficiencies in the range of 20% to 30% and find application in some technological areas requiring the cooling of small volumes to very low temperatures.

During the latter part of the 20th century, many useful applications of ultrasound were developed. Ultrasound is used in motion detectors that turn on the lights when a person enters a room or that set off an alarm when an intruder enters an area. It is also used to control rodents and insects; clean jewelry and intricate mechanical and electronic components; weld plastics; sterilize medical instruments; enhance certain chemical reactions; and measure the speed of the wind (see the Meteorological Application at the end of this section).

Ivan Kuzmin/Shutterstock.com

Figure 6.32 Bats use ultrasonic echolocation to navigate in the dark.

Figure 6.33 Ultrasonic image of author Don Bord's son Jeffrey at 13 weeks' gestation.

Ultrasound can also be used to produce light. First discovered in the 1930s, **sonoluminescence** (from the Latin words for sound and light) has been the subject of intense research in recent years, yet still defies a complete and universally accepted explanation. A bubble inside water emits flashes of light as pressure oscillations caused by sound waves with frequencies around 26,000 Hz make the bubble expand and collapse. The temperature inside the bubble rises to more than 10,000 K during collapse—hotter than the surface of the Sun—and the light pulse lasts less than a billionth of a second. Recent research suggests that the light-producing process is similar to that occurring inside x-ray tubes (discussed in Chapter 8).

Ultrasound has several uses in medicine. It is routinely used to form images of internal organs and fetuses (**Figure 6.33**). High-frequency ultrasound, typically 3.5 million hertz, is sent into the body and is partially reflected as it encounters different types of tissue. These reflections are analyzed and used to form an image on a video monitor. Some sophisticated ultrasonic scanning devices also use the Doppler effect. The beating heart of a fetus and the flow of blood in arteries can be monitored by detecting the frequency shift of the reflected ultrasound.

Recently developed *acoustic surgery* uses ultrasound for tasks such as destroying tumors. Focused, high-intensity sound causes heating that destroys tissue. The precision of such an "acoustic scalpel" can exceed that of a conventional knife.

Another use of ultrasound in medicine is *ultrasonic lithotripsy*, a procedure that breaks up kidney stones that have migrated to the bladder. (This is not the same as the ESWL described at the beginning of this chapter.) A large-amplitude 23,000- to 25,000-Hz sound wave travels through a steel tube inserted into the body and placed in contact with the stone. The ultrasound breaks the stone into small pieces, somewhat like a singer breaking a wineglass. A procedure similar to ultrasonic lithotripsy has recently been developed to break up blood clots.

Learning Check

1. (True or False.) Sound cannot travel through solid steel.
2. We are able to hear because our ears respond to the changes in _____ associated with a sound wave.
3. Which of the following is *not* one of the ways sound can be used?
   (a) to make a bubble in water produce light
   (b) to relay information to orbiting satellites
   (c) to form images of internal organs
   (d) to cool small volumes to very low temperatures

ANSWERS: 1. False **2.** pressure **3.** (b)

METEOROLOGICAL APPLICATIONS Putting Sound to Work

Perhaps you would like to know if it is windy enough to go sailing or too windy to launch a hot-air balloon. Or maybe you want to see if the snow is deep enough for skiing or if the water is high enough in a river to go kayaking. So you go online or use your cell phone to get the information, but who or what actually takes the measurement of the wind speed or snow depth? It may be an automatic device that uses sound.

The desire for such information is not limited to those with recreation in mind. An electric utility may want to monitor the wind a hundred meters above the ground at some location, day and night for months, to see if it would be a good place to install a wind-energy "farm." The stakes are much higher for a pilot coming in for a landing, wanting to know if there will be a drastic change in wind speed or direction, called **wind shear**, that could bring a jumbo jet crashing to the ground. Meteorologists whose job might be to forecast what the weather will be like tomorrow or to predict how Earth's climate might change over the next century need all of the information they can get about what the atmosphere is doing, both near Earth's surface and extending upward hundreds or thousands of meters. Increasingly, measurements of wind velocity, air temperature, and other quantities are being made by instruments that utilize sound.

Probably the simplest such device is a **sonic depth gauge**, basically a sonar unit pointed downward. It sends out a pulse of ultrasound that reflects off snow or water and uses the time it takes for the reflected wave to return to compute the distance to the reflecting surface— similar to Figure 6.20—and from that, the depth. (The air temperature must be measured and used in the calculation because the speed of sound changes as the air temperature changes.)

Measuring wind speed and direction with sound is a bit more complicated. Traditional **anemometers**—wind meters—use propellers or "cups" on spokes that are spun by the wind. But these devices wear out, and they can be frozen in ice in cold weather. Hundreds of weather stations spread across the United States are replacing these with **sonic anemometers**, which have no moving parts. They are also critical components of wind-shear alert systems at some major airports, as well as in security systems designed to monitor airflow in major transportation centers in the event of emergencies. Sound is carried along as the air moves, so it takes sound less time to travel a given distance downwind than it takes it to travel the same distance upwind. (Just like a boat motoring down a river versus up a river.) A sonic anemometer sends a pulse of ultrasound in one direction and records the time it takes to travel a short distance (roughly a foot or less). It repeats this with a pulse sent in the opposite direction (**Figure 6.34**). The component of the wind's velocity in the given direction is computed using these two times. One benefit of using the times for the sound going both ways is that the actual speed of sound cancels out: the anemometer gives the correct wind speed regardless of the air temperature.

Direct measurement of the air temperature and wind velocity above the ground is done twice a day, at about 1,000 sites worldwide. A lightweight instrument pack called a **radiosonde** is carried upward tens of kilometers by a hydrogen- or helium-filled balloon. As it rises, it radios back to a ground station the measurements it takes. This system has been in place for decades and is still one of the most important sources of vital information about the atmosphere. But the instruments are expensive, and they are usually not recovered for reuse, so only two "profiles" of the atmosphere are measured every 24 hours at each site.

There are two types of sound-based, remote-sensing instruments that measure wind velocity or air temperature at many levels far above the ground, day and night: **sodar** (<u>so</u>nic <u>d</u>etection and <u>r</u>anging) and **RASS** (<u>r</u>adio <u>a</u>coustic <u>s</u>ounding <u>s</u>ystem). Both are essential tools in atmospheric research programs, and a few units are in regular use to supplement data from radiosondes.

Sodar functions somewhat like the sonic depth meter, but it sends sound upward. The returning sound that it detects is that scattered by the air itself. The atmosphere is generally filled with turbulent eddies, like the small vortices seen in smoke rising from burning incense, except the eddies are usually invisible. The twinkling of stars and the apparent shimmering of things seen far in the distance on a hot day are caused by these temperature-induced eddies affecting light traveling through them. When sound encounters these eddies, some of it is scattered in all directions, including back toward the sound source. Sodar uses a very sensitive receiver to detect this returning sound. Any motion of the air toward or away from the sodar will cause a Doppler shift in the sound frequency, which is used to measure the wind speed. A typical sodar instrument will send sound pulse "beams" in three directions (vertical and two near vertical) and use the Doppler shifts of the return sound to compute the wind speed and direction. By waiting a specific period of time before accepting the return sound—half a second, for example—the unit can select which elevation above the ground to measure the wind velocity. A common hourly wind report from a sodar unit consists of wind speeds and directions at 10, 20, and 30, and so on meters above the ground and up to perhaps several hundred meters, depending on the system and the conditions.

RASS measures air temperatures aloft using the Doppler effect and an interaction between sound waves and high-frequency radio waves (**Figure 6.35**). A sound wave and a radio wave are emitted upward, with the radio wave's wavelength twice that of the sound wave's. (This makes the frequency of the radio wave more than 400,000 times that of the sound wave. Why?) The radio wave is scattered by the higher-density air in the sound-wave compressions, and some of this scattered energy is detected back on the ground. Because the radio-wave signal that is received is reflecting off sound-wave compressions moving upward, the decrease in the radio wave's frequency because of the Doppler effect can be used to determine the speed of the sound wave. Then the equation for the speed of sound given in Section 6.1 is used to calculate the air temperature. Incidentally, the two-to-one ratio of the wavelengths of the two waves is chosen so that interference enhances the return signal: radio waves reflected downward from two successive sound-wave compressions are in phase and undergo constructive interference.

Bryan Smith/ZUMA Press/Newscom

Figure 6.34 A sonic anemometer suspended in Grand Central Terminal in New York City is among the instruments deployed by the U.S. Department of Homeland Security to study the movement of air in the event of an emergency. Three ultrasonic transducers located at the corners of the two triangular devices serve as transmitters and detectors of sound pulses sent between them. Analysis of the pulse travel times between the various pairs of transducers can be used to establish the air speed and direction in the vicinity of the anemometer.

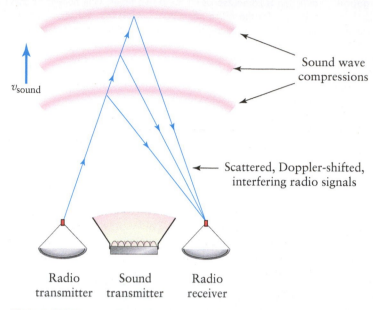

Research on new ways to use sound to probe the atmosphere continues. For example, studies have shown that severe tornado-producing thunderstorms generate infrasound that might be detected and used to warn of approaching tornadoes. The devices described here and others like them indicate that, just as from one person to another, sound turns out to be a useful way for the atmosphere to communicate information to us about what it is doing.

■ **QUESTION**

1. Name three modern meteorological devices that use sound waves to make weather-related measurements in the atmosphere. Describe how sound is employed in each instrument to facilitate the measurements.

v_{sound}

Sound wave compressions

Scattered, Doppler-shifted, interfering radio signals

Radio transmitter   Sound transmitter   Radio receiver

Figure 6.35 Simplified diagram of an RASS system. Radio waves (blue) are scattered downward by the compressions of the upward-moving sound wave (pink), and undergo constructive interference. The Doppler shift of the radio signal is used to compute the speed of sound, and from that, the air temperature.

6.4 Production of Sound

In the remaining sections of this chapter, we will take a brief look at the three P's of acoustics: the production, propagation, and perception of sound.

The sounds that we hear range from simple pure tones such as a steady whistle to complicated and random waveforms like those found on a noisy street corner. Most of the sound we hear is a combination of many sounds from different sources. The loudness usually fluctuates, as do the frequencies of the component sounds.

Sound is produced when vibration causes pressure variations in the air. Any flat plate, bar, or membrane that vibrates produces sound. The tuning fork shown in Figure 6.30 is a nice example. A dropped garbage-can lid, a vibrating speaker cone, and a struck drumhead produce sound the same way. The tuning fork executes simple harmonic motion and produces a pure tone. The garbage-can lid and the drumhead have more complicated motions and so produce complex tones or noise.

The various musical instruments represent some of the basic types of sound producers. Drums, triangles, xylophones, and other percussion instruments produce sound by direct vibration. Each is made to vibrate by a blow from a mallet or drumstick.

Guitars, violins, and pianos use vibrating strings to produce sound (**Figure 6.36**). By itself, a vibrating string produces only faint sound because it is too thin to compress and expand the air around it effectively. These instruments employ "soundboards" to increase the sound production. (The electric versions of guitars and violins pick up and amplify the string vibrations electronically.) One end of the string is attached to a wooden soundboard, which is made to vibrate by the string. The vibrating soundboard, in

Figure 6.36 A young guitarist makes the strings vibrate by plucking them. Most of the sound that we hear comes from the front plate of the guitar, which is made to vibrate by the strings.

istockphoto.com/Ryan Lane

turn, produces the sound. **Figure 6.37** shows a simplified diagram of the process used in pianos. When a note is played, a hammer strikes the piano wire and produces a wave pulse. The pulse travels back and forth on the wire, being reflected each time at the ends. The soundboard receives a "kick" each time the pulse is reflected at that end. This makes the soundboard vibrate at a frequency equal to the frequency of the pulse's back-and-forth motion. The sound dies out because the pulse loses energy to the soundboard during each cycle.

The strings on guitars are plucked instead of struck, giving the pulses a different shape. This is partly why the sound of a guitar is different from that of a piano. Violin strings are bowed, resulting in even more complicated wave pulses. In all three instruments, the frequency of the pulse's motion depends on the speed of waves on the string and on the length of the string. When a string is tuned by being tightened, the wave speed is increased. The pulse moves faster on the string and makes more "round-trips" each second—the frequency of the sound is raised. Different notes are played on the same guitar or violin string by using a finger to hold down the string some distance from its fixed end. The pulse travels a shorter distance between reflections, makes more round-trips per second, and produces a higher-frequency sound.

A similar process is used in flutes, trumpets, and other wind instruments. Here it is a pressure pulse in the air inside a tube that moves back and forth (**Figure 6.38**). Initially, a sound pulse is produced at one end of the tube by the musician. This pulse travels down the tube and is partially reflected and partially transmitted at the other end. The transmitted part spreads out into the air, becoming the sound that we hear. The reflected part returns to the mouthpiece end, where it is reflected again and reinforced by the musician. (The sound pulse is also inverted at each end: it goes from a compression to an expansion, and vice versa.) The musician must supply pressure pulses at the same frequency that the pulse oscillates back and forth in the tube. This, of course, is the frequency of the sound that is produced.

Different notes are played by changing the length of the tube—by opening side holes in woodwinds and by using valves or slides in brasses. The speed of the pulses is determined by the temperature of the air. This is one reason why musicians "warm up" before a performance. The air inside the instrument is warmed by the musicians' breath and hands. Hence the frequencies of the notes are higher than when the air inside is cool.

The human voice uses several types of sound production and modification mechanisms. Some consonant sounds like "*sss*" and "*fff*" are technically noise: they are hissing sounds produced by air rushing over the teeth and lips. The randomly swirling air produces sounds with random, changing frequencies. The vocal cords, located inside the Adam's apple in the throat, are the primary sound producers for singing and for spoken vowel sounds (**Figure 6.39**). When air is blown through the vocal cords, they vibrate and produce pressure pulses (sound) much like the reed of a saxophone. This sound is modified by the shapes of the air cavities in the throat, mouth, and nasal region. Muscles in the throat are used to tighten and loosen the vocal cords, thereby changing the pitch of the sound. Moving the tongue or jaw changes the shape of the mouth's air cavity and allows for different sounds to be produced. A sinus cold can change the sound of one's voice because swelling changes the configuration of the nasal cavity.

Figure 6.37 Sound production in a piano. The hammer creates a wave pulse that oscillates back and forth on the piano wire. (The amplitude is not to scale.) The pulse causes the soundboard to vibrate at the frequency of the pulse's oscillation.

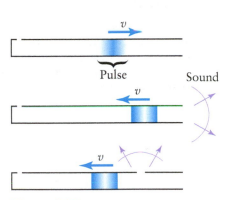

Figure 6.38 Sound production in a flute. The musician causes a sound (pressure) pulse to oscillate back and forth in the tube. Each time the pulse reaches the right end, it sends out a compression—part of the sound that we hear. Opening a side hole shortens the path of the pulse, thereby increasing the frequency.

Figure 6.39 All singers produce sound by making their vocal cords vibrate. The position of the jaw, the size of the mouth opening, and other factors also influence the sound.

Physics To Go 6.6

Two simple exercises illustrate how complicated human speech is.

1. Using a mirror, look carefully at your jaw, tongue, and lips as you recite the alphabet slowly. Note just how much their positions change as you speak different sounds.
2. Speak the first four vowel sounds (a, e, i, o). Now speak one of these sounds but "lock" your jaw, tongue, and lips in place (don't let them move) and then try to make the other three sounds. What happens?

Perhaps you've heard someone speak who had inhaled helium. (This is *not* a recommended exercise. It is possible to suffocate because of lack of oxygen in the lungs.) The speed of sound in helium is nearly three times that in air (refer to Table 6.1). This raises the frequencies of the sounds and gives the speaker a falsetto voice.

Sound waves carry energy, as do all waves. This means that the source of the sound must supply energy. Speaking loudly or playing an instrument for extended periods can tire you out for this reason. For continuous sounds, it is more relevant to consider the power of the source, because the energy must be supplied continuously. Most instruments, including the human voice, are very inefficient; typically, only a small percentage of the energy output of the performer is converted into sound energy.

Learning Check

1. Increasing the speed of waves on a guitar string increases the _____ of the sound that it produces.

2. (True or False.) The vocal cords are used to make all sounds that are produced in human speech.

ANSWERS: 1. frequency   2. False

6.5 Propagation of Sound

Once a sound has been produced, what factors affect the sound as it travels to our ears? The general aspects of wave propagation discussed in Section 6.2 of course apply to sound waves. Of these, reflection, diffraction, and the reduction of amplitude with distance from the sound source are most important in influencing the sound that actually reaches us.

The simplest situation is a single source of sound in an open space—such as a person talking in an empty field. The sound travels in three dimensions, and its amplitude decreases as the wave fronts expand. In particular, the amplitude is inversely proportional to the distance from the sound source.

$$\text{amplitude} \propto \frac{1}{d}$$

When you move to twice as far away from a steady sound source, the amplitude of the sound is decreased by one-half. The sound becomes quieter as you move away from the sound source.

6.5a Reverberant Sound

Sound propagation is more complicated inside rooms and other enclosures. First, diffraction and reflection of sound allow you to hear sound from sources that you can't see because they are around a corner. We are so accustomed to this phenomenon that it doesn't seem mysterious.

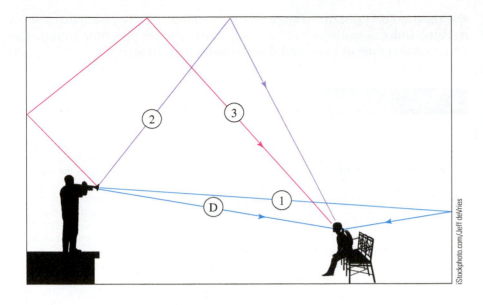

Figure 6.40 Some possible pathways for the sound in a room to travel from the source to your ears; D represents the sound that reaches your ears directly. Rays 1 and 2 are reflected once. Ray 3 is reflected twice. In most rooms, the sound can be reflected more than a dozen times and still be heard.

Second, even when the source is inside the room with you, most of the sound that you hear has been reflected one or more times off the walls, ceiling, floor, and any objects in the room. This has a large effect on the sound that you hear.

Figure 6.40 shows that sound emitted by a source in a room can reach your ears in countless ways. Consider a single sound pulse like a hand clap. In an open field, you would hear only a single momentary sound as the pulse moves by you. A similar pulse produced in a room is heard repeatedly: you hear the sound that travels directly to your ears; then sound that reflects off the ceiling, floor, or a wall before reaching your ears; then sound that is reflected twice, three times, and so on. These reflected sound waves travel greater and greater distances before reaching your ears and are heard successively later and generally with smaller amplitudes than the direct sound. This process of repeated reflections of sound in an enclosure is called **reverberation**. The single hand clap is heard as a continuous sound that fades quickly.

Figure 6.41 compares the sound from a hand clap as it is heard in an open field and in a room. Each graph shows the amplitude of the sound that is heard versus the time after the sound pulse is produced. The reverberation causes the sound to "linger" in the room. The indirect sound that one hears after the initial direct pulse is called the *reverberant sound*. The amount of time it takes for the reverberant sound to fade out depends on the size of the room and the materials that cover the walls, ceiling, and floor. Sound is never completely reflected by a surface: Some percentage of the energy in an incoming wave is absorbed by the surface, leaving the reflected wave with a reduced amplitude. (Concrete absorbs only about 2 percent of the incident sound's energy, whereas carpeting and acoustical ceiling tile can absorb around 90 percent.) A room with a large amount of sound-absorbing materials in it will have little reverberation. After a few reflections, the sound loses most of its energy and cannot be heard.

6.5b Reverberation Time

The *reverberation time* is used to compare the amount of reverberation in different rooms. It is the time it takes for the amplitude of the reverberant sound to decrease by a factor of 1,000. It varies from a small fraction of a second for small rooms with high sound absorption to several seconds for large, brick-walled gymnasiums and similar

Figure 6.41 Comparison of the sound heard from a hand clap in (a) an open field and in (b) a room. A single pulse is heard in the field, whereas a continuous, fading sound is heard in the room.

Figure 6.42 The marble used in the building of the Taj Mahal in India is a very poor absorber of sound energy. This results in a reverberation time of more than 10 seconds in its large central dome.

enclosures. The Taj Mahal (**Figure 6.42**), a breathtakingly beautiful mausoleum in Agra, India, is made of solid marble, which absorbs very little sound. The reverberation time of its central dome is more than 10 seconds.

> ### Physics To Go 6.7
>
> 1. You can make a rough measurement of the reverberation time of a large room or a racquetball court with a stopwatch. Have a friend produce a single loud handclap while you start the stopwatch at the same instant. Stop it the moment that the reverberant sound can no longer be heard. You may want to do the exercise several times and compute the average of the values.
> 2. If the room has a great deal of reverberation, you can easily illustrate the impact it has on speech communication. Stand next to a friend near the middle of the room and begin a normal conversation (or simply repeat a sentence to each other). Move a few steps apart and exchange sentences again. What is different? Continue moving farther apart and conversing until you can no longer understand each other.
> 3. Repeat the process while speaking more slowly. What is different about your experiences this time? Explain.
> 4. Find a room that is about the same size but with carpeting, drapes, or other sound absorbing material. Repeat steps 1 to 3. What differences do you find now?

When a steady sound is produced in a room, such as a trumpet playing a long note in an auditorium, the sound that one hears is affected by reverberation in a number of ways:

1. The sound is louder than it would be if you had heard it at the same distance in an open field.

2. The sound "surrounds" you; it comes from all directions, not just straight from the source (Figure 6.40).

3. Beyond a short distance from the sound source, the loudness does not decrease as rapidly with distance as it would in an open field. That is why one can often hear as well near the back of an auditorium as in the middle.

4. Not only does the sound fade gradually when the source stops, but the sound also "builds" when the source starts.

Figure 6.43 In Zankel Auditorium at Carnegie Hall, violinist Leila Josefowicz rehearses with pianist John Novacek, using reverberation to enhance the sound quality of her performance.

Moderate reverberation has an overall positive effect on the sound that we hear, particularly music (**Figure 6.43**). However, excessive reverberation adversely affects the clarity of both speech and music. Speech and music are a series of short, steady sounds interspersed with short moments of silence. Each note, word, or syllable is followed by a brief pause. If we again graph the amplitude of sound versus time, we can see the effect of reverberation (**Figure 6.44**). In an open field, one hears each syllable or note as a distinct, separate sound. In a room, the individual sounds begin to merge. As a new note is played or a new word is spoken, the reverberant sound from the preceding one can still be heard. The longer the reverberation time, the more the sounds overlap each other and the harder it is to understand speech. Racquetball courts have hard, smooth walls and very high reverberation times; that is why it is very difficult for players to converse unless they are close to each other. It is recommended that the reverberation time of rooms used for oral presentations and lectures should be around 0.5 to 1.0 second. For concert halls, it should be from 1 to 3 seconds, depending on the type of music being performed.

(a)

Reverberant sound

(b)

Figure 6.44 A series of steady sounds, such as spoken syllables, is heard (a) in an open field and (b) in a room. Reverberation in the room causes the sounds to merge. This tends to blend musical notes and makes speech more difficult to understand.

Many other factors besides reverberation time must be taken into account by architects and building designers. For example, a balcony must be high above the main floor and not extend out too far, or little reverberant sound will reach the seats below it. Also, sound is focused by concave walls and ceilings. Building an elliptical auditorium could result in the sound being concentrated in a small area of the room (Figure 6.17).

Learning Check

1. A stationary lawnmower produces a steady sound in an open field. If a listener moves to a distance one-half as far away from the mower, the amplitude of the sound would be
   (a) one-half as large
   (b) one-fourth as large
   (c) twice as large
   (d) four times as large

2. (True or False.) Reverberation can both inhibit conversation and enhance music.
3. The _____ of a room with walls that are good reflectors of sound is longer than that of a similar room with walls that are good absorbers of sound.

ANSWERS: 1. (c) **2. True 3.** reverberation time

6.6 Perception of Sound

In this section, we consider some aspects of sound perception—how the physical properties of sound waves are related to the mental impressions we have when we hear sound. We will be comparing psychological sensations, which can be quite subjective, to measurable physical quantities. (A similar situation: "hot" and "cold" are subjective perceptions that are related to temperature, which is a measurable physical quantity.) To make things simple, we will limit ourselves to steady, continuous sounds. This frees us from having to include such effects as reverberation in a room.

The main categories that we use to describe sounds subjectively are **pitch, loudness,** and **tone quality**.

The *pitch* of a sound is the perception of highness or lowness. The sound of a soprano voice has a high pitch and that of a bass voice has a low pitch. The pitch of a sound depends primarily on the frequency of the sound wave.

The *loudness* of a sound is self-descriptive. We can distinguish among very quiet sounds (difficult to hear), very loud sounds (painful to the ears), and sounds with loudness somewhere in between. The loudness of a sound depends primarily on the amplitude of the sound wave.

The *tone quality* of a sound is used to distinguish two different sounds even though they have the same pitch and loudness. A note played on a violin does not sound quite like the same note played on a flute. The tone quality of a sound (also referred to as the *timbre* or *tone color*) depends primarily on the waveform of the sound wave.

6.6a Pitch

Pitch is perhaps the most accurately discriminated of the three categories, particularly by trained musicians. It depends almost completely on the frequency of the sound wave: the higher the frequency, the higher the pitch. Noise does not have a definite pitch, because it does not have a definite frequency.

Pitch is essential to nearly all music. There is a great deal of arithmetic in the musical scale; each note has a particular numerical frequency. **Figure 6.45** shows the frequencies of the notes on the piano keyboard. There are seven octaves on the piano, each consisting of 12 different notes. Within each octave, the notes are designated A through G, plus five sharps and flats. Each note in a given octave has exactly twice the frequency of the corresponding note in the octave below. For example, the frequency of the lowest note on the piano, an A, is 27.5 hertz; for the A in the next octave, it is 55 hertz; it is 110 hertz for the third A, and so on. The frequency of middle C is 261.6 hertz.

Figure 6.45 The frequencies of the notes on a piano are indicated and compared with the approximate frequency ranges of different singing voices and musical instruments. This frequency scale, and therefore the piano keyboard, is a logarithmic scale: the interval between 50 and 100 hertz is the same as the interval between 500 and 1,000 hertz. (Adapted from *Physics in Everyday Life* by Richard Dittman and Glenn Schmieg. Used by permission.)

Certain combinations of notes are pleasing to the ear, whereas others are not. Nearly 2,500 years ago, Pythagoras indirectly discovered that two different notes are in harmony when their frequencies have a simple whole-number ratio. For example, a musical fifth is any pair of notes with frequencies in the ratio 3 to 2. Any E and the first A below it have this ratio of frequencies, as do any G and the first C below it.

Figure 6.45 also shows the approximate ranges of singing voices and some instruments. For normal speech, the ranges are approximately 70 to 200 hertz for men and 140 to 400 hertz for women. When whispering, you do not use your vocal cords, and you produce much higher-frequency "hissing" sounds.

6.6b Loudness

The loudness of a sound is determined mainly by the amplitude of the sound wave. The greater the amplitude of the sound wave that reaches your eardrums, the greater the perceived loudness of the sound. The actual pressure amplitudes of normal sounds are extremely small, typically around one-millionth of 1 atmosphere. This causes the eardrum to vibrate through a distance of around 100 times the diameter of a single atom. An extremely faint sound has an amplitude of less than one-billionth of 1 atmosphere, and it makes the eardrum move *less* than the diameter of an atom. The ear is an amazingly sensitive device.

There is a specially defined physical quantity that depends on the amplitude of sound but is more convenient for relating amplitude to perceived loudness. This is the sound pressure level or simply the **sound level**.

The standard unit of sound level is the decibel (dB). The range of sounds that we are normally exposed to has sound levels from 0 decibels to about 120 decibels. Figure 6.46 shows some representative sound levels along with their relative perceived loudness. Sound level does not take into account the irritation of the sound. Your favorite music played at 100 decibels may not sound as loud as an annoying screech of fingernails on a blackboard with a sound level of 80 decibels.

The relationship between the amplitude of a steady sound and its sound level is based on factors of 10. A sound with 10 times the amplitude of another sound has a sound level that is 20 decibels higher. A 90-decibel sound has 10 times the amplitude of a 70-decibel sound.

The following five statements describe how the perceived loudness of a sound is related to the measured sound level. These are general trends that have been identified by researchers after testing large numbers of people. For a particular person, the actual numerical values of the sound levels can vary somewhat from those listed. Also some of the values given, particularly

Jet takeoff (60 m)		120 dB	
Construction site		110 dB	*Intolerable*
Shout (1.5 m)		100 dB	
Heavy truck (15 m)		90 dB	*Very loud*
Urban street		80 dB	
Automobile interior		70 dB	*Noisy*
Normal conversation (1 m)		60 dB	
Office, classroom		50 dB	*Moderate*
Living room		40 dB	
Bedroom at night		30 dB	*Quiet*
Broadcast studio		20 dB	
Rustling leaves		10 dB	*Barely audible*
		0 dB	

Figure 6.46 The decibel scale with representative sounds that produce approximately each sound level. Sound levels above about 85 decibels pose a danger to hearing. Levels above 120 decibels cause ear pain and the potential for permanent hearing loss.

Figure 6.47 Ten equivalent sources sound about twice as loud as one.

in numbers 1 and 3 below, depend on the frequencies of the sounds.

1. The sound level of the quietest sound that can be heard under ideal conditions is 0 decibels. This is called the *threshold of hearing*.

2. A sound level of 120 decibels is called the *threshold of pain*. Sound levels this high cause pain in the ears and can result in immediate damage to them.

3. The minimum increase in sound level that makes a sound noticeably louder is approximately 1 decibel. For example, if a 67-decibel sound is heard and after that a 68-decibel sound, we can just perceive that the second sound is louder. If the second sound had a sound level of 67.4 decibels, we could not notice a difference in loudness.

4. A sound is judged to be twice as loud as another if its sound level is about 10 decibels higher. A 44-decibel sound is about twice as loud as a 34-decibel sound. A 110-decibel sound is about twice as loud as a 100-decibel sound. This is a cumulative factor: a 110-decibel sound is about four times as loud as a 90-decibel sound and so on.

5. If two sounds with equal sound levels are combined, the resulting sound level is about 3 decibels higher. If one lawn mower causes an 80-decibel sound level at a certain point nearby, starting up a second identical lawn mower next to the first will raise the sound level to about 83 decibels. It turns out that 10 similar sound sources are perceived to be twice as loud as a single source (**Figure 6.47**).

The loudness of pure tones and, to a lesser degree, of complex tones, also depends on the frequency. This is because the ear is inherently less sensitive to low- and high-frequency sounds. The ear is most sensitive to sounds in the frequency range of 1,000 to 5,000 hertz. For example, a 50-hertz pure tone at 78 decibels, a 1,000-hertz pure tone at 60 decibels, and a 10,000-hertz tone at 72 decibels all sound equally loud. (At very high sound levels, 80 decibels and above, the ear's sensitivity does not vary as much with frequency as it does at lower sound levels.) One reason for this variation in sensitivity is that a considerable amount of low-frequency sound is produced inside our bodies by flowing blood and flexing muscles. The ear is less sensitive to low-frequency sounds, so these internal sounds do not "drown out" the external sounds that we need to hear.

Sound levels are measured with sound-level meters (**Figure 6.48**). Most sound-level meters are equipped with a special weighting circuit (called the A scale) that allows them to respond to sound much as the ear does. The response to low and high frequencies is diminished. The readings on the A scale are designated dBA. When operating in the normal mode (called the C scale) a sound-level meter measures the sound level in decibels, treating all frequencies equally. When in the A scale mode, a sound-level meter responds like the human ear and therefore indicates the relative loudness of the sound.

Loud sounds can not only damage your hearing, but also affect the physiological and psychological balance of your body. Since the beginning of humankind, the sense of hearing has been used as a warning device: loud sounds often indicate the possibility of danger, and the body automatically reacts by becoming tense and apprehensive. Constant exposure to loud or annoying sounds puts the body under stress for long periods of time and consequently jeopardizes the physical and mental well-being of the individual.

The Occupational Safety and Health Administration (OSHA) has established standards designed to protect workers from excessive sound levels. Workers must be supplied with sound-protection devices such as earmuffs and earplugs if they are exposed to sound levels of 90 dBA or higher (**Table 6.2**). Some communities also have noise ordinances designed to reduce the sound levels of traffic and other activities.

Figure 6.48 A sound-level meter.

Table 6.2 OSHA Noise Limits

Sound Level (dBA)	Daily Exposure (hours)
90	8
92	6
95	4
97	3
100	2
102	1.5
105	1
110	0.5
115	0.25

6.6c Tone Quality

The tone quality of a sound is not as easily described as loudness or pitch. Comparisons such as full versus empty, harsh versus soft, or rich versus dry are sometimes used. The tone quality of a sound is very important to our ability to identify what produced the sound. The sound of a flute is different from the sound of a clarinet, and we notice this even if they produce the same note at the same sound level. The tone quality of a person's voice helps us identify the speaker.

The tone quality of a sound depends primarily on the waveform of the sound wave. If two sounds have different waveforms, we usually perceive different tone qualities. The simplest waveform is that of a pure tone: sinusoidal. Pure tones have a soft, pleasant tone quality (unless they are very loud or high pitched). Complex tones with waveforms that are nearly sinusoidal share the same characteristics. Unlike frequency and sound level, a waveform cannot be expressed as a single numerical factor.

What determines the waveform of a sound wave? The following mathematical principle gives us a way to comparatively analyze waveforms.

> **PRINCIPLES** Any complex waveform is equivalent to a combination of two or more sinusoidal waveforms with definite amplitudes. These component waveforms are called *harmonics*. The frequencies of harmonics are whole-number multiples of the frequency of the complex waveform.

For our purposes, this means that any complex tone is equivalent to a combination of pure tones. These pure tones, called **harmonics**, have frequencies that are equal to 1, 2, 3, . . . times the frequency of the complex tone. For example, **Figure 6.49** shows the waveform of a complex tone with a frequency of, let's say, 100 hertz. It is equivalent to a combination of the three pure tones shown of frequencies 100, 200, and 300 hertz. This means that we could artificially produce this complex tone by carefully playing the individual pure tones simultaneously. This is one method that electronic music synthesizers often use to create sounds.

The tone quality of a complex tone depends on the number of harmonics that are present and on their relative amplitudes. These two factors give us a quantitative way of comparing waveforms. A spectrum analyzer is a sophisticated electronic instrument that indicates which harmonics are present in a complex tone and what their amplitudes are. In general, complex tones with a large number of harmonics have a rich tone quality. Notes played on a recorder or a flute contain only a couple of harmonics, so they sound similar to pure tones. Violins and clarinets, on the other hand, have more than a dozen harmonics in their notes and consequently have richer tone qualities.

The waveform of noise is a random "scribble" that doesn't repeat. This is because noises are composed of large numbers of frequencies that are not related

Figure 6.49 The complex-tone waveform on the left is a combination of the three pure-tone waveforms on the right. The pure tones (harmonics) have frequencies that are 1, 2, and 3 times the frequency of the complex tone.

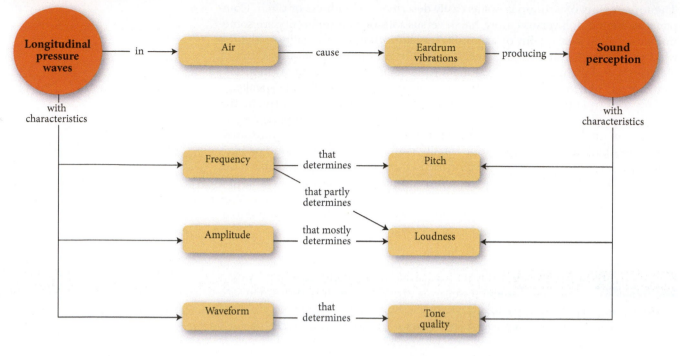

to each other: they are not harmonics. "White noise" contains equal amounts of all frequencies of sound. (It is so named because one way to produce "white light" is to combine equal amounts of all frequencies of light.) The sound of rushing air approximates white noise.

▶ Physics To Go 6.8

For this one, you have to be able to whistle.

1. Do as in Physics to Go 6.5 steps 1 and 2, but with a vowel sound or a sung note. Is the tone quality different when your ears are blocked?
2. Repeat while whistling a steady tone. Is the tone quality different when your ears are blocked? Is the effect as pronounced as in Step 1?

The existence of higher-frequency harmonics in musical notes explains why high-fidelity sound-reproduction equipment must respond to frequencies up to about 20,000 hertz. Even though the frequencies of musical notes are generally less than 4,000 hertz, the frequencies of the harmonics *do* go up to 20,000 hertz and higher. (Since our ears can't hear any harmonic with a frequency above 20,000 hertz, it isn't necessary to reproduce them.) To accurately reproduce a complex tone, each higher-frequency harmonic must be reproduced. Today's iPhones, which respond to frequencies from about 80 Hz up to around 20,000 Hz, do a good job in this regard.

The study of sound perception brings together the fields of physics, biology, and psychology. The mechanical properties of sound waves interact with the physiological mechanisms in the ear to produce a psychological perception. One of the challenges is to relate subjective descriptions of different sound perceptions to measurable aspects of the sound waves. The hearing apparatus itself is one of the most remarkable in all of Nature. It responds to an amazing range of frequencies and amplitudes and still captures the beauty and subtlety of music. At the same time, it is very fragile. We need to learn more about how our sense of hearing works and how it can be protected.

Concept Map 6.2 summarizes the general aspects of sound perception.

1. Choose the *correct* statement.
   (a) The perceived pitch of a sound usually doesn't depend on the sound's frequency.
   (b) The frequency of each musical note is 10 hertz greater than that of the next lower note.
   (c) The decibel is the standard unit of measure of tone quality.
   (d) The perceived loudness of a sound depends somewhat on its frequency.
2. (True or False.) The sound from two identical sound producers will sound twice as loud as that from just one.
3. A complex tone can be produced by combining two or more _____.
4. (True or False) Telephones in use as part of the North American Public Switched Network do not transmit sounds with frequencies above 3,400 Hz. But all musical notes with frequencies below 3,400 Hz should sound normal when they are heard over a telephone.

ANSWERS: 1. (d) 2. False 3. pure tones (harmonics) 4. False

Profiles in Physics   Acoustics Pioneers

The struggle to understand the nature of waves and wave propagation was historically waged on two fronts: optics and acoustics. Optics is the formal study of light, and acoustics is the study of sound and vibration. Progress in both areas over the last four centuries or so has produced a nearly complete picture of wave phenomena. The two areas of investigation complemented each other well because some wave properties are more easily exhibited with light, whereas others are more easily illustrated with sound. For example, although diffraction of sound is a very common occurrence, the diffraction of light was "discovered" first by Italian mathematician Francesco Maria Grimaldi (1618–1663). A beam of light passing through a tiny hole in a plate produces a projected circle of light on a screen that is larger than the size of the hole. This is because the light spreads out as it passes through the hole. Isaac Newton also experimented with light diffraction.

Similarly, the Doppler effect was first investigated with sound. Predicted for both light and sound by Austrian physicist Christian Johann Doppler (1803–1853) in 1842, the Doppler effect is not easily observed with light because of its extreme speed. However, in 1845, a Dutch meteorologist named Christoph Heinrich Dietrich Buys-Ballot (1817–1890) illustrated the Doppler effect using trains to provide the motion for a sound source, trumpets. He also used the trumpets as a stationary source and put listeners on the moving train.

Sound

The earliest discoveries in acoustics were apparently made around 550 BCE by Pythagoras, a Greek philosopher and mathematician whose name is immortalized by a theorem in geometry. Although Pythagoras left no record of his work, the accounts of his successors indicate that he used a stretched string to discover that two pitches sound good together (are in harmony) if the ratio of their frequencies is simple: 2 to 1, 3 to 2, and so on. A treatise written by Aristotle more than 200 years later indicates that he had a good understanding of the nature of sound and sound production.

A rare contribution to science by the Romans was made by the architect Vitruvius around 50 BCE. In a treatise on the design of theaters, he discussed the effects of reverberation and echoes on sound clarity.

The next notable discoveries in acoustics were made by the father of experimental science, Galileo. He completed the analysis of the vibrations of strings and correctly determined the relationship between the frequency of vibrations and the string's length, mass density, and tension. Galileo also showed that the pitch of a sound is determined by its frequency. At about this same time, a French Franciscan friar named Marin Mersenne (1588–1648; **Figure 6.50**) made fairly good measurements of the speed of sound in air and of the actual frequencies of musical notes.

MARIN   MERSENNE
1588–1648

Figure 6.50 Marin Mersenne.

In his *Principia*, published in 1687, Isaac Newton showed how the properties of air can be used to theoretically predict the speed of sound. However, his predicted value was markedly lower than the measured value because he did not take into account the fact that the air is rapidly heated and cooled by the compressions and expansions associated with a sound wave. Newton then "fudged" his analysis by introducing erroneous compensating factors that raised his theoretical value so as to agree with the measured one. Even the great master was not correct in all matters.

A major step toward understanding the nature of complex tones was made by French mathematician Joseph Fourier (1768–1830; **Figure 6.51**). He discovered that any complex waveform is actually a composite of simple, sinusoidal waveforms (Section 6.6). Fourier did not apply this principle to sound. German physicist Georg Ohm (1787–1854), famous for his work on the conduction of electricity, used Fourier's discovery to distinguish pure tones and complex tones and to show that complex musical tones are combinations of pure tones.

Perhaps the greatest contributions to the science of acoustics, particularly sound perception, were made by Hermann von Helmholtz (1821– 1894; **Figure 6.52**). Helmholtz was a man of prodigious analytical talent, which he applied as a surgeon, physiologist, physicist, and mathematician. In addition to his work in acoustics, Helmholtz made important contributions to electricity and magnetism and to the principle of the conservation of energy. It has been said that as a doctor he began to study the eye and the ear but found that he needed to learn more physics. Then he found that he needed to study mathematics to learn physics. In this way,

Figure 6.52 Hermann von Helmholtz.

he became a noted physiologist, a noted physicist, and a noted mathematician.

Helmholtz used his medical understanding of the ear to make several discoveries about sound perception. He distinguished the sensations of loudness, pitch, and tone quality in a sound. He showed that the tone quality of a complex tone depends on the harmonics that it contains. To detect these harmonics, he designed a set of "Helmholtz resonators"—hollow spheres tuned to particular frequencies that would amplify an individual pure tone in a complex tone. Most of his discoveries are beyond the level of this text. Helmholtz's book, *Sensations of Tone*, was perhaps *the* musical acoustics textbook for nearly a century. Only with the advent of modern electronic instrumentation in the mid- to late 1900s has it been possible to improve on Helmholtz's findings about sound perception.

In recent decades, there has been a resurgence of interest in the field of musical acoustics. One example of this is the establishment of the Institut de Recherche et Coordination Acoustique/Musique (IRCAM) in Paris in the late 1970s. Here musicians and scientists work together, using computers and other sophisticated electronic instruments, to study and create musical sounds. Many physicists find musical acoustics to be a particularly gratifying field of research because it combines the art and beauty of music with the elegance and depth of physical science.

■ QUESTIONS

1. Describe some of the contributions to our understanding of acoustical phenomena made by scientists and mathematicians during the 16th, 17th, and 18th centuries, including Galileo, Newton, and Fourier.

2. A *polymath* is someone who knows a great deal about many things, a person with encyclopedic knowledge. Give some evidence to support the contention that Hermann von Helmholtz could be called a polymath.

Figure 6.51 Joseph Fourier.

SUMMARY

» Waves are everywhere. They can be classified as **transverse** or **longitudinal** according to the orientation of the wave oscillations relative to the direction of wave propagation.

» The speed of a wave depends on the properties of the medium through which it travels. For continuous periodic waves, the product of the **frequency** and the **wavelength** equals the wave speed.

» Once waves are produced, they are often modified as they propagate. **Reflection** and **diffraction** cause waves to change their direction of motion when they encounter boundaries.

» **Interference** of two waves produces alternating regions of larger-amplitude and smaller or zero-amplitude waves.

» The **Doppler effect** and the formation of **shock** waves are phenomena associated with moving wave sources. The former also occurs when the receiver of the waves is moving.

» Although sound can refer to a broad range of mechanical waves in all types of matter, we often restrict the term to the longitudinal waves in air that we can hear. Sound waves in air are generally represented by the air-pressure fluctuations associated with the **compressions** and **expansions** in the wave.

» Sound with frequencies too high to be heard by humans, above about 20,000 hertz, is called **ultrasound**. Ultrasound is used for echolocation by bats and for a variety of procedures in medicine. Sound with frequencies too low to hear, below about 20 hertz, is called **infrasound** and is used by some large mammals for communication.

» The sounds that we can hear in the range 20 Hz to 20,000 Hz can be divided into **pure tones**, **complex tones**, and **noise** according to the shapes of their **waveforms**.

» Sound is produced in many different ways, all resulting in rapid pressure fluctuations that travel as a wave. Different musical instruments employ diverse and sometimes multiple sound-production mechanisms, including vibrating plates or membranes, vibrating strings, or vibrating columns of air in tubes.

» Sound propagation inside rooms and other enclosures is dominated by repeated reflection, called **reverberation**. This causes individual sounds to linger after they are produced. Moderate reverberation causes a positive blending of successive sounds, such as musical notes, but too much reverberation can adversely affect the clarity of speech.

» We use three main characteristics to classify a steady sound we perceive: **pitch**, **loudness**, and **tone quality**. The pitch of a sound depends mainly on the frequency of the sound wave.

» The loudness of a sound depends mainly on the amplitude (or **sound level**) of the sound, but it is also affected by the frequency.

» The tone quality of a sound depends on the **waveform** of the sound wave. The waveform of a complex tone depends on the number and amplitudes of the separate pure tones, called **harmonics**, that comprise it.

IMPORTANT EQUATIONS

Equation	Comments	Equation	Comments
$\rho = \dfrac{m}{l}$	Linear mass density of rope, wire, string, Slinky, etc.	$v = f\lambda$	Relates frequency, wavelength, and speed for continuous waves
$v = \sqrt{\dfrac{F}{\rho}}$	Speed of waves on linear media (a rope, wire, string, Slinky, etc.)	$v = H_0 d$	Hubble relation for expansion of the universe
$v = 20.1 \times \sqrt{T}$	Speed of sound waves in air (SI units)	amplitude $\propto \dfrac{1}{d}$	Dependence of sound amplitude on distance

MAPPING IT OUT!

1. Review Section 6.2 on wave propagation. Identify at least five new concepts introduced in this section and devise a way to integrate these concepts in a meaningful way into Concept Map 6.1 at the end of Section 6.1.

QUESTIONS

(■ **Indicates a review question, which means it requires only a basic understanding of the material to answer. Questions without this designation typically require integrating or extending the concepts presented thus far.**)

1. Take a close look at the pulse traveling on the rope in Figure 6.2. It was produced by moving the hand quickly to the left (up in the photo), then right. Notice that the rope is blurred ahead of the peak of the pulse and behind it but that the peak itself is not. Explain why.

2. ■ Give an example of a wave that does not need a medium in which to travel and a wave that does need a medium.

3. ■ What is the difference between a longitudinal wave and a transverse wave? Give an example of each.

4. A popular distraction in large crowds at sporting events since the 1980s is the "wave." People in one section quickly

stand up and raise their arms and then sit down; people in the neighboring sections follow suit in succession, resulting in a visible pattern in the crowd that travels around the stadium. Which of the two types of waves is this? Compared to the waves described in this chapter, how is the stadium wave different?

5. A long row of people are lined up behind one another at a service window. Joe E. Clumsy stumbles into the back of the person at the end and pushes hard enough to generate a wave in the people waiting. What type of wave is produced?

6. A person attaches a paper clip to each coil of a Slinky—about 90 in all—in such a way that waves will still travel on it. What effect, if any, will the paper clips have on the speed of the waves on the Slinky?

7. The speed of an aircraft is sometimes expressed as a Mach number: Mach 1 means that the speed is equal to the speed of sound. If you wish to determine the speed of an aircraft going Mach 2.2, for example, in meters per second or miles per hour, what additional information do you need?

8. Based on information given in Sections 6.1 and 4.3, does there seem to be a relationship between the speed of sound in a gas and its density? Explain.

9. If you were actually in a battle fought in space like the ones shown in science fiction movies like *Star Wars VII*, would you hear the explosions that occur? Why or why not?

10. ■ Explain what the amplitude, frequency, and wavelength of a wave are.

11. ■ A low-frequency sound is heard and then a high-frequency sound is heard. Which sound has a longer wavelength?

12. The rays that could be used to represent the steady sound emitted by a warning siren on the top of a pole would look similar to the gravitational field lines (Section 2.7) representing the field around an object. Explain why they are alike and point out the one important difference.

13. When trying to hear a faint sound from something far away, we sometimes cup a hand behind an ear. Explain why this can help.

14. ■ What useful thing can happen to a wave when it encounters a concave reflecting surface?

15. ■ A person riding on a bus hears the sound from a horn on a car that is stopped. What is different about the sound when the bus is approaching the car compared to when the bus is moving away from the car?

16. ■ Explain the process of echolocation. How is the Doppler effect sometimes incorporated?

17. In the past, ships often carried small cannons that were used when approaching shore in dense fog to estimate the distance to the hidden land. Explain how this might have been done.

18. You can buy a measuring device billed as a "sonic tape measure." Describe how a device equipped with an (ultrasonic) speaker, a microphone, and a precision timer could be used to measure the distance from the device to a wall (for example).

19. While a shock wave is being generated by a moving wave source, is the Doppler effect also occurring? Explain.

20. Describe some of the things that would happen if the speed of sound in air suddenly decreased to, say, 20 m/s. What would it be like living next to a freeway?

21. ■ If a boat is producing a bow wave as it moves over the water, what must be true about its speed?

22. When a wave passes through two nearby gaps in a barrier, interference will occur, provided that there is also diffraction. Why must there be diffraction?

23. A recording of a high-frequency pure tone is played through both speakers of a portable stereo placed in an open field. A person a few meters in front of the stereo walks slowly along an arc around it. How does the sound that is heard change as the person moves?

24. As a loud, low-frequency sound wave travels past a small balloon, the balloon's size is affected. Explain what happens. (The effect is too small to be observed under ordinary circumstances.)

25. ■ Describe the waveforms of pure tones, complex tones, and noise.

26. ■ What is ultrasound? Give two examples of what it can be used for.

27. ■ Describe how sound is produced in string instruments. Why does tightening a string change the frequency of the sound it makes?

28. A conditioning drill consists of repeatedly running from one end of a basketball court to the other, turning around and running back. Sometimes the drill is changed and the runner turns around at half court, or perhaps at three-fourths of the length of the court. Describe how the number of round-trips a runner can do each minute changes when the distance is changed and how this is related to a guitarist changing the note generated by a string by pressing a finger on it at some point.

29. A special room contains a mixture of oxygen and helium that is breathable. Two musicians play a guitar and a flute in the room. Does each instrument sound different from when it is played in normal air? Why or why not?

30. ■ An audio speaker producing a steady sound at an outdoor concert is 25 ft away from you. If you move to a position where the speaker is 75 ft distant, by what factor will the amplitude of the sound change?

31. ■ What is reverberation? How does reverberation affect how we hear sounds?

32. ■ What are the three categories used to describe our mental perception of a sound? On what physical properties of sound waves does each depend?

33. ■ A 100-Hz pure tone at a 70-dB sound level and a 1,000-Hz pure tone at the same sound level are heard separately. Do they sound equally loud? If not, which is louder, and why?

34. A sound is produced by combining three pure tones with frequencies of 200 Hz, 400 Hz, and 600 Hz. A second sound is produced using 200 Hz, 413 Hz, and 600 Hz pure tones. What important difference is there between the two sounds?

35. ■ The highest musical note on the piano has a frequency of 4,186 Hz. Why would a tape of piano music sound terrible if played on a tape player that reproduces frequencies only up to 5,000 Hz?

36. Normal telephones do not transmit pure tones with frequencies below about 300 Hz. But a person whose speaking voice has a frequency of 100 Hz can be heard and understood over the phone. Why is that?

37. Six transverse waves move through different media. They all have the same frequency, but their amplitudes, A, and wavelengths, λ, vary as indicated. Rank these waves according to their propagation speeds, from largest to smallest. That is, rank the wave that is traveling fastest first, and place the wave moving with the slowest speed last. If two (or more) waves have the same speed, give them the same ranking. Explain the reasoning you used in establishing your rankings.

Wave I: $A = 6$ units; $\lambda = 3$ m
Wave II: $A = 8$ units; $\lambda = 2$ m
Wave III: $A = 4$ units; $\lambda = 6$ m
Wave IV: $A = 6$ units; $\lambda = 6$ m
Wave V: $A = 8$ units; $\lambda = 3$ m
Wave VI: $A = 16$ units; $\lambda = 2$ m

38. Repeat Question 37, but now rank the waves according to the energy they carry in the mediums through which they travel from highest to lowest. Again, if two (or more) waves have the same energy, give them the same ranking. Explain the reasoning you used to arrive at your rankings.

39. Six strings, all having the same lengths, are stretched taut between two fixed points. The tensions, F, and masses, m,

of the strings vary as shown. A transverse pulse with a given amplitude is initiated near one end of each of the strings and propagates toward the other end. Rank the speeds of the pulses along their respective strings from highest to lowest. If two (or more) strings have the same speed, give them the same ranking. Explain the reasoning you employed to arrive at your rankings.

String A: $F = 2.0$ N; $m = 0.050$ kg
String B: $F = 1.5$ N; $m = 0.075$ kg
String C: $F = 4.0$ N; $m = 0.100$ kg
String D: $F = 1.5$ N; $m = 0.025$ kg
String E: $F = 2.0$ N; $m = 0.100$ kg
String G: $F = 6.0$ N; $m = 0.150$ kg

PROBLEMS

1. Two children stretch a jump rope between them and send wave pulses back and forth on it. The rope is 3 m long, its mass is 0.5 kg, and the force exerted on it by the children is 40 N.
   (a) What is the linear mass density of the rope?
   (b) What is the speed of the waves on the rope?
2. The force stretching the D string on a certain guitar is 150 N. The string's linear mass density is 0.005 kg/m. What is the speed of waves on the string?
3. What is the speed of sound in air at the normal boiling temperature of water?
4. The coldest and hottest temperatures ever recorded in the United States are −83°F (210 K) and 134°F (330 K), respectively. What is the speed of sound in air at each temperature?
5. A 4-Hz continuous wave travels on a Slinky. If the wavelength is 0.5 m, what is the speed of waves on the Slinky?
6. A 500-Hz sound travels through pure oxygen. The wavelength of the sound is measured to be 0.65 m. What is the speed of sound in oxygen?
7. A wave traveling 80 m/s has a wavelength of 3.2 m. What is the frequency of the wave?
8. What frequency of sound traveling in air at 20°C has a wavelength equal to 1.7 m, the average height of a person?
9. Verify the figures given in Section 6.1 for the wavelengths (in air) of the lowest and highest frequencies of sound that people can hear (20 and 20,000 Hz, respectively).
10. What is the wavelength of 3.5 million Hz ultrasound as it travels through human tissue?
11. The frequency of middle C on the piano is 261.6 Hz.
    (a) What is the wavelength of sound with this frequency as it travels in air at room temperature?
    (b) What is the wavelength of sound with this frequency in water?
12. A steel cable with total length 30 m and mass 100 kg is connected to two poles. The tension in the cable is 3,000 N, and the wind makes the cable vibrate with a frequency of 2 Hz. Calculate the wavelength of the resulting wave on the cable.

13. In a student laboratory exercise, the wavelength of a 40,000-Hz ultrasound wave is measured to be 0.868 cm. Find the air temperature.
14. A 1,720-Hz pure tone is played on a stereo in an open field. A person stands at a point that is 4 m from one of the speakers and 4.4 m from the other. Does the person hear the tone? Explain.
15. A person stands directly in front of two speakers that are emitting the same pure tone. The person then moves to one side until no sound is heard. At that point, the person is 7 m from one of the speakers and 7.2 m from the other. What is the frequency of the tone being emitted?
16. Ultrasound probes can resolve structural details with sizes approximately equal to the wavelength of the ultrasound waves themselves. What is the size of the smallest feature observable in human tissue when examined with 20-MHz ultrasound? The speed of sound in human tissue is 1,540 m/s.
17. A sonic depth gauge is placed 5 m above the ground. An ultrasound pulse sent downward reflects off snow and reaches the device 0.03 seconds after it was emitted. The air temperature is −20°C.
    (a) How far is the surface of the snow from the device?
    (b) How deep is the snow?
18. The huge volcanic eruption on the island of Krakatoa, Indonesia, in 1883 was heard on Rodrigues Island, 4,782 km (2,970 miles) away. How long did it take the sound to travel to Rodrigues?
19. A baseball fan sitting in the "cheap seats" is 150 m from home plate (**Figure 6.53**). How much time elapses between the instant the fan sees a batter hit the ball and the moment the fan hears the sound?
20. A geologist is camped 8,000 m (5 miles) from a volcano as it erupts.
    (a) How much time elapses before the geologist hears the sound from the eruption?
    (b) How much time does it take the seismic waves produced by the eruption to reach the geologist's camp, assuming the waves travel through granite as sound waves do?

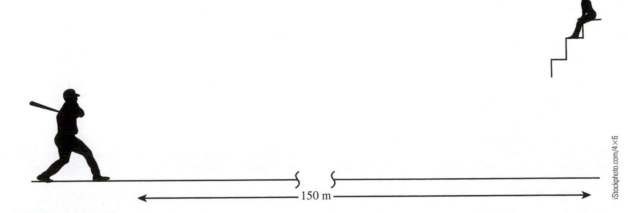

Figure 6.53 Problem 19.

iStockphoto.com/4×6

21. A person stands at a point 300 m in front of the face of a sheer cliff. If the person shouts, how much time will elapse before an echo is heard?

22. A sound pulse emitted underwater reflects off a school of fish and is detected at the same place 0.01 s later (**Figure 6.54**). How far away are the fish?

Figure 6.54 Problem 22.

23. The sound level measured in a room by a person watching a movie on a home theater system varies from 65 dB during a quiet part to 95 dB during a loud part. Approximately how many times louder is the latter sound?

24. Approximately how many times louder is a 100-dB sound than a 60-dB sound?

25. What are the frequencies of the first four harmonics of middle C (261.6 Hz)?

26. The frequency of the highest note on the piano is 4,186 Hz.
    (a) How many harmonics of that note can we hear?
    (b) How many harmonics of the note one octave below it can we hear?

CHALLENGES

1. Redraw Figure 6.15 for a convex boundary (one with the curve in the opposite direction). What happens to the amplitude of the reflected wave?

2. Perform the Physics to Go 6.3 exercise and measure the distance to the wall when you could no longer distinguish the echo. Use this distance to determine the amount of time between the direct sound that you heard and the echo.

3. Jack and Jill go for a walk along an abandoned railroad track. Jack puts one ear next to a rail, while Jill, 200 m away, taps on the rail with a stone (**Figure 6.55**). How much sooner does Jack hear the sound through the steel rail than through the air?

4. A stationary siren produces a sound with a frequency of 500 Hz. A person hears the sound while riding in a car traveling 30 m/s toward the siren. Compute the frequency of the sound that is heard by using the relative speed of the listener and the sound wave. Assume the air temperature is 20°C.

5. An entrepreneur decides to invent and market a device that will fool the Doppler radar units used to detect cars that are speeding. The device would be placed at the very front of the car and would detect the radar signal, determine its frequency, and transmit back its own radar signal that would make the radar unit register a legal speed. What would the frequency of the "fake" signal have to be in comparison to the original? If a second unit were designed to be placed at the back end of the car, what would be different about the frequency it would have to use compared to that used by the unit at the front?

6. The guitar string described in Problem 2 is approximately 0.6 m long. What is the frequency of oscillation of a pulse on the string? (*Hint:* The frequency equals 1 divided by the period, the amount of time it takes the pulse to go back and forth once.)

7. Use the relationship between amplitude and sound level described in Section 6.6 to determine how many times larger the amplitude of a 120-dB sound is compared to the amplitude of a 0-dB sound.

8. The frequency of the lowest note played on a flute at room temperature (20°C) is 262 Hz. What would be the frequency of the same note when the flute is played outside in the winter and the air temperature in the flute is 0°C?

9. In Western music, each of the seven base notes is assigned a letter A through G and the so-called *half-step* intervals between them have the following pattern for a major scale: 2-1-2-2-1-2-2. There are two half-steps (or one whole step) between A and B, one half-step between B and C, one whole step between C and D, etc. The frequencies of the notes in the most commonly used tuning scheme are based on the 12th root of 2: $\sqrt[12]{2} = 1.05946$. The frequency of each note equals that of the note n half-steps lower multiplied by $\left(\sqrt[12]{2}\right)^n$. Thus, going up the scale one octave (eight notes: A to the next highest A, D to the next highest D, etc.) comprises 12 half-steps and yields a note with a frequency that is $\left(\sqrt[12]{2}\right)^{12} = 2$ times that of the initial one (see Section 6.6a). Middle C on the piano keyboard has a frequency of 261.6 Hz. What is the frequency of a C# ("C-sharp") key that is one half-step above middle C? What about the frequency of the F above middle C? Verify that the A above middle C has a frequency of 440 Hz (Figure 6.45).

Figure 6.55 Challenge 3.

200 m

7

ELECTRICITY

Figure CO-7 Twenty-first-century iProducts find their basis in 19th-century physics of electric fields.

CHAPTER INTRODUCTION: iProducts—iPods, iPads, and iPhones

Within a few years of its introduction in 2001, the iPod revolutionized the way people purchase and enjoy music. Later models and derivative products such as the iPhone and iPad have brought the same portability to images, video, and the Internet, and they now offer thousands of "apps" that you can use to locate the nearest coffee shop, monitor your blood pressure, manage your bank accounts, find the most recent results from the world of sports, follow the latest "tweets" from your favorite celebrities, and much, much more. The iPod and its relatives represent 21st-century triumphs in the application of electricity, magnetism, materials science, and optics (**Figure CO-7**). Several key components of these *iProducts* function by exploiting electric fields, one of the principal topics in this chapter.

For example, the click wheel on an iPod or the screen-scroll feature of an iPad detects the presence and motion of the user's finger by sensing the changes the finger induces in the electric field maintained by a fine conducting grid embedded in the device. Similarly, the liquid crystal displays (LCDs) on these devices, like those on iPhones and laptop computers, create images by using electric fields to selectively activate individual picture elements (*pixels*) embedded in the screens. What makes these "iProducts" so powerful, of course, is the incredible quantity of digital information stored on them and precisely converted into audio or video signals ultimately detected by the users' ears and eyes. The enormous computational task required to carry out this conversion is itself accomplished by millions of miniaturized transistors embedded in integrated circuit chips, each of which is subject to electric fields that control the flow of electricity through them. It is no exaggeration to say that it is the electric field that serves as the workhorse for some of the most desirable and ubiquitous electronic devices now in use by you and millions of your fellow human beings around the world.

In this chapter, we consider some of the basic aspects of electricity, starting with electrostatic phenomena involving stationary electric charges. We also introduce the important quantities of voltage, resistance, and electric current and show how they are involved in many devices around us, as well as in living things. The final two sections of the chapter deal with electric power and energy and the two different types of electric current: AC and DC.

7.1 Electric Charge

How many electrical devices have you used so far today? How many are operating around you right now? Do you realize that as you read these words, the information sent to your brain and processed there also relies on electric charges and electric signals? When you turn or scroll down this page, your brain will communicate to the muscles in your hand and eyes in the same way. Electricity governs your life in ways you probably rarely think about. Even the properties of all matter that surrounds you—the air you breathe, the water you drink, the clothes that protect and insulate your body—are largely determined by electrical forces acting in and between atoms. From the latest electronic gizmo you've just acquired to the "glue" that holds matter together, electricity is inextricably woven into your life. Indeed, most of the material in the remainder of this text is connected to electricity to one degree or another. So let's take a closer look at what electricity is.

Figure 7.1 The amber effect. The comb attracts bits of paper because it was charged by rubbing.

GIPhotoStock/Science Source

> ▶ Physics To Go 7.1
>
> Caveat: Exercises like this sometimes do not work well when the relative humidity is high.
> You need a Styrofoam cup (or hard plastic comb or inflated balloon), some short pieces of thread, and a collection of fingernail-sized pieces of paper. "Charge" the cup (or comb or balloon) by rubbing it back and forth through your hair or over other furlike material. Lower the cup to a few centimeters above the pieces of paper and thread. What happens? Think about what you've seen in terms of forces and Newton's laws of motion. (The cup can be "recharged" as needed. To "discharge" a cup, you can rub its outside on a metal water faucet.)

The word *electricity* comes from *electron*, which itself is based on the Greek word for amber. Amber is a fossil resin that attracts bits of thread, paper, hair, and other things after it has been rubbed with fur. You may have noticed that a plastic comb can do the same after you run it through your hair (**Figure 7.1**). This phenomenon, known as the *amber effect*, was documented by the ancient Greeks, but its cause remained a mystery for more than two millennia. The results of numerous experiments, some conducted by American scientist and statesman Benjamin Franklin, indicated that matter possessed a "new" property not connected to mass or gravity. This property was eventually traced to the atom and is called **electric charge.**

DEFINITION	**Electric Charge**

An inherent physical property of certain subatomic particles that is responsible for electrical and magnetic phenomena. Charge is represented by q, and the SI unit of measure is the *coulomb* (C).

Early on, we stated that there are three fundamental things that physicists can quantify or measure: space, time, and the properties of matter (Section 1.1). Until now, mass has been the only fundamental property of matter that we have used. Electric charge is another basic property of matter, but it is intrinsically possessed only by electrons, protons, and certain other subatomic or "elementary" particles (more on this in Chapter 12). Unlike mass, there are two different (and opposite) types of electric charge, *positive* and *negative.* One coulomb of positive charge will "cancel" 1 coulomb of negative charge. In other words, the *net* electric charge would be zero ($q = -1$ C $+ 1$ C $= 0$ C).

Recall that every atom is composed of a nucleus surrounded by one or more electrons (Section 4.1). The nucleus itself is composed of two types of particles: protons and neutrons (**Figure 7.2**). Every electron has a charge of -1.6×10^{-19} C, and every proton has a charge of $+1.6 \times 10^{-19}$ C. (To have a total charge of -1 C, more than 6 billion billion electrons are needed.) Neutrons are so named because they are neutral; they have no net electric charge. Normally, an atom will have the same number of electrons as it

Figure 7.2 Simplified model of an atom of the element helium. The nucleus contains two protons; that is why helium's atomic number is 2. Also in the nucleus are uncharged neutrons—two in this case. Orbiting the nucleus are electrons. As long as the number of electrons equals the number of protons, the atom has no net charge.

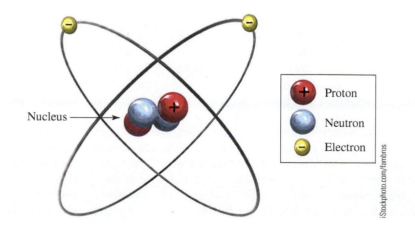

Nucleus →

⊕	Proton
⬤	Neutron
⊖	Electron

iStockphoto.com/fambros

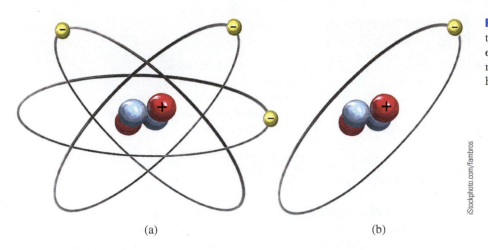

(a)   (b)

Figure 7.3 An atom is ionized when the number of electrons does not equal the number of protons. (a) A negative helium ion. (b) A positive helium ion.

has protons, which means the atom as a whole is neutral. The number of protons in the nucleus is the element's *atomic number*, which determines the atom's identity. For example, an atom of the element helium has two protons in the nucleus and two electrons in orbit around the nucleus. The positive charge possessed by the two protons is exactly balanced by the negative charge possessed by the two electrons, so the net charge is zero. Most of the substances that we normally encounter are electrically neutral simply because the total number of electrons in all of the atoms is equal to the total number of protons.

A variety of physical and chemical interactions can cause an atom to gain one or more electrons or to lose one or more of its electrons. In these cases, the atom is said to be *ionized*. For example, if a helium atom gains one electron, it has three negative particles (electrons) and two positive particles (protons). This atom is said to be a **negative ion** because it has a net negative charge (**Figure 7.3a**). The value of its net charge is just the charge on the "extra" electron, $q = -1.6 \times 10^{-19}$ C. Similarly, if a neutral helium atom loses one electron, it becomes a **positive ion** because it has two positive particles and only one negative particle. Its net charge is $+1.6 \times 10^{-19}$ C (**Figure 7.3b**).

In many situations, ions are formed on the surface of a substance by the action of friction. When a piece of amber, plastic, or hard rubber is rubbed with fur, negative ions are formed on its surface. The contact between the fur and the material causes some of the electrons in the atoms of the fur to be transferred to some of the atoms on the surface of the solid. The fur acquires a net positive charge because it has fewer electrons than protons. Similarly, the amber, plastic, or hard rubber acquires a net negative charge because it has an excess of electrons (**Figure 7.4**). Combing your hair can charge the comb in the same way. Rubbing glass with silk causes the glass to acquire a net positive charge. Some of the electrons in the surface atoms of the glass are transferred to the silk, which becomes negatively charged. Ion formation by friction is a complicated phenomenon that is still not completely understood. It is affected by many factors, including what materials are used and how high the relative humidity is.

Figure 7.4 The contact between fur and plastic when they are rubbed together causes some of the electrons in the fur to be transferred to the plastic. This leaves the plastic with a net negative charge and the fur with a net positive charge.

▶ Physics To Go 7.2

The "popcorn" exercise. Caveat: Exercises like this sometimes do not work when the relative humidity is high.

For this activity, you'll need a Styrofoam cup, several fingernail-sized pieces of aluminum foil wadded up into little balls, and a horizontal metal surface (a baking pan or a cookie sheet work well). Place the foil wads on the metal surface and charge the cup (as in Physics to Go 7.1). Slowly lower the cup to within a few centimeters above the foil wads. What happens? (If all goes well, things will happen fast—so you'll have to watch carefully.) Speculate on what the causes are that might produce the observed results.

1. The physical property possessed by certain particles that is responsible for electrical and magnetic phenomena is _____.
2. (True or False.) All three principal types of particles found in atoms are electrically charged.

3. A positive ion is formed when
   (a) a neutral atom gains an electron.
   (b) a neutral atom loses an electron.
   (c) a negative ion loses an electron.
   (d) All of the above.

ANSWERS: 1. electric charge   2. False   3. (b)

7.2 Electric Force and Coulomb's Law

The original amber effect illustrates that electric charges can exert forces. You may have noticed hair being drawn toward a charged comb or "static cling" between items of clothes removed from a dryer. These are among the most common situations illustrating this effect—two objects with opposite charges attracting each other (**Figure 7.5**). The negatively charged comb exerts an attractive force on the positively charged hair. In addition, two objects with the same kind of charge (both positive or both negative) repel each other. When two similarly charged combs are suspended from threads, they push each other apart. Just remember this simple rule: *like charges repel, unlike charges attract*.

> ### Physics To Go 7.3
>
> Caveat: Exercises like this sometimes do not work when the relative humidity is high.
>   Attach a thread roughly 0.5-m long to a Styrofoam cup. Charge the cup following the procedure used in Physics to Go 7.1 and then let it dangle from the thread. Charge a second Styrofoam cup, and then bring it close to the hanging one. What happens? The size of the net force on the hanging cup is related to how far the thread is angled from vertical. Experiment by holding the second cup at various distances from the hanging cup. Does the force on the suspended cup seem to be affected by how far apart the two cups are?

7.2a Coulomb's Law

This force between charged objects is extremely important in the physical world, particularly at the atomic level. It is this interaction that holds atoms together and makes it possible for them to exist. In each atom, the positively charged protons in the nucleus exert attractive forces on the negatively charged electrons. The electric force on each electron provides the centripetal force that keeps it in its orbit, much as the gravitational force exerted by the Sun keeps Earth in its orbit. The forces between the atoms in many compounds arise because opposite charges attract. For example, when salt is formed from the elements sodium and chlorine, each sodium atom gives up an electron to a chlorine atom. The resulting ions exert attractive forces on one another because they are oppositely charged (**Figure 7.6**). Matter as we know and experience it would not exist without the electrical force.

Figure 7.5 Objects with opposite electric charges exert attractive forces on one another. Objects with the same kind of charge repel each other.

Recall Newton's law of universal gravitation in Section 2.7, which gives the size of the force acting between two masses. The corresponding law for the electrical force, **Coulomb's law,** is very similar.

LAWS   **Coulomb's Law**   The force acting on each of two charged objects is directly proportional to the net charges on the objects and inversely proportional to the square of the distance between them:

$$F \propto \frac{q_1 q_2}{d^2}$$

The constant of proportionality in SI units is 9×10^9 N-m^2/C^2. Therefore, in SI units,

$$F = \frac{(9 \times 10^9) q_1 q_2}{d^2}$$

with F in newtons, q_1 and q_2 in coulombs, and d in meters.

The force on q_1 is equal and opposite to the force on q_2 by Newton's third law of motion. Note that if both objects have the same kind of charge (both positive or both negative), then the force F is positive. This indicates a repulsive force. If one charge is negative and the other positive, then the force F is negative, indicating an attractive force. If the distance between two charged objects is doubled, then the forces are reduced to one-fourth their original values (**Figure 7.7**).

Perhaps it is not a surprise that Coulomb's law has the same form as Newton's law of universal gravitation. After all, mass and charge are both fundamental properties of the particles that comprise matter. We must remember, however, that the (gravitational) force between two bodies because of their masses is *always* an attractive force, whereas the (electrostatic) force between two bodies from their electric charges can be attractive or repulsive, depending on whether or not they have opposite charges. Also, all matter has mass and so experiences and exerts gravitational forces, whereas the electrostatic force normally acts between objects only when there is a net charge on one or both of them. Generally, when two objects have electric charges, the electrostatic force between them is much stronger than the gravitational force. For example, the electrostatic force between an electron and a proton is about 10^{39} times as large as the gravitational force between them. (For more information on the properties and relative strengths of the forces in Nature, see Chapter 12.)

Figure 7.6 Ordinary table salt consists of positive sodium ions and negative chlorine ions. The strong attractive forces between the oppositely charged ions hold them in a rigid crystalline array.

Figure 7.7 Coulomb's law. Doubling the charge on one object, (b), doubles the force on both. Doubling the distance between the charges, (c), reduces the forces to one-fourth their original values.

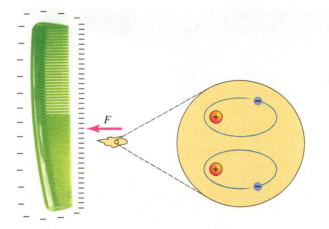

Figure 7.8 A charged comb exerts a net attractive force on a neutral piece of paper because the electrons are displaced slightly away from the negatively charged comb. (The distorted electron orbits are exaggerated in this drawing.) The attractive force on the closer nuclei is stronger than the repulsive force on the electrons.

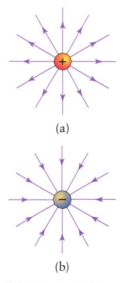

(a)

(b)

Figure 7.9 Electric field lines in the space around a positive charge (a) and around a negative charge (b). The arrows show the direction of the force that would act on a positive charge placed in either field.

Electric field

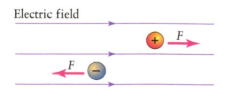

Figure 7.10 In an electric field, the force on a positively charged body is parallel to the field, whereas the force on a negatively charged body is opposite to the direction of the field.

It is possible for a charged object to exert a force of attraction on a second object that has no net charge. This is what happens when a charged comb is used to pick up bits of paper or thread. Here the negatively charged comb attracts the nuclei of the atoms and repels the electrons. The orbits of the electrons are distorted so that the electrons are, on the average, farther away from the charged comb than the nuclei (**Figure 7.8**). This results in a net attractive force because the repulsive force on the slightly more distant, negatively charged electrons is smaller than the attractive force on the closer, positively charged protons. The process of inducing a small charge separation (or displacement) between the nucleus of an atom and its electrons is called *polarization*.

Some molecules are naturally polarized—that is, they have a net negative charge displaced to one side of the net positive nuclear charge. They are called *polar molecules*. Water molecules have this property. (As we will see in Chapter 8, that's why microwave ovens work so effectively.) If a polar molecule is free to rotate—as in a liquid—it will be attracted to a charged object. Its side with the charge opposite that on the object will turn toward the object, and the attractive force on that side will be stronger than the repulsive force on the other side, as with the atoms previously discussed.

> ▶ Physics To Go 7.4
>
> Caveat: Exercises like this sometimes do not work when the relative humidity is high.
> Turn on a water faucet just enough to produce a very small, but steady, stream trickling out. Bring a charged Styrofoam cup near the stream. What happens?

7.2b The Electric Field

The electrostatic force is another example of "action at a distance." As with gravitation, the concept of a field is useful. In the space around any charged object, there is an *electric field*. This field is the "agent" of the electrostatic force: it will cause any charged object to experience a force. The electric field around a charged particle is represented by lines that indicate the direction of the force that the field would exert on a <u>positive</u> charge. Thus the *electric field lines* around a positively charged particle point radially outward, and the field lines around a negatively charged particle point radially inward (**Figure 7.9**).

The strength of an electric field at a point in space is equal to the size of the force that it would exert on a given charged object placed at that point, divided by the size of the charge on the object.

$$\text{electric field strength} = \frac{\text{force on a charged object}}{\text{charge on the object}}$$

Where the field is strong, a charged object will experience a large force. The strength of the electric field is indicated by the spacing or density of the field lines: where the lines are close together, the field is strong. The electric field around a charged particle is clearly weaker at greater distances from it. As defined here, the SI units for the electric field are newtons per coulomb: N/C.

Any time a positive charge is in an electric field, say due to other charged objects in its vicinity, it experiences a force in the *same* direction as the field lines. A negative charge in an electric field feels a force in the *opposite* direction of the field lines (**Figure 7.10**). Remember that the electric field exerts the force on a charged object. In Chapter 8, we will show that electric fields can be produced in other ways without using only electric charges.

Perhaps you have had the experience of walking across a carpeted floor and receiving a shock when you touched a metal doorknob. This is more likely to happen in winter than in summer because the relative humidity is usually lower then, and electrostatic charging of your body as you move over (rub against) the carpet takes place more readily. The shock results from charges flowing between you and the doorknob, and it may be accompanied by a visible spark. Air normally does not allow charges to flow through it. A spark occurs when there is an electric field strong enough to ionize atoms in the air. Freed electrons accelerate in a direction opposite to the direction of the electric field, and positive ions accelerate in the same direction as the field. The electrons and ions pick up speed and collide with other atoms and molecules, ionizing them or causing them to emit light (**Figure 7.11**). Lightning is produced in this same way on a much larger scale, as Benjamin Franklin demonstrated using kites, keys, and metal rods in the middle of the 18th century (see the application feature at the end of this section). In Chapter 10, we will discuss how atomic collisions cause the emission of light.

Figure 7.11 A large spark created in a lab. The spark is produced because there is a very strong electric field in the space between the two shiny, spherical charged objects.

Adam Hart-Davis/SPL/Photo Researchers

> **Physics To Go 7.5**

Caveat: Exercises like this sometimes do not work when the relative humidity is high.

For this you'll need a source of smoke such as a smoldering twig or a stick of burning incense, in a room where the air is still. Bring a charged Styrofoam cup near the rising stream of smoke. What happens? Compare this phenomenon to what was observed in Physics to Go 7.4.

Although most of the electrical devices we rely on make use of electric currents, some depend primarily on electrostatics. One important example of the latter is the *electrostatic precipitator*, an air-pollution control device (**Figure 7.12**). Tiny particles of soot, ash, and dust are major components of the airborne emissions from power plants that burn fossil fuels and from many industrial processing plants. Electrostatic precipitators can remove nearly all of these

Precipitator operating

Precipitator not operating

Photographs courtesy of Bateman Engineering N.V.

(a)

(b)

Figure 7.12 A large refining facility with its electrostatic precipitator in operation (left). The same plant with the precipitator not in operation (right). Note the difference in the stack emissions in the two cases.

Figure 7.13 Schematic of an electrostatic precipitator. (a) The emissions (flue gas) flow around negatively charged wires hanging between positively charged plates. (b) Top view of one channel. The particles are charged negatively by the strong electric field around the wires. Consequently, they are attracted to, and collect on, the plates.

Wires (−) Soot, ash, dust

Plate (+)

Flue gas

Collection hopper

Flue gas

(a) (b)

particles from the emissions. The flue gas containing the particles is passed between a series of positively charged metal plates and negatively charged wires (**Figure 7.13**). The strong electric field around the wires creates negative ions in the particles. These negatively charged particles are attracted by the positively charged plates and collect on them. Periodically, the plates are shaken so the collected soot, ash, and dust slide down into a collection hopper. This "fly ash" must then be disposed of, but sometimes it has its own uses—for example, as a filler in concrete.

Electronic signs that behave like electronically erasable paper make use of electric fields to form letters and other images. One type, called SmartPaper, consists of millions of tiny beads between two thin plastic sheets (**Figure 7.14**). One side of each bead is a particular color and negatively charged, and the other side is a contrasting color and positively charged. An electric field exerts opposite forces on the two sides (recall Figure 7.10), causing the beads to rotate until they are aligned with the field. (As we will see in Chapter 8, this is just like what a compass needle does in a magnetic field.) Letters are formed on the electronic paper by selectively applying upward and downward electric fields at different places so that parts of the display are one color and the rest are the other color.

The type of transistor used most widely in computers and similar devices is the field-effect transistor (FET). In FETs, an electric field controls the flow of electricity through the transistor. Electric fields play crucial roles in the operation of liquid crystal displays (LCDs), the touchpads on laptop computers, and, as noted in the chapter introduction, the touch-sensitive screens on iPads and iPhones.

Concept Map 7.1 summarizes the ideas presented in these first two sections of Chapter 7.

Figure 7.14 Simplified edge view of a segment of SmartPaper electronic paper. One half of each tiny bead has a negative charge and is some color (red in this example), and the other half has a positive charge and a contrasting color (blue here). An electric field exerts opposite forces on the two sides of each bead, so the bead rotates and aligns with the field. Letters are formed by using electric fields to turn the red sides up in parts of the display and the blue sides up elsewhere.

Transparent film Electric field

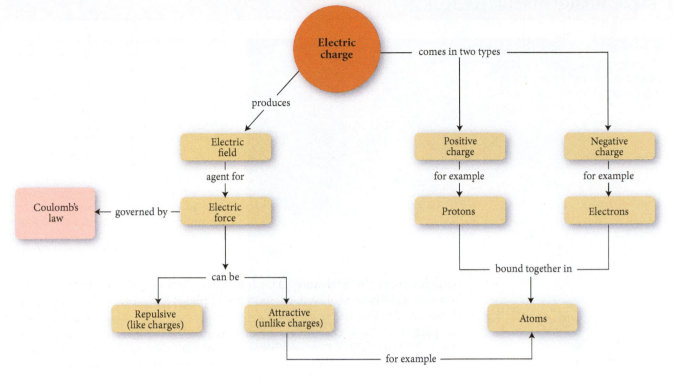

■ Learning Check

1. Like charges _____ ; unlike charges _____ .

2. Two charged objects exert forces on each other. The sizes of both forces will increase if
   (a) the objects are brought closer together.
   (b) the charge on one of them is increased.
   (c) the charges on both of them are increased.
   (d) All of the above.

3. A charged particle creates an electric _____ in the space around it.

4. (True or False.) An electrostatic precipitator uses an electric field to make it rain.

ANSWERS: 1. repel, attract 2. (d) 3. field 4. False

METEOROLOGICAL APPLICATIONS Electrifying Sights and Sounds—A Thunderstorm Primer

At one time or another, most of us have witnessed the fearsome power and awesome beauty of a thunderstorm, complete with torrential rains and hail, earsplitting claps of thunder, and brilliant—sometimes deadly—strokes of lightning (**Figure 7.15**). At any given time, there are roughly 2,000 thunderstorms in progress around the world, producing some 30 to 100 cloud-to-ground lightning flashes each second, totaling about 5 million lightning strikes a day. But what are these lightning strikes? How are they produced and what makes them so dangerous? The answers, as Benjamin Franklin first discovered in the early 1750s (cf. Profiles in Physics at the end of this chapter) are rooted in the physics of electricity.

Lightning is akin to the short sparks we experience when we touch a metal doorknob after walking across a wool rug in a dry environment in the winter—only on a much grander scale! Cumulonimbus clouds, the most common thunderstorm cloud, and Earth effectively acquire opposite electrical charges (like your finger and the

doorknob), with the air between them serving as an insulating material. When the separated charge grows sufficiently large, the strong electric field between the cloud and Earth, typically some 300,000 to 400,000 newtons/coulomb, causes an electrical breakdown of the insulating air and creates an ionized path between the cloud and the ground along which charge can flow—that is, a lightning discharge. Usually during this discharge, negative charge is transferred from the lower part of the cloud to neutralize the induced positive charge on Earth below. The average maximum current in such lightning flashes is about 30,000 amperes, lasting for only about 30 microseconds and delivering about a coulomb of charge. It is the large currents carried by lightning discharges that make them so dangerous to human beings (see the medical applications feature in Section 7.4). Every year an average of 200 persons in the United States die from injuries sustained after being struck by lightning. Lightning is the leading cause of weather-related personal injuries. Thus, although the overall

Figure 7.15 Lightning and sparks are caused by very strong electric fields.

James "BO" Insogna/Shutterstock.com

wherein the discharge occurs along an ionized track connecting a positive portion of a cloud with a negatively charged region of the same cloud.) The characteristic lightning flash is produced by atmospheric ions recombining with electrons and relaxing to their normal energy states by emitting light (Section 10.3). What appears as a single lightning stroke may actually be composed of several individual strokes, the average for most displays being about three or four, each one separated by 40 to 80 milliseconds. The average peak power delivered by such lightning flashes is about 10^{12} watts, equivalent to 10 billion 100-watt lightbulbs!

Like lightning, thunder, which accompanies most displays, is another natural phenomenon about which humans have speculated for centuries. The consensus view today, however, remains that described by M. Hirn in 1888, namely, that air along the lightning train is heated to a high temperature ($\sim 20{,}000\,°\text{C}$) and rapidly expands to form a shock wave, which soon decays to a sound

functional explanation for lightning strikes is well understood, many of the details have only been elucidated relatively recently.

For example, one major puzzle surrounding cloud-to-ground lightning, the most important type of lightning in terms of its impact on human life, has been how the cloud acquires charge—or more properly, a charge separation—in the first place. One popular theory takes advantage of the mixed phases of water typically found in large multikilometer-sized thunderstorm clouds: liquid water droplets, ice crystals, and *graupel* or soft hail. It is believed that, as the faster-falling graupel collides with the smaller ice crystals, positive charge is transferred from the porous hail to the ice crystals so that the former acquire a net negative charge and the latter a net positive charge. (This is a bit like rubbing a plastic rod with a piece of fur; Figure 7.4.) As the graupel continues to move downward relative to the ice crystals, a charge separation develops within the cloud, with the negative charges concentrated in the lower portions of the cloud and the positive charges residing higher up (as much as a kilometer or more) in the cloud (**Figure 7.16**). Although this scenario for producing a charge separation shows promise in explaining lightning in common terrestrial thunderstorms, it is not the only theory of thundercloud electrification, and it may not be able to explain other types of discharges called *warm cloud* lightning.

The average lightning stroke is between 6 and 8 miles long and consists of a large surge of current usually moving upward from the ground into the cloud along a path of ionized air. (*Intracloud lightning* is also quite common

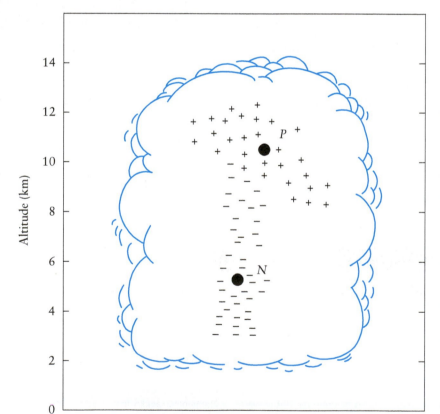

Figure 7.16 Schematic drawing showing the distribution of the main concentrations of positive and negative charge in a large thundercloud. The solid black dots indicate the effective centers of charge for each distribution. If each cloud of charge were replaced by an equivalent point charge located at these centers of charge, then the size of the charge at P would be about $+40$ C, whereas that at N would be -40 C. [After M. A. Uman, *Lightning*, p. 3 (New York, Dover, 1984).]

wave we hear. However, even under the best of conditions, thunder generally cannot be heard at all at distances of more than about 25 kilometers from the lightning channel that produced it. This is because the upward refraction of the sound waves resulting from vertical temperature variations and wind shear. But that's another story. (As noted in Physics to Go 1.3 you can often estimate the distance to a lightning strike from the time difference between the visible flash and the audible thunder associated with it.)

So, next time you find yourself in the middle of one of Nature's fantastic fireworks displays, as you contemplate its majesty and beauty, think too about the awesome physics that makes it all possible!

■ **QUESTIONS**

1. Describe one promising explanation for how thunderclouds become electrified, that is, acquire a charge separation that often leads to cloud-to-ground lightning strikes.
2. Loud thunder claps frequently accompany lightning discharges. Give a description of how scientists believe thunder is produced during a lightning storm.

7.3 Electric Currents—Superconductivity

An **electric current** is a flow of charged particles. The cord on an electrical appliance encloses two separate metal wires covered with insulation. When the appliance is plugged in and operating, electrons inside each wire move back and forth. Inside older television picture tubes and computer monitors, free electrons are accelerated from the back of the unit to the screen at the front. There is a near vacuum inside the video tube so the electrons can travel without colliding with gas molecules (**Figure 7.17**). When salt is dissolved in water, the sodium and chlorine ions separate and can move about just like the water molecules. If an electric field is applied to the water, the positive sodium ions will flow one way (in the direction of the field), and the negative chlorine ions will flow the other way.

7.3a Electric Current

Regardless of the nature of the moving charges, the quantitative definition of electric current is as follows.

Figure 7.17 Examples of electric current. (a) Electrons flowing inside a metal wire. (b) Electrons moving through the near vacuum inside a television picture tube. (c) Positive ions and negative ions, from dissolved salt, flowing through water.

Light bulb

(a)

(b)

Wire

(c)

Salt water

iStockphoto.com/Don Nichols; iStockphoto.com; iStockphoto.com/KateLeigh

Table 7.1 Typical Currents in Common Devices

Device	Current (A)
Calculator	0.0001
Spark plug	0.001
iPhone (while charging)	0.003
60-watt conventional lightbulb	0.5
LCD television	1.9
1,800-watt hair dryer	15

DEFINITION **Current** The rate of flow of electric charge. The amount of charge that flows by per second.

$$\text{current} = \frac{\text{charge}}{\text{time}} \qquad I = \frac{q}{t}$$

The SI unit of current is the *ampere* (A or amp), which equals 1 coulomb per second. Current is measured with a device called an *ammeter*.

A current of 5 amperes in a wire means that 5 coulombs of charge flow through the wire each second. (Table 7.1 lists some representative currents.) Either positive charges or negative charges can comprise a current. The effect of a positive charge moving in one direction is the same as that of an equal negative charge moving in the opposite direction. Formally, an electric current is represented as a flow of *positive* charge. This is because it was originally believed that positive charges moved through metals. Even after it was discovered that it is negatively charged electrons that flow in a wire to comprise the current, the convention of defining the direction of current flow as that which would be associated with positive charges was retained. If positive ions are flowing to the right in a liquid, then the current is to the right. If negative charges (like electrons) are flowing to the right, then the direction of the current is to the left. In Figure 7.17, the current is to the left in (a) and (b) and to the right in (c).

The ease with which charges move through different substances varies greatly. Any material that does not readily allow the flow of charges through it is called an electrical *insulator*. Substances such as plastic, wood, rubber, air, and pure water are insulators because the electrons are tightly bound in the atoms, and electric fields are usually not strong enough to rip them free so they can move. Our lives depend on insulators: the electricity powering the devices in our homes could kill us if insulators, like the covering on power cords, didn't keep it from entering our bodies.

An electrical *conductor* is any substance that readily allows charges to flow through it. Metals are very good conductors because some of the electrons are only loosely bound to atoms and so are free to "skip along" from one atom to the next when an electric field is present. In general, solids that are good conductors of heat are also good conductors of electricity. As mentioned before, liquids such as water are conductors when they contain dissolved ions. Most drinking water has some natural minerals and salts dissolved in it and so conducts electricity. Solid insulators can become conductors when wet because of ions in the moisture. The danger of being electrocuted by electrical devices increases dramatically when they are wet.

Semiconductors are substances that fall in between the two extremes. The elements silicon and germanium, both semiconductors, are poor conductors of electricity in their pure states, but they can be modified chemically ("doped") to have very useful electrical properties. Transistors, solar cells, and numerous other electronic components are made out of such semiconductors. The electronic revolution that began in the second half of the 20th century and continues today, including the development of inexpensive calculators, computers, sound-reproduction systems, and other devices, came about because of semiconductor technology (Figure 7.18).

Figure 7.18 Enlarged image of a micro-electronic circuit etched on a semiconductor chip. The pin lying on top of the chip provides an idea of the extremely small scale of the circuit.

Science Faction/SuperStock

DEFINITION **Resistance** A measure of the opposition to current flow. Resistance is represented by R, and the SI unit of measure is the *ohm* (Ω).

7.3b Resistance

What makes a 100-watt lightbulb brighter than a 60-watt bulb? The size of the current flowing through the filament determines the brightness. That, in turn, depends on the filament's **resistance.**

In general, a conductor will have low resistance and an insulator will have high resistance. The actual resistance of a particular piece of conducting material—a metal wire, for example—depends on four factors:

Composition. The particular metal making up the wire affects the resistance. For example, an iron wire will have a higher resistance than an identical copper wire.

Length. The longer the wire is, the higher its resistance.

Diameter. The thinner the wire is, the higher its resistance.

Temperature. The higher the temperature of the wire, the higher its resistance.

The filament of a 100-watt bulb is thicker than that of a 60-watt bulb, so its resistance is lower. As we will see in later sections, this means a larger current normally flows through the 100-watt bulb, so it is brighter.

Resistance can be compared to friction. Resistance inhibits the flow of electric charge, and friction inhibits relative motion between two substances. In metals, electrons in a current move among the atoms and in the process collide with them and give them energy. This impedes the movement of the electrons and causes the metal to gain internal energy. The consequence of resistance is the same as that of kinetic friction—heating. The larger the current through a particular device, the greater the heating.

7.3c Superconductivity

In 1911, Dutch physicist Heike Kamerlingh Onnes made an important discovery while measuring the resistance of mercury at extremely low temperatures. He found that the resistance decreased steadily as the temperature was lowered, until at 4.2 K (−452.1°F) it suddenly dropped to zero (**Figure 7.19**). Electric current flowed through the mercury with *no* resistance. Onnes named this phenomenon *superconductivity* for good reason: mercury is a perfect conductor of electric current below what is called its *critical temperature* (referred to as T_c) of 4.2 K. Subsequent research showed that hundreds of elements, compounds, and metal alloys become superconductors, but only at very low temperatures. Until 1985, the highest known T_c was 23 K for a mixture of the elements niobium and germanium.

Superconductivity seems too good to be true: electricity flowing through wires with *no* loss of energy to heating. Once a current is made to flow in a loop of superconducting wire, it can flow *for years* with no battery or other source of energy because there is no energy loss from resistance. A great deal of the electrical energy that is wasted as heat in wires could be saved if conventional conductors could be replaced with superconductors. But the superconducting state for a given material has limitations. Resistance returns if the temperature is raised above the superconductor's T_c, if the current through the substance becomes too large, or if it is placed in a magnetic field that is too strong.

Practical superconductors were developed in the 1960s and are now widely used in science and medicine. Most of them are copper oxide compounds that contain calcium, barium, yttrium, and other rare-earth elements. *Superconducting electromagnets*, the strongest magnets known, are used to study the effects of magnetic fields on matter and to direct high-speed charged particles. The Large Hadron Collider (LHC), an enormous particle accelerator located near Geneva, Switzerland (Figure 8.16), uses superconducting electromagnets to guide and focus protons as they are accelerated to nearly the speed of light. An entire experimental passenger train was built that levitated by superconducting electromagnets. *Magnetic resonance imaging* (MRI) uses superconducting electromagnets to form incredibly detailed images of the body's interior. (Magnetism and electromagnets are discussed in Chapter 8.)

Figure 7.19 Graph of the resistance of a sample of mercury versus temperature, showing the transition to superconductivity.

Widespread practical use of these superconductors is severely limited because they must be kept cold using liquefied helium. Helium is very expensive and requires sophisticated refrigeration equipment to cool and to liquefy. Once a superconducting device is cooled to the temperature of liquid helium, bulky insulation equipment is needed to limit the flow of heat into the helium and the superconductor. These factors combine to make the so-called low-T_c superconductors unwieldy or uneconomical except in certain special applications when there are no alternatives.

But hope for wider use of superconductivity blossomed beginning in 1987 when a new family of "high-T_c" superconductors was developed with critical temperatures that now reach as high as about 140 K. This was an astounding breakthrough because these materials can be made superconducting through the use of liquid nitrogen (boiling point 77 K). Liquid nitrogen is widely available, is inexpensive to produce compared to liquid helium, and can be used with much less-sophisticated insulation. However, the new high-T_c superconductors are handicapped by a couple of unfortunate properties: they are brittle and consequently are not easily formed into wires, and they aren't very tolerant of strong magnetic fields or large electric currents. If these problems can be overcome, a new revolution in superconducting technology will occur.

In 2008, superconducting technology took a significant step forward toward this revolution with the inauguration of the $60 million *Holbrook Superconductor Project*, the world's first production superconducting transmission cable. Funded by the U.S. Department of Energy and operated as part of the Long Island Power Authority, a 138-kV suburban power station is fed via a 0.6-kilometer-long underground tunnel containing nearly 100 miles of high-T_c superconducting wire cooled with liquid nitrogen. The superconducting material is a bismuth strontium calcium copper oxide with a critical temperature of 108 K. An extensive support system is required for producing and storing the liquid nitrogen used to cool the ribbon-like cables. In 2012, the tolerance of the cable system to magnetic fields was improved, and the system continues to function today.

In 2014, an even larger 40-MW project called *AmpaCity* was initiated in Essen, Germany, to connect two substations in the heart of the city separated by ~1 kilometer. Although the start-up costs for such a system far exceed those of a comparable conventional wire grid supply, the AmpaCity project is based around the premise that, over a 40-year lifetime, the superconducting cable will be cheaper than a conventional alternative even before including any savings due to efficiencies associated with the lossless energy transmission afforded by the superconducting technology. The question now becomes one of scaling: Can the superconducting grids be feasibly expanded to meet the current power consumption needs of the most populous regions of the globe? Only time will tell. For the record, the highest-T_c superconductor yet discovered is hydrogen sulfide gas compressed to a solid state using pressures of up to 150 GPa; it has a critical temperature of 203 K ($-70°C$).

Learning Check

1. An electric current can consist of
   (a) positive ions flowing in a liquid.
   (b) negative ions flowing in a liquid.
   (c) electrons moving through a wire.
   (d) All of the above.
2. A substance that is neither a good conductor nor a good insulator is called a _____.

3. What is the distinction between an excellent conductor, such as a copper wire used in an electronic device, and a superconductor?

ANSWERS: 1. (d) **2.** semiconductor. **3.** small resistance versus zero resistance

7.4 Electric Circuits and Ohm's Law

An electric current will flow in a lightbulb, a laptop, or other such device only if an electric field is present to exert a force on the charges. A conventional flashlight works because the batteries produce an electric field that forces electrons to flow through the lightbulb. An *electric circuit* is any system consisting of a battery or other electrical *power supply*, some electrical device such as a lightbulb, and wires or other conductors to carry the current to and from the device (**Figure 7.20**). The power supply acts like a "charge pump": it forces charges to flow out of one terminal, go through the rest of the circuit, and flow into the other terminal. Electrons typically move through a circuit quite slowly, about 1 millimeter per second. In this respect, an electric circuit is much like the cooling system in a car in which the water pump forces coolant to flow through the engine, radiator, and the hoses connecting them.

Figure 7.20 A simple electric circuit. The battery supplies the energy or "pressure" needed to move the charges through the circuit. Charge does not build up anywhere; for each coulomb of charge that leaves the positive terminal, one coulomb enters the negative terminal.

7.4a Voltage and Ohm's Law

The concepts of energy and work are used to quantify the effect of a power supply in a circuit. In a conventional flashlight, for instance, the batteries cause electrons to flow through the bulb's filament. Because a force acts on the electrons and causes them to move through a distance, *work is done* on the electrons *by the batteries*. In other words, the batteries give the electrons energy. This energy is converted into internal energy and light as the electrons go through the lightbulb and heat the filament. This leads to the concept of electric **voltage.**

> **DEFINITION** **Voltage** The work that a charged particle can do divided by the size of the charge. The energy per unit charge that is provided to charged particles by a power supply.
>
> $$V = \frac{\text{work}}{q} \quad V = \frac{E}{q}$$
>
> The SI unit of voltage is the *volt* (V), which is equal to *1 joule per coulomb*. Voltage is measured with a device called a *voltmeter*.

A 12-volt battery gives 12 joules of energy to each coulomb of electric charge that it moves through a circuit. Each coulomb does 12 joules of work as it flows through the circuit. (**Table 7.2** lists some typical voltages.)

If we return to the analogy of a battery as a charge pump, the voltage plays the role of pressure. A high voltage causing charges to flow in a circuit is similar to a high pressure causing a fluid to flow (**Figure 7.21**). Even when the circuit

Table 7.2 Examples of Common Voltages

Description	Voltage (V)
Nerve impulse	0.1
D-cell battery	1.5
Smoke detector battery	9.0
Standard car battery	12
Electric vehicle battery pack	300
Power plant generator	24,000
Dental x-ray machine (typical)	80,000
High-voltage transmission line (typical)	345,000

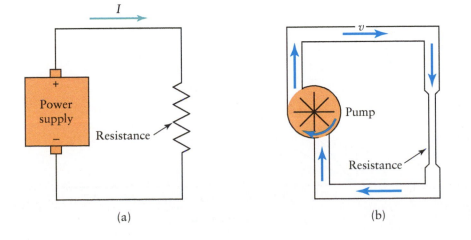

(a) (b)

Figure 7.21 (a) The flow of charge in an electric circuit is much like (b) the flow of water through a closed pipe. The power supply corresponds to the water pump, and the resistance corresponds to the narrow segment of pipe. The pressure on the output side of the pump is much like the voltage on the "+" terminal of the power supply. The electric current corresponds to the rate of flow of the water.

is disconnected from the power supply and there is no charge flow, the power supply still has a voltage. In this case, the electric charges have potential energy. (Voltage is also referred to as electric *potential*.)

The size of the current that flows through a conductor depends on its resistance and on the voltage causing the current. **Ohm's law,** named after its discoverer, Georg Simon Ohm, expresses the exact relationship.

LAWS **Ohm's Law** The current in a conductor is equal to the voltage applied to it divided by its resistance:

$$I = \frac{V}{R} \quad \text{or} \quad V = IR$$

The units of measure are consistent in the two equations: if I is in amperes and R is in ohms, then V will be in volts.

By Ohm's law, the higher the voltage for a given resistance, the larger the current. The larger the resistance for a given voltage, the smaller the current. By applying different-sized voltages to a given conductor, one can produce different-sized currents. A graph of the voltage versus the current will be a straight line with a slope that is equal to the conductor's resistance (**Figure 7.22**). Reversing the polarity of the voltage (switching the "+" and "−" terminals) will cause the current to flow in the opposite direction.

EXAMPLE 7.1 A lightbulb used in a 3-volt flashlight has a resistance equal to 6 ohms. What is the current in the bulb when it is switched on?

SOLUTION By Ohm's law,

$$I = \frac{V}{R} = \frac{3 \text{ V}}{6 \text{ }\Omega}$$

$$= 0.5 \text{ A}$$

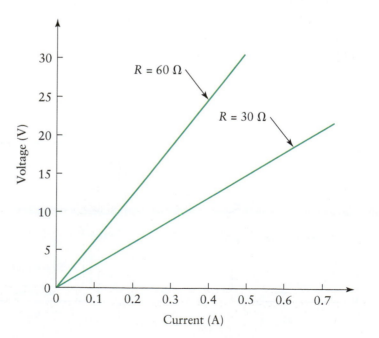

Figure 7.22 Graph of voltage versus current for two conductors with different resistances (R). For each resistance, the voltage needed to produce a given current is directly proportional to the current.

EXAMPLE 7.2 A small electric heater has a resistance of 15 ohms when the current in it is 2 amperes. What voltage is required to produce this current?

SOLUTION

$$V = IR = 2\,\text{A} \times 15\,\Omega$$

$$= \boxed{30\,\text{V}}$$

Not all devices remain "ohmic"—that is, obey Ohm's law—as the voltage applied to them changes. Often, instead of remaining constant, the resistance of a conductor changes when the voltage changes. At higher voltages, a larger current flows through the filament of a lightbulb, so its temperature is also higher. The resistance of the hotter filament is consequently greater (**Figure 7.23**). Some semiconductor devices, called *diodes*, are designed to have very low resistance when current flows through them in one direction but very high resistance when a voltage tries to produce a current in the other direction. Water with salt dissolved in it generally has lower resistance when higher voltages are applied to it: doubling the voltage will more than double the current. A graph of *V* versus *I* for ordinary tap water is less steep at higher voltages.

Many electrical devices are controlled by changing a resistance. The rotary or slide dimmer control on a room lighting circuit simply varies the resistance in the circuit. Turning up the dimmer switch reduces the resistance, so more current flows in the circuit, resulting in brighter illumination. A volume slider on a studio audio mixer used to change the sound level during a recording session works the same way, as do the speed setting switches on a kitchen blender.

7.4b Series and Parallel Circuits

In many situations, several electrical devices are connected to the same electrical power supply. A house may have a hundred different lights and appliances all connected to one cable entering the house. An automobile has dozens of devices connected to its battery. There are two basic ways in which more than one device can be connected to a single electrical power supply—by a series circuit and by a parallel circuit.

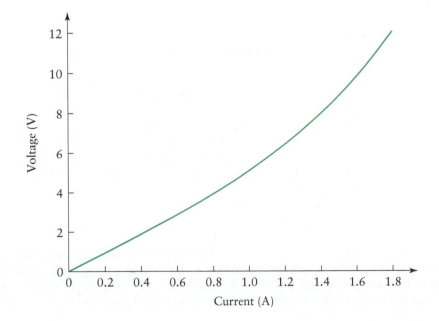

Figure 7.23 Graph of voltage versus current for a conventional brake lightbulb on a car. The graph curves upward because the resistance as measured by the slope is higher with larger currents. This is because the filament is hotter.

Figure 7.24 A simple series circuit. The current is the same in each of the bulbs.

In a *series circuit*, there is only one path for the charges to follow, so the same current flows in each device (**Figure 7.24**). In such a circuit, the voltage is divided among the devices: the voltage on the first device plus the voltage on the second device, and so on, equals the voltage of the power supply. For example, if three lightbulbs with the same resistance are connected in series to a 12-volt battery, the voltage on each bulb is 4 volts. If the bulbs had different resistances, each one's "share" of the voltage would be proportional to its resistance.

Notice that the current in a series circuit is stopped if any of the devices breaks the circuit (**Figure 7.25**). A series circuit is not normally used with, say, a number of lightbulbs because if one of them burns out, the current stops and all of the bulbs go out. A string of Christmas lights that flash at the same time uses a series circuit so that all the bulbs go on and off together.

In a *parallel circuit*, the current through the power supply is "shared" among the devices while each has the same voltage (**Figure 7.26**). The current flowing in the first device plus the current in the second device, and so on, equals the current output by the power supply. There is more than one path for the charges to follow—in this case, three. If one of the devices burns out or is removed, the others still function. The lightbulbs in multiple-bulb light fixtures are in parallel so that if one bulb burns out, the others remain lit. Often, the two types of circuits are combined: one switch may be in series with several lightbulbs that are in parallel.

EXAMPLE 7.3 Three lightbulbs are connected in a parallel circuit with a 12-volt battery. The resistance of each bulb is 24 ohms. What is the current produced by the battery?

SOLUTION The voltage of each bulb is 12 volts. Therefore, the current in each bulb is

$$I = \frac{V}{R} = \frac{12\ \text{V}}{24\ \Omega}$$

$$= 0.5\ \text{A}$$

The total current supplied by the battery equals the sum of the currents in the three bulbs.

$$I = 0.5\ \text{A} + 0.5\ \text{A} + 0.5\ \text{A}$$

$$= 1.5\ \text{A}$$

Burned-out bulb

Figure 7.25 If one device in a series circuit fails, such as a bulb burning out, the current stops and all of the devices go off.

The concept of voltage is quite general and is not restricted to electrical power supplies and electric circuits. Whenever there is an electric field in a region of space, a voltage exists because the field has the potential to do work on electric charges. The strength of an electric field can be expressed in terms of the voltage change per unit distance along the electric field lines. As defined in Section 7.2, the SI unit for the electric field is the newton per coulomb. Here we see an alternative, equivalent unit for the electric field: volt/m. $1\ \text{N/C} = 1\ \text{V/m}.$

For example, air conducts electricity when the electric field is strong enough to ionize atoms in the air. The minimum electric field strength required for this to happen is between 10,000 and 30,000 volts per centimeter, depending on the conditions. This means that if there is a spark one-fourth of an inch

Figure 7.26 A simple parallel circuit. Each bulb has the same voltage (the voltage of the power supply). Because there are three separate pathways for the current, should one of the bulbs burn out, the others will remain lit.

iStockphoto.com/Don Nichols; iStockphoto.com

long between your finger and a doorknob, the voltage that causes the spark is at least 7,500 volts.

As transistors and other components on integrated circuit chips (ICs) are made smaller, even the low voltages that are used to make them operate (typically around 1 volt) produce very strong electric fields. Inside modern ICs, electric field strengths can reach 400,000 V/cm. Designers of ICs must keep this in mind because electric fields only about 25 percent stronger than this can disrupt circuit processes.

■ Learning Check

1. (True or False.) A battery stores electric charge in the same way that a bottle stores water.
2. Which of the following is not true about voltage?
   (a) It is analogous to fluid pressure.
   (b) It equals the energy per unit charge.
   (c) It equals the work that can be done per unit charge.
   (d) Its unit of measure is equivalent to a joule per ampere.

3. The size of the current flowing through a conductor equals the _____ divided by the _____.

4. (True or False.) In a parallel circuit, the size of the current flowing through each device may be different.

ANSWERS: 1. False **2.** (d) **3.** voltage, resistance **4.** True

---

MEDICAL APPLICATIONS   Electricity and the Human Body

As in all living organisms from tiny single-celled protozoans to huge whales and sequoia trees, electricity plays vital roles in the normal functioning of the human body. Our nervous system, sketched in **Figure 7.27**, is a highly sophisticated internal information network composed of different types of *neurons* (nerve cells). Sensory neurons provide input to the brain, associative neurons process information in the brain, and motor neurons carry signals from the brain to muscles and some glands. There are approximately 12 billion nerve cells in the human body, some as long as 1 meter.

A signal is transmitted along an individual neuron in the form of a small voltage change between the inside and the outside of the cell. A neuron at rest has an excess of negative ions at its inner surface, resulting in a voltage of about −0.07 volts. A nerve signal involves a momentary flow of positive sodium ions into the neuron until the polarity is reversed and the voltage is about +0.03 volts. That part of the neuron quickly returns to the resting state, but the small voltage "pulse" travels along the neuron at up to 100 m/s (about 225 mph).

The specialized *sensory cells* associated with touch, taste, sight, and the other senses produce electrical signals when they are stimulated by the outside environment. These signals are passed on from neuron to neuron, through the spinal cord, and into various regions of the brain. Command signals from the brain are transmitted by motor neurons in the opposite direction to muscles, which contract after the electrical signal is received.

Recently, researchers have developed ways to control computers using electrical signals associated with muscle contraction. One goal is to provide physically disabled individuals full use of computers without the need of keyboards and pointing devices.

The brain is constantly swarming with neural signals as it processes information and formulates response signals. An *electroencephalogram* (EEG) is a record of the periodic voltage changes associated with brain activity. EEGs are made by monitoring the minute voltages that appear on the scalp near different parts of the brain. EEGs show characteristic patterns for different levels of mental

Figure 7.27 The human nervous system.

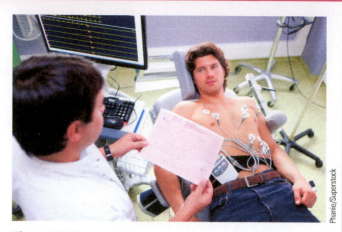

Figure 7.28 An electrocardiogram being recorded.

activity (fully alert wakefulness, sleeping, at rest with the eyes closed, and so on) and can indicate certain disorders such as epilepsy.

The heart pumps blood by coordinated muscle contraction followed by relaxation. Unlike other muscles, such as those in your arms, the heart does not rely on a signal from the brain to contract. A small group of cells in the heart produces electrical impulses spontaneously at a rate of around 70 per minute. (The actual frequency of the signal, and therefore the heart rate, is regulated by the brain and autonomic nervous system. For example, the heart rate is increased during exercise.) This impulse initiates a wave of contraction of heart muscle in the required coordinated fashion. An *electrocardiogram* (ECG, more commonly called EKG) is similar to an EEG in that it is a record of the voltages produced by various parts of the heart as it beats (**Figure 7.28**). EKGs are routinely used to detect irregularities in the heart's electrical conduction system.

Often, with age, the intrinsic ability of the heart to control its beating diminishes. A common problem is that different parts of the heart start to beat at different rates because the controlling signal is too weak. A pacemaker can often be implanted near the heart to remedy this problem. This device supplies a small electrical shock to the heart at regular intervals to reestablish coordinated heart contraction.

Because the body relies so heavily on small voltages for its normal function, it is extremely vulnerable to electricity from the outside. Interestingly, it is the *current* that actually flows through the body, not the voltage, that is most important in determining the physiological effect. A single shock of static electricity received from a doorknob may be caused by several thousand volts, but the effect is localized (a sharp pain in your finger) because only a small amount of charge flows through the interior of your body.

The effect of a steady current on your body depends on the size of the current and on the part of the body through which it flows. Currents less than about 1 milliampere usually cannot be felt. A current of around 7 milliamperes or above through your hand (for example) would block signals to your muscles, and you would not be able to move your fingers. A current of 100 milliamperes or more through the heart can induce uncoordinated contractions, called *fibrillation,* leading to death. Steady currents above 300 milliamperes through the heart are usually fatal.

You should realize that a current cannot flow through your body unless it is part of a complete circuit. A person hanging by the arms from a 240-volt power line will have no current flowing through his or her body. But touching the ground or a different wire would provide a complete circuit for the charge to flow through.

In some cases, sending a current through part of the body is beneficial—even life-saving. The pacemaker is one example. Another is the defibrillator. For more than 60 years, defibrillators have been used to save lives by sending a momentary current through fibrillating hearts, thereby restoring normal heartbeat. Some individuals with high risk of fibrillation can have a miniature, automatic defibrillator surgically implanted. Within the last 15 years, the Food and Drug Administration (FDA) has approved the implantation of a neurostimulator (sometimes referred to as a "brain pacemaker") to send electrical impulses to specific parts of the brain for the treatment of movement and affective disorders. This *deep brain stimulation* has provided therapeutic benefits for patients suffering from Parkinson's disease, multiple sclerosis, chronic pain, major depression, and obsessive-compulsive disorder.

From the very "glue" that holds together atoms and molecules to the vehicle for thought and communication in the human body, electricity is essential to the existence of matter and life itself.

■ QUESTION

1. What does the term *fibrillation* mean in connection with heart function? Give two important causes of heart fibrillation in human beings. How do pacemakers and defibrillators help to prolong life for patients suffering from this condition?

7.5 Power and Energy in Electric Currents

Because a battery or other electrical supply must continually put out energy to cause a current to flow, it is important to consider the **power output**—the rate at which energy is delivered to the circuit. The power is determined by the voltage of the power supply and the current that is flowing. Think of it this way: the power output is the amount of energy expended per unit amount of time. The power supply gives a certain amount of energy to each coulomb of charge that flows through the circuit. Consequently, the energy output per unit time equals the *energy given to each coulomb of charge* multiplied by the *number of coulombs that flow through the circuit per unit time:*

energy per unit time = energy per coulomb × number of coulombs per unit time

These three quantities are just the power, voltage, and current, respectively. Consequently, the power output of an electrical power supply is

$$power = voltage \times current$$

$$P = VI$$

The units work out correctly in this equation also: joules per coulomb (volts) multiplied by coulombs per second (amperes) equals joules per second (watts).

The power output of a battery is proportional to the current that it is supplying: the larger the current, the higher the power output.

EXAMPLE 7.4   In Example 7.1, we computed the current that flows in a flashlight bulb. What is the power output of the batteries?

SOLUTION Recall that the batteries produce 3 volts and that the current in the lightbulb is 0.5 amperes. The power output is

$$P = VI = 3\,\text{V} \times 0.5\,\text{A}$$

$$= 1.5\,\text{W}$$

The batteries supply 1.5 joules of energy each second.

What happens to the energy delivered by an electrical power supply? In a lightbulb, less than 5 percent is converted into visible light, and the rest becomes internal energy. Even the visible light emitted by a lightbulb is absorbed eventually by the surrounding matter and transformed into internal energy. (Interior lighting is actually used to heat some buildings.) Electric motors in hair dryers, vacuum cleaners, and the like convert about 60 percent of their energy input into mechanical work or energy, while the remainder goes to internal energy. The mechanical energy is generally dissipated as internal energy through friction. In a similar way, we can trace the energy conversions in other electrical devices and the outcome is the same: most electrical energy eventually becomes internal energy.

Ordinary metal wire converts electrical energy into internal energy whenever there is a current flowing. You may have noticed when using a hair dryer that its cord becomes warm. This heating, called **ohmic heating**, occurs in any conductor that has resistance, even when the resistance is quite small. The huge cables used to conduct electricity from power plants to cities are heated by this effect. This heating represents a loss of usable energy.

The temperature that a current-carrying wire reaches from ohmic heating depends on the size of the current and on the wire's resistance. Increasing the current in a given wire will raise its temperature. Many devices utilize this effect. The resistances of heating elements in toasters and electric heaters are chosen so that the normal operating current is large enough to heat them until

Figure 7.29 The current causes the thin filament wire to become white hot, even as the larger wires connected to it stay much cooler.

they glow red hot and can toast bread or heat a room. The filament in an incandescent lightbulb is made so thin that ohmic heating causes it to glow white hot and emit enough light to illuminate a room (**Figure 7.29**).

Ohmic heating is a major consideration in the design of sophisticated integrated circuit chips. Even though the currents flowing through the tiny transistors are extremely small, there are so many circuits in such a small space that special steps must be taken to make sure the heat produced is conducted away. Because a superconductor has zero resistance, there is no ohmic heating. The overall efficiencies of most electrical devices could be improved if regular wires could be replaced by superconductors. Superconducting transmission lines would allow electricity to be carried from a power plant to a city with no loss of energy. The limitations of currently known superconductors, however, make such uses impracticable (see the discussion in Section 7.3c).

A sufficiently large current in any conventional wire can cause it to become very hot—hot enough to melt any insulation around it or to ignite combustible materials nearby. Fuses and circuit breakers are put into electric circuits as safety devices to prevent dangerous overheating of wires. If something goes wrong or if too many devices are plugged into the circuit and the current exceeds the recommended safe limit for the size of wire used, the fuse or circuit breaker will automatically "break" the circuit and the current will stop. (A fuse is a fine wire or piece of metal inside a glass or plastic case. When the current exceeds the fuse's design limit, the metal melts away, and the circuit is broken; **Figure 7.30**.) Designers of electric circuits in cars, houses, and other buildings must choose wiring that is large enough to carry the currents needed without overheating. They must also include fuses or circuit breakers that will disconnect a circuit if it is overloaded.

Most electrical devices are rated by the power that they consume in watts. The equation $P = VI$ can be used to determine how much current flows through the device when it is operating.

EXAMPLE 7.5 An electric hair dryer is rated at 1,875 watts when operating on 120 volts. What is the current flowing through it?

SOLUTION

$$P = VI$$

$$1{,}875 \text{ W} = 120 \text{ V} \times I$$

$$\frac{1{,}875 \text{ W}}{120 \text{ V}} = I$$

$$I = \boxed{15.6 \text{ A}}$$

The wires in the electric cord must be large enough to allow 15.6 amperes to flow through them without becoming dangerously hot.

Figure 7.30 Two 20-ampere fuses used in automobiles. At some point, the current in the fuse on the left exceeded 20 amperes, so it burned out.

The highest current that can flow in a particular wire without causing excessive heating depends on the size of the wire. This is one reason why electric utilities use high voltages in their electrical power supply systems. The electricity delivered to a city, subdivision, or individual house must be transmitted with wires. Because $P = VI$, using a large voltage makes it possible to transmit the same power with a smaller current. If low voltages were used, say, 100 volts instead of the more typical 345,000 volts, much larger cables would have to be required to handle the larger currents.

Customers pay for the electricity supplied to them by electric companies based on the amount of energy they use. An electric meter keeps track of the total energy used by monitoring the power (rate of energy use) and the amount of time each power level is maintained (**Figure 7.31**). Recall the equation used to define power:

$$P = \frac{E}{t}$$

Therefore,

$$E = Pt$$

The amount of energy used is equal to the power times the time elapsed. If P is in watts and t is in seconds, then E will be in joules.

EXAMPLE 7.6   If the hair dryer discussed in Example 7.5 is used for 3 minutes, how much energy does it use?

SOLUTION The power is 1,875 watts. To get E in joules, we must convert the 3 minutes into seconds.

$$t = 3 \text{ min} = 3 \times 60 \text{ s} = 180 \text{ s}$$

So,

$$E = Pt = 1{,}875 \text{ W} \times 180 \text{ s}$$

$$= \boxed{340{,}000 \text{ J}}$$

This is a large quantity of energy—about the same as the kinetic energy of a small car going 60 mph (Example 3.6). Another comparison: a 150-pound person would have to climb 1,700 feet (about 170 floors) to gain 340,000 joules of potential energy.

Figure 7.31 Modern, digital *smart meter* equipped with a radio frequency transmitter and capable of providing automated meter reading (as well as other services) as part of local monitoring network.

Jeff Wilber/Shutterstock.com

A typical household can consume more than a billion joules of electrical energy each month. For this reason, a more appropriately sized unit of measure is employed for electrical energy—the kilowatt-hour (kWh). Energy in kilowatt-hours is computed by expressing power in kilowatts (kW) and time in hours. The conversion factor between joules and kilowatt-hours is

$$1 \text{ kWh} = 1 \text{ kW} \times 1 \text{ h}$$

$$= 1{,}000 \text{ W} \times 3{,}600 \text{ s}$$

$$= 3{,}600{,}000 \text{ J}$$

In Example 7.6, the energy used by the hair dryer is about 0.1 kilowatt-hours. The cost of electricity varies from region to region, but it is typically around 10 cents per kilowatt-hour. This means that it costs about 1 cent to run the hair dryer 3 minutes. (Would you climb 1,700 feet for 1 cent?)

Perhaps you have wondered why a common 1.5-volt D-cell battery is larger than a 9-volt battery used in smoke alarms and other common electrical devices. The voltage of a battery really has nothing to do with its physical size. Different 1.5-volt batteries range from the size of a button (for wristwatches) to larger than a beer can. The size is more an indication of the amount of electrical energy stored in the battery. A large battery can supply the same current (and the same power) for a longer time than a small battery with the same voltage. In applications with small power requirements, such as in hearing aids and calculators, even a small battery can provide enough stored electrical energy to operate the device for a year or more.

1. An automobile headlight and a dashboard light use the same voltage, but the power input to the headlight is much larger because
   (a) its resistance is lower.
   (b) the current in it is larger.
   (c) it uses energy faster.
   (d) All of the above.
2. Name one or more devices in which ohmic heating is desirable.

3. (True or False.) A fuse is designed to stop the flow of charge in a circuit if the current is too large.
4. The standard unit of energy used in electric utility bills is the _____.

ANSWERS: 1. (d) **2.** toaster, electric heater **3.** True **4.** kilowatt-hour

7.6 AC and DC

The electric current supplied by a battery is different from the current supplied by a normal household wall socket. Batteries supply **direct current (DC)**, and household outlets supply **alternating current (AC)**. A DC power supply, such as a battery, causes a current to flow in a fixed direction in a circuit (**Figure 7.32**). The current flows out of the positive (+) terminal of the power supply, moves through the circuit, and flows into the negative (−) terminal of the power supply. If the total resistance in the circuit doesn't change, the size of the current remains constant (as long as the battery doesn't run down). A graph of the current I versus time t is simply a horizontal line.

In an AC power supply, the polarity of the two output terminals switches back and forth—the voltage alternates. This causes the current in any circuit connected to the power supply to alternate as well. It flows counterclockwise, then clockwise, then back to counterclockwise, and so on. All the while, the size of the current is increasing, then decreasing, and so forth. A graph of the current in an AC circuit shows this variation in the size and direction of the current. (When I goes below zero, it means that the direction has reversed; **Figure 7.33**).

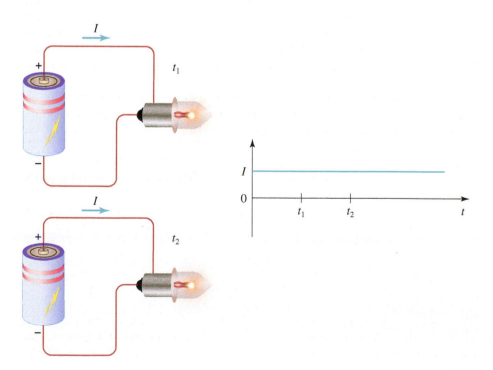

Figure 7.32 Direct current. The current flows in one direction and doesn't increase or decrease, as shown in the graph of current versus time.

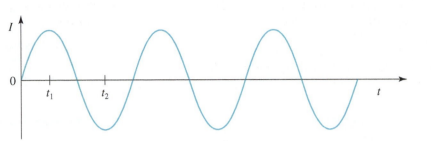

Figure 7.33 Alternating current. The direction of the current switches back and forth, and the size of the current varies continuously.

▶ Physics To Go 7.6

For this experiment, you will have to be in a room lit only by tube-type fluorescent lights (compact fluorescent lights don't work as well) and have a pen, pencil, or other thin object that is white or light colored. Position yourself so that the fluorescent light is behind or above you. Hold the pen in such a way that the light shines on it and there is a dark background behind it. Move the pen rapidly back and forth (sideways) with your hand. You should see faint lines parallel to the pen. What causes this? (*Hint:* AC is involved.) Why doesn't this work nearly as well with incandescent lightbulbs?

We have seen this kind of oscillation before in Section 2.5 and throughout Chapter 6. One can even think of AC as a kind of "wave" causing the charges in a conductor to oscillate back and forth. Almost all public electric utilities in the United States supply 60-hertz AC. The voltage between the two slots in a wall outlet oscillates back and forth 60 times per second. (In Europe, the standard frequency of AC is 50 hertz. This is why, when traveling abroad, you often have to bring along a plug-in converter.)

Some electric devices (such as lightbulbs) can operate on AC or DC, whereas others require one or the other. Electric motors and generators must be designed to operate on or produce either AC or DC. There are devices that can convert an AC voltage to a DC voltage and vice versa. Batteries can produce direct current only. For this reason, automobiles have DC electrical systems. (The "alternator" in an automobile generates AC, which is then converted into DC to be compatible with the battery.)

Alternating current has one distinct advantage over DC: simple, highly efficient devices called *transformers* can "step up" or "step down" AC voltages. This makes it possible to generate AC at a power plant at some intermediate voltage, step it up to a very high voltage (typically more than 300,000 volts) for economical transmission, and then step it down again to lower voltages for use in homes and industries. There is no counterpart of the transformer for DC. Another important use of AC is in electronic sound equipment. One example: if a 440-hertz tone is recorded on tape and then played back, the "signal" going to the speaker will be an alternating current with a frequency of 440 hertz. (We will discuss transformers and sound reproduction in more detail in Chapter 8.)

■ **Learning Check**

1. Direct current (DC)
   (a) is produced by batteries.
   (b) is used by most electric utilities.
   (c) carries the audio signal to speakers.
   (d) oscillates.

2. (True or False.) An incandescent lightbulb normally used with AC would not work if connected to DC (with the proper voltage).

3. AC voltages can be easily increased or decreased by using a _____ .

ANSWERS: 1. (a) **2.** False **3.** transformer.

Profiles in Physics   Founders of Electrical Science

The first person to undertake a successful systematic analysis of electric as well as magnetic effects was William Gilbert (1544–1603; **Figure 7.34**), a contemporary of Galileo. Born into an English family of comfortable means, Gilbert studied medicine at Cambridge and was later appointed physician to Queen Elizabeth I. This position left him with sufficient time and financial resources to pursue his studies of electricity and magnetism. Gilbert showed that the two types of effects are distinct, and he dispensed with a number of misconceptions about them. He carefully tested various substances to see which exhibited the amber effect. Some of the terminology that we use today, such as "pole," originated with Gilbert.

Figure 7.34 William Gilbert, English physician and scientist, was born in Colchester and served as personal physician to Elizabeth I from 1601 until his death in 1603.

After Gilbert published his findings in the book *De Magnete* in 1600, nothing new was discovered for some 60 years. Otto von Guericke, famous for his experiments with atmospheric pressure (see Profiles in Physics for Chapter 4), constructed a huge electrostatic machine to illustrate the amber effect on a large scale. It consisted of a large, rotating sulfur ball that was charged by friction. It could attract feathers, bits of paper, and other things from considerable distances. He was also the first to record electrostatic repulsion. During the next 200 years, demonstrations of electrostatic effects with such machines became very popular as parlor amusements and as topics of public lectures. Benjamin Franklin's interest in electricity began after he viewed such a demonstration in Boston in 1746.

The findings of two other experimenters, Englishman Stephen Gray (1666–1736) and Frenchman Charles Dufay (1698–1739), are also noteworthy. Gray discovered, somewhat by accident, that the electrostatic charge could flow through some substances but not through others. He thus discovered electrical conductors and insulators. Materials that are good at showing the amber effect turn out to be insulators. Conductors such as iron cannot be charged in the usual way because the charge will simply flow into the person holding it. Conductors can only be charged by keeping them out of contact with other conductors. Gray showed that the human body is itself a conductor.

Dufay studied electrostatic attraction and repulsion and concluded that there must be two different kinds of electric charge to account for them. The two types, which he called *vitreous* and *resinous*, obeyed the rule "like charges repel and unlike charges attract."

Our current names for the two types of charges, positive and negative, come from Benjamin Franklin (**Figure 7.35**). Franklin performed his electrical experiments during the late 1740s and early 1750s. In describing his results, he imagined that electricity (or "electrical fire") is a single fluid found in all objects and that it is capable of being circulated among objects. The flow of electrical fluid is initiated by rubbing (friction), a process in which one object, say B, acquires an excess of electrical fire while another object, say A, receives a deficit. Franklin described this circumstance as follows: "Hence have arisen some new terms among us: we say, B . . . is electrised positively; A negatively. Or rather, B is electrised plus; A, minus." Thus Franklin introduced the terms *positive* and *negative* to describe electricity—not to distinguish different types of charges, but to describe the states of objects having a greater or lesser amount of electrical fire in proportion to their normal share.

World History Archive/Alamy Stock Photo

Figure 7.35 Benjamin Franklin (1706–1790).

Alessandro Graf Volta.

Figure 7.36 Alessandro Volta, inventor of the battery.

Coulomb's law, the force law for static electric charges, seems to have been discovered independently by three different men: Charles Coulomb (1736–1806), John Robison (1739–1805), and Henry Cavendish (1731–1810). (Cavendish, whom we encountered in Section 2.7, was a noted recluse who amassed a wealth of experimental findings that were not published until long after his death.) But the Frenchman Coulomb is credited with discovering the law. His experience as an engineer, including a 9-year stint in Martinique, provided him with the skills he needed to construct precision force balances.

The investigation of electric currents was triggered by a chance observation of an Italian biologist, Luigi Galvani (1737–1798). Frog legs that he was preparing twitched when touched by a charged scalpel. A careful experimenter, Galvani undertook an extensive study of the phenomenon and thus made the first discoveries of the role of electricity in living systems. However, Galvani clung to the mistaken notion that the electricity in living organisms was somehow different from ordinary electricity.

Galvani's work came to the attention of the physicist Alessandro Volta (1745-1827; Figure 7.36) in northern Italy. Volta was already an accomplished experimenter in electricity: he was the first to devise a way to measure what was later named *voltage*. In the process of verifying and extending the discoveries of Galvani, Volta invented a type of battery. This was a pivotal discovery: batteries were able to supply much larger currents than were possible with electrostatic generators. The invention brought immediate acclaim to Volta. In 1801, Napoleon viewed a demonstration of the device and was very impressed.

Experimenters throughout Europe quickly began building their own "Voltaic cells," and the understanding of moving electricity advanced rapidly. The foremost researcher in this area was a German schoolteacher named Georg Simon Ohm (1787–1854).

Working in isolation from the more famous physicists of the time, Ohm introduced the important quantities of voltage, current, and resistance. He carefully studied the factors that affect the resistance of conductors and discovered the relationship that bears his name. Ohm's discoveries, unlike Volta's, were first met with skepticism and even contempt, perhaps because he was not a member of the traditional scientific community. It was only a few years before his death that Ohm was finally given the respect and recognition that he deserved.

The large currents that could be produced by batteries were a boon to science. Some of the larger batteries built had power outputs of several thousand watts. It was soon discovered that sending a current through water decomposed it into hydrogen and oxygen. This process, known as *electrolysis,* was applied to many other substances and led to the discovery of several new elements. It also showed that electricity is involved in the very structure of matter.

The proliferation of electrical devices that started in the late 19th century, and that continues to this day, revolutionized nearly every aspect of the lives of billions of people worldwide. Our dependence on electricity is perhaps best illustrated by what happens when it is unexpectedly cut off. The great blackouts that occurred in the northeastern United States in 2003, 1977, and 1965 brought commerce, transportation, science—nearly every aspect of modern society—almost to a complete halt. One could argue that the discoveries of researchers into the basics of electricity affect how we live more than those in any other field of physics.

■ QUESTIONS
1. For what discovery is Alessandro Volta most remembered? Why was his invention so important to the development of electrical science in the 19th century?
2. Describe Benjamin Franklin's main contribution to the way we conceptualize and talk about electricity today.

SUMMARY

» Electrons, protons, and certain other subatomic particles possess a physical property called **electric charge** that is the basic source of electrical and magnetic phenomena.

» Forces act between any objects that possess a net electric charge, positive or negative. Objects with like charges experience repulsive forces; objects with unlike charges experience attractive forces.

» The electrostatic force, expressed by **Coulomb's law**, is responsible for binding electrons to the nucleus in atoms, for the amber effect (such as static cling), and for a number of other natural or technological phenomena.

» Electric charges produce electric fields in the space around them. This field is the agent for the electrostatic force, just as the gravitational field is the agent of the gravitational attraction between objects.

» Most useful applications of electricity involve electric currents. They frequently involve the flow of electrons through metal wires driven by an electrical power supply, such as a battery.

» The flow of charge is analyzed using the physical concepts **voltage**, **current**, and **resistance**.

» **Ohm's law** states that the current in a circuit equals the voltage divided by the resistance.

» The power consumption in a circuit depends on the product of the voltage and the current.

» The electrical energy needed to cause a current to flow through a resistive element is converted into internal energy. This **ohmic heating** is usefully exploited directly by space heaters and toasters, as well as by incandescent lightbulbs to produce light.

» Fuses and circuit breakers are used to automatically disconnect a circuit if the current is large enough to cause excessive ohmic heating.

» At extremely low temperatures, many materials become superconductors—they have zero resistance. Consequently, no energy is lost to heating when electric currents flow through them. Superconductors now in use are still primarily limited to special-purpose scientific and medical instruments.

» There are two types of electric current: **alternating (AC)** and **direct (DC)**. Because batteries produce DC, battery-powered devices generally employ DC.

» Transformers can be built to "step up" or "step down" an AC voltage from one value to another. This makes AC particularly convenient for electrical supply networks such as electric utilities.

IMPORTANT EQUATIONS

Equation	Comments	Equation	Comments
$F = \dfrac{(9 \times 10^9)\, q_1 q_2}{d^2}$	Coulomb's law (in SI units)	$I = \dfrac{V}{R}$	Ohm's law
$I = \dfrac{q}{t}$	Definition of current	$V = IR$	Ohm's law
		$P = VI$	Electrical power consumption
$V = \dfrac{E}{q} = \dfrac{\text{work}}{q}$	Definition of voltage	$E = Pt$	Energy consumed during time t

MAPPING IT OUT!

1. Sections 7.3 and 7.4 introduce several important concepts for understanding electric circuits. These include, but are not limited to, *current, resistance,* and *voltage.* Using these fundamental concepts as a starting point, develop a concept map that captures your understanding of them and that properly establishes their relationships to one another and to electrical circuits (another concept). Is there one main concept that unifies your map? If so, which one is it? If not, how did or does that fact influence the structure of your concept map?

2. Concept maps can be used as substitutes for traditional note taking when you read articles in magazines and newspapers. This question illustrates the point.

   Chapter 7 describes a number of interesting devices based on the concepts of electric charge and force, electric currents, and electric circuits. Some of these applications of the principles of electricity include semiconductor devices, electrostatic precipitators, superconducting devices (such as MRI machines), transformers, and so on. Using the Internet or resources available at your library, locate and read an article on one of these devices. After reading the article, look back through it and circle or identify the key concepts in it. Then, using only the concepts you have identified, try to construct a concept map that accurately relates the major ideas or issues raised in the article. After completing the map, reflect for a moment on the degree of difficulty of this exercise. Are some major "concepts" missing from the article that you, the reader, had to "add," based on your own knowledge, to make sense of the article and, hence, your map? What does this suggest to you about the quality or level of sophistication of the piece you read?

(■ **Indicates a review question, which means it requires only a basic understanding of the material to answer. Questions without this designation typically require integrating or extending the concepts presented thus far.)**

1. ■ All matter contains both positively and negatively charged particles. Why do most things have no net charge?
2. A particular solid is electrically charged after it is rubbed, but it is not known whether its charge is positive or negative. How could you determine which charge it has by using a piece of plastic and fur?
3. ■ What is a positive ion? A negative ion?
4. What remains after a hydrogen atom is positively ionized?
5. ■ Describe the similarities and the differences between the gravitational force between two objects and the electrostatic force between two charged objects.
6. When clothes are removed from a hot dryer, they often cling together, but when two identical articles of clothing, such as two matched socks, are taken out, they usually repel one another. Explain this difference in behavior.
7. In otherwise empty space, what would happen if the size of the electrostatic force acting between two positively charged objects was exactly the same as that of the gravitational force acting between them? What would happen if they were moved closer together or farther apart?
8. (a) A negatively charged iron ball (on the end of a plastic rod) exerts a strong attractive force on a penny even though the penny is neutral. How is this possible?
   (b) The penny accelerates toward the ball, hits it, and then is immediately repelled. What do you think causes the sudden change from an attractive to a repulsive force on contact between the penny and the ball?
9. ■ What is an electric field? Sketch the shape of the electric field around a single proton.
10. ■ At one moment during a storm, the electric field between two clouds is directed toward the east. What is the direction of the force on any electron in this region? What is the direction of the force on a positive ion in this region?
11. ■ Explain what an electrostatic precipitator is and how it works.
12. ■ If electrons are flowing clockwise around an electric circuit, which way is the conventional current flowing in the circuit?
13. Saltwater contains an equal number of positive and negative ions. When saltwater is flowing through a pipe, does it constitute an electric current?
14. ■ Materials can be classified into four categories based on the ease with which charges can flow through them. Give the names of these categories and describe each one.
15. A solid metal cylinder has a certain resistance. It is then heated and carefully stretched to form a longer, thinner cylinder. After it cools, will its resistance be the same as, greater than, or less than what it was before? Explain your choice.
16. A student using a sensitive meter that measures resistance finds that the resistance of a thin wire is changed slightly when it is picked up with a bare hand. What causes the change in the resistance, and does it increase or decrease?
17. If a new material is found that is a superconductor at all temperatures, what parts of some common electric devices would definitely *not* be made out of it? Explain.
18. ■ Explain what current, resistance, and voltage are.
19. ■ Make a sketch of a simple electric circuit and label the key elements in the circuit.
20. ■ Describe Ohm's law.

21. A power supply is connected to two bare wires that are inserted into a glass of saltwater. The resistance of the water decreases as the voltage is increased. Sketch a graph of the voltage versus the current in the water showing this type of behavior.
22. ■ There are two basic methods for connecting more than one electrical device in a circuit. Name, describe, and give the advantages of each.
23. Make a sketch of an electric circuit that contains a switch and two lightbulbs connected in such a way that if either bulb burns out, the other still functions, but if the switch is turned off, both bulbs go out.
24. Two 1.5-volt batteries are connected in series in an electric circuit. Use the concept of energy to explain why this combination is equivalent to a single 3-volt battery. When connected in parallel, what are two 1.5-volt batteries equivalent to?
25. An electrical supply company sells two models of 100-watt power supplies (the maximum power output is 100 W), one with an output of 12 V and the other 6 V. What can you conclude about the maximum current that the two power supplies can produce?
26. A simple electric circuit consists of a constant-voltage power supply and a variable resistor. What effect does reducing the resistance have on the current in the circuit and on the power output of the power supply?
27. ■ What is the purpose of having fuses or circuit breakers in electric circuits? How should they be connected in circuits so they will be effective?
28. A 20-A fuse in a household electric circuit burns out. What catastrophe could occur if it is replaced by a 30-A fuse?
29. ■ Why is it economical to use extremely high voltages for the transmission of electrical power?
30. ■ Explain what AC and DC are. Why is AC used by electric utilities? Why is DC used in flashlights?
31. If AC in a circuit can be thought of as a wave, which kind is it, longitudinal or transverse?
32. If the electric utility company where you live suddenly changed the frequency of the AC to 20 Hz, what problems might this cause?
33. Rank the force on each of the point charges q_1 and q_2 in the following pairings from most attractive to most repulsive. The separation, d, between each pair is given. Explain the reasoning you used to arrive at your rankings.
    Pair 1: $q_1 = +2\ \mu C$; $q_2 = -4\ \mu C$; and $d = 10$ cm
    Pair 2: $q_1 = +2\ \mu C$; $q_2 = -2\ \mu C$; and $d = 5$ cm
    Pair 3: $q_1 = +3\ \mu C$; $q_2 = +3\ \mu C$; and $d = 10$ cm
    Pair 4: $q_1 = -5\ \mu C$; $q_2 = -2\ \mu C$; and $d = 20$ cm
34. A simple circuit of the form shown in **Figure 7.37** is constructed using a battery and three lightbulbs. The bulbs may have different resistances as given in the six circumstances **A** through **F**. For each situation, however, the battery in the circuit is always the same, as are the wire connections. Rank these circuits from greatest to smallest, according to the voltage across the bulb shown in red in the list. If two circuits have the same voltage associated with the selected bulb, give them the same ranking. Justify your rankings using your knowledge of the physics of such circuits.
    A: Bulb 1: 10 Ω; bulb 2: 20 Ω; bulb 3: 30 Ω
    B: Bulb 1: 10 Ω; bulb 2: 10 Ω; bulb 3: 20 Ω
    C: Bulb 1: 10 Ω; bulb 2: 20 Ω; bulb 3: 10 Ω
    D: Bulb 1: 10 Ω; bulb 2: 20 Ω; bulb 3: 30 Ω
    E: Bulb 1: 10 Ω; bulb 2: 10 Ω; bulb 3: 20 Ω
    F: Bulb 1: 10 Ω; bulb 2: 20 Ω; bulb 3: 10 Ω

Figure 7.37 Question 34.

35. Six resistors, all having the same temperatures and compositions, are each attached to a battery with voltage V to form a simple circuit. The resistors have different lengths, l, but the same diameters. Rank the circuits according to the current that flows through the resistor from smallest to greatest. If two (or more) circuits have the same current flowing, give them the same ranking. Explain the rationale you used to establish your rankings.

Circuit 1: $l = 2$ cm; $V = 3$ V
Circuit 2: $l = 2$ cm; $V = 6$ V
Circuit 3: $l = 1$ cm; $V = 3$ V
Circuit 4: $l = 1$ cm; $V = 6$ V
Circuit 5: $l = 3$ cm; $V = 2$ V
Circuit 6: $l = 3$ cm; $V = 6$ V

■ PROBLEMS

1. Two charged particles exert an electric force of 16 N on each other. What will the magnitude of the force be if the distance between the particles is reduced to one-half of the original separation? If the original distance between the particles is now restored, how much larger would the charge on either one of the particles have to become to yield the same force that was present when the particle separation was only one-half its original value?

2. What would the separation between two identical objects, one carrying 1 C of positive charge and the other 1 C of negative charge, have to be if the electrical force on each was precisely 1 N?

3. A particle having a positive charge of 2.0×10^{-6} C experiences an upward force of 8 N when placed at a certain location in space. What is the magnitude and direction of the electric field at that point? What would be different about your answer if the particle possessed a negative charge of 2.0×10^{-6} C?

4. During 30 seconds of use, 250 C of charge flow through a microwave oven. Compute the size of the electric current.

5. A lightning stroke lasts 0.05 s and involves a flow of 100 C. What is the current?

6. A current of 0.7 A goes through an electric motor for 1 min. How many coulombs of charge flow through it during that time?

7. A calculator draws a current of 0.0001 A for 5 min. How much charge flows through it?

8. A current of 12 A flows through an electric heater operating on 120 V. What is the heater's resistance?

9. A 120-V circuit in a house is equipped with a 20-A fuse that will "blow" if the current exceeds 20 A. What is the smallest resistance that can be plugged into the circuit without causing the fuse to blow?

10. The resistance of each brake lightbulb on an automobile is 6.6 Ω. Use the fact that cars have 12-V electrical systems to compute the current that flows in each bulb if they are connected in series.

11. The lightbulb used in a computer projector has a resistance of 80 Ω. What is the current through the bulb when it is operating on 120 V?

12. The resistance of the skin on a person's finger is typically about 20,000 Ω. How much voltage would be needed to cause a current of 0.001 A to flow into the finger? (This is about enough current to just be felt by most people.)

13. A 150-Ω resistor is connected to a variable-voltage power supply.

(a) What voltage is necessary to cause a current of 0.3 A in the resistor?
(b) What current flows in the resistor when the voltage is 18 V?

14. Compute the power consumption of the electric heater in Problem 8.

15. An electric eel can generate a 400-V, 0.5-A shock for stunning its prey. What is the eel's power output?

16. An electric train operates on 750 V. What is its power consumption when the current flowing through the train's motor is 2,000 A?

17. All of the electrical outlets in a room are connected in a single parallel circuit (**Figure 7.38**). The circuit is equipped with a 20-A fuse, and the voltage is 120 V.

(a) What is the maximum total power that can be supplied by the outlets without blowing the fuse?
(b) How many 1,200-W appliances can be plugged into the sockets without blowing the fuse?

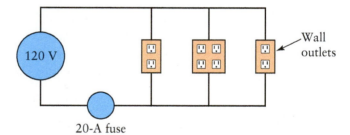

Figure 7.38 Problem 17.

18. Your cell phone typically consumes about 400 mW of power when you text a friend. If the phone is operated using a lithium-ion battery with a voltage of 3.6 V, what is the current flowing through the cell-phone circuitry under these circumstances? Compare your result with the data given in Table 7.1.

19. A car's headlight consumes 40 W when on low beam and 50 W when on high beam.
(a) Find the current that flows in each case ($V = 12$ V).
(b) Find the resistance in each case.

20. Find the current that flows in a 40-W bulb used in a common household circuit ($V = 120$ V). Compare this with the answer to the first part of Problem 19.

21. An electric clothes dryer is rated at 4,000 W. How much energy does it use in 40 min?

22. A clock consumes 2 W of electrical power. How much energy does it use each day?

23. Which costs more, running a 1,200-W hair dryer for 5 min or leaving a 60-W lamp on overnight (10 h)?

24. A representative lightning strike is caused by a voltage of 200,000,000 V and consists of a current of 1,000 A that flows for a fraction of a second. Calculate the power.

25. A toaster operating on 120 V uses a current of 9 A.
    (a) What is the toaster's power consumption?
    (b) How much energy does it use in 1 min?
    (c) What is the total resistance of the heating element wires?

26. A certain electric motor draws a current of 10 A when connected to 120 V.
    (a) What is the motor's power consumption?
    (b) How much energy does it use during 4 h of operation? Express the answer in joules and in kilowatt-hours.

27. The generator at a large power plant has an output of 1,000,000 kW at 24,000 V.
    (a) If it were a DC generator, what would be the current in it?
    (b) What is its energy output each day—in joules and in kilowatt-hours?
    (c) If this energy is sold at a price of 10 cents per kilowatt-hour, how much revenue does the power plant generate each day?

28. A lightbulb is rated at 60 W when connected to 120 V.
    (a) What current flows through the bulb in this case?
    (b) What is the bulb's resistance?
    (c) What would be the current in the bulb if it were connected to 60 V, assuming the resistance stays the same?
    (d) What would be its power consumption in this case?

29. About 40,000 J of energy is stored in a typical 1.5-V D-cell battery. If two such batteries are joined to produce a total of 3.0 V in a flashlight circuit and cause 2.0 A of current to flow through the flashlight bulb, how long will the batteries be able to deliver power to the flashlight at this level?

30. An electric car is being designed to have an average power output of 4,000 W for 2 h before needing to be recharged. (Assume there is no wasted energy.)
    (a) How much energy would be stored in the charged batteries?
    (b) The batteries operate on 30 V. What would the current be when they are operating at 4,000 W?
    (c) To be able to recharge the batteries in 1 h, how much power would have to be supplied to them?

31. The resistance of an electric heater is 10 Ω when connected to 120 V. How much energy does it use during 30 min of operation?

CHALLENGES

1. Compute the electric force acting between the electron and the proton in a hydrogen atom. The radius of the smallest orbit of the electron around the proton is about 5.3×10^{-11} m.

2. Use the result from Challenge 1 and the equation for centripetal force from Chapter 2 to compute the speed of the electron as it moves around the proton. The electron's mass is 9.1×10^{-31} kg.

3. Compute the number of electrons that flow through a wire each second when the current in the wire is 0.2 A.

4. Using your understanding of the nature of internal energy and temperature, explain why we might expect the resistance of a solid to increase if its temperature increases.

5. The current that flows through an incandescent lightbulb immediately after it is turned on is higher than the current that flows moments later. Why?

6. An electrical device called a *diode* is designed to have very low resistance to current flowing through it in one direction but very large resistance to current flow in the other direction. Sketch a graph of the voltage versus current for such a device.

7. Imagine a company offering a line of hair dryers that operate on different voltages, say, 12, 30, 60, and 120 V. Assume that all of the dryers are rated at 1,200 W and find the current that would flow in each as it operates. What would be different about the heating filament wires and the motors in the various hair dryers?

8. Perform the calculation referred to in the last sentence of Example 7.6.

9. Combine Ohm's law and the equation for power consumption to derive the equation that gives the power in terms of current and resistance. Use the result to answer the following question. A cable carrying electrical energy wastes 10 kWh of energy each day because of ohmic heating. If the current in the cable is doubled but the cable's resistance remains the same, how much energy will it waste each day?

10. A defibrillator sends approximately 0.1 C of charge through a patient's chest in about 2 ms. The average voltage during the discharge is approximately 3,000 V. Compute the average current that flows, the average power output, and the total energy consumed (*Note:* Only a small portion of this current actually flows through the heart in a typical defibrillating procedure. Much of the charge is diffused throughout the patient's thoracic cavity.).

8

ELECTROMAGNETISM AND EM WAVES

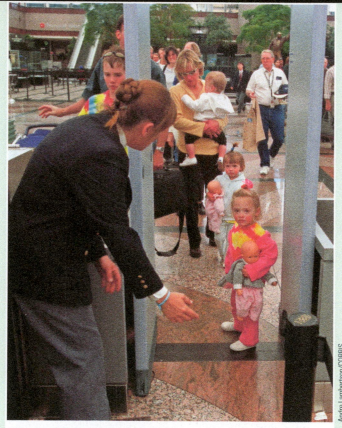

Figure CO-8 Airport metal detector.

Andre Lambertson/CORBIS

CHAPTER INTRODUCTION: Metal Detectors

Metal detectors (**Figure CO-8**) are the first line of defense against persons trying to smuggle weapons onto passenger planes or into schools, government buildings, and many other places. Metal detectors probe your clothing and body without physically touching you, looking for metal that could be part of a gun, a knife, or other dangerous object. A device that can find hidden items on a person walking through an arch seems like something from science fiction. But in today's world, it is routine.

How do these devices work their magic? Although metal detectors operate on electricity, it is magnetism that probes you. Brief magnetic pulses are sent around and through you, typically at a rate of about 100 times a second. The device carefully monitors how swiftly each magnetic pulse dies out. Any metal object encountered by a pulse is induced to produce its own magnetic pulse, which affects how rapidly the total pulse dies out. Sophisticated electronics in the metal detectors sense this change and signal that metal is present. They detect iron and other metals that ordinary magnets attract as well as metals such as aluminum and gold that do not respond to magnets.

Magnetism and its relationship to electricity are the subjects of this chapter. As we shall see, the inextricable linkage between electric and magnetic fields explains the operation of many common devices like motors, generators, and metal detectors, and predicts the existence of electromagnetic (EM) waves upon which all remote control, Wi-Fi, and modern communication and entertainment devices depend. The properties and uses of the different types of EM waves are the main topics of the latter half of this chapter.

8.1 Magnetism

Magnetism was first observed in a naturally occurring ore called *lodestone*. Lodestones were fairly common around Magnesia, an ancient city in Asia Minor. Small pieces of iron, nickel, and certain other metals are attracted by lodestones, much as pieces of paper are attracted by charged plastic (**Figure 8.1**). The Chinese were probably the first to discover that a piece of lodestone will orient itself north and south if suspended by a thread or floated on water on a piece of wood. The *compass* revolutionized navigation because it allowed mariners to determine the direction of north even in cloudy weather. It was also one of the few useful applications of magnetism up to the 19th century.

8.1a Magnetic Materials and Fields

Magnets today are made into a variety of sizes and shapes out of special alloys that exhibit much stronger magnetism than lodestone. All simple magnets exhibit the same compass effect—one end or part of it is attracted to the north, and the opposite end or part is attracted to the south. The north-seeking part of a magnet is called its **north pole**, and the south-seeking part is its **south pole**. All magnets have both poles. If a magnet is broken into pieces, each part will have its own north and south poles. The south pole of one magnet exerts a mutually attractive force on the north pole of a second magnet. The south poles of two magnets repel each other, as do the north poles (**Figure 8.2**). Simply put: *like poles repel, unlike poles attract* just as with electric charges.

Metals that are strongly attracted by magnets are said to be *ferromagnetic*. Such materials have magnetism induced in them when they are near a magnet. If a piece of iron is brought near the south pole of a magnet, the part of the iron nearest the magnet has a north pole induced in it, and the part farthest away has a south pole induced in it (**Figure 8.3**). Once the iron is removed from the vicinity of the magnet, it loses most of the induced magnetism. Some ferromagnetic metals actually retain the magnetism induced in them—they become *permanent magnets*. Common household magnets and compass needles are made of such metals. Ferromagnetism is also the basis of magnetic data recording, but more on this later.

As with gravitation and electrostatics, it is useful to employ the concept of a field to represent the effect of a magnet on the space around it. A *magnetic field* is produced by a magnet and acts as the agent of the magnetic force. The poles of a second magnet experience forces when in the magnetic field: its north pole has a force in the same direction as the magnetic field, but its south pole has a force in the opposite direction. A compass can be thought of as a "magnetic field detector" because its needle will always try to align itself with a magnetic field (**Figure 8.4**). The shape of the magnetic field produced by a magnet can be "mapped" by noting the orientation of a compass at various places nearby. Magnetic *field lines* can be drawn to show the shape of the field, just as electric field lines are used to show the shape of an electric field (**Figure 8.5**). The direction of a field line at a particular place is the direction that the north pole of a compass needle at that location points.

8.1b Earth's Magnetic Field

Because magnets respond to magnetic fields, the fact that compass needles point north indicates that Earth itself has a magnetic field. The shape of Earth's field has been mapped carefully over the course of many centuries because of the importance of compasses in navigation. Earth's magnetic field has the same general

Figure 8.1 Lodestone, a natural magnet, attracting pieces of metallic ore.

Figure 8.2 The poles of two magnets exert forces on each other. Like poles repel each other, and unlike poles attract each other.

Figure 8.3 When a piece of ferromagnetic material (like iron) is brought near a magnet, it has magnetism induced in it. That is why it is attracted by the magnet.

Figure 8.4 The forces on the poles of a compass placed in a magnetic field are in opposite directions. This causes the needle to turn until it is aligned with the field.

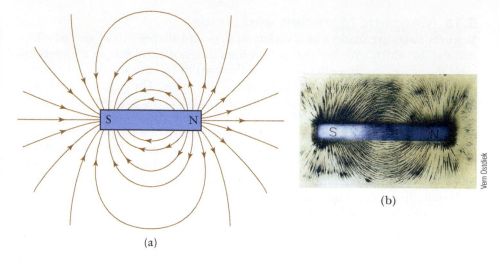

Figure 8.5 (a) Sketch of the magnetic field in the space around a bar magnet. Note that the field lines point toward the south pole and away from the north pole. (b) Photograph of iron filings around a magnet. Each tiny piece of iron becomes magnetized and aligns itself with the magnetic field.

(a)

(b)

Vern Ostdiek

shape as the field around a bar magnet, with its poles tilted about 11° with respect to the axis of rotation (**Figure 8.6**). The direction of "true north" shown on maps is determined by the orientation of Earth's axis of rotation. (The axis is aligned closely with Polaris, the North Star.)

Because of the tilt of Earth's "magnetic axis," at most places on Earth compasses do not point to true north. For example, in the western two-thirds of the United States, compasses point to the right (east) of true north, whereas in New England compasses point to the left (west) of true north. The difference, in degrees, between the direction of a compass and the direction of true north varies from place to place (and with time as well) and is referred to as the *magnetic declination.* In parts of Alaska, the magnetic declination is as high as 25° east. This must be taken into account when navigating with a compass.

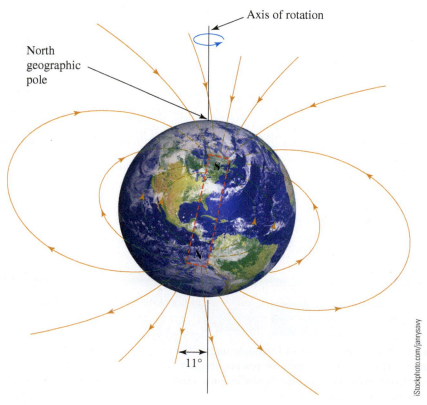

Axis of rotation

North geographic pole

11°

iStockphoto.com/jamysavy

Figure 8.6 Earth's magnetic field. It is shaped as if there were a huge bar magnet deep inside Earth tilted 11° relative to its axis of rotation.

▶ Physics To Go 8.1

Determine what the magnetic declination is where you live. You can do this by finding it on a topographic map or other high-precision navigation map or by searching for a Web site that finds it for you based on your zip code or latitude and longitude. If its magnitude is a large number, say, 10° or so, get a compass and find a place where you can see Polaris (the North Star) at night. Compare the direction the compass points to the direction of Polaris. Is this difference in reasonable agreement with the reported magnetic declination?

Incidentally, Earth's field is responsible for the magnetism in lodestone. This naturally occurring ferromagnetic ore is weakly magnetized by Earth's magnetic field. Another thing to note about Earth's magnetic field: Earth's *north* magnetic pole is at (near) its *south* geographic pole, and vice versa. Why?

1. The north pole of a magnet is attracted to the south pole of a second magnet.

2. The north pole of a compass needle points to the north.

Therefore, a compass's north pole points at Earth's south magnetic pole. This is not a physical contradiction: it is a result of naming the poles of a magnet after directions instead of, say, + and −, or A and B.

Some organisms use Earth's magnetic field to aid in navigation. Although the biological mechanisms that they employ have not yet been fully identified, certain species of fish, frogs, turtles, birds, newts, and whales are able to sense the strength of Earth's field or its direction (or both). The former allows the animal to determine its approximate latitude (how far north or south it is) because Earth's magnetic field is stronger near the magnetic poles (Figure 8.6). Some migratory species travel thousands of miles before returning home, guided—at least in part—by sensing Earth's magnetic field.

Superconductors, so named because of their ability to carry electric current with zero resistance (recall the discussion in Section 7.3), react to magnetic fields in a rather startling fashion. In the superconducting state, the material will expel any magnetic field from its interior. This phenomenon, known as the *Meissner effect,* is why strong magnets are levitated when placed over a super-conductor (**Figure 8.7**). When trying to determine whether a material is in the superconducting state, it is easier to test for the presence of the Meissner effect than it is to see if the resistance is exactly zero.

You have probably noticed that magnetism and electrostatics are very similar: there are two kinds of poles and two kinds of charges. Like poles repel, as do like charges. There are magnetic fields and electric fields. However, there are some important differences. Each kind of charge can exist separately, whereas magnetic poles always come in pairs. (Modern theory indicates the possible existence of a particular type of subatomic "elementary particle" that has a single magnetic pole. As of this writing, such a "magnetic monopole" has not been found.) Furthermore, all conventional matter contains positive and negative charges (protons and electrons, for example) and can exhibit electrostatic effects by being "charged." But, with the exception of ferromagnetic materials, most matter shows very little response to magnetic fields.

We should also point out that the electrostatic and magnetic effects described so far are completely independent. Magnets have no effect on pieces of charged plastic, for instance, and vice versa. This is the case as long as there is no motion of the objects or changes in the strengths of the electric and magnetic fields. As we shall see in the following sections, a number of fascinating and useful interactions between electricity and magnetism take place when motion or change in field strength occurs.

Concept Map 8.1 summarizes the similarities and differences between electrostatics and magnetism.

Figure 8.7 The Meissner effect. A magnet levitating above a high-T_c superconductor cooled with liquid nitrogen.

David Parker/Science Source

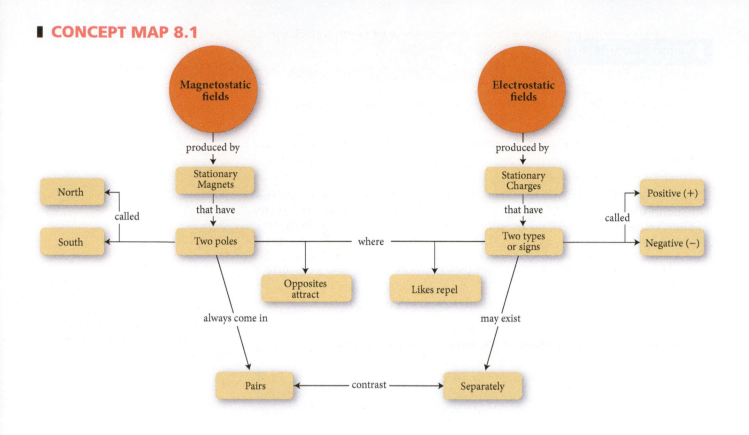

Learning Check

1. An iron nail is brought near a magnet. Which of the following is *not* true?
   (a) The nail has both a north pole and a south pole.
   (b) The nail becomes either a north pole or a south pole.
   (c) The magnet exerts an attractive force on the nail.
   (d) The nail exerts an attractive force on the magnet.

2. (True or False.) The shape of the magnetic field around a bar magnet is almost exactly the same as the shape of the electric field around a positive charge.

3. Magnetic _____ indicates how far away from true north a compass needle points.

ANSWERS: 1. (b) **2.** False **3.** declination

8.2 Interactions between Electricity and Magnetism

Consider the following items that we usually take for granted: electric motors in hair dryers, automobiles, computer disc drives, elevators, and countless other devices; generators that produce most of the electricity we use; speakers, audio and videotape recorders, and high-fidelity microphones; and the waves that make satellite audio receivers, cell phones, radar, microwave ovens, medical x-rays, and our eyes work. What do all of these have in common? They all are possible because electricity and magnetism interact with each other in basic—and very useful—ways. The word *electromagnetic,* which appears dozens of times in this chapter, is perhaps the best indication of just how intertwined these two phenomena are.

Before we delve into these interactions, let's summarize and review the key aspects of electrostatics and magnetism presented in Sections 7.1, 7.2, and 8.1:

- Electric charges produce electric fields in the space around them (see Figure 7.9).

- An electric field, regardless of its origin, causes a force on any charged object placed in it (see Figure 7.10).
- Magnets produce magnetic fields in the space around them (see Figure 8.5).
- A magnetic field, regardless of its origin, causes forces on the poles of any magnet placed in it (see Figure 8.4).

These statements have been worded in a particular way because, as we shall see, it is the electric and magnetic *fields* that are involved in the interplay between electricity and magnetism. In this section, we describe three basic observations of these interactions and discuss some useful applications of them. In Section 8.3, we summarize the underlying concepts in the form of two organizing principles. We emphasize the ways in which electricity and magnetism interact and how these help us understand how many electrical devices work and what light and other electromagnetic waves are. Fortunately, we can do this without having to go into the complex underlying causes of these interactions.

8.2a Electromagnets

The first of the three observations is the basis of electromagnets.

> **Observation 1:** A moving electric charge produces a magnetic field in the space around it. An electric current produces a magnetic field around it.

A single charged particle creates a magnetic field only when it is moving. The magnetic field produced is in the shape of circles around the path of the charge (**Figure 8.8**). For a steady (DC) current, which is basically a succession of moving charges in a wire, the field is steady, and its strength is proportional to the size of the current and inversely proportional to the distance from the wire. (The field is quite weak unless the current is large. A current of 10 amperes or more will produce a field strong enough to be detected with a compass; **Figure 8.9**.) Reversing the direction of the current in the wire will reverse the directions of the magnetic field lines.

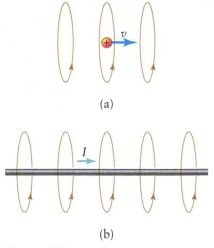

(a)

(b)

Figure 8.8 Magnetic field produced by (a) a moving charge and by (b) a wire carrying DC. The field lines are circles concentric with the path of the charges. (The power source for the current is not shown.)

> ### ▶ Physics To Go 8.2
>
> For this, you will need a car or similar vehicle and a compass. Open the hood of the car (when the engine is not running), and locate the battery and the large cables that carry current from it to the starter. Close the hood and hold a compass over the hood just above where a cable is located. Have a friend start the car, and watch the compass while the starter is engaged. What happens? What causes this?

(a)

(b)

Figure 8.9 (a) The compass needles align with Earth's magnetic field when no current is in the wire. (b) The compass needles show the circular shape of the magnetic field produced by a large current (15 amperes) flowing in the wire.

(a)   (b)

Figure 8.10 (a) The magnetic field produced by a current in a coil of wire. The field has the same shape as that produced by a bar magnet. When the direction of the current is reversed (b), the polarity of the magnetic field is also reversed.

Most applications of this phenomenon use coils—long wires wrapped in the shape of a cylinder, often around an iron core. The magnetism induced in the iron greatly enhances the magnetic field of the coil. The magnetic field of such a coil (when carrying a direct current) has the same shape as the field around a bar magnet. (**Figure 8.10** and compare it to Figure 8.5.) This device is an *electromagnet*. It behaves just like a permanent magnet as long as there is a current flowing. One end of the coil is a north pole, and the other is a south pole. Electromagnets have an advantage over permanent magnets in that the magnetism can be "turned off" simply by switching off the current.

A coil with a length that is much greater than its diameter is called a *solenoid*. If an iron rod is partially inserted into a solenoid with a hollow core, the rod will be pulled in when the current is switched on: the magnetic pole associated with the coil's field nearest the rod induces a magnetic field with the opposite polarity in the rod, thus exerting an attractive force on it (**Figure 8.11**). Solenoids are used in common devices for striking doorbell chimes, opening valves to allow water to enter and to leave washing machines, withdrawing deadbolts in electric door locks, and engaging starter motors on car and truck engines.

Electromagnets are used to produce the strongest magnetic fields on Earth. Two factors contribute to stronger fields: wrapping more coils around the cylinder and using a larger electric current. The former suggests the use of thinner wire so that more coils can fit into the same amount of space. But smaller wire requires smaller electric current so the wire does not overheat and melt. This limitation is overcome in *superconducting electromagnets* (**Figure 8.12**). When the wire used in an electromagnet is a superconductor, it can carry huge electric currents with no ohmic heating because there is no resistance. Very small superconducting electromagnets can generate very strong magnetic fields while using much less electrical energy than a conventional electromagnet. (We described some uses of superconducting electromagnets in Section 7.3 and will introduce others later in this section.)

Superconducting electromagnets do have limitations, though. The superconducting state is lost if the temperature, electric current, or magnetic field strength exceeds certain values. Most superconducting electromagnets now

Figure 8.11 (a) The iron rod is pulled into the coil (solenoid) when the current flows. (b) The solenoid (red) in this doorbell chime pulls in the iron rod when a current flows in it. The rod strikes the black bar at the left, producing sound. When the current is shut off, the spring retracts the rod, and it returns to strike the bar at the right. If the bars are tuned to different frequencies, together they can produce the familiar "ding-dong" sound.

(a)   (b)

Figure 8.12 A technician stands in front of the ATLAS detector, part of the Large Hadron Collider (LHC) near Geneva, Switzerland. This detector uses two large superconducting magnets to produce enormous fields that exert forces on high speed charged particles like protons. These forces alter the particles' trajectories and make it possible to measure their linear momenta.

David Parker/Science Source

found in laboratories throughout the world must be kept cold with liquid helium ($T = 4$ K). The added cost of the liquid helium system is offset by the high magnetic fields achieved and the great reduction in use of electric energy compared to conventional electromagnets.

The polarity of an electromagnet is reversed if the direction of the current is reversed (Figure 8.10b). An alternating current in a coil will produce a magnetic field that oscillates: it increases, decreases, and switches polarity with the same frequency as the current. Such an oscillating magnetic field will cause a nearby piece of iron to vibrate. The oscillating magnetic field of a coil with AC in it is used in many common devices, as we shall see in the following sections.

This first interaction not only explains how electromagnets work, but also gives us new insight into permanent magnets as well. Because electrons in atoms are charged particles in motion about the nucleus, they produce magnetic fields. Also, the electrons have their own magnetic fields associated with their spin (more on this in Chapter 12). In any unmagnetized material, the individual magnetic fields of the electrons are randomly oriented and cancel each other out (**Figure 8.13**). In ferromagnetic materials, these fields can be aligned with one another by an external magnetic field; the material then produces a net magnetic field. So we can conclude that moving electric charges are the causes of magnetic fields even in ordinary bar and horseshoe magnets.

This brings us back to a statement made at the beginning of Chapter 7: electric charges are the cause of both electrical and magnetic effects. We might regard electricity and magnetism as two different manifestations of the same thing—charge.

8.2b Electric Motors

The second observation helps us understand how things such as electric motors and speakers work.

> **Observation 2:** A magnetic field exerts a force on a moving electric charge. Therefore, a magnetic field exerts a force on a current-carrying wire.

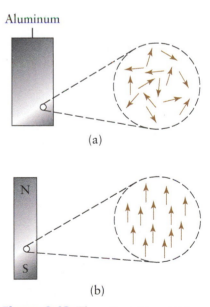

Figure 8.13 The arrows represent the magnetic fields of electrons in individual atoms. (a) The fields remain randomly oriented in nonferromagnetic materials. (b) Inside ferromagnetic material that is magnetized, the individual magnetic fields are aligned.

(a)   (b)   (c)

Figure 8.14 (a) The force on a current-carrying wire in a magnetic field. (b) When the direction of the current is reversed, the direction of the force is also reversed. (c) A current-carrying wire levitating in a magnetic field.

A stationary electric charge is not affected by a magnetic field, but a moving charge usually is. Note that this second observation is a logical consequence of the first: anything that produces a magnetic field will itself be affected by other magnetic fields. Caveat: If a charge's velocity or the direction of a current is parallel to that of the magnetic field or in the opposite direction, the magnetic field does not exert this force.

A curious characteristic of electromagnetic phenomena is that the effects are often perpendicular to the causes. The direction of the magnetic field from a current-carrying wire is perpendicular to the direction the current is flowing (Figure 8.8). Similarly, the force that a magnetic field exerts on a moving charge or on a current-carrying wire is perpendicular to both the direction of the magnetic field and the direction the charge is flowing. For example, if a horizontal magnetic field is directed away from you and a wire is carrying a current to your right, the force on the wire is *upward* (**Figure 8.14**). If the direction of the current is reversed, the direction of the force is reversed (downward). An alternating current would cause the wire to experience a force that alternates up and down.

Electric motors—like those in power drills and elevators—exploit this electromagnetic interaction. The simplest type of electric motor consists of a coil of wire mounted so that it can rotate in the magnetic field of a horseshoe-shaped magnet (**Figure 8.15**). A direct current flows through the coil, the magnetic field causes forces on the sides of the coil, and the coil rotates. Once the coil has completed half of a rotation, a simple mechanism reverses the direction of the current. This reverses the force on the coil, causing it to rotate another half-turn. This process is repeated, and the coil spins continuously. Motors designed to run on AC can exploit the fact that the direction of the current is automatically reversed 120 times each second (60-cycles-per-second AC, with two reversals each cycle).

Liquid metals, such as the molten sodium used in certain nuclear reactors, can be moved through pipes using an electromagnetic pump that has no moving parts. If the metal has to be moved in a pipe that is oriented north–south, for example, a large electric current can be sent across the pipe—east to west perhaps. Then, if a strong magnetic field is directed downward through the same section of pipe, the current-carrying metal will be forced to move southward.

Figure 8.15 Simplified sketch of an electric motor. The loop of wire rotates because of the forces on its sides. Each time the loop becomes horizontal, the direction of the current is reversed, and the rotation continues.

Figure 8.16 Aerial view of the Large Hadron Collider near Geneva, Switzerland. Inside a subterranean tunnel (outlined in red here) 8.6 kilometers in diameter, charged particles such as protons are accelerated to nearly the speed of light. Strong magnetic fields are used to keep the charges traveling in a circle.

Several large-scale devices used in experimental physics make use of the effect of magnetic fields on moving charged particles. High-temperature plasmas cannot be confined in any conventional metal or glass container because the container would melt. Because plasmas are composed of charged particles, magnetic fields can be used to contain them in what is known as a *magnetic bottle*. This is one approach being employed in the attempt to harness nuclear fusion as an energy source (Section 11.7).

In the absence of other forces, a charged particle moving perpendicularly to a magnetic field will travel in a circle: the force on the particle is always perpendicular to its velocity and is therefore a centripetal force. An electron, proton, or other charged particle can be forced to move in a circle by a magnetic field and then gradually accelerated during each revolution. Particle accelerators used for experiments in atomic, nuclear, and elementary particle physics, as well as those used for producing radiation for cancer treatments at some large hospitals, operate on this principle. The world's highest-energy particle accelerator is the Large Hadron Collider (LHC) (**Figure 8.16**) located along the border between France and Switzerland near the city of Geneva. This device comprises a circular tunnel with a diameter of 8.6 kilometers (5.3 miles) buried 50 to 175 meters beneath Earth's surface in which two counter-rotating beams of charged particles travel in a vacuum guided by superconducting magnets. The head-on collisions between the particles in these oppositely moving beams yield information about the fundamental forces and interactions in Nature. More will be said on this topic in Chapter 12.

8.2c Electromagnetic Induction and Electric Generators

The third observed interaction between electricity and magnetism is used by electric generators. Recall that the first observation tells us that moving charges create magnetic fields. The third one is a similar statement about moving magnets.

(a)

v

I

(b)

(c)

(d)

iStockphoto.com; iStockphoto.com/Dale Robins; Vern Ostdiek/University of Michigan–Dearborn; Vern Ostdiek/University of Michigan–Dearborn

Figure 8.17 (a) The electric field produced by a moving magnet is circular. (The magnetic field of the magnet is not shown.) (b) A magnet will induce a current in a coil as it passes through. (c) The ammeter shows there is no current in the coil of wire (yellow) when the horseshoe magnet (blue) is not moving. (d) A current flows when the magnet is moving.

Observation 3: A moving magnet produces an electric field in the space around it. A coil of wire moving through a magnetic field has a current induced in it.

The electric field around a moving magnet is in the shape of circles around the path of the magnet. This circular electric field will force charges in a coil of wire to move in the same direction—as a current (**Figure 8.17**). The process of inducing an electric current with a magnetic field is known as **electromagnetic induction**. All that is required is that the magnet and coil move relative to each other. If the coil moves and the magnet remains stationary, a current is induced. If the motion is steady in either case, the induced current is in one direction. If either the coil or the magnet oscillates back and forth, the current alternates with the same frequency—it is AC.

Electromagnetic induction is used in the most important device for the production of electricity: the generator. The simplest generator is basically an electric motor. When the coil is forced to rotate, it moves relative to the magnet, so a current is induced in it (see **Figure 8.18**). We might call this device a "two-way energy converter." When electrical energy is supplied to it, it is a motor. It converts this electrical energy into mechanical energy of rotation. When it is mechanically turned (by hand cranking, by a drive belt on a car engine, or by a turbine in a power plant), it is a generator. It converts mechanical energy into electrical energy.

This motor–generator duality is used in dozens of *pumped-storage* hydroelectric power stations. During the night when surplus electrical energy is available from other power stations, the motor mode is used to pump water from one reservoir to another that is at a higher elevation. Most of the electrical energy is converted into "stored" gravitational potential energy. During the peak time of electric use the next day, water flows in the opposite direction, and the generator mode is used as the moving water turns the pumps (now acting as turbines) that turn the motors (now acting as generators), thereby producing electricity. You might say that the system functions like a rechargeable gravitational battery.

N

I (output)

S

Figure 8.18 Simplified sketch of a generator. As the loop of wire rotates relative to the magnetic field, a current is induced in it.

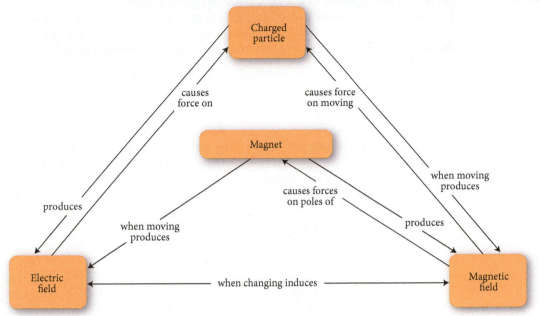

Another application of this technology is *regenerative braking*, which is used in electric and hybrid vehicles. While accelerating and cruising, electric motors turn the wheels using electricity from batteries. During braking, the motors function as generators: the wheels turn them, and the electricity that is generated can partially recharge the batteries. Instead of all of the vehicle's kinetic energy being converted into wasted heat—the case with conventional friction brakes—some of it is saved for reuse.

In summary, when electric charges or magnets are in motion, electricity and magnetism are no longer independent phenomena. The three observations given here are statements of experimental facts that illustrate this interdependence. They can be demonstrated easily using a battery, wires, a compass, a large magnet, and a sensitive ammeter. The fact that electricity and magnetism interact only when there is motion (and then the effects are perpendicular to the causes) is somewhat startling when compared to, say, gravitation and electrostatics. As we saw in Chapters 2 and 7, gravitational and electrostatic forces are always toward or away from the objects causing them, and they act whether or not anything is moving or changing. These basic yet surprising interactions between electricity and magnetism are crucial to our modern electrified society.

Concept Map 8.2 summarizes the interactions between electricity and magnetism.

■ Learning Check

1. (True or False.) A stationary electric charge produces a magnetic field around it.
2. How does one reverse the polarity (switch the north and south poles) of an electromagnet?
3. A magnetic field exerts a force on an electric charge if the charge is _____.
4. A current can be made to flow in a coil of wire if
   (a) it is connected to a battery.
   (b) it is put into motion near a magnet.

(c) a magnet is put into motion near it.
(d) All of the above.

5. When the coil in a simple electric motor is turned—by a crank, for example—the motor functions as a _____.

ANSWERS: 1. False **2.** reverse the current **3.** moving **4.** (d) **5.** generator.

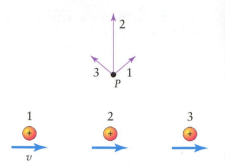

Figure 8.19 The electric field at point P changes as the charged particle moves by. The upper violet arrows indicate the magnitude and direction of the electric field for three different locations of the particle. The magnetic field that is induced at point P is directed straight out of the paper.

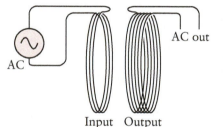

Figure 8.20 Simplified diagram of a transformer. The alternating magnetic field produced by the AC in the input coil induces an alternating current in the output coil. In this case, the output voltage would be higher than the input voltage.

8.3 Principles of Electromagnetism

The interactions between electricity and magnetism described in the previous section, along with other similar observations, suggest the following two general statements. We might call these the **principles of electromagnetism**:

1. An electric current or a changing electric field induces a magnetic field.

2. A changing magnetic field induces an electric field.

These two statements summarize the previous observations and also emphasize the symmetry that exists. In both cases, a "changing" field means that the strength or the direction (or both) of the field is changing. The first principle can be used to explain the first observation: as a charge moves past a point in space, the strength of the electric field increases and then decreases. All the time the direction of the field is changing as well (**Figure 8.19**). The effect of this is to cause a magnetic field to be produced. Similarly, the second principle explains electromagnetic induction.

As mentioned in Section 7.6, a transformer is a device used to step up or step down AC voltages. It represents one of the most elegant applications of electromagnetism. In essence, a transformer consists of two separate coils of wire in close proximity. An AC voltage is applied to one of the coils called the *input* or *primary* coil, and an AC voltage appears at the other coil called the *output* or *secondary* coil (**Figure 8.20**). The AC in the primary coil produces an oscillating magnetic field through both coils. Most transformers have both coils wrapped around a single ferromagnetic core to intensify the magnetic field and guide it from one coil to the other. This oscillating (and therefore changing) magnetic field induces an AC current in the output coil. Note that a DC input would produce a steady magnetic field that would not induce a current in the output coil. Transformers do not work with DC.

Now, how can the voltage of the output be different from the voltage of the input? Each "loop" or "turn" of the output coil has the same voltage induced in it. The voltages in all of the turns add together so that the more turns there are in the output coil, the higher the total voltage. The ratio of the number of turns in the two coils determines the ratio of the input and output voltages. In particular,

$$\frac{\text{voltage of output}}{\text{voltage of input}} = \frac{\text{number of turns in output coil}}{\text{number of turns in input coil}}$$

$$\frac{V_o}{V_i} = \frac{N_o}{N_i}$$

If there are twice as many turns in the output coil as in the input coil, then the output voltage will be twice the input voltage. If there are one-third as many turns in the output coil, then the output voltage will be one-third the input voltage. Thus, the AC voltage can be stepped up or stepped down by any desired amount by adjusting the ratio of the number of turns in the two coils.

EXAMPLE 8.1 A transformer is being designed to have a 600-volt output with a 120-volt input. If there are to be 800 turns of wire in the input coil, how many turns must there be in the output coil?

SOLUTION

$$\frac{V_o}{V_i} = \frac{N_o}{N_i}$$

$$\frac{600 \text{ V}}{120 \text{ V}} = \frac{N_o}{800}$$

$$800 \times 5 = N_o$$

$$N_o = 4{,}000 \text{ turns}$$

Vern Ostdiek/University of Michigan-Dearborn

Figure 8.21 The transformer in this lamp converts 120 volts AC into 14 volts AC ("Hi") or 12 volts AC ("Lo").

In addition to being used to change voltages in electrical distribution systems, transformers are found in a wide variety of electrical appliances. Most electrical components used in audio systems, calculators, and the like require voltages that are much smaller than 120 volts. Appliances designed to operate on household AC must include transformers to reduce the voltage accordingly. High-intensity desk lamps also use transformers (**Figure 8.21**). The spark used to ignite gasoline in automobile engines is generated using a type of transformer called a "coil." The number of turns in the output coil is many times the number of turns in the input coil. A spark is produced by first sending a brief current into the input. A magnetic field is produced that quickly disappears. This induces a very high voltage (around 25,000 volts) in the output, which is conducted to the spark plugs to ignite the fuel.

Understanding electromagnetism allows us to better appreciate how the metal detector introduced at the start of the chapter work. The magnetic pulses are produced by sending an electric current through a coil of wire for a short period of time. When the current stops, the magnetic field that was created dies out quickly, and this decreasing field induces an electric current in the coil. This current is used to monitor how swiftly the magnetic pulse dies out.

Metals are detected because the rapidly changing magnetic field of each pulse induces electrons in the metal to move—as in the secondary coil in a transformer—and this current produces an opposite magnetic pulse. This change in the total magnetic field affects the current induced in the coil. The electronics are designed to detect any such change and signal an alarm.

Learning Check

1. A magnetic field will be produced at some point in space if the electric field at that point
   (a) gets stronger.
   (b) gets weaker.
   (c) changes direction.
   (d) All of the above.
2. (True or False.) A transformer works with AC but not DC.

ANSWERS: 1. (d) 2. True

8.4 Applications to Sound Reproduction

A hundred years or so ago, the only people who listened to music performed by world-class musicians were those few who could attend live performances. Today, people in the most remote corners of the world can hear concert-quality sound from large home entertainment systems, pocket-sized or smaller MP3 players, and many devices in between. The first Edison phonographs were strictly mechanical and did a fair job of reproducing sound. It was the invention of electronic recording and playback machines that brought true high fidelity to sound reproduction, however. The sequence that begins with sound in a recording studio and ends with the reproduced sound coming from a speaker in your home, headphones or earbuds, or car includes components that use electromagnetism.

8.4a Microphones and Magnetic Tape Devices

The key to electronic sound recording and playback is first to translate the sound into an alternating current and then later retranslate the AC back into sound. The first step requires a microphone, and the second step requires a speaker. Although there are several different types of microphones, we will take a look at what is called a *dynamic* microphone. It consists of a magnet surrounded by a coil of wire attached to a diaphragm (**Figure 8.22**). The coil and diaphragm are free to oscillate relative to the stationary magnet. When sound waves reach the microphone, the pressure variations in the wave push the diaphragm back and forth, making it and the coil oscillate. Because the coil is moving relative to the magnet, an oscillating current is induced in it. The frequency of the AC in the coil is the same as the frequency of the diaphragm's oscillation, which is the same as the frequency of the original sound. That is all it takes. This type of dynamic microphone is also referred to as a *moving coil* microphone. The alternative is to attach a small magnet to the diaphragm and keep the coil stationary—a *moving magnet* microphone.

Let's skip ahead now to when the sound is played back. The output of the CD player, radio, or other audio component is an alternating current that has to be converted back into sound by a speaker. The basic speaker is quite similar to a dynamic microphone. In this case, the coil (called the *voice coil*) is connected to a stiff paper cone instead of to a diaphragm (**Figure 8.23**). Recall from Section 8.2 that an alternating current in the voice coil in the presence of the magnet will cause the coil to experience an alternating force. The voice coil and the speaker cone oscillate with the same frequency as the AC input. The oscillating paper cone produces a longitudinal wave in the air—sound.

The tiny speakers built into earbud earphones now commonly used with iPods and other portable music devices operate by these same principles. They are made possible by the use of small but powerful permanent magnets made of an alloy of the elements neodymium, iron, and boron (NIB). The extreme strength of NIB magnets (which in some cases can approach that of large medical MRIs) makes them capable of reproducing a very broad range of frequencies with exceptional fidelity. Coupled with their small size, this has made them indispensable in the design of compact earphones.

Microphones and speakers are classified as *transducers:* they convert mechanical oscillation from sound into AC (microphone), or they convert AC into mechanical oscillation and sound (speaker). They are almost identical. In fact,

Figure 8.22 Simplified diagram of one type of microphone. Sound waves cause the diaphragm and coil to oscillate relative to the magnet. This induces AC in the coil.

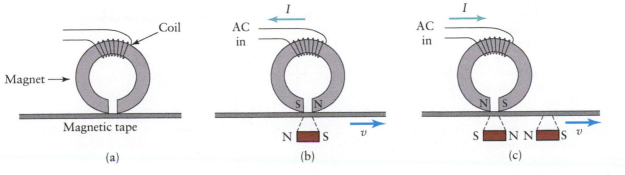

(a)

(b)

a microphone can be used as a speaker, and a speaker can be used as a microphone. But, as with motors and generators, each is best at doing what it is designed to do.

Figure 8.23 (a) Simplified sketch of a speaker. AC in the voice coil forces the cone in and out, thereby producing sound. (b) Photograph of a speaker with the speaker cone (between fingers) and voice coil (red band) assembly removed. The voice coil fits around the magnet (labeled "S").

▶ Physics To Go 8.3

For this, you will need a pair of headphones and a stereo or tape recorder with a microphone input jack that matches the headphone plug. Plug the headphones into the microphone jack and talk into them as if they were microphones. Does it work? How good is the quality of the sound that is produced?

Even now, much sound recording, particularly that done in studios and on sound sets, is done on magnetic tape. The tape is a plastic film coated with a thin layer of fine ferromagnetic particles that retain magnetism. Sound is recorded on the tape using a recording head, a ring-shaped electromagnet with a very narrow gap (**Figure 8.24**). During recording, an AC signal (from a microphone, for example) produces an alternating magnetic field in the gap of the recording head. As the tape is pulled past the gap, the particles in each part of the tape are magnetized according to the polarity of the head's magnetic field at the instant they are in the gap. The polarity of the particles changes from north–south to south–north, and so on, along the length of the tape, constituting a binary-like encoding of the tape in (N,S) or (+,−) "digits" instead of (0,1) bits used by computers.

Figure 8.24 (a) Simplified sketch of a tape head. During recording, (b) and (c), AC in the coil induces alternating magnetism in the tape.

(a)

(b)

(c)

To play back the recording, the tape is pulled past a playback head, often the same head used for recording. The magnetic field of the particles in the tape oscillates back and forth and induces an oscillating magnetic field in the tape head (**Figure 8.25**). This oscillating magnetic field induces an oscillating current (AC) in the coil—electromagnetic induction again.

Magnetic recording is not limited to sound reproduction. Older television videocassette recorders (VCRs) record both sound and visual images on magnetic tape. Computers store information magnetically on tapes, disks, and hard drives.

The AC signals produced by microphones, CD players, and tape playback heads are quite weak. Amplifiers are used to increase the power of these signals before they are sent to speakers. Amplifiers also allow the listener to modify the sound by adjusting its loudness with the volume control and its tone quality with the bass and treble controls.

8.4b Digital Sound

A revolution in sound reproduction occurred in the 1980s with the advent of digital sound reproduction, the method used in *compact discs* (CDs) and various computer sound file formats, including MP3. In a process known as analog-to-digital conversion, the sound wave to be recorded is measured and stored as numbers. For CDs, the actual voltage of the AC signal from a microphone is measured 44,100 times *each second* (**Figure 8.26**). Note that this frequency is

Figure 8.26 Digital sound reproduction in CDs. (a) To record the sound, the voltage of the waveform is measured 44,100 times each second. The resulting numbers are stored on magnetic media for later use. (b) During playback, the voltage of the output at each time is set equal to the numerical value that was stored originally. The reconstructed waveform is then "smoothed" using an electronic filter. The resulting waveform is an almost perfect reproduction of the original.

Pits   Label   Lacquer

1.2 mm

Aluminum

Plastic

Air   Laser beam

(a)   (b)

Andrew Syred/Science Source

Figure 8.27 (a) Information is stored on CDs and DVDs in digital form as microscopic pits used to represent 0s and 1s. (b) The data are read from the spinning disc by the beam from a tiny laser.

more than twice the highest frequency that people can hear. The waveform of the sound is "chopped up" into tiny segments and then recorded as numerical values. These numbers are stored as binary numbers using 0s and 1s, just as information is stored in computers. To play back the sound, a digital-to-analog conversion process reconstructs the sound wave by generating an AC signal whose voltage at each instant in time equals the numerical value originally recorded. After being "smoothed" with an electronic filter, the waveform is an almost perfect copy of the original. (Refer to the discussion of waveforms in Section 6.3.)

A huge amount of data is associated with digital sound reproduction— millions of numbers for each minute of music. CDs (and DVDs) store these data in the form of microscopic pits in a spiral line several miles long (**Figure 8.27**). A tiny laser focused on the pits reads them as 0s and 1s. The amount of information stored on a 70-minute CD is equivalent to more than a dozen full-length encyclopedias. A standard DVD can store about seven times as much data, and new Blu-ray discs (which employ special blue lasers to scan the pits) can handle as much as 40 times more. Little wonder that CDs and DVDs have also been embraced by the personal-computer industry as a way to store huge amounts of information in durable, portable form.

The superior quality of digital sound comes about because the playback device looks only for numbers. It can ignore such things as imperfections in the disc or tape, the weak random magnetization in a tape that becomes tape hiss on cassettes, and the mechanical vibration of motors that we hear as a rumble on phonographs. A sophisticated error-correction system can even compensate for missing or garbled numbers. Because the pickup device in a CD player does not touch the disc, each CD can be played over and over without the slow deterioration in quality that results from a needle moving in a phonograph groove or from the constant unwinding and rewinding of a cassette tape over the recorder heads. This combination of high fidelity and disc durability made the CD system an immediate hit with consumers.

This is just a glimpse of some of the factors in state-of-the-art high-fidelity sound reproduction. Perhaps we are all now so accustomed to it that we cannot appreciate how much of a technological miracle it really is. The next time you listen to high-quality recorded music, remember that it is all possible because of the basic interactions between electricity and magnetism described in Section 8.2.

Learning Check

1. Which of the following does *not* use electromagnetic induction?
   (a) speaker
   (b) transformer
   (c) dynamic microphone
   (d) cassette tape player

2. (True or False.) A speaker can be used as a crude microphone.

3. The process of making a digital recording of a sound wave involves
   (a) digital-to-analog conversion.
   (b) analog-to-digital conversion.
   (c) electromagnetic induction.
   (d) All of the above.

ANSWERS: 1. (a) **2.** True **3.** (b)

DEFINITION **Electromagnetic Wave** A transverse wave consisting of a combination of oscillating electric and magnetic fields.

8.5 Electromagnetic Waves

Eyes, radios, televisions, radar, x-ray machines, microwave ovens, heat lamps. . . . What do all of these things have in common? They all use **electromagnetic waves (EM waves)**. EM waves occupy prominent places both in our daily lives and in our technology. These waves are also involved in many natural processes and are essential to life itself. In the rest of this chapter, we will discuss the nature and properties of electromagnetic waves and look at some of their important roles in today's world.

As the name implies, EM waves involve both electricity and magnetism. The existence of these waves was first suggested by 19th century physicist James Clerk Maxwell while he was analyzing the interactions between electricity and magnetism. Consider the two principles of electromagnetism stated in Section 8.3. Let's say that an oscillating electric field is produced at some place. The electric field switches back and forth in direction while its strength varies accordingly. This oscillating electric field will induce an oscillating magnetic field in the space around it. But the oscillating magnetic field will then induce an oscillating electric field. This will then induce an oscillating magnetic field and so on in an endless "loop": the principles of electromagnetism tell us that a continuous succession of oscillating magnetic and electric fields will be produced. These fields travel as a wave—an EM wave.

Most electromagnetic waves are transverse waves because the oscillation of both of the fields is perpendicular to the direction the wave travels, although some EM waves in plasmas can be longitudinal. **Figure 8.28** shows a "snapshot" of a transverse EM wave traveling to the right. (The three axes are perpendicular to each other.) In this particular case, the electric field is vertical. As the wave travels by a given point in space, the electric field oscillates up and down, the way a floating petal oscillates on a water wave. The magnetic field at the point oscillates horizontally, in and out.

Figure 8.28 should remind you of the transverse waves we described in Chapter 6 (Figure 6.5, for example). Electromagnetic waves do differ from mechanical waves in two important ways, however. First, they are a combination of two waves in one: an electric field wave and a magnetic field wave. These cannot exist separately. Second, EM waves do not require a medium in which to travel. They can travel through a vacuum: the light from the Sun does this. They can also travel through matter. Light through air and glass, and x-rays through your body are common examples.

Electromagnetic waves travel at an extremely high speed. Their speed in a vacuum, called the "speed of light" because it was first measured using light, is represented by the letter c. Its value is

$$c = 299{,}792{,}458 \text{ m/s} \qquad \text{(speed of light)}$$

or

$$c = 3 \times 10^8 \text{ m/s} \qquad \text{(approximately)}$$
$$= 300{,}000{,}000 \text{ m/s}$$
$$= 186{,}000 \text{ miles/s} \qquad \text{(approximately)}$$

All of the parameters introduced for waves in Chapter 6 apply to EM waves. The wavelength can be readily identified in Figure 8.28. The amplitude is the maximum value of the electric field strength. The equation $v = f\lambda$ holds with v replaced by c. There is an extremely wide range of wavelengths of EM waves, from the size of a single proton, about 10^{-15} meters, to almost 4,000 kilometers for one type of radio wave. The corresponding frequencies of these extremes are about 10^{23} hertz and 76 hertz, respectively. Most EM waves used in practical applications have extremely high frequencies compared to sound.

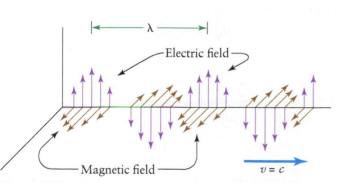

Figure 8.28 Representation of a portion of an EM wave. The electric field is always perpendicular to the magnetic field. The entire pattern moves to the right at the speed of light.

EXAMPLE 8.2 An FM radio station broadcasts an EM wave with a frequency of 100 megahertz. What is the wavelength of the wave?

SOLUTION The prefix *mega* stands for 1 million. Therefore, the frequency is 100 million hertz.

$$c = f\lambda$$
$$300{,}000{,}000 \text{ m/s} = 100{,}000{,}000 \text{ Hz} \times \lambda$$
$$\frac{300{,}000{,}000 \text{ m/s}}{100{,}000{,}000 \text{ Hz}} = \lambda$$
$$\lambda = 3 \text{ m}$$

Electromagnetic waves are named and classified according to frequency. In order of increasing frequency, the groups, or "bands," are **radio waves**, **microwaves**, **infrared radiation**, **visible light**, **ultraviolet radiation**, **x-rays**, and **gamma rays** (γ-rays). (Use of the word *radiation* instead of *waves* is not significant here.) **Figure 8.29** shows these groups along with frequency and wavelength scales. This is called the **electromagnetic spectrum**. Notice that the groups overlap. For example, a 10^{17}-hertz EM wave could be ultraviolet radiation or an x-ray. In cases of overlap, the name applied to an EM wave depends on how it is produced.

We will briefly discuss the properties of each group of waves in the electromagnetic spectrum—how they are produced, what their uses are, and how they can affect us. The great diversity of uses of EM waves arises from the variety of ways in which they can interact with different kinds of matter. All matter around us contains charged particles (electrons and protons), so it seems logical that EM waves can affect and be affected by matter. The oscillating electric field can cause AC currents in conductors; it can stimulate vibration of molecules, atoms, or individual electrons; or it can interact with the nuclei of atoms. Which sort of interaction occurs, if any, depends on the frequency (and wavelength) of the EM wave and on the properties of the matter through which it is traveling—its density, molecular and atomic structure, and so on.

Figure 8.29 The electromagnetic (EM) spectrum.

Figure 8.30 A radio is an EM wave detector that can select a single-frequency radio wave.

In principle, an electromagnetic wave of any frequency could be produced by forcing one or more charged particles to oscillate at that frequency. The oscillating field of the charges would initiate the EM wave. The "lower-frequency" EM waves (radio waves and microwaves) are produced this way: a transmitter generates an AC signal and sends it to an antenna. At higher frequencies, this process becomes increasingly difficult. Electromagnetic waves above the microwave band are produced by a variety of processes involving molecules, atoms, and nuclei. Note that charged particles are present in all of these processes.

There is one other factor to keep in mind: electromagnetic waves are a form of energy. Energy is needed to produce EM waves, and energy is gained by anything that absorbs EM waves. The transfer of heat by way of radiation is one example.

8.5a Radio Waves

Radio waves, the lowest frequency EM waves, extend from less than 100 hertz to about 10^9 Hz (1 billion hertz or 1,000 megahertz; **Figure 8.30**). Within this range are a number of frequency bands that have been given separate names—for example, ELF (extremely low frequency), VHF (very high frequency), and UHF (ultrahigh frequency). Most frequencies are given in kilohertz (kHz) or megahertz (MHz). Sometimes radio waves are classified by wavelength: long wave, medium wave, or short wave.

As mentioned earlier, radio waves are produced using AC with the appropriate frequency. Radio waves propagate well through the atmosphere, which makes them practical for communication. Lower-frequency radio waves cannot penetrate the upper atmosphere, so higher frequencies are used for space and satellite communication. Only the very lowest frequencies can penetrate ocean water.

By far the main application of radio waves is in communication. The process involves broadcasting a certain frequency of radio wave with sound, video, or other information "encoded" in the wave. The radio wave is then picked up by a receiver, which recovers the information. Sometimes, this is a one-way process (commercial AM and FM radio and broadcast [or terrestrial] television), but in most other applications, it is two-way: each party can broadcast as well as receive. Narrow frequency bands are assigned for specific purposes. For example, frequencies from 88 to 108 megahertz (88 million hertz to 108 million hertz) are reserved for commercial FM radio. There are dozens of bands assigned to government and private communication.

8.5b Microwaves

The next band of EM waves, with frequencies higher than those of radio waves, is the microwave band. The frequencies extend from the upper limit of radio waves to the lower end of the infrared band, about 10^9–10^{12} hertz. The wavelengths range from about 0.3 m to 0.3 mm.

One use of microwaves is in communication. Bluetooth and WiFi signals that interconnect computers, cell phones, and other devices are microwaves. Early experiments with microwave communication led to the most important use of microwaves, *radar* (*ra*dio *d*etection *a*nd *r*anging), after the discovery that microwaves are reflected by the metal in ships and aircraft. As we discussed in Section 6.2, radar is echolocation using microwaves. The time it takes microwaves to make a round-trip from the transmitter to the reflecting object and back is used to determine the distance to the object. Radar systems are quite sophisticated: Doppler radar can determine the speed of an object moving toward or away from the transmitter by measuring the frequency shift of the reflected wave. Such radars are essential tools for air traffic control and monitoring severe weather. During its initial four-year mission to explore Saturn and its environs, the *Cassini* spacecraft used imaging radar to penetrate the dense,

NASA - digital version copyright Science Faction/Getty Images

Figure 8.31 This radar image of the surface of Saturn's moon Titan was obtained by the *Cassini* spacecraft on Sept. 7, 2005. The bright, rough region on the left side of the image is topographically high terrain that is cut by channels and bays. The boundary of the bright region and the dark, smooth region appears to be a shoreline. The patterns in the dark area indicate that it may once have been flooded, with the liquid having at least partially receded. The image is 175 km (109 mi) by 330 km (205 mi).

perpetual smog that envelopes Titan, Saturn's largest moon, and to map its surface topology (**Figure 8.31**). Similar radar equipment placed in orbit around Earth is used to form images of its surface, for such purposes as monitoring changes in the global environment and searching for geological formations (ancient craters, for example) and archaeological sites.

Microwaves have gained wide acceptance as a way to cook food. The goal of cooking is to heat the food, in other words, increase the energies of the molecules in the food. Conventional ovens heat the air around the food and rely on conduction (in solids) or convection (in liquids) to transfer the heat throughout the food. Microwave ovens send microwaves (typically with $f = 2,450$ megahertz and $\lambda = 0.122$ meters) into the food. The microwaves penetrate the food and raise the energies of the molecules directly. Recall from Section 7.2 that water consists of polar molecules—they have a net positive charge on one side and a net negative charge on the other side (**Figure 8.32**). The electric field of a microwave exerts forces on the two sides of the water molecules in food. These forces are in opposite directions and twist the molecule. Because the electric field is oscillating, the molecules are alternately twisted one way and then the other. This process increases the kinetic energy of the molecules and thereby raises the temperature of the food. Cooking with microwaves is fast because energy is given directly to all of the molecules. It does not rely primarily on the conduction of heat from the outside to the inside of the food—a much slower process.

8.5c Infrared

Infrared radiation (IR; also called *infrared light*) occupies the region between microwaves and visible light in the electromagnetic spectrum. The frequencies are from about 10^{12} hertz to about 4×10^{14} hertz (400,000,000 megahertz). The wavelengths of IR range from approximately 0.3 to 0.00075 millimeters.

Infrared radiation is ordinarily the main component of heat radiation (introduced in Section 5.4). Everything around you is both absorbing and emitting infrared radiation, just as you are. The warmth you feel from a fire or heat

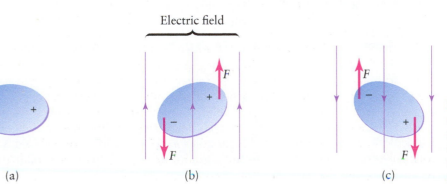

Figure 8.32 (a) Simplified sketch of a water molecule showing the net charges on its sides. (b, c) The oscillating electric field of a microwave twists the molecule back and forth, giving it energy.

Figure 8.33 This remote-control unit uses infrared radiation to transmit signals to other electronic devices to control their operation.

lamp is the result of your skin absorbing the IR. Infrared radiation is constantly emitted by atoms and molecules because of their thermal vibration. Absorption of IR by a cooler substance increases the vibration of the atoms and molecules, thus raising the temperature. We will take a closer look at heat radiation and its uses in Section 8.6.

Infrared radiation is commonly used in wireless remote-control units for televisions and for short-distance wireless data transfer between such devices as garage door openers, video game controllers, audio receivers and laptop computers (Figure 8.33). These units emit coded IR that is detected by other devices. In this capacity, IR is used much like radio waves. Another use of IR is in lasers; some of the most powerful ones in use emit infrared light (see Section 10.8).

8.5d Visible Light

Visible light is a very narrow band of frequencies of EM waves that happens to be detectable by human beings. Certain specialized cells in the eye, called *rods* and *cones*, are sensitive to EM waves in this band. They respond to visible light by transmitting electrical signals to the brain, where a mental image is formed. (The visible ranges of some animals such as hummingbirds and bees extend into the ultraviolet band. Some flowers that seem plain to humans are quite attractive to these nectar eaters.)

Visible light is a component of the heat radiation emitted by very hot objects. About 44 percent of the Sun's radiation is visible light: it glows white hot. Incandescent lightbulbs produce visible light in the same way. Fluorescent and neon lights use excited atoms that emit visible light. In Chapter 10, we will discuss this process and describe how infrared and ultraviolet light and even x-rays are emitted by excited atoms.

Within the narrow band of visible light, the different frequencies are perceived by people as different colors. The lowest frequencies of visible light, next to the infrared band, are perceived as the color red. The highest frequencies are perceived as violet. Table 8.1 shows the approximate frequencies and wavelengths of the six main colors in the rainbow.

Note how narrow the frequency band is: the highest frequency of light we can see is less than twice the lowest. By comparison, the range of frequencies of sound that can be heard is huge: the highest is 1,000 times the lowest.

Most colors that you see are combinations of many different frequencies. White represents the extreme: one way to produce white light is to combine equal amounts of all frequencies (colors) of light. Rainbow formation involves reversing the process: white light is separated into its component colors. When no visible light reaches the eye, we perceive black.

In our daily lives, visible light is the most important of all electromagnetic waves. The entire next chapter is dedicated to optics, the study of light and its interaction with matter.

Table 8.1 Approximate Frequencies and Wavelengths of Different Colors

Color	Frequency Range ($\times 10^{14}$ Hz)	Wavelength Range ($\times 10^{-7}$ m)
Red	4.0–4.8	7.5–6.3
Orange	4.8–5.1	6.3–5.9
Yellow	5.1–5.4	5.9–5.6
Green	5.4–6.1	5.6–4.9
Blue	6.1–6.7	4.9–4.5
Violet	6.7–7.5	4.5–4.0

8.5e Ultraviolet Radiation

Ultraviolet (UV) radiation, also called *ultraviolet light*, is a band of EM waves that begins just above the frequency of violet light and extends to the x-ray band. The frequency range is from about 7.5×10^{14} hertz to 10^{18} hertz.

Ultraviolet light is also part of the heat radiation emitted by very hot objects. About 7 percent of the radiation from the Sun is UV. This part of sunlight is responsible for suntans and sunburns. Ultraviolet radiation does not warm the skin as much as IR, but it does trigger a chemical process in the skin that results in tanning. Overexposure leads to sunburn as a short-term effect, and repeated overexposure during a person's lifetime increases the chance of developing skin cancer. In Section 8.7, we describe how the ozone layer protects us from excessive UV in sunlight.

Some substances undergo *fluorescence* when irradiated with UV: they emit visible light. The inner surfaces of fluorescent lights are coated with such a substance. The UV emitted by excited atoms in the tube strikes the fluorescent coating, and visible light is produced. The same process is used in plasma TVs. Some fluorescent materials appear to be colorless under normal light and can be used as a kind of invisible ink. They can be seen under a UV lamp but are invisible otherwise.

UV radiation has many practical applications. For example, it is used as an investigative tool at crime scenes to help identify bodily fluids such as blood and bile. Ultraviolet lights are used by entomologists to attract and collect nocturnal insects for cataloging and study. Ultraviolet lamps are used to sterilize workspaces and tools used in biology laboratories and medical facilities, and in barber and styling salons. And, increasingly, ultraviolet lasers (see Chapter 9) are finding use in many fields from metallurgy (engraving) to medicine (dermatology and optical keratectomy) and computing (optical data storage).

8.5f X-Rays

The next higher frequency electromagnetic waves are x-rays. They extend from about 10^{16} to 10^{20} hertz. An important feature of x-rays is that their range of wavelengths (about 10^{-8} to 10^{-11} meters) includes the size of the spacing between atoms in solids. X-rays are partially reflected by the regular array of atoms in a crystal and so can be used to determine the arrangement of the atoms. X-rays also travel much greater distances through most types of matter compared to UV, visible light, and other lower-frequency EM waves.

X-rays are produced by smashing high-speed electrons into a "target" made of copper, tungsten or some other metal (**Figure 8.34**). The electrons spontaneously emit x-rays as they are rapidly decelerated on entering the metal. X-rays are also emitted by some of the atoms excited by the high-speed electrons.

Medical and dental "x-ray" photographs are made by sending x-rays through the body. Typically, x-rays with frequencies between 3.6×10^{18} hertz and 12×10^{18} hertz are used. As x-rays pass through the body, the degree to which they are absorbed depends on the material through which they pass. Tissue containing elements with relatively large atomic numbers (Z), such as calcium ($Z = 20$), tend to absorb x-rays more effectively than those that contain predominantly light elements such as carbon ($Z = 6$), oxygen ($Z = 8$), or hydrogen ($Z = 1$). Lead, with atomic number 82, is a particularly good shield for blocking x radiation. Bones, which are rich in calcium, absorb x-rays better

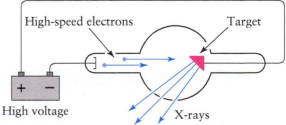

Figure 8.34 Simplified sketch of an x-ray tube. Electrons are accelerated to a very high speed by the high voltage. X-rays are emitted as the electrons enter the metal target.

Figure 8.35 X-ray photograph of a human hand. In this image, areas that appear white are those that strongly absorbed the incident x radiation. Bones are much more efficient at absorbing x-rays because of their calcium content. Elements like gold and silver found in most jewelry are even better absorbers of x-rays because of their higher atomic numbers.

than soft tissue such as muscle or fat, and hence they show up more clearly on x-rays (**Figure 8.35**).

X-rays (and gamma rays) can be harmful because they are **ionizing radiation**— radiation that produces ions as it passes through matter. Such radiation can "kick" electrons out of atoms, leaving a trail of freed electrons and positive ions. This process can break chemical bonds between atoms in molecules, thereby altering or destroying the molecule. Living cells rely on very large, sophisticated molecules for their normal functioning and reproduction. Disruption of such molecules by ionizing radiation can kill the cell or cause it to mutate, perhaps into a cancer cell. The human body can (and does) routinely replace dead cells, but massive doses of x-rays or other ionizing radiation can overwhelm this process and cause illness, cancer, or death. Because medical x-rays are the largest source of artificially produced radiation in the United States, comprising about 10 percent of the total annual radiation dose for the average resident, it is little wonder that protecting the public from unnecessary exposure to damaging radiation in diagnostic radiology is one of the greatest challenges to health and radiological physicists.

8.5g Gamma Rays

The highest-frequency EM waves are gamma rays (γ-rays). The frequency range is from about 3×10^{19} hertz to beyond 10^{23} hertz. The wavelength of higher-frequency gamma rays is about the same distance as the diameter of individual nuclei. Gamma rays are emitted in a number of nuclear processes: radioactive decay, nuclear fission, and nuclear fusion, to name a few. We will study these processes in detail in Chapter 11.

This concludes our brief look at the electromagnetic spectrum. Even though the various types of waves are produced in different ways and have diverse uses, the only real difference in the waves themselves is their frequency and, therefore, their wavelength.

■ Learning Check

1. What simple thing can you do with a charged object to make it generate an electromagnetic wave?

2. Which of the following is *not* a common use of microwaves?
   (a) cooking
   (b) radar
   (c) medical imaging
   (d) communication

3. If we see two objects that have different colors, the light waves coming from them must have different _____.

4. Tanning and sunburn are caused by the _____ component of sunlight.

5. (True or False.) Bones show up in x-ray images because they don't absorb x-rays as effectively as muscle and other tissue does.

ANSWERS: 1. make it oscillate **2.** (c) **3.** frequencies **4.** ultraviolet **5.** False

---

BEHAVIORAL SCIENCES APPLICATION N Rays – "C'est une erreur."

The years before the beginning of the 20th century were truly revolutionary in the history of modern physical science. Starting in 1893 with Sir William Crookes's experiments with cathode-ray tubes (the forerunners of television picture tubes) and continuing through 1895 with Wilhlem Roentgen's discovery of x-rays, the world of physics was turned topsy-turvy. Virtually every physicist in Europe was conducting experiments on these new phenomena. The period from 1895 to 1905

was one of intense activity, intense excitement, and intense rivalry— among individuals and nations—in physics.

One of the early experimenters with x-rays was René Blondlot (**Figure 8.36a**), a professor of physics at the University of Nancy in France. Blondlot had already established a solid reputation as a physicist when, in 1903, in connection with his research on x-rays, he reported the discovery of a new type of radiation, which he

Figure 8.36 (a) René Prosper Blondlot (1849–1930). (b) Title page from Blondlot's book on N rays.

dubbed "N rays" after the place of their discovery, Nancy. Using first a small spark and later a low-intensity gas flame, which increased in brightness when N rays were present, Blondlot reported that these rays were emitted from a variety of sources: x-ray tubes, ring-shaped gas burners (but not ordinary Bunsen burners), sheets of iron and silver heated to glowing, and the Sun. Originally thought to be similar to infrared radiation, N rays possessed a number of unique qualities: they passed through plates made of platinum but not chunks of rock salt; they passed through dry but not wet paper; they could be produced by objects on which stresses were applied (like a stick bent by the hands) or by hardened metal (as in a steel file). The marvels, mysteries, and means of manufacturing and manifesting these rays were described by Blondlot in great detail in 26 articles and a book (Figure 8.36b).

Once the discovery of N rays was announced, experimenters around the world rushed to reproduce and extend Blondlot's work. The results were mixed. Some workers reported success, but most reported failure, including such renowned physicists as Rayleigh, Langevin, and Rubens, a German physicist who pioneered studies in infrared radiation. Suspicion about the reality of N rays began to grow, reaching a high point in the summer of 1904 when a group of concerned scientists decided to send an envoy to Blondlot's laboratory in Nancy to investigate the activities there firsthand. The spy was Professor Robert L. Wood of Johns Hopkins University, a well-known expert in optical phenomena and debunker of numerous spiritualist scams.

Upon arriving in Nancy, Wood was treated to the complete gamut of N-ray phenomena by none other than Blondlot himself. And having witnessed the demonstrations, Wood parted on good terms with his host and published his report. In it Wood recounted how, when asked to hold a steel file (a well-known emitter of N rays) near Blondlot's forehead so as to enhance the latter's ability to see a dimly lit clock face, he instead substituted a piece of wood (one of the few objects known *not* to emit N rays), with no adverse effects on the results of the experiment as reported by Blondlot. Wood himself reported no improvement in the clock's visibility when the file was placed near his line of vision. If this were not enough, Wood noted that during a critical experiment in which the spectrum of N rays was to be produced by diffracting them through an aluminum prism, Wood pocketed the prism in the dark with no apparent alteration of the successful outcome of the demonstration as described by Blondlot. After Wood's exposé, the issue of N rays was dead.

To this day, explanations of the N-ray affair are unsatisfying. Probably the best that can be said is that problems associated with the subjective observation of low-intensity sources with varying energy output, coupled with the failure to perform a well-controlled experiment and the desire to achieve personal prestige and to foster national pride, led to spurious results. What does seem clear is that the story of N rays is not a tale of deliberate fraud on the part of Blondlot or his associates. Moreover, it was not a hoax. It was, quite simply, a mistake. In the words of Professor Josef Bolfa, when interviewed on the subject, "C'est une erreur."

Perhaps one point to take from all of this is that physics is a human endeavor carried out by human beings, and, as in all areas of human activity, mistakes are made. Physicists, like all scientists, are not infallible. Nor are they always as honest, unbiased, or objective as we might wish. In the first three years of the current century, scandals involving two physicists, one at Lucent Technologies' Bell

Laboratories and another at the Lawrence Berkeley National Laboratory, showed all too clearly that physicists are subject to the same temptations and have the same character flaws as any other mortal. In each of these instances, an expert panel of investigators at each of the two affected institutions concluded that the physicists had fabricated or falsified data relating to the alleged discovery of superconductivity in buckyballs (Section 4.1c) on the one hand and the existence of element 118 in the periodic table (Section 4.1b) on the other. Reports issued by these review committees led to the dismissal of the two individuals and to the retraction by coauthors or publishers of many of the scientific articles based on the fraudulent data.

A second lesson to emerge from sagas like these is that physics, as a scientific discipline, is largely self-correcting. The inability of experimenters at other, independent laboratories around the world to duplicate or corroborate the results reported by these miscreants ultimately led to their discovery and downfall. Thus, although a few physicists may be guilty of cheating—either subconsciously as in the case of N rays or deliberately as in the more recent cases— to obtain the answers they desire, the collective action of many physicists over a period of years generally produces results that are reliable, unbiased, and reflective of how the natural world truly operates. As Pier Oddone, deputy director of the Berkeley Lab, has said, "In the end, Nature is the checker. Experiments have to be reproducible."

■ **QUESTION**

1. Give two lessons that may be learned from the unfortunate N-ray affair and other more recent incidents of uncorroborated and ultimately erroneous scientific reports that have been uncovered.

8.6 Blackbody Radiation

Every object emits electromagnetic radiation because of the thermal motion of its atoms and molecules. We have already discussed how this radiation offers one method of transferring heat (see Section 5.4). Without radiation from the Sun, Earth would be a frozen rock. In this section, we take a closer look at heat radiation and consider some of its uses.

The nature of the radiation emitted by a given object—the range of frequencies or wavelengths of EM waves present and their intensities—depends on the temperature of the object and on the characteristics of its surface (for example, its color). A hypothetical object that is perfectly black—one that absorbs all EM waves that strike it—would actually be the best at emitting heat radiation. Referred to as a *blackbody*, it would emit radiant energy at a higher rate than any other object at the same temperature, and the intensities of all of the wavelengths of EM waves emitted could be predicted quite accurately. The heat radiation emitted by such an object is referred to as **blackbody radiation (BBR)**.

Blackbody radiation, an idealized representation of heat radiation, has been analyzed thoroughly and is well understood. The actual radiation emitted by real objects usually is not too much different from BBR, so we can use it as a model of heat radiation.

8.6a BBR Laws

The heat radiation emitted by any object (such as your own body, the Sun, a blackbody) is a broad band of electromagnetic waves. Within this band, some wavelengths are emitted more strongly than others: the intensity of the different wavelengths, the amount of energy released per square meter of emitting surface per second, varies with wavelength. For example, the heat radiation from the Sun contains more energy in each wavelength of visible light than in each wavelength of IR radiation. The intensity of the visible wavelengths is higher than that of the IR wavelengths.

A graph showing the intensity of each wavelength of radiation emitted by a blackbody is called a *blackbody radiation curve* (**Figure 8.37**). The size and shape of the graph change with the object's temperature. The graph for a real object (not a blackbody) would be similar.

The total amount of radiation emitted per second by an object obviously depends on how large it is. A 100-watt lightbulb is brighter (it emits more light) than a 10-watt lightbulb because its filament is larger. Aside from this factor, it is the object's *temperature* that has the greatest influence on the amount and

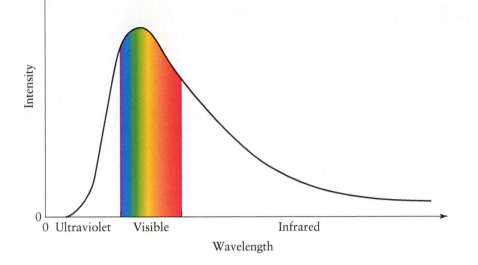

Figure 8.37 Typical blackbody radiation curve. It indicates the amount of energy emitted at each wavelength of the EM spectrum.

types of radiation emitted. Three aspects of heat radiation are affected by the object's temperature.

1. The amount of each type of radiation (such as microwave and IR) emitted increases with temperature.

An incandescent lightbulb fitted with a dimmer control illustrates this well. Dimming the light causes the filament temperature to decrease. This reduces the amount of visible light emitted: the bulb's brightness is decreased. The amount of infrared is also reduced.

2. The total amount of radiant energy emitted per unit area per unit time increases rapidly with any increase in temperature. For a blackbody with surface area A, the total radiant energy emitted per second (power) is proportional to the temperature (in kelvins) raised to the fourth power. Specifically, in SI units,

$$P = (5.67 \times 10^{-8})AT^4$$

Doubling the Kelvin temperature of an object will cause it to emit 16 times as much radiant energy each second. The human body at 310 K (98.6°F) radiates about 25 percent more power than it would at room temperature, 293 K (68°F). If you could see infrared with your eyes, humans would appear to glow more brightly than their cooler surroundings.

3. At higher temperatures, more of the power is emitted at successively shorter wavelengths (higher frequencies) of electromagnetic radiation. For a blackbody, the wavelength that is given the maximum power (the peak of the blackbody radiation curve) is inversely proportional to its temperature.

$$\lambda_{max} = \frac{0.0029}{T} \quad (\lambda_{max} \text{ in meters, } T \text{ in kelvins})$$

Objects cooler than about 700 K (about 800°F) emit mostly IR with smaller amounts of microwaves, radio waves, and visible light. There is not enough visible light to be detected by the human eye. Above this temperature, objects emit enough visible light to glow. They appear red hot because more red (longer wavelength) radiation is emitted than any other visible wavelength. The peak wavelength is still in the infrared region. (Several factors influence the minimum temperature needed to cause an object to glow. These include the size and color of the object, the brightness of the background light, and the acuity of the observer's eyesight.)

The filament of an incandescent lightbulb can be as hot as 3,000 K. The peak wavelength of its heat radiation curve is in the infrared region, not far from the visible band. The visible light emitted is a bit stronger in the longer wavelengths, so it has a slightly reddish tint. At 6,000 K, the Sun's surface emits heat radiation that peaks in the visible band (the wavelength of the peak is one-half that of the lightbulb's). It appears to be white hot (**Figure 8.38**). When comparing the curves, keep in mind that the Sun's is higher *not* because the Sun is bigger than a lightbulb but because it is hotter.

Figure 8.38 Blackbody radiation curves for the Sun and an ordinary incandescent lightbulb.

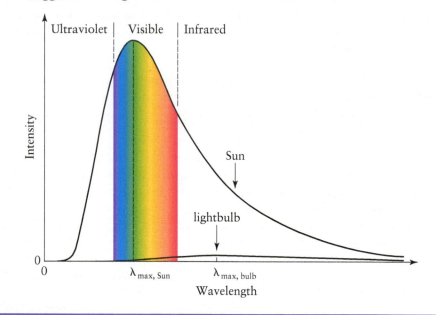

EXAMPLE 8.3 A blackbody in the shape of a sphere of radius 0.25 m has a surface area of 0.79 m². If the temperature of the object is 1,500 K, how much energy does it emit each second?

SOLUTION

$$P = (5.67 \times 10^{-8}) \, AT^4$$
$$= (5.67 \times 10^{-8}) \, (0.79 \text{ m}^2) \, (1{,}500 \text{ K})^4$$
$$= 2.27 \times 10^5 \text{ W}$$

EXAMPLE 8.4 Assuming that the Sun is a blackbody with a temperature of 6,000 K, at what wavelength does it radiate the most energy?

SOLUTION

$$\lambda_{max} = \frac{0.0029}{T}$$
$$= \frac{0.0029}{6{,}000 \text{ K}}$$
$$= 4.8 \times 10^{-7} \text{ m}$$

Table 8.1 shows this to be in the blue-green part of the visible band (see Figure 8.38).

Some stars are hot enough to appear bluish. Sirius and Vega are two examples. The peaks of their radiation curves are in the UV, so they emit more of the shorter wavelength visible light (blue) than the longer wavelengths (see Physics to Go 8.5).

The temperature dependence of blackbody radiation is responsible for a number of interesting phenomena, and it is used in some ingenious ways. The following are examples.

8.6b Temperature Measurement

The temperature of an object can be determined by examining the radiation that it emits. This is particularly useful when very high temperatures are involved, as in a furnace, because nothing has to come into contact with hot matter. Special devices called *pyrometers* measure the amount and types of radiation emitted and use the rules mentioned above to determine the temperature. A similar process is used to measure the temperature of the Sun and other stars. The electronic ear thermometer works in a similar way: it determines the patient's body temperature by measuring the intensity of infrared radiation emitted by an eardrum.

8.6c Detection of Warm Objects

Most things on Earth have temperatures that cause them mainly to emit infrared light. Anything that can detect IR can use this fact to locate warmer-than-average objects, because they will emit more IR. Rattlesnakes and certain other snakes use IR to hunt mice and other warm-blooded animals at night. These snakes have sensitive organs that detect the higher-intensity infrared emitted by objects warmer than their surroundings.

Infrared-sensitive photographic film, video cameras, and other detection devices have many practical uses. For instance, IR photographs, called *thermograms,* can show where heat is escaping from a poorly insulated house and can detect ohmic heating caused by a short circuit in an electrical substation (**Figure 8.39a**). IR detectors can locate warmer areas on the human body that might be caused by tumors. The military uses IR detectors to locate soldiers at night, and healthcare practitioners use them to determine patients' temperatures (**Figure 8.39b**). Heat-seeking missiles automatically steer themselves toward the hot exhaust of aircraft.

Figure 8.39 (a) Thermograms—infrared photographs—show regions that are warmer or cooler than the surroundings. (b) An infrared thermometer measures the IR energy emitted by a human body and converts it to a temperature.

altrendo images/Getty Images

(a)

© Cengage Learning/Edward L. Dodd Jr.

(b)

Next time you're outside on a clear night in a place well removed from city lights and free of obstructions, take a careful look at the sky. Do the stars you see all appear equally bright? Do they all appear the same color? The color (and to some extent the brightness) of a star is directly related to its surface temperature, as described earlier in this section. You can establish the relative temperatures for many of the brightest stars by assessing their colors and applying the physics of blackbodies that you have learned thus far.

To get started, use the star chart in **Figure 8.40** to locate the prominent wintertime constellation of Orion. Identify the bright stars Rigel and Betelgeuse. Notice the differences in their colors. Betelgeuse has a surface temperature of only about 3,200 K (quite low by stellar standards but hot enough to melt iron). It should appear distinctly red to you. By contrast, Rigel is a much hotter star ($T = 10,000$ K) and should look bluish-white to your eyes.

Scan the sky for other bright stars in the vicinity of Orion, such as Sirius in Canis Major or Aldebaran in Taurus. Based on their colors, how do their surface temperatures compare to those of Rigel and Betelgeuse? How would you describe the colors of the majority of the brightest-appearing stars in the sky? Suppose you were told that most of the stars in our Galaxy near the Sun have rather low surface temperatures and should appear red-orange in color. Is this statement at odds with your answer to the previous question? Can you offer an explanation for any discrepancy between your observations and the true state of stars in the solar neighborhood?

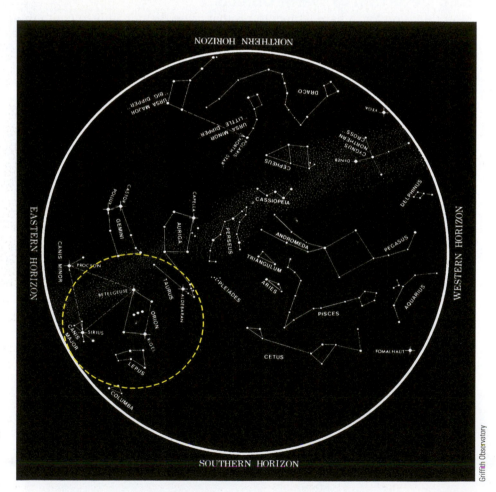

Figure 8.40 Winter constellations in the evening at midnorthern latitudes. Dashed circle highlights the area around the constellation Orion.

1. The total radiant power emitted by a blackbody is proportional to its _____ raised to the fourth power.
2. (True or False.) A glowing blackbody that appears bluish in color is hotter than one that appears reddish.
3. The blackbody radiation emitted by something as hot as the Sun consists of
   (a) infrared waves.
   (b) visible waves.
   (c) ultraviolet waves.
   (d) All of the above.
4. (True or False.) Visible light is the only type of EM wave that living organisms can detect.

ANSWERS: 1. temperature 2. True 3. (d) 4. False

ASTRONOMICAL APPLICATION   Cosmic Background Radiation (CBR)—A Relic of the Big Bang

Not too long ago, when broadcast (or terrestrial) television was the principal source of programming, you could tune your TV set to a channel that was not used by your local network affiliates and observe the "snow" on the screen. A part of the "signal" you were receiving was from radiation that was produced during the formation of the universe in a catastrophic event popularly referred to as the *Big Bang*. Indeed, our belief that the universe was in fact born in a fiery explosion of space and time nearly 14 billion years ago is intimately connected to the existence and character of the weak microwave radiation picked up by your TV as "noise."

The description of what is now called the *cosmic background radiation* as "noise" is an apt one, for it was the search for the persistent source of radio interference in their 20-foot horn antenna (**Figure 8.41**) in 1965 that eventually led Bell Laboratory researchers Arno Penzias and Robert Wilson to the startling conclusion that the sky is filled with microwaves. Regardless of the direction in which a suitably tuned receiver is pointed, it will detect such radiation and with nearly the same intensity. The equivalent temperature of the interfering radiation discovered by Penzias and Wilson was found to be about 3.5 K. Because of the limitations of their original equipment, they were not able to fully describe the wavelength dependence of this "noise." But, in the years following their report, measurements were made by other groups at other wavelengths, and all yielded temperatures of between about 2.7 and 3.0 K. The spectrum was thus shown to be that of a blackbody, characteristic of matter in thermal equilibrium with radiation at a temperature of a bit less than 3 K. The best demonstration of the blackbody nature of this cosmic radiation comes from data obtained by the *Cosmic Background Explorer* (COBE) satellite; **Figure 8.42a** shows the results of measurements made by this instrument, which yield an excellent fit to a blackbody radiation curve for a temperature of 2.726 K. Using the method of Example 8.3, this temperature gives a wavelength of 1.06×10^{-3} meters (1.06 millimeters) for the peak in the cosmic blackbody radiation curve; this is squarely in the microwave region of the EM spectrum (Figure 8.29). Nobel Prizes went to Penzias

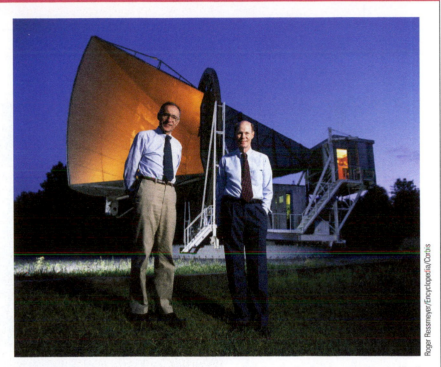

Figure 8.41 Arno Penzias (left) and Robert Wilson. In the background is the Bell Labs microwave antenna that was used to detect the cosmic background radiation.

and Wilson in 1978 and to John Mather and George Smoot, COBE team leaders, in 2006 for their investigations of the CBR.

Just exactly *how* are the existence and character of this microwave radiation connected to the creation of the universe, and *what* can it tell us about the conditions present during this birthing event? One of the implications of the discovery that the galaxies are rushing away from one another with speeds that are proportional to their separations is that at earlier times in the history of the universe, the galaxies were all closer together than they are now (see the applications section on the Hubble relation at the end of Section 6.2). The earlier the epoch, the more densely packed were the galaxies, until at some point all the matter and energy in the universe were concentrated in an infinitely small volume. Insofar as the universe comprises everything, nothing can exist outside it; thus this period of infinite density marks not only the temporal but also the spatial beginning of the universe. Space and time were created simultaneously in an

Wavelength (millimeters)
(a)

(b)

NASA Goddard Space Flight Center (NASA-GSFC)

Figure 8.42 (a) COBE data showing the blackbody spectrum of the cosmic background radiation. (b) WMAP all-sky image showing microkelvin temperature variations in the background radiation. Regions shown in red are slightly warmer than average, and those in blue-violet are slightly cooler.

There came a time, however, when the continued expansion of the universe rendered the density and temperature of matter low enough that the interactions between the radiation and the matter became far less pronounced. At this stage, believed to have occurred some 400,000 to a million years after the initial explosion, the radiation "decoupled" from the matter in the universe and was free to move through it virtually unimpeded. The universe became transparent to radiation. This radiation then filled the universe nearly uniformly and isotropically. Of course, just after its separation from matter, the temperature of the radiation was still quite high. Current theories place it at about 3,000 K. As the universe inexorably expanded and cooled, though, this temperature dropped until at the present time we expect it to have reached about 3 K. Because the radiation was originally in equilibrium with the matter in the universe, it remained so after decoupling and exhibits today the spectrum of a blackbody. Thus, one prediction of the Big Bang model arising from the expansion of the universe from an infinitely hot, dense state is the existence of a low-temperature cosmic background radiation with a blackbody distribution. And this is exactly what Penzias and Wilson observed.

The discovery of the microwave background radiation was one of the most important pieces of physical evidence leading to the widespread acceptance by scientists of the Big Bang model as the correct description of the formation of the universe. Continuing studies of the cosmic microwave background radiation by

explosion of mass and energy from this singular condition. This *is* the Big Bang, a term coined by astronomer Fred Hoyle in 1950.

If we treat the early universe like a highly compressed gas, then we expect it to have a high temperature according to the laws of thermodynamics discussed in Chapter 5. As this hot, dense "gas" expanded, it thinned out and cooled off. In the process, matter, initially in the form of elementary particles (see Chapter 12), began to condense out of the sea of pure radiant energy that was the universe at that time. Once this condensation took place, matter began to interact strongly with the radiation, absorbing, scattering, and emitting it profusely. The distance traveled by any given beam of radiation was very small before it encountered particles that delayed and deflected it. In this way, the matter acted as a very effective dam to the free propagation of radiation at this phase in the development of the universe.

COBE and more recently by the Wilkinson Microwave Anisotropy Probe (WMAP) launched by NASA in June 2001 have now produced highly detailed maps of the sky revealing microkelvin fluctuations in temperature (**Figure 8.42b**). Such displays are the "baby pictures" of the universe when it was only 379,000 years old. The size and scale of these minute temperature variations are helping scientists understand how galaxy formation, driven by such fluctuations, began in the early universe. But that's another story.

■ **QUESTIONS**

1. Describe the principal features or characteristics of the cosmic microwave background radiation.
2. What is believed to be the origin or source of this radiation and why do astronomers hold this to be true?

8.7 EM Waves and Earth's Atmosphere

Many substances in the atmosphere surrounding Earth interact with EM waves in important ways. Some of these interactions are crucial to the existence of life on this planet, another helps us communicate, and some add to the beauty that characterizes life on Earth. The visible phenomena—rainbows, for example—are described in the next chapter dedicated to visible light. Some of the others are discussed here.

8.7a Ozone Layer

The sunlight that keeps Earth warm and provides the energy for plants to grow also contains ultraviolet radiation that is harmful to living things. But life has evolved on this planet because the atmosphere has protected it from much of this UV. In the region between about 20 and 40 kilometers above Earth known as the *ozone layer*, there is a comparatively high concentration of *ozone* (O_3), the form of oxygen with three atoms in each molecule. This ozone absorbs most of the harmful UV in sunlight. The ozone layer has been a shield protecting living things on Earth.

In 1974, it was reported that *chlorofluorocarbons* (CFCs), chemical compounds such as Freon used in refrigerators, air conditioners, and as an aerosol propellant in spray cans, could be depleting the ozone layer. The CFCs that are released into the air drift upward to the ozone layer and chemically break up the ozone molecules with alarming efficiency. (The 1995 Nobel Prize in chemistry was awarded to the discoverers of this effect.) Because of this, CFCs were banned for use as aerosol propellants in the United States. Then in 1985 it was discovered that a "hole" developed in the ozone layer over Antarctica during the later part of each year. The concentration of ozone in a huge section of the atmosphere was reduced by about one-half. A review of old satellite measurements revealed that this hole had developed in previous years as well and that the one in 1982 was twice the area of the United States (**Figure 8.43**). Scientists now believe that the ozone hole is produced by a complex set of processes involving chlorine that originates in CFCs. During winter in the southern hemisphere, chlorine molecules (Cl_2) are released by chemical reactions that take place in extremely high clouds, called *polar stratospheric clouds* (PSCs), over the sunless South Pole. The return of sunlight in spring then triggers the reactions that cause chlorine to break up ozone molecules.

Until recently it was thought that a similar ozone hole would not form over the North Pole. But in March 1997 the ozone level over the Arctic dropped to a record low for that time of year. The depletion was not as severe as in the southern ozone hole, but it suggests that the damage to the ozone layer is increasing.

Global monitoring of the ozone layer revealed an overall decline during the latter part of the 20th century and not just over the poles. Continued reduction in ozone levels could have tragic consequences. Rates of occurrence of skin cancer could rise and crop yields could decrease because increased UV adversely affects many plant species. But unprecedented international cooperation has led to the banning of CFCs throughout the developed countries. Levels of CFCs in the atmosphere seem to be declining, but the extremely high chemical stability of these compounds means they will continue to do damage for many years to come.

8.7b Greenhouse Effect

The **greenhouse effect** is so named because it is partly responsible for keeping greenhouses warm in cold weather. Glass and certain other materials allow visible light to pass through them

Figure 8.43 False-color image of the ozone hole over the southern hemisphere, based on satellite measurements taken in October 2015. The region of greatly reduced ozone is shown in blue. The "hole" spans 28.2 million square kilometers (10.9 million square miles)—the fourth largest area measured since satellite records began in 1979.

while they absorb or reflect the longer-wavelength infrared radiation. A glass wall or roof on a building allows visible light to enter and to warm the interior. As the temperature of things inside increases, they emit more IR. Without the glass, this IR would escape from the enclosure and carry away the added energy. But the glass blocks the IR, so the added internal energy is trapped and the enclosure is warmed. (The glass not only reduces heat loss by radiation but also eliminates convective losses: the heated air that rises cannot leave and take internal energy with it.) A car parked in the Sun with the windows up is much warmer than the outside air because of this heating. Windows on the sunny side of a building help to keep it warm in the same way.

The greenhouse effect occurs naturally in Earth's atmosphere. Water vapor, carbon dioxide (CO_2), and other gases in the air act somewhat like glass in that they allow visible light from the Sun to pass through to Earth's surface while they absorb part of the infrared radiation emitted by the warmed surface (**Figure 8.44a**). The atmosphere is heated by the IR that is absorbed; it is about 35°C (65°F) warmer than it would be without this effect.

From 1958 to 2008, the carbon-dioxide content of the atmosphere rose 21 percent (**Figure 8.44b**). Evidence points to human activity as the likely cause. During the last century, huge quantities of fossil fuels—coal, oil, and natural gas—have been burned. This has released vast amounts of carbon dioxide into the air. At the same time, forests have been cut down for building materials and cleared for farming and human occupation. This contributes to the problem because trees and other plants take carbon dioxide out of the air. The concentration of methane is also increasing. Methane gas, released into the air by

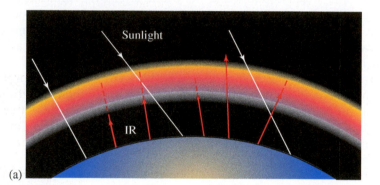

(a)

Atmospheric CO$_2$ at Mauna Loa Observatory

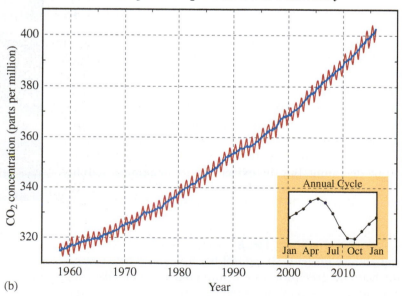

(b)

Figure 8.44 (a) Earth's atmosphere produces a greenhouse effect. Increased carbon dioxide content in the air could cause too much heating. (b) Graph of CO_2 content in the air versus time, measured at the observatory atop Mauna Loa in Hawaii. The concentration varies throughout each year because plants in the northern hemisphere take in more CO_2 in the summer when they are most active.

some animals and by rice as it grows, and the ozone-threatening chlorofluoro-carbons are also so-called greenhouse gases like water vapor and CO_2. The net result of these modifications of Earth's atmosphere is the possibility of global warming—the entire atmosphere heating up.

Evidence to sustain the view that global warming is occurring is highly persuasive. For example, global sea levels rose approximately 17 cm (6.7 in.) in the last century, with the rate of increase during the last decade being nearly double the average over that period. Global surface temperature models show Earth has steadily warmed since 1880, with the bulk of the effect occurring during the 1970s; all ten of the warmest years have been measured in the last 12 years. This effect has also been seen in the oceans, where the temperature of the upper 700 m (2300 ft) has increased by 0.167°C (0.302°F) since 1969. One result of this rise in ocean temperature has been the rapid shrinking of ice sheets in Greenland and Antarctica and the decline in sea ice in the Arctic region.

Some have argued that these changes are merely part of a long-term cyclical variation in atmospheric conditions, whereas most scientists see this as foreshadowing of a potentially irreversible trend leading to a runaway greenhouse effect unless steps are taken soon to rein in humankind's environmentally destructive practices. A runaway greenhouse effect did occur naturally on the planet Venus, where the surface temperature is now 460°C. Conditions on Venus prevented excessive atmospheric carbon dioxide from being trapped in carbonate rocks or dissolved in oceans, as is the case on Earth.

The atmospheric greenhouse effect is such a complex phenomenon that it is difficult to predict accurately what will happen. Different scenarios have been proposed: as Earth heats up, huge amounts of CO_2 dissolved in the oceans could be released and could exacerbate the global warming. Increased evaporation of water would raise the water-vapor content of the atmosphere and possibly cause more heating. Or the higher humidity could lead to more cloud cover, which might cool Earth as less sunlight reaches its surface. About all we can be sure of is that we are altering the life-sustaining blanket in which we live even though we cannot predict precisely what the long-term effects may be.

8.7c The Ionosphere

About 50–90 kilometers above Earth's surface, in a region of the atmosphere known as the *ionosphere*, there is a relatively high density of ions and free electrons. Radio waves from surface transmitters travel upward into the ionosphere. Higher-frequency radio waves, such as those used in FM radio and terrestrial television, pass through the ionosphere and out into space (**Figure 8.45a**). Lower-frequency radio waves, such as the 500–1,500 kilohertz waves used in AM radio, reflect off the ionosphere and return to Earth (**Figure 8.45b**), allowing such radio stations to be picked up over distances of several hundred kilometers. The range for high-frequency radio waves is limited to "line-of-sight" reception with distances of about 80 kilometers or less. The curvature of Earth eventually blocks the signal from the transmitting tower. Low-frequency radio waves can skip off the ionosphere and travel farther around the planet, a fact that amateur radio enthusiasts called DXer's, after the telegraphic shorthand *DX* for "distant" or "distance," take full advantage of. The radio waves used to communicate with spacecraft must have high frequency to pass through the ionosphere.

Figure 8.45 (a) High-frequency radio waves pass through the ionosphere. Because of this, transmission of commercial FM and television is limited to less than about 80 km. (b) Low-frequency radio waves are bent back to Earth's surface by the ionosphere. This allows commercial AM radio to be transmitted hundreds of kilometers.

The ionosphere is also the home of the *auroras*—the northern lights and the southern lights. Charged particles from the Sun excite atoms and molecules in the ionosphere, causing them to emit light. (More on this in Chapter 10.)

8.7d Astronomy

Stars, galaxies, and other objects in space emit all types of electromagnetic radiation. Astronomers were originally limited to studying only visible light through optical telescopes. Now they examine the entire spectrum of EM radiation to gain a more complete understanding about the universe. The atmosphere is a hindrance to some of these investigations. Ultraviolet light from stars and galaxies is absorbed by ozone and other gases. Infrared light is also absorbed, mainly by water vapor. Lower-frequency radio waves are absorbed in the ionosphere. Even visible light is affected by the random swirling of the air, which causes stars to twinkle and degrades the images formed in large telescopes. Microwaves and higher-frequency radio waves are about the only EM waves from space that are not affected by the atmosphere.

Since the early 1960s, dozens of telescopes and other astronomical instruments have been placed in orbit to overcome the deleterious effects of Earth's atmosphere. Among the most important and sophisticated of recent space missions are the four that comprise NASA's Great Observatories Program. Each was designed to examine different parts of the EM spectrum from infrared light for studying cool stars and interstellar dust to gamma radiation emitted in high-energy processes associated with supernova explosions and neutron-star collisions. The most famous of these, the Hubble Space Telescope (HST), was launched in 1990 and is equipped with instruments that analyze visible, ultraviolet, and shorter-wavelength infrared light (**Figure 8.46**). After a shaky start, HST has achieved enormous success: research astronomers, as well as millions of people worldwide, have been captivated by the stunning images it has returned. (There is more on HST in Section 9.2.) The other spacecraft in the program, along with the year in which each was launched, are the Compton Gamma-Ray Observatory (1991), the Chandra X-Ray Observatory (1999), and the Spitzer Space Telescope (infrared, 2003). Astronomy textbooks for years to come are likely to contain illustrations and findings supplied by these observatories. Space exploration continues to give astronomers access to the entire electromagnetic spectrum.

Figure 8.46 (Left) The Hubble Space Telescope during deployment from the space shuttle *Discovery*. (Right) Hubble Space Telescope image of the Cat's Eye nebula showing gas and dust ejected from a dying star.

1. Why would a decrease in the amount of ozone in the atmosphere increase the likelihood of people getting sunburn?
2. Of the following gases, which is the most important contributor to Earth's greenhouse effect?
   (a) water vapor
   (b) ozone
   (c) CFCs
   (d) methane
3. (True or False.) The Hubble Space Telescope and devices that detect nonvisible segments of the EM spectrum are sometimes placed in orbit to overcome the distortion and absorption caused by Earth's atmosphere on EM waves passing through it.

ANSWERS: 1. Increases solar UV **2.** (a) **3.** True

Profiles in Physics   19th Century Leaders in Electromagnetism

The first recorded investigation of magnetism was carried out by William Gilbert (see Chapter 7 Profiles in Physics) at the same time he was studying electricity. In the course of his work, Gilbert built a sphere out of lodestone to model Earth's effects on a compass needle. With this *terrella* (little Earth), he accounted for a number of phenomena, such as magnetic declination, that had been observed by navigators as they traveled around the globe. One of these was an effect observed by Columbus on his famous voyage: the declination angle changes with longitude.

As recently as 200 years ago, electricity and magnetism seemed to be similar but independent phenomena. Reports that compasses were affected by lightning storms suggested that some kind of interaction between the two existed, but nothing concrete was observed until 1820. In that year, Hans Christian Oersted (1777–1851; **Figure 8.47**), a physics professor at the University of Copenhagen, announced that he had observed a compass needle being deflected by an electrical current in a nearby wire. Oersted had begun experimenting with electricity after he heard of Volta's invention of the battery. His first attempts to observe the effect failed because, understandably, he was not expecting the deflection to be perpendicular to the current-carrying wire.

Within months, news of the discovery had spread throughout the European scientific community. The exact relationship between an electric current and the strength and configuration of the magnetic field that it produced were quickly deduced. Most of this work was done by French physicist André Marie Ampère (1775–1836).

Within a week of hearing about Oersted's discovery, Ampère had thoroughly investigated it and prepared his own paper on the topic. Ampère showed that the magnetic field of a current could be concentrated by bending the wire into a loop or wrapping it up as a solenoid. He correctly suggested that the magnetism in permanent magnets is caused by tiny currents in the molecules. He made the second important observation of electromagnetism: a magnetic field exerts a force on a current-carrying wire. Ampère's most famous finding is that two current-carrying wires exert forces on each other (**Figure 8.48**). The force results from magnetism even though no permanent magnets are involved. This simple setup elegantly illustrates how magnetism is an inextricable part of moving electric charges.

An American physicist, Joseph Henry (1797–1878), is notable for his research on electromagnets. He improved their efficiency to the point of making an electromagnet capable of lifting 50 times its own weight when connected to a small battery. Henry also had the idea of using electromagnets for long-distance communication—the telegraph. He freely mentioned this idea to others, who later patented the process and became wealthy because of it.

The final major discoveries in electromagnetism were made by two of the greatest names in physics, Michael Faraday and James Clerk Maxwell.

Faraday (1791–1867), son of a blacksmith, grew up near London. His apprenticeship as a bookbinder afforded him the opportunity to read extensively about the great discoveries in electricity. He first experimented with electricity in 1821, but his greatest work was

Bettmann/Corbis

Figure 8.47 Hans Christian Oersted.

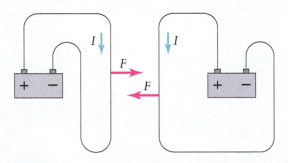

Figure 8.48 Two parallel current-carrying wires exert forces on each other. The magnetic field produced by the wire on the left exerts a force on the wire on the right, and vice versa.

Figure 8.49 Apparatus used by Faraday to discover electromagnetic induction. A current is induced in the right coil only as the direct current in the left coil is either switched on or switched off. The compass is used to detect the induced current.

done in the 1830s. Like many other experimenters, he assumed that magnetism could be used to produce electricity because the reverse was true. His initial attempts failed because he used steady magnetic fields. But on 29 August 1831, using a device similar to a transformer, Faraday noted that a current was momentarily induced in one coil as the other was connected to or disconnected from a battery (**Figure 8.49**). Electromagnetic induction occurred in the coil only if the magnetic field in it was changing.

In the following months, Faraday (**Figure 8.50**) developed a keen understanding of electromagnetic induction. He constructed primitive electric motors, generators, and transformers. In addition to these practical inventions, he made important contributions to the theoretical understanding of electricity and magnetism. He introduced the concept of "lines of force" (field lines) to model what were later called *electric* and *magnetic fields*. Faraday's successors

Figure 8.51 James Clerk Maxwell in 1855, while at Cambridge University.

marveled at his ability to understand sophisticated phenomena without the aid of higher mathematics.

One of Faraday's greatest admirers was the young James Clerk Maxwell (1831–1879; **Figure 8.51**). Maxwell's work in electricity and magnetism began in 1855, about the time he accepted a physics professorship in Aberdeen, Scotland. He synthesized the findings of Faraday and others and provided a concrete mathematical framework for describing electrical and magnetic phenomena. He solidified Faraday's concept of fields, which was later applied to gravity as well as electricity and magnetism, and introduced an important clarification of the idea of current. He is most famous for a set of equations, now called *Maxwell's equations*, that summarize the basics of electromagnetism. These equations are on the same level of importance in physics as Newton's laws.

Maxwell saw that his mathematical description of electricity and magnetism indicated that traveling waves of electric and magnetic fields could exist. That in itself was not particularly startling, but when he computed the speed of these waves, it turned out to be nearly equal to the measured speed of light. The conclusion was inescapable: light was one type of "electromagnetic wave." Maxwell not only had integrated electricity and magnetism but also brought light into the same realm.

Figure 8.50 Michael Faraday delivering a public lecture on electricity and magnetism at the Royal Institution, London, in 1856. Faraday devoted a great deal of effort to his lectures and drew large crowds.

Some nine years after Maxwell's death, his prediction of electromagnetic waves was proved correct. German physicist Heinrich Hertz (1857–1894), a former assistant to Hermann von Helmholtz, produced radio waves and showed that they had properties similar to light. In the more than a century that has followed, our technological society has filled the skies with electromagnetic waves.

■ **QUESTIONS**

1. Explain why two wires, each with a current flowing in it, exert forces on each other even when they are not touching (see Figure 8.48).
2. Compare and contrast the lives, scientific careers, and approaches to scientific discovery of Michael Faraday and James Clerk Maxwell. Name one important contribution to our understanding of electromagnetism made by each of these 19th-century scientists.

SUMMARY

» Every simple magnet has a **north pole** and a **south pole**, so named because the two parts of the magnet are naturally attracted to the north and to the south, respectively.

» The magnetic field produced by one magnet causes forces to act on the poles of any other magnet in its vicinity. When two magnets are near each other, the like poles repel and the unlike poles attract.

» A compass is simply a small magnet that is free to rotate when in the presence of a magnetic field. It can be used to determine the direction of the magnetic field at any point in space.

» Earth has its own magnetic field that causes compasses to point toward the north. Earth's magnetic poles do not coincide with its geographic poles, so compasses do not point exactly north at most places on Earth.

» Many phenomena depend on the interactions between electricity and magnetism. These interactions occur only when there is some kind of change taking place, such as motion of charges or a magnet.

» The basic electromagnetic interactions can be stated in the form of three simple observations:

(1) Moving charges produce magnetic fields.
(2) Magnetic fields exert forces on moving charges.
(3) Moving magnets induce currents in coils of wire (**electromagnetic induction**).

These processes are exploited in a variety of useful devices from electromagnets and electric motors to microphones and speakers.

» The **principles of electromagnetism** express the fundamental relationship between electricity and magnetism: "an electric current or a changing electric field induces a magnetic field," and "a changing magnetic field induces an electric field."

» These principles of electromagnetism summarize the three observations and predict the existence of **electromagnetic (EM) waves**—traveling combinations of oscillating electric and magnetic fields. All EM waves in a vacuum travel at the same speed, c, the "speed of light."

» EM waves can be classified according to frequency. From low to high, the bands are **radio waves, microwaves, infrared radiation, visible light, ultraviolet radiation, x-rays,** and **gamma rays**. The different waves are involved in a diverse number of natural processes and technological applications.

» **Blackbody radiation** is a broad band of electromagnetic waves emitted by an object because of the thermal motion of atoms and molecules. The amount of radiation emitted and the intensities of different wavelengths depend on an object's temperature.

» The ozone layer in Earth's atmosphere absorbs most of the harmful UV present in sunlight. Compounds known as CFCs drift upward and reduce the concentration of ozone.

» Carbon dioxide, water vapor, methane gas, and CFCs contribute to a **greenhouse effect** in Earth's atmosphere. They allow sunlight to pass through and warm Earth's surface while they absorb much of the IR emitted by the heated surface. The result is a warming of the atmosphere that may increase because of heightened concentrations of these gases.

» The ionosphere, a region in the upper atmosphere containing ions and free electrons, reflects lower-frequency radio waves back to Earth, but allows higher frequency radio waves to pass out into space. This greatly increases the range of radio communication for the lower frequencies.

» Astronomers have overcome the absorption of different bands of EM waves by the atmosphere by placing telescopes and other instruments in space.

IMPORTANT EQUATIONS

Equation	Comments
$\dfrac{V_o}{V_i} = \dfrac{N_o}{N_i}$	Input and output voltages of a transformer
$c = f\lambda$	Relates speed, wavelength, and frequency of EM waves
$\lambda_{max} = \dfrac{0.0029}{T}$	Peak wavelength of BBR curve (SI units)
$P = (5.67 \times 10^{-8})\, AT^4$	Power radiated by a blackbody with emitting area A (SI units)

1. Reread Section 8.5 on electromagnetic waves. Make a list of the main concepts introduced in this section and then, taking *electromagnetic waves* as the organizing concept, develop a concept map that includes the basic properties of EM waves, their relationship to electric and magnetic fields, as well as the principal types of EM waves that scientists have distinguished. How could your map be integrated or connected with Concept Map 8.2? That is, how and where would you link your map to Concept Map 8.2?

2. As mentioned in Exercise 2, in Mapping It Out! in Chapter 7, concept maps can often take the place of more traditional note taking when reading magazine or newspaper articles. Here is another opportunity for you to practice the advantages of this technique.

   Using the Internet or the resources of your local or campus library, locate and read *two* articles on either global warming *or* the ozone-layer problem. After reading them, review both again and circle what you believe to be the key concepts introduced in each. Then, using *only* the concepts you circled, construct a concept map for each that represents the major points or positions discussed in the articles. After finishing the task, consider the difficulty of this exercise as you did in Chapter 7. Are some important concepts missing from the articles that you had to supply from your own background or knowledge in order to make sense of the articles? Evaluate how conceptually complete the articles were. Consider how your answer to the previous question might affect your judgment of the quality of the articles. Then use your concept maps to reconsider the degree to which you might be inclined to trust the accuracy or fairness of the articles. Comment on these issues as part of your report.

QUESTIONS

(■ **Indicates a review question, which means it requires only a basic understanding of the material to answer. Questions without this designation typically require integrating or extending the concepts presented thus far.**)

1. Three bar magnets are placed near each other along a line, end to end, on a table. The net magnetic force on the middle one is zero, and its north pole is to the left. Make two sketches showing the possible arrangements of the poles of the other magnets.
2. ■ Sketch the shape of the magnetic field around a bar magnet.
3. ■ What happens to a ferromagnetic material when it is placed in a magnetic field?
4. What causes magnetic declination? Is there a place where the magnetic declination is 180° (a compass points south)? If so, approximately where?
5. ■ Describe the three basic interactions between electricity and magnetism.
6. ■ Explain what *superconducting electromagnets* are. What advantages do they have over conventional electromagnets? What disadvantages do they have?
7. ■ Name five different *basic* devices that use at least one of the electromagnetic interactions.
8. ■ In many cases, the effect of an electromagnetic interaction is perpendicular to its cause. Describe two different examples that illustrate this.
9. To test whether a material is a superconductor, a scientist decides to make a ring out of the material and then to see whether a current will flow around in the ring with no steady energy input.
   (a) Explain how a magnet could be used to initiate the current.
   (b) At some later time, how could the scientist check to see whether the current is still flowing in the ring without touching the ring?
10. A coil of wire has a large alternating current flowing in it. A piece of aluminum or copper placed near the coil becomes warm even if it does not touch the coil. Explain why.
11. In the particular accelerator shown in Figure 8.16, what are the possible directions of the magnetic field that keeps the particles traveling around the circular path?
12. ■ What is the "motor-generator duality"? Explain how it is used.
13. ■ Explain why a transformer doesn't work with DC.
14. What would an audio speaker like that shown in Figure 8.23 do at the instant a low-voltage battery is connected to it?
15. The type of microphone described in Section 8.4 can be thought of as a speaker "operating in reverse." Explain.
16. ■ What is *analog-to-digital conversion* and how is it used in sound reproduction?
17. ■ Explain how electromagnetic waves are a natural outcome of the principles of electromagnetism.
18. An electromagnetic wave travels in a region of space occupied only by a free electron. Describe the resulting motion of the electron.
19. ■ List the main types of electromagnetic waves in order of increasing frequency. Give at least one useful application for each type of wave.
20. ■ Alternating current with a frequency of 1 million Hz flows in a wire. What in particular could be detected traveling outward from the wire?
21. ■ What are the main uses of microwaves? Explain how each process works.
22. A liquid compound is not heated by microwaves the way water is. What can you conclude about the nature of the compound's molecules?
23. Aircraft equipped with powerful radar units are forbidden from using them when parked on the ground near people. Explain why this is so.
24. ■ Which type of EM wave does your body emit most strongly?
25. ■ What is different about our perceptions of the different frequencies within the visible light band of the EM spectrum?
26. A heat lamp is designed to keep food and other things warm. Would it also make a good tanning lamp? Why or why not?
27. ■ How are x-rays produced?
28. ■ Why are x-rays more strongly absorbed by bones than by muscles and other tissues?
29. ■ What is *blackbody radiation*? In what ways does the radiation emitted by a blackbody change as its temperature increases?
30. A lightbulb manufacturer makes bulbs with different "color temperatures," meaning that the spectrum of light they emit is similar to a blackbody with that temperature. What would

be the difference in appearance of the light from bulbs with color temperatures of 2,000 K and 4,000 K?

31. Referring to the lightbulbs described in Question 30 and assuming the emitting areas of the filaments in the two bulbs are the same, which of the two is the brighter? About how much brighter is this bulb than its companion?

32. Explain how infrared light can be used to detect some types of animals in the dark. Can you think of a situation in which this would not work?

33. Stars emit radiation whose spectrum is very similar to that of a blackbody. Imagine two stars identical in size, each of which is at the same distance from us. One of the stars appears reddish in color, while the other one looks distinctly bluish. Based on this information, what can you say about the relative temperatures of the two stars? Which is the hotter? How do you know?

34. ■ What effect does the ozone layer have on the EM waves from the Sun? What is currently threatening the ozone layer?

35. ■ Describe the greenhouse effect that is occurring in Earth's atmosphere.

36. ■ How does the ionosphere affect the range of radio communications?

37. Five transformers, all with the same basic construction (Figure 8.52), have different numbers of turns of wire as their input, N_{in}, and output, N_{out}, coils according to the circumstances given below. The input voltages, V_{in}, for the five transformers also vary as indicated. Rank the output voltages, V_{out}, for each transformer from greatest to smallest. If two or more transformers have the same output voltage, give them the same ranking. Explain the reasoning you used in making your rankings.

Transformer A: V_{in} = 25 V; N_{in} = 100 turns; N_{out} = 250 turns
Transformer B: V_{in} = 25 V; N_{in} = 200 turns; N_{out} = 250 turns
Transformer C: V_{in} = 50 V; N_{in} = 100 turns; N_{out} = 50 turns
Transformer D: V_{in} = 50 V; N_{in} = 50 turns; N_{out} = 200 turns
Transformer E: V_{in} = 12.5 V; N_{in} = 50 turns; N_{out} = 100 turns

Figure 8.52 Question 37.

38. Six blackbody radiators have temperatures and emitted powers as given. Rank these blackbodies according to their emitting surface areas from smallest to largest. Justify your rankings based on your knowledge of the physics of blackbodies.
Blackbody A: P = 6000 W; T = 50 K
Blackbody B: P = 6000 W; T = 10 K
Blackbody C: P = 20,000 W; T = 30 K
Blackbody D: P = 50,000 W; T = 50 K
Blackbody E: P = 10,000 W; T = 10 K
Blackbody F: P = 50,000 W; T = 50 K

PROBLEMS

1. The charger cord used to recharge a cell phone contains a transformer that reduces 120 V AC to 5 V AC. For each 1,000 turns in the input coil, how many turns are there in the output coil?

2. The generator at a power plant produces AC at 24,000 V. A transformer steps this up to 345,000 V for transmission over power lines. If there are 2,000 turns of wire in the input coil of the transformer, how many turns must there be in the output coil?

3. Compute the wavelength of the carrier wave of your favorite radio station.

4. What is the wavelength of the 60,000-Hz radio wave used by "radio-controlled" clocks and wristwatches?

5. Compute the frequency of an EM wave with a wavelength of 1 in. (0.0254 m).

6. The wavelength of an electromagnetic wave is measured to be 600 m.
(a) What is the frequency of the wave?
(b) What type of EM wave is it?

7. Determine the range of wavelengths in the UV radiation band.

8. A piece of iron is heated with a torch to a temperature of 900 K. How much more energy does it emit as blackbody radiation at 900 K than it does at room temperature, 300 K?

9. The filament of a light bulb goes from a temperature of about 300 K up to about 3,000 K when it is turned on.

How many times more radiant energy does it emit when it is on than when it is off?

10. A rectangular metal plate measures 0.15 m long and 0.1 m wide. The plate is heated to a temperature of 1,200 K by passing a current through it. Assuming that it behaves like a blackbody, how much power does the plate radiate under these conditions?

11. A copper cylinder with a temperature of 450 K is found to emit 175 J of energy each second. If the cylinder radiates like a blackbody, what is its surface area?

12. What is the wavelength of the peak of the blackbody radiation curve for the human body (T = 310 K)? What type of EM wave is this?

13. What wavelength EM radiation would be emitted most strongly by matter at the temperature of the core of a nuclear explosion, about 10,000,000 K? What type of wave is this?

14. What is the frequency of the EM wave emitted most strongly by a glowing element on a stove with temperature 1,500 K?

15. The blackbody radiation emitted from a furnace peaks at a wavelength of 1.2×10^{-6} m (0.0000012 m). What is the temperature inside the furnace?

16. What is the lowest temperature that will cause a blackbody to emit radiation that peaks in the infrared?

1. Earth's magnetic field lines are not parallel to its surface except in certain places: they actually "dip" downward at some angle to the ground. (An ordinary compass does not show this.)
   (a) At what places on Earth would the "magnetic dip" be the greatest?
   (b) Where on Earth would the dip be zero?

2. A *cyclometer* is a device mounted on a bicycle that records and displays trip information such as elapsed time, distance traveled, and average speed. Modern versions employ small magnets attached to a spoke on either the front or rear wheel and a sensor located on the front fork or rear frame. As the magnet is carried past the detector once each wheel revolution, it produces an electrical pulse that is recorded and logged by an integrated circuit in the device. The distance traveled equals the total number of revolutions multiplied by the circumference of the wheel.
   (a) A typical mountain bike has a wheel diameter of 26 in. How far will a cyclist have traveled in both miles and kilometers if her cyclometer indicates that her front wheel has made 25,000 revolutions during her trip?
   (b) If the cyclometer clock shows that the elapsed time for the trip was 2 hours and 40 minutes, what was the rider's average speed during this period?

3. A solenoid connected to a 60-Hz AC source will produce an oscillating magnetic field, as we have seen. If a permanent magnet is inserted into the solenoid, it will oscillate, but *not* with the same frequency that an unmagnetized piece of iron would. Why? (This is why the "hum" or "buzz" from electrical devices is sometimes 60-Hz sound and sometimes 120-Hz sound.)

4. The *right-hand rule* is a way to determine the direction of the magnetic field produced by moving charges. Imagine wrapping your right hand around the path of the charges so that the positive charges (or the current) flow from the little finger side of your fist to the thumb side (**Figure 8.53**). Then your fingers circle the path in the same direction as the magnetic field lines. Use this rule to verify the directions of the magnetic fields shown in Figures 8.8 and 8.10. How would you use the rule to find the direction of the magnetic field lines around a moving negative charge?

5. Sketch a diagram of an atom with one electron orbiting the nucleus. Use the right-hand rule (see Challenge 4) to determine the direction of the magnetic field produced by the electron.

6. If a coil of wire is connected to a very sensitive ammeter and then waved about in the air, a current will be induced in it even if there are no magnets around. Why?

7. Even though the output voltage of a transformer can be much larger than the input voltage, the power output is nearly the same as the power input. (There is some energy loss from ohmic heating.) Use this to determine the relationship between the input and output currents and the number of turns in the input and output coils.

8. The highest frequency sound that can be recorded by a tape recorder depends on the size of the gap in the recording head. Why? Would a wider or a narrower gap be capable of recording higher frequencies?

9. Describe what happens to an electric charge as an electromagnetic wave passes through the region around it. Explain why the charge will produce another electromagnetic wave. (This process occurs in the ionosphere.)

10. Would an electromagnetic pump be able to move a liquid superconductor? Why or why not?

11. The nuclei of carbon atoms that are found in Nature come in two main "varieties" called *isotopes* (more on this in Chapter 11). The carbon-14 nuclei have a greater mass than the carbon-12 nuclei. The two types of atoms can be separated from each other by ionizing them, accelerating them all to some uniform velocity, and then passing them between the poles of a magnet. Why does this separate the two isotopes? Describe the motion of the two ion types after passing out of the magnetic field. What would happen if all the atoms did not have the same speed?

Figure 8.53 Challenge 4.

9

OPTICS

Figure CO-9 Sundogs.

CHAPTER INTRODUCTION: Doggone It!

When the "dog days" of summer give way to the "three dog nights" of winter, inhabitants of colder climates are often treated to spectacular apparitions like that shown in the accompanying photo (**Figure CO-9**). On brisk days when the Sun is near the horizon and the sky is filled with thin clouds of ice crystals, one often sees bright patches of light flanking the Sun at an angular distance of about 22°. These patches frequently exhibit rainbow-like coloration and are variously called *mock suns, parhelia,* or *sundogs.* Next to rainbows themselves, sundogs are among the most common atmospheric optics effects seen at midlatitudes and above.

This phenomenon, like that of rainbows and halos, touches us with its beauty, but it also stirs our intellectual curiosity about what naturally occurring circumstances conspire to produce such an exquisite display. What are the underlying physical principles governing the processes that lead to such striking visualizations? How, if at all, might they be related to other optical phenomena often seen in the sky such as rainbows?

The wonder and fascination afforded to us by Nature in the form of sundogs can be enhanced by finding answers to these questions. A primary ingredient in the mix that provides the answers we seek is the law of refraction.

In this chapter, we will investigate the law of refraction together with its counterpart, the law of reflection, to discover that they comprise two important elements in the study of light and its interaction with matter—the field of *optics.* These laws also provide the basis for many practical devices, such as cameras, telescopes, and liquid-crystal displays, as well as for many of the most striking natural phenomena, such as rainbows, soap-bubble iridescence, and, of course, sundogs.

9.1 Light Waves

Light generally refers to the narrow band of electromagnetic (EM) waves that can be seen by human beings. These are transverse waves with frequencies from about 4×10^{14} hertz to 7.5×10^{14} hertz (see Table 8.1 in Section 8.5). The corresponding wavelengths are so small that we will find it useful to express them in nanometers (nm). One *nanometer* is one-billionth of a meter:

$$1 \text{ nanometer} = 10^{-9} \text{ meter} = 0.000000001 \text{ meter}$$

$$1 \text{ nm} = 10^{-9} \text{ m}$$

The wavelengths of visible light (in a vacuum or in air) thus range from about 750 nanometers for low-frequency red light to about 400 nanometers for high-frequency violet light. Keep in mind two important points from Section 8.5: (1) different frequencies of light are perceived as different colors, and (2) white light is typically a combination of all frequencies in the visible spectrum.

The various properties of waves described in the first part of Chapter 6 also apply to light. (You may find it useful to review Sections 6.1 and 6.2 at this time.) As with sound and water ripples, we will use both wave fronts and rays to represent light waves. Recall that a wave front shows the location in space of one particular phase (peak or valley, for example) of the wave. For a lightbulb or other spherical light source, the wave fronts are spherical shells (not unlike balloons) expanding outward at

Figure 9.1 The light from a lightbulb represented (a) by wave fronts and (b) by light rays.

the speed of light (**Figure 9.1**). A *light ray* is a line drawn in space representing a "pencil" of light that is part of a larger beam. Rays are represented as arrows and indicate the direction the light is traveling. A laser beam can often be thought of as a single light ray. The light from a lightbulb can be represented by light rays radiating outward in all directions. (Be careful not to confuse these light rays with the electric and magnetic field lines discussed in previous chapters.)

Some of the general characteristics of wave propagation, such as reflection, are readily observed with light waves. But other phenomena are more rare in everyday experience because of two factors:

1. The speed of light is extremely high (3×10^8 m/s in a vacuum).

2. The wavelengths of light are extremely short.

For example, we must turn to distant galaxies moving away from us at high speeds to easily observe the Doppler effect with light (see the special feature, "The Hubble Relation—Expanding Our Horizons," at the end of Section 6.2). As we saw in Section 6.2, the Doppler effect with sound, on the other hand, is quite common because the speed of sound is only about 350 m/s and the wavelengths of sound (in air) are in the centimeter to meter range.

In these first two sections, we will describe some of the phenomena that can occur when light encounters matter. The remaining sections of the chapter deal with important things that occur after light has traveled inside transparent material.

9.1a Reflection

Reflection of light waves is extremely common: except for light sources such as the Sun and lightbulbs, everything we see is reflecting light to our eyes. There are two types of reflection: specular and diffuse.

Specular reflection is the familiar type that we see in a mirror or in the surface of a calm pool of water. A mirror is a very smooth, shiny surface, usually made by coating glass with a thin layer of aluminum or silver. Specular reflection occurs when the direction the light wave is traveling changes (**Figure 9.2**).

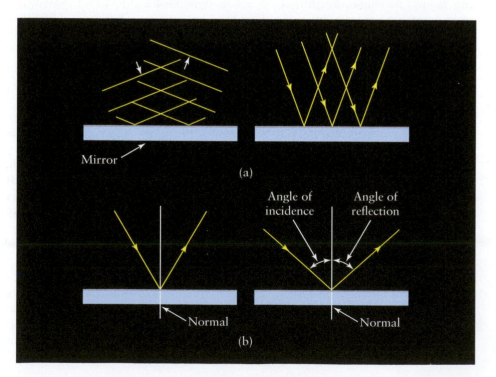

Figure 9.2 (a) Specular reflection using wave fronts and light rays. (b) Specular reflection of a single light ray with different angles of incidence.

By changing the angle of the incident (incoming) light ray and observing the reflected ray, we see that the light behaves somewhat like a billiard ball bouncing off a cushion on a pool table.

Figure 9.2b shows an imaginary line drawn perpendicular to the mirror and touching it at the point where the incident ray strikes it. This line is called the **normal**. The angle between the incident ray and the normal is called the **angle of incidence**, and the angle between the reflected ray and the normal is called the **angle of reflection**. Our observations indicate that these angles are always equal. The following law, first described in a book titled *Catoptrics* and thought to have been written by Euclid in the third century BCE, states this formally.

LAWS	**Law of Reflection** The angle of incidence equals the angle of reflection.

So specular reflection of light is much like sound echoing off a cliff, as described in Chapter 6.

Don Bord

> ### ▷ Physics To Go 9.1
>
> Stand one edge of a small plane mirror on a flat, horizontal surface (a piece of cardboard or foam board will do nicely) so that its reflecting side is upright (vertical). Position four straight pins so that they form an upside-down "V" with its "corner" approximately in the middle of the mirror (**Figure 9.3**). Look into the mirror and move your head until you can align the pair of pins on one side with the images of the other pair in the mirror. Devise a way to use this arrangement to verify the law of reflection. In other words, how might you show that the angle of incidence equals the angle of reflection by this demonstration?

Figure 9.3 By placing a pair of pins in front of a plane mirror and then aligning their images with another pair of pins, you can easily convince yourself of the validity of the law of reflection.

The other type of reflection, **diffuse reflection**, occurs when light strikes a surface that is not smooth and polished but uneven like the bottom of an aluminum pan or the surface of a piece of paper. The light rays reflect off the random bumps and nicks in the surface and scatter in all directions (**Figure 9.4**). The law of reflection still applies, but the rays encounter segments of the irregular surface oriented at different angles and therefore leave the surface with different directions. That is why you can shine a flashlight on the aluminum and see the reflected light from different angles around the pan. With specular reflection from a mirror, you could see the reflected light from only one direction. (Try it!)

Except for light sources and smooth, shiny surfaces such as mirrors, every object we see is reflecting light diffusely. This diffuse reflection causes light to radiate outward from each point on a surface. You can see every point on your hand as you turn it in front of your face because each point on your skin is reflecting light in all directions.

Things can have color because light actually penetrates into the material and is partially reflected and partially absorbed along its way into and out of the material. The reflected light that leaves the surface will have color if pigments in the material absorb some frequencies (colors) more efficiently than others. A white surface reflects all frequencies of light nearly uniformly. If you shine just red light on it, it will appear red. With just blue light, it will appear blue. A colored surface, like that of a red fire extinguisher, "removes" some frequencies of the light. A red surface reflects the lower-frequency light (red) most effectively and absorbs much of the rest (**Figure 9.5**). If you shine red light on it, it will appear red. With blue or any other single color, it will appear black: very little of the light will be reflected.

Figure 9.4 Diffuse reflection of light from a rough surface.

Figure 9.5 A red surface reflects the red contained in white light much more effectively than it does blue, green, and other colors.

▶ Physics To Go 9.2

Check out your supply of holiday wrapping materials. You're likely to find some transparent, colored-plastic gift wrap in the collection. Common hues include red, green, yellow, and blue. (If you don't have a ready supply of gift wrap, plastic food wrap now comes in several colors and may be purchased at your local grocery store.) Cover a flashlight with one color of plastic, say, red, and a second flashlight with another, say, green. Shine the two beams at the same spot on a white piece of paper. What color do you see? Try the experiment with two filters of other colors. Again, what color results from adding the two beams together? This type of color production is called *additive mixing*. Now look at a brightly lit window through one of the colored sheets, say, the red one. What do you see? Without changing your perspective, interpose a second filter—say, the green one—along your sight line. What do you see now? Is the effect different from what you saw when you combined the two flashlight beams with the same two filters? If so, how has it changed? Can you account for the differences based on your understanding of how color is produced in common objects? Try sandwiching together two of the other colored sheets. Do you find a similar result? This type of color production is called *subtractive mixing* because, typically, all but one color is subtracted (absorbed) from an incident beam of white light, leaving the residual color to be passed to the eye.

9.1b Diffraction

As with all waves, diffraction of light as it passes through a hole or slit is observable only when the width of the opening is not too much larger than the wavelength of the light. This means that light doesn't spread out after passing through a window nearly as much as sound does, but diffraction is observed when a very narrow slit (about the width of a human hair) is used (**Figure 9.6**). The narrower the slit, the more the light spreads out.

9.1c Interference

In Section 6.2, we described how waves can undergo interference. Recall that when two identical waves arrive at the same place, they add together. If the

Figure 9.6 Diffraction of light. (a) Light passing through a narrow slit spreads out as shown on the screen. (b) Photograph of laser light projected onto a screen after passing through a slit 0.008 centimeter wide. The screen was 10 meters from the slit.

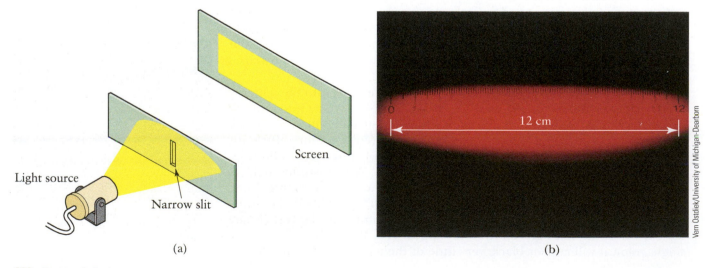

(a)

(b)

two waves are "in phase"—peak matches peak—the resulting amplitude is doubled. This is called **constructive interference**. At any point where the two waves are "out of phase"—peak matches valley—they cancel each other. This is **destructive interference**.

Interference of light waves is an important phenomenon for two reasons. First, in experiments conducted around 1800, British physician Thomas Young used interference to prove that light is indeed a wave. Second, interference is routinely used to measure the wavelength of light. We will consider two types of interference: *two-slit* interference and *thin-film* interference.

When a light wave passes through two narrow slits that are close together, the two waves emerging from the slits diffract outward and overlap. If the light consists of a single frequency (color), a screen placed behind the slits where the two light waves overlap will show a pattern of bright areas alternating with dark areas (**Figure 9.7**). At each bright area, the two waves from the slits are completely in phase and undergo constructive interference. Conversely, at each dark area the two waves are completely out of phase and undergo destructive interference—they cancel each other. There is a bright area at the center of this *interference pattern* because the two waves travel exactly the same distance in getting there, so they are in phase. At the first bright area to the left of center, the wave from the slit on the right has to travel a distance exactly equal to one wavelength farther than the wave from the slit on the left. This puts them in phase as well. Similarly, at each successive bright area on the left side, the wave from the right slit has to travel 2, 3, 4, . . . wavelengths farther than the wave from the left slit. At each bright area on the right side of the pattern, it is the wave from the left slit that has to travel a whole number of wavelengths farther.

At the first dark area to the left of the center of the pattern, the wave from the right slit travels one-half wavelength farther than the wave from the left slit. The two waves are out of phase and interfere destructively. At the next dark area on the left, the additional distance is $1\frac{1}{2}$ wavelengths, then $2\frac{1}{2}$ wavelengths at the next, and so on. True constructive and destructive interference actually occurs only at the centers of the bright and dark areas. At points in between, the waves are neither exactly in phase nor exactly out of phase, so they partially reinforce or partially cancel each other.

The distance between two adjacent bright or dark areas is determined by the distance between the two slits, the distance between the screen and the slits, and the wavelength of the light. Because the first two can be measured easily, their values can be used to compute the wavelength of the light. Specifically, the separation, Δx, between two adjacent bright areas in the interference pattern is determined by the distance, a, between the slits; the distance, S, between the plane of the slits and the screen on which the pattern is projected;

Figure 9.7 Two-slit interference. (a) Light passing through two narrow slits forms an interference pattern on the screen. The bright areas occur at places where the light waves from the two slits arrive in phase. (b) Photograph of an interference pattern formed by laser light passing through two narrow slits 0.025 centimeter apart. The screen was 10 meters from the slits.

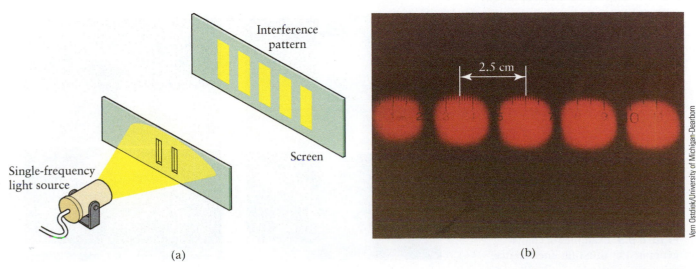

(a)

(b)

Vern Ostdiek/University of Michigan-Dearborn

and the wavelength, λ, of light used in the experiment. In mathematical terms, the relationship among these variables is given by:

$$\Delta x = (S/a)\lambda$$

EXAMPLE 9.1 For the experimental setup used to produce Figure 9.7b, the slit separation, a, is 0.025 cm; the slit-to-screen distance, S, is 10 m; and λ = 633 nm, the wavelength of red light from an He-Ne laser. Verify that the separation between adjacent bright areas in the resulting interference pattern is 2.5 cm.

SOLUTION

$$\Delta x = \left(\frac{S}{a}\right)\lambda$$

$$= \left(\frac{10 \text{ m}}{0.025 \text{ cm}}\right)(633 \text{ nm})$$

$$= \left(\frac{10 \text{ m}}{2.5 \times 10^{-4}\,\text{m}}\right)(633 \times 10^{-9}\,\text{m})$$

$$= (4 \times 10^4)(633 \times 10^{-9}\,\text{m})$$

$$= 2{,}532 \times 10^{-5}\,\text{m} = 2{,}532 \times 10^{-3}\,\text{cm}$$

$$= \boxed{2.53 \text{ cm}}$$

The swirling colors you see in oil or gasoline spills floating on wet pavement are caused by *thin-film interference*. Part of the light striking a thin film of oil is reflected from it, and part passes through to be reflected off the water (**Figure 9.8**). The light wave that passes through the film before being reflected travels a greater distance than the wave that reflects off the upper surface of oil. If the two waves emerge in step, there is constructive interference. If they emerge out of step, there is destructive interference.

The wavelength of the light, the thickness of the film, and the angle at which the light strikes the film combine to determine whether the interference is constructive, destructive, or in between. With single-color (one wavelength) light, one would see bright areas and dark areas at various places on the film. With white light, one sees different colors at different places on the film. At some places, the film thickness and angle of incidence will cause constructive interference for the wavelength of red light, at other places for the wavelength of green light, and so on.

Interference in thin films in hummingbird and peacock feathers is the cause of their iridescent colors. Soap bubbles are also colored by interference of light reflecting off the front and back surfaces of the soap film (**Figure 9.9**).

Figure 9.8 Interference of light striking a thin film of oil. The light reflecting from the upper surface of the film undergoes interference with the light that passes through to, and is reflected by, the lower surface. Part of the light continues on through the lower surface.

Figure 9.9 Examples of colors generated by thin film interference.

G.C. Kelly/Science Source

David Parker/SPL/Science Source

9.1d Polarization

The fact that light could undergo diffraction and interference convinced Young and other scientists of his time that light can behave like a wave. The other model of light elaborated by Newton (see Profiles in Physics at the end of this chapter) held that light is a stream of tiny particles, but this approach could not account for these distinctively wavelike phenomena. **Polarization** reveals that light is a *transverse* wave rather than a longitudinal wave like sound.

A rope secured at one end can be used to demonstrate polarization. If you pull the free end tight and move it up and down, a wave travels on the rope that is *vertically polarized*. Each part of the rope oscillates in a vertical plane (**Figure 9.10**). In a similar manner, moving the free end horizontally produces a *horizontally polarized* wave on the rope. Moving the free end at any other angle with the vertical will also produce a polarized wave. Polarization is possible only with transverse waves.

The fact that light can be polarized reveals its transverse nature. A *Polaroid filter*, like the lenses of Polaroid sunglasses, absorbs light passing through it unless the light is polarized in a particular direction. This direction is coincident with the *transmission axis* of the filter. Light polarized in this direction passes through the Polaroid largely unaffected, light polarized perpendicular to this direction is blocked (absorbed), and light polarized in some direction in between is partially absorbed (**Figure 9.11**). (For simplicity, we will assume that our Polaroid filters are 100-percent efficient in absorbing light polarized in a direction perpendicular to the transmission axis.)

> **▶ Physics To Go 9.3**
>
> Go to the sunglasses display in a department or drug store and choose a polarized pair. Examine an overhead light fixture through one lens of the glasses. Rotate the lens as you look at the light. Does the intensity (brightness) of the source change? Now find a place where bright light is reflected off the top of a display case or a highly polished floor. Again view the light through one lens of the glasses, and rotate the lens through at least 90°. Is there any change in the image brightness now? Can you relate the image intensity to the angle of rotation of the lens? Pick up another pair of polarized sunglasses. Arrange them so that overhead light passes through the right lens of one and the left lens of the other before reaching your eye. Rotate one pair with respect to the other. What happens? Can you relate what you see now to what you saw with the reflected light?

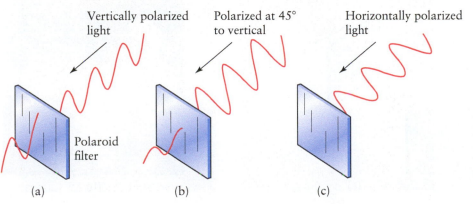

Figure 9.11 Light polarized in different directions encountering a Polaroid filter with its transmission axis vertical. The lines are not actually visible in a Polaroid filter but are drawn here to show the direction of the filter's transmission axis.

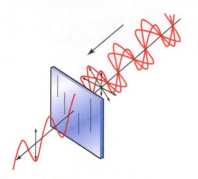

Figure 9.12 Unpolarized light being vertically polarized by a Polaroid filter. The star-shaped cluster of arrows represents a combination of light waves with all different directions of polarization. The filter blocks all but the vertically polarized components in the light.

The light that we get directly from the Sun and from ordinary light fixtures is a mixture of light waves polarized in all different directions. The light is said to be "natural" or "unpolarized" because it has no preferred plane of vibration. When natural light encounters a Polaroid filter, it emerges polarized along the transmission axis (**Figure 9.12**). The filter allows only that portion of the incident light that oscillates along this direction to pass through; the rest of the radiation is absorbed.

Now, if the emergent light encounters a second Polaroid filter, the amount of light that passes through will depend on the orientation of the transmission axis of the second filter. If the axis of the second filter is aligned with that of the first, then the light will continue on unimpeded. If the axis of the second is perpendicular to that of the first, then all of the light will be blocked by the second filter. This is referred to as *crossed Polaroids* (**Figure 9.13**). When the angle between the transmission axes of the two filters is other than zero or 90°, some of the light will pass through, with the intensity becoming progressively less for angles closer to 90°.

Polaroid sunglasses are very useful because light is polarized to some extent when it reflects off a smooth surface such as that of water, asphalt, or the paint on the hood of a car. In particular, reflected sunlight is partially polarized horizontally (**Figure 9.14**). This reflected light, called *glare*, is usually bright and annoying. Sunglasses using Polaroid lenses with their transmission axes vertical will block most of this reflected light, which makes it easier to see the surface itself.

> **Physics To Go 9.4**

Grab those polarizing sunglasses again. This time examine a liquid-crystal display (LCD) on a watch or a calculator or the gas pump the next time you fill 'er up. Rotate the polarizing lens as you did in the previous Physics to Go exercise. Describe what happens. What does this tell you about the light emanating from the LCD? Try the same experiment with light from the sky on a clear sunny day. When the Sun is near the horizon, look through the polarizing sunglasses at the blue sky in directions north or south (approximately 90° away from the Sun). Rotate the glasses as before. What do you see? Can you draw any conclusions about the nature of the scattered sunlight in these directions?

LCDs used in calculators, digital watches, laptop computers, and video games also use polarization. The liquid crystal is sandwiched between crossed Polaroids, and this assembly is placed in front of a mirror. Without the liquid crystal present, the display would be dark: light passing through the first

Figure 9.13 Crossed Polaroids. (a) Light emerging from the first Polaroid is polarized vertically. The second Polaroid, with its transmission axis horizontal, blocks the light. (b) A photograph of crossed Polaroid sheets demonstrating the effect visually.

(a)

(b)

No polaroid · Polaroid lens

GIPhotoStock/Science Source · GIPhotoStock/Science Source

(a) · (b)

Figure 9.14 (a) Sunlight reflected from the windshield and the hood of a car is partially polarized horizontally. (b) Polaroid sunglasses with vertical transmission axes block this glare.

Polaroid is polarized vertically and would be totally absorbed when it reached the second Polaroid. No light would reach the mirror, so none would be reflected back from the display. The specially chosen liquid-crystal material between the Polaroids is arranged so that in its normal state its molecules change the polarization of the light passing through it. The liquid crystal twists the polarized light 90°. The polarization is changed from vertical to horizontal for light passing through from the front and from horizontal to vertical for light passing through from the rear. The light that was vertically polarized by the first Polaroid is made horizontally polarized by the liquid crystal, passes through the second Polaroid, is reflected back through the display, and emerges polarized vertically (**Figure 9.15a**).

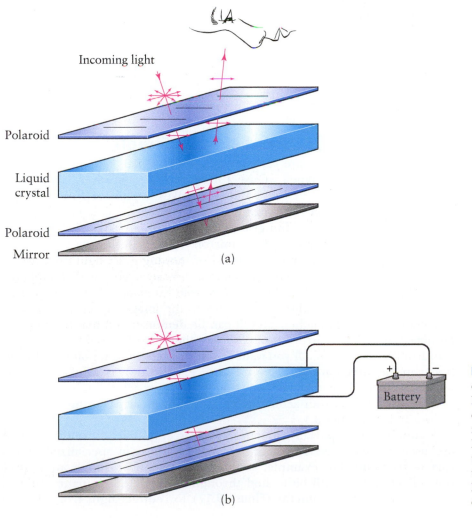

Figure 9.15 Expanded view of part of a liquid-crystal display. (a) The liquid crystal changes the polarization of the light so that it can make a round-trip through the crossed Polaroids. (b) The voltage neutralizes this part of the liquid crystal so the light is blocked by the second Polaroid. This part of the display is darkened.

To make images on the display, segments of it are darkened. Here is where the liquid crystal property is used. An electric field is switched on in the parts of the display to be darkened. This causes the molecules of the liquid crystal to rotate and become aligned in one direction. As a result, they no longer change the polarization of light passing through. Light going through this segment is absorbed by the second Polaroid, and that part of the display is dark (**Figure 9.15b**). By the way, the transmission axis of the first Polaroid is commonly oriented vertically so that you can see the display while wearing Polaroid sunglasses. If you rotate an LCD—say, like that on a digital watch—and look at it while wearing Polaroid sunglasses, at one point all of it will be dark.

It has been discovered that some color-blind animals such as squid can sense the polarization of light. This helps squid see plankton, which are transparent. Foraging honey bees use their ability to detect polarized light to help orient their *waggle dances*, which are used to communicate the locations of pollen sources to other bees. Even the human eye is weakly sensitive to polarized light, although to detect polarization without the use of intervening filters takes practice.

Learning Check

1. (True or False.) According to the law of reflection, if the angle of incidence of a ray striking a mirrored surface is 35°, then the angle of reflection of the ray leaving the surface is 55°.
2. The two types of reflection are _____ and _____.
3. Two otherwise identical light waves arriving at the same point with peak matching valley undergo
   (a) constructive interference.
   (b) destructive interference.

   (c) polarization.
   (d) reflection.
   (e) None of the above.
4. Light is totally absorbed when sent through two Polaroid filters if their respective transmission axes are _____.

ANSWERS: 1. False **2.** specular, diffuse **3.** (b) **4.** perpendicular.

Mirror

Image

Figure 9.16 Image formed by a plane mirror. The rays reaching the viewer's eyes form an image that appears to be on the other side of the mirror. (The normal for the lower ray is shown as a dashed line to illustrate that the law of reflection applies.)

9.2 Mirrors: Plane and Not So Simple

Most mirrors that we use are *plane* mirrors: they are flat, smooth, almost perfect reflectors of light. When we use a mirror to "see ourselves," light that is diffusely reflected off our clothes and face strikes the mirror and undergoes specular reflection. Some of the rays leaving the mirror are going in the proper direction to enter our eyes and give us an image of ourselves. The image appears to be on the other side of the mirror. **Figure 9.16** shows a person viewing her image in a plane mirror. Instead of showing every light ray traveling outward from every point on the person, we show selected rays that happen to enter the person's eyes. The dashed lines from the image show the apparent paths taken by the rays when traced back to the image. Applying the law of reflection and a little mathematics, it can be demonstrated that the image of an object in a plane mirror is as far behind the mirror's surface as the object is in front of the mirror. (If you want to test this conclusion, point your index finger in the direction of a mirrored surface, and slowly move your hand toward the mirror. The image of your finger will approach the mirror's surface at the same rate that your actual finger does and will arrive at the surface just as your finger touches it.)

A number of important, practical devices employ plane mirrors. They are used as *optical levers* to amplify small rotations in specialized laboratory instruments. For example, as the mirror rotates through an angle θ, the reflected beam will be turned through an angle 2θ. A plane mirror is used in many reflex cameras (**Figure 9.17**) to redirect light from the lens

Penta prism
viewfinder

Incoming
light rays

Camera lens

Film or electronic detector

Flip mirror

Figure 9.17 A typical reflex camera design. Light enters through the lens, strikes the mirror, travels to the prism and then to the photographer's eye. When the shutter is released, the mirror pops up, allowing light to pass directly to the film or electronic detector. The mirror then drops down after the exposure.

to the viewfinder. When the shutter is pressed, the mirror tilts up, allowing the light to reach the film. More recently, *micromirrors*, small enough to fit through the eye of needle (0.5 mm or less), have become indispensable parts of modern telecommunications networks, where they are used in high-speed optical switches to control the flow of information through optical fibers (see Section 9.3).

❯ Physics To Go 9.5

Hold up your right hand in front of a plane mirror. What do you see? (Be cautious now! Describe the image carefully.) If you're having trouble, place a tube of toothpaste or other object with writing on it in front of the mirror. Now what can you say? The process you're witnessing is called *inversion* and is one of the common characteristics of plane mirrors.

9.2a *"One-Way Mirror"*

A "one-way mirror" is made by partially coating glass so that it reflects some of the light and allows the rest to pass through. This is called a *half-silvered mirror* (**Figure 9.18**). When used as a window or wall between two rooms, it will function as a one-way mirror if one of the rooms is brightly lit and the other is dim. It will appear to be an ordinary mirror to anyone in the bright room, but it will appear to be a window to anyone in the dim room. This is because, in the bright room, the light reflected off the half-silvered mirror is much more intense than the light that passes through from the other room. In the dim room, the transmitted light from the bright room dominates (**Figure 9.19**). A person in the dim room can see what is happening in the bright room without being seen by anyone in the bright room.

This device is often used in interview and interrogation rooms (**Figure 9.20**) and as a means of observing customers in stores and gambling casinos. Note that if a bright light is turned on in the dimmer room, the one-way effect is destroyed. Ordinary window glass is a crude one-way mirror because it does reflect some of the light (about 4 percent) that strikes it. At night, one can see into a brightly lit room through a window, but anyone in the room has difficulty seeing out because room light is reflected by the windowpane.

Figure 9.18 Light striking a half-silvered mirror. Part of the light is reflected, and part passes through.

Figure 9.19 A half-silvered mirror used as a one-way mirror between two rooms. In the room on the left, a person sees mostly reflected light (a mirror). In the room on the right, a person sees mostly transmitted light (a window).

Darkened room

> ## Physics To Go 9.6

Stand a small, rectangular plane mirror vertically on a table. Now place another similar plane mirror adjacent to the first with one of its vertical edges running alongside that of the other. Adjust the angle between the mirrors to be about 45°. Place a small object (a coin or a die will do nicely) midway between the mirrors. How many images of the object do you see reflected in the mirrors? Make the angle between the mirrors roughly 60°. How many reflected images do you see now? Set the mirror angle to 30°. Count the number of images in this case. Do you see a pattern developing between the total number of objects (actual plus images) arrayed around the cluster and the angle separating the mirrors? How do you think your observations might be applied to the construction of a kaleidoscope?

9.2b Curved Mirrors

As we saw in Section 6.2, reflectors—mirrors in this case—that are curved have useful properties. Parallel light rays that reflect off a properly shaped *concave mirror*—a mirror that is curved inward—are focused at a point called the **focal point** (Figure 9.21a). The energy in the light is concentrated at that point. Sunlight focused by a concave mirror can heat things to very high temperatures (Figure 9.21b). Even when a mirror's surface is curved, the law of reflection still holds at each point that a ray strikes the mirror. If a normal line is drawn at each point (as was done in Figure 9.2b), the angle of incidence equals the angle of reflection. One such normal line is shown in Figure 9.21a.

Figure 9.20 (Left) In a well-lit room, the half-silvered mirror appears to be an ordinary mirror. (Right) In a dark room on the other side, the half-silvered mirror appears to be a window.

Charles D. Winters

Charles D. Winters

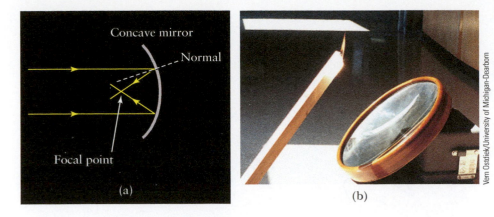

Figure 9.21 (a) Parallel rays reflecting off a concave mirror converge to a point called the *focal point*. (The normal for the upper ray shows that the law of reflection applies.) (b) Sunlight focused by a concave mirror can generate high temperatures. This piece of wood was in flames a few seconds after being placed at the mirror's focal point.

A concave mirror can be used to form images that are enlarged—magnified. Magnifying makeup and shaving mirrors are concave mirrors, as are the large mirrors used in astronomical telescopes. **Figure 9.22b** shows a magnified image seen in a concave mirror.

A *convex mirror* is one that is curved outward. The image formed by a convex mirror is *reduced*—it is smaller than the image formed by a plane mirror (**Figure 9.22c**). The advantage of a convex mirror is that it has a wide field of view—images of things spread over a wide area can be viewed in it. **Figure 9.23** shows the fields of view for a convex mirror and a plane mirror of the same size. One glance at a well-placed convex mirror on a bike path allows quick surveillance of a large area (**Figure 9.24**). Passenger-side rearview mirrors on cars and auxiliary "wide-angle" rearview mirrors on trucks and other vehicles are convex so the driver can view a large region to the rear. Care must be taken when using such a mirror because the reduced image makes any object appear to be farther away than it actually is.

> ▶ Physics To Go 9.7
>
> A shiny metal spoon can serve as either a concave mirror or a convex mirror, although its shape is usually far from ideal. The front side of the spoon is a concave mirror. Place your finger close to the spoon and look at its image in the spoon. Describe what you see. Now hold the same side of the spoon a comfortable distance in front of your face. What is different about the image of your face that you see? The back side of a spoon is a convex mirror. Describe the image of your face and surrounding environment seen in this side of the spoon.

Figure 9.22 (a) Plane mirror image (no magnification). (b) Enlarged image in a concave mirror. (c) Reduced image in a convex mirror. (Note the wide field of view of this mirror that allows the image of the photographer to be seen.)

(a)

(b)

(c)

Figure 9.23 The field of view of a convex mirror is much larger than that of a plane mirror.

Figure 9.24 This convex mirror on a bike path at the University of Minnesota allows quick surveillance of a large area that would normally be hidden from view.

Vern Ostdiek/University of Michigan-Dearborn

9.2c **Astronomical Telescope Mirrors**

The largest telescopes used by astronomers to examine stars, galaxies, and other celestial objects make use of curved mirrors. **Figure 9.25** shows a common design for such telescopes. Light from the distant source enters the telescope and reflects off a large concave mirror called the *primary mirror*. The reflected rays converge onto a much smaller convex mirror called the *secondary mirror*. The rays are reflected back toward the primary mirror, pass through a hole in its center, and converge to form an image at the focal point *F*. The primary mirror is the key component of the telescope.

Telescope mirrors have as their basic functions the gathering of light and the concentration of that light to a point. The ability of a mirror to collect light increases with its surface area. To acquire enough radiation to study faint objects adequately, astronomers have sought to build instruments with larger and larger apertures (openings) and, hence, larger light-collecting areas.

The quality of the images produced by telescopes is greatly affected by the shapes of the mirrors. The easiest curved mirror to make is one that has a surface in the shape of portion of a sphere. But such a *spherical mirror* is not perfect for the task of focusing light rays. **Figure 9.26a** shows that parallel light rays reflecting off a spherical mirror are not all focused at the same point. An image formed using such a mirror will be somewhat blurred. This phenomenon is called **spherical aberration**. We will see in Section 9.4 that the same thing happens with lenses.

As the name implies, spherical aberration is a defect associated with spherical surfaces. A concave mirror in the shape of a parabola does not have this aberration. (You may recall that we saw the parabola in Section 2.7.) A *parabolic mirror* will concentrate all the rays coming from a distant source at the same point (**Figure 9.26b**). Thus, the ideal surface for a telescope mirror (or for that matter, reflectors in auto headlamps and household flashlights)

Figure 9.25 This is a basic design of a large astronomical telescope. The large, concave primary mirror and the small, convex secondary mirror combine to focus incoming light at the focal point *F*.

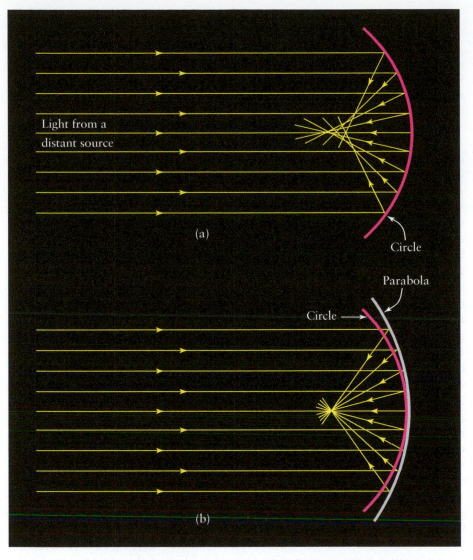

Light from a
distant source

(a)

Circle

Parabola

Circle →

(b)

Figure 9.26 (a) A spherical mirror does not reflect all incoming rays from a distant source to the same point. Rays located well off the symmetry axis of the mirror are brought to a focus closer to the mirror than those rays nearer the axis. This effect is called *spherical aberration*. (b) A parabolic mirror focuses all the light from a distant source to a single point. It is the ideal shape for a telescope mirror.

is one shaped like a parabola. Fabricating very large mirrors, some as big as 10 meters (33 feet) in diameter, with the precise parabolic shape is an enormous technical challenge. One technique called *spin casting* exploits the fact that the surface of a liquid rotating at a steady rate has the required parabolic shape.

> ### ◗ Physics To Go 9.8
>
> If you have a phonograph turntable, you can make your own spinning parabolic mirror. Place a lightweight soup bowl—dark in color if possible—on the turntable so that *its center matches that of the platter.* (You may have to remove the short spindle from the center or place a couple of useless records on first so that the spindle doesn't hold up the center of the bowl.) Carefully pour cooking oil into the bowl until it is about half full (water works, but it sloshes a lot). Turn on the turntable, and use the $33\frac{1}{3}$ revolutions per minute speed. If there is enough light in the room, place your hand about a foot above the bowl and to one side. Place your head above the bowl and off to the other side, and look for the image of your hand in the oil's surface. How does it appear? If the room is not well lit, replace your hand with a dim lamp or a low-wattage frosted lightbulb. (Be *very careful* when using a lamp, because its magnified image can be very bright.) Move your hand or the lamp up and down. How does the image change? Turn the speed up to 45 revolutions per minute. What is different? Watch the image as you turn off the turntable and it slows to a stop. Describe what you see.

Since the 1980s, several telescopes have been constructed that employ rotating liquid mirrors, the most ambitious being the 6-meter Large Zenith Telescope (LZT) in British Columbia. In the simplest designs, a large bowl of mercury is spun at the proper rate to produce a surface with the desired parabolic shape. Although such liquid mirror telescopes can't be tilted and can only examine the sky nearly directly above them, they are relatively cheap to build: the 6-meter LZT cost only about $1 million, 100 times less than a comparable conventional glass mirror telescope.

Shortly after the Hubble Space Telescope (HST; Figure 8.46) was placed in orbit on 25 April 1990, scientists discovered that its primary mirror was afflicted with a type of spherical aberration. At the edge of the 2.4-meter-diameter mirror, its surface is misshapened by 0.002 millimeters from what it is supposed to be. This seemingly minuscule error drastically reduced the telescope's ability to form sharp images. In December 1993, space shuttle astronauts installed corrective optics on the instrument platform of the Space Telescope to alleviate this problem and allow the observatory to perform as designed. **Figure 9.27** shows the dramatic improvement in the ability of the Space Telescope to resolve fine detail as a result of the repairs. Since this first service mission, four others have been launched, the latest in May 2009, all designed to improve the optical performance and extend the life of the HST. For example, in this latest—and last—service mission, a new wide-field camera (WFC3) and cosmic origins spectrograph (COS) were installed. The COS instrument operates primarily in the UV portion of the EM spectrum and increases the sensitivity of HST in this wavelength region by more than a factor of 10. Mission scientists also made important repairs on two existing instruments: the Advanced Camera for Surveys (ACS) and the Space Telescope Imaging Spectrograph (STIS), restoring them to full functionality. In addition, all of Hubble's 18-year-old batteries were replaced and six new gyroscopes and a new fine guidance system were installed. To make room for the new instrumentation, two older units were removed, including the corrective optics package put into service in 1993. This instrument was no longer needed because all recently added scientific equipment has contained optical components that compensate for the originally flawed primary mirror.

From the ground, recent efforts to combat the blurring effects of Earth's atmosphere (see the discussion in Section 8.7d) and thereby to increase the resolution of optical instruments, have involved the use of "adaptive optics" (AO). Here, elaborate wave-front sensors, fast computers, and deformable mirrors are used to produce images so sharp that it's as though the atmosphere had disappeared entirely. **Figure 9.28a** shows a typical design of an AO system, similar to those in use with the two 10-m Keck Telescopes in Hawaii, the 8.2-m Gemini North (Hawaii) and South (Chile) instruments, and the Very Large Telescope array (Chile). The key optical element is the "rubber mirror"—a thin glass mirror that has a surface that can be slightly deformed by up to a hundred

Figure 9.27 Hubble Space Telescope images of the core of the galaxy M100 taken before (right) and after (left) the installation of corrective optics in December 1993.

(a)

(b)

Figure 9.28 (a) An adaptive optics (AO) system. The distorted wave fronts Σ_1 are analyzed and reshaped by the use of a deformable "rubber" mirror. The corrected wave fronts Σ_2 are transmitted to the scientific instruments and to a wave-front sensor that provides feedback for controlling the deformable mirror. (b) Robo-AO in operation at the Palomar Observatory 1.5-m telescope. The laser beam is not visible to the human eye; a camera sensitive to ultraviolet light was used to capture this image. (Image credit: Christoph Baranec/University of Hawaii.)

tiny actuators attached to the back. If the tiny deformations in the rubber mirror can be made to counteract the deformations in the wave fronts from distant sources caused by turbulence in Earth's atmosphere, then the original wave fronts can be restored and the blurring in the image removed. Fast computers are required to monitor continuously the rapidly changing conditions in the atmosphere, compute the necessary corrections, and signal the accuators to move appropriately.

This technique normally uses a bright star in the field of view of the telescope as the reference beacon for the wave-front assessment. In the absence of such a star, an artificial one can be created using a high-power, highly focused laser beam projected up through the telescope. Called "laser guidestars," the light from the beam that is scattered downward either from air molecules at altitudes of 10 to 40 km or from sodium atoms residing as high as 90 km produces a faint but measurable target for the wave-front sensors.

In 2011, an autonomous laser guidestar AO instrument called *Robo-AO* was installed on the 1.5-m telescope at the Palomar Observatory (**Figure 9.28b**). This device robotically operates the telescope, laser, AO system, and science camera to observe a variety of different classes of astronomical objects. In its first three years of operation, Robo-AO logged more than 18,000 observations with much higher efficiency than any other AO system currently in operation.

As indispensable as AO systems are for optimizing the light-gathering and resolving power of large, 10-m class telescopes, still greater advances will be needed in this field to harness the full capabilities of the *really* large telescopes with 20-m apertures or larger planned for the future, like the 24.5-m Giant Magellan Telescope under construction at the Las Campanas Observatory in Chile, the Thirty Meter Telescope now being built at the Mauna Kea Observatory in Hawaii, and the nearly 40-m European Extremely Large Telescope that has broken ground at the Cerro Armazones Observatory also in Chile. To meet the challenges, new initiatives described as "atmospheric tomography"—similar to "medical tomography" where a 3-D view of a patient is produced—will be required.

Concept Map 9.1 summarizes the types and uses of mirrors.

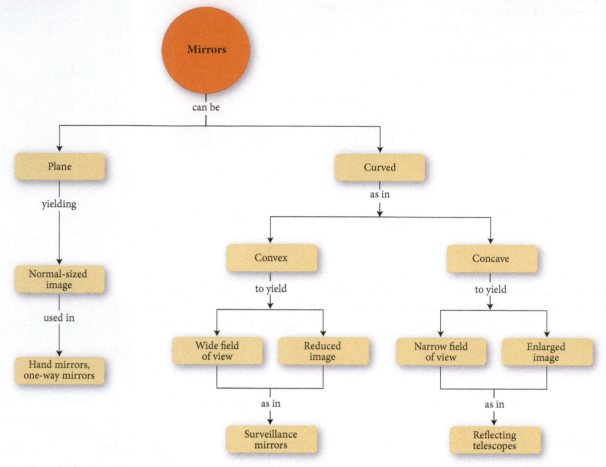

■ Learning Check

1. An object is placed 20 cm in front of a plane mirror. The image of the object formed by the mirror is located
   (a) 10 cm in front of the mirror.
   (b) 10 cm behind the mirror's surface.
   (c) at the mirror's surface.
   (d) 20 cm in front of the mirror.
   (e) 20 cm behind the mirror's surface.

2. Parallel light rays from a distant object strike the surface of a concave mirror. After leaving the mirror's surface, the rays will converge toward the _____ of the mirror.

3. (True or False.) A curved mirror that produces an enlarged image will also have a greater field of view than a plane mirror.

4. The primary mirror in an astronomical telescope must be a
   (a) half-silvered mirror.
   (b) convex mirror.
   (c) concave mirror.
   (d) plane mirror.
   (e) hexagonal mirror.

ANSWERS: 1. (e) **2.** focal point **3.** False **4.** (c)

9.3 Refraction

In this section, we describe what happens when light interacts with glass and other transparent material. Imagine a light ray from some source in air arriving at the surface of a second transparent substance like glass. The boundary between the air and glass is called the *interface*. As mentioned earlier, some of the incident light is reflected back into the air, but the rest of it is transmitted across the interface into the glass (**Figure 9.29a**). The law of reflection gives us the direction of the reflected light ray, but what about the transmitted light ray?

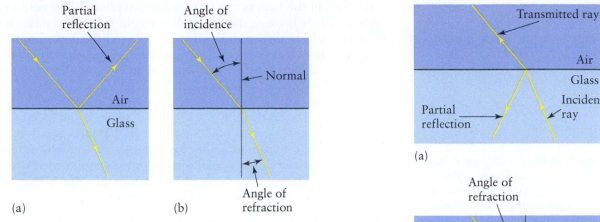

Figure 9.29 (a) Refraction of light as it enters glass, with partial reflection shown. (b) The normal and angles of incidence and refraction. Note that the ray is bent toward the normal.

9.3a The Law of Refraction

The light that passes into the glass is *refracted*—the transmitted ray is bent into a different direction than the incident ray. (Only if the incident ray is perpendicular to the interface is there no bending.) This bending is caused by the fact that light travels slower in glass than in air. We again draw in the normal, a line perpendicular to the interface. The angle between the transmitted ray and the normal, called the **angle of refraction**, is smaller than the angle of incidence (**Figure 9.29b**). We can also have the reverse process: a light ray traveling in glass reaches the interface and is transmitted into air (**Figure 9.30a**). In this case, the angle of refraction is larger than the angle of incidence (**Figure 9.30b**). The following law summarizes these observations.

> **LAWS** **Law of Refraction** A light ray is bent toward the normal when it enters a transparent medium in which light travels more slowly. It is bent away from the normal when it enters a medium in which light travels faster.

Notice the symmetry in Figures 9.29 and 9.30. This is one example of the *principle of reversibility:* the path of a light ray that is refracted at an interface is reversible. The path that a ray takes when going from air into glass is the same path it would take if it turned completely around and went from glass into air. In all of the figures in this and the following sections, the arrows on the light rays could be reversed and the paths would be the same.

Refraction of light affects how you see things that are in a different medium, such as under water. **Figure 9.31a** shows a coin at the bottom of an empty glass; **Figure 9.31b** shows the same coin at the bottom of a glass full of water. Note the

Figure 9.30 (a) Refraction of light as it goes from glass into air. (b) The path of the refracted ray is the reverse of the path shown in Figure 9.29b. (c) The apparent discontinuity of a pencil's length when it is partially submerged in a glass of water is due to the refraction of light from the submerged portion of the pencil as it emerges into the air. Note also the magnifying effects (see Section 9.4b).

Figure 9.31 A coin under water (b) appears closer and larger to an observer looking down from above than does an identical one in air (a).

Air

Water

Apparent location of coin

Figure 9.32 Ray diagram showing why the object in Figure 9.31b appears magnified to the observer when seen through water. The object looks larger because its image lies closer to the observer. This is similar to the effect shown in Figure 9.30c.

difference in the sizes of the coin in the two photos. The coin in Figure 9.31b appears larger because the image of the coin formed in water is closer to the observer than the image formed in air. **Figure 9.32** displays a ray diagram for this circumstance revealing how the bending of the rays of light from the edges of the coin as they enter the air from the water causes the coin to appear to be closer to the surface than it is.

The speed of light in any transparent material is less than the speed of light c in a vacuum. For example: $c = 3 \times 10^8$ m/s, whereas the speed of light in water is 2.25×10^8 m/s, and the speed of light in diamond is only 1.24×10^8 m/s. **Table 9.1** gives the speed of yellow light in selected transparent media. (Critical angle and the variation of speed with color will be discussed later.)

At this point, you may have two questions in mind. Why is the speed of light lower in transparent media such as water and glass, and why does this cause refraction? To answer the first question, we must remember that light is an electromagnetic wave—a traveling combination of oscillating electric and magnetic fields. As light enters glass, the oscillating electric field of the wave causes electrons in the atoms of the glass to oscillate at the same frequency. These oscillating electrons emit secondary EM waves (light) that travel outward to neighboring electrons. However, the wave emitted by each electron is not exactly "in step" with the primary incident wave that is causing the electron to oscillate: the secondary wave lags the primary wave slightly. This means, for example, that a secondary wave crest arrives at a given point in the medium a little bit later in time than the corresponding primary wave crest. As the process repeats from electron to electron, these phase lags accumulate, yielding a net EM disturbance formed from the addition of the primary and secondary waves whose speed of propagation through the medium is reduced.

As to the second question, we can see why a change in the speed of light causes bending by carefully following wave fronts as they cross an interface. A good analogy to this is a marching band, lined up in rows, crossing (at an angle) the boundary between dry ground and a muddy area (**Figure 9.33**). Each person in the band is slowed as his or her feet slip and sink in the mud. To remain aligned and properly spaced, each person has to turn a bit, and the whole band ends up marching in a slightly different direction. A similar change in direction occurs for light entering glass. Adding together the slower moving

Table 9.1 Speed of (Yellow) Light and Critical Angles for Selected Materials

Phase	Substance	Speed	Critical Angle[a]
Gases at 0°C and 1 atm	Carbon dioxide	2.9966×10^8 m/s	
	Air	2.9970	
	Hydrogen	2.9975	
	Helium	2.9978	
Liquids at 20°C	Benzene	1.997×10^8 m/s	41.8°
	Ethyl alcohol	2.203	47.3°
	Water	2.249	48.6°
Solids at 20°C	Diamond	1.239×10^8 m/s	24.4°
	Glass (dense flint)	1.81	37.0°
	Glass (light flint)	1.90	39.3°
	Glass (crown)	1.97	41.1°
	Table salt	2.00	41.8°
	Fused silica	2.05	43.2°
	Ice	2.29	49.8°

[a] Critical angles are for air as the second medium.

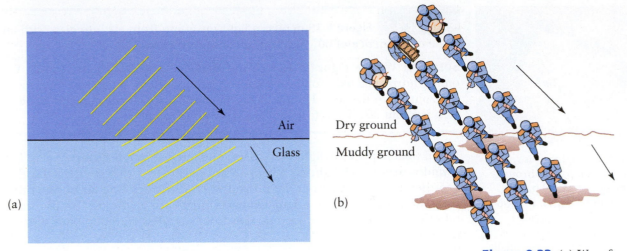

(a)

(b)

wave fronts emitted by the electron oscillators located at different distances from the interface in a way that respects their relative phases produces a net (total) wave front that propagates in a different direction compared to that of the incident wave. Note that the reduction in speed also makes the wavelength *shorter*. This has to be the case because the frequency of the wave remains the same (why?) and we must have $v = f\lambda$.

The law of refraction was discovered in 1621 by the Dutch physicist Willebrord Snel and later expressed in its common form by French mathematician René Descartes (see Challenge 4 at the end of the chapter). It gives the exact value of the angle of refraction when the angle of incidence and the speeds of light in the two media are known.

The graph in **Figure 9.34** shows the relationship between the two angles for an air–glass boundary. (A similar graph could be constructed for other different pairs of media.) The angle between the normal and the light ray in air is plotted on the horizontal axis, and the corresponding angle for the ray in glass is plotted on the vertical axis. Notice that the angle in glass is smaller than the angle in air. For a ray going from air into the glass, this means that the angle of refraction is smaller than the angle of incidence, as shown in Figure 9.29. For the largest possible angle of incidence, 90°, the angle of refraction is about 43°.

Figure 9.34 Graph showing the angles between a light ray and the normal when it passes through an air–glass interface. For a ray passing from air into glass, the angle of incidence is the "angle in air." When the ray passes from glass into air, the angle of incidence is the "angle in glass."

Figure 9.35 The light ray is refracted twice as it passes through the block of glass. Its final path is parallel to its initial path if the two surfaces of the glass are parallel. (Note the partial reflection at the upper surface where the ray initially enters the glass.)

EXAMPLE 9.2 **Figure 9.35** depicts a light ray going from air into glass with an angle of incidence of 60°. Find the angle of refraction.

SOLUTION We use Figure 9.34 with the angle in air equal to 60°. Locate 60° on the horizontal axis, follow the dashed line up to the curve, and then move horizontally to the left. The point on the vertical scale indicates the angle in glass is about 36°. Therefore, the angle of refraction is 36°.

But what happens to light that passes completely through a sheet of glass like a windowpane? Each light ray is bent toward the normal when it enters the glass and then is bent away from the normal when it reenters the air on the other side. No matter what the original angle of incidence is, the two bends exactly offset each other, and the ray emerges from the glass traveling parallel to its original path (Figure 9.35).

9.3b Total Internal Reflection

Consider now the following experiment shown schematically in **Figure 9.36**. Rays of light originating in glass strike the boundary (separating it from the surrounding air) at ever-increasing angles. In this case, because the speed of light in glass is smaller than the speed of light in air, we see that the ray is bent *away* from the normal. In other words, the angle of refraction is *larger* than the angle of incidence, as shown in Figure 9.36b. Moreover, the angle of refraction increases as the angle of incidence does.

Because of reversibility we can use the same graph, Figure 9.34, to find the angle of refraction when the angle of incidence is known. In this situation, the angle

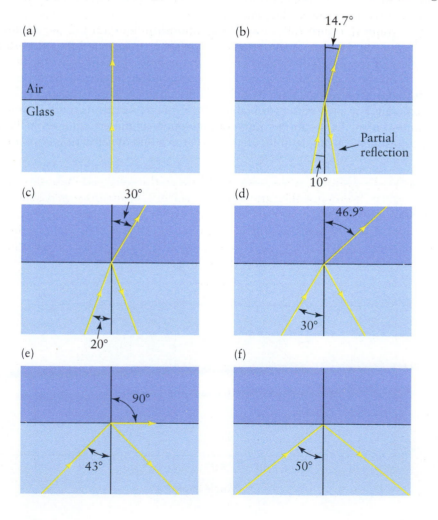

Figure 9.36 A light ray traveling in glass strikes a glass–air interface with different angles of incidence. The angle of incidence in (e) is the *critical angle*, for which the angle of refraction is 90°. Total internal reflection (e and f) occurs when the angle of incidence is equal to, or greater than, the critical angle. No light enters the air.

Vern Ostdiek/University of Michigan-Dearborn

of incidence is the "angle in glass," whereas the angle of refraction is the "angle in air." For example, when the angle of incidence is 20°, we locate 20° on the vertical scale, move horizontally to the curve, and then drop down to the axis where we see that the angle of refraction would be about 30°. This is shown in Figure 9.36c.

What distinguishes this case from the previous one, in which the incident ray was in air, is the existence of a **critical angle** of incidence for which the angle of refraction reaches 90°. When the angle of incidence equals this critical angle, the transmitted ray travels out *along* the interface between the two media (Figure 9.36e). For angles of incidence greater than the critical angle, the formerly transmitted ray is bent back into the incident medium and does not travel appreciably into the second medium. When this happens, we have a condition called **total internal reflection**. This is shown in Figures 9.36e and f.

▶ Physics To Go 9.9

Fill a large, clear glass beaker or other similar vessel (a clear glass coffee pot works well) about halfway with water. Place the container on a sheet of white paper on which you have drawn a circle with a diameter that is about two-thirds that of the bottom of the beaker. Make sure the circle and beaker are concentric with one another. Look through the side of the beaker obliquely at the water's surface. Can you see the entire circle? If not, what portion of the circle cannot be observed? Shift your gaze up and down a bit. How does the image of the circle seen through the water's surface change? What do you think is happening to prevent you from seeing the light rays originating from the part of the circle that is hidden from view?

The graph in Figure 9.34 indicates that the critical angle for a glass–air interface, the angle in glass that makes the angle in air equal to 90°, is about 43°. In general, different media have different critical angles. Table 9.1 includes the values of the critical angle for light rays entering air from different transparent media. Notice that the critical angle is smaller for materials in which the speed of light is lower. Diamond, with the lowest speed of light in the list, has a critical angle of less than 25°—little more than half that of glass.

EXAMPLE 9.3   A homeowner wishes to mount a floodlight on a wall of a swimming pool under water so as to provide the maximum illumination of the surface of the pool for use at night (**Figure 9.37**). At what angle with respect to the wall should the light be pointed?

SOLUTION   To illuminate the surface of the water, the refracted ray at the water–air interface should just skim the water's surface. This means that the angle of refraction must be 90°, so the incident angle must be the critical angle.

Table 9.1 shows that the critical angle for water is about 49°. The homeowner should direct the floodlight upward so that the beam makes an angle of roughly 49° with the vertical side wall of the pool.

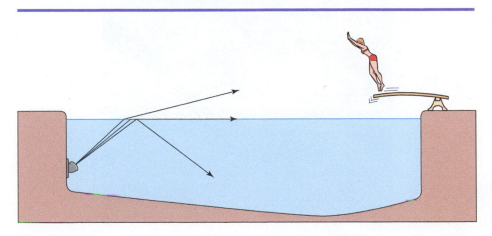

Figure 9.37 Maximizing illumination of the surface of a swimming pool.

Figure 9.38 Multiple internal reflections within an optical fiber.

Optical fibers are flexible, coated strands of glass that utilize total internal reflection to channel light. In a sense, an optical fiber does to light what a garden hose does to water. **Figure 9.38** shows the path of a particular light ray as it enters one end of an optical fiber. When this ray strikes the wall of the fiber, it does so with an angle of incidence that is greater than the critical angle, so the ray undergoes total internal reflection. This is repeated each time the ray encounters the fiber's wall, so the light is trapped inside until it emerges from the other end. You may have seen decorative lamps with light bursting from the ends of a "bouquet" of optical fibers. Fiber-optic cables consisting of dozens or even hundreds of individual optical fibers are now in common use to transmit information (**Figure 9.39**). Telephone conversations, audio and video signals, and computer information are encoded ("digitized") and then sent through fiber-optic cables as pulses of light produced by tiny lasers. A typical fiber-optic cable can transmit thousands of times more information than a conventional wire cable that is much larger in diameter. For example, the first fiber-optic transatlantic cable, which went into service in 1988, was designed to carry 40,000 simultaneous conversations over just two pairs of glass fibers. By contrast, the last of the large copper bundles installed for overseas communications (1983) could handle only about 8,000 calls.

As noted above, one significant advantage of using optical fibers to conduct light from one place to another is their flexibility. If bundles of free fibers are bound together in a way that preserves the relative positions of adjacent fibers from one end to the other, the bundle may transmit images as well as light. Devices incorporating this technology are routinely used to examine hard-to-reach places like the insides of a nuclear reactor, a jet engine, or the

Spencer Grant/Science Source

Figure 9.39 A fiber-optic light-guide cable (below) can typically carry many more telephone conversations than a much larger wire cable (above).

human body. When used in the latter capacity, the instrument is generally called an *endoscope*. Specific examples of endoscopes include bronchoscopes (for examining lung tissue), gastroenteroscopes (for checking the stomach and digestive tract), and colonoscopes (for surveying the bowel).

■ Learning Check

1. Upon entering a medium in which the speed of light is 1.5×10^8 m/s from a medium in which the speed of light is 2.0×10^8 m/s, an oblique ray will
   (a) be bent away from
   (b) be bent toward
   (c) travel along
   (d) travel perpendicular to
   (e) remain undeviated with respect to the normal to the interface between the two media.
2. In air, a source of light has a wavelength of 590 nm. The same source under water will have (a) a longer wavelength, (b) a shorter wavelength, or (c) the same wavelength.
3. The critical angle is that angle of incidence for which the angle of refraction equals
   _____.
4. Name an important communication system component that makes use of total internal reflection.

ANSWERS: 1. (b) **2.** (b) **3.** 90° **4.** optical fiber

9.4 Lenses and Images

In Section 9.2, we described how curved mirrors are used in astronomical telescopes and other devices to redirect light rays in useful ways. Microscopes, binoculars, cameras, and many other optical instruments use specially shaped pieces of glass called *lenses* to alter the paths of light rays. As was the case with mirrors, the key to making glass or other transparent substances redirect light is the use of curved surfaces rather than flat (planar) ones.

Suppose we grind a block of glass so that one end takes the shape of a segment of a sphere as shown in cross section in **Figure 9.40**. Let parallel rays strike the convex spherical surface at various points above and below the line of symmetry (called the *optical axis*) of the system. If one applies the law of refraction at each point to determine the angle of refraction of each ray, the results shown in Figure 9.40 are found. In particular, rays traveling along the optical axis emerge from the interface undeviated. Rays entering the glass at points successively above or below the optical axis are deviated ever more strongly toward the optical axis. The result is to cause the initially parallel bundle of rays to gradually *converge* together—to become focused—into a small region behind the interface. This point is called the *focal point* and is labeled F in the figure. (Compare this to Figure 9.21a in which a concave mirror causes light rays to converge.)

Figure 9.41 shows the behavior of parallel rays refracted across a spherical interface that, instead of bowing outward, curves inward. In this case, the emergent rays *diverge* outward as though they had originated from a point F' to the

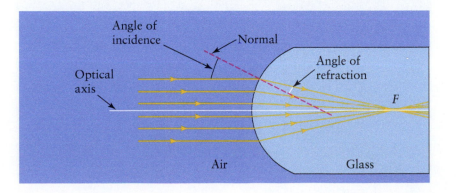

Figure 9.40 Refraction at a convex spherical surface showing the convergence of light rays. F is the focal point—that is, the point at which the light rays are concentrated after having passed through the surface.

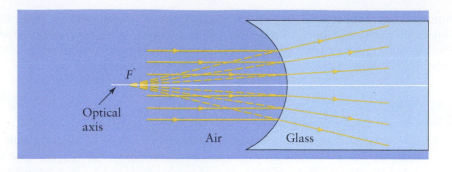

left of the interface. The ability to either bring together or spread apart light rays is the basic characteristic of lenses, be they camera lenses, telescope lenses, or the lenses in human eyes.

Although the situation shown in Figure 9.40 does occur in the eye, in most devices the light rays must enter and then leave the optical element (lens) that redirects them. Common lenses have two refracting surfaces instead of one, with one surface typically in the shape of a segment of a sphere and the second either spherical as well or flat (planar). The effect on parallel light rays passing through both surfaces is similar to that in the previous examples with one refracting surface. A *converging lens* causes parallel light rays to converge to a point, called the *focal point* of the lens (**Figure 9.42**). The distance from the lens to the focal point is called the **focal length** of the lens. A more sharply curved lens has a shorter focal length. Conversely, if a tiny source of light is placed at the focal point, the rays that pass through the converging lens will emerge parallel to each other. This is the principle of reversibility again.

> ### ▶ Physics To Go 9.10
>
> Look around your home, office, or dorm room for a simple, hand-held magnifying lens. Stand under an overhead light, and holding the lens in one hand, project the image of the source onto your other hand. Move the lens up or down to produce a clearly focused image (not just a blur of light) on your palm. Now estimate the distance from the lens to the image. Congratulations! You've just found the focal length of the lens. This same technique works for any converging lens as long as the source is a good deal farther from the lens than its focal length. Project the image on a white index card or a piece of white paper instead of your hand. Describe the image of the source produced by the lens.

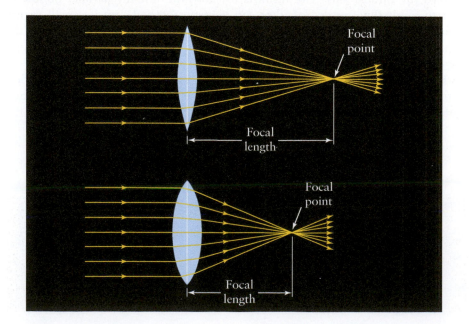

Figure 9.42 Converging lenses focusing parallel light rays at their focal points. A more sharply curved lens has a shorter focal length.

A *diverging lens* causes parallel light rays to diverge after passing through it. These emergent rays appear to be radiating from a point on the other side of the lens. This point is the focal point of the diverging lens (**Figure 9.43**). The distance from the lens to the focal point is again called the focal length, but for a diverging lens it is given as a negative number, −15 centimeters, for example. If we reverse the process and send rays converging toward the focal point into the lens, they emerge parallel.

For both types of lenses there are two focal points, one on each side. Clearly, if parallel light rays enter a converging lens from the right side in Figure 9.45, they will converge to the focal point to the left of the lens. Whether a lens is diverging or converging can be determined quite easily: if it is *thicker* at the middle than at the edges, it is a converging lens; if it is *thinner* at the center, it is a diverging lens (**Figure 9.44**).

9.4a Image Formation

The main use of lenses is to form images of things. First, let's consider the basics of image formation when a symmetric converging lens (see, for example, Figure 9.44a left) is used. Our eyes, most cameras (both still and video), slide projectors, movie projectors, and overhead projectors all form images this way. **Figure 9.45** illustrates how light radiating from an arrow, called the *object*, forms an *image* on the other side of the lens. One practical way of demonstrating this would be to point a flashlight at the arrow so that light would reflect off the arrow and pass through the lens. The image could be projected onto a piece of white paper placed at the proper location to the right of the lens.

Although each point on the object has countless light rays spreading out from it in all directions, it is simpler to consider only three particular rays from a single point—the arrow's tip. These rays, shown in Figure 9.45, are called the **principal rays**.

1. The ray that is initially parallel to the optical axis passes through the focal point (*F*) on the other side of the lens.

2. The ray that passes through the focal point (*F'*) on the same side of the lens as the object emerges parallel to the optical axis.

3. The ray that goes exactly through the center of the lens is undeviated because the two interfaces it encounters are parallel.

Figure 9.43 Parallel light rays diverging after passing through a diverging lens. The rays appear to radiate from the focal point, to the left of the lens.

Converging   Diverging   Converging   Diverging   Converging   Diverging

(a)   (b)   (c)

Figure 9.44 Examples of different types of lenses: (a) bi-convex (left), bi-concave (right); (b) plano-convex (left), plano-concave (right); (c) meniscus-convex (left), meniscus-concave (right).

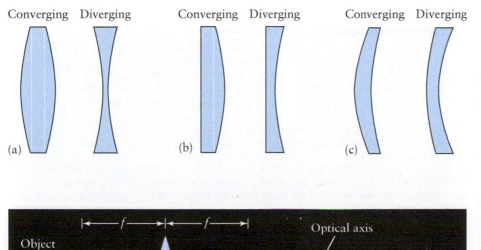

Figure 9.45 Arrangement of a simple converging lens showing the object and image positions as well as the focal points, *F* and *F'* and the focal lengths. Here *s* and *p* are on opposite sides of the lens and are both considered positive. The three principal rays from the arrow's tip are shown.

Figure 9.46 Image formation in a camera. The central principal ray from a point at the top and from a point at the bottom of the object are shown. (Figure not to scale.)

Image

Note that the image is *not* at the focal point of the lens. Only parallel incident light rays converge to this point (assuming an ideal lens).

We could draw principal rays from each point on the object, and they would converge to the corresponding point on the image. This kind of image formation occurs when you take a photograph (**Figure 9.46**) or view a PowerPoint slide on a projection screen. In the latter case, light radiating from each point on the slide converges to a point on the image on the screen. Note that the image is inverted (upside down). (Digital images are electronically manipulated to compensate for this effect so they appear properly oriented to the viewer when projected on a screen.)

The distance between the object and the lens is called the **object distance**, represented by s, and the distance between the image and the lens is called the **image distance**, p. By convention (with the light traveling from left to right), s is positive when the object is to the left of the lens, and p is positive when the image is to the right of the lens. If we place the object at a different point on the optical axis, the image would also be formed at a different point. In other words, if s changes, then so does p. Using a lens with a different focal length for fixed s would also cause p to change. For example, the image would be closer to the lens if the focal length were shorter.

▶ Physics To Go 9.11

Let's experiment some more with the magnifying lens. This time use a small electric light as a source; a high-intensity desk lamp works well. Stand as far away from the lamp as possible and again project a focused image of it on a white index card or piece of white paper (the screen; **Figure 9.47**). Move the lens toward the source, adjusting the position of the screen continuously to produce a clear, focused image. How does the distance of the screen from the lens (the image distance) change as the distance between the source and the lens (the object distance) is reduced? Describe the nature and change in the image as you gradually shorten the object distance. What happens when the lens gets closer to the lamp than about one focal length? At this point, remove the screen and look back through the lens at the source. (Be sure to turn down the lamp intensity or otherwise shield your eyes so that you can view the source comfortably.) How does the lamp appear now? Repeat the experiment, keeping especially careful track of the appearance of the image when the object distance is greater than about two times the focal length of the lens, about equal to two focal lengths, between one and two focal lengths, and, finally, less than one focal length. Record your observations for later study.

Vern Ostdiek/University of Michigan-Dearborn

Figure 9.47 Single-lens image formation.

The following equation, known as the *lens formula*, relates the image distance p to the focal length f and the object distance s:

$$p = \frac{sf}{s - f} \qquad \text{(lens formula)}$$

The following example shows how the lens formula can be used.

EXAMPLE 9.4 In a slide projector, a slide is positioned 0.102 meter from a converging lens that has a focal length of 0.1 meter. At what distance from the lens must the screen be placed so that the image of the slide will be in focus?

SOLUTION The screen needs to be placed a distance p from the lens, where p is the image distance for the given focal length and object distance. So,

$$p = \frac{sf}{s - f} = \frac{0.102 \text{ m} \times 0.1 \text{ m}}{0.102 \text{ m} - 0.1 \text{ m}}$$

$$= \frac{0.0102 \text{ m}^2}{0.002 \text{ m}}$$

$$= 5.1 \text{ m}$$

If the slide-to-lens distance is increased to 0.105 meter (about a 3% change), the distance to the screen (p) would have to be reduced to 2.1 meters (more than a factor of 2.4 reduction in distance).

If the lens is replaced by one that has a shorter focal length, the distance to the screen would have to be reduced as well.

The images formed in the manner just described are called **real images**. Such images can be projected onto a screen. We see the image on the screen because the light striking the screen is diffusely reflected to our eyes. A simple magnifying glass is a converging lens, but the image that it forms under normal use is *not* a real image—it can't be projected onto a screen. We see the image by looking *into* the lens, just as we see a mirror image by looking *into* the mirror. This type of image is called a **virtual image**. **Figure 9.48** shows how an image is formed in a magnifying glass. In this case, the object is between the focal point F' and the lens, so the object distance s is *less than* the focal length f of the lens. Note that the image is enlarged and that it is upright. It is also on the same side of the lens as the object, which means that p is negative. This situation is very much like the image formation with a concave mirror (Figure 9.22b).

EXAMPLE 9.5 A converging lens with focal length 10 centimeters is used as a magnifying glass. When the object is a page of fine print 8 centimeters from the lens, where is the image?

SOLUTION

$$p = \frac{sf}{s - f} = \frac{8 \text{ cm} \times 10 \text{ cm}}{8 \text{ cm} - 10 \text{ cm}}$$

$$= \frac{80 \text{ cm}^2}{-2 \text{ cm}}$$

$$= -40 \text{ cm}$$

The negative value for p indicates that the image is on the same side of the lens as the object. Therefore, it is a virtual image and must be viewed by looking through the lens.

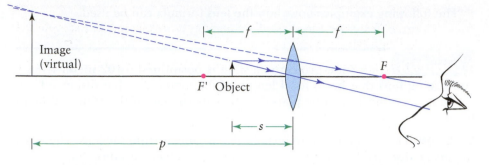

Figure 9.48 Image formation when the object is between the focal point and the lens. A virtual image of the arrow is formed, seen by looking through the lens toward the object. (Only two of the principal rays are shown here.)

A virtual image is also formed when you look at an object through a diverging lens (**Figure 9.49**) for all values of s. In this case, the image is smaller than the object, as it is with a convex mirror.

9.4b Magnification

In Figures 9.45 through 9.49, the size of the image is not the same as the size of the object. This is one of the most beneficial properties of lenses: they can be used to produce images that are enlarged (larger than the original object) or reduced (smaller than the original object). In either case, the **magnification**, M, of a particular configuration is the height of the image divided by the height of the object.

$$M = \frac{\text{image height}}{\text{object height}}$$

If the image is twice the height of the object, the magnification is 2. If the image is upright, the magnification is positive. If the image is inverted, the magnification is negative (because the image height is negative).

The magnification that one gets with a particular lens changes if the object distance is changed. Because of the simple geometry, the magnification also can be written in an alternate, equivalent form as *minus* the image distance divided by the object distance.

$$M = \frac{-p}{s}$$

From this we can conclude that:

If p is positive (image is to the right of the lens and real), M is negative; the image is inverted (Figures 9.45 and 9.46).

If p is negative (image is to the left of the lens and virtual), M is positive; the image is upright (Figures 9.48, 9.49, and **9.50**).

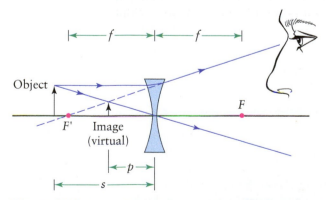

Figure 9.49 Image formation with a diverging lens. The image is virtual, so it must be viewed by looking through the lens. (Again, only two of the principal rays are shown.)

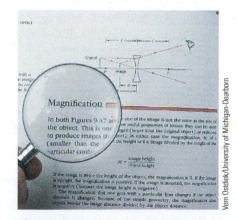

Figure 9.50 The image seen through the converging lens is magnified, upright, and virtual.

Vern Ostdiek/University of Michigan–Dearborn

EXAMPLE 9.6 Compute the magnification for the projector in Example 9.4 and for the magnifying glass in Example 9.5.

SOLUTION In the first case in Example 9.4, $s = 0.102$ meters and $p = 5.1$ meters. Therefore:

$$M = \frac{-p}{s} = \frac{-5.1 \text{ m}}{0.102 \text{ m}}$$

$$= -50$$

The image is 50 times as tall as the object, but it is inverted (because M is negative). A slide that is 35 millimeters tall has an image on the screen that is 1,750 millimeters (1.75 meters) tall. When s is 0.105 meters, p is 2.1 meters, and the magnification is -20.

In Example 9.5, $s = 8$ centimeters and $p = -40$ centimeters. Consequently:

$$M = \frac{-p}{s} = \frac{-(-40 \text{ cm})}{8 \text{ cm}}$$

$$= +5$$

The image of the print seen in the magnifying glass is five times as large as the original, and it is upright (because M is positive).

Telescopes and microscopes can be constructed by using two or more lenses together. **Figure 9.51** shows a simple telescope consisting of two converging lenses. The real image formed by lens 1 becomes the object for lens 2. The light that could be projected onto a screen to form the image for lens 1 simply passes on into lens 2. In essence, lens 2 acts as a magnifying glass and forms a virtual image of the object. In this telescope, the image is magnified but inverted. Replacing lens 2 in Figure 9.51 with a diverging lens yields a telescope that produces an upright image.

9.4c Aberrations

In real life, lenses do not form perfect images. Suppose we carefully apply the law of refraction to a number of light rays, all initially parallel to the optical axis, as they pass through a real lens that has a surface shaped like a segment of a sphere (**Figure 9.52**). We would find that the lens exhibits the same flaw we saw in Section 9.2 with spherically shaped curved mirrors: *spherical aberration*. Figure 9.52 shows that rays striking the lens at different points do not cross the optical axis at the same place (compare Figure 9.26a). In other words, there is no single focal point. This causes images formed by such lenses to be somewhat blurred. Lens aberrations of this type can be corrected, but this process is complicated and often necessitates the use of several simple lenses in combination.

Another type of aberration shared by all simple lenses even when used under ideal conditions is *chromatic aberration*. A lens affected by chromatic aberration,

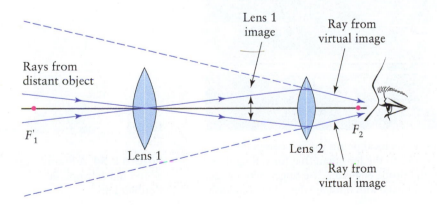

Figure 9.51 Simple telescope. Only the third principal rays from the top and from the bottom of the distant object are shown. The dashed lines show that the virtual image subtends a larger angle than the incoming rays from the object do—the image is magnified.

Figure 9.52 Spherical aberration for a convex lens illuminated by a beam of parallel light. Rays 1 and 1′ are brought to a focus at $F_{11'}$, while rays 2 and 2′, and 3 and 3′ are focused at $F_{22'}$, and $F_{33'}$, respectively.

when illuminated with white light, produces a sequence of more or less overlapping images, varying in size and color. If the lens is focused in the yellow-green portion of the EM spectrum where the eye is most sensitive, then all the other colored images are superimposed and out of focus, giving rise to a whitish blur or fuzzy overlay (**Figure 9.53**). For a converging lens, the blue images would form closer to the lens than the yellow-green images, whereas the reddish images would be brought to a focus farther from the lens than the yellow-green ones.

The cause of chromatic aberration has its roots in the phenomenon of *dispersion*, to be discussed in Section 9.6. The remedy for this problem, originally thought to be insoluble by none other than Newton himself, was discovered around 1733 by C. M. Hall and later (in 1758) developed and patented by John Dolland, a London optician. It involves using two different types of glass mounted in close proximity. **Figure 9.54** shows a common configuration called a *Fraunhofer cemented achromat* (meaning "not colored"). The first lens is made of crown glass, the second of dense flint glass (see Table 9.1). These materials are chosen because they have nearly the same dispersion. To the extent to which this is true, the excess convergence exhibited by the first lens at bluish wavelengths is compensated for by the excess divergence produced by the second lens at these same wavelengths. Similar effects occur at the other wavelengths in the visible spectrum, permitting cemented doublets of this type to correct more than 90 percent of the chromatic aberration found in simple lenses.

> ### ▶ Physics To Go 9.12
>
> Most households have large, thick magnifying glasses. Using this handy device and a burning candle, you can easily demonstrate some of the basics of chromatic aberration. With the candle as your source of illumination, locate the lens so that it forms a real image of the candle flame on a white card or piece of paper (the screen). (If you need help in determining where to place the lens relative to the candle and the screen to accomplish this task, see Physics to Go 9.11.) Move the screen closer to the lens. What color is the outside edge of the now-blurred image of the candle flame? Next, move the screen away from the lens past the position of best focus of the candle image. What color does the perimeter of the blurred image appear now? Try examining the source directly by looking back through the lens at the candle. The chromatic effects should be more apparent.

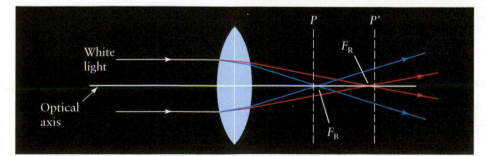

Figure 9.53 Diagram showing the positions of the focal points for two colors for a simple lens. The focal point for blue light is closer to the lens than that for red light. A screen placed at point P will show an image with outside edges that are tinged red-orange, whereas a screen placed at point P' will reveal an image with outside edges that are bluish.

Figure 9.54 A Fraunhofer cemented achromatic doublet lens.

1. (True or False.) A light ray can pass through a diverging lens without being deflected.

2. _____ lenses are thicker at their centers than at their edges, whereas _____ lenses are thinner at their centers than at their edges.

3. An image formed using a converging lens
   (a) is always real.
   (b) is always virtual.
   (c) can always be projected on a screen.
   (d) can have a negative magnification.
   (e) can never be upright.

4. The magnification of an object by a certain lens is equal to −0.5. The image formed by the lens must be
   (a) upright and enlarged.
   (b) upright and reduced.
   (c) inverted and enlarged.
   (d) inverted and reduced.
   (e) virtual and unchanged in size.

ANSWERS: 1. True **2.** Converging, diverging **3.** (d) **4.** (d)

OPTICAL ENGINEERING APPLICATIONS Fresnel, Pharos, and Physics

For coastal dwellers, be it along one of the Great Lakes or the eastern or western seaboards, the sight of a lighthouse (or *pharos*) with its sweeping beam warning mariners miles away of the dangers lurking near the shore is not an uncommon one. The measure of safety afforded to seafarers depends on the distance to which these beacons can be seen. This, in turn, depends upon the intensity and degree of focus of the beams. We all owe a debt of gratitude to a 19th-century physicist for developing the principles that underlie the successful function of lighthouses, as well as traffic lights, classroom overhead projectors, stage spotlights, recreational vehicle rear-viewing screens, spacecraft solar panels, and many other practical, everyday devices.

Augustin Jean Fresnel (1788–1827; **Figure 9.55**) was a French engineer and physicist who made many important contributions to our understanding of the nature of light. He was a strong proponent of the wave theory and the first to argue persuasively that light was a transverse wave based on its polarization. Using the wave theory, he developed mathematical formulas that accurately predicted the diffraction patterns produced by apertures of various

Stefano Bianchetti/Historical/CORBIS

Figure 9.55 Augustin Jean Fresnel. French engineer and physicist whose work in optics included both practical (lighthouse lenses) and theoretical (mathematical description of diffraction) innovations.

shapes and sizes. But Fresnel was a practical man, and much of his effort throughout his adult life was devoted to public works projects, like road construction, designed to improve the infrastructure in his native France.

In 1824, Fresnel merged his interests in optics with his desire to serve his country by accepting a commission with the Lighthouse Board. In this capacity, Fresnel set about improving the science of pharos by developing compound lenses to capture and focus light instead of using mirrors, as had been done previously. The result was the much-heralded *Fresnel lens*; its modern usage has spread far beyond the lighthouse for which it was originally intended.

Fresnel realized that the refraction of light by a lens occurs only at the two curved surfaces of the lens. The material between serves no purpose in focusing the light and may, if thick enough, actually reduce the light intensity significantly by absorption. So, to preserve the refractive effects of the surfaces but eliminate the bulk and absorption of the lens, Fresnel subdivided his lens into prism-like segments and then cut out the middle part of each prism (**Figure 9.56**). The resulting wedges have the same refracting properties as the prisms and, hence, the original lens. The main advantages that derive from this design are ones of light weight, low absorption, and, by suitable arrangement of the wedges, high light-collecting efficiency.

When employed in a lighthouse, the various wedge segments are arrayed along an arc. In this configuration, the top and bottom prisms serve to both refract and internally reflect the rays from the source. Moreover, the central wedge is frequently replaced by a very thin, plano-convex lens. Originally, Fresnel lenses for lighthouses were ranked in six main sizes called *orders*; the order is determined by the distance from the light source to the lens. A first-order Fresnel lens might have included up to 1000 prisms, stood nearly 10 feet tall, measured 6 feet in diameter, and weighed up to 3 tons. The first Fresnel lens was installed in the Cardovan Tower Lighthouse on the Gironde River in France. It produced a beam that was five times more powerful than the reflector system used previously and was visible to the horizon more than 20 miles away. Although Fresnel lenses were introduced in the United States in 1841 for service in lighthouses in New York harbor, it was not until 1851 that the

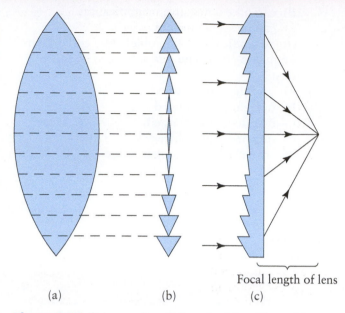

Focal length of lens

(a) (b) (c)

Figure 9.56 Cut-away view of the principle of Fresnel lenses. An ordinary converging lens may be thought of as a stacked set of prisms (a). If the central portions of each prism are removed and the remaining wedges aligned, the resulting structure (b) has the same refracting characteristics without the added weight or light absorption. Modern Fresnel lenses (c) display a circular symmetric array of grooves on one side of a piece of clear plastic cut to match the shapes of the wedges. This device works to focus light from a distant source just like the original lens.

government authorized the use of this technology throughout its national system of warning beacons (**Figure 9.57**).

Modern Fresnel lenses are usually made out of plastic with rotary symmetric grooves on one surface only for ease of production. Lenses of this type can be found in many applications where the need to create bright, focused beams of light exists. Two good examples are the lenses in spotlights illuminating performance venues and in many common traffic lights at busy intersections. Because Fresnel lenses also have a wide field of view, they are ideal for imaging panoramic scenes that often lie hidden behind large vehicles. Perhaps you've seen such lenses in the rear windows of large recreational vehicles. So once again, the theme of safety threads the application of Fresnel's design. One cannot help wondering how many lives have been saved

Courtesy of The Michigan Lighthouse Conservancy

Figure 9.57 A second-order Fresnel lighthouse lens from Standard's Rock lighthouse on Lake Superior. This is one of only a few lenses of this size to be used on the Great Lakes; most were much smaller. For example, fourth-order Fresnel lenses were typical in lighthouses along the shores of Lake Michigan. The bull's-eye lenses in the center created a characteristic flash pattern as the device rotated.

at sea and on land during the past nearly two centuries through the innovation of a young Frenchman with an eye for detail and a commitment to service.

■ **QUESTIONS**

1. Give three advantages that the Fresnel lens design has over that of a conventional lens with the same size and focal length.
2. List several examples of the use or application of Fresnel lenses in today's world.

9.5 The Human Eye

The eye may be the most sophisticated of our sense organs. But up to the point when the light rays are absorbed and the signal to the brain is formed, the eye is a fairly simple optical instrument.

Figure 9.58 shows a simplified cross section of the human eye. Light enters from the left and is projected onto the retina at the rear of the eyeball. The iris, the part of the eye that is colored, controls the amount of light that enters the eye. Its circular opening, the pupil, is large in dim light and small in bright light. As light enters the cornea, it converges because of the cornea's convex shape (Figure 9.40 shows this effect). The eye's *auxiliary lens,* simply called the

Figure 9.58 The human eye. Light passes through the opening in the iris, the pupil, and forms an image on the retina. The cornea and the lens act as a single converging lens.

lens, makes the light converge even more. This lens is used by the eye to ensure that the image is in focus on the retina. The effect of the cornea and the lens is the same as if the eye were equipped with a single converging lens (albeit one with a variable focal length), so the image formation process shown in Figure 9.45 applies.

9.5a Defects of the Eye

For an object to be seen clearly, the image must be in focus on the retina. Within the eye, the image distance is always the same—the diameter of the eyeball. We are able to focus on objects near and far (with small and large object distances) because the focal length of the lens can be varied by changing its shape. When the eye is focused on a distant object (farther than, say, 5 meters away) the lens is thin and has a long focal length (**Figure 9.59**). For a near object, special muscles make the lens thicker. This shortens the focal length of the lens so that the image of the near object is focused on the retina. Unlike many cameras and other optical devices, the eye has a constant value for p, the image distance, but it accommodates different values for s, the object distance, by changing its focal length f. (Humans, like most mammals, accommodate by

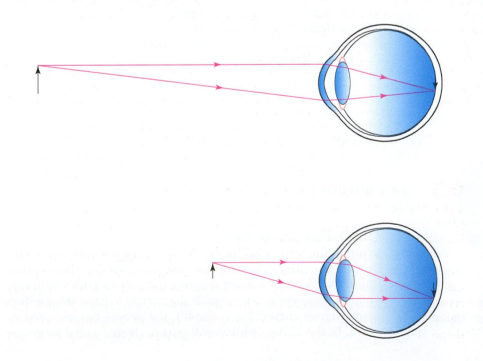

Figure 9.59 The eye can form images on the retina of either distant or near objects by changing the thickness (and therefore the focal length) of the lens.

Figure 9.60 (a) A nearsighted eye causes rays from a distant object to converge too quickly. (b) A diverging lens corrects the problem.

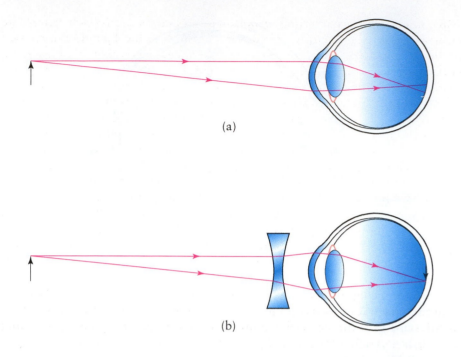

(a)

(b)

changing the shape of the eye lens. Fish, however, move the lens itself toward or away from the retina, thus changing the value of p.)

The two most common types of poor eyesight, nearsightedness and farsightedness, are the result of improper focusing. *Nearsightedness*, or myopia, occurs when close objects are in focus but distant objects are not. The light rays from a distant object are brought into focus before they reach the retina. When the rays do reach the retina, they are out of focus (**Figure 9.60**). The problem is remedied by placing a properly chosen *diverging lens* in front of the eye. The corrective lens is either held in place by a frame (glasses) or placed in direct contact with the cornea (contact lenses). This lens makes the light rays diverge slightly before they enter the eye. This moves the point where the rays meet back to the retina.

Farsightedness, or hyperopia, is the opposite: distant objects are in focus, but near objects are not. The cornea and the lens do not make the rays from a near object converge enough. The rays reach the retina before they meet, and the image is out of focus (**Figure 9.61**). The remedy for this condition is a properly chosen *converging lens* placed in front of the eye. This lens makes the light rays converge slightly before they enter the eye, thereby bringing the light rays into focus on the retina.

Another common problem is *astigmatism*, which occurs when the cornea is not symmetric. For example, the cornea's focal length might be shorter for two parallel rays that enter the eye one above the other compared to the focal length for two parallel rays that enter side by side. Objects appear distorted. This condition can often be corrected by using a specially shaped lens that has the opposite asymmetry.

9.5b Eye Surgery

During the second half of the 20th century, several types of corrective eye surgery became commonplace. As described earlier in this section, the convex shape of the cornea causes light rays entering the eye to converge. If the cornea's shape is imperfect (too sharply curved, for example), the person has impaired vision. In the 1970s, ophthalmologists perfected *radial keratotomy*

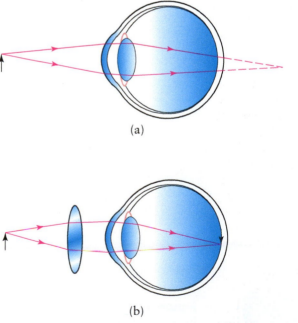

(a)

(b)

Figure 9.61 (a) A farsighted eye does not cause light rays from a near object to converge enough. (b) A converging lens corrects the problem.

(RK), a method for correcting myopia. The strategy is to precisely flatten the cornea—make it less sharply curved—so that the eye's focal length is increased to the correct value. A surgeon makes several radial (spokelike) incisions in the cornea. As the cornea heals, it becomes flatter, so that the patient no longer needs corrective lenses in most cases.

In the 1990s, several forms of laser surgery were developed that offer alternatives to RK. *Photorefractive keratectomy* (PRK) is a procedure that reshapes the cornea using an ultraviolet laser controlled by a computer. (For more on lasers, see Section 10.8.) The laser selectively vaporizes (removes) tissue on the cornea's surface until the desired shape is achieved. PRK is used primarily to correct myopia and astigmatism. Another method, called *laser-assisted in situ keratomileusis* (LASIK), is similar to PRK but more complicated. Here, the surgeon cuts a flap of corneal tissue, uses a laser to remove tissue beneath it, and then replaces the flap. In a similar fashion, during *laser-assisted in situ epithelial keratomileusis* (LASEK) procedures, a thin flap of corneal epithelial cells is created using an alcohol solution. The folded layer of cells is similar to the corneal flap made with LASIK but much thinner. The laser is then applied to reshape the cornea. LASEK eye surgery is recommended for patients with thin corneas because it saves more of the corneal tissue than LASIK, and it holds especially good promise for treating patients suffering from hyperopia.

9.6 Dispersion and Color

Most of you at one time or another have seen decorative hanging glass pendulums through which sunlight is streaming. If so, you probably noticed patches of bright, rainbow-hued light playing about as the pendulums slowly turned in response to air currents. Did you ever wonder how such beauty was produced? Sir Isaac Newton did, and he performed several experiments in an attempt to answer this question. He concluded that sunlight—white light—was a mixture of all the colors of the rainbow and that upon being refracted through transparent substances like glass, it could be *dispersed*, or separated, into its constituent wavelengths (colors). The process of refraction was seen to be color dependent!

This color dependence of refraction comes about because the speed of light in any medium is slightly different for each color. In glass, diamond, ice, and most other common transparent materials, *shorter* wavelengths of light travel slightly *slower* than longer wavelengths. Violet travels a little slower than blue, blue travels slower than green, and so on. In the case of common glass, the speed of violet light is about 1.95×10^8 m/s, whereas the speed of red light is about 1.97×10^8 m/s. The speeds of the other colors lie between these two values. This slight difference in the speeds of different colors causes dispersion. In diamond, the difference in speeds of violet and red light is comparatively larger—about 2 percent, compared to about 1 percent for glass—so the dispersion is greater. That is why one sees such brilliant colors in diamonds. The difference in speeds of violet and red light in water is comparatively smaller than it is in glass.

Up until this time, we have ignored the fact that the speed of light in a medium depends on wavelength. (You can think of our previous study of refraction as having been done using light of only a single color or wavelength, something known as *monochromatic* light.) What effect does this now have on how violet light rays are refracted at an air–glass interface relative to how red rays are refracted?

Consider **Figure 9.62**, showing an incoming ray of light that we will assume is a mixture of only blue and red wavelengths. Because the speeds of both blue light *and* red light are lower in glass than they are in air, we expect that

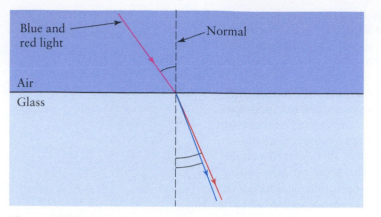

Figure 9.62 Dispersion at an air–glass interface produces a separation of red and blue rays. The angle of refraction of the blue ray is smaller than the angle of refraction of the red ray.

Figure 9.63 Dispersion of white light into a spectrum of colors as it passes through an equilateral prism. Each color is bent a different amount because the speed is slightly different for each wavelength.

both rays will be bent toward the normal based on our analysis in Section 9.3. But we know that the speed of blue light in glass is lower than the speed of red light in the same medium, so that the blue light will be bent slightly *more* toward the normal than the red light. In Figure 9.62, this results in the angle of refraction for blue light being a bit smaller than the angle of refraction for red light. Thus, although both rays are bent toward the normal upon passing into the glass, the blue ray is refracted more strongly and emerges from the interface along a different path than the red ray. The colors have been dispersed, or separated, as a result of the refraction process because of the wavelength dependence of the speed of light.

If the incoming beam is now allowed to contain the remaining colors between red and blue, the emergent rays for each will fall between the limits set by the red and blue rays. What is produced is a *spectrum*—the different colors spread over a range of angles. It is the process of dispersion, acting to sort out the different colors, that causes the chromatic aberration in simple lenses described at the end of Section 9.4.

The difference between the angles of refraction for the red and blue rays above amounts to less than 0.5°. This may not sound like much, but it is some 30 times the minimum angular separation between rays that the human eye can detect under bright conditions and, therefore, would certainly be noticeable. If additional air–glass surfaces are introduced, more refractions may occur, and the angular spread in the emerging rays may be increased. The light is said to be more highly dispersed in this case.

A *prism* is a common device used to disperse light and form a spectrum (**Figure 9.63**). Prisms were well known and highly prized by the Chinese from the early 1600s for their ability to generate color. Today, they are highly valued by scientists for much the same reason. For example, one can analyze the radiation emitted by a source of light by dispersing the light into a spectrum and measuring the intensity (amount) of radiation coming off in the various wavelengths (colors). If the source radiates like a blackbody (see Section 8.6), this information might be used to determine the temperature of the source. As we shall see in Chapter 10, it is also possible to determine the chemical composition of a source by examining its spectrum. **Figure 9.64** shows two common configurations of dispersing prisms and the path a single-color light ray follows through each.

Figure 9.64 Two common forms of dispersing prisms, with their angular deviations shown for monochromatic light. (a) The Pellin-Broca prism. (b) The Abbe prism. Used as shown, these prisms are very convenient because a light source and viewing system can be set up at a fixed angle—90° in (a)—and then the prisms rotated slightly to look at a particular wavelength (color). In each case, at the point labeled P, total internal reflection of the light ray occurs.

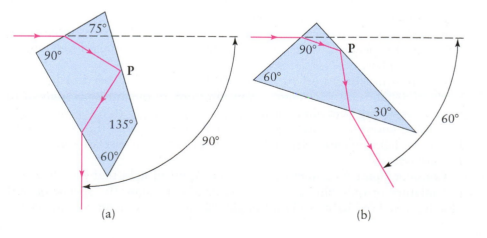

(a) (b)

Learning Check

1. A _____ lens is used to compensate for hyperopia.
2. When myopia is corrected surgically, the shape of the _____ is changed.
3. (True or False.) Blue light travels more slowly through glass than red light does.
4. A ray of violet light and a ray of yellow light enter a block of glass from air with the same angle of incidence. The angle of refraction for the violet ray will be

   (a) much larger than that of the yellow ray.
   (b) a little larger than that of the yellow ray.
   (c) equal to that of the yellow ray.
   (d) a little smaller than that of the yellow ray.
   (e) much smaller than that of the yellow ray.

ANSWERS: 1. converging 2. cornea 3. True 4. (d)

9.7 Atmospheric Optics: Rainbows, Halos, and Blue Skies

9.7a Rainbows

"My heart leaps up when I behold a rainbow in the sky." This is how the poet Wordsworth described his reaction to a rainbow, and it is probably not too bad a description of how many of us feel upon seeing a dazzling, colored arc stretching across the sky. Rainbows are both beautiful and puzzling. How do the elements of water and sunlight combine to produce such spectacles? Armed with the information in Sections 9.3 and 9.6, we are in a position to find out.

> **Physics To Go 9.13**

Most of us have seen rainbows while washing our cars or watering our lawns or gardens on sunny days, but have you ever considered what goes into creating a rainbow? The next time you're outside using a hose and see a rainbow, ask yourself some of these questions, and experiment to find the answers if they're not obvious to you. Where is the Sun in the sky relative to you and the direction in which the rainbow is seen? How does the production of a rainbow depend upon the size of the water droplets in the mist provided by the hose? How are the colors of the rainbow ordered from inside to outside? How large is the rainbow arc? (To help you with the assessment of the dimensions of the rainbow, you can use the information in **Figure 9.65**, which shows some angle-measuring techniques or tricks long used by astronomy students to establish the angular separations of celestial objects in the sky.) Keep the answers to these questions in mind to reinforce and/or confirm what you read in the remainder of this section.

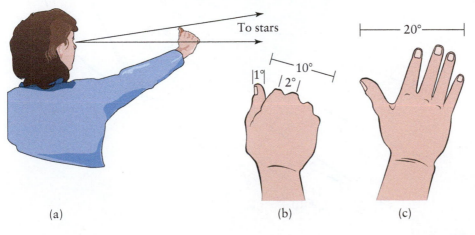

(a)　　(b)　　(c)

Figure 9.65 Using the human hand to roughly measure angles. (a) Extend the hand at arm's length and sight along it. (b) The thumb subtends about 1°, two knuckles about 2°, and the fist about 10°. (c) The open hand subtends about 20°.

Before doing so, however, we need to point out some rainbow basics. First, rainbows consist of arcs of colored light (spectra) stretching across the sky, with the red part of the spectrum lying on the outside of the bow and the blue-violet part lying on the inside. Second, rainbows are always seen against a background of water droplets with the Sun typically at our backs. These two basic characteristics of rainbows are what we seek to understand.

Imagine a beam of light from the Sun striking a raindrop. For simplicity, we will assume that raindrops are spherical, although real falling raindrops are more oblate in shape. If we apply the law of refraction, concentrating only on those rays that are internally reflected at the back of the drop and return in the general direction of the Sun (to be consistent with the second rainbow basic above), we find the result shown in **Figure 9.66**. The ray striking the drop at its center (ray 1) returns directly back along its incident direction and defines the axis of the drop. Rays entering above the axis exit below the axis, and vice versa. The farther above the axis the ray enters, the greater its emergent angle up to a point defined by ray 7. This ray is called the *Descartes ray*, after René Descartes, who first suggested this explanation for rainbows in 1637.

For rays entering above the Descartes ray, the exit angles are *less* than that of the Descartes ray. Thus rays entering the drop on either side of the Descartes ray emerge at about the same angle as the Descartes ray itself, leading to a concentration of rays leaving the droplet at a maximum angle corresponding to that of the Descartes ray. This angle is about 41° for rays 6 through 10 in Figure 9.66.

This concentration of sunlight at exit angles near 41° produces rainbows. The Descartes model predicts that rainbows should consist of circles of light of angular radii equal to 41°, centered on a point opposite the Sun in the sky—the *antisolar point*. Thus, to see a rainbow, we need to look for these concentrated rays in a direction about 41° from the "straight back" direction with the Sun behind us (**Figure 9.67**). Notice, if the Sun is above the horizon, the antisolar point

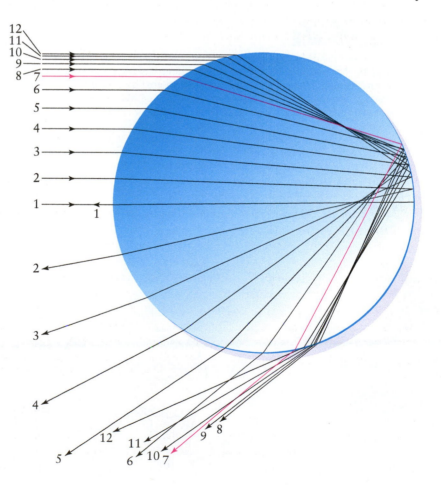

Figure 9.66 Paths of light rays through a water drop. Ray 7 is the Descartes ray.

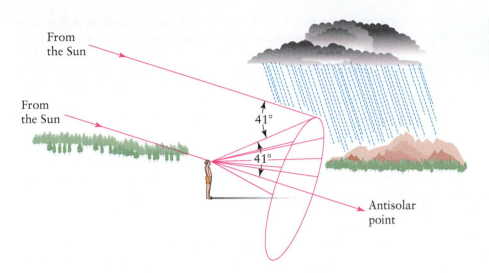

Figure 9.67 From Descartes' construction, the rainbow is predicted to be a circle of angular radius 41°, centered on the antisolar point.

From the Sun

From the Sun

41°

41°

Antisolar point

will be below the horizon along the direction of your shadow. In this case, the rainbow circle intersects the horizon, and we see only an arc of the circle. For earthbound observers, the best rainbow apparitions occur when the Sun is on the horizon, for then we see half of the rainbow circle. If the Sun is higher in the sky than about 41° above the horizon, then no rainbow can be seen from the ground because the antisolar point lies 41° or more below the horizon, and the rainbow circle never reaches above the horizon. This is why observers throughout most of the continental United States rarely see rainbows at noon. When viewed from an aircraft, a rainbow can form a complete circle, however.

So far, we have addressed several aspects of the shape and location of rainbows but not their colors. To do so, we must include the phenomenon of dispersion. Recall from Section 9.6 that blue light is more strongly deviated in passing through transparent media than is red light. This means that the maximum emergent angle from the raindrop for blue light will be smaller than the maximum emergent angle for red light (**Figure 9.68**). Therefore, the blue light is concentrated at slightly smaller angles than is the red light. Calculations show that blue-violet light is concentrated in a circle of angular radius of about 40°, whereas red light is concentrated at an angle of about 42°. The other colors of the rainbow fall in between. A more detailed model, including dispersion, thus predicts that real rainbows should consist of bands of color in the sky a total of some 2° or so wide, with blue-violet colors on the inside and red-orange colors on the outside. And this is precisely what is seen.

The application of the laws of reflection and refraction (including dispersion) to falling raindrops gives us an explanation for what is called the *primary* rainbow. The primary results from *one* internal reflection of the rays in the drop at the rear surface. Higher-order rainbows may be produced by rays executing two or more internal reflections before leaving the droplet. Some of you no doubt have seen *secondary* rainbows lying outside the primaries along arcs of circles having angular radii of approximately 51° (**Figure 9.69**). The ordering of the colors of these rainbows is reversed from that of the primaries. All of these properties are explicable in terms of the laws of optics with which we have become familiar (**Figure 9.70**).

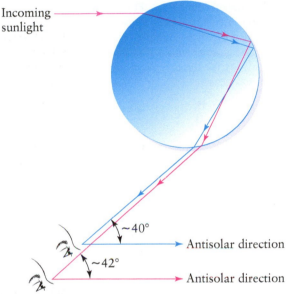

Incoming sunlight

~40°

Antisolar direction

~42°

Antisolar direction

Figure 9.68 Dispersion of sunlight by a spherical raindrop.

9.7b Halos

Halos, circular arcs of light, often with reddish inner edges, surrounding the Sun or full Moon might be considered winter's answer to rainbows. When the temperature in the upper atmosphere drops below freezing, ice crystals form.

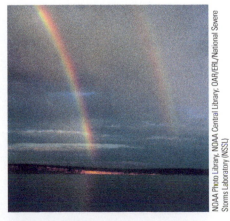

Figure 9.69 Primary and secondary rainbows.

NOAA Photo Library, NOAA Central Library; OAR/ERL/National Severe Storms Laboratory (NSSL)

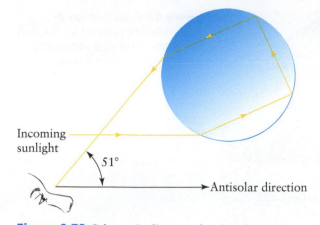

Figure 9.70 Schematic diagram showing the production of a secondary rainbow from two internal reflections in a raindrop. The path shown is for a typical ray of yellow light. The ray emerges at an angle of about 51° with respect to the antisolar direction. Red rays emerge with slightly smaller angles, whereas blue rays emerge with somewhat larger angles, thus accounting for the reversal of the color-ordering in the secondary bows.

At high elevations in the temperate regions of Earth, one common shape exhibited by such crystals is that of a hexagon, similar to a short, unsharpened pencil (**Figure 9.71a**). When seen in cross section, such crystals may be considered as pieces of equilateral (60°) prisms, and they deviate light like them (**Figure 9.71b**).

As in the case of rays entering raindrops, if one traces the paths of rays entering such a crystal at various incident angles, one finds that there is a concentration of exiting rays with deviation angles near 22°. Thus, when light from the Sun or the Moon enters a cloud of such ice crystals having all possible orientations, the emergent rays tend to be clustered into circular arcs having angular radii of 22° centered on the source of illumination.

To see a ray of light forming part of a halo, we should look in a direction 22° away from the Sun or Moon (**Figure 9.72**). When doing so, you may notice that the inner edge of the halo circle is tinted red. This is again the result of dispersion. At each refraction, the blue component of sunlight (or moonlight, which is merely reflected sunlight) is more strongly refracted than is the red component. Consequently, the angle of concentration for the blue light is somewhat greater than it is for red light, and the latter piles up preferentially at the inner edge of the halo, as indicated in **Figure 9.73**.

What we have described is the well-known 22° halo. There are also 46° halos, which result from light entering one face of the pencil crystal and leaving through one end. These halos are much fainter than the 22° halos and are much harder to see—partly because they occupy such a large portion of the sky, having angular diameters of more than 90°! The sundogs mentioned at the beginning of the chapter also arise by refraction and dispersion in ice crystals, but not randomly oriented pencil-like ones. Instead, plate-like crystals (Figure 9.71a lower) with their large flat sides parallel to the ground are responsible for concentrating (and coloring) the light at angular positions along the horizon 22° ahead of and/or behind the Sun. And these are but a few of the many, many other phenomena associated with ice crystal reflection and refraction. Such magnificence surrounds us almost daily if only we allow our eyes to be open to it. A knowledge of physics can help us to appreciate these natural wonders more deeply, making apparitions like the sundogs shown in Figure CO-9 even more impressive and inspiring.

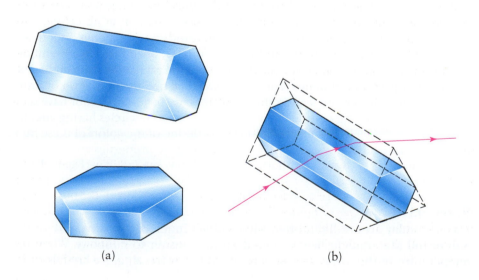

Figure 9.71 (a) Two simple ice crystal forms: top, a columnar or pencil crystal; bottom, a plate crystal. (b) A light ray passing through a pencil crystal is refracted as if it were passing through an equilateral (60°) prism.

(a)

(b)

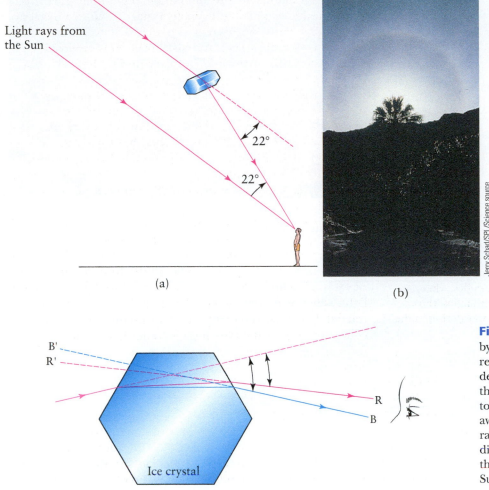

Figure 9.72 (a) Schematic representation of how a 22° halo is produced, showing an enlarged image of the ice crystals typically responsible for this phenomenon. To see sunlight that is refracted through an angle of 22° by the ice crystals to form part of the halo, one looks to an angle of 22° away from the sun in the sky. (b) Actual 22° halo around the sun (itself partially obscured by the intervening tree).

Jerry Schad/SPL/Science source

Figure 9.73 Dispersion of sunlight by ice crystals produces halos with reddish inner edges. Because the deviation of blue rays is larger than that of red rays, the blue rays appear to originate along the line $B'B$, farther away from the Sun than the red rays, which appear to come from the direction $R'R$. Thus, the inner edge of the halo—that is, the part nearest the Sun—is tinged reddish.

9.7c Blue Skies

The most common of all atmospheric optical phenomena is the blue sky. It is caused by air molecules *scattering* sunlight in all directions. As a light wave travels through the atmosphere, the wave's oscillating electric field causes the electrons in air molecules to oscillate with the same frequency. From the discussion in Chapter 8, we know that oscillating electric charges (electrons, in this case) emit electromagnetic radiation. This emitted light, which travels outward in all directions (hence the use of the term *scattered*), is what we see filling the sky.

But why is it blue instead of white like the incident sunlight? It turns out that the electrons in air molecules are much more efficient at absorbing and radiating higher frequencies of light. When blue light makes an electron in an air molecule oscillate, it absorbs and scatters much more of the incident radiant energy than when red light makes it oscillate. In particular, the amount of radiant energy scattered per second by air molecules is inversely proportional to the wavelength raised to the fourth power:

$$\frac{E_{scattered}}{t} \propto \frac{1}{\lambda^4}$$

The wavelength of blue light is about 0.7 times that of red light (from Table 8.1). Consequently, there is about 4.2 ($= 1/0.7^4$) times as much blue in scattered sunlight as there is red. The sky is pale blue rather than "pure" blue because all frequencies of light still remain present to some extent in the scattered sky light, diluting the blue to some extent.

Not only does molecular scattering give us beautiful blue skies, it is also responsible for the stunning orange and red colors often seen in clouds shortly

Figure 9.74 (a) Photo of clouds shortly after sunset. (b) Sunlight on its way to a cloud at sunset has most of the blue removed by scattering.

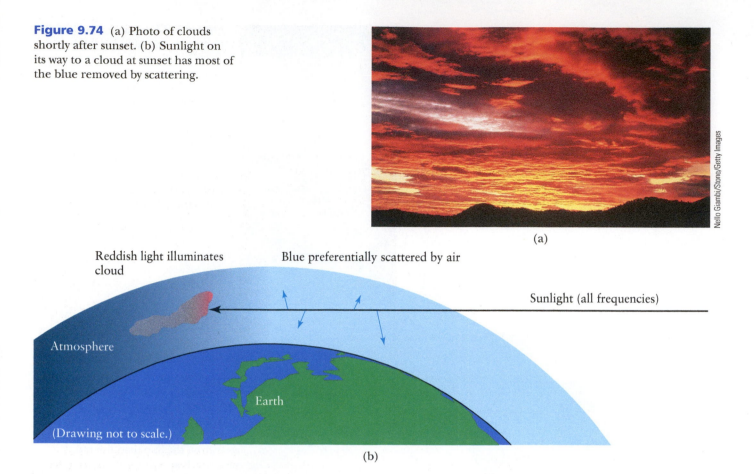

(a)

Reddish light illuminates cloud

Blue preferentially scattered by air

Sunlight (all frequencies)

Atmosphere

Earth

(Drawing not to scale.)

(b)

after sunset and shortly before sunrise (**Figure 9.74a**). The sunlight that reaches such clouds must travel hundreds of kilometers through the atmosphere. Along the way, more and more of the radiant energy is removed from the sunlight and transferred to scattered light. The higher frequencies (like blue) are much more strongly attenuated by this process than are the lower frequencies (like red). Consequently, the light that reaches the cloud is reddish in color because comparatively less of the red in the original sunlight has been removed (**Figure 9.74b**). The same scattering process that produces reddish-hued clouds near sunrise and sunset is also responsible for the fiery red-orange appearance of the Sun itself when located near the horizon.

Learning Check

1. Which of the following is *not* involved in the formation of a typical 41° rainbow? (Indicate all that apply.)
   (a) diffraction
   (b) dispersion
   (c) internal reflection
   (d) refraction
   (e) interference

2. If the Sun were just rising in the east at the time a rainbow is seen, then in what part of the sky would the rainbow be located?
   (a) directly overhead at the zenith point
   (b) toward the northern horizon
   (c) toward the southern horizon
   (d) toward the eastern horizon
   (e) toward the western horizon

3. Which of the following is responsible for producing halos around the Sun?
   (a) water droplets
   (b) dust particles
   (c) ice crystals
   (d) carbon-dioxide molecules
   (e) ozone molecules

4. (True or False.) In the atmosphere, the intensity of scattered sunlight with wavelength 480 nm is less than that of light with wavelength 590 nm.

ANSWERS: 1. (a), (e) **2.** (e) **3.** (c) **4.** False

Although much knowledge about the way light propagates in transparent media had been amassed before 1600, the developments in optics that occurred in the 17th century quickly eclipsed all that had been done during the previous 1,500 years. The first decade of that century saw the invention of the microscope and the refracting telescope, the latter being effectively used by Galileo to discover craters on the Moon, the phases of Venus, and four of the moons of Jupiter (**Figure 9.75**). By 1611, Johannes Kepler (1571–1630) had discovered total internal reflection. In or about 1621, the law of refraction was found by Willebrord Snel (or in its anglicized form Snell), a professor of mathematics at the University of Leiden, after many years of experimentation. Unfortunately, Snel's results were not publicized until years after his death, and by that time Descartes had already succeeded in getting his version of the law of refraction into print.

René Descartes (1596–1650; **Figure 9.76**), "father of modern philosophy" and famous for his statement *Cogito, ergo sum* ("I

Figure 9.76 René Descartes (right foreground) at the court of Queen Christina of Sweden.

physical quantity. These results appeared in 1637 at the beginning of Descartes' *La Dioptrique*.

The laws of reflection and refraction reappear in another of Descartes' works entitled *Météores*. Here he presents a mathematical explanation of primary and secondary rainbows, accounting for their angular dimensions and locations but not for their colors. It remained for Isaac Newton to produce the correct explanation for color in rainbows.

At the end of Chapter 2, several of the accomplishments of Sir Isaac Newton in the field of optics were mentioned, including his invention of the reflecting telescope and his explanation of the phenomenon of dispersion. These two items are very closely connected, the former being a direct consequence of the latter. In about 1666, Newton performed an *experimentum crucis* (a "critical experiment") in which, by a clever arrangement of two prisms, he was able to demonstrate that light of a single color undergoes no further dispersion upon being refracted through a prism (**Figure 9.77**). From this and similar experiments, Newton concluded, "Light itself is a heterogeneous mixture of differently refrangible rays," asserting that there is an exact correspondence between color and the "degree of

Figure 9.75 Galileo demonstrating his discovery of the satellites of Jupiter to the counselors of Venice in 1610.

think, therefore I am"), was also an accomplished mathematician and physicist. He founded the study of analytical geometry and was the first to develop the use of the hypothetical model as a tool of research. For example, his model for the propagation of light may be compared to that of a tennis ball moving uniformly. The law of reflection can be deduced by imagining an elastic collision of the ball with a stationary, impenetrable surface and applying the principle of conservation of linear momentum. Interestingly, Descartes was one of the first to recognize the momentum of a particle as an important

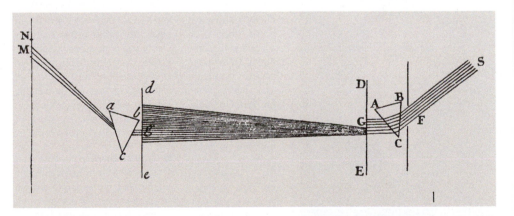

Figure 9.77 Newton's drawing of his two-prism experiment. Sunlight *S* enters the prism *ABC* at the right and is dispersed into a spectrum. Slit *G* selects a small portion of the spectrum with essentially a single color which is then projected on the screen *de* where slit *g* permits the light to enter a second prism *abc*. The light emerging from the second prism and striking the screen at *M* is found to consist only of the single color that entered it, not a complete spectrum. This demonstrates that a prism does not somehow "add" color to the light.

refrangibility," the least refrangible rays being "disposed to exhibit a red colour."

Newton's prismatic experiments seem to have convinced him of two things: first, that "light"—white light or sunlight—was not a "body" unto itself but an aggregate of corpuscles (tiny particles). "Each colour is caused by uniformly moving globuli," he wrote. "The uniform motion which gives the sensation of one colour is different from the motion which gives the sensation of any other colour." In particular, Newton believed the particles producing the sensation of blue light moved faster in glass than those associated with red light and that it was for this reason that blue rays were more strongly refracted. Newton's corpuscular theory of light permitted him to account for many of the properties of light, notably its straight-line propagation. But Newton recognized that certain phenomena, for example, diffraction and interference, could best be explained by treating light as a wave. Thus, even before the beginning of the 18th century, we see an apparent duality (particle versus wave) emerging in the nature of light. We will have more to say on this matter in Chapter 10.

The second thing of which Newton became convinced as a result of his color experiments was the impossibility of producing refracting telescopes (ones using only combinations of lenses) that were free from chromatic aberrations. It was for this reason that Newton turned his attention to the development of reflecting telescopes.

Newton's discoveries about light and color were published in 1704 in his *Opticks*. In one of Newton's investigations, he extended Descartes' model of rainbows, explicitly taking into account dispersion, and calculated the size of the rainbow arcs for each of the different colors in the primary and secondary rainbows. This book ends with a series of "Queries"—questions about the nature of light and related phenomena in need of further explanation and study. It has been argued that these questions formed the most important part of the book insofar as they strongly influenced the course of research on light for the next two centuries.

Newton's particle model of light stood for less than a century before it was dealt a nearly mortal blow by Thomas Young (1773–1829; **Figure 9.78**). Young was born into an English family of Quakers and was quickly seen to be a genius. By the age of 14, he knew several languages and had written an autobiography in Latin. Although Young studied medicine and became a physician, his intellectual pursuits spanned an incredible variety of disciplines, from physics to hieroglyphics (he worked on the Rosetta Stone) to insurance.

Figure 9.78 Thomas Young. His work with interference demonstrated that light possesses wavelike properties.

Karl Friedrich Gauß.

Figure 9.79 Karl Friedrich Gauss.

Young championed the wave model of light and showed that it could account for interference and diffraction of light. In his most famous experiment, Young passed light through two pinholes and produced an interference pattern similar to the one in Figure 9.7. From this pattern he was able to measure the wavelengths of different colors of light. Young was also the first to suggest that light could be polarized and was therefore a transverse wave. Young's success, along with that of other physicists, including Fresnel, in the early 1800s, caused the wave model of light to replace Newton's particle model—no small feat considering Newton's great stature. But as we shall see in Section 10.2, the particle model of light was successfully reincarnated a century later.

Optical science in the 19th century also profited from the work of another polymath, Karl Friedrich Gauss (1777–1855; **Figure 9.79**). Gauss was a child prodigy whose accomplishments in mathematics, physics, and astronomy mark him as one of the greatest intellects of all time. In the field of optics, Gauss invented the heliotrope, an instrument used in surveying and triangulation work, but only *after* he fully developed the necessary optical theory. In 1841, he published *Dioptrische Untersuchungen*, in which he analyzed the path of light through a system of lenses. In this work, in addition to determining how image distance, object distance, and the focal length of a lens are related to one another (the lens formula), he showed that any system of lenses is equivalent to a properly chosen single lens. This was Gauss's last significant contribution to science, and one of his biographers has called it his greatest work.

Advancements in optics by no means came to an end in the mid-1800s. Beginning shortly after World War II, applied optics began to flourish as the mathematical methods of communication theory and high-speed digital computers were brought into the discipline. Optical technology, driven by the development of the laser in 1960 (see Section 10.8), continues to advance, bringing improvements in our lives and in our understanding of the universe. But if rapid progress has recently occurred, it has happened because modern scientists have, in the words of Newton, "stood on the shoulders of giants," including Descartes, Gauss, and Newton himself.

■ QUESTIONS

1. Isaac Newton made significant contributions to the science of optics in addition to those that he developed in the field of mechanics. Give three examples of key discoveries or explanations that emerged from Newton's investigations into optical phenomena.
2. Thomas Young's conception of the fundamental nature of light differed from that of Newton. Compare these two views of light, and give some experimental or observational evidence to sustain each of these representations.

SUMMARY

» Visible light consists of electromagnetic waves with wavelengths between about 400 and 750 nanometers.

» When two or more light waves overlap in any region of space, **interference** occurs. If the waves are in phase, they reinforce one another, yielding **constructive interference**. If the waves are out of phase, they tend to cancel one another to some degree, producing **destructive interference**.

» Because light is a transverse wave, it can be **polarized**, that is, made to oscillate along some preferred direction. Polaroid sunglasses and LCDs make use of this property of light.

» The **law of reflection** states that the **angle of incidence** equals the **angle of reflection**.

» Mirrors make use of the law of reflection to modify and control the direction of propagation of light rays: plane mirrors produce undistorted virtual images of the object before them; concave mirrors can concentrate light rays and yield magnified images; convex mirrors disperse light rays and have a wide field of view.

» **Refraction** occurs when light crosses a boundary between two different transparent media. The **law of refraction** connects the size of the **angle of refraction** to that of the angle of incidence: The refraction angle is smaller than the incident angle if the wave speed in the transmitted medium is lower than that in the incident medium. Conversely, it is larger than the incident angle if the wave speed in the transmitted medium is higher than that in the incident medium.

» When the wave speed in the medium in which the transmitted ray travels is higher than that in the medium in which the incident ray travels, there exists a **critical angle** of incidence for which the angle of refraction will be 90°. For angles of incidence larger than the critical angle, incident rays undergo **total internal reflection** and remain trapped in the incident medium. This effect produces the brilliant sparkle in diamonds and underlies the operation of optical fibers.

» Lenses are optical devices that use refraction to control the paths of light rays and to form images. Simple, thin lenses may be either diverging or converging, according to whether their **focal lengths** are negative or positive, respectively. Knowing the focal length of a lens and the position of an object relative to the lens, the location of the image formed by the lens may be determined using the lens formula.

» The **magnification** of the image provides information about the size and orientation of the image. It equals the negative ratio of the image distance to the object distance.

» **Dispersion** describes the process whereby individual wavelengths (colors) comprising a beam of light are separated upon passing across the boundary between two transparent media. Dispersion occurs because the speed of light in any medium other than a vacuum is different for different wavelengths. Dispersion is responsible for unwanted chromatic aberrations in simple lenses, but when controlled by the use of prisms, it can be employed to establish the temperature and composition of luminous sources.

» Application of the laws of reflection and refraction, including the effects of dispersion, can lead to an increased understanding and appreciation of many natural phenomena in the area of atmospheric optics, like rainbows and halos.

IMPORTANT EQUATIONS

Equation	Comments
$\Delta x = (S/a)\lambda$	Fringe spacing, double slit interference
$p = \dfrac{sf}{s-f}$	Lens formula
$M = \dfrac{\text{image height}}{\text{object height}}$	Magnification of a lens system
$M = \dfrac{-p}{s}$	Magnification using image distance and object distance

MAPPING IT OUT!

1. The "shell" of a concept map dealing with lenses and their properties is shown in **Concept Map 9.2**. Most of the concepts and all of the linking phrases needed to form meaningful propositions have been left out. Complete this map by selecting the appropriate concepts and linkages from the lists that follow. Some connecting words will have to be used more than once to finish the map correctly. If you need to, be sure to review Sections 9.3 through 9.5 for help in properly determining and relating the relevant concepts.

2. Sections 9.6 and 9.7 deal with the phenomena of dispersion and its applications to atmospheric optics. One such application is the explanation of rainbows. Develop a concept map that you could use to teach a friend or family member about the formation and basic characteristics of rainbows. Your map should include links to the fundamental physical processes of reflection, refraction, and dispersion and should distinguish between primary and secondary rainbows. It should also indicate the conditions that must be satisfied to observe rainbows.

Concepts	Linking Words
Altered size	which can be
Real ($p > 0$)	come in
Upright ($M > 0$)	for example
Diverging	form
Ray diagram	using
Altered orientation	can be combined in
Eye lens	located by
Inverted ($M < 0$)	
Reduced ($|M| < 1$)	
Two principal rays	
Converging	
Telescope	
Enlarged ($|M| > 1$)	
Microscope	
Virtual ($p < 0$)	

▌**CONCEPT MAP 9.2** Mapping It Out! Exercise 1.

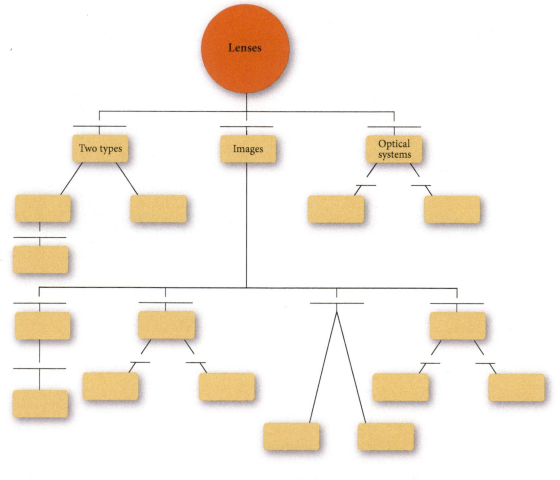

▌ QUESTIONS

(■ **Indicates a review question, which means it requires only a basic understanding of the material to answer. Questions without this designation typically involve integrating or extending the concepts presented thus far.**)

1. ■ Why are the Doppler effect and diffraction not as commonly experienced with light as they are with sound?
2. ■ Distinguish between specular reflection and diffuse reflection.

3. ■ The law of reflection establishes a definite relationship between the angle of incidence of a light ray striking the boundary between two transparent media and its angle of reflection. Describe this relationship.
4. The cover of a book appears blue when illuminated with white light. What color will it appear in blue light? Red light? Explain.
5. A ballet dancer is on stage dressed in a green body suit. How could you light the stage so that the dancer's costume looked black to the audience?

6. ■ Describe how light passing through two narrow slits can produce an interference pattern.

7. A person looking straight down on a thin film of oil on water sees the color red. How thick could the film be to cause this? How thick could it be at another place where violet is seen? (Some useful information is given at the beginning of Section 9.1.)

8. If the wavelength of visible light were around 10 cm instead of 500 nm and we could still see it, what effect would this have on our ability to see diffraction and interference effects?

9. An interference pattern is formed by sending red light through a pair of narrow slits. If blue light is then used, the spacing of the bright areas (where constructive interference takes place) won't be the same. How will it be different? Why?

10. For a given slit spacing, a, and wavelength, λ, if the screen distance, S, is increased, what happens to the fringe spacing—that is, the distance between adjacent bright regions in a typical double slit interference pattern?

11. If the separation, a, between the slits in a double slit interference experiment is reduced, then, for a fixed value of screen distance, S, and wavelength of light, λ, how will the separation, Δx, between adjacent bright areas in the interference pattern change?

12. ■ What is polarized light? How do Polaroid sunglasses exploit polarization?

13. Describe how you could use two large, circular Polaroid filters in front of a circular window as a kind of window shade.

14. Just before the Sun sets, a driver encounters sunlight reflecting off the side of a building. Will Polaroid sunglasses stop this glare? Why or why not?

15. You get a new pair of sunglasses as a birthday gift from a friend. When wrapping the present, your friend has removed the price tag and all the other labels from the sunglasses. How could you determine whether the pair of shades has polarizing lenses in them or not?

16. Compared to a person's height, what is the minimum length (top to bottom) of a mirror that will allow the person to see a complete image of themselves from head to toe?

17. An observer O stands in front of a plane mirror as shown in **Figure 9.80**. Which of the numbered locations 1 through 5 best represents the location of the image of the source S seen by the observer? Justify your choice by appealing to the appropriate law(s) of optics.

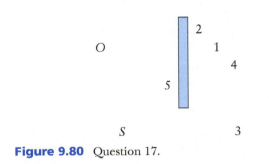

Figure 9.80 Question 17.

18. Light traveling along direction SO in **Figure 9.81** strikes the surface of a plane mirror. Which of the paths OP, OQ, OR, or OT best describes the path of the light reflected from the mirror? Defend your choice by appealing to the appropriate law(s) of optics.

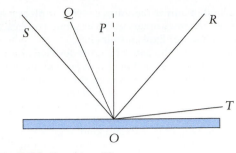

Figure 9.81 Question 18.

19. Consider **Figure 9.82** below. At which of the lettered positions would an observer be able to see the image of the dot in the mirror?

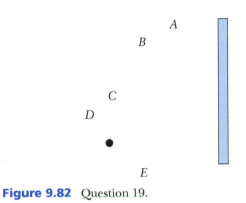

Figure 9.82 Question 19.

20. Two plane mirrors are hinged along one edge and set at right angles to one another as shown in **Figure 9.83**. A light ray enters the system, striking mirror 1 with an angle of incidence of 45°. Draw a diagram showing the direction in which the ray will exit the system. This device is called a *corner reflector* and is the basis for the design of bicycle reflectors and some highway signs.

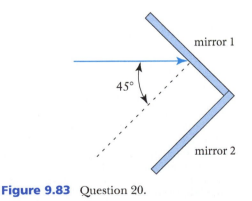

Figure 9.83 Question 20.

21. ■ What is different about an image (of a nearby object) formed with a convex mirror compared to an image formed with a concave mirror? What are the advantages of each type of mirror?

22. ■ What is the ideal shape of concave mirrors used in telescopes?

23. What kind of mirror (concave, convex, or plane) do you think dental care workers use to examine patients' teeth and gums for disease? Explain your choice.

24. ■ Describe how the path of a ray is deviated as it passes (at an angle) from one medium into a second medium in which the speed of light is lower. Contrast this with the case when the speed of light in the second medium is higher.

25. How would Figure 9.29a be different if the glass were replaced by water or by diamond?

26. The speed of light in a certain kind of glass is exactly the same as the speed of light in benzene—a liquid. Describe what happens when light passes from benzene into this glass, and vice versa.

27. A piece of glass is immersed in water. If a light ray enters the glass from the water with an angle of incidence greater than zero, in which direction is the ray bent?

28. A light ray in air enters a block of clear plastic as shown in **Figure 9.84**. Which of the numbered paths represents the correct one for the ray in the plastic? Justify your choice by appealing to the appropriate law(s) of optics.

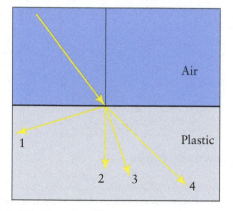

Figure 9.84 Question 28.

29. After hitting a ball into a water trap, a golfer looks into the pond and spies the ball within apparent easy reach. Reaching in to retrieve the ball, the golfer is surprised to find that it cannot be grasped even with a fully extended arm. Explain why the golfer was deceived into thinking that the location of the ball was close at hand.

30. One sometimes hears the expression, "It was like shooting fish in a barrel!" This usually is taken to mean that the task, whatever it was, was easy to complete. But is it really easy to shoot fish in a barrel? Only if you know some optics! Suppose you're in a boat and spy a large fish a few meters away. If you want to shoot the fish, how should you aim? Above the image of the fish? Below it? Directly at the image? Explain your choice. (You may assume that the path of the projectile you fire will not be deviated from a straight line upon entering the water, unlike light.)

31. Rank (from smallest to largest) the angle of refraction for a light ray in air entering each of the following substances with an angle of incidence equal to 30°: (a) water, (b) benzene, (c) dense flint glass, (d) diamond.

32. ■ What is total internal reflection, and how is it related to the critical angle?

33. Would the critical angle for a glass–water interface be less than, equal to, or greater than the critical angle for a glass–air interface? Explain your choice.

34. Explain why images seen through flat, smooth, uniform, plate-glass windows are undistorted.

35. ■ For transparent solids, distinguish between effects of surfaces that curve inward and those that curve outward on the paths of a parallel bundle of light rays incident on each.

36. ■ Distinguish converging lenses from diverging lenses, and give examples of each type.

37. If you were lost in the forest and wanted to start a small fire to keep warm or cook a meal using sunlight and a small lens, what type of lens—converging or diverging—would you use and why?

38. Of the three converging lenses shown in Figure 9.44, which would you expect to have the shortest focal length? Why?

39. ■ Describe the three principal rays used to locate an image.

40. ■ Contrast real images with virtual images in as many ways as you can.

41. Indicate whether each of the following is a real image or a virtual image.
(a) Image on the retina in a person's eye
(b) Image one sees in a rearview mirror
(c) Image one sees on a movie screen
(d) Image one sees through a magnifying lens

42. A convex lens forms a clear, focused image of some small, fixed object on a screen. If the screen is moved closer to the lens, will the lens have to be (a) moved closer to the object, (b) moved farther from the object, or (c) left at the same location to produce a clear image on the screen at its new location? Explain your answer.

43. ■ How is the magnification of a lens related to the object distance? How is it related to the image distance? What is the significance of the sign of the magnification? What is the significance of its magnitude (size)?

44. Estimate the values of the magnification in Figures 9.45, 9.48, and 9.49.

45. ■ What is chromatic aberration? How can it be remediated?

46. ■ How is the eye able to form focused images of objects that are different distances away?

47. ■ When a person is nearsighted, what happens in the eye when the person is looking at something far away? How is this condition commonly corrected?

48. ■ Describe the phenomenon of dispersion, and explain how it leads to the production of a spectrum.

49. Two light waves that have wavelengths of 700 and 400 nm enter a block of glass (from air) with the same angle of incidence. Which has the larger angle of refraction? Why? Would the answer be different if the light waves were going from glass into air?

50. Suppose a 20-m-long tube is filled with benzene and the ends sealed off with thin disks of glass. If pulses of red and blue light are admitted simultaneously at one end of the tube, will they emerge from the opposite end together, that is, at the same time? If not, which pulse will arrive at the far end of the tube first? Why?

51. The difference in speed between red light and violet light in glass is smaller than the difference in speed between the same two colors in a certain type of plastic. For which material, glass or plastic, would the angular spread of the two colored rays after entering the material obliquely from air be the largest? Why?

52. ■ What is a prism? Why are such devices useful to scientists?

53. Would a prism made of diamond be better at dispersing light than one made of glass? Why or why not?

54. Referring to Figure 9.64, why doesn't the ray inside each of the two prisms get refracted out of the prisms at the second interface instead of turned back into the glass? Similarly, referring to Figure 9.17, why don't the rays that undergo two reflections inside the pentaprism get refracted out of the prism at one or both of these boundaries instead of being turned back into the glass? Explain what must be happening at the glass–air interfaces in these prisms to produce the ray paths shown.

55. ■ Describe the accepted model of rainbows. Specifically, discuss how the model accounts for the size, shape, location, and color ordering of primary rainbows.

56. Suppose an explosion at a glass factory caused it to "rain" tiny spheres made of glass. Would the resulting rainbow

be different from the normal one? If so, how might it be different and why?

57. ■ Compare the primary and secondary rainbows as regards their angular size, color ordering, and number of internal reflections that occur in the rain droplets.

58. Suppose you are told by close friends that they had witnessed a glorious rainbow in the west just as the Sun was setting. Would you believe them? Why or why not?

59. ■ How is a 22° halo formed? Describe a measurement technique you might use to distinguish a genuine 22° halo from some other "halo-like" phenomenon seen surrounding the Sun or the Moon.

60. Water droplets in clouds scatter sunlight in ways that are similar to those of air molecules. Because clouds are white (during the day), what can you conclude about the frequency dependence of scattering by cloud particles?

PROBLEMS

1. Suppose a beam of red light from an He-Ne laser ($\lambda = 633$ nm) strikes a screen containing two narrow slits separated by 0.200 mm. The fringe pattern is projected on a white screen located 1.00 m away. Find the distance in millimeters between adjacent bright areas near the center of the interference pattern that is produced.

2. In a double-slit interference experiment, a special lamp emitting yellow light from heated sodium atoms is used to produce an interference pattern on a screen located 1.50 m from a pair of slits separated by 0.10 mm. If the distance between adjacent bright regions in the resulting pattern is 8.84 mm, what is the wavelength of the sodium light?

3. A light ray traveling in air strikes the surface of a slab of glass at an angle of incidence of 50°. Part of the light is reflected and part is refracted. Find the angles the reflected and refracted rays make with respect to the normal to the air–glass interface.

4. A ray of yellow light crosses the boundary between glass and air, going from the glass into air. If the angle of incidence is 20°, what is the angle of refraction?

5. Using Figure 9.34, find the angles of refraction for a light ray passing from air into glass with the following angles of incidence: 5°, 10°, and 20°. Do you notice a trend in the resulting values? If so, describe it. Based on your observations, what would you predict the angle of refraction to be for an angle of incidence of 40°? How does your value compare with that inferred from Figure 9.34?

6. A fish looks up toward the surface of a pond and sees the entire panorama of clouds, sky, birds, and so on, contained in a narrow cone of light, beyond which there is darkness. What is going on here to produce this vision, and how large is the opening angle of the cone of light received by the fish?

7. A camera is equipped with a lens with a focal length of 30 cm. When an object 2 m (200 cm) away is being photographed, how far from the film should the lens be placed?

8. A 2.0-cm-tall object stands in front of a converging lens. It is desired that a virtual image 2.5 times larger than the object be formed by the lens. How far from the lens must the object be placed to accomplish this task, if the final image is located 15 cm from the lens?

9. When viewed through a magnifying glass, a stamp that is 2 cm wide appears upright and 6 cm wide. What is the magnification?

10. A person looks at a statue that is 2 m tall. The image on the person's retina is inverted and 0.005 m high. What is the magnification?

11. What is the magnification in Problem 7?

12. A small object is placed to the left of a convex lens and on its optical axis. The object is 30 cm from the lens, which has a focal length of 10 cm. Determine the location of the image formed by the lens. Describe the image.

13. If the object in Problem 12 is moved toward the lens to a position 8 cm away, what will the image position be? Describe the nature of this new image.

14. (a) In a camera equipped with a 50-mm focal-length lens, the maximum distance that the lens can be from the film is 60 mm. What is the smallest distance an object can be from the camera if its image on the film is to be in focus? What is the magnification?

    (b) An extension tube is added between the lens and the camera body so that the lens can be positioned 100 mm from film. How close can the object be now? What is the magnification?

15. The focal length of a diverging lens is negative. If $f = -20$ cm for a particular diverging lens, where will the image be formed of an object located 50 cm to the left of the lens on the optical axis? What is the magnification of the image?

16. The equation connecting s, p, and f for a simple lens can be employed for spherical mirrors, too. A concave mirror with a focal length of 8 cm forms an image of a small object placed 10 cm in front of the mirror. Where will this image be located?

17. If the mirror described in the previous problem is used to form an image of the same object now located 16 cm in front of the mirror, what would the new image position be? Assuming that the magnification equations developed for lenses also apply to mirrors, describe the image (size and orientation) thus formed.

18. Compute the approximate ratio of the amount of blue light in scattered sky light to the amount of orange.

19. If the wavelength of light is doubled, by what fraction does the amount of the light scattered by Earth's atmosphere change? Is the amount of light scattered at the new wavelength larger or smaller than the amount of light scattered at the old wavelength?

CHALLENGES

1. When white light undergoes interference by passing through narrow slits, dispersion occurs. Why? What is the ordering of the various colors, starting from the middle of the pattern?

2. In Section 9.6, we described how the speed of light varies with wavelength (or frequency) for transparent solids. But the speed of light in matter is also a function of temperature and pressure. This dependence is most marked for gases and is instrumental in producing such things as mirages and atmospheric refraction, the latter phenomenon being the displacement of an astronomical object (like the Sun

or another star) from its true position because of the passage of its light through the atmosphere. Because Earth's atmosphere is a gaseous mixture and easily compressed, its density is highest near Earth's surface and gradually declines with altitude. (Refer to the discussion in Section 4.4 and Figure 4.29.) Thus, the speed of light in the atmosphere is lowest near the surface and gradually gets higher, approaching c as one goes farther and farther into space. Using this fact and the law of refraction, sketch the path a light ray from the Sun would follow upon entering Earth's atmosphere, and predict the apparent position of the Sun relative to its true position (**Figure 9.85**). What does this tell you about the actual location of the Sun's disk relative to your local horizon when you see it apparently setting brilliantly in the west in the evening?

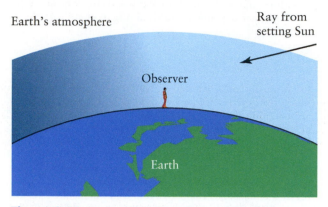

Earth's atmosphere

Ray from setting Sun

Observer

Earth

Figure 9.85 Challenge 2. (Drawing not to scale.)

3. Would the critical angle for a glass–water interface be less than, equal to, or greater than the critical angle for a glass–air interface? Why?

4. Every transparent material can be characterized by what is called its *index of refraction, n,* given by the equation

$$n = \frac{c}{v}$$

Here c is the speed of light in a vacuum and v is the speed of light in the transparent material. Although normal air is not quite a vacuum, its density is small enough that v_{air} is approximately equal to c, so that we may take $n_{air} = 1.00$ without too much error.

The precise mathematical relationship between the angles of incidence and refraction is given by what is known as *Snell's law,*

$$n_1 \sin A_1 = n_2 \sin A_2$$

where n_1 and A_1 are the index of refraction and the angle between the light ray and the normal, respectively, in medium 1, and n_2 and A_2 are the corresponding quantities for medium 2.

(a) Using the information in Table 9.1, compute the index of refraction for water and for fused silica (glass).

(b) Verify the result in Example 9.2 using Snell's law explicitly. Show, using the same relationship, that the values for the angles of refraction in air (medium 2) in panels (b), (c), (d), and (e) of Figure 9.36 are correct for the given angles of incidence in glass (medium 1).

5. While above water, a swimmer's eyes are focused on a nearby boat. When the swimmer submerges, the underwater part of the boat will not be in focus even though it is the same distance away and the swimmer's eyes have not been refocused. Why? Watertight goggles correct this. Why?

6. The form of the lens formula used most commonly in physics is

$$\frac{1}{f} = \frac{1}{s} + \frac{1}{p}$$

Use this to derive an equation that gives s in terms of p and f.

7. For $f = 10$ cm, use the lens formula (in Challenge 6) to compute the image distance, p, for the object distances, s, given in the accompanying table. A few values of p have already been included so you can check your work. Make a graph of p versus s using your results. What does the sign of p signify? For what values of s does p change the most? What happens to p as s gets larger and larger? Can you relate your conclusions to the operation of a simple camera or your eye? (*Note*: If you are familiar with the use of computer spreadsheet programs, the calculations required in this exercise are particularly easy to complete. Even the plot may be done using the graphics capabilities supported by most such software.)

s (cm)	p (cm)	s (cm)	p (cm)
2.0		25.0	
4.0	−6.67	50.0	12.50
8.0		100	
9.999		200	10.53
12.0		500	
14.0	35.00	1000	
18.0		10,000	
20.0			

8. A camera is used to photograph a kitten. The camera is equipped with a standard lens that has a focal length of 50 mm. The camera is focused by moving the lens closer to or farther away from the film at the back of the camera.

(a) First the kitten is photographed when it is 350 mm (13.8 in.) in front of the camera lens. Use the lens formula to compute the distance that the lens must be from the film for the image to be in focus.

(b) Compute the magnification and the height of the image, assuming that the kitten is 100 mm tall.

(c) The kitten is then photographed from a distance of 3,500 mm (11.5 ft). Compute the lens-to-film distance, the magnification, and the image height.

9. A camera is equipped with a telephoto lens that has a focal length of 200 mm. Repeat part (c) of Challenge 8 for this situation.

10. Why can't you go to the end of a rainbow? What happens to the rainbow when you walk toward one of its ends?

10

ATOMIC PHYSICS

Figure CO-10 Plasma display panel (PDP).

Photo folks

CHAPTER INTRODUCTION: Something Old, Something New

Fluorescent light and plasma TV. One is a familiar, trusted, and often taken-for-granted device that's been providing efficient lighting for decades. The other became one of the first status symbols of the 21st century (**Figure CO-10**). One is typically hidden in fixtures and noticed only if it doesn't work correctly (it flickers or buzzes). The other can cost a thousand dollars or more and is often the center of attention—even becoming part of the studio set for TV news programs and military press briefings. Its inventors shared an Emmy Award in 2002. One is mundane, the other glamorous.

And yet, they both produce light using the same physical process. In fact, a plasma display is basically millions of tiny fluorescent lamps emitting light in a coordinated fashion. It is a three-step process: electrons and ions move through a plasma and give energy to atoms. These atoms in turn emit ultraviolet (UV) light, which strikes a special material called a *phosphor*. The phosphor's atoms absorb the UV, which we can not see, and emit the visible light that we do see.

Fluorescent lights and plasma TVs are examples of the many devices that take advantage of modern atomic

physics and the quantum nature of light. The first part of the chapter describes how the efforts of early 20th-century physicists to understand three puzzling physical processes led to the concept of photons and the Bohr model of the atom. This is followed by a description of the wave nature of atomic particles and the emergence of the revolutionary new physics known as *quantum mechanics*. The remainder of the chapter demonstrates how the quantum-mechanical model of the atom successfully explains the production of atomic spectra and x-rays and the operation of lasers.

10.1 The Quantum Hypothesis

By the end of the 19th century, physicists were quite satisfied with the great progress that had been made in the study of physics. Many questions that had been puzzling scientists for thousands of years had been answered. Advances made possible by Newton's great treatise on mechanics and the elucidation of electromagnetism by Maxwell gave physicists cause to celebrate the deep understanding of the physical world that they had acquired. In fact, some physicists worried that the field might be dying; they feared that *all* of their questions might soon be answered.

But some problems defied solution, and advancements in experimental equipment and techniques led to new discoveries. The implications arising from attempts to explain these phenomena were so revolutionary that by the early 1900s many of these same physicists were bewildered and wondered just how much they really did know.

In the first five sections of this chapter, we take an historical approach to three of these problems. We describe how investigations into them led to a reinterpretation of the very nature of light and the way it interacts with matter. The phenomena are (1) **blackbody radiation**, (2) the **photoelectric effect**, and (3) **atomic spectra**.

Physicists knew a great deal about these phenomena, but the fundamental understanding of their *causes* had eluded them. It was much like the period before Newton: astronomers knew a lot about the shapes of the orbits of the Moon and the planets, but they didn't know what caused these particular shapes. Newton's mechanics and his law of universal gravitation provided the answer. Similarly, scientists some two centuries later were seeking the theoretical basis for these three phenomena.

10.1a Blackbody Radiation

Everything around you, as well as your own body, is constantly emitting electromagnetic (EM) radiation. At normal room temperature, mostly infrared (IR) radiation is emitted. Very hot bodies that are actually glowing, such as the Sun and the heating elements in a toaster, emit visible light and ultraviolet (UV) radiation as well as copious amounts of infrared. Bodies that are very cold, on the other hand, emit very weak, long-wavelength infrared and microwaves. This radiation, the characteristics of which we have already discussed at some length in Section 8.6, had been studied carefully by scientists in the 19th century, and its properties were well known but not understood. For example, it had been determined that a perfectly "black" body, one that would absorb all light and other EM radiation incident on it, would also be a perfect emitter of EM radiation. Such an ideal *blackbody* was said to emit *blackbody radiation* (BBR).

> ### ▶ Physics To Go 10.1
>
> The glowing filament of a standard incandescent lamp acts roughly like a blackbody with a temperature of about 2,500 to 3,000 K. You can view the visible part of the radiation it emits with a compact disc. The spiral line of data on the CD causes light to undergo dispersion—the CD acts somewhat like a prism. (The light undergoes diffraction and interference as it reflects off the closely spaced segments of the line, and different colors (frequencies) of light undergo constructive interference in different directions.)
>
> Use the "shiny" side of the CD (usually the bottom), and position it so you can see the reflection of a glowing lightbulb near the hole in the center. (It's OK if the bulb is in a light fixture.) Slowly tilt the CD so the image of the bulb moves to one side of the hole in the center, and note the appearance of colors on the opposite side of the hole. What do the colors look like? Does it appear that all colors of the visible spectrum are present?

The well-known characteristics of the BBR emitted by a given blackbody at a particular temperature can be conveniently illustrated with a graph. Imagine examining each wavelength of radiation in turn, from very short wavelength ultraviolet to much longer wavelength microwaves, and measuring the energy that is emitted each second (the power) from the blackbody. The graph of the resulting data, plotted as the intensity of the radiation versus wavelength, is called a *blackbody radiation curve.* **Figure 10.1** shows two BBR curves for blackbodies at two different temperatures. These graphs illustrate two important ways that BBR changes when the temperature of the body is increased:

1. More *energy* is emitted per second *at each wavelength* of EM radiation.
2. The *wavelength* at which the *most energy* is emitted per second (in other words, the *peak* of the BBR curve) *shifts to smaller values.* That is why toaster elements glow red hot and extremely hot stars glow blue hot.

Why does a blackbody emit different amounts of EM radiation at different wavelengths in precisely this way? As a start, using the principles of electromagnetism, we can easily see why EM waves are emitted. Atoms and molecules are continually oscillating, and they contain charged particles—electrons and protons. We saw in Chapter 8 that this kind of system will produce EM waves. But the exact mechanism involved, one that would account for the two features above, was a mystery to scientists at the end of the 19th century.

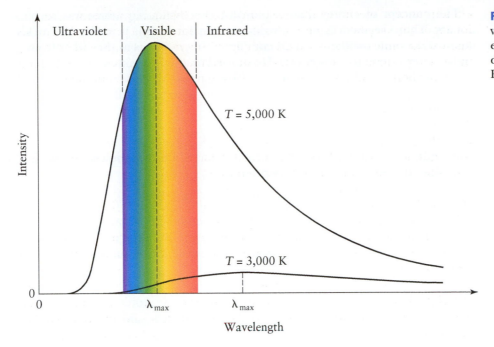

Figure 10.1 Graph of intensity versus wavelength for the EM waves emitted by blackbodies at two different temperatures (compare with Figure 8.38).

10.1b Quantized Oscillators

The clue to solving the puzzle was discovered by German physicist Max Planck (Figure 10.2) in the year 1900. By trial and error, Planck derived a mathematical equation that fit the shape of the BBR curves. (This is a common first step in theoretical physics. The mathematical shapes of the planetary orbits—ellipses—were known about a century before Newton explained *why* they were ellipses.) The equation did little to increase the understanding of the fundamental process, but Planck also developed a model that would account for his equation.

Planck proposed that an oscillating atom in a blackbody can have only certain fixed values of energy. It can have zero energy or a particular energy *E*, or an amount equal to 2*E*, or 3*E*, or 4*E*, and so on. In other words, the energy of each atomic oscillator is **quantized**. The energy *E* is called the fundamental **quantum** of energy for the oscillator. The allowed values of energy for the atom are integral multiples of this quantum:

allowed energy = *nE*, where *n* = 0, 1, 2, 3, . . .

This is quite different from an ordinary oscillator, such as a mass hanging from a spring or a child on a swing, which can have a continuous range of energies, not just certain values.

We can illustrate the difference between quantized and continuous energy values by comparing a stairway to a ramp. A cat lying on a stairway has quantized potential energy: it can be on only one of the steps, and each step corresponds to a particular *PE*. A cat lying on a ramp does not have quantized potential energy. It can be anywhere on the ramp. Its height above the ground can have any value within a certain range, and therefore its *PE* is one of a continuous range of values (Figure 10.3).

Figure 10.2 Max Planck (1858–1947), whose work started what may be called the *quantum revolution*.

Figure 10.3 (a) A "quantized" cat. Its potential energy is restricted to certain values, one for each step. (b) On the ramp the cat can be anywhere, so its potential energy is not quantized.

The concept of energy quantization for the oscillating atoms was revolutionary. There seemed to be no logical reason for it. But it worked. Planck's quantized atomic oscillators could emit light only in bursts as they went from a higher energy level to a lower one. He showed that light emitted in this fashion by a blackbody resulted in the correct BBR curves. Moreover, he determined that the basic quantum of energy was proportional to the oscillator's frequency: $E \propto f$. In particular,

$$E = (6.63 \times 10^{-34}) f = hf$$

The constant h is called *Planck's constant* and has SI units of joule-seconds (J-s). The allowed energies can now be written as:

$$\text{allowed energy} = nhf, \text{ with } n = 0, 1, 2, 3, \ldots$$

Planck was unsure of the implication of his model. He regarded it mainly as a helpful construct that gave him the correct result. But it turned out to be the first of many scientific discoveries showing the existence of quantized effects at the atomic level.

■ Learning Check

1. The temperature of blackbody A is 2,000 K and that of blackbody B is 1,000 K. Regarding the electromagnetic radiation emitted by the two blackbodies:
   (a) A emits more energy per second at any given wavelength than B.
   (b) The peak of A's blackbody radiation curve is at a longer wavelength than that of B's.
   (c) B emits infrared but A does not.
   (d) All of the above.

2. (True or False.) The energy of an oscillating atom in a blackbody can be any value within a certain limited range.

3. When something is restricted to having only certain numerical values, it is said to be _____.

ANSWERS: 1. (a) **2.** False **3.** quantized

10.2 The Photoelectric Effect and Photons

The second phenomenon that puzzled physicists at the beginning of the 20th century was the photoelectric effect. It seemed to be unrelated to blackbody radiation, other than that it also involved light, but the concept of quantization turned out to be the key to explaining it, too. The photoelectric effect was accidentally discovered by Heinrich Hertz during his experiments with electromagnetic waves. Hertz was producing EM wave pulses by generating a spark between two conductors. The EM wave would travel out in all directions and induce a spark between two metal knobs used to detect the wave. Hertz noticed that when these knobs were illuminated with ultraviolet light, the sparks were much stronger. The UV was somehow increasing the current in the spark. This is one example of the photoelectric effect.

10.2a Photoelectric Physics

The photoelectric effect is exhibited by metals when exposed to x-rays, UV light, or (for some metals, including sodium and potassium) high-frequency visible light. By some means, the EM waves give energy to electrons in the metal, and the electrons are ejected from the surface (Figure 10.4). (It was these freed electrons that enhanced the spark in Hertz's apparatus.) Because of the nature of EM waves, we shouldn't be too surprised that this sort of thing can happen. But again, some of the characteristics of the photoelectric effect that scientists observed persuaded them that the fundamental process was not

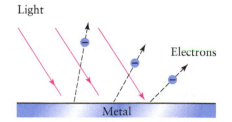

Figure 10.4 The photoelectric effect. Light striking a metal surface causes electrons to be ejected.

entirely understood. The biggest puzzle was the relationship between the speed or energy of the electrons and the incident light. It might seem reasonable to suppose that brighter light would cause the electrons to gain more energy and be ejected from the surface with higher speeds. Yet even though it was found that brighter light caused *more* electrons to be ejected each second, it did not increase their energies. Even more of a surprise was the finding that only the frequency (color) of the light affected the electron energies. Higher-frequency light ejected electrons with higher energy, and even extremely dim light with high enough frequency would immediately cause electrons to be ejected.

The explanation of the photoelectric effect was supplied in 1905 by Albert Einstein (**Figure 10.5**). Planck had suggested that light is emitted in discrete "bursts" or "bundles." Einstein took this idea one step further and proposed that the light itself remains in bundles or "packets" and is absorbed in this form. The electrons in a metal can only take in light energy by absorbing one of these discrete quanta of radiation. The amount of energy in each quantum of light depends on the frequency. In particular,

$$E = hf \qquad \text{(energy of a quantum of EM radiation)}$$

This is the same equation for the energy of Planck's quantized atomic oscillators. Einstein suggested that light and other electromagnetic waves are quantized, just like the energy of oscillating atoms.

This idea, that the energy of an EM wave is quantized, allows us to picture the radiation as being made up of individual particles now called **photons**. (The word was coined by G. N. Lewis in 1926.) Each photon carries a quantum of energy, $E = hf$, and propagates at the speed of light in empty space. The total energy in an EM wave is just the sum of the energies of all of the photons associated with it. Notice how different this picture is from the wave picture of light developed in Chapter 8. Both pictures are correct and present mutually complementary aspects of EM radiation. Under certain circumstances, such as those involving refraction or interference of light, the wave nature of light is manifested. In other circumstances, including those involving the emission or absorption of light, the particle aspects of light are demonstrated. These two different sides of light are like the front and back sides of a person (**Figure 10.6**). Sometimes, we see only the front of an individual. We recognize the individual by facial structure, eye color, hair color, and so on. At other times, we see only the back of the person. Here again we may recognize the person, but now by virtue of body structure and gait. In each case, we see different aspects of

Figure 10.5 Albert Einstein (1879–1955) enjoying one of his favorite pastimes—sailing.

(a)

(b)

Figure 10.6 (a) Recognition of an individual as she approaches is often made using such qualities as facial structure, eye color, hair color and style, and so on. (b) Recognition of the same person as she moves away from us can often be made using different qualities, such as body structure, gait, and so on.

the same person, but we are still able to recognize the person, although for the most part the clues leading to recognition are not the same in the two instances. The situation with light is analogous. Different experiments reveal different aspects of what we recognize to be the same type of EM radiation.

With Einstein's proposal, the observed aspects of the photoelectric effect could be synthesized together. Higher-frequency light ejects electrons with more energy because each photon has more energy to give. Brighter light simply means that more photons strike the metal each second. This results in more electrons being ejected each second, but it does not increase the energy of each electron. In truth, for each metal, there does exist a threshold (or cutoff) frequency below which no photoelectrons will be emitted, regardless of how intense the light is. Photons with frequencies less than this threshold possess too little energy to eject any electrons from the metal. The electrons are too tightly bound to the metal atoms, and the available energy from the photons is too small to overcome the binding energy. Einstein's explanation of the photoelectric effect earned him the 1921 Nobel Prize in physics. This new understanding of the nature of light and the way it interacts with matter profoundly altered the course of 20th-century physics.

Just how much energy does a typical photon have? Not very much. The energy of a photon of visible light is only about 3×10^{-19} joules. On this minute scale, it is convenient to use a much smaller unit of energy called the *electronvolt* (eV). One electronvolt is the potential energy of each electron in a 1-volt battery. Because voltage is energy per charge, the charge of an electron multiplied by voltage equals energy:

$$1 \text{ electronvolt} = 1.6 \times 10^{-19} \text{ coulomb} \times 1 \text{ volt}$$

$$= 1.6 \times 10^{-19} \text{ coulomb} \times 1 \text{ joule/coulomb}$$

$$1 \text{ eV} = 1.6 \times 10^{-19} \text{ J}$$

The energy of visible photons is on the order of 2 electronvolts, a much easier number to deal with. In terms of the electronvolt, the value of Planck's constant is

$$h = 6.63 \times 10^{-34} \text{ J-s} = 4.14 \times 10^{-15} \text{ eV/Hz}$$

Figure 10.7 shows the photon energies corresponding to different types of EM waves.

EXAMPLE 10.1 Compare the energies associated with a quantum of each of the following types of EM radiation.

$$\text{red light: } f = 4.3 \times 10^{14} \text{ Hz}$$
$$\text{blue light: } f = 6.3 \times 10^{14} \text{ Hz}$$
$$\text{x-ray: } f = 5 \times 10^{18} \text{ Hz}$$

SOLUTION For each one, we use Planck's original equation for a quantum of energy.
$$E = hf$$

$$= 4.136 \times 10^{-15} \text{ eV/Hz} \times 4.3 \times 10^{14} \text{ Hz}$$

$$= 1.78 \text{ eV} \qquad \text{(red light)}$$

Using the same equation for blue light and x-rays, we get

$$E = 2.61 \text{ eV} \qquad \text{(blue light)}$$

$$E = 20{,}700 \text{ eV} \qquad \text{(x-ray)}$$

Notice how much greater the energy of a quantum of x radiation is compared to that of either red or blue light. Little wonder then that high doses of x-rays can be harmful to the human body.

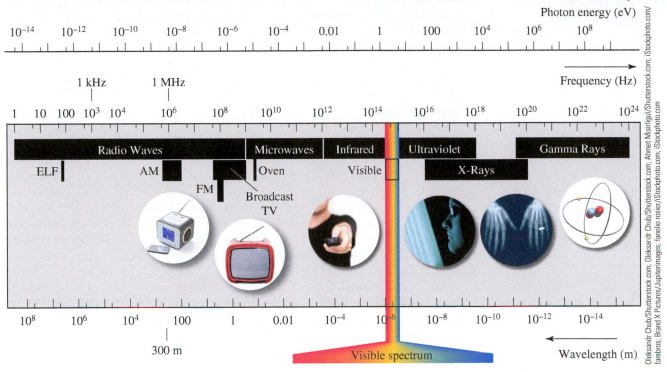

Oleksandr Chub/Shutterstock.com; Oleksandr Chub/Shutterstock.com; Ahmet Misirligul/Shutterstock.com; iStockphoto.com/
fambros; Brand X Pictures/Jupiterimages; tanelie rosier/iStockphoto.com; iStockphoto.com

Figure 10.7 Photon energies in the EM spectrum. Visible light photons range from approximately 1.7 to 3.1 electronvolts (red light to violet light).

10.2b Applications of the Photoelectric Effect

Before we move on to the third puzzle that faced scientists in the early 1900s, let's look at some applications of the photoelectric effect. This phenomenon is the key to "interfacing" light with electricity. Just as the principles of electromagnetism make it possible to convert motion into electrical energy and vice versa, the ability of electrons to absorb the energy in photons makes it possible to detect, measure, and extract energy from light. **Figure 10.8** shows a schematic of a device that can detect light. When no light strikes the metal, no current flows in the circuit because there is nothing to carry the charge through the tube. When light strikes the metal, electrons are ejected and attracted to the positive terminal of the tube. The result is a current flowing in the circuit. This sort of detector could be used to automatically count people entering a building. The detector is placed on one side of the door, and a light source is placed on the other side, pointed toward the detector. When something blocks the light going to the detector, the current stops. A counter connected to an

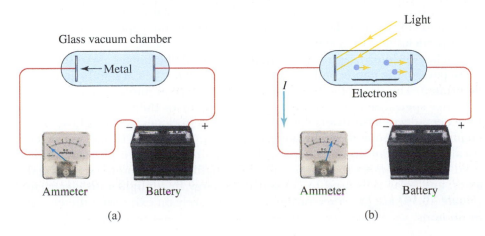

Figure 10.8 Schematic of a light detector. (a) No current flows in the circuit because there is nothing to carry charge through the tube. (b) Light releases electrons from the metal, allowing a current to flow. The size of the current indicates the brightness of the light.

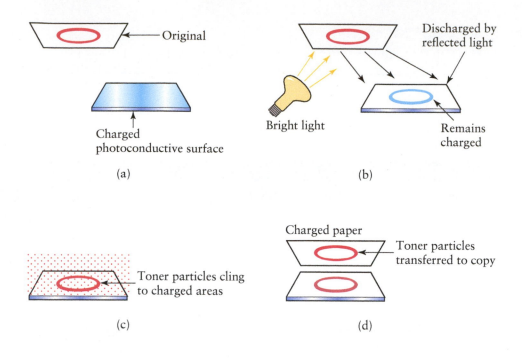

Figure 10.9 The main steps in the photocopying process.

Original

Charged
photoconductive surface

(a)

Discharged by
reflected light

Bright light

Remains
charged

(b)

Toner particles cling
to charged areas

(c)

Charged paper

Toner particles
transferred to copy

(d)

ammeter in the circuit then automatically tallies one count. A similar setup could be used to automatically open and close a door.

Photocopying machines and laser printers use the interplay between electricity and light in a process known as *electrophotography*. The key part of the operation is a special *photoconductive* surface. It is normally an insulating material, but it becomes a conductor when exposed to light. Electrons bound to atoms are freed when they absorb photons in the incident light.

The process of forming an image begins when the photoconductive surface is charged electrostatically. The surface retains the charge until light strikes it, and the freed electrons allow the charge to flow off the surface. A mirror image of the material to be printed is formed on the photoconductive surface using light. In the case of a photocopying machine, a bright light shines on the original, and the reflected light strikes the charged surface (**Figure 10.9**). White areas on the original reflect most of this light onto the corresponding areas of the photoconductive surface, which are consequently discharged by the large number of incident photons. Dark parts of the original—printed letters, for example—reflect very little light. Consequently, the corresponding regions on the photoconductive surface retain their electrical charge. Then fine particles of toner, somewhat like a solid form of ink, are brought near the surface. The toner particles are attracted to the charged regions and collect on them, while the discharged areas remain clear. A blank piece of paper, also electrically charged, is placed in contact with it. The toner particles are attracted to the paper and collect on it. The final image is "fused" on the paper by melting the toner particles into the paper.

In a laser printer, a laser under computer control illuminates and discharges those parts of the photoconductive surface that will not be dark in the image to be printed. The rest of the process parallels the steps in photocopiers.

Other specialized *photosensitive* materials, many of them semiconductors, have been devised for diverse uses. Light meters in cameras, the light-sensing elements in video and digital cameras that convert optical images into electrical signals, scanning elements in fax machines, sensitive photodetectors used by astronomers to measure the faint light from distant stars and galaxies, and solar cells used to convert the energy in sunlight into electricity (**Figure 10.10**) are just some of the devices that rely on extracting the energy in photons.

Vern Ostdiek/University of Michigan-Dearborn

Figure 10.10 Author Don Bord (left) and students on top of Emory Peak in Big Bend National Park, Texas. The solar cells above them absorb photons in sunlight to generate electrical energy that powers a radio transmitter for park personnel.

Learning Check

1. Is the energy associated with a photon of blue light
   (a) greater than
   (b) less than
   (c) equal to
   the energy associated with a photon of red light? Why?
2. (True or False.) In the photoelectric effect, increasing the frequency of the light being used will increase the energy of the electrons that are ejected.
3. (True or False.) In the photoelectric effect, increasing the brightness of the light being used will increase the energy of the electrons that are ejected.
4. Name two common devices that make use of the photoelectric effect or similar process in which electrons absorb photons.

ANSWERS: 1. (a), higher frequency **2.** True **3.** False **4.** photocopier, laser printer, digital camera, solar cell, . . .

10.3 Atomic Spectra

The third problem that defied explanation at the beginning of the 20th century involved the spectra produced by the various chemical elements. Suppose we use a prism to examine the light produced by the heated filament of an ordinary incandescent bulb. The light from the bulb will be dispersed on passing through the prism, as described in Chapter 9, and a spectrum will be produced (Figure 9.63 and Physics to Go 10.1). The spectrum will appear as a continuous band of the colors of the rainbow, one color smoothly blending into the next. Such a spectrum is called a **continuous spectrum**, and it is characteristic of the radiation emitted by a hot, luminous solid.

Imagine now a sample of some gas confined in a glass tube and induced to emit light, as by heating (**Figure 10.11**). If we examine the light from the luminous gas with a prism, we do not see a continuous spectrum. Instead we see an **emission-line spectrum** consisting of a few, isolated, discrete lines of color. In this case, the source is not emitting radiation at all wavelengths (colors) but only at certain selected wavelengths. If a different type of gas is investigated, one finds that the resultant line spectrum is different. Each type of gas has its own unique set of spectral lines (**Figure 10.12**).

> ### Physics To Go 10.2
>
> This is just like Physics to Go 10.1—only this time you use a compact fluorescent lamp. (The type shown in Figure 4.1 works well.) It's best if the CD is a couple of meters from the lamp and there are no other light sources around. What is different about the spectrum of colors? (The spectrum of light emitted by a compact fluorescent lamp is not a simple emission-line spectrum, but it is very similar to one.) You might also try this with sodium or mercury streetlights and neon lights.

Figure 10.11 The spectrum of a hot gas consists of several discrete colors (three in this example).

Figure 10.12 Emission spectra of selected elements.

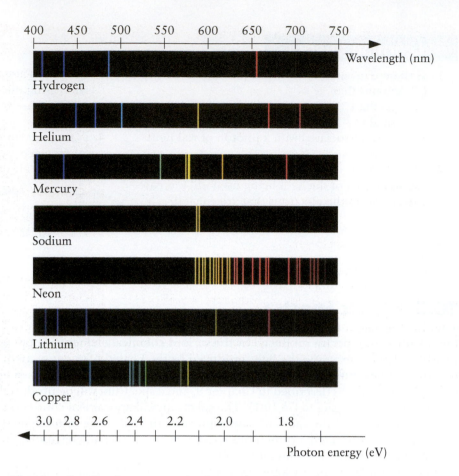

The fact that luminous, vaporized samples of material produce line spectra when their light is dispersed (as by a prism) was discovered in the 1850s. Once chemists recognized that each element has its own special spectrum, it became possible for them to determine the compositions of substances in the laboratory by examining their spectra. Two German scientists, Robert Bunsen (of Bunsen burner fame) and Gustav Kirchhoff, were pioneers in this new field of *spectroscopy*, the study of spectra, during the last half of the 19th century (see the applications feature at the end of this section). The problem that remained was to understand why luminous gases produce line spectra and not continuous spectra and how it is that each element has its own unique spectral "fingerprint" by which it can be identified.

Spectroscopy has grown to be one of the most useful tools for chemical analysis in fields as diverse as law enforcement and astronomy. Suspected poisons or substances found at a crime scene can be identified by comparing their line spectra to those in a catalog of known elements and compounds. Astronomers can determine what chemicals exist in the atmospheres of stars by examining spectra of the starlight with the aid of telescopes.

The explanation of atomic spectra inspired one of the most crucial periods of advancement in physics ever. In the remainder of this chapter, we describe how a picture of the structure of the atom was created and refined to account for atomic spectra. This effort spawned a "new" physics for dealing with matter on the scale of atoms—quantum mechanics.

Learning Check

1. A hot luminous solid emits a _____ spectrum, whereas a hot luminous gas produces a(n) _____ spectrum.

2. (True or False.) Two different gases can have the same emission-line spectrum.

ANSWERS: 1. continuous, emission-line **2.** False

In 1844, French philosopher Auguste Comte published the following view of the prospects of the science of astronomy:

> The stars are only accessible to us by a distant visual exploration. This inevitable restriction therefore not only prevents us from speculating about life on all these great bodies, but also forbids the superior inorganic speculations relative to their chemical or even their physical natures.

Like many other predictions about the future course of scientific discovery, Comte's was soon proved false. Indeed, within 20 years, two German scientists would pioneer techniques that would allow the composition of substances to be discovered from a long-distance examination of the light they emit.

Beginning in 1859, Robert Bunsen (1811–1899) and Gustav Kirchhoff (1824–1887; **Figure 10.13**) undertook a series of experiments at the University of Heidelberg that laid the foundations for what is now known as the field of spectroscopy. A key ingredient to their success was the burner developed by Bunsen a few years earlier. The Bunsen burner, which produces a flame with a very high temperature but low luminosity (brightness), is ideal for examining the distinctive colors imparted by chemical salts to flames containing them (**Figure 10.14**). Bunsen had originally used filters to distinguish the various hues emitted by different substances, but Kirchhoff suggested that a much surer distinction between different chemicals might be obtained by examining the spectra of the colored flames. Together they designed and constructed the first laboratory spectroscope using a 60° hollow prism filled with carbon disulfide and then carefully examined the light emitted by various hot gases (Figure 10.11). What they found was that each element produced its own special set of bright colored lines, the line patterns being as unique as a person's fingerprints. The importance of this observation was not lost on Bunsen and Kirchhoff. "We can base a method of qualitative analysis on these lines," Bunsen wrote, "that greatly broadens the field of chemical research and leads to the solution of problems previously beyond our grasp."

University of Michigan-Dearborn

(a)

University of Michigan-Dearborn

(b)

University of Michigan-Dearborn

(c)

University of Michigan-Dearborn

(d)

Figure 10.14 Characteristic colors for Bunsen burner flame tests for the elements (a) lithium, (b) sodium, (c) copper, and (d) potassium. In each case, the flame color reflects the wavelengths of the dominant emission lines in the spectrum of each element (Figure 10.12).

One problem thought to be beyond the grasp of scientists in the mid-19th century (as evidenced by Comte's remark) was the determination of the chemical compositions of the stars. Bunsen and Kirchhoff's efforts to do just that are thought to have originated from a fire in the city of Mannheim, some 10 miles west of Heidelberg. Bunsen and Kirchhoff apparently observed the fire from their laboratory window in the evening and turned their spectroscope toward the flames. There they were able to detect barium and strontium. Some time later, while reflecting on this event, Bunsen remarked that if he and Kirchhoff could analyze a fire in Mannheim, might they not do the same for the Sun? "But," he added, "people would think we were mad to dream of such a thing."

But dreams sometimes have a way of coming true, and such was the case here. The realization of the dream

AIP Neils Bohr Library

Bettman/Corbis

Figure 10.13 (left) Chemist Robert Bunsen, inventor of the Bunsen burner and codeveloper of the science of spectroscopy. (right) Physicist Gustav Kirchhoff, who worked with Bunsen, left his mark in many areas of physics.

began with Kirchhoff's demonstration that a conspicuous dark line (labeled with the letter *D* by Fraunhofer in 1814) in the spectrum of the Sun was unambiguously from the element sodium. Further experiments convinced Kirchhoff that a substance capable of emitting a certain spectral line has a strong absorptive power for the same line. He concluded that the dark *D* line in the solar spectrum was produced by absorption by sodium in the atmosphere of the Sun. The Sun, like Earth, contained sodium! By 1862, Kirchhoff had detected calcium, magnesium, iron, chromium, nickel, barium, copper, and zinc in the Sun, in addition to sodium. The analysis of the composition of celestial bodies had begun.

The excitement felt by Bunsen and Kirchhoff during this period may be best conveyed by quoting from a letter written by Bunsen to an English colleague on 15 November 1859:

> At present, Kirchhoff and I are engaged in an investigation that doesn't let us sleep. Kirchhoff has made a wonderful, entirely unexpected discovery in finding the cause of the dark lines in the solar spectrum, and he can increase them artificially in the Sun's spectrum or produce them in a continuous spectrum and in exactly the same position as the corresponding Fraunhofer lines. Thus, a means has been found to determine the composition of the Sun and the fixed stars with the same accuracy as we determine strontium chloride, etc. with our chemical reagents.

No sooner had Kirchhoff and Bunsen unlocked the door to the chemical secrets of the universe beyond Earth than they announced (in 1860) a new terrestrial metal, cesium (from the Latin *caesius*, "sky blue"), so named because of its brilliant blue spectral lines. The following year they published the discovery of yet another element, named rubidium (from the Latin *rubidus*, "dark red") for its strong red emission lines. In the years that followed, several other elements were identified spectroscopically: thallium (1861), indium (1863), gallium (1875), scandium (1879), and germanium (1886). In addition, studies of the spectra of the Sun and stars led to the discovery of helium and to the recognition that hydrogen is the most abundant element in the universe.

Although Bunsen and Kirchhoff did not understand *how* the spectra of the different elements were produced, they were highly successful at exploiting the uniqueness of such spectra for the purpose of identifying elements in various sources. The ultimate explanation for atomic spectra did not come until the work of Max Planck and Niels Bohr at the beginning of the 20th century. Nonetheless, it is hardly an understatement to say that the birth of what might be called "cosmic chemistry" occurred in Heidelberg in 1859 as a result of the blending of the talents of a skilled chemist and a mathematical physicist who dared to dream.

■ **QUESTION**

1. Describe at least three major contributions (e.g., inventions, discoveries, new techniques, etc.) made by Bunsen and Kirchhoff to the field of spectroscopy.

10.4 The Bohr Model of the Atom

At the beginning of the 20th century, little was known about the atom. In fact, many doubted that atoms existed at all. This made the origin of atomic spectra a mystery, in spite of the fact that spectroscopy was a booming field. In 1911, an important experiment performed by Ernest Rutherford in England revealed that the positive charge in an atom is concentrated in a tiny core—the nucleus (more on this in Chapter 11). This result was quickly followed in 1913 by a model of the atom put forth by the great Danish physicist Niels Bohr (**Figure 10.15**). But like Planck's explanation of blackbody radiation, the **Bohr model** of the atom was based on assumptions that did not seem sensible at the time. The basic features of Bohr's model are the following:

1. The atom forms a miniature "solar system," with the nucleus at the center and the electrons moving about the nucleus in well-defined orbits. The nucleus plays the role of the Sun and the electrons are like planets.

2. The electron orbits are quantized—that is, electrons can only be in certain orbits around a given atomic nucleus. Each allowed orbit has a particular energy associated with it, and the larger the orbit, the greater the energy. Electrons do not radiate energy (emit light) while in one of these stable orbits.

3. Electrons may "jump" from one allowed orbit to another. In going from a low-energy orbit to one of higher energy, the electron must *gain* an amount of energy equal to the difference in energy it has in the two orbits. When passing from a high-energy orbit to a lower-energy one, the electron must *lose* the corresponding amount of energy.

Figure 10.15 Niels Bohr (1885–1962), one of the giants in the field of atomic physics.

Figure 10.16 shows the Bohr model for the simplest atom, that of the element hydrogen (atomic number = 1). A lone electron orbits the nucleus, in this case a single proton. The electron can be in any one of a large number of orbits (four are shown). In each case, the electrical force of attraction between the oppositely charged electron and proton supplies the centripetal force needed to keep the electron in orbit.

When in orbit 1, the electron has the lowest possible energy. The electron has more energy in each successively larger orbit (**Figure 10.17**). The electron's energy while in any of the orbits is negative because it is bound to the nucleus (see the end of Section 3.5). To go to a larger orbit, the electron must gain energy. The maximum energy the electron can have and still remain bound to the proton is called the **ionization energy**. If the electron acquires more than this energy, it breaks free from the nucleus, and the atom is ionized. The resulting positive ion in this case is just a bare proton.

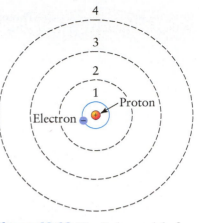

Figure 10.16 The Bohr model of the hydrogen atom. The electron can have only certain orbits. Four of these are shown. The figure is not drawn to scale: the fourth orbit is actually 16 times the size of the first orbit.

10.4a Atomic Spectra Explained

How does the Bohr model account for the characteristic spectra emitted by luminous hydrogen gas? When an electron in a larger (higher-energy) orbit "jumps" to a smaller orbit, it loses energy. One of the ways it can lose this energy is by emitting light (photons). This process is the origin of atomic spectra.

Imagine that the electron in our hydrogen atom is in the sixth allowed orbit, with its energy represented by E_6. Suppose the electron makes a transition (a jump) to an inner orbit—say, the second one—where its energy is E_2. To do so, the electron must lose an amount of energy, ΔE, equal to

$$\Delta E = E_6 - E_2$$

In what is called a *radiative transition*, the electron loses this energy through the emission of a photon that has an energy just equal to ΔE (**Figure 10.18**). The photon, spontaneously created during the transition of the electron from orbit 6 to orbit 2, carries off the excess energy into space in the form of EM radiation. This is much like Planck's atomic oscillators emitting photons. Because the energy of a photon equals h times its frequency, we have

$$\text{photon energy} = \Delta E = E_6 - E_2$$

and

$$\text{photon energy} = hf$$

Therefore,

$$hf = E_6 - E_2$$

(a)   (b)

Figure 10.17 A hydrogen atom with its electron (a) in orbit 2 and (b) in orbit 3. The electron has more energy when it is in orbit 3.

Figure 10.18 Photon emission. The electron makes a transition from orbit 6 to orbit 2 and emits a photon. The photon's energy equals the energy lost by the electron.

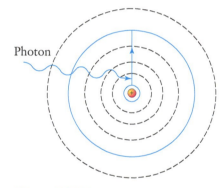

Figure 10.19 Photon absorption. The electron makes a transition from orbit 1 to orbit 5 by absorbing a photon.

Figure 10.20 Absorption spectrum of a gas. The gas absorbs photons in the light passing through it, so at those frequencies (colors) the spectrum is fainter.

The frequency of the emitted light is directly proportional to the difference in the energy of the orbits between which the electron jumped. The larger the energy difference, the higher the frequency of the light given off in the process. A downward transition from orbit 3 to orbit 2 will produce a photon with lower energy and lower frequency, because the energy difference between orbit 3 and orbit 2 is smaller:

$$\Delta E = E_3 - E_2 \qquad \text{is smaller than} \qquad \Delta E = E_6 - E_2$$

For the 6-to-2 transition in hydrogen, a violet-light photon is emitted. For the 3-to-2 transition, it is a red-light photon.

Each possible downward transition from an outer orbit to an inner orbit results in the emission of a photon with a particular frequency. An appropriately heated sample of hydrogen gas will emit light with these different frequencies but no other radiation. This is the line spectrum of hydrogen. (We will discuss this more in Section 10.6.)

A downward electron transition can also occur without the emission of light through what is called a *collisional transition*. In this case, a collision between a hydrogen atom and another particle (perhaps another hydrogen atom) can induce the electron in the outer orbit to spontaneously jump to an inner orbit. The energy that the electron loses can be transferred to the other particle, or it can be converted into increased kinetic energy of both colliding particles. Collision-induced transitions are generally important in dense gases where the numbers of atoms or molecules per unit volume of gas are large and the likelihood of two or more gas particles colliding is therefore quite high.

So far, we have focused on how an electron in an outer orbit may lose energy by jumping to an inner orbit. But the reverse of this process also occurs: an electron in an inner orbit can gain just the right amount of energy and jump to an outer orbit. For example, a hydrogen atom in the lowest energy state might gain the amount of energy needed for its electron to jump from orbit 1 to, say, orbit 5. To do this, the electron would have to acquire energy:

$$\Delta E = E_5 - E_1$$

One way to do this would be through a collision: the hydrogen atom could collide with another atom with more energy.

Another way an electron can jump to an outer orbit is by absorbing a photon with the proper energy (**Figure 10.19**). For example, an electron in orbit 1 can jump to orbit 5 if it absorbs a photon with energy:

$$\text{photon energy} = \Delta E = E_5 - E_1$$

If a sample of hydrogen gas is irradiated with a broad band of EM waves (like blackbody radiation), many such transitions to outer orbits will occur. Some of the photons in the incident radiation will have just the right energies to induce transitions from inner orbits to outer orbits. In the process, the number of photons with these particular energies will be reduced, and the intensity of the EM radiation at the corresponding frequencies will decrease. This reduction of intensity of light at only certain frequencies after it passes through a gas results in an **absorption spectrum** (**Figure 10.20**). For example, if white light is passed

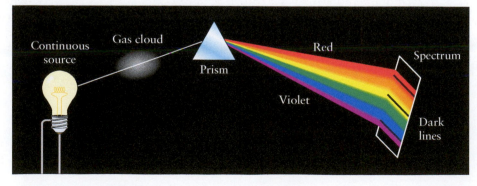

through hydrogen gas and then dispersed with a prism, dark bands appear at certain frequencies. They are exactly the same frequencies that are in the emission spectrum of luminous hydrogen.

This is how the element helium was discovered. In 1868, some of the absorption lines in the spectrum of sunlight were found not to correspond with those of any elements known at that time. The existence of a new element in the Sun's atmosphere was suggested to account for these lines. This new element was named after the Greek word (*helios*) for Sun.

10.4b Model Assumptions

The Bohr model was successful at explaining the origin of atomic spectra. But Bohr's model of electron orbits rested on two unexplained assumptions. First, there are only certain allowed orbits. Although one can place a satellite in orbit around Earth with any radius, the electron orbits were restricted to specific radii. These radii were determined by a seemingly arbitrary but nonetheless effective condition: that the angular momentum of the electron in its orbit is quantized. Much like the energy of Planck's quantized atomic oscillators, the orbital angular momentum of Bohr's electrons could only be integer multiples of Planck's constant divided by 2π; that is,

$$\text{allowed angular momentum} = n\left(\frac{h}{2\pi}\right), \text{ with } n = 1, 2, 3, \dots$$

So quantization was popping up everywhere: in the energy of oscillating atoms, the energy of EM waves, and now in the orbital angular momentum of atomic electrons.

The second assumption that physicists of the time found objectionable was that as long as an electron remained in one of its allowed orbits, it did not emit EM radiation. Maxwell's work had indicated that whenever a charged object undergoes acceleration, including centripetal acceleration, it will radiate. Put another way, an electron in a periodic orbit is much like a charge oscillating back and forth, and the latter results in the production of an EM wave. By this logic, an orbiting electron should be continually radiating, losing energy in the process and spiraling into the nucleus. According to the physical laws known at the time, atoms could not remain stable and should collapse in a fraction of a second! Obviously, something was missing.

Even though Bohr had a model that worked to account for atomic spectra, clearly the physics behind it was not fully elucidated. What was needed was a revolution in our understanding of how Nature works at the atomic level.

Concept Map 10.1 summarizes the Bohr model of the atom.

Learning Check

1. Which of the following is *not* true in the Bohr model of the atom?
   (a) The electrons orbit the nucleus.
   (b) The orbits of the electrons are quantized.
   (c) While moving in a stable orbit, electrons can emit photons.
   (d) Electrons can jump to a higher-energy orbit if they gain the right amount of energy.
2. In the Bohr model of the atom, a photon may be _____ when an electron jumps from a large, high-energy orbit to a smaller, low-energy orbit.

3. An atom is said to have been _____ when it absorbs a photon with sufficient energy to free an electron.
4. (True or False.) There is always a direct correlation between the absorption spectrum of a particular gas and its emission spectrum.

ANSWERS: 1. (c) **2.** emitted **3.** ionized **4.** True

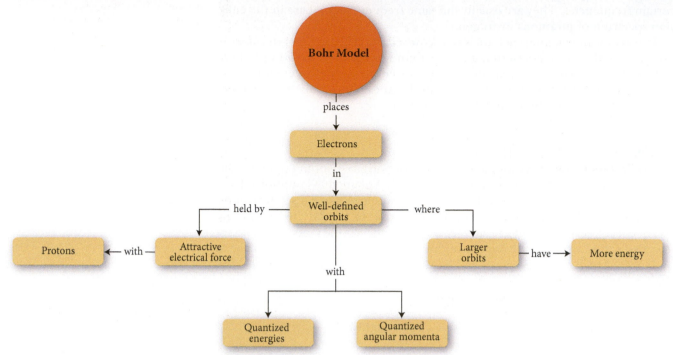

10.5 Quantum Mechanics

10.5a de Broglie's Hypothesis

The success of Bohr's model of the atom, even though it was at odds with accepted principles of physics at the time, indicated that perhaps a new physics was needed to describe what goes on at the atomic level. The first step in this direction came during the summer of 1923. While working on his doctorate in physics, a French aristocrat named Louis Victor de Broglie (rhymes with "Troy") proposed that electrons and other particles possess wavelike properties (**Figure 10.21**). Einstein had shown that light has both wavelike and particle-like properties, so why not electrons too? The wave associated with any moving particle has a specific wavelength (called the **de Broglie wavelength**) that depends on the particle's momentum, mv:

$$\lambda = \frac{h}{mv} \quad \text{(de Broglie wavelength)}$$

The higher the momentum of a particle, the shorter its wavelength. High-speed electrons have shorter wavelengths than low-speed electrons.

Once again, Planck's constant shows up. Because h is such a tiny number, de Broglie wavelengths are extremely small. This means that the wave properties of particles are only manifested at the atomic and subatomic level.

Figure 10.21 Louis de Broglie (1892–1987) was the first to suggest that particles can behave like waves.

AIP Neils Bohr Library

EXAMPLE 10.2 What is the de Broglie wavelength of an electron with speed 2.19×10^6 m/s? (This is the approximate speed of an electron in the smallest orbit in hydrogen.)

SOLUTION Using the electron mass given in the Table of Conversion Factors, the electron's momentum is

$$mv = 9.11 \times 10^{-31} \text{ kg} \times 2.19 \times 10^6 \text{ m/s}$$

$$= 1.995 \times 10^{-24} \text{ kg-m/s}$$

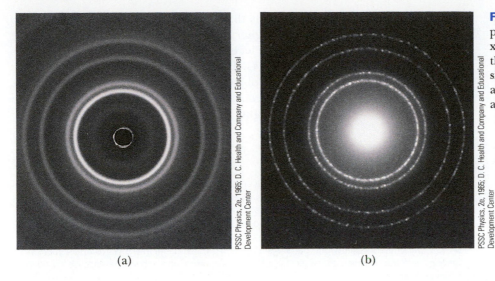

(a)   (b)

PSSC Physics, 2e, 1965; D. C. Health and Company and Educational Development Center

PSSC Physics, 2e, 1965; D. C. Health and Company and Educational Development Center

Figure 10.22 These patterns were produced by sending (a) a beam of x-rays and (b) a beam of electrons through an aluminum target. The similarity exists because the electrons act like waves as they interact with the aluminum atoms.

Using the value of h in SI units (Section 10.1):

$$\lambda = \frac{h}{mv} = \frac{6.63 \times 10^{-34} \text{J-s}}{1.99 \times 10^{-24} \text{ kg-m/s}}$$

$$= 3.32 \times 10^{-10} \text{ m} = 0.332 \text{ nm}$$

This distance is in the same range as the diameters of atoms.

Another radical theory had entered the arena of physics. Although submitted by a newcomer, de Broglie's hypothesis was not completely rejected by the physics community because Einstein himself found it plausible. Then, in 1925, the puzzling results of some experiments were interpreted as proof of the existence of de Broglie's matter waves. In a series of experiments, American physicist Clinton Davisson (with various collaborators) showed that a beam of high-speed electrons underwent diffraction when sent into a nickel target. The electrons were behaving just like waves. In fact, one gets the same kind of scattering pattern using x-rays or electrons (**Figure 10.22**).

Other experiments have verified the wave properties of electrons, protons, and other particles. Young's classic two-slit experiment, by which he proved that light is a wave, can be used to make particles undergo interference (**Figure 10.23**). The *electron microscope* exploits the wave properties of electrons. Instead of using ordinary light like a conventional microscope, an electron

Stream of particles

Particles detected at these points

Figure 10.23 Particles, like electrons, undergo interference when passed through narrow slits. They exhibit a property of waves (compare to Figure 9.7).

Proceedings of the National Academy of Sciences of the United States of America, PNAS October 18, 2005 vol. 102 no. 42 14952-14959

Figure 10.24 A false-color electron microscope image of an algae cell.

microscope uses a beam of electrons that acts like a beam of electron waves. The magnification can be much higher with an electron microscope because the de Broglie wavelength of the electrons is much shorter than the wavelengths of visible light (**Figure 10.24**). Diffraction and interference of waves passing around and between small objects such as cells strongly affect image clarity. Shorter-wavelength waves (electrons) diffract and interfere much less so than do longer-wavelength waves (light).

▌ Physics To Go 10.3

You can investigate in more detail the diffraction and interference of electrons passing through slits using a suite of programs developed at Kansas State University by Professor Dean Zollman and collaborators. Visit the Visual Quantum Mechanics Web site at http://phys.educ.ksu .edu/ and select one of the available simulations (for example, Double Slit Diffraction). Follow the instructions and explore the dependence of the resulting electron interference patterns on such parameters as slit width, particle energy, and particle-beam intensity. Compare your findings with the observed characteristics of light-wave diffraction and interference discussed in Chapters 6 and 9.

A more recently developed microscope uses the wavelike nature of electrons to form tantalizing images of individual atoms on the surfaces of solids. The *scanning tunneling microscope* (STM) uses an extremely fine pointed needle that scans back and forth over a surface. A positive voltage maintained on the needle attracts the electrons at the surface of the sample. If electrons were simply particles, they could not traverse the gap—less than 1 nanometer wide—between the surface and the needle. But their wave nature allows the electrons to penetrate or to "tunnel" through the gap and reach the needle (**Figure 10.25**). The tiny current of electrons that flows decreases rapidly if the gap widens. As the needle scans back and forth over the surface (like someone mowing a lawn), a feedback mechanism moves the needle up and down to keep this tunneling current constant. The varying height of the needle is recorded and used to draw a contour map of the surface that clearly shows individual molecules and atoms (**Figure 10.26** and Figure 4.7c). The 1986 Nobel Prize in physics was shared by the inventor of the electron microscope and the codevelopers of the STM (see Appendix A).

Figure 10.25 Simplified sketch of a scanning tunneling microscope (STM). The wavelike nature of the electrons allows them to cross the gap from the surface to the needle.

Figure 10.26 False-color scanning tunneling micrograph (STM) of DNA. The main feature of this image is a double-stranded DNA molecule, which appears as the row of orange/yellow peaks. The magnification of these features is about 1.6 million times.

10.5b The Bohr Atom Explained

Another major triumph for de Broglie's wave hypothesis was its ability to explain Bohr's quantized orbits. De Broglie reasoned that because an orbiting electron acts like a wave, its wavelength has to affect the circumference of its orbit. In particular, the electron's wave wraps around on itself, and in doing so it must interfere with itself. Only if the wave interferes constructively (peak matches peak) can the electron's orbit remain stable (**Figure 10.27**). This means that the circumference of its orbit must equal exactly one de Broglie wavelength, or two, or three, and so on.

$$\text{circumference of orbit} = n\lambda, \text{ with } n = 1, 2, 3, \ldots$$

If r is the radius of a circular orbit, then the circumference is $2\pi r$. Thus, the radius is

$$r = n\left(\frac{\lambda}{2\pi}\right), \text{ with } n = 1, 2, 3, \ldots$$

This distance sets the scale for atomic dimensions and is often used as a standard of measure in comparing the relative sizes of atoms.

Figure 10.27 Simplified wave representation of an electron orbiting a nucleus. The orbit in (a) is not possible because the wave interferes destructively with itself. The orbit in (b) is allowed. This corresponds to orbit 5, because it has 5 wavelengths fitted into the orbit.

EXAMPLE 10.3 Using the results of Example 10.2, find the radius of the smallest orbit in the hydrogen atom.

SOLUTION In de Broglie's model, the circumference of the smallest orbit ($n = 1$) must equal the de Broglie wavelength of the electron. We calculated this wavelength to be 0.332 nanometers for the smallest orbit. So,

$$r = \frac{\lambda}{2\pi}$$

$$r = \frac{0.332 \text{ nm}}{2\pi} = \frac{0.332 \text{ nm}}{2 \times 3.14} = \frac{0.332 \text{ nm}}{6.28}$$

$$= 0.0529 \text{ nm}$$

The allowed values for the circumference (or radius) of an electron's orbit leads to Bohr's allowed values for the electron's angular momentum. Because the de Broglie wavelength is given by

$$\lambda = \frac{h}{mv}$$

the radius can have the following values:

$$r = \frac{n}{2\pi}\left(\frac{h}{mv}\right), \text{ with } n = 1, 2, 3, \ldots$$

When we multiply both sides by mv, we get

$$mvr = n\left(\frac{h}{2\pi}\right), \text{ with } n = 1, 2, 3, \ldots$$

But the quantity mvr is the angular momentum of the electron (see Section 3.8), and we have recovered the same relationship that served as one of the assumptions in Bohr's theory of the atom at the end of Section 10.4. De Broglie's condition on the size of the electron's orbit turns out to be identical to Bohr's condition on the angular momentum of the electron.

The success of the concepts of quantization and wave-particle duality at the atomic level could not be overlooked. In the later half of the 1920s, a flurry

Figure 10.28 German physicist Werner Heisenberg (1901–1976), codeveloper of the field of quantum mechanics.

of activity resulted in a formal mathematical model that incorporated these ideas—**quantum mechanics**. The two principal founders were Werner Heisenberg (**Figure 10.28**) and Erwin Schrödinger.

One of the main contributions of Heisenberg was the **uncertainty principle**. Because electrons and other particles on the atomic scale have wavelike properties, we can no longer think of them as being like very tiny marbles or ball bearings. They are a bit spread out in space, more like tiny, fuzzy cotton balls. In the old particle model of the electron, it was possible, at least in theory, to state exactly where an electron is and exactly what its momentum is at any instant in time. Heisenberg stated that the wave nature of particles makes this impossible. One cannot specify *both* the position *and* the momentum of an electron to arbitrarily high precision. The more precisely you know the position of the electron, the less precisely you can determine its momentum and vice versa. If we let Δx represent the uncertainty in the position of a particle and Δmv represent the uncertainty in the momentum of the particle, then

$$\Delta x \, \Delta mv \geq h \qquad \text{(uncertainty principle)}$$

The product of the two uncertainties is greater than or equal to Planck's constant. No matter how good the experimental apparatus, the best precision we can achieve in the measurement of a particle's position and momentum is limited by this principle. On the atomic scale, particles cannot be localized in the same way that they can be on a large scale.

Schrödinger established a mathematical model for the waves associated with particles. In his model, a quantum system like a hydrogen atom can be described by a *wave function*. The wave function contains all the information needed to predict the characteristics and future evolution of the system once the nature of the interactions in which the system is involved are known. For example, knowledge of the wave function allows one to calculate the most probable position of the electron in a hydrogen atom with respect to the nucleus, as well as the electron's average energy. Although often difficult to determine in practice, the wave function of a system is the key ingredient to understanding its behavior on an atomic scale.

During the early decades of the 20th century, the view of what matter is like at the submicroscopic level changed dramatically. Electromagnetic waves have particle-like properties, and electrons and other particles have wavelike properties. Clearly, Nature is very different at this level compared to what we commonly experience on a macroscopic scale.

Learning Check

1. If the speed of an electron increases, its de Broglie wavelength
   (a) increases.
   (b) decreases.
   (c) stays the same.
   (d) may increase or decrease.
2. Name a device that makes use of the wavelike nature of electrons.
3. The wavelike aspect of particles is responsible for which of the following?
   (a) The diffraction of electrons as they pass through aluminum
   (b) the quantized orbits of the electron in a hydrogen atom
   (c) The ability of electrons to tunnel through a gap
   (d) All of the above.
4. (True or False.) At any instant in time, it is possible to specify simultaneously the position and the momentum of an electron to arbitrarily high precision.

ANSWERS: 1. (b) **2.** electron microscope, STM **3.** (d) **4.** False

10.6 Atomic Structure

The findings of de Broglie, Heisenberg, and Schrödinger force us to revise the simple Bohr model of the atom with its planetary structure. We can no longer represent electrons as little particles with sharp boundaries moving in precisely defined circular orbits. In reality, each electron is described by a wave that determines the probability of finding it at different locations. The atom is pictured as a tiny nucleus surrounded by an "electron cloud" (**Figure 10.29**). The density of the cloud at each point in space indicates the likelihood of finding the electron there. The different allowed orbits of the electrons appear as clouds with different sizes and shapes. So the simple drawings of atoms that we used earlier in this text (such as Figure 7.2) are not strictly correct. They did, however, serve the purpose of presenting the basic structure of the atom without complicating the picture with the wave nature of the electrons.

Courtesy of Forinash, Kyle. Hydrogen W Simulation. Indiana University Southeast. Retrieved on 2008-12-18. Tokita, Sumio; Sugiyama, Takao; Noguchi, Fumio; Fujii, Hidehiko; Kobayashi, Hidehiko (2006). "An Attempt to Construct an Isosurface Having Symmetry Elements". *Journal of Computer Chemistry, Japan* 5 (3): 159–164 DOI:10.2477/jcc 5.159.

Figure 10.29 Computer-generated plots of the "electron clouds" for a few atomic orbits in hydrogen. These are cross-sections of the electron probability density and are color-coded with black representing zero probability density and white representing the highest probability density. The principal quantum number n in the Bohr theory is marked to the right of each row. The letters s, p, and d, shown along the top of the display, indicate different angular momentum states of the electron, s being the lowest and d the highest. The probability density in three-dimensional space is obtained by rotating each graph shown here around a vertical axis through the center of each plot. The electron distribution is spherical for the lowest angular momentum state (the s-state) and develops an increasingly complicated lobe-like pattern as the angular momentum of the electron increases. Note the narrow "forbidden" (black) regions where the chances of finding the electron are vanishing small.

10.6a Atomic Energy Levels

Because we can't say exactly where a bound electron is with respect to the nucleus at any given moment, it is more useful to concentrate on the electron's average energy which can be precisely predicted. After all, when determining the frequency of radiation emitted or absorbed by an atom, the important quantity is the difference between the electron's initial and final energy. In the new model, we describe the electrons as being in certain allowed energy states or **energy levels**. For the hydrogen atom, the lowest energy level (corresponding to the innermost orbit in the Bohr theory) is called the **ground state**. The higher energy states are referred to as **excited states**. We now represent the structure of the atom schematically using an **energy-level diagram**, such as the one for hydrogen shown in **Figure 10.30**. Each energy level is labeled with a **quantum number**, n, beginning with the ground state having $n = 1$ and continuing on up. As the quantum number

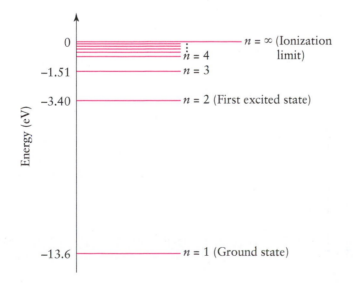

Figure 10.30 Energy-level diagram for hydrogen. Each level corresponds to one of the allowed electron orbits.

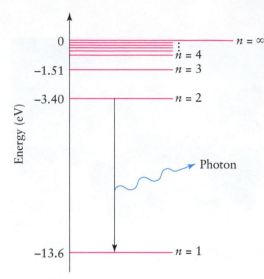

Figure 10.31 Downward electronic transition from level $n = 2$ to level $n = 1$ in hydrogen. The transition is accompanied by the emission of a photon with energy 10.2 electronvolts.

increases, so does the energy associated with the state. Moreover, as n gets larger, the difference in energy between the adjacent states becomes smaller. The difference in energy between the $n = 4$ and the $n = 3$ states is smaller than the difference in energy between the $n = 3$ and the $n = 2$ states. Finally, we again note the existence of a maximum allowed energy above which the electron is no longer bound to the nucleus. The state is designated $n = \infty$, and its energy is the ionization energy.

The numbers on the left in the energy-level diagram are the electron energies for each state. (The negative values indicate that the electron is bound to the nucleus, much like the golf ball trapped in the hole in Section 3.5.) The transition of an electron from one orbit to another corresponds to the atom going from one energy level to another. The change in energy of the electron as a result of the "energy-level transition" is found by comparing the energies of the two states.

Using the energy-level diagram, we can give a more complete picture of the emission spectrum of hydrogen. Suppose an atom in the $n = 2$ state undergoes a transition to the $n = 1$ state by emitting a photon. (Typically, an atom will remain in an excited state for only about a billionth of a second.) The transition is represented by an arrow drawn from the initial level to the final level (**Figure 10.31**). The photon that is emitted has an energy equal to the difference in energy between the two levels. Because electron transitions may occur only from one allowed energy state to another, the arrows representing such transitions *must* begin and end on allowed levels within the energy-level diagram. In this case,

$$\text{photon energy} = \Delta E = E_2 - E_1 = -3.4 \text{ eV} - (-13.6 \text{ eV})$$

$$= 10.2 \text{ eV}$$

This is a photon of UV light (refer to Figure 10.7).

Similarly, we can represent all possible downward transitions from higher energy levels to lower energy levels. **Figure 10.32** is an enlarged energy-level

Figure 10.32 Energy-level diagram for hydrogen showing different possible energy-level transitions. The number with each arrow is the wavelength (in nanometers) of the photon that is emitted.

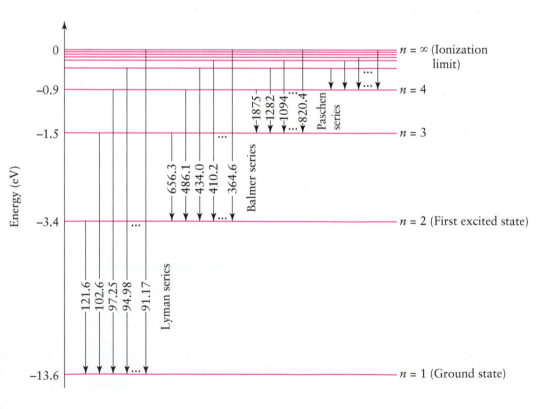

diagram for hydrogen showing various possible energy-level transitions from higher levels to lower ones. The number with each arrow is the wavelength (in nanometers) of the photon that is emitted.

EXAMPLE 10.4 Find the frequency and wavelength of the photon emitted when a hydrogen atom goes from the $n = 3$ state to the $n = 2$ state.

SOLUTION We first find the energy of the photon and then use that to determine its frequency. The wavelength we get from the equation $c = f\lambda$:

$$\text{photon energy} = hf = E_3 - E_2$$
$$= -1.51 \text{ eV} - (-3.4 \text{ eV})$$
$$= 1.89 \text{ eV}$$

Therefore,

$$f = \frac{1.89 \text{ eV}}{h} = \frac{1.89 \text{ eV}}{4.136 \times 10^{-15} \text{ eV/Hz}}$$

$$= 4.57 \times 10^{14} \text{ Hz}$$

For the wavelength,

$$\lambda = \frac{c}{f} = \frac{3 \times 10^8 \text{ m/s}}{4.57 \times 10^{14} \text{Hz}}$$

$$= 6.56 \times 10^{-7} \text{ m} = 656 \text{ nm}$$

From Figure 10.7, we see that this is a photon of visible light. (To be more precise, Table 8.1 shows that it is red light.)

By doing similar calculations for the other transitions, we can draw the following conclusions about the light that can be emitted by excited hydrogen atoms:

1. Transitions from higher energy levels to the *ground* state ($n = 1$) result in the emission of *ultraviolet* photons. This series of emission lines is referred to as the Lyman series.

2. Transitions from higher energy levels to the $n = 2$ state result in the emission of *visible* photons. This series of emission lines is referred to as the Balmer series (Figure 10.12 and Example 10.4).

3. Transitions from higher energy levels to the $n = 3$ state result in the emission of infrared photons. This series of emission lines is referred to as the Paschen series.

4. Downward transitions to other states with n greater than 3 result in the emission of other, lower-energy photons.

An atom in the $n = 3$ state or higher can make transitions to intermediate energy levels instead of jumping directly to the ground state. For example, an atom in, say, the $n = 4$ state may jump to the $n = 1$ state, or it may go from $n = 4$ to $n = 2$ and then from $n = 2$ to $n = 1$. In the latter case, two different photons would be emitted (**Figure 10.33**). Atoms in the higher energy levels can undergo different "cascades" in returning to the ground state.

We can envision what happens when hydrogen gas in a tube is heated to a high temperature, as in Figure 10.11, or is excited by passing an electric current through it. The billions and

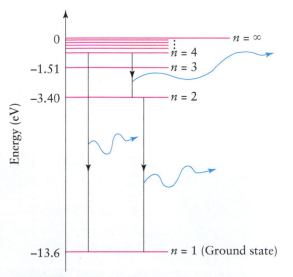

Figure 10.33 Two different ways for a hydrogen atom in the $n = 4$ state to return to the ground state. The left arrow represents a direct transition with the emission of one photon. The two arrows on the right show a transition to the $n = 2$ level, followed by a transition to the ground state. Two lower-energy (longer-wavelength) photons are emitted.

billions of atoms will be excited, some to each of the possible higher energy levels. The excited atoms then undergo spontaneous transitions to lower levels, with different atoms "stopping" at different levels according to a complex set of quantum-mechanical transition probabilities. The hydrogen gas continuously emits photons corresponding to all of the possible energy-level transitions, although some transitions are more favored than others, depending in part on the gas temperature. This is hydrogen's emission spectrum. (We have assumed here that the gas density is low enough or the temperature is high enough that interactions between the gas atoms are negligible. Under such conditions, we may ignore electron transitions brought about by collisions in which no photons are emitted.)

Upward energy-level transitions occur when the atom gains energy by absorption of a photon with the proper energy or perhaps by collision. For example, a hydrogen atom in the ground state can jump to the $n = 2$ state by absorbing a photon with energy 10.2 electronvolts. When white light—containing photons with many different energies— passes through hydrogen, those photons with just the right energy can be absorbed, leading to the observed absorption spectrum (Figure 10.20).

An atom is ionized if it absorbs enough energy to make the electron energy greater than zero. For example, a hydrogen atom in the ground state is ionized if it absorbs a photon with energy greater than 13.6 electronvolts. This process of *photoionization* is essentially the same as the photoelectric effect. Any excess energy the electron has appears as kinetic energy.

10.6b Some Applications

We've presented a fairly complete picture of the hydrogen atom, but what of the other elements? The presence of more than one electron complicates things, but the general principles of atomic structure are much the same. Each element has its own atomic energy-level diagram, with correspondingly different energy values for each level. The various downward transitions between these levels produce the element's characteristic emission spectrum. Atoms with larger atomic numbers have more protons in the nucleus, so the force on the inner electrons is stronger. This means that the electrons are more tightly bound and their energies have larger magnitudes but are still negative.

One very common device that uses the emission spectra of elements is the neon sign (**Figure 10.34a**)—so named because neon is one of the most commonly used elements in them. These signs are made by placing low-pressure gas in a sealed glass tube. A high-voltage alternating current power supply connected across the ends of the tube causes electrons to move back and

Figure 10.34 Two examples of emission spectra from excited gases. (a) A neon sign. (b) An auroral display as seen from the International Space Station.

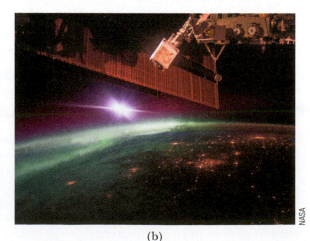

(a)   (b)

forth through the tube. The electrons excite the atoms of the gas by collision, and the atoms emit photons as they return to lower energy states. Because different elements have different emission spectra, signs can be made to emit different colors by being filled with different gases. Neon-filled signs are red because there are several bright red lines in neon's emission spectrum (Figure 10.12).

Fluorescent lights and plasma displays (TVs) make use of ultraviolet emission lines (see the chapter introduction). As in neon lights, the atoms of a gas (mercury in the former and xenon or neon in the latter) are collisionally excited by a current. The photons that are emitted are mostly ultraviolet, so a second substance called a *phosphor* is used to convert these to the visible-light photons that we see. Atoms in the phosphor, which is coated onto the inside surface of the gas-containing chamber, absorb the UV photons and jump to higher energy levels. They return to the ground state via two steps. First, they jump to intermediate energy levels, which gives energy to neighboring atoms; second, they return to the ground state as they emit visible-light photons. Different phosphors produce different frequencies of light. Fluorescent lights use phosphors that emit several different colors, which combine to approximate white light. Each pixel in a plasma display consists of three separate subpixel cells, one with a red phosphor, another with green, and the third with blue. Different colors are produced by varying the brightness of the red, green, and blue light. This is done by varying the current in each cell.

Nature provides us with a spectacular example of emission spectra: the *aurora borealis* (northern lights; **Figure 10.34b**) and the *aurora australis* (southern lights). These light displays, seen at night mainly in the far north and the far south, arise from a fascinating combination of physical processes. It begins at the Sun, where ionized atoms and free electrons are flung out into space as part of the solar wind. After a journey of several days they reach Earth, where the charged particles are guided by the magnetic field surrounding our planet (recall Figure 8.6). The lightweight electrons spiral around the magnetic field lines and follow them toward the north and south magnetic poles. If the conditions are right, these energetic electrons enter the upper atmosphere (down to about 100 kilometers above Earth) and collide with atoms, molecules, and ions, thereby exciting them to emit spectra. The most common emission is a whitish-green glow from oxygen atoms. Energy-level transitions in excited nitrogen molecules are responsible for a pink that is often seen. Much of the emitted radiation is in the ultraviolet and infrared bands, which we can't see.

> ## ▶ Physics To Go 10.4
>
> Go back to the Visual Quantum Mechanics site on the Web at http://phys.educ.ksu.edu/ and select the simulation called Gas Lamps—Emission. This activity offers you the chance to construct a set of energy levels that can produce (match) the emission spectrum of one of several available elements. Start with a simple spectrum, like that of hydrogen, which you know something about, to get the hang of the "game." Then try some of the less familiar and more complicated examples. After you've explored this simulation to your satisfaction, return to the program menu and select Gas Lamps—Absorption and complete the equivalent exercise. What conclusions can you draw about the relationship between the energy level structure of an atom and the appearance of the absorption and emission spectra it is capable of producing?

10.6c The Pauli Exclusion Principle

The structure of atoms with more than one electron is governed by a principle formulated by Wolfgang Pauli in 1925. Pauli, who was awarded the 1945 Nobel Prize in physics for his work, carefully analyzed the emission spectra of different elements noting that some expected transitions did not occur when the lower energy level was already occupied by a certain number of electrons. From this and other evidence, Pauli concluded that only a limited, fixed number of electrons could populate each energy level in an atom. (By analogy, we can't sit comfortably on a sofa if it is already full of people.) Pauli stated his conclusion in the **exclusion principle**:

PRINCIPLES | **Pauli Exclusion Principle** Two electrons cannot occupy precisely the same quantum state at the same time.

For each energy level, there exists a set number of quantum states available to the electrons. Once all of the quantum states corresponding to a given energy are filled, any remaining electrons in the atom must occupy other energy levels that have vacancies. The number of quantum states associated with each energy level is related to the electron's orbital angular momentum and spin (see Section 12.3). In the $n = 1$ level, the number of distinct quantum states is 2, so the maximum number of electrons that can occupy this level is 2. For the $n = 2$ energy level, the maximum number of allowed states and, hence, electrons is 8; for the $n = 3$ level, it is 18. The general rule is, for level n, the occupation limit of electrons is $2n^2$. The ground state of a multi-electron atom is one in which all electrons are in the lowest energy levels consistent with the Pauli exclusion principle. If any one electron is in a higher energy level, the atom is in an excited state. Table 10.1 shows the ground-state energy-level populations for several atoms.

The properties of each element are determined to a great extent by the ground-state configuration of its atoms, particularly the number of electrons in the highest energy level that is occupied. Table 10.1 shows that helium in the ground state has the $n = 1$ energy level filled and that neon has the $n = 2$ energy level filled. Because of this, helium and neon have similar properties: they are both gases at normal room temperature and pressure and are very stable. They don't burn or react chemically in other ways except under special circumstances. Hydrogen, lithium, and sodium have similar properties because they all have one electron in the highest occupied energy level (when in their respective ground states).

The periodic table of the elements (back inside cover of the print edition) was developed by Dmitri Mendeleev in 1869. He arranged the elements known at the time according to their properties: elements with similar properties were placed in the same column. After Pauli's discovery, it was determined that the elements in each column have similar ground-state configurations, and that is why they are alike. (For example, hydrogen, lithium, and sodium are in the first column, called Group 1A, because each has one electron in its

Table 10.1 Ground-State Configurations of Several Atoms

Element	Atomic Number	Number of Electrons in Level		
		$n = 1$	$n = 2$	$n = 3$
Hydrogen	1	1	0	0
Helium	2	2	0	0
Lithium	3	2	1	0
Carbon	6	2	4	0
Oxygen	8	2	6	0
Neon	10	2	8	0
Sodium	11	2	8	1

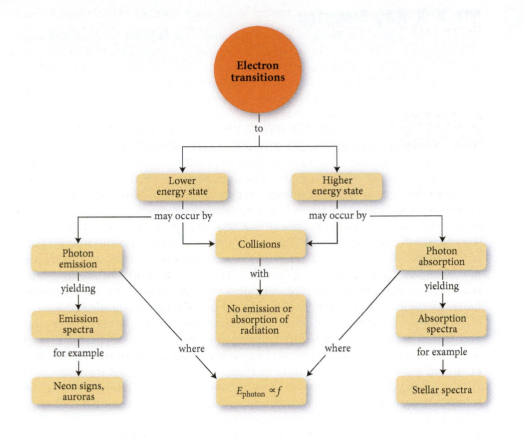

highest occupied energy level.) This is one indication of how understanding quantum mechanics can play a crucial role in other scientific disciplines such as chemistry.

Concept Map 10.2 summarizes the types of transitions atomic electrons can undergo.

■ Learning Check

1. Indicate whether the following transitions within the energy-level diagram of hydrogen are accompanied by the *emission* or *absorption* of photons.
    (a) $n = 5$ to $n = 3$
    (b) $n = 1$ to $n = 8$
    (c) $n = 8$ to $n = 9$
    (d) $n = 3$ to $n = 5$
    (e) $n = 6$ to $n = 5$

2. Assuming only one photon is emitted or absorbed in the course of each transition given in Question 1, rank order (from highest to lowest) the frequencies of the photons involved in each transition.

3. The Balmer series is
    (a) a set of play-off matches in golf named after the famous professional golfer Arnold Balmer.

(b) a sequence of emission lines in the ultraviolet portion of the electromagnetic spectrum from hydrogen.
(c) a sequence of emission lines in the visible portion of the electromagnetic spectrum from hydrogen.
(d) a sequence of emission lines from hydrogen involving transitions from higher levels down to the $n = 3$ energy level.

4. What exactly is (or are) "excluded" by the Pauli exclusion principle?

5. (True or False.) The chemical properties of the elements in the periodic table are primarily determined by the electronic ground-state configuration of their atoms.

ANSWERS: 1. Emission (a), (e) Absorption (b), (c), (d). **2.** (b), (a) & (d), (e), (c) **3.** (c) **4.** Electrons in the same quantum state. **5.** True.

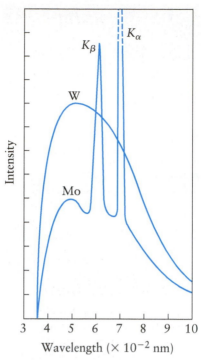

Figure 10.35 X-ray spectra of tungsten (W) and molybdenum (Mo) resulting from the bombardment of targets made of each element by electrons having energies of 35,000 electronvolts. Over the wavelength range shown, the spectrum of tungsten is a continuous one produced by bremsstrahlung processes. The spectrum of molybdenum includes two very strong characteristic peaks.

10.7 X-Ray Spectra

We have seen that the Bohr model of the atom (as refined by quantum mechanics) had great success in explaining the spectrum produced by hydrogen. Another early triumph of the Bohr atom came in connection with the study of x-rays. Soon after Bohr published his model, a young English physicist named H. G. J. Moseley used it to explain the characteristic spectra of x-rays.

In Section 8.5, we described how x-rays are produced. Electrons are accelerated to a high speed and then directed into a metal target (Figure 8.34). A band of x-rays is emitted, with different wavelengths having different intensities. **Figure 10.35** shows the x-ray spectra when the elements tungsten (W) and molybdenum (Mo) are used as targets. These graphs are like blackbody radiation curves in that they are plots of intensity versus wavelength. Most of the x-rays are produced as the electrons are rapidly decelerated on entering the target. (This is one example of Maxwell's finding that accelerated charges emit EM waves.) This *bremsstrahlung* (German for "braking radiation") appears as the smooth part of the spectra covering the full range of wavelengths. Bremsstrahlung spectra are much the same for different elements. But the two sharp peaks for the molybdenum target are unique to this element and constitute its characteristic spectrum. Other elements exhibit characteristic spectra but at different wavelengths.

Moseley compared the characteristic x-ray spectra of different elements and found a simple relationship between the frequencies of the peaks and the atomic number of the element. For each peak, a graph of the square root of the frequency versus the atomic number of the element was a straight line (**Figure 10.36**). This was a very practical discovery because it allowed him to determine the atomic number, and therefore, the identity, of unknown elements. All he had to do was determine the wavelengths of the characteristic x-ray peaks and use his graph. This method was a key factor in the discovery of several new elements, including promethium (atomic number 61) and hafnium (atomic number 72).

After Bohr developed his model of the atom, Moseley (**Figure 10.37**) quickly realized that the characteristic x-rays are much like the emission lines of hydrogen. He showed that they are produced when one of the innermost electrons jumps from one orbit (energy level) to another. Different elements have x-ray peaks with different wavelengths because the electron energy levels have different energy values. In particular, atoms with higher atomic numbers have more protons in the nucleus and bind the inner electrons more tightly; their energies are greater.

Figure 10.36 Moseley's diagram showing the relationship between the square root of the frequency for two characteristic x-ray peaks and the atomic number (Z) of the emitting element.

Moseley proposed the following scenario to account for the x-ray spectra of the elements. If a bombarding electron collides with, say, a molybdenum atom and knocks out an electron from the $n = 1$ level, electrons in the upper levels will cascade down to fill the vacancy left by the ejected electron, and photons will be emitted. The lowest-frequency (longest-wavelength) x-ray photon corresponds to the transition from $n = 2$ to $n = 1$. Because the lowest energy level was called the "K shell" by x-ray experimenters, this is called the K_α (K-alpha) peak. Electrons jumping from $n = 3$ to $n = 1$ emit photons that form the K_β (K-beta) peak. Notice that the K_α and K_β peaks correspond to the two lowest-frequency lines in hydrogen's Lyman series.

With this knowledge, we can see why high-speed electrons are needed to produce the characteristic x-rays of heavy elements. The inner electrons are so tightly bound that it takes very-high-energy electrons to knock them out.

10.8 Lasers

The word *laser* is an acronym derived from the phrase "light amplification by stimulated emission of radiation" (Figure 10.38). By examining this phrase one piece at a time, we will describe how the laser operates. In doing so, we will find yet another example of how our theory of atomic structure provides the basis for understanding a process whose practical application has touched nearly every aspect of modern life.

Imagine that an electron in an atom has been excited to some higher energy level by either a collision or the absorption of a photon with the proper energy, E. Generally, such an electron will remain in this excited state for only a short time (around a billionth of a second) before returning to a lower energy level. If this decay to a lower state occurs spontaneously by radiation (which is usually the case in low-density gases where collisions are rare), a photon will be emitted in some random direction.

It turns out, however, that an electron can be *stimulated* to return to its original energy level through the intercession of a second photon with the same energy, E (Figure 10.39). If a group of atoms all having their electrons in this same excited state is "bathed" in light consisting of photons with energy E, the atoms will be stimulated to decay by emitting additional photons with the same

Figure 10.37 H. G. J. Moseley (1887–1915), whose work with x-ray spectra provided another confirmation of Bohr's model of the atom. Moseley's brilliant career was cut short when he enlisted during World War I and was killed in action.

Figure 10.38 *Laser* acronym.

Figure 10.39 Stimulated emission. (a) Photon absorption places an atom in an excited state. (b) A photon with the same energy stimulates the excited atom to return to the lower state and emit an identical photon.

(a) Incoherent radiation
     from excited atoms

(b) Coherent radiation
     from excited atoms

Figure 10.40 (a) Excited atoms in incoherent sources like ordinary lightbulbs emit photons independently. The directions of propagation and the relative phases of emitted light are random. (b) Excited atoms in a laser emit coherent light. Photons produced by stimulated emission all travel in the same direction in phase with one another.

Components of the first ruby laser

Figure 10.41 Schematic of a ruby laser system. The pumping radiation is provided by a high-intensity quartz flash tube. The stimulating photons are reflected back and forth between the parallel end mirrors to build up a beam of high intensity. The laser beam consists of photons that escape through the partially transmitting end reflector at the right. Lasers of this type must be pulsed to avoid overheating and possibly cracking the ruby rod.

energy E. By this process, the intensity of a beam of light is increased or amplified as a result of stimulated emission from atoms in the region through which the radiation passes. We have a laser.

To achieve this amplification, we must first arrange for the majority of the atoms through which the stimulating radiation travels to have their electrons in the same excited state. Otherwise, the unexcited atoms will just absorb the radiation. This is not an easy situation to arrange because the excited atoms normally don't stay that way for long.

The problem can be overcome because many atoms possess excited states that are said to be *metastable*. Once in such a state, the electrons tend to remain there for a relatively long time (perhaps a thousandth of a second instead of a billionth) before spontaneously decaying to a lower state. During the time it takes to excite more than half of the atoms to that state, the ones already excited don't jump back down. This process, called *pumping*, can result in the majority of the atoms being in the same metastable, excited state. This condition is called a *population inversion* because there are more atoms with electrons populating an upper energy level than a lower one, the reverse of the usual situation. This condition is necessary for the generation of laser light.

If we now irradiate the population-inverted atoms with photons of the correct energy, a chain reaction can be established that greatly amplifies the incident light beam. One photon stimulates the emission of another identical one, these two stimulate the emission of two more, these four stimulate the emission of four more, and so on. In the end, an avalanche of photons is produced, all having the same frequency (color). The beam is said to be *monochromatic*. The light amplification can produce a very high-intensity beam of light of a single wavelength.

Besides being intense and monochromatic, laser light has an additional property that makes it extremely useful: *coherence*. This means that the stimulating radiation and the additional emitted laser radiation are in phase: the crests and troughs of the EM waves at a given point all match up (**Figure 10.40b**). An ordinary light source such as an incandescent bulb emits incoherent light (different parts of the beam are not in phase with one another) because the excited atoms giving off the radiation do so independently of each other. The emitted photons in this case may be considered to be individual, short *wave trains* that bear no constant phase relation to one another (**Figure 10.40a**). By contrast, stimulated emission of radiation by excited atoms produces photons that are in phase with the stimulating radiation. The coherence of laser light contributes to its high intensity. Coherent sources like lasers are important for producing interference patterns and *holograms* (three-dimensional images described at the end of this section).

Although not in widespread use except for some specialized scientific applications, the ruby laser is simple in concept and illustrates all the important design characteristics of most other types of lasers. This device is shown schematically in **Figure 10.41**. It consists of a ruby rod with ends that are polished and silvered to become mirrors, one of which is partially (a few percent) transparent. The ruby rod is composed of aluminum oxide (Al_2O_3) in which some of the aluminum atoms have been replaced by chromium atoms. The chromium atoms produce the lasing effects. The rod is surrounded by a flash tube capable of producing a rapid sequence of short, intense bursts containing green light with a wavelength of 550 nanometers. The chromium atoms

Figure 10.42 Energy-level transitions used by a ruby laser. The chromium atoms are excited by green pumping radiation. A transition to the metastable state follows. Stimulated emission from this level produces the red laser light.

are excited by these flashes from their lower state E_0 to state E in a process referred to as *optical pumping*. (The high-intensity flash tube "pumps" energy into the chromium atoms.) The chromium atoms quickly spontaneously decay back to level E_0, or, in some cases, to the metastable level E_1 (**Figure 10.42**). With very strong pumping, a majority of the atoms can be forced into state E_1, and a population inversion is produced. Eventually, a few of the chromium atoms in state E_1 decay to the state E_0, thus emitting photons that stimulate other excited chromium atoms to execute the same transition. When these photons strike the end mirrors, most of them are reflected back into the tube. As they move back in the opposite direction, they cause more stimulated emission and increased amplification. A small fraction of the photons oscillating back and forth through the rod is transmitted through the partially silvered end and makes up the narrow, intense, coherent laser beam. The beam produced by a ruby laser has a wavelength of 694.3 nanometers, a deep red color.

The ruby laser is an example of a pulsed solid-state laser: it produces a single, short pulse of laser light each time the flash tube fires, and the lasing material (the chromium atoms) is distributed in a solid matrix. Neodymium:yttrium-aluminum garnet (Nd:Yag) lasers are also of this type. Another common type of laser is the gas laser, of which the helium-neon (He-Ne) laser is a good example. Here, the lasing material is mixture of about 15% helium gas and 85% neon gas. In most applications, the neon gas produces coherent radiation of wavelength 632.8 nanometers (red) as a result of stimulated emission from a metastable level to which it has been excited by collisions with the helium atoms. Helium-neon lasers can also be designed to yield green light with a wavelength of 543.5 nanometers. The He-Ne laser is a continuous laser in that it produces a steady beam of laser light.

Semiconductor lasers, sometimes called *diode lasers*, are not solid-state lasers but small, layered electronic devices that rely on a population inversion that is achieved electronically by injecting charges into the active, lasing medium (often gallium arsenide [GaAs]) from adjacent semiconducting layers ("cladding"). The cladding, which has an index of refraction that is lower than that of the active medium, also serves to confine the laser output from the active medium by total internal reflection (see Section 9.3). The laser light in these devices is amplified by multiple reflections between the end facets of the semiconductor crystal that have been carefully cleaved and act as mirrors. Diode lasers typically operate in the near-infrared portion of the EM spectrum between about 750 and 880 nanometers. They are widely used in the telecommunications industry: most long-distance telephone calls are carried by diode-laser–generated light signals transmitted over fiber-optic cables. The recent development of compact blue-emitting gallium nitride semiconducting lasers operating at wavelengths near 400 nanometers will likely lead to new applications of these devices in areas such as optical data storage and printing where shorter wavelengths mean higher spatial resolution.

Two other types of lasers, excimer lasers and dye lasers, are also in use. Excimer lasers use reactive gases like chlorine and fluorine mixed with inert gases (argon, krypton, or xenon) to produce laser light in the ultraviolet range. Dye

Table 10.2 Characteristics of Some Common Lasers

Active Medium	Wavelength (nm)	Type
Argon fluoride	193 (UV)	Pulsed
Nitrogen	337.1	Pulsed
Helium-cadmium	441.6	Continuous
Argon	476.5, 488.0, 514.5	Continuous
Krypton	476.2, 520.8, 568.2, 647.1	Continuous
Helium-neon	543.5, 632.8	Continuous
Rhodamine 6G dye	570–650	Pulsed
Ruby	694.3	Pulsed
Gallium arsenide	780–904* (IR)	Continuous
Neodymium:Yag	1060 (IR)	Pulsed
Carbon dioxide	10,600 (IR)	Continuous

*Depends on temperature.

lasers employ complex organic dyes in a liquid solution to produce laser light with wavelengths that can be varied ("tuned") over a range of about 100 nanometers. Table 10.2 summarizes the characteristics of some of the most common types of lasers.

> ### ▶ Physics To Go 10.5
>
> Reach into your wallet or purse and take out one of your credit cards. It's a pretty good bet that, if your card isn't more than a few years old, it will show a silvery region about 2 centimeters or so in size containing a reflection hologram. Examine the hologram carefully under strong (white) light. Slowly tilt the card to change the angle of incidence of the illumination. Pay close attention to the subtle alteration in the color and appearance of the hologram image (a bird taking flight, a world map, etc.) as the card is rotated. Use a hand magnifying lens to see the image in greater detail if necessary. Notice also the apparent depth (third dimension) of the image. The presence of such complex holographic images makes credit cards carrying them very difficult to counterfeit. Of course, the same level of complexity makes holograms suitable elements for incorporation in graphic design projects. If you are interested in exploring the full range of commercially available transmission and reflection holograms, try a Web search using these terms as key words. You will be amazed at the beauty and variety of holographic art.

Employing lasers to create holograms (see the accompanying Optical Application at the end of this section) is only one of many ways that these extraordinary devices are exploited in our society. Today, lasers are used to perform surgery to correct certain vision defects (see Section 9.5) and to remove cancerous tumors of the skin, to cut metals and other materials precisely, to induce nuclear fusion reactions, to measure accurately the distances separating objects (Earth and the Moon or two mirrors on opposite sides of a geological fault), to transmit telephonic information along optical fibers, to faithfully replay recorded music, movies, and data (CD, CD-ROM, DVD and Blu-ray players; see Figure 8.27), and to determine the price of goods at the supermarket checkout counter (Figure 10.43). Regardless of the location or application, lasers all operate according to the same basic physics—physics entirely accessible and comprehensible once the structure of the atom is understood.

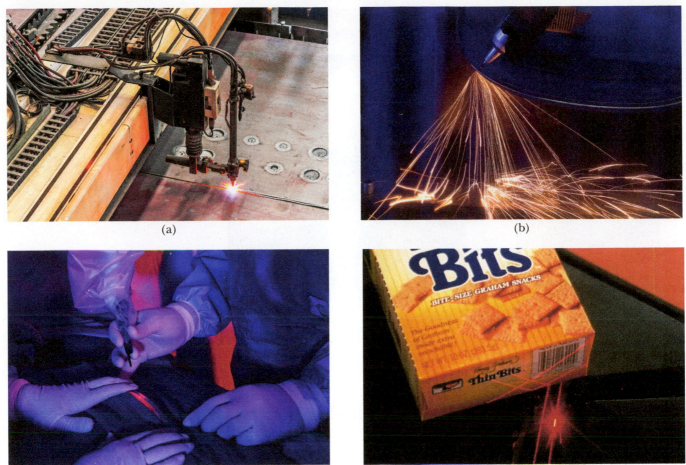

Figure 10.43 Applications of laser technology. (a) Laser welding. (b) Laser drilling. (c) Laser surgery. (d) Laser price scanning.

◼ **Learning Check**

1. (True or False.) An x-ray can be emitted only when an electron undergoes an allowed energy-level transition.

2. Rank (from lowest to highest) the frequencies of the characteristic K_α radiation emitted by the following elements:
   (a) iron ($Z = 26$)
   (b) zinc ($Z = 30$)
   (c) titanium ($Z = 22$)
   (d) copper ($Z = 29$)

3. Which of the following processes or phenomena is *not* involved in the operation of lasers?
   (a) stimulated emission
   (b) population inversion
   (c) metastable state
   (d) blackbody radiation

4. Which of the following common devices does *not* employ a laser in its operation?
   (a) DVD player
   (b) supermarket checkout price scanner
   (c) telephonic fiber-optics communication system
   (d) plasma TV

5. _____ are three-dimensional images made with the aid of lasers.

ANSWERS: 1. False **2.** (c), (a), (d), (b) **3.** (d) **4.** (d) **5.** Holograms

Figure 10.44 One method of hologram production. Laser beams reflected from the mirror and from the object combine at the photographic film to produce an interference pattern. The developed film is the hologram.

Increasingly, 3-D imagery is becoming the norm in commercial and home movie theaters, video gaming, and Web-based displays for everything from action sequences and architectural renderings to Martian landforms to zoological and biochemical nanostructures. For the most sophisticated of these applications, a headset using high-speed shuttered LCD lenses and costing hundreds of dollars is required, but for more common 3-D displays, only an inexpensive ($2 to $3) pair of glasses with one red lens and one blue one is sufficient to provide a sense of depth—a third dimension—when viewing a properly prepared photograph or cinematic image.

We enjoy a sense of depth or three dimensions largely because each eye sees a slightly different view of the same scene. 3-D *anaglyphic* images are created by superimposing two photographs, one taken in red light and the other in blue, which record a particular scene from two slightly different perspectives. When viewing such a composite picture with a pair of red–blue glasses, the red image is seen by one eye while the blue one is seen by the other. The brain combines these two different images together to produce a 3-D image.

True 3-D objects, however, generally change appearance as you move around them. You see different parts of an object depending on where you sit or stand. Such is not the case with typical, simple, non-interactive "3-D" images. The same aspects of the scene are observed from the left side of the image as from the right—glasses and all. The images are basically two-dimensional because ordinary photographs and video frames capture only the intensity of the light striking the film and nothing about its phase.

The production of true 3-D images that faithfully reproduce all aspects of the original objects was first accomplished in 1947 by Dennis Gabor. Gabor developed a way to preserve information about the relative phases of light beams emitted by or reflected from different points on an object in a complex *interference pattern*. A simple version of the experimental setup is shown in **Figure 10.44**. A beam of coherent radiation, in modern applications usually from a laser, is split in two, one beam traveling directly to a piece of photographic film and the second reaching the film after being reflected off some object. At the position of the film, the two beams recombine, interfering with one another to produce a complex pattern of bright and dark areas (see "Interference" in Section 9.1.). This pattern is recorded on the film and contains information not only about the relative intensities of the object and reference beam but also about their phase differences.

To see the image (to reconstruct the object), light of the same type as that used originally is shone on the film. Depending on the details of the original exposure process, the transmitted or reflected light is viewed obliquely to reveal the image. The resulting hologram (from the Greek word *holos*, meaning "whole") is a true 3-D image of the original object. Its aspect or color (or both) changes as you move your head; some parts of the object become visible and others disappear. The apparent shape, size, and brightness of the image change just as they would if you were examining the genuine article (**Figure 10.45**). Gabor's research earned him the Nobel Prize in physics in 1971.

Dr. Pierre-Alexandre Blanche

Figure 10.45 Holograms of four difference vehicles (shown in the far right panel) made using a diode laser. The holographic images to the left of each vehicle reveal their three-dimensional character insofar as different aspects of the vehicles are displayed depending on the viewer's perspective on the scene.

When first perfected during the 1960s, holography—the science of producing holograms—was viewed as little more than a curiosity, a clever way to generate interesting 3-D images and to illustrate simultaneously some basic physical concepts. In recent years, however, the ability of holograms to store huge quantities of information has been recognized by those in the computer industry involved in data storage and retrieval. Other applications involve component testing and analysis in which two holograms of an object taken at different times are superimposed. Any changes in the object, like deformations produced by, say, forces exerted during industrial stress tests or by tumor growth beneath the surface of the object, show up immediately as interference patterns. Recently, banks and credit unions have begun to place embossed holograms on their membership cards to reduce fraud. The presence of these highly detailed holographic images makes card duplication nearly impossible and attempts at counterfeiting readily apparent (see Physics to Go 10.5). Finally, researchers at the University of Michigan have developed holographic filmstrips that can display art objects, rare artifacts, and old musical instruments in full three-dimensionality for interested students, something ordinary photographs simply cannot do, 3-D glasses notwithstanding.

■ **QUESTION**

1. What is a hologram? Give several ways in which a holographic image differs from an ordinary photographic image.

Profiles in Physics   The Solvay Conferences

It is extremely difficult to describe in a few paragraphs very much of the history associated with the rapid and revolutionary developments that occurred in physics during the 40-year period from 1895 to 1935. It is equally difficult, if not impossible, in the same space to do justice to the many brilliant men and women physicists and chemists who contributed their talents to bring about these developments. The list of Nobel Prize recipients in physics and their citations for this period (Appendix A) may help to establish the significance of the work done by some of the "giants" of this era, but it does not begin to capture the excitement and spirit of adventure that existed within the scientific community at this time. Nowhere was this spirit more in evidence than at the Solvay Conferences held between 1911 and 1933.

Ernest Solvay (1838–1922) was a Belgian industrial chemist who perfected a commercial process for producing sodium carbonate and became extremely wealthy as the use of the "Solvay process" spread across the world. Solvay had long been interested in the fundamental structure of matter and by 1910 had become especially intrigued with the developing crisis between classical physics and the recently introduced quantum theories. In an effort to promote discussion and to resolve the crisis, Solvay held an international conference in October 1911 to which he invited many of the most prominent physicists of the period, including Planck, Einstein, and Marie Curie. The first conference proved to be so successful that Solvay established a foundation to sponsor similar conventions from time to time in the years that followed, right up to the present time.

The first Solvay Conference had as its topic radiation theory and quanta, and the second conference in 1913 dealt with the structure of matter. Five additional conferences were held between 1921 and 1933 on topics ranging from the electrical conductivity of metals, through magnetism, to the structure and properties of the nucleus. **Figure 10.46** is a group portrait of the participants in the 1927 Solvay Conference. The participants in the other conferences held during this period were no less luminary than those shown in this photograph.

The progress made in the understanding and interpretation of natural phenomena during these early Solvay conferences was considerable. In what follows, we will mention some of the highlights of these sessions in an attempt to convey the importance of the work done during these meetings and the intense, spirited, and often humorous manner in which it was carried out.

During the third conference (1921), the two major subjects discussed were the nuclear model of the atom, as proposed and elaborated upon by Ernest Rutherford (more on this in Chapter 11), and Bohr's theory. During the discussion following Rutherford's report, three remarkable notions emerged—all of which were later confirmed as being substantially correct. (1) Jean Perrin suggested that the mechanism by which the Sun derives its energy is the transmutation of hydrogen into helium, whereby radiation equivalent to the mass difference between the reactants and the product is produced according to Einstein's equation $E = mc^2$ (Section 11.5). (2) Rutherford, after agreeing with Perrin that huge quantities of energy could be released in the combination of hydrogen nuclei to form helium, then commented:

> It has occurred to me that the hydrogen atoms of the (solar) nebula might consist of particles which one might call the "neutrons," which would be formed by a positive nucleus with an electron at a very short distance. These neutrons would hardly exercise any force in penetrating into matter. They will serve as intermediaries in the assemblage of nuclei of elements of higher atomic weights.

The neutron was later discovered in 1932 by James Chadwick. (3) Marie Curie argued that the stability of the nucleus could not be accounted for on the basis of electrostatic forces and that such stability required strong but very short-range, attractive forces of another type. These forces are now referred to as "nuclear" forces (Section 11.1).

The 1927 Solvay Conference focused on the new quantum mechanics and initiated what has been called the Einstein-Bohr dialogues concerning the implications of the probabilistic nature of this theory. Briefly, Bohr believed that the best one could do at the subatomic level was to calculate the probabilities of finding a system in a given set of allowed states and that the very act of measurement involved interactions with the system that forced it into one of these allowed states where it is "observed." Numerous repeated measurements of the same type would reveal the system, on the average, to be in that state having the highest quantum-mechanical probability.

Einstein was not at all happy with the lack of determinism (see the special feature on "Chaos" at the end of Section 2.5) in the quantum theory and argued that the quantum-mechanical

Benjamin Couprie, Institut International de Physique Solvay/Photo Researchers, Inc

Figure 10.46 Participants of the 1927 Solvay Conference. Front row: I. Langmuir, M. Planck, M. Curie, H. A. Lorentz, A. Einstein, P. Langevin, Ch. E. Guye, C. T. R. Wilson, O. W. Richardson. Middle row: P. Debye, M. Knudsen, W. L. Bragg, H. A. Kramers, P. A. M. Dirac, A. H. Compton, L. de Broglie, M. Born, N. Bohr. Back row: A. Piccard, E. Henriot, P. Ehrenfest, Ed. Herzen, Th. de Donder, E. Schrödinger, E. Verschaffelt, W. Pauli, W. Heisenberg, R. H. Fowler, L. Brillouin.

description of Nature may not have exhausted all the possibilities of accounting for observable phenomena. He believed that, if carried further, the analysis would reveal a means of knowing precisely, with 100-percent surety, the outcome of an experiment of a particular type on a given quantum system. Einstein's attitude toward quantum mechanics, as regards its being a complete description of Nature on a microscopic level, may be summarized in his own words: "To believe this is logically possible without contradiction: but it is so very contrary to my scientific instinct that I cannot forgo the search for a more complete conception."

The Einstein–Bohr dialogues produced some lively debates during the 1927 Solvay meeting. The chairman of the session, H. A. Lorentz, himself an eminent physicist and master of three languages (English, German, and French), tried to maintain order but found it exceedingly difficult to do so, as one speaker after another joined in the fray, each in his own language. At one point, things became so confused that one of the participants, Paul Ehrenfest, a friend and colleague of both Einstein and Bohr, went to the blackboard and wrote: "The Lord did there confound the language of all the Earth." And, at a later time, during one of the lectures, Ehrenfest passed a note to Einstein

that read: "Don't laugh! There is a special section in purgatory for professors of quantum theory, where they will be obliged to listen to lectures on classical physics ten hours every day." Present-day students of quantum theory might still find solace in Ehrenfest's remark!

The significance of the events at these meetings for later generations of physicists has been great indeed. The traditions of excellence established during these early conferences continue even to the present. As Werner Heisenberg, one of the founders of modern quantum theory and a Solvay participant in 1927, 1930, and 1933, wrote in 1974: "There can be no doubt that in those years (1911–1933) the Solvay Conferences played an essential role in the history of physics . . . the Solvay Meetings have stood as an example of how much well-planned and well-organized conferences can contribute to the progress of science. . . ."

■ **QUESTION**

1. What are the Solvay Conferences? In what ways did these meetings contribute to the development of quantum mechanics during the years 1911–1933?

SUMMARY

» Near the beginning of the 20th century, developments in experimental physics, particularly those related to **blackbody radiation**, the **photoelectric effect**, and **atomic spectra**, forced theoretical physicists to introduce revolutionary new concepts that defied classical physics as articulated by Newton and Maxwell.

» Key characteristics of these three processes were explained by assuming that physical quantities on the atomic scale are **quantized**. Light quanta (**photons**) have energies that are proportional to their frequencies.

» The **Bohr model** of the atom has electrons restricted to certain orbits in which the angular momentum is quantized. Photons are emitted and absorbed when electrons jump from outer orbits to inner orbits or vice versa.

» During the 1920s and 1930s, the field of **quantum mechanics** emerged as it became clear that particles possess wavelike properties. This allowed a refinement of Bohr's model of the atom and led to a new probabilistic interpretation of physics on the atomic scale. No longer simply tiny spheres, electrons are now represented as probability "clouds" surrounding nuclei.

» The atomic theory put forth by Bohr and his colleagues has been widely applied since its inception in 1913. One early use of the model was to account for the characteristic x-ray spectra from heavy elements.

» More recently, the establishment of the energy-level structure of such elements as neon and chromium, including the identification of relatively long-lived metastable states, has been instrumental in the development of laser systems.

» Today, quantum mechanics forms the foundation of our understanding of physics on the microscopic scale and stands as a monument to the combined genius of the men and women who participated in the early Solvay conferences where the foundations of this branch of physics were laid.

IMPORTANT EQUATIONS

Equation	Comments
$E = hf$	Energy of atomic oscillator with frequency f. Energy of a photon with frequency f.
$\Delta E = hf$	Energy of photon emitted or absorbed in energy-level transition
allowed angular momentum $= n\left(\dfrac{h}{2\pi}\right)$	Quantization of electron orbital angular momentum (Bohr model)
$\lambda = \dfrac{h}{mv}$	de Broglie wavelength of material particle
$\Delta x\, \Delta mv \geq h$	Heisenberg's uncertainty principle
maximum number of allowed quantum states $= 2n^2$	Permitted number of electrons in each energy level (Pauli exclusion principle)

MAPPING IT OUT!

1. Quantum mechanics is one of the most daunting of all areas in physics. To students encountering this subject for the first time at the introductory level, quantum mechanics can seem mysterious, even contradictory, because it is so counterintuitive to what we are used to experiencing in the macroscopic domain where Newton's laws apply. Some of the confusion can often be dispelled by organizing the material in a way that highlights some of the major connections between the principal elements of the subject. Concept maps can be of value in this regard.

   Review the reading in Sections 10.5 and 10.6. Then make a list of the main ideas or concepts introduced in these sections. Your list should include some, if not all, of the following items: uncertainty principle, exclusion principle, wave–particle duality, de Broglie wavelength, and wave function. Try to organize the concepts you have identified around the central theme of *quantum mechanics,* linking the various concepts appropriately by connecting words or short phrases to form meaningful propositions. When you have completed your concept map, compare your result with that of a fellow student. What similarities exist between the two maps? What distinctions can be seen in them? If there are significant differences in the linkages between concepts held in common in the two maps, discuss them with your colleague and try to resolve them. Consult your instructor for assistance in this effort, if necessary.

2. Section 10.8 introduces the laser as a device whose development followed closely from our understanding of quantum mechanics and atomic structure. Make a list of the basic concepts related to the development, function, and application of the laser in modern society. Reexamine Section 10.8 if necessary to ensure that your list is as complete as possible. Next, prioritize the items in your list from most important to least important. If some concepts are of equal importance or significance with respect to others, be sure to identify them as such. Now exchange your list with another student in your class. Does your list contain the same elements as that of your neighbor? Are his or her items prioritized in the same manner as yours? Discuss the differences in the two lists with your colleague and attempt to reconcile them. Ask your instructor or another classmate for help if you need it.

(■ **Indicates a review question, which means it requires only a basic understanding of the material to answer. Questions without this designation typically involve integrating or extending the concepts presented thus far.**)

1. ■ What does it mean when we say that the energy of something is "quantized"?
2. Of the things that a car owner has to purchase routinely—gasoline, oil, antifreeze, tires, spark plugs, and so on—which are normally sold in quantized units, and which are not?
3. How might you explain the concept of quantization to a younger child (brother, sister, niece, etc.), using money as the quantized entity?
4. Two common controls in a car are the steering wheel and the windshield wiper speed selector. Indicate whether each is controlling something in a quantized or a continuous fashion.
5. ■ What assumption allowed Planck to account for the observed features of blackbody radiation?
6. ■ What is a photon? How is its energy related to its frequency? To its wavelength?
7. ■ Describe the photoelectric effect. Name some devices that make use of this process.
8. If Nature suddenly changed and Planck's constant became a much larger number, what effect would this have on things such as solar cells, atomic emission and absorption spectra, lasers, and so on?
9. Sodium undergoes the photoelectric effect, with one electron absorbing a photon of violet light and another absorbing a photon of ultraviolet light. What is different about the two electrons afterward?
10. Can you think of a reason why metals exhibit the photoelectric effect most easily? (*Hint:* Do you see any connection between this phenomenon and the properties of a good electrical or thermal conductor?)
11. What aspects of the photoelectric effect can be explained without recourse to the concept of the photon? What aspects of this phenomenon require the existence of photons for their explanation?
12. Based on what you learned about image formation in Chapter 9, describe how you might design a photocopying machine that could make a copy that is enlarged or reduced compared to the size of the original.
13. If sunlight can be conceived of as a beam of photons, each of which carries a certain amount of energy and momentum, why don't we experience (or feel) any recoil as these particles collide with our bodies when, say, we're at the beach on a sunny day?
14. ■ What is the difference between a continuous spectrum and an emission-line spectrum?
15. A mixture of hydrogen and neon is heated until it is luminous. Describe what is seen when this light passes through a prism and is projected onto a screen.
16. ■ What are the basic assumptions of the Bohr model? Describe how the Bohr model accounts for the production of emission-line spectra from elements like hydrogen.
17. ■ Discuss what is meant by the term *ionization*. Give two ways by which an atom might acquire enough energy to become ionized.
18. ■ Compare the emission spectra of the elements hydrogen and helium (Figure 10.12). Which element emits photons of red light that have the higher energy?
19. If an astronomer examines the emission spectrum from luminous hydrogen gas that is moving away from Earth at a high speed and compares it to a spectrum of hydrogen seen in a laboratory on Earth, what would be different about the frequencies of spectral lines from the two sources? (See Section 6.2 if you need a little help in answering this question.)

20. The spectrum of light from a star that is observed with the Hubble Space Telescope is not exactly the same as that star's spectrum observed by a telescope on Earth. Explain why this is so.
21. A high-energy photon can collide with a free electron and give it some energy. (This is called the *Compton effect*.) How are the photon's energy, frequency, and wavelength affected by the collision?
22. ■ What is the de Broglie wavelength? What happens to the de Broglie wavelength of an electron when its speed is increased?
23. ■ An electron and a proton are moving with the same speed. Which has the longer de Broglie wavelength? (You may want to consult the Table of Conversion Factors for some useful information.)
24. ■ In an electron microscope, electrons play the role that light does in optical microscopes. What makes this possible?
25. ■ What is the uncertainty principle?
26. Explain why the Bohr model of the atom is incompatible with the Heisenberg uncertainty principle.
27. ■ What does it mean when a hydrogen atom is said to be in its "ground state"?
28. ■ Explain why a hydrogen atom with its electron in the ground state cannot absorb a photon of just any energy when making a transition to the second excited state ($n = 3$).
29. Describe the spectrum produced by ionized hydrogen—that is, a sample of hydrogen atoms all of which have lost one electron.
30. ■ Will the energy of a photon that ionizes a hydrogen atom from the ground state be larger than, smaller than, or equal to, the energy of a photon that ionizes another hydrogen atom from the first excited state ($n = 2$)? Explain.
31. What would an energy-level diagram look like for the quantized cat in Figure 10.3?
32. ■ In what part of the EM spectrum does the Lyman series of emission lines from hydrogen lie? The Balmer series? The Paschen series? Describe how each of these series is produced. In what final state do the electron transitions end in each case?
33. Radioactive strontium (Sr) tends to concentrate in the bones of people who ingest it. Why might one expect that strontium would behave like calcium (Ca) chemically and thus be preferentially bound in bone material, which is predominantly calcium in composition?
34. ■ The x-ray spectrum of a typical heavy element consists of two parts. What are they? How is each produced?
35. ■ Will the frequency of the K_α peak in the x-ray spectrum of copper (Cu) be higher or lower than the frequency of the K_α peak of tungsten (W)? Explain how you arrived at your answer.
36. ■ Describe how the Bohr model may be used to account for characteristic x-ray spectra in heavy atoms.
37. ■ What is the origin of the word *laser*?
38. ■ Distinguish laser light from the light emitted by an ordinary lightbulb in as many ways as you can.
39. ■ Distinguish between a metastable state and a normally allowed energy state within an atom. Discuss the role of metastable states in the operation of laser systems.
40. ■ Define what is meant by the term *population inversion*. Why must this condition be achieved before a system can successfully function as a laser?
41. ■ Describe the operation of a pulsed ruby laser.
42. A friend tells you that physicists have just invented a new pulsed laser that is pumped with yellow light and produces laser light in the ultraviolet portion of the EM spectrum. Why might you be a little skeptical of this claim? Explain.

PROBLEMS

1. Find the energy of a photon with a frequency of 1×10^{16} Hz.

2. Things around you are emitting infrared radiation that includes the wavelength 9.9×10^{-6} m. What is the energy of these IR photons?

3. In what part of the EM spectrum would a photon of energy 9.5×10^{-25} J be found? What is its energy in electronvolts?

4. Gamma rays (γ-rays) are high-energy photons. In a certain nuclear reaction, a γ-ray of energy 0.511 MeV (million electronvolts) is produced. Compute the frequency of such a photon.

5. Electrons striking the back of a conventional TV screen travel at a speed of about 8×10^7 m/s. What is their de Broglie wavelength?

6. In a typical electron microscope, the momentum of each electron is about 1.6×10^{-22} kg-m/s. What is the de Broglie wavelength of the electrons?

7. During a certain experiment, the de Broglie wavelength of an electron is 670 nm = 6.7×10^{-7} m, which is the same as the wavelength of red light. How fast is the electron moving?

8. If a proton were traveling the same speed as electrons in Problem 5, what would its de Broglie wavelength be? The mass of a proton is 1.67×10^{-27} kg.

9. (a) A hydrogen atom has its electron in the $n = 4$ level. The radius of the electron's orbit in the Bohr model is 0.847 nm. Find the de Broglie wavelength of the electron under these circumstances.
   (b) What is the momentum, mv, of the electron in its orbit?

10. (a) A small ball with a mass of 0.06 kg moves along a circular orbit with a radius of 0.5 m at a speed of 3.0 m/s. What is the angular momentum of the ball?
    (b) If the angular momentum of this ball were quantized in the same manner as the angular momentum of electrons in the Bohr model of the atom, what would be the approximate value of the quantum number n in such a case?

11. A hydrogen atom has its electron in the $n = 2$ state.
    (a) How much energy would have to be absorbed by the atom for it to become ionized from this level?
    (b) What is the frequency of the photon that could produce this result?

12. A hydrogen atom initially in the $n = 3$ level emits a photon and ends up in the ground state.
    (a) Compute the energy of the emitted photon?
    (b) If this atom then absorbs a second photon and returns to the $n = 3$ state, what must the energy of this photon be?

13. **Figure 10.47** is the energy-level diagram for a particularly simple, fictitious element, Vernium (Vn). Indicate by the use of arrows all allowed transitions leading to the emission of

photons from this atom and order the frequencies of these photons from highest (largest) to lowest (smallest).

14. A neutral calcium atom ($Z = 20$) is in its ground state electronic configuration. How many of its electrons are in the $n = 3$ level? Explain how you arrived at your answer.

15. An atom of neutral zinc possesses 30 electrons. In its ground configuration, how many fundamental energy levels are required to accommodate this number of electrons? That is, what is the smallest value of n needed so that all 30 of zinc's electrons occupy the lowest possible quantum energy states consistent with the Pauli exclusion principle?

16. Referring to Figure 10.35, notice that the bremsstrahlung x-ray spectrum of both Mo and W cut off (have zero intensity) at about 0.035 nm. X-rays with this wavelength are produced by the target element when bombarding electrons are promptly stopped in a single collision and give up all their energy in the form of EM waves. Confirm that electrons having energies of 35,000 eV will produce x-ray photons with wavelengths near 0.035 nm by this process.

17. The characteristic K_α and K_β lines for copper have wavelengths of 0.154 nm and 0.139 nm, respectively. What is the ratio of the energy difference between the levels in copper involved in the production of these two lines?

18. Referring to Figure 10.36, we see that the atomic number Z is proportional to $f^{1/2}$ or that Z^2 is proportional to f. Because the frequency of the characteristic x-ray lines is itself proportional to the energy of the associated x-ray photon, we are led to conclude that ΔE and hence the energies of the atomic levels also scale as Z^2. Based on this analysis (and ignoring any differences from the masses of their nuclei), how much greater is the energy associated with the ground state of helium ($Z = 2$) than that of hydrogen ($Z = 1$)? Make a similar comparison between the ground state energies for sodium ($Z = 11$) and hydrogen.

19. Characteristic x-rays emitted by molybdenum have a wavelength of 0.072 nm. What is the energy of one of these x-ray photons?

20. In a helium-neon laser, find the energy difference between the two levels involved in the production of red light of wavelength 632.8 nm by this system.

21. The carbon-dioxide laser is one of the most powerful lasers developed. The energy difference between the two laser levels is 0.117 eV.
    (a) What is the frequency of the radiation emitted by this laser?
    (b) In what part of the EM spectrum is such radiation found?

22. If you bombard hydrogen atoms in the ground state with a beam of particles, the collisions will sometimes excite the atoms into one of their upper states. What is the *minimum* kinetic energy the incoming particles must have if they are to produce such an excitation?

23. (a) Which of the following elements emits a K-shell x-ray photon with the highest frequency?
    (b) Which of the following emits a K-shell photon with the lowest frequency?
       (i) Silver (Ag)
       (ii) Calcium (Ca)
       (iii) Iridium (Ir)
       (iv) Tin (Sn)

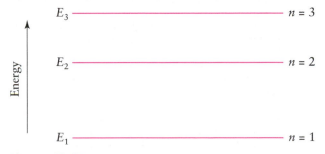

Figure 10.47 Problem 13.

1. One serious problem with sending intense laser beams long distances through the atmosphere to receiving targets is that the beams spread out. This effect, called *thermal blooming,* is caused by the fact that the beam heats the air, thereby changing the speed that the light travels. Explain how such a change in the speed of light in the air could produce the observed effect. Will the speed increase or decrease upon heating? Why?

2. The muon is a negatively charged particle that is much like an electron but with a mass about 200 times larger. A "muonic" hydrogen atom forms when a muon orbits a proton. The muonic atom's orbits and energy levels follow the basic rules of the Bohr model and the quantum mechanical model of the atom. Using the analysis in Section 10.5, explain why the size of each Bohr orbit in the muonic atom is much smaller than the corresponding orbit in an ordinary hydrogen atom. How would the energies of each of the energy levels in the muonic atom be different from those in the regular hydrogen atom? What would be different about the emission spectrum?

3. The rate at which solar wind particles enter the atmosphere is higher during the day than at night, yet the intensity of the auroral emissions remains high well after the Sun has set. Can you suggest a means by which the atmospheric molecules might be able to radiate long after the period of collisions with charged particles has ended? (*Hint:* How long does it take a typical atom to radiate from a normal allowed energy state? How could this time be lengthened?)

4. Compute a rough estimate of the number of photons emitted each second by a 100-W lightbulb. You might make the simplifying assumptions that (1) all of the electrical energy is converted into radiant energy and (2) the "average" photon that is emitted has an energy corresponding to the peak of the BBR curve for an object at 3,000 K (refer to Section 8.6).

5. In the Bohr model for hydrogen, the radius of the nth orbit can be shown to be n^2 times the radius of the first Bohr orbit $r_1 = 0.05$ nm (see Example 10.3). Similarly, the energy of an electron in the nth orbit is $\frac{1}{n^2}$ times its energy when in the $n = 1$ orbit. What is the *circumference* of the $n = 100$ orbit? (This is the distance the electron has traveled after having revolved around the proton once.) For such large-n states, the orbital frequency is about equal to the frequency of the photon emitted in a transition from the nth level to an adjacent level with $n + 1$ or $n - 1$. Given this, find the frequency and corresponding period of the electron's orbit by computing the frequency associated with the transition from $n = 100$ to $n = 101$. Using your values for the electron's orbital size (distance) and travel time (period), calculate the approximate speed of the electron in the 100th orbit. How does this speed compare to the speed of light?

6. Long before the advent of quantum mechanics, physicists had developed an empirical formula to predict the wavelengths of the emission lines of hydrogen. Specifically, for an electron transition from the mth energy level down to the nth level ($m > n$), the wavelength of the emitted photon is given by

$$1/\lambda = (10{,}973{,}731.57)(1/n^2 - 1/m^2) \text{ m}^{-1}$$

Use this equation to confirm that the first member of the Balmer series has a wavelength of 656.3 nm. What is the wavelength of a photon emitted by hydrogen during a transition from the ninth excited state ($m = 10$) to the second excited state ($n = 3$)? In what part of the EM spectrum does this spectral line reside? What wavelength photon would be required to ionize a hydrogen atom from the $n = 8$ level?

11

NUCLEAR PHYSICS

Figure CO-11 Ionization-type smoke detector.

Peter Mukherjee/Getty Images

CHAPTER INTRODUCTION: Radioactive Sentinel

You probably have a radioactive device in your home. Put there deliberately, its job is to save your life and those of your family. Ionization smoke detectors (**Figure CO-11**) have been standing guard in homes for decades, ready to detect the tiny particles that comprise smoke and sound an alarm. Although an electrical device, this type of smoke detector also makes use of a small amount of radioactive material.

The heart of an ionization smoke detector is a chamber through which air can pass. Ions are continually created in the air in the chamber, and a battery causes them to carry a small electrical current through the air. This current is closely monitored by the device's electronics.

When smoke particles enter the chamber, the ions are attracted to them and become attached. This reduces the current, which triggers the alarm. The device is very sensitive, as anyone can attest who has burned toast or food while cooking and was startled by the sound of the smoke alarm going off.

The ions that contribute to the current in these devices are produced when nuclear radiation released by the decay of radioactive material in the alarm chamber knocks electrons off nitrogen and oxygen atoms in the air. This phenomenon, known as *radioactive decay*, is one of the main subjects of this chapter, along with the basic properties of nuclei, key applications of nuclear processes, and the promises and perils of nuclear energy produced via nuclear fission and nuclear fusion.

11.1 The Nucleus

Unimaginably tiny yet incredibly dense, the nucleus occupies the very center of the atom. More than 99.9 percent of the atom's mass is compressed into roughly *one-trillionth* of its total volume to yield a density of approximately 2×10^{17} kg/m³. This is 2×10^{14} times the density of water. If an atom could be enlarged until it were 2,000 feet (0.38 mi) across, its nucleus would only be about the size of a pea. The nucleus is impervious to the chemical and thermal processes that affect the atom's electrons. But it is seething with energy—energy, for example, that makes the Sun and other stars shine.

The nucleus contains two kinds of particles: protons and neutrons. These particles have nearly the same mass, about 1,840 times that of an electron. The masses are extremely small, so it becomes convenient to introduce an appropriate unit of mass, the *atomic mass unit*, u:

$$1 \text{ u} = 1.66 \times 10^{-27} \text{ kg}$$

Using this measure, the masses of a proton and a neutron are each about 1 u. The proton carries a positive

● Proton
○ Neutron

Figure 11.1 The different possible nuclei of helium atoms. All have two protons, but the number of neutrons varies. The electrons in orbit around the nuclei are not shown. With this scale, the radius of such an orbit would be about 1,000 feet (305 m).

DEFINITION **Neutron Number** The number of neutrons contained in a nucleus.

DEFINITION **Mass Number** The total number of protons and neutrons in a nucleus.

DEFINITION **Isotopes** Isotopes of a given element have the same number of protons in the nucleus but different numbers of neutrons.

Table 11.1 Properties of the Particles in the Atom

Particle	Mass	Charge
Electron	9.109×10^{-31} kg = 0.00055 u	-1.602×10^{-19} C
Proton	1.67262×10^{-27} kg = 1.00730 u	1.602×10^{-19} C
Neutron	1.67493×10^{-27} kg = 1.00869 u	0

charge, and the neutron is uncharged (Table 11.1). The number of protons in the nucleus is the **atomic number** Z of the atom. There is one proton in the nucleus of each hydrogen atom ($Z = 1$), two in that of helium ($Z = 2$), eight in oxygen ($Z = 8$), and so on. This number determines the identity of the atom.

The uncharged neutrons have much less influence on the properties of the atom: their main effect is on the atom's mass. In fact, the number of neutrons in the nuclei of a particular element can vary. Most helium atoms have two neutrons in the nucleus, but some have one, three, four, or even six (Figure 11.1). Each is still a helium atom; the only difference is their mass. This allows us to associate another number with each nucleus, the **neutron number.**

The symbol N is used to represent the neutron number. For all helium atoms, $Z = 2$, but N can be 1, 2, 3, 4, or 6.

The mass of an atom is determined primarily by how many protons and neutrons there are in the nucleus. (The electrons are so light that they contribute only a negligible amount to the total mass.) Therefore, it is useful to define yet a third number for each nucleus.

The **mass number** is represented by the letter A. Note that

$$A = Z + N$$

The mass number indicates what the mass of a nucleus is, just as the atomic number indicates the amount of electric charge in the nucleus. Protons and neutrons are collectively referred to as **nucleons.** (The terms *nucleon number* and *atomic mass number* are sometimes used instead of mass number.) The mass number is just the total number of nucleons in the nucleus. The possible mass numbers for helium are $A = 3$, 4, 5, 6, or 8. Each different possible "type" of helium is called an **isotope.**

The different isotopes of an element have essentially the same atomic properties: these are determined by the atomic number Z. For example, most of the carbon atoms in your body have six neutrons in their nuclei, but small numbers of the isotopes with seven and eight are also present. The three different carbon isotopes are indistinguishable as far as chemical processes (such as burning) are concerned. The different isotopes of an element often *do* have vastly different nuclear properties. For example, most nuclear power plants rely on the splitting of uranium-235 nuclei—that is, uranium atoms with mass number $A = 235$. Uranium-238 will not work.

Most of the 118 different elements have several isotopes. Some have only a few (hydrogen has 3), and others have more than 20 (iodine, silver, and mercury, to name a few). More than 3,100 different isotopes have been identified and studied. Of these, only about 340 occur naturally. The rest are produced in certain nuclear processes, such as nuclear explosions. Of the known isotopes, only 256 are long-lived. The remaining 2,800 or so are unstable; the nuclei eventually transform into different nuclei through a process known as *radioactive decay.* (More on this in Section 11.2.)

Each different isotope of an element is designated by its mass number. The most common isotope of helium has $A = 4$ ($Z = 2$ and $N = 2$) and is called helium-4. The other helium isotopes are helium-3, helium-5, helium-6, and helium-8. The three isotopes of carbon in your body are carbon-12, carbon-13, and carbon-14. Two of the isotopes of hydrogen have been given special names:

hydrogen-2 is called *deuterium* and hydrogen-3 is called *tritium.* (The prefixes come from the Greek words for *second* and *third.*)

Freezing, boiling, burning, crushing, and other chemical and physical processes do not affect the nuclei of atoms. These processes are influenced by the forces between different atoms, forces that involve only the outer electrons. But nuclei are not indestructible: a number of nuclear processes do affect them. Nuclei can lose or gain neutrons and protons, absorb or emit gamma rays, split into smaller nuclei, or combine with other nuclei to form larger ones. These processes are called *nuclear reactions.* Some occur around us naturally, but most are produced artificially in laboratories. Most of the remainder of this chapter deals with the nature of nuclear reactions and their applications.

To diagram nuclear reactions, we use a special notation to represent each isotope. It consists of the atom's chemical symbol with a subscript and a superscript on the left side. The *subscript* is the atom's atomic number Z, and the *superscript* is the atom's mass number A. Some examples:

Helium-4	$^{4}_{2}\text{He}$	Carbon-14	$^{14}_{6}\text{C}$
Carbon-12	$^{12}_{6}\text{C}$	Uranium-235	$^{235}_{92}\text{U}$

The two numbers indicate the relative mass and charge that the nucleus possesses. The neutron number N can be found by subtracting the lower number (Z) from the upper number (A). For example, each uranium-235 nucleus contains 143 neutrons ($235 - 92 = 143$). Note that both the chemical symbol and the subscript indicate what the element is, so they must always agree. Regardless of what the superscript is, if the subscript is 6, the element is carbon and the symbol must be C. There is a certain redundancy in this notation insofar as the atomic number and the proper chemical symbol really convey the same information about the isotope. For example, we could eliminate the use of the atomic number and designate carbon-14 as simply ^{14}C. This simplification is adopted in many books, but we will continue to include the atomic number in our notation throughout this chapter because it is helpful in verifying charge conservation in the course of nuclear reactions (see Section 11.2).

This notation can be extended to represent individual particles as well. The designations for neutrons, protons, and electrons are

Neutron	$^{1}_{0}\text{n}$
Proton	$^{1}_{1}\text{p}$
Electron	$^{0}_{-1}\text{e}$

For the electron, 0 is used for the mass number because its mass is so small. The -1 indicates that it has the same size charge as a single proton, except it is negative.

Perhaps you've wondered how protons can be bound together in a nucleus. After all, like charges do repel each other. There is indeed a large electrostatic force acting to push the protons apart, but another force many times stronger acts to hold them together inside a nucleus. This is called the *strong nuclear force,* one of the four fundamental forces in Nature. (The *weak nuclear force* is involved in certain nuclear processes, but it is not important in holding the nucleus together. More on this in Chapter 12.) The strong nuclear force is an attractive force that acts between nucleons irrespective of their charge: every proton and neutron in a nucleus exerts an attractive force on every other proton and neutron. Compared to the gravitational and electromagnetic forces, the strong nuclear force is rather strange. It is much stronger than the others, but it has an extremely short range: the attractive nuclear force between particles in the nucleus begins to weaken appreciably when the particles are more than about 3×10^{-15} meters apart, and it effectively disappears if the particles become separated by more than 10^{-14} meters. This puts an upper limit on the size that a nucleus can have and still be stable. In a large nucleus, protons on opposite

Figure 11.2 Plot showing the number of neutrons and the number of protons in the isotopes that are stable. Each small square represents one stable isotope. The green line indicates the trend for nuclei that would have the same number of protons and neutrons. The points curve away from this line, indicating that larger stable nuclei have successively more neutrons than protons.

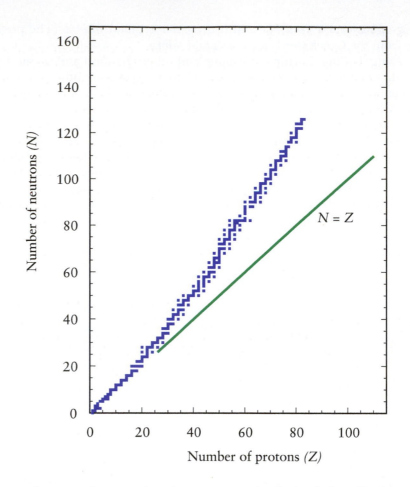

sides are far enough apart that they are near the limit of the effective range of the strong attractive nuclear force; in these instances, the repulsive electric force between the protons becomes important. No known stable isotope has an atomic number larger than 83.

The effectiveness of the nuclear force at holding a nucleus together also depends on the relative numbers of neutrons and protons. If there are too many or too few neutrons compared to the number of protons, the nucleus will not be stable. For example, carbon-12 and carbon-13 are stable, but carbon-11 and carbon-14 are not. The ratio of N to Z for stable nuclei is about 1 for small atomic numbers and increases to about 1.5 for large nuclei (**Figure 11.2**). The stable isotope lead-208 has 126 neutrons and 82 protons in each nucleus.

In summary, the nucleus is a collection of neutrons and protons held together by the strong nuclear force. This force is limited in its ability to hold nucleons together. Nuclei that are too large or that do not have the proper ratio of neutrons to protons are unstable. They eject particles and release energy. There are several different mechanisms used by various unstable nuclei to accomplish this. The main ones are discussed in the following section.

Learning Check

1. Different isotopes of an element have the same number of _____ in the nucleus but different numbers of _____ .

2. Which of the following is *not* true about a nucleus of the isotope $^{18}_{8}O$?
   (a) It contains 8 protons.
   (b) It contains 18 nucleons.
   (c) It contains 8 electrons.
   (d) It contains 10 neutrons.

3. (True or False.) Protons can stay close together inside a nucleus because the repulsive electric force between them disappears at short range.

ANSWERS: 1. protons, neutrons **2.** (c) **3.** False

11.2 Radioactivity

Radioactivity, also called *radioactive decay,* is a process wherein an unstable nucleus emits particles or EM radiation (or both). Isotopes with unstable nuclei are called **radioisotopes.** The majority of all isotopes are radioactive. For the moment, we will put aside the question of where radioisotopes come from and concentrate on the manner in which radioactive decay occurs.

When it was first investigated, nuclear radiation was found to be similar to x-rays. For example, it exposes photographic film. Soon it was determined that there were actually three different types of nuclear radiation named **alpha** (α), **beta** (β), and **gamma** (γ) **radiation.** Alpha rays and beta rays are actually high-speed charged particles. They can be deflected with magnetic and electric fields (**Figure 11.3**). Gamma rays are extremely high-frequency electromagnetic waves (high-energy photons). It is now known that there are many different *decay channels* that unstable nuclei may use when undergoing radioactive decay. For simplicity, we will focus here on only a few of the most common mechanisms of radioactivity: alpha, beta, and gamma decay.

Several different devices are used to detect radiation, the most common being the *Geiger counter.* It exploits the fact that nuclear radiation is ionizing radiation. A gas-filled cylinder is equipped with a fine wire running along its axis (**Figure 11.4**). A high voltage is maintained between the cylinder wall and the wire, producing a strong electric field. When an alpha, beta, or gamma ray enters the cylinder and ionizes some of the gas atoms, the freed electrons are accelerated by the electric field. These in turn ionize other atoms, and an avalanche of electrons reaches the wire and causes a current pulse. Most Geiger counters emit an audible click each time a ray is detected. They also keep track of how many are detected each second. In other words, they indicate the count rate—the number of alpha, beta, or gamma rays detected each second.

The emission of each type of radiation has a different effect on the nucleus. Both alpha decay and beta decay alter the identity of the nucleus (the atomic number is changed) as well as release energy. Gamma-ray emission does not in itself change the nucleus: it simply carries away excess energy in much the same way that photon emission carries away excess energy from excited atoms. Gamma-ray emission often accompanies alpha decay and beta decay because these two processes often leave the nucleus in an excited state with excess energy that is quickly lost through the emission of a high-energy photon. **Table 11.2** lists several radioisotopes and their modes of decay.

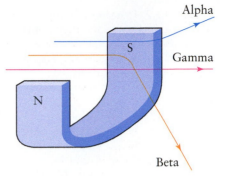

Figure 11.3 The three common types of nuclear radiation are affected differently by a magnetic field. Alpha and beta rays are deflected as they go through the field because they are charged particles. Gamma rays (high-energy photons) are not deflected.

11.2a Alpha Decay

An alpha particle is really four particles tightly bound together: two protons and two neutrons. It is identical to a nucleus of helium-4. For this reason, an alpha particle can be represented as

$$\text{alpha particle:} \quad \alpha \quad \text{or} \quad {}^{4}_{2}\text{He}$$

Figure 11.4 (Left) Simplified sketch of a Geiger counter. Nuclear radiation ionizes the gas in the cylinder. The freed electrons are accelerated to the wire and produce a current pulse. (Right) Geiger counter detecting beta rays emitted by an isotope in a lead cup (gray). The count rate is enhanced by the magnet (blue), since beta rays are deflected by the magnetic field into the detector (marked G).

Table 11.2 Properties of Selected Isotopes[a]

Element	Isotope	Decay Mode(s)	Relative Abundance (%)
Hydrogen	$_1^1\text{H}$	. . . (stable)	99.985
	$_1^2\text{H}$(deuterium)	. . .	0.015
	$_1^3\text{H}$(tritium)	beta	. . .
Helium	$_2^3\text{He}$	. . .	0.00014
	$_2^4\text{He}$	. . .	99.9999
	$_2^5\text{He}$	alpha	. . .
	$_2^6\text{He}$	beta	. . .
	$_2^8\text{He}$	beta, gamma	. . .
Carbon	$_6^{12}\text{C}$	. . .	98.90
	$_6^{13}\text{C}$	. . .	1.10
	$_6^{14}\text{C}$	beta	trace
	$_6^{15}\text{C}$	beta	. . .
Silver	$_{47}^{107}\text{Ag}$*	gamma, then stable	51.84
	$_{47}^{108}\text{Ag}$	beta, gamma	. . .
	$_{47}^{109}\text{Ag}$*	gamma, then stable	48.16
	$_{47}^{110}\text{Ag}$	beta, gamma	. . .
Uranium	$_{92}^{232}\text{U}$	alpha, gamma	. . .
	$_{92}^{233}\text{U}$	alpha, gamma	. . .
	$_{92}^{234}\text{U}$	alpha, gamma	0.0055
	$_{92}^{235}\text{U}$	alpha, gamma	0.72
	$_{92}^{236}\text{U}$	alpha, gamma	. . .
	$_{92}^{237}\text{U}$	beta, gamma	. . .
	$_{92}^{238}\text{U}$	alpha, gamma	99.27
	$_{92}^{239}\text{U}$	beta, gamma	. . .

[a]Isotopes in excited nuclear states are marked with asterisks (*) in the table. Not every known isotope for the elements carbon, silver, and uranium is included here.

A nucleus that undergoes alpha decay *does not* contain a helium-4 nucleus. This just happens to be a particularly stable combination of nuclear particles that can be ejected as a group from the nucleus.

The emission of an alpha particle reduces both the atomic number Z and the neutron number N by 2. The mass number A is reduced by 4. **Figure 11.5** is a diagram of the alpha decay of a plutonium-242 nucleus. The atomic number of the nucleus decreases from 94 to 92: the nucleus is transformed from plutonium to uranium. The original plutonium-242 nucleus is called the *parent*, and the resulting uranium-238 nucleus is called the *daughter*.

Figure 11.5 shows why the isotopic notation introduced earlier is so convenient. When used to represent a nuclear process, such as alpha decay, it clearly shows how the mass and the charge of the nucleus are affected. Because both conservation of charge and conservation of mass must hold in nuclear

Figure 11.5 A nucleus of plutonium-242 undergoing alpha decay. (Because of the large mass difference between the daughter nucleus and the alpha particle, we have ignored the recoil velocity of the former in this figure.)

transformations (more on this in Chapter 12), the total electric charge and the total number of protons and neutrons must be the same before and after the process. This means that the sum of the subscripts on the right side of the arrow must equal the sum of those on the left. The same is true for the superscripts. This allows one to determine what the decay product (the daughter) is when a given nucleus undergoes alpha decay.

EXAMPLE 11.1 The isotope radium-226 undergoes alpha decay. Write the reaction equation and determine the identity of the daughter nucleus.

SOLUTION From the periodic table of the elements, we find that the atomic number of radium is 88 and its chemical symbol is Ra. So the reaction will appear as follows:

$$^{226}_{88}\text{Ra} \rightarrow ? + {}^{4}_{2}\text{He}$$

The mass number A of the daughter nucleus must be 222 for the superscripts to agree on both sides of the arrow. By the same reasoning, the atomic number Z must be 86. From the periodic table, we find the element radon (Rn) has $Z = 86$. The daughter nucleus is radon-222.

$$^{226}_{88}\text{Ra} \rightarrow {}^{222}_{88}\text{Rn} + {}^{4}_{2}\text{He}$$

Because alpha decay causes such a drastic change in the mass of the nucleus, it generally occurs only in radioisotopes with high atomic numbers. The alpha particle is ejected with very high speed (typically around 1/20th the speed of light). Alpha particles are quickly absorbed when they enter matter: even something as thin as a sheet of paper can stop a significant fraction of them.

11.2b Beta Decay

Beta decay is easily the oddest of the three kinds of radioactivity. A beta particle is simply an electron ejected from a nucleus. This means that a beta particle has the same symbol as an electron.

$$\text{Beta particle:} \quad \beta \quad \text{or} \quad {}^{0}_{-1}\text{e}$$

But wait a minute: there aren't any electrons in a nucleus. During beta decay, one of the neutrons is spontaneously converted into an electron (the beta particle) and a proton. The electron is ejected with very high speed, and the proton remains in the nucleus. We can represent this process as follows:

$$\text{neutron} \rightarrow \text{proton} + \text{electron}$$

$$^{1}_{0}\text{n} \rightarrow {}^{1}_{1}\text{p} + {}^{0}_{-1}\text{e}$$

The total electric charge remains the same—zero.

Another particle, a *neutrino* (Italian for "little neutral one"), is also emitted in beta decay. Neutrinos are very strange little beasts—they have no charge, their mass is extremely small, and they rarely interact with matter. Neutrinos routinely pass through the entire Earth without being absorbed, deflected, or otherwise affected. For our purposes, the neutrino can be regarded merely as part of the energy that is released during beta decay. (Chapter 12 has more to say about neutrinos.)

A nucleus that undergoes beta decay loses one neutron and gains one proton. **Figure 11.6** shows the beta decay of a carbon-14 nucleus. Note that the mass number A of the nucleus is unchanged.

Like many nuclear processes, the decay of a neutron is reversible. In other words, it is possible for a proton to combine with an electron to produce a neutron. This is what happens during an *electron capture* process: an atomic electron

Figure 11.6 A nucleus of carbon-14 undergoing beta decay. (As in Figure 11.5, because the mass of the nitrogen-14 nucleus is so much larger than that of the electron, we have ignored the small recoil velocity of the former in this drawing.)

$$^{14}_{6}\text{C} \quad ---\!\!\rightarrow \quad ^{14}_{7}\text{N} \quad + \quad ^{0}_{-1}\text{e}$$

penetrates the nucleus and interacts with a bound proton to form a bound neutron in what might be thought of as an *inverse beta decay*. In such circumstances, the nucleus loses a proton and gains a neutron. As with ordinary beta decay, the mass number of the nucleus remains the same, but now the atomic number is reduced by one unit.

EXAMPLE 11.2 The isotope iodine-131 undergoes beta decay. Write the reaction equation and determine the identity of the daughter nucleus.

SOLUTION From the periodic table, we find that iodine's chemical symbol and atomic number are I and 53, respectively. Therefore,

$$^{131}_{53}\text{I} \rightarrow ? + ^{0}_{-1}\text{e}$$

The mass number stays the same, and the atomic number is increased by 1 to 54. Appealing again to the periodic table, we find that this is the element xenon (Xe). The daughter nucleus is xenon-131.

$$^{131}_{53}\text{I} \rightarrow ^{131}_{54}\text{Xe} + ^{0}_{-1}\text{e}$$

Often the daughter nucleus in both alpha and beta decay is itself radioactive and decays into another isotope. Plutonium-242 undergoes alpha decay to uranium-238. This is a radioisotope that undergoes alpha decay into thorium-234. This process continues until, after a total of nine alpha decays and six beta decays, the stable isotope lead-206 is reached. This is called a **decay chain.** When Earth was formed from the debris of exploded stars, hundreds of different radioisotopes were present in varying amounts. Geological formations that were originally rich in uranium-238 now contain large amounts of lead-206 as well.

11.2c Gamma Decay

Because a gamma ray has no mass or electric charge, gamma-ray emission has no effect on the mass number A or the atomic number Z of a nucleus. Nuclei can exist in excited states in much the same way that atomic electrons can. Gamma rays are emitted when the nucleus makes a transition to a lower energy state. Figure 11.7 shows the gamma decay of a strontium-87 nucleus. The identity of the nucleus is not changed during the process.

Figure 11.7 A nucleus of strontium-87 undergoing gamma decay. The asterisk (*) on the symbol for the parent nucleus indicates that it is in an excited state. (As in previous drawings of this type, the recoil velocity of the strontium-87 nucleus has been ignored.)

$$^{87}_{38}\text{Sr}^* \quad ---\!\!\rightarrow \quad ^{87}_{38}\text{Sr} \quad + \quad \gamma$$

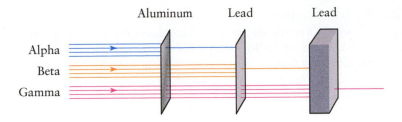

Gamma-ray photons, as indicated in Figure 10.7, have energies from about 100,000 electronvolts to more than 1 billion electronvolts. Most gamma-ray photons emitted in gamma decay are around 1 million electronvolts. (The unit of energy that is used most often in nuclear physics is the megaelectronvolt [MeV], which is 1 million electronvolts.)

One way to compare the different decay processes is to focus on three properties of the nucleus: mass, electric charge, and energy. Alpha decay alters all three: an alpha particle carries away considerable mass, two charged protons, and a great deal of kinetic energy. Beta decay has little effect on the mass of the nucleus, but it does increase the positive charge of the nucleus, and it takes away energy in the form of the beta particle's kinetic energy. Gamma decay only takes energy away from the nucleus.

The three types of nuclear radiation differ considerably in their ability to penetrate solid matter. All three are ionizing radiation. They collisionally ionize atoms as they pass through matter. Alpha particles are the least penetrating. Their positive charge causes them to interact strongly with atomic electrons and nuclei, and they are absorbed after traveling only a short distance. Beta particles are more penetrating than alpha particles. For comparison, a thin sheet of aluminum that will block essentially all alpha particles will stop only a small fraction of beta particles. Gamma rays are the most penetrating of the three. Because they have no electric charge and travel at the speed of light, they interact with atoms much less frequently. It requires several centimeters of lead to block gamma rays (**Figure 11.8**).

11.2d Radioactivity and Energy

What becomes of the energy of nuclear radiation as it is absorbed? Most of it goes to heat the material. Early experimenters with radioactivity noticed that highly radioactive samples were physically warmer than their surroundings. As long as the substance continued to emit radiation, it stayed warm. Because of this, radioactivity has been used as an energy source on spacecraft. Many interplanetary space probes, including the *Cassini* spacecraft sent to probe the planet Saturn and its moons (see Section 3.6b), were equipped with *radioisotope thermoelectric generators* (RTGs). The heat from the radioactive decay of an isotope such as plutonium-238 is converted into electricity to operate cameras, radio transmitters, and other onboard equipment.

Earth's interior is so hot that much of it is molten. Some of this heat reaches the surface in volcanoes, in geysers, and through other geothermal processes. Geologists are not certain why Earth's interior is so hot, but they know what keeps it from cooling off: heat from radioactive decay. In Earth's interior, relatively small amounts of radioisotopes are still present. The energy released as these radioisotopes decay is enough to compensate for the transfer of heat to Earth's surface. Because much of this radioactive material is concentrated in a layer just below Earth's outer crust, this layer is kept in a partially liquid state by heat from the radioactivity. This in turn allows the crustal plates (continental land masses) to slowly slide around over Earth in a process called *continental drift*.

11.2e Applications

Radiation from radioactive decay is used routinely for many purposes. Nuclear medicine makes use of radioisotopes for diagnosis and treatment (see the Medical Applications feature at the end of Section 11.3). Dozens of industrial facilities around the world use gamma radiation from cobalt-60 and other isotopes to sterilize disposable medical products such as syringes and healthcare workers' gloves. The gamma rays can penetrate deep into the interiors of product containers and kill bacteria and other microorganisms that may be present there. Many of these facilities also irradiate food items, from spices and herbs to fresh meat, with gamma rays and other ionizing radiation to prevent spoilage, prolong shelf life, and kill harmful bacteria such as salmonella.

Because gamma rays are absorbed in a predictable way as they pass through solid material, they can be used to measure the thickness and integrity of manufactured items. Flaws in jet engine parts and cracks in large cables can be detected because more gamma rays will pass through items with gaps in them than through solid material. The thickness of metal sheets as they are manufactured can be measured by monitoring how much gamma radiation passes through them.

Another way in which radioactivity is put to good use is in the most common type of smoke detector (see the Chapter Introduction). The alpha radiation from a radioisotope (americium-241) ionizes the air between two plates connected to the terminals of a battery (**Figure 11.9**). Consequently, an electric current flows through the circuit. When smoke enters the detection chamber, the ions are attracted to the smoke particles and attach to them. This reduces the flow of charge between the plates. An ammeter built into the electronic circuitry detects the resulting decrease in the current and triggers an audible alarm. Thousands of lives have been saved by these simple, inexpensive devices.

Analysis of radioactive decay is an important tool in nuclear physics. The type of radiation emitted by a particular radioisotope, along with the amount of energy released, provide clues to the structure of the nucleus. Because nuclei are much too small for us to examine with our eyes, we have to use indirect information, such as that carried by radiation when it leaves a nucleus, to learn about them.

Concept Map 11.1 summarizes the common types of radioactive decay.

Figure 11.9 Simplified diagram of a smoke detector. (a) A current flows between the plates because the radiation ionizes the air. (b) Ions are attracted to the smoke particles (gray ovals) and attach to them. This reduces the current, which triggers the alarm.

(a)

(b)

Learning Check

1. Nuclear radiation can be detected by using a
   _____ .

2. Of the three types of nuclear decay, only
   _____ does not affect the atomic
   number of the nucleus.

3. (True or False.) Alpha decay causes the atomic
   number of the nucleus to increase.

4. Radioactive decay is involved in which of the
   following?
   (a) Keeping Earth's interior hot.
   (b) Making smoke detectors work.
   (c) Supplying energy for interplanetary space probes.
   (d) All of the above.

ANSWERS: 1. Geiger counter **2.** gamma decay
3. False **4.** (d)

11.3 Half-Life

We now address an aspect of radioactive decay that we have avoided so far—
time. If a nucleus is radioactive, when will it decay? In truth, there is no way
of knowing precisely when a particular nucleus will decay: it may wait a billion
years or only a millionth of a second. Radioactive decay is a random process,
much like throwing dice (**Figure 11.10**). One can't predict exactly what will
come up each throw, but one can analyze the results of hundreds of throws
and establish how likely it is that each possible value will come up. Similarly,
we can predict how much time will elapse, on the average, before a nucleus of
a given radioisotope will decay. There is wide variation in decay times among
the thousands of different radioisotopes. Nearly all nuclei of some isotopes

Figure 11.10 The exact value of one roll of a pair of dice can't be predicted, but the number of times that a total of seven will turn up in 1,000 rolls can be predicted quite closely. Similarly, the exact time at which a given nucleus will decay can't be predicted, but the fraction of 1 million nuclei that will decay during some time interval can be established very accurately.

Vern Ostdiek

will decay in less than a second, whereas only a fraction of the nuclei of other isotopes will decay in 4.5 billion years, the age of Earth. This leads us to the concept of **half-life.**

DEFINITION **Half-Life** The time it takes for half the nuclei in a sample of a radioisotope to decay. The time interval during which each nucleus has a 50-percent probability of decaying.

11.3a Quantifying Half-Lives

The half-lives of radioisotopes range from a tiny fraction of a second to billions of years (Table 11.3). During the span of one half-life, approximately half of the nuclei in a sample will decay—that is, emit their radiation. Half of the remaining nuclei will decay during the span of a second half-life, leaving only one-fourth of the original nuclei. After three half-lives, one-eighth remain, and so on. After n half-lives, one-half raised to the power n of the original nuclei will remain undecayed—that is,

$$N = N_0 (1/2)^n$$

Here N_0 is the original number of radioactive nuclei present in the sample, and N is the number of radioactive nuclei remaining after the elapse of n half-lives.

Table 11.3 Half-Lives of Isotopes in Table 11.2

Element	Isotope	Half-Life
Hydrogen	^1_1H	...
	^2_1H (deuterium)	...
	^3_1H (tritium)	12.3 yr
Helium	^3_2He	...
	^4_2He	...
	^5_2He	2×10^{-21} s
	^6_2He	0.805 s
	^8_2He	0.119 s
Carbon	$^{12}_6C$	...
	$^{13}_6C$	...
	$^{14}_6C$	5,730 yr
	$^{15}_6C$	24 s
Silver	$^{107}_{47}Ag^*$	44.2 s
	$^{108}_{47}Ag$	2.42 min
	$^{109}_{47}Ag^*$	39.8 s
	$^{110}_{47}Ag$	24.6 s
Uranium	$^{232}_{92}U$	70 yr
	$^{233}_{92}U$	159,000 yr
	$^{234}_{92}U$	245,000 yr
	$^{235}_{92}U$	704,000,000 yr
	$^{236}_{92}U$	23,400,000 yr
	$^{237}_{92}U$	6.75 d
	$^{238}_{92}U$	4,470,000,000 yr
	$^{239}_{92}U$	23.5 min

Note: Asterisks (*) indicate that the nuclei are in an excited state.

For example, let's say that we start with 8 million nuclei of a radioisotope with a half-life of 5 minutes. About 4 million of the nuclei will decay in the first 5 minutes. Half of the remaining 4 million will decay during the next 5 minutes, leaving 2 million. After 15 minutes, there will be 1 million left undecayed. After 50 minutes (10 half-lives), there will be about 7,800 nuclei left (8 million $\times \left(\frac{1}{2}\right)^{10} = 7,800$).

EXAMPLE 11.3 The accumulation of naturally occurring radioactive radon gas in the basements of homes after entering from the surrounding soil and rocks through cracks and fissures in the foundations is a nationwide health hazard. The radon gas decays into daughter nuclei that are themselves radioactive, and these radioactive nuclei can attach to dust and smoke particles that can be inhaled, where they remain in the lungs and release tissue-damaging radiation. Suppose that 5.5×10^7 atoms of radon are trapped in a basement after which the space is sealed against any further accumulation of the gas. If the half-life of radon is 3.83 days, how many radon atoms remain after one month's time (31 days)?

SOLUTION In one month, there are (31 days)/(3.83 days/half-life), or a total of 8.1 elapsed radon half-lives. In 8 half-lives, the original number of radon atoms is reduced by a factor of $(1/2)^8$ or $1/256$. Disregarding the small difference between 8.0 and 8.1 half-lives, the number of radon atoms remaining after one month's time is approximately

$$\frac{(5.5 \times 10^7 \, \text{atoms})}{256} = 2.15 \times 10^5 \, \text{atoms}$$

In reality, it isn't possible to count how many nuclei are left undecayed every 5 minutes. But with a Geiger counter, one can keep track of how the rate of decay—that is, the number of nuclei that decay each minute—decreases. In the previous example, 4 million decay during the first 5 minutes, 2 million during the next 5 minutes, and so on. A Geiger counter placed nearby might show an initial count rate of 10,000 counts per minute. (The actual, observed count rate would depend on how close the counter is to the sample, as well as how efficient the counter is in detecting the type of nuclear radiation emitted.) Five minutes later, the count rate would be one-half that, about 5,000 counts per minute. In other words, the count rate also is halved during each half-life (**Figure 11.11**). Thus, by monitoring the count rate of the nuclear radiation emitted by a radioisotope, one can determine the isotope's half-life.

Is it useful to know the half-lives of radioisotopes? Yes, for a number of reasons. Knowing the half-life of a radioisotope allows us to estimate over what period of time a sample will emit appreciable amounts of radiation. The

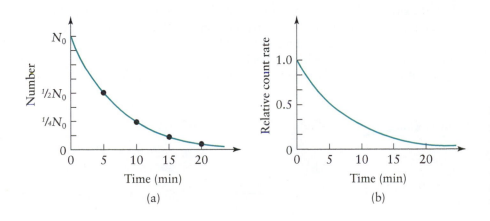

Figure 11.11 (a) Graph of the number of remaining nuclei versus time for a radioisotope with a 5-minute half-life. (N_0 is the initial number of nuclei.) (b) Graph of the relative count rate versus time for the same radioisotope.

isotope americium-241 is used in smoke detectors partly because its long half-life (432 years) ensures that it will continue to produce the ions needed for the detector's operation throughout the device's lifetime.

Small amounts of certain radioisotopes are sometimes used medically in the human body (see the Medical Applications feature at the end of this section). The flow of the isotope through the bloodstream (for example) can be monitored with a Geiger counter. The radioisotope that is used must have a long enough half-life to remain detectable during the time that it takes to move through the body. Its half-life must not be too long, however. To minimize any possible harmful effects, the body should not be subjected to the radiation any longer than is necessary to complete the diagnostic test or therapeutic treatment. One of the most commonly used radioisotopes in nuclear medicine is technetium-99, a gamma emitter with a half-life of 6 hours.

> ## ▶ Physics To Go 11.1
>
> You can simulate the decay of radioactive nuclei with a large number of pennies or similar small, disc-like objects with differently marked sides. It's best to use 50 or more.
>
> 1. Place the pennies in a small box or bag and shake them thoroughly. Dump them out on a flat surface like a desktop.
> 2. Treat each penny that shows "tails" as a nucleus that has decayed. Push these off to the side and record how many pennies are left. Collect these "undecayed" pennies and repeat the entire procedure several times.
> 3. Each penny has a 50 percent chance of turning up "tails" after each throw. Thus each throw represents one half-life for the pennies. On the average, half of them "decay" during each throw.
> 4. Notice how the number of pennies left undecayed decreases. Make a simple graph of these numbers. You should get plots similar to those in Figure 11.11.
> 5. Think about the probabilities. After, say, the fifth throw, any surviving penny has turned up "heads" five times in a row, a sequence of events with a likelihood of $(1/2)^5 = 1/32 = 0.0313 = 3.13\%$. What is its chance of turning up "heads" on the *next* throw?

11.3b Radioactive Dating

The regular rate of decay of radioisotopes can be used like a clock. **Carbon-14 dating** is a good example. Carbon is a key element in all living things on Earth. Carbon enters the food chain as plants take in carbon dioxide from the air. The complex carbon-based molecules formed by plants are ingested by animals and people. About 99 percent of this carbon is the stable isotope carbon-12. About one out of every trillion carbon atoms is radioactive carbon-14. In the upper atmosphere, carbon-14 is constantly being produced as cosmic rays (high-energy charged particles) from outer space collide with atoms in air molecules. Some of the fragments are free neutrons that cause the formation of carbon-14 when they collide with nitrogen-14. The reaction goes like this:

$$\ce{^1_0n} + \ce{^{14}_7N} \rightarrow \ce{^{14}_6C} + \ce{^1_1p}$$

(**Figure 11.12.**) The carbon-14 atoms can combine with oxygen atoms to form carbon-dioxide molecules, then mix in the atmosphere and enter the food chain. A small percentage of all the carbon atoms in plants, animals, and people is carbon-14. (You are slightly radioactive! However, the amount of carbon-14 is so small that the radiation—beta particles—is not a hazard.) The half-life of carbon-14 is about 5,700 years, so very little of it decays during the lifetime of most organisms. An exception would be certain trees that can live thousands of years.

Figure 11.12 The formation of carbon-14 from nitrogen-14. This process occurs naturally in the upper atmosphere.

(a)

(b) $\frac{1}{2}$ as much $^{14}_{6}C$ 5,700 years

(c) $\frac{1}{4}$ as much $^{14}_{6}C$ 11,400 years

Figure 11.13 (a) As soon as a tree dies, the amount of carbon-14 in it begins to decrease. (b) After 5,700 years (one half-life), it contains one-half as much as it did originally. (c) After 11,400 years, it contains one-fourth as much.

After an organism dies, no new carbon-14 is added, so the percentage of carbon-14 decreases as the nuclei undergo radioactive decay. This can be used to measure the age of the remains. For example, if a tree is cut down and used to build a shelter, 5,700 years later there will be one-half as much carbon-14 in the wood as in a live tree (**Figure 11.13**). After 11,400 years, there would be one-fourth as much, and so on. Thus, by measuring the carbon-14 content in ancient logs, charcoal, bones, fabric, or other such artifacts, archaeologists can estimate their age. This process has become an invaluable tool to archaeologists and is quite accurate for material ranging up to about 40,000 years old.

Geologists use radioisotopes to estimate the ages of rock formations and Earth itself. The ratio of the amount of the parent radioisotope to the amount of the daughter is used. For example, about 28 percent of the naturally occurring atoms of the element rubidium are the radioisotope rubidium-87. Its half-life is so long—49 billion years—that only a small percentage of it has decayed since Earth was formed. Rubidium-87 undergoes beta decay into the stable isotope strontium-87. The age of rock formations that contain rubidium-87 can be estimated by measuring the relative amount of strontium-87 that is present. The more strontium-87 present, the older the rock.

Because of radioactive decay, most of the radioisotopes present when Earth formed have long since decayed away. Only radioisotopes with very long half-lives, such as rubidium-87 and uranium-238, have survived to this day. A few radioisotopes with short half-lives occur naturally because they are constantly being produced, carbon-14 being a good example. We humans have added a large number of radioisotopes to the environment through processes like nuclear explosions and nuclear power production. Concern over the amount of radioactive fallout in the atmosphere produced by nuclear weapons testing led to the limited Test Ban Treaty of 1963. A Comprehensive Nuclear Test-Ban Treaty (CTBT) that bans all nuclear explosions in all environments was adopted by the United Nations in 1996. Although signed by more than 180 nations, including the United States, the treaty has yet to enter into force, lacking the ratification of key states possessing nuclear capability. In October 1999, the U.S. Senate rejected ratification of the CTBT.

■ Learning Check

1. A certain radioisotope has a half-life of 3 hours. If there are initially 10,000 nuclei, after 6 hours have elapsed the number of undecayed nuclei will be
   (a) 10,000.
   (b) 5,000.
   (c) 2,500.
   (d) 0.

2. If a piece of wood is suspected to be several thousand years old, its true age can likely be determined by using the isotope _____ .

3. (True or False.) Earth is so old that all radioisotopes present when it was formed have decayed away.

ANSWERS: 1. (c) **2.** carbon-14 **3.** False

As pointed out in Section 8.5, x-rays, gamma rays, and other ionizing radiation can be harmful to life. They can break up the large complex molecules in living cells, causing them to die or mutate. But life on this planet thrives in an environment where there has always been small amounts of nuclear radiation present. There are cosmic rays coming from outer space, radioactive nuclei in the soil and rocks, and even radioisotopes in living tissue, such as carbon-14 and potassium-40. Such radiation does kill cells, but the human body routinely replaces millions of dead cells each day, so it can repair the damage caused by low radiation levels.

In the early decades of the nuclear era, it became apparent that high levels of nuclear radiation can be harmful, even fatal. But it was also realized that there are a number of ways that radiation could be medically beneficial. Today, nuclear medicine is a major branch of the medical field. Thousands of hospitals use nuclear medicine daily. It has been estimated that one-third of all hospitalized patients benefit from the use of radioisotopes for diagnosis or treatment.

Diagnosing an ailment often involves finding abnormally high or low activity in some part of the body or determining if substances are circulating as they should. Radioactive nuclei make excellent probes for these tasks because their location in the body can be determined by detecting the telltale radiation that they emit. A small quantity of a radioactive material can be placed in the bloodstream. Then its movement through the circulatory system can be monitored by placing a Geiger counter near a vein or using a camera that responds to radiation instead of light.

Some chemical elements naturally accumulate in specific places. For example, the body uses calcium to "build" bones and teeth. The element iodine accumulates in the thyroid gland at a rate affected by such conditions as congestive heart failure or improper thyroid function. In one standard test, a patient ingests a pill that includes a radioactive isotope of iodine such as iodine-123 (half-life = 13.1 hours). The iodine enters the bloodstream and begins to accumulate in the thyroid after a few hours. A radiation detector placed by the patient's neck indicates if the accumulation rate is normal or not.

The fact that radiation can kill cells makes it a powerful tool for fighting cancer. The strategy is to irradiate and kill the cancer cells but at the same time limit the damage to surrounding normal tissue. External radiation therapy commonly involves sending multiple beams of radiation into the body from different directions, all focused on the cancerous tissue. Gamma rays are usually used for this purpose to treat brain tumors in a procedure called the *gamma knife*. First developed in 1967, modern applications of the gamma knife utilize as many as 200 cobalt-60 sources arrayed in a circular fashion in a heavily shielded helmet that is rigidly fixed to the patient's skull (**Figure 11.14**). The device aims the gamma radiation emitted by each of the cobalt-60 sources at the location of the tumor in the brain. The individual beams are of relatively low intensity, so the radiation has little effect on the intervening healthy tissue, but the concentration of the totality of the beams produces radiation dosages capable of killing the cancer cells. Most gamma knife procedures are done in an outpatient setting without the use of a general anesthetic and require no incisions, thus avoiding the long hospital stays, lengthy rehabilitation, and potential physical trauma and disruption of normal brain function that commonly accompanies traditional brain surgery.

There are also therapies that place the source of radiation inside the body. In *brachytherapy*, radioactive isotopes are formed into small seeds and implanted directly into malignant tumors in a preset pattern. Seeds containing iridium-192 are used to treat many cancers, whereas those containing palladium-103 are often used on prostate tumors.

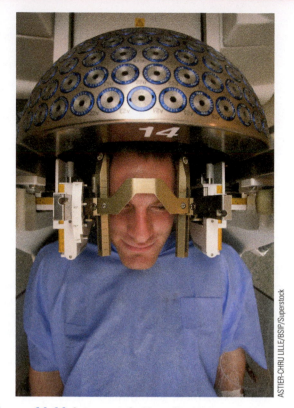

ASTIER-CHRU LILLE/BSIP/Superstock

Figure 11.14 In gamma knife radiosurgery, a protective metal helmet is placed over the patient's head. The numerous holes serve to focus the gamma rays on the tumor within the brain.

This technique allows high doses to be administered to very specific locations. An oral dose of the radioisotope iodine-131 (half-life = 8 days) can be used to irradiate the thyroid selectively. The beta and gamma radiation it emits destroys cancerous tissue in the thyroid where the iodine accumulates. A more recent innovation is to incorporate radioactive nuclei in antibodies that naturally seek out cancer cells and attach to them. The radiation from the radioisotope can then kill the cells.

Radiation therapy is not limited to fighting cancer. It is used to treat certain blood and thyroid disorders. In Europe, it is used to treat arthritis.

Nuclear medicine relies heavily on the knowledge that has been amassed in the field of nuclear physics. For example, gamma-emitting radioisotopes are preferred for diagnostic procedures that require placing the material inside the body. Because gamma rays interact less with matter than some other forms of radiation, they do less damage to tissue and are more likely to make it out of the body where they can be detected. The half-life of an implanted radioisotope is a critical parameter for determining dosages, because it indicates over what period of time the irradiation will be appreciable.

The evolution of our attitude toward nuclear radiation during the last century is remarkable and complex. In the early 1900s, nuclear radiation was a scientific curiosity, its potential for causing harm largely unknown or unappreciated. The bombing of Hiroshima and Nagasaki near the end of World War II gave the world stark images of thousands of victims suffering from massive doses of radiation. For decades afterward, the nuclear arms race raised the specter of most of Earth's inhabitants being exposed to radioactive fallout if all-out war occurred. Accidents such as the explosion of the Chernobyl nuclear power plant in 1986 and the discovery that naturally occurring radioactive radon gas can seep into our homes showed that weapons are not the only way that radiation can reach us. But by the time the threat of global nuclear war had receded in the 1990s, nuclear medicine had quietly established itself as a true lifesaving discipline. We can hope that nuclear radiation will continue to be far more widely used as a tool for saving lives than as a weapon for taking them.

■ **QUESTION**

1. Describe several ways in which nuclear radiation is used in the diagnosis and treatment of human diseases.

11.4 Artificial Nuclear Reactions

Radioactivity and the formation of carbon-14 from nitrogen-14 are examples of natural nuclear reactions. With the exception of gamma decay, each of these results in *transmutation*—the conversion of an atom of one element into an atom of another (the medieval alchemist's dream come true). Many other types of nuclear reactions can be induced artificially in laboratories. A common example of this is the bombardment of nuclei with alpha particles, beta particles, neutrons, protons, or other nuclei. If a nucleus "captures" the bombarding particle, it will become a different element or isotope. (This is called *artificial transmutation*. Note that here something is added to the nucleus, whereas in radioactive decay something leaves the nucleus.)

One example of a useful artificial reaction involves bombardment of uranium-238 nuclei with neutrons. The result is uranium-239:

$$^{238}_{92}U + ^{1}_{0}n \rightarrow ^{239}_{92}U$$

This new nucleus undergoes two beta decays, resulting in plutonium-239:

$$^{239}_{92}U \rightarrow ^{239}_{93}Np + ^{0}_{-1}e$$

$$^{239}_{93}Np \rightarrow ^{239}_{94}Pu + ^{0}_{-1}e$$

The plutonium-239 can be used directly in nuclear reactors, but the original uranium-238 can't. A breeder reactor is a nuclear reactor designed to use neutrons to produce, or "breed," reactor fuel, in this case plutonium-239. This process also shows how elements with Z greater than 92 can be produced artificially.

Similarly, neutron bombardment can be used to produce hundreds of other isotopes, most of which are radioactive. Some of the radioisotopes used in nuclear medicine are produced this way. **Neutron activation analysis** is an accurate method for determining what elements are present in a substance. The material to be tested is bombarded with neutrons. Many of the nuclei in the substance are transformed into radioisotopes. By monitoring the intensity and energy of the nuclear radiation that is then emitted, it is possible to determine which elements were originally present (**Figure 11.15**). For example, the only naturally occurring isotope of sodium, sodium-23, becomes radioactive sodium-24 by neutron absorption:

$$^{23}_{11}Na + ^{1}_{0}n \rightarrow + ^{24}_{11}Na$$

The sodium-24 emits both gamma and beta rays with specific energies. If sodium is present in an unknown substance, it can be detected by bombarding the substance with neutrons and then looking to see whether gamma and beta rays with the proper energies are released.

Neutron activation analysis played a key role in the discovery that a large asteroid or comet struck Earth 65 million years ago, resulting in the extinction of dinosaurs. The element iridium is rare on Earth, but in the 1970s neutron

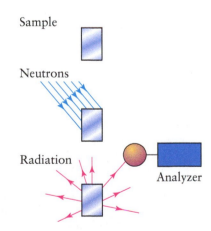

Figure 11.15 In neutron activation analysis, the sample is irradiated with neutrons. This produces radioisotopes in the sample. Analysis of the radiation that is emitted makes it possible to determine the original composition of the sample.

activation analysis revealed a high concentration of it in ancient sediments. Scientists concluded that a massive impact threw iridium-laden debris into the atmosphere, inducing adverse climatic conditions that eventually proved fatal for all but the smallest land animal species. In time, the iridium-rich material settled on Earth's surface and became incorporated into the rock record.

Neutron activation analysis has a number of uses related to law enforcement. For example, it can reveal the presence and concentration of toxins in a single hair of a poisoning victim. Likewise, an art forgery can be detected by using neutron activation analysis to determine if the paint or other materials conform to what was known to be in use when the piece was supposedly created.

This procedure can also be readily used to find explosives and illegal drugs—even if they are inside a locked vehicle. Most chemical explosives contain nitrogen, and some illegal drugs, including cocaine, contain chlorine. A truck-mounted neutron activation analysis system can be parked next to a suspicious vehicle to scan it for the presence of either element. The vehicle is bombarded with neutrons, thereby producing radioisotopes in the various substances present inside. Gamma rays that are emitted in the decay of these radioisotopes pass into the detecting system and are analyzed to see if their energies match those of activated nitrogen or chlorine nuclei.

Learning Check

1. (True or False.) It is possible to change a nucleus that is stable into one that is not stable.

2. _____ activation analysis is used to determine which elements are present in a sample.

ANSWERS: 1. True   2. Neutron

11.5 Nuclear Binding Energy

A very important and useful characteristic of a nucleus is its binding energy. Imagine the following experiment. A nucleus is dismantled by removing each proton and neutron one at a time, and the total amount of work done in the process of overcoming the strong nuclear force that binds the nucleons is measured. If the process is now reversed, and these protons and neutrons are allowed to reassemble under their mutual strong force attraction to form the original nucleus, an amount of energy equal to the previous work done would be released. This energy is called the **binding energy** of the nucleus. It indicates how tightly bound a nucleus is. A useful quantity for comparing different nuclei is the average binding energy for each proton and neutron; this is called the *binding energy per nucleon*. It is just the total binding energy divided by the total number of protons and neutrons in the nucleus—that is, the mass number. If we measure the binding energy per nucleon for all of the elements, we find that it varies considerably from about 1 MeV to nearly 9 MeV (**Figure 11.16**).

Nuclei with mass numbers around 50 have the highest binding energies per nucleon: the protons and neutrons are more tightly bound to the nucleus than in larger or smaller nuclei, so these nuclei are the most stable. The relatively high stability of the helium-4 nucleus (alpha particle) is indicated by the small peak in the graph at $A = 4$.

A proton or neutron bound to a nucleus is similar to a ball resting in a hole in the ground; it's total energy is negative. The ball's binding energy is the amount of energy that would have to be provided (the amount of work that would have to be done) to lift it out of the hole, thus releasing it from its confinement. A deeper hole means a higher binding energy.

The graph of binding energy per nucleon suggests how one can tap nuclear energy. Imagine taking a large nucleus with A around 200 and splitting it into

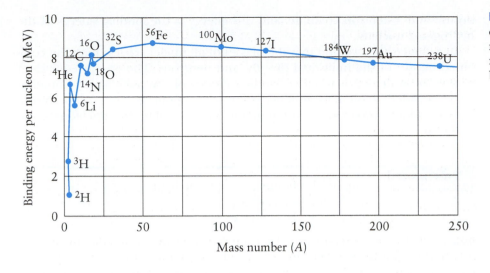

Figure 11.16 Graph of the binding energy per nucleon versus mass number. A higher binding energy means the nucleus is more tightly bound.

two smaller nuclei. The graph shows that each of the smaller nuclei—say, with A around 100—has a higher binding energy per nucleon than the original nucleus. All of the neutrons and protons have become more tightly bound together and have *released* energy in the process. (This is like moving the ball to a deeper part of the same hole. Its potential energy is decreased [made more negative], so it has given up some energy.) The act of splitting a large nucleus, referred to as *nuclear fission,* releases energy.

Similarly, we can combine two very small nuclei into one larger nucleus and release energy. If a hydrogen-1 nucleus and a hydrogen-2 nucleus are combined to form helium-3, the binding energy of each proton and neutron is increased, and energy is again released. This process is called *nuclear fusion.*

Clearly, the graph of binding energy per nucleon versus mass number is very important in nuclear physics because it shows why nuclear fission and nuclear fusion release energy. Fission is exploited in nuclear power plants and atomic bombs, and fusion is the source of energy in the Sun, the stars, and hydrogen bombs. But how does one actually go about measuring the binding energy of a nucleus? It is not practical to actually dismantle a nucleus and keep track of the amount of work done. (Remember that the largest nuclei contain more than 200 nucleons.) The best way to measure binding energy is to use the equivalence of mass and energy.

One prediction of Einstein's special theory of relativity, presented in Section 12.1, is that the mass of an object increases when it gains energy. The exact relationship between the energy E of a particle and its mass m is the famous equation

$$E = mc^2$$

When a particle is given energy, its mass increases; and when it loses energy, its mass decreases. In other words, mass can be converted into energy and vice versa. The amount of mass "lost" (converted into energy) by things around us when we take energy from them is generally much too small to be measured. But the quantities of energy involved in nuclear processes are so great that mass–energy conversion can be measured.

For example, the mass of a hydrogen-1 atom (one proton and one electron) is 1.00785 u, and the mass of one neutron is 1.00869 u (see Table 11.1). The total mass of a hydrogen atom and a free neutron is 2.01654 u. But careful measurements of the mass of a hydrogen-2 (deuterium) atom give the value 2.01410 u. When the neutron and the proton are bound together in the hydrogen-2 nucleus, their combined mass is 0.00244 u *less* than when they are apart (**Figure 11.17**). This much mass is converted into energy when a

Proton   Neutron

Mass is
0.00244 u less

Figure 11.17 A proton and a neutron bound together in a nucleus have *less mass* than a proton and a neutron separated. When the two combine, some of their mass is converted into energy—the binding energy of the nucleus. (The electron is not shown.)

proton and a neutron combine, and this energy is the binding energy of the hydrogen-2 nucleus.

In a similar way, we find that the masses of all nuclei are less than the combined masses of the individual protons and neutrons. The energy equivalent of this "mass defect" is the total binding energy of the nucleus. The binding energy divided by the mass number is the binding energy per nucleon of the nucleus. This is a good way to compute the binding energy per nucleon for the different isotopes.

EXAMPLE 11.4 Compute the binding energy for hydrogen-2 (deuterium) in joules and MeV. Find the binding energy per nucleon for this species and compare the result with that given in Figure 11.16.

SOLUTION The mass defect for deuterium is 0.00244 u, and the energy equivalent to this mass is given by

$$E = mc^2$$
$$= (0.00244 \text{ u} \times 1.66 \times 10^{-27} \text{ kg/u}) \; c^2$$
$$= 4.0504 \times 10^{-30} \text{ kg} \times (3 \times 10^8 \text{ m/s})^2$$
$$= 4.0504 \times 10^{-30} \text{ kg} \times (9 \times 10^{16} \text{ m}^2/\text{s}^2)$$
$$= 3.6454 \times 10^{-13} \text{ J}$$

In MeV, the result is

$$E = (3.6454 \times 10^{-13} \text{ J})/(1.602 \times 10^{-19} \text{ J/eV})$$
$$= 2,275,506 \text{ eV} = 2.276 \text{ MeV}$$

The ^{2}H nucleus contains two nucleons (one proton and one neutron). The binding energy per nucleon for deuterium is thus 1.1375 MeV, which agrees well with the value shown in Figure 11.16.

Incidentally, the atomic mass unit, u, is defined to be exactly 1/12th of the mass of one atom of carbon-12. In other words, u was defined so that the mass of an atom of carbon-12 is exactly 12 u. Other isotopes could have been used for establishing the size of u, but carbon-12 is convenient because it is very common and plays an important role in the life cycle of most organisms on Earth.

Fission and fusion are processes that convert matter into energy. In both cases, the total mass of all the nucleons afterward is less than the total mass before. Some of the original mass, typically from 0.1 percent to 0.3 percent, is converted into energy. In the next two sections, we take a closer look at fission and at fusion.

Concept Map 11.2 summarizes the concepts presented in this section.

■ Learning Check

1. The _____ of a nucleus is a measure of how tightly bound it is.

2. (True or False.) The average mass of each neutron in the nucleus of an iron atom is not the same as the average mass of each neutron in the nucleus of a gold atom.

ANSWERS: 1. binding energy 2. True

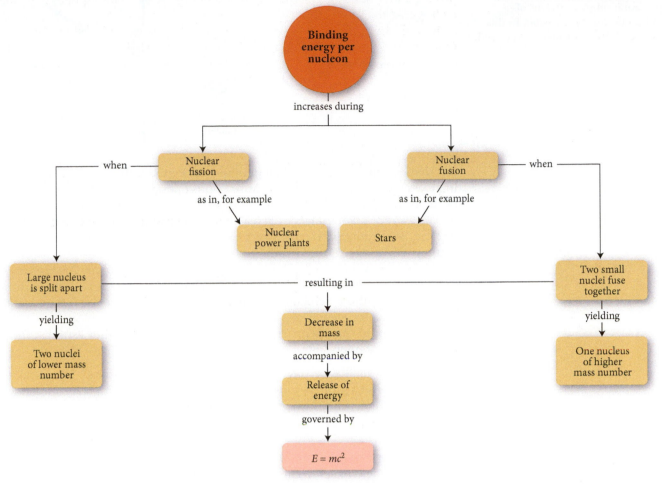

11.6 Nuclear Fission

In the 1930s, a discovery was made during neutron bombardment experiments that was one of the most fateful in human history: the nuclei of certain isotopes actually split when they absorb neutrons. Uranium-235 and plutonium-239 are the two most important nuclei that do this. Energy is released in the process along with several free neutrons. This process is called **nuclear fission.**

The two resulting nuclei are called *fission fragments*. For a particular type of fissioning isotope, the set of fission fragments produced in the decay of the activated species will not always be the same. There are dozens of different ways that a nucleus can split, and there are dozens of different possible fission fragments.

Fission is most commonly induced by bombarding nuclei with neutrons. Protons, alpha particles, and gamma rays also have been used. After absorbing a neutron, the nucleus becomes highly unstable and quickly divides (**Figure 11.18**). Two of the many possible fission reactions of uranium-235 are

$$\,^{1}_{0}n + \,^{235}_{92}U \rightarrow \,^{236}_{92}U^* \rightarrow \,^{141}_{56}Ba + \,^{92}_{36}Kr + 3\,^{1}_{0}n$$

$$\,^{1}_{0}n + \,^{235}_{92}U \rightarrow \,^{236}_{92}U^* \rightarrow \,^{140}_{54}Xe + \,^{94}_{38}Sr + 2\,^{1}_{0}n$$

The asterisk (*) indicates that the uranium-236 is in an excited, unstable state—so much so that it splits immediately. In both cases, energy is released during the fission event. The total mass of the fission fragments and the neutrons is

DEFINITION **Nuclear Fission**
The splitting of a large nucleus into two smaller nuclei. Free neutrons and energy are also released.

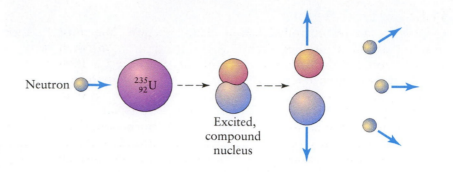

Figure 11.18 A schematic representation of a nucleus of uranium undergoing fission. In this example, three neutrons are released.

Neutron

$^{235}_{92}U$

Excited, compound nucleus

less than the mass of the original uranium nucleus and the neutron. The missing mass is converted into energy. Most of this energy appears as kinetic energy of the fission fragments: they emerge from the interaction with high speeds.

The average amount of energy released by the fissioning of one uranium-235 nucleus is 215 million electronvolts (3.4×10^{-11} joules). By comparison, the amount of energy released during chemical processes such as burning, metabolism in your body, and chemical explosions is typically about *10 electronvolts* for each molecule involved. This is why nuclear-powered ships and submarines can go years without refueling, whereas ships that use diesel or oil must take on tons of fuel for each trip.

In addition to the energy released during fission, two other aspects of this process are extremely important.

1. Almost all fission fragments are themselves radioactive. The ratio of neutrons to protons in most fission fragments is too high for the nuclei to be stable, and they undergo beta decay. These radioactive fission fragments are important components of the radioactive fallout from nuclear explosions and of the nuclear waste produced in nuclear power plants.

2. The neutrons that are released during fission can strike other nuclei and cause them to split. This process leads to what is called a *chain reaction*.

For uranium-235, 2.5 neutrons are released on the average by each fission reaction. If just one of the neutrons from each fission induces another fission, a *stable* chain reaction results. The number of nuclei that split each second remains constant. Energy is therefore released at a steady rate. This type of reaction is used in nuclear power plants. If, on the other hand, more than one of the neutrons causes other fissions, an *unstable* chain reaction results—a nuclear explosion. One fission could trigger two, these two could trigger four more, then eight, sixteen, and so on. Energy is released at a rapidly increasing rate. This is the process used in atomic bombs.

We can make a comparison between a nuclear chain reaction (fission) and a chemical chain reaction (burning). A stable chain reaction is similar to the burning of natural gas in a furnace or on a stove top. The number of gas molecules that burn each second is kept constant, and the energy is released at a steady rate. An unstable chain reaction is like the explosion of gunpowder in a firecracker. The energy released at the start quickly causes the rest of the powder to burn rapidly until all of it is consumed in a violent, uncontrolled fashion.

In the remainder of this section, we look at some of the details involved in constructing atomic bombs and nuclear power plants. Keep in mind that there are two different types of nuclear bombs: atomic bombs, which use nuclear fission, and hydrogen bombs (also called *thermonuclear bombs*), which use both fission and nuclear fusion. The latter will be discussed in the next section.

The key raw material for both atomic bombs and nuclear power plants is uranium. Naturally occurring uranium is approximately 99.3 percent uranium-238 and 0.7 percent uranium-235. The uranium-235 fissions readily; the uranium-238 does not (unless irradiated with extremely high-speed neutrons), but it can be "bred" into fissionable plutonium-239.

11.6a Atomic Bombs

To produce an uncontrolled fission chain reaction, one must ensure that the fission of each nucleus leads to more than one additional fission. This is accomplished by using a high density of fissionable nuclei so that each neutron emitted in a fission is likely to encounter another nucleus and cause it to fission (**Figure 11.19**). In other words, nearly pure uranium-235 or plutonium-239 must be used. Uranium-235 can be extracted from uranium ore through a complicated and costly series of processes referred to as *enrichment*, or *isotope separation*. Because all uranium isotopes have essentially the same chemical properties, the separation processes rely on the difference in the masses of the nuclei.

Not only is nearly pure uranium-235 or plutonium-239 necessary but also there must be a sufficient amount of it and it must be put into the proper configuration—a sphere, for example. This increases the probability that each neutron will collide with a nucleus before it escapes through the surface of the containment chamber holding the fissionable material. When these conditions are met, there is a *critical mass* of fissionable material. If there are too few fissionable nuclei present or if they are spread out too far (the density is low), too many of the neutrons will escape without causing other fissions. Also, the critical mass must be held together long enough for the chain reaction to cause an explosion. If not, the initial energy and heat from the first fissions will blow the critical mass apart and stop the chain reaction: a "fizzle." A firecracker is tightly wrapped with paper for a similar reason.

Another important consideration is timing: a premature explosion is highly undesirable, to say the least. This is prevented by keeping the fissionable material out of the critical mass configuration before the explosion. Two different techniques are used. In a gun-barrel atomic bomb, two subcritical lumps of uranium-235 are placed at opposite ends of a large tube (**Figure 11.20**). The explosion is triggered by forcing the two lumps together into a critical mass. This type of bomb was dropped on Hiroshima, Japan.

The other type of bomb, the implosion bomb, is used with plutonium-239. A subcritical sphere of the plutonium is surrounded by a shell of specially shaped conventional explosives (**Figure 11.21**). These explosives squeeze the plutonium-239 so much that it becomes a critical mass, and an explosion occurs. This type of bomb was dropped on Nagasaki, Japan.

A nuclear explosion releases an enormous amount of energy in the form of heat, light, and other electromagnetic and nuclear radiation. The temperature at the center of the blast reaches millions of degrees. Even a mile away, the heat radiation is intense enough to instantly ignite wood and other combustible materials. The air near the explosion is heated and expands rapidly, producing a shock wave followed by hurricane-force winds. In addition to the radioactive fission fragments, large quantities of radioisotopes are produced by the neutrons and other radiation that bombard the air, dust, and debris. As this material settles back to the ground, it is collectively referred to as *fallout*.

The amount of energy released by an atomic bomb explosion is generally expressed in terms of the number of tons of TNT (trinitrotoluene), a conventional high explosive, that would release the same amount of energy. One ton of TNT releases about 4.5 billion joules when it explodes. Most atomic bombs are in the 10–100 kiloton range: they are equivalent to between 10,000 and 100,000 tons of TNT. For comparison, an entire 100-unit freight train can carry only about 10,000 tons of commercial TNT. Some fission-based nuclear weapons are small and lightweight enough to be carried by one person.

Figure 11.19 Fission chain reaction in pure uranium-235. Because all of the nuclei are fissionable, each neutron emitted by one fission event is likely to strike another nucleus and cause it to undergo fission. This leads to an explosive chain reaction. Note: For clarity, the neutrons produced in subsequent fission reactions have been omitted from the figure.

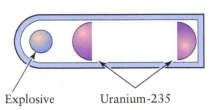

Explosive   Uranium-235

Figure 11.20 Gun-barrel atomic bomb. The two lumps of uranium are forced together by a conventional explosive to form a critical mass.

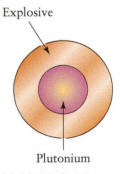

Explosive

Plutonium

Figure 11.21 Implosion atomic bomb. The plutonium is subcritical until it implodes under the force of the surrounding explosives. This brings it to a critical mass.

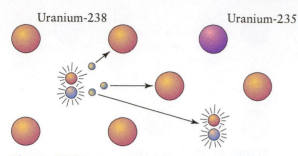

Uranium-238   Uranium-235

Figure 11.22 A controlled fission chain reaction in enriched uranium. Because only a small percentage of the nuclei are fissionable, not all of the neutrons induce other fissions. Some are absorbed by uranium-238 nuclei, which do not fission. Note: For clarity, the neutrons produced in subsequent fission reactions have been omitted from the figure.

11.6b Nuclear Power Plants

About one-eighth of the world's electricity is generated by nuclear power plants. As of July 2015, 438 nuclear plants were in operation around the globe, 99 of them in the United States alone. The U.S. sites produce approximately 20 percent of America's electricity, or ~8.5 percent of its total energy output. The reactors in these plants release the energy from nuclear fission in a controlled manner. The key to preventing a fission chain reaction from escalating to an explosion is controlling how the neutrons induce other fissions.

There are dozens of different designs for nuclear reactors. Most of them use only slightly enriched uranium fuel, typically about 3 percent uranium-235. This keeps the density of fissionable nuclei low so that not all neutrons from each fission are likely to strike other U-235 nuclei and induce other fissions, and it makes it impossible for a critical mass configuration to arise: a nuclear reactor cannot explode like an atomic bomb. (**Figure 11.22**). This is also economical because the enrichment process is very expensive.

Another important factor in producing and controlling a fission chain reaction in a nuclear reactor is the fact that slow neutrons are much better at causing uranium-235 nuclei to undergo fission than fast neutrons. Slowing down the neutrons released in the fissioning of uranium-235 makes it much more likely that they will be captured by other uranium-235 nuclei, causing them to split. This again makes it feasible to use a low concentration of uranium-235 in a reactor.

How are the neutrons slowed down? This is done through the use of a moderator, a substance that contains a large number of small nuclei. Small nuclei are more effective at removing kinetic energy from the neutrons in the same way that a golf ball will lose more kinetic energy when it collides with a baseball at rest than with a bowling ball at rest. As neutrons pass through the moderator, they collide with the nuclei and lose some of their energy. They slow down. After many collisions, they have the same average kinetic energy as the surrounding nuclei. (For this reason, they are called *thermal neutrons* because their energies are determined by the temperature of the moderator.) Almost all nuclear power plant reactors in the United States use water as the moderator because of the hydrogen nuclei in the water molecules. Many reactors in England and the former Soviet Union use carbon in the form of graphite as the moderator.

The uranium fuel is shaped into long rods that are kept separate from each other. Large reactors have tens of thousands of individual fuel rods. Each rod relies on thermal neutrons from neighboring rods to sustain the chain reaction. This makes it possible to control the chain reaction by lowering *control rods* between the fuel rods (**Figure 11.23**). Control rods are made of materials that absorb neutrons effectively, such as the elements cadmium and boron. The rate of fissioning in a reactor and, therefore, the rate of energy release, is regulated by the number of control rods that are withdrawn.

The energy released in a nuclear reactor is mainly in the form of heat. This heat is used to boil pressurized water into high-temperature steam. The rest of a nuclear power plant's electricity generating system is much the same as that found in coal-fired power plants (refer to Figure 5.35). The steam turns a turbine that turns a generator that produces electricity. In most nuclear reactors, water flows around the fuel rods and serves both as a moderator slowing the released neutrons and as a coolant carrying away the heat in much the same way that the liquid in a car's radiator carries away heat from hot engine parts. If, for some reason, the reactor loses its water (called a "loss of coolant accident"), the reactor can become dangerously overheated.

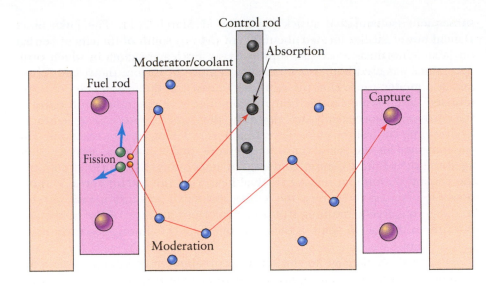

Figure 11.23 Simplified diagram of a portion of the core of a nuclear reactor. Neutron-absorbing control rods can be inserted between the fuel rods to control the chain reaction. The control rods prevent neutrons released in one fuel rod from inducing fissions in another fuel rod.

11.6c Power Plant Accidents

Nuclear power plants are equipped with a sophisticated array of safety features and emergency backup systems. Millions of gallons of water are poised to flood the reactor if the temperature becomes too high or if the coolant system leaks. The control rods are designed to drop into place at the slightest sign of trouble. The entire nuclear reactor and supporting systems are housed inside a huge steel-reinforced concrete containment building. This building is typically about 200 feet high, dome shaped, and has walls that are more than 1 foot thick. It is designed to have an airtight seal if there is a possibility that radiation has leaked from the reactor. All of this is needed because there is far more radioactivity inside a nuclear reactor than is released by a typical nuclear bomb.

Three notable accidents at nuclear power plants point out the potential for disaster that exists when human operators and designers make mistakes. In 1979, an unlikely series of errors and equipment malfunctions caused the core of a reactor at Three Mile Island, Pennsylvania, to lose most of its cooling water for 3 hours. More than half of the core melted. The containment building served its purpose, however, and very little radiation escaped from the plant. But the costly reactor was ruined, and the core was reduced to a pile of intensely radioactive rubble. The billion-dollar cleanup lasted for more than a decade.

The second major nuclear disaster occurred in April 1986 when a graphite-moderated power plant at Chernobyl (near Kiev, Ukraine) blew itself apart. Operators performing tests on the generator pushed the reactor beyond its limits. Because the plant did not have the kind of containment building used at Three Mile Island, tons of radioactive debris were released into the atmosphere and then carried by winds literally around the world. The huge mass of graphite was ignited by the heat and burned for 10 days, adding to the severity of the accident. Thirty-one people were killed by the immediate effects of the accident, and more than 300,000 were required to move out of the contaminated area and resettle elsewhere. Early predictions that thousands might eventually die from radiation exposure now appear to have been wrong. However, a United Nations report published in 2011 indicated that more than 6,000 cases of thyroid cancer, mostly in individuals who were children at the time of exposure, were caused by the accident. Estimates of the total economic loss to the economies of Ukraine and neighboring countries run into hundreds of billions of dollars.

The third major disaster involving nuclear power plants took place very recently as a result of a severe earthquake (9.0 on the Richter scale) and subsequent tsunami that struck Japan on 11 March 2011. The Fukushima Daiichi power facility, located about 90 km (54 mi) south of the city of Sendai on Japan's northeastern coast, supported six nuclear reactors in which ordinary water was used both as a coolant to remove heat from the reactor cores and as a moderator to slow the neutrons released in the fission reactions. Automated safety systems in the four reactors that were operating on the day of the earthquake successfully shut down the plants as they were designed to do by dropping control rods into place and thereby killing further reactions. The tsunami that hit the island an hour after the quake, however, severely damaged the site's infrastructure, disabling the generators that provided power to the coolant pumps that remove residual heat from the reactor cores. Without power to maintain the pumps, the temperatures and pressures in reactors rose to dangerous levels, resulting in major explosions at three of the plants and two fires at the fourth plant in the days immediately following the earthquake and tsunami.

As a result of damage to the reactor containment buildings, radioactive material was released from the Fukushima Daiichi site, and the evacuation of nonessential personnel within a 30 km (19 mi) radius of the plant in which radiation levels were significantly higher than normal was ordered on 25 March 2011; more than 160,000 residents were forced to flee the area as a result. Sample monitoring of drinking water and vegetables in the vicinity of the reactor site revealed measurable levels of radioactive ^{131}I and ^{137}Cs, but the majority of the tests done during the weeks after the disaster showed contamination levels below regulatory limits. Additional contamination of seawater along the adjacent coast also occurred as a result of the discharge of radioactive water from the basements of three turbine buildings into the sea.

After more than five years, monitoring activities and cleanup operations involving nearly 8,000 workers continue at the site, with predictions that a full restoration of the area may require up to 40 years. Current efforts focus on preventing groundwater from flowing into the basement areas of the four affected reactors, becoming contaminated, and then leaching into the Pacific Ocean. As of this writing, work continues on construction of an *ice wall* consisting of more than 60 buried cylindrical containers surrounding the reactors. Refrigerants will be fed through a network of 1.5 km of piping and used to freeze the deep subsurface soil to form an impervious barrier around which the groundwater will then flow.

Although it is still too early to predict the long-term consequences of this tragic event for the human population and the natural environment along the coastal area neighboring the Fukushima Daiichi facility, it worth noting that compared with the Chernobyl disaster, which released an estimated 10 times as much radioactive material into the atmosphere, the aftereffects of the Fukushima meltdown, although serious, are not expected to be as severe as the 1986 event in Ukraine. The incident does point out, however, the vulnerability of nuclear power stations to natural disasters such as earthquakes, hurricanes, and tornadoes, and the importance of providing adequate redundancy in safety mechanisms in the design of such facilities.

Eventually, the fissionable isotopes in the fuel rods in the nuclear reactors are consumed, and the spent rods have to be replaced. Spent fuel rods are highly radioactive because they contain a large quantity of fission fragments in the form of more than 200 different radioisotopes. So much radiation is emitted that the rods must be kept underwater for months to remain cool. Even after the radioisotopes with short half-lives have decayed away and the rods have cooled, they remain dangerously radioactive for thousands of years. Finding a safe way to dispose of spent fuel rods and other nuclear waste is a major concern of government regulatory officials and nuclear power plant authorities.

Learning Check

1. Which of the following is *not* the usual outcome of nuclear fission?
   (a) The binding energy per nucleon of the fission fragments is smaller than that of the original nucleus.
   (b) Energy is released.
   (c) Neutrons are released.
   (d) The fission fragments are radioactive.
2. (True or False.) Of the thousands of different isotopes known to exist, only a few can be used to make an atomic bomb.
3. In a nuclear reactor, _____ are used to limit the number of neutrons passing from one fuel rod to another.
4. (True or False.) In a worst-case accident at a nuclear power plant, the core could explode like an atomic bomb.

ANSWERS: 1. (a) **2.** True **3.** control rods **4.** False

11.7 Nuclear Fusion

As incredible as it may seem, another source of nuclear energy exists that dwarfs nuclear fission: **nuclear fusion.** It is the source of energy for the Sun and the stars and for hydrogen bombs.

> **DEFINITION** **Nuclear Fusion**
> The combining of two nuclei to form a larger nucleus with the accompanying release of energy.

Fusion is like the reverse of fission, although it usually involves very small nuclei. One example of a fusion reaction is shown in **Figure 11.24**. In this case, two hydrogen nuclei, one hydrogen-1 and the other hydrogen-2 (deuterium), fuse to form a nucleus of helium-3. There are many other possible fusion reactions that result in a release of energy. Some of these reactions, including the amount of energy released, are

$$^2_1\text{H} + {^2_1}\text{H} \rightarrow {^3_2}\text{He} + {^1_0}\text{n} + 3.3 \text{ MeV}$$

$$^2_1\text{H} + {^3_1}\text{H} \rightarrow {^4_2}\text{He} + {^1_0}\text{n} + 17.6 \text{ MeV}$$

$$^2_1\text{H} + {^3_2}\text{He} \rightarrow {^4_2}\text{He} + {^1_1}\text{p} + 18.3 \text{ MeV}$$

In each case, energy is released because the total mass of the nucleons after the fusion is less than the total mass before. As with fission, the missing mass is converted into energy. This energy appears as kinetic energy of the fused nucleus and the energy of the released proton, neutron, or gamma ray.

Fusion can occur only when the two nuclei are close enough for the short-range nuclear force to pull them together. This turns out to be a major problem in trying to induce a fusion reaction. Nuclei are positively charged. When brought close together, they exert strong repulsive forces on each other and resist fusion.

How can this difficulty be overcome? The most common way is with extremely high temperatures. If nuclei can be given enough average kinetic energy, their inertia will carry them close enough together to fuse when they collide (**Figure 11.25**). This process is called *thermonuclear fusion*.

$$^1_1\text{H} \quad + \quad {^2_1}\text{H} \quad \longrightarrow \quad {^3_2}\text{He} \quad + \quad \gamma$$

Figure 11.24 Schematic representation of the fusion of a hydrogen-1 nucleus and a hydrogen-2 nucleus. A gamma ray is emitted in the process.

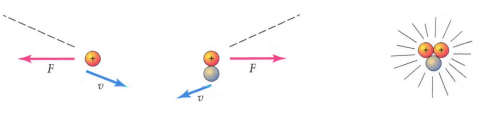

Figure 11.25 In a high-speed collision of two nuclei, the repulsive force between them is overcome, and fusion occurs. At extremely high temperatures, nuclei possess such high speeds because of their thermal motions. This is thermonuclear fusion.

When we say that the temperatures are high, we really mean it: the hydrogen-1 plus hydrogen-2 fusion reaction requires a temperature of about *50 million* degrees Celsius. Some fusion reactions require several hundred million degrees Celsius. These unearthly temperatures occur in the interiors of stars and at the centers of nuclear fission explosions.

11.7a Fusion in Stars

The Sun glows white hot and emits enormous amounts of energy. This energy originates in a natural fusion reaction in the Sun's interior. The Sun is composed mostly of hydrogen in a dense, high-temperature plasma. At the Sun's core, the temperature is around 15 million degrees Celsius, and the pressure is more than 1 billion atmospheres. Under these conditions, the hydrogen undergoes a series of fusion reactions that results in the formation of helium. Each second, hundreds of millions of tons of hydrogen deep in the Sun's interior fuse, and more than 4 million tons of matter are converted into energy.

Stable stars get their energy from nuclear fusion. Some produce energy by reactions like those in the Sun, whereas others that are larger and have hotter cores use other fusion reactions. In large stars with core temperatures in excess of 100 million degrees Celsius, the helium fuses to form larger nuclei such as carbon and oxygen, which in turn may fuse to form silicon and, eventually, iron. In this way, the elements in the periodic table are built up from the basic raw material, hydrogen. The heaviest elements are produced during *supernova events*—the gigantic explosions of massive stars at the ends of their life cycles (**Figure 11.26**)—by successive neutron capture followed by beta decay. The majority of the elements contained in Earth and everything on it (including you) were formed in the cataclysms of supernovas that occurred billions of years ago.

Life as we know it would be impossible on this planet without the Sun. Directly or indirectly, its energy supports the entire web of life. It is ironic that solar energy—that familiar, reliable, placid, and natural form of energy—originates from a violent nuclear process on a scale nearly beyond imagination.

11.7b Thermonuclear Weapons

The most destructive weapons in the world's arsenals are *thermonuclear warheads*, also called *hydrogen bombs*. These weapons get most of their explosive energy from the fusion of hydrogen. To produce the high temperatures necessary for the fusion reactions to occur, a nuclear fission explosion is used as a trigger. The fission explosion, an incredibly huge blast in itself, is but a primer for the monstrous fusion explosion. In terms of the relative amount of energy released, a hydrogen bomb is to an atomic bomb what a stick of dynamite is to a small firecracker.

Truly enormous amounts of energy are released. Many thermonuclear devices are in the *megaton* range; their energy output is given in terms of *millions* of tons of TNT. The largest weapon ever tested was rated at more than 50 megatons. To put this in perspective, this one blast released more energy than the total of all of the explosions in all of the wars in history, including the two atomic bomb blasts in World War II. However, it wasn't the most powerful blast to have ever occurred on Earth. Some volcanic eruptions, like the one on the island of Krakatoa in 1883, released even more energy.

Related to fission–fusion thermonuclear weapons is the *neutron bomb*, also called an *enhanced radiation weapon*. This device is

Figure 11.26 Remnant of Tycho's supernova, which erupted in November 1572 and was carefully observed and reported on by the famous Danish astronomer Tycho Brahe. This bubble of expanding gas and dust is located in the constellation of Cassiopeia at a distance of 7,500 light years. The image is a composite made from observations taken in the x-ray region of the spectrum by the orbiting Chandra X-Ray Observatory, in the infrared by the Spitzer Space Telescope, and in the optical using the 3.5-m Calar Alto telescope in Spain.

NASA/JPL-Caltech/CXC/Calar Alto O. Krause (Max Planck Institute for Astronomy)

designed to release a significant portion of its energy (as much as 50 percent) as nuclear radiation, specifically high-speed neutrons. Unlike typical hydrogen bombs in which the containment material is made of neutron-absorbing elements such as uranium or lead, the case material in a neutron bomb is made of chromium or nickel through which the neutrons produced in the fusion reactions can easily pass. Although the blast and heat effects of the detonation of a neutron bomb are not inconsequential, it is the neutron radiation that poses the greatest threat and produces the greatest number of casualties. Such radiation is so terribly lethal because of its ability to penetrate buildings and military vehicles and to kill or incapacitate individuals who would otherwise be protected from the explosion.

Neutron bombs also differ from other fusion devices in that they require substantial amounts of tritium (hydrogen-3) for their function; the ignition of such devices typically involves the fusion of deuterium with tritium to form helium-4 and a high-energy neutron:

$$\ce{^2_1H + ^3_1H -> ^4_2He + ^1_0n} + 17.6 \text{ MeV}$$

However, as Table 11.2 shows, tritium is radioactive and undergoes beta decay with a half-life of about 12.3 years. Thus, the "shelf life" of a neutron bomb is relatively short and, to remain effective, such devices must have their tritium components replaced periodically. Primarily developed as a tactical weapon intended for use against armored vehicles, the last of the U.S.-built neutron warheads was dismantled in 2003, and no nation is currently believed to have such weapons deployed for offensive use.

11.7c Controlled Fusion

Soon after fusion was discovered, scientists looked for ways to harness it as an energy source. The initial success of nuclear fission reactors gave them hope. But fusion presents technical challenges that have resisted solutions for decades. For a thermonuclear fusion reaction to occur, two conditions must be met:

1. The nuclei must be raised to an extremely high temperature to ignite the fusion reaction.

    It is possible to produce temperatures in the millions of degrees, but the problem is containing (referred to as *confining*) a plasma at this high temperature. Containers made of conventional materials would melt long before such temperatures would be reached.

2. There must be a high enough density of nuclei for the probability of collisions and, therefore, fusions, to remain high.

    If energy is to be released at a usable rate, a large number of fusions must occur each second. This means that the density of the plasma must be kept sufficiently high for the nuclei to collide often.

Research on controlled fusion is being pursued along a number of avenues. Several of these employ *magnetic confinement* of the plasma. Specially shaped magnetic fields are used to keep the plasma confined without letting it come into contact with other matter. This is possible because the nuclei are charged particles and consequently experience a force when moving in a magnetic field. A "magnetic bottle" is formed out of magnetic fields, and the plasma is injected into it. One of the most promising designs, called a *tokamak*, has the plasma confined inside a toroid—the shape of a doughnut (**Figure 11.27**). Fusion reactions have been produced, but so far (c. January 2016) the amount of energy released has been less than the amount used to produce the reactions. Hopes for greater success with this design

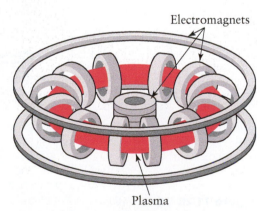

Figure 11.27 Schematic of a tokamak fusion device. The plasma containing hydrogen nuclei is trapped by the magnetic field produced by the various electromagnets. [Adapted from *Scientific American* 249, no. 4 (October 1983):63.]

rest primarily with the ITER (Latin for "the way") project, which began construction in the south of France in 2013. Operated by an international seven-member consortium, ITER is the largest tokamak facility in the world; if the construction schedule is maintained, plasma experiments are expected to begin in 2020.

Another approach to controlled fusion is to use extremely intense bursts of laser light to produce miniature fusion explosions. A small cylindrical capsule, often made of gold, called a *hohlraum* (German for "hollow room") containing deuterium (D) and tritium (T) is blasted from several different directions by powerful lasers and heated to temperatures high enough to emit intense x-rays. This radiation, in turn, heats and compresses the D-T fuel inside the capsule to bring about the desired fusion reactions. The National Ignition Facility (NIF) at Lawrence Livermore National Laboratory is the newest and largest such facility (**Figure 11.28**). At a cost of more than $4 billion, the NIF, which was certified by the Department of Energy to begin testing in March 2009, will use short pulses from 192 laser beams to deliver 5×10^{14} watts of power to a fuel capsule. Tests of the laser systems conducted through September 2012 yielded temperatures and pressures less than 30 percent of those needed for ignition of the D-T fuel, and more recent efforts up through the present have been refocused on using plutonium targets and on increasing the laser shot rate to achieve values closer to the design estimates of six per day.

A third technology has also achieved fusion and shows promise of one day producing a self-sustaining fusion reaction. The *pulsed power*, or "Z," *machine* makes use of an electromagnetic process similar to that shown in Figure 8.48. Hundreds of very thin wires are arranged symmetrically around, and parallel to, a central axis (the "z-axis"). A huge momentary current (millions of amperes) is sent simultaneously through each wire. The wires are quickly vaporized by the ohmic heating and form a plasma. The huge magnetic field produced by the cumulative current exerts forces on the current-carrying plasma and compresses it violently; the rapid, inward acceleration of the charged particles produces an intense pulse of x-rays. These x-rays trigger fusion in a BB-sized deuterium pellet. This route to sustained fusion is being vigorously pursued at the Sandia National Laboratories in Albuquerque, New Mexico. The newest Z machine (c. 2013) can now produce currents of up to 27 million amperes in as short a time as 95 nanoseconds. The radiated power

Figure 11.28 The 10-meter-diameter target chamber (blue) during construction of the National Ignition Facility. An assembly of 192 high-powered laser beams will converge on a fuel capsule at the center of the chamber.

Lawrence Livermore National Laboratory

has been raised to 350 terawatts and the x-ray energy output to 2.7 megajoules. A major problem yet to be overcome includes producing fusion energies in a single Z-pinch shot and then quickly reloading the reactor after each shot. If this impediment can be removed, early estimates suggest that an implosion of a fuel capsule every 10 seconds could economically produce 300 MW of fusion energy. That said, it remains to be seen whether this avenue of research, or any of the others being pursued, will ever lead to commercial power from nuclear fusion.

There are important reasons why controlled fusion would make a good energy source. The main one is that the oceans contain an enormous supply of the fuel: hydrogen nuclei. Compared to fission reactors, much less radioactive waste would be generated. Chief by-products would be helium, which has a number of uses, and tritium, which can be used as fuel. Fusion chain reactions would be easier to control than fission chain reactions. But because of the huge technical challenges, it does not appear likely that controlled fusion will provide a viable source of energy in the near future.

11.7d Cold Fusion

There are ways to induce fusion without using extremely high temperatures. Referred to as *cold fusion,* these processes use other means to bring the fusing nuclei close together. The largest nuclei known, "superheavy" elements with atomic numbers of more than 100, are produced in the laboratory using one type of cold fusion. Smaller nuclei are accelerated to high speeds and collide with large nuclei. Under proper conditions, the nuclei fuse to form larger nuclei. Between 1958 and 1974, scientists in the United States and in the Soviet Union accelerated very small nuclei to synthesize elements 102 through 106. In the 1980s, a team in Germany specialized in creating even larger nuclei by colliding not-so-small nuclei with large nuclei. For example, in 1996, the team produced element 112 by bombarding lead-208 with zinc-70. The reaction equation is

$$^{70}_{30}\text{Zn} + ^{208}_{82}\text{Pb} \rightarrow ^{277}_{112}\text{Cn} + ^{1}_{0}\text{n}$$

Cn is the symbol for copernicium, the proposed name for element 112.

The kinetic energy of the incoming nucleus must be just large enough to overcome the electrostatic repulsion between the two nuclei but not so large that the resulting nucleus has enough excess energy to undergo immediate fission. Researchers are optimistic that, using this technique and fine tuning the choices of the targets, projectiles, and their relative energies, prospects for creating increasingly larger and more massive nuclei will remain promising.

Another form of cold fusion seen in the laboratory is produced with the aid of an exotic elementary particle known as the *muon.* (Muons and other elementary particles are discussed in Chapter 12.) Hydrogen atoms normally combine in pairs to form H_2 molecules. If one of the two electrons is removed, the result is two hydrogen nuclei bound together by their mutual attraction to the electron. But the nuclei in this case are too far apart to undergo fusion. The negatively charged muon is basically an overweight electron: its mass is about 200 times larger. Consequently, a "muonic atom" can exist that is just a muon in orbit about a hydrogen nucleus. Now, if the electron in the molecule described above is replaced by a muon, the two nuclei will be about 200 times closer together, and it is possible for them to fuse. This type of cold fusion has been observed, but it is not likely to be used as a source of energy. Muons are unstable, with a half-life of only 2.2×10^{-6} seconds. Too much energy would be needed to create a constant supply of muons.

Learning Check

1. What is the origin of solar energy?
2. A _____ weapon makes use of both nuclear fission and nuclear fusion.
3. Which of the following is *not* a method of producing nuclear fusion?
   (a) Heating a magnetically confined plasma.
   (b) Forcing nuclei together using lasers.
   (c) Cooling nuclei to very cold temperatures.
   (d) Forcing nuclei together with a nuclear fission explosion.
4. (True or False.) Fusion is used to produce the largest nuclei known to exist.

ANSWERS: 1. fusion of hydrogen **2.** thermonuclear **3.** (c) **4.** True

Profiles in Physics A Gallery of Nuclear Scientists

The history of nuclear physics began more than 100 years ago, at a time when the very existence of atoms was still disputed. It is a remarkable history for many reasons. The pace of discovery, compared to the days of Galileo and Newton, was swift. The way in which physics research was done changed during the period. Instead of lone scientists working in small laboratories, most of the work was done by research groups using increasingly sophisticated equipment. The classification of physicists into experimentalists and theoreticians, starting in the 19th century with Michael Faraday and James Clerk Maxwell, shaped the growing physics community.

The first discoveries in nuclear physics were made before the structure, or even the existence, of the nucleus was known. Henri Becquerel (1852–1908) was the third in a line of four generations of prominent French physicists (**Figure 11.29**). Like his father before him, Becquerel studied fluorescence—the emission of visible light from a substance when it is irradiated with ultraviolet light. (This process is exploited in fluorescent lights.)

Becquerel heard of the new x-rays, discovered by Wilhelm Roentgen in 1895, and in early 1896 he began to test fluorescent substances to see if they too emitted x-rays. In one experiment, he placed uranium on a photographic film plate that was wrapped to keep out visible light. After exposing the uranium to sunlight, he found that the film had been irradiated with what he thought to be x-rays emitted by the uranium because of the sunlight. He tried to repeat the experiment, but the late winter weather in Paris turned cloudy for several days. Becquerel checked the film and expected very little exposure because of the weak sunlight. To his surprise, he found that the radiation from the uranium had been just as strong as before. During the following weeks, he determined that the uranium constantly emitted the radiation and needed no external stimulation. Clearly these were not x-rays. Nuclear radiation, dubbed "Becquerel rays," had been discovered.

At this point, another famous family of physicists arrived on the scene, the Curies (**Figure 11.30**). Marie Curie (1867–1934), originally Maria Sklodowska, grew up in Russian-occupied Warsaw. Her scientific career began when she joined her sister in Paris in 1891. While working toward her doctorate, Marie met Pierre Curie (1859–1906), a French chemist. They were married in July 1895. Two years later, a daughter, Irène, was born. She would follow in her mother's footsteps and become a renowned scientist.

At about this time, Marie, at the suggestion of Pierre, undertook an investigation of "Becquerel rays" as a thesis project. She tested all of the known elements and found that only uranium and thorium emitted the radiation. Upon testing ore samples from a museum, she found that some minerals were more radioactive than could be accounted for by uranium and thorium. She suspected that some new substance was emitting the radiation. Her husband joined the investigation, and in July of 1898 they announced the discovery of a new radioactive element, which Marie patriotically named *polonium*, for her native Poland.

Later that same year, in September, the Curies discovered another radioactive element, radium. The task of isolating pure samples of the new elements was enormous. To get a tiny sample of radium for analysis, the Curies spent years extracting it from a ton of ore. The conditions in their crude laboratory were primitive, and no one realized the danger of radiation. Marie eventually died from a condition probably caused by radiation poisoning. Even her research notes were found to be radioactive.

Marie and Pierre Curie shared the 1903 Nobel Prize in physics with Henri Becquerel. Three years later, Pierre was killed by a runaway carriage. Marie carried on her research and received a second Nobel Prize in 1911 in chemistry.

Pantheon/Superstock

Figure 11.29 French physicist Henri Becquerel, the discoverer of radioactivity.

Figure 11.30 Pierre and Marie Curie shortly after their marriage.

radioactivity causes transmutation of an element. They began to discover other radioactive substances that were later identified as different isotopes of radium. During this time, Rutherford reached the controversial conclusion that alpha particles are ionized helium atoms.

In 1907, Rutherford returned to England and took a position at the University of Manchester. The following year he won the Nobel Prize in chemistry. (The fields of chemistry and physics overlap a great deal, particularly in the study of the nucleus.) At this time, he experimentally verified the identity of alpha particles. His laboratory flourished, and much of his experimentation was carried out with the help of students and assistants. One such experiment led Rutherford to his most noted discovery, the nuclear model of the atom.

The most prominent figure in the early years of nuclear physics was Ernest Rutherford (1871–1937; Figure 11.31). Born into a Scottish family in New Zealand, Rutherford developed an interest in physics at an early age. He attended college in New Zealand and won a scholarship to Cambridge. There his brilliance was soon recognized. Rutherford was an excellent experimenter with a reputation for simplicity and elegance. In 1897, he began to study the newly discovered nuclear radiation. The next year he made his first discovery: two different types of radiation are emitted by uranium. These he named *alpha* and *beta*. The third type of nuclear radiation, *gamma*, was discovered later in France.

In 1898, Rutherford took a position at McGill University in Montreal and continued his work. In 1900, he began collaborating with a chemist, Frederick Soddy, and together they established that

At this time, the structure of the atom was a topic of speculation. The accepted model held that the electrons were embedded in some kind of positively charged sphere, somewhat like raisins in a cake. In an experiment performed by a student in Rutherford's laboratory, it was observed that some alpha particles were deflected as they traveled through a thin gold foil. The amount of deflection was much larger than would occur if the positive charge of an atom were spread evenly throughout its volume. Rutherford concluded, and then verified with experiments, that the positive charge is confined to a small region at the center of the atom, which he called the *nucleus*. This was in 1911, 15 years after the discovery of nuclear radiation.

The work in nuclear physics continued, with an understandable lull during World War I. Rutherford found that alpha particles could be used to disintegrate nuclei of nitrogen. He gave the proton its name and speculated that a similar particle that had no charge might exist in the nucleus. The eventual identification of the neutron was made in 1932 by James Chadwick (1891–1974), a former student of Rutherford.

Two of the most prominent experimenters in the 1930s were Irène Curie (1897–1956) and her husband Frédéric Joliot (1900–1958; Figure 11.32). They met when they both worked for Irène's mother, Marie. Early in 1932, Joliot and Curie reported results from experiments with a newly discovered type of radiation that they could not explain. They missed their chance: the radiation was high-speed neutrons. Chadwick quickly performed the necessary experiments and verified their identity. The discovery of the neutron completed Rutherford's nuclear model of the atom and solved some important mysteries in nuclear physics. Chadwick received the 1935 Nobel Prize in physics for his work.

The greatest discovery of Joliot and Curie, which they did interpret correctly, was artificially induced radioactivity. In a one-page paper dated 10 February 1934, they announced that bombarding

Figure 11.31 Ernest Rutherford.

Figure 11.32 Husband-and-wife team Frédéric Joliot and Irène Curie, daughter of Marie and Pierre.

undergone fission, was made by Lise Meitner (1878–1968; Figure 11.34) and her nephew, Otto Frisch (1904–1979). (Meitner was a colleague of Hahn but had fled to Sweden because of the rise of the Nazis in Germany.)

The news spread to physicists around the world, and many groups quickly confirmed the result. Soon uranium fission was recognized as a potential source of energy and as a process that could be exploited in bombs or in some other controlled process. Coming as it did, just as the world was being plunged into the worst war in history, one has to marvel at the timing of this discovery. The incentive to develop nuclear weapons was great.

Volumes have been written about the great atomic bomb project, called the *Manhattan Project,* during World War II. Some of the greatest physicists of the time, many of them refugees from Hitler's Germany, were gathered in top-secret laboratories around the United States. The technical challenges were enormous, but the thought of a German atomic bomb was a powerful motivator.

certain kinds of nuclei with alpha particles resulted in their transmutation into radioactive nuclei. This earned them a Nobel Prize.

One of the first to exploit the discovery of artificial radioactivity was Italian physicist Enrico Fermi (1901–1954; **Figure 11.33**). Fermi was a brilliant student and earned his doctorate at the age of 20. He became the key figure in a push to restore Italy's greatness in physics. Already famous at the time of Joliot and Curie's discovery, Fermi decided to use neutrons to induce artificial radioactivity. Alpha particles are positively charged and are therefore repelled by nuclei. The uncharged neutrons are much more likely to enter a nucleus. Fermi and his group in Rome started irradiating all of the known elements and soon produced about 40 new radioisotopes (see Section 11.4). This brought Fermi the 1938 Nobel Prize in physics.

Fermi and others missed an opportunity to make a critical discovery when the element uranium was irradiated with neutrons. Although they reported strange results, it was Otto Hahn (1879–1968; **Figure 11.34**), once a student of Rutherford in Canada, and Fritz Strassman (1902–1980) who discovered barium and other midsized elements in an irradiated uranium sample. The correct interpretation of this finding, that some of the uranium had

Figure 11.33 Nuclear physicist Enrico Fermi.

Figure 11.34 Otto Hahn and Lise Meitner, two of the discoverers of nuclear fission.

Three bombs were exploded: one was tested near Alamogordo, New Mexico, and the other two were used to destroy the Japanese cities of Hiroshima and Nagasaki. The hundreds of thousands of casualties caused by these explosions closed the final chapter on a war that saw civilians killed by the millions around the world.

The impact of the Manhattan Project on physics was enormous. On the one hand, its result was a triumph in nuclear physics and engineering, but it left a cruel legacy for all involved. More clearly than ever before, it showed how the work of scientists can be used to destroy life. This vivid demonstration of the destructive aspects of applied physics spurred many of the principal scientists to seek careful control of nuclear weapons after the war. But Pandora's box had been opened, and the nuclear arms race was on.

■ QUESTIONS

1. For each of the following nuclear scientists, give one major discovery, achievement, or contribution that they made to their field of study: (a) Marie Curie; (b) Ernest Rutherford; (c) Irène Curie and Frédéric Joliot; and (d) Enrico Fermi.

2. For what major discovery in nuclear physics was James Chadwick awarded the Nobel Prize?

SUMMARY

» The nucleus is the tiny, extremely dense core of an atom. It is composed of protons and neutrons held together by the short-range **strong nuclear force.**

» The different **isotopes** of a given element have the same number of protons in their nuclei but different numbers of neutrons.

» Isotopes are designated by the name of the element and the **mass number,** A, the total number of protons and neutrons in the nucleus, as in uranium-235.

» The nuclear force cannot hold a nucleus together if the ratio of protons to neutrons is too high or too low or if the nucleus is too large. In such cases, the nuclei are **radioactive** and emit nuclear radiation.

» There are three common types of radioactive decay—**alpha, beta,** and **gamma**—and each has a different effect on a nucleus.

» The **half-life** of a **radioisotope** is the time it takes for one-half of the nuclei in a given sample to decay. There are more than 2,000 different radioisotopes with half-lives that range from tiny fractions of a second to billions of years.

» Some radioisotopes, most notably carbon-14, are quite useful for determining the ages of organic matter and of geological formations.

» Measurements of the **binding energy** per nucleon for different elements indicate that large and small nuclei are not as tightly bound as those with A around 50 to 60. This explains why splitting large nuclei and fusing small nuclei release energy.

» Energy is so concentrated in nuclear processes that Einstein's equivalence of mass and energy can be observed.

» Bombarding nuclei with neutrons and other particles can induce a variety of nuclear reactions.

» **Nuclear fission,** the splitting of a nucleus into two smaller nuclei, occurs when certain large nuclei are struck by neutrons. The energy released is exploited in atomic bombs and in nuclear power plants.

» **Nuclear fusion** is the combining of two small nuclei to form a larger nucleus. The Sun, the stars, and thermonuclear warheads use energy released by nuclear fusion.

» Efforts to harness fusion as a source of energy are hampered by the extreme temperatures and densities that are required to induce fusion.

IMPORTANT EQUATIONS

Equation	Comments
$N = N_0 (1/2)^n$	Radioactive decay
$E = mc^2$	Einstein's mass-energy relation

MAPPING IT OUT!

1. Reread the material presented in Section 11.3 on the concept of *half-life* and, following the principles for concept mapping that we have applied previously throughout the book, extend Concept Map 11.1 to include the notion of the half-life of a radioactive decay process. Be sure to incorporate the application of nuclear half-life to the dating of rocks and artifacts of early civilizations. Try to link the half-life concept and its ancillary properties not only to the main concept of radioactive decay but also to the particular types of decay processes already expressed in Concept Map 11.1.

QUESTIONS

(■ **Indicates a review question, which means it requires only a basic understanding of the material to answer. Questions without this designation typically involve integrating or extending the concepts presented thus far.)**

1. ■ Why do different isotopes of an element have the same chemical properties?
2. The atomic number of one particular isotope is equal to its mass number. Which isotope is it?
3. A mixture of two common isotopes of oxygen, oxygen-16 and oxygen-18, is put in a cylindrical chamber that is then spun around its long axis at a very high speed. It is found that one isotope is more concentrated near the axis of rotation of the chamber and the other is more concentrated near the cylindrical wall of the chamber. Why is that, and which isotope is located near the axis? near the wall?
4. A material is known to consist entirely of an isotope of calcium, although the particular isotope is not known. From such limited information, which of the following quantities can you specify for the isotope? (a) Its atomic number? (b) Its neutron number? (c) Its atomic mass (or nucleon) number?
5. Two nuclei have different mass numbers A_1 and A_2. Are the two nuclei necessarily isotopes of the same element? Explain.
6. ■ What is the name of the force that holds protons and neutrons together in the nucleus?
7. ■ What aspects of the composition of a nucleus can cause it to be unstable?
8. ■ Describe the common types of radioactive decay. What effect does each have on a nucleus?
9. A nuclear explosion far out in space releases a large amount of alpha, beta, and gamma radiation. Which of these would be detected first by a radiation detector on Earth?
10. The deflection of an alpha particle as it passes through a magnetic field is much less than the deflection of a beta particle (Figure 11.3). Explain why.
11. A standard treatment for some cancers inside the body is to use nuclear radiation to kill cancer cells. If the radiation has to pass through normal tissue before reaching the cancer, why would alpha radiation not be a good choice for such treatment?
12. A concrete wall in a building is found to contain a radioactive isotope that emits alpha radiation. What could be done to protect people from the radiation (short of razing the building)? What if it were gamma radiation that was being emitted?
13. ■ Explain the concept of nuclear half-life.
14. One-half of the nuclei of a given radioisotope decays during one half-life. Why doesn't the remaining half decay during the next half-life?
15. A large number of regular six-sided dice are shaken together in a box, then dumped onto a table. Those showing 1 or 2 are removed, and the process is repeated with the remaining

ones. (See Physics to Go 11.1.) Is the half-life of the dice greater than, equal to, or less than one throw?
16. ■ How is carbon-14 used to determine the ages of wood, bones, and other artifacts?
17. One cause of uncertainty in carbon-14 dating is that the relative abundance of carbon-14 in atmospheric carbon dioxide is not always constant. If it is discovered that during some era in the past carbon-14 was more abundant than it is now, what effect would this have on the estimated ages of artifacts dated from that period?
18. The half-life of plutonium-238, the isotope used to generate electricity on the *Voyager* spacecraft, is about 88 years. What effect might this have on the spacecraft's anticipated useful lifetime?
19. ■ The half-lives of most radioisotopes used in nuclear medicine range between a few hours and a few weeks. Why?
20. ■ What are the principal steps in neutron activation analysis? Why are neutrons used in this technique to probe the composition of objects instead of, say, protons or alpha particles?
21. During the normal operation of nuclear power plants and nuclear processing facilities, machinery, building materials, and other things can become radioactive even if they never come into physical contact with radioactive material. How can this happen?
22. If the binding energy per nucleon (see the graph in Figure 11.16) increased steadily with mass number instead of peaking around $A = 56$, would nuclear fission and nuclear fusion reactions work the same way they do now? Explain.
23. ■ How can a nucleus of uranium-235 be induced to fission? Describe what happens to the nucleus.
24. ■ What aspect of nuclear fission makes it possible for a chain reaction to occur? What is the difference between a chain reaction in a bomb and one in a nuclear power plant?
25. ■ Explain how materials that absorb neutrons are used to control nuclear fission chain reactions.
26. ■ What are fission fragments, and why are they so dangerous?
27. ■ There is much more uranium-235 in a typical nuclear power plant than there was in the bomb that destroyed the city of Hiroshima. Why can't the reactor explode like an atomic bomb?
28. After a fuel rod in a fission reactor reaches the end of its life cycle (typically 3 years), most of the energy that it produces comes from the fissioning of plutonium-239. How can this be?
29. ■ Why is a nuclear fusion reaction so difficult to induce?
30. If the strong nuclear force had a longer range than it does, what effect (if any) would that have on efforts to harness controlled fusion as an energy source?
31. ■ Why are extremely high temperatures effective at causing fusion? What is used to produce such temperatures in a thermonuclear warhead?
32. ■ Why is magnetic confinement being used in fusion research?
33. ■ What is meant by the term *cold fusion*?

PROBLEMS

(Note: In some of these, you may need to use a Periodic Table of the Elements like that included on the back inside cover of the print edition of the book.)

1. Determine the nuclear composition (number of protons and neutrons) of the following isotopes.
   (a) carbon-14
   (b) calcium-45
   (c) silver-108
   (d) radon-225
   (e) plutonium-242

2. The isotope helium-6 undergoes beta decay. Write the reaction equation and determine the identity of the daughter nucleus.
3. The isotope silver-110 undergoes beta decay. Write the reaction equation and determine the identity of the daughter nucleus.
4. A nucleus of oxygen-15 undergoes electron capture. Write out the reaction equation and determine the identity of the daughter nucleus.
5. The isotope polonium-210 undergoes alpha decay. Write the reaction equation and determine the identity of the daughter nucleus.

6. The isotope plutonium-239 undergoes alpha decay. Write the reaction equation and determine the identity of the daughter nucleus.
7. The isotope silver-107* undergoes gamma decay. Write the reaction equation, and determine the identity of the daughter nucleus.
8. The following is a possible fission reaction. Determine the identity of the missing nucleus.

$$^{1}_{0}n + {}^{235}_{92}U \rightarrow {}^{236}_{92}U^* \rightarrow {}^{95}_{39}Y + ? + 2{}^{1}_{0}n$$

9. The following is a possible fission reaction. Determine the identity of the missing nucleus.

$$^{1}_{0}n + {}^{235}_{92}U \rightarrow {}^{236}_{92}U^* \rightarrow {}^{143}_{57}La + ? + 3{}^{1}_{0}n$$

10. Two deuterium nuclei can undergo two different fusion reactions. One of them is given at the beginning of Section 11.7. In the second possible reaction, two deuterium nuclei fuse to form a new nucleus plus a lone proton. Write the reaction equation, and determine the identity of the resulting nucleus.
11. At the centers of some stars, nitrogen-15 can undergo a fusion reaction with a proton (a hydrogen-1 nucleus) to produce two different nuclei. One of the products is an alpha particle (a helium-4 nucleus). What is the other? Write down the reaction equation for this transformation.
12. Iron-58 and bismuth-209 fuse into a large nucleus plus a neutron. Write the reaction equation and determine the identity of the resulting nucleus.
13. A Geiger counter registers a count rate of 4,000 counts per minute from a sample of a radioisotope. Twelve minutes later, the count rate is 1,000 counts per minute. What is the half-life of the radioisotope?

14. Iodine-131, a beta emitter, has a half-life of 8 days. A 2-gram sample of initially pure iodine-131 is stored for 32 days. How much iodine-131 remains in the sample afterward?
15. An accident in a laboratory results in a room being contaminated by a radioisotope with a half-life of 3 days. If the radiation is measured to be eight times the maximum permissible level, how much time must elapse before the room is safe to enter?
16. The amount of carbon-14 in an ancient wooden bowl is found to be one-half that in a new piece of wood. How old is the bowl?
17. When the plutonium bomb was tested in New Mexico in 1945, approximately 1 gram of matter was converted into energy. How many joules of energy were released by the explosion?
18. A nucleus of element 112—copernicium—is formed using the reaction equation given near the end of Section 11.7. It then undergoes six successive alpha decays. Give the identity of the isotope that results after each step of this process.
19. A nucleus of element 114—flerovium—is produced by fusing calcium-48 with plutonium-244. Write the reaction equation assuming three neutrons are also released.
20. Suppose that the mass of the proton and the mass of the neutron were each exactly 1.0 u. What would the mass defect for a copper-63 nucleus be under these circumstances if its atomic mass is 62.5 u?
21. The most abundant isotope of lithium, Li-7, contains three protons and four neutrons and has a mass of 7.01600 u. What is the mass defect for this nucleus in atomic mass units? In kilograms?
22. Careful measurements of the mass of a hydrogen-3 nucleus (tritium) yield a value of 3.01605 u. Compute the binding energy in MeV and the binding energy per nucleon for this radioactive nuclide. Compare your value with that for ^{3}H given in Figure 11.16.

CHALLENGES

1. Geiger counters are not very accurate when the count rates are very high; they indicate a count rate lower than the actual value. Explain why this is so.
2. As a general rule, the radioactivity from a particular radioisotope is considered to be reduced to a safe level after 10 half-lives have elapsed. (Obviously, the initial quantity of the isotope is also important.) By how much is the rate of emission of radiation reduced after 10 half-lives? Plutonium-239 is considered to be one of the most dangerous radioisotopes. Its half-life is about 25,000 years. How long would a sample of plutonium-239 have to be kept isolated before it could be considered safe?
3. The naturally occurring radioisotopes uranium-238 and uranium-235 have decay chains that end with the stable isotopes lead-206 and lead-207, respectively. Natural minerals such as zircons contain these uranium and lead isotopes. Careful measurements of the relative amounts of the isotopes can be used to estimate the ages of such minerals.
   (a) How many separate alpha decays are there in each decay chain?
   (b) For a mineral sample that is 4.5 billion years old, what would be the approximate ratio of the number of uranium-238 nuclei to the number of lead-206 nuclei? (Assume the initial decay of uranium-238 is the slowest in the decay chain.)
   (c) Repeat part (b) for uranium-235 and lead-207.
   (d) If you could only measure the ratio of the number of lead-206 nuclei to the number of lead-207 nuclei, what would the approximate value be, assuming the normal relative abundance of uranium-238 and uranium-235?
4. For nuclei with $A > 20$ that are not in highly excited states, the protons and neutrons are generally clustered together in

an approximately spherical volume whose radius, r, has been empirically found to scale as $A^{1/3}$. Specifically,

$$r \approx (1.2 \times 10^{-15} \text{ m}) \, A^{1/3}$$

Compare the sizes of the following isotopes to the size of the first Bohr orbit in hydrogen computed in Example 10.3 to be 5.29×10^{-11} m: (a) aluminum-27; (b) argon-40; (c) copper-63; (d) cesium-134; and (e) bismuth-211.
5. Using the information given in Section 11.5 and the mass–energy conversion equation, compute the binding energy (in MeV) and the binding energy per nucleon for hydrogen-2 (deuterium).
6. The Q-value of a nuclear reaction is a measure of the kinetic energy that the reaction products carry away as a result of the mass that is converted into energy by the nuclear transformation prescribed by the reaction. For example, in Section 11.7, the fusion of two deuterium nuclei produces a He-3 nucleus and a free neutron and yields 3.3 MeV of kinetic energy in the process. The Q-value for this reaction is thus 3.3 MeV, which is equal to the energy equivalent of the difference between the mass of the reactants (the two deuterium nuclei) and the mass of the products (the ^{3}He and the neutron). Using the data given in Section 11.5 and Einstein's equation, verify the Q-value for this reaction. The mass of a helium-3 nucleus is 3.01603 u.
   Repeat this calculation for the third reaction in the sequence involving the conversion of a deuterium nucleus and a helium-3 nucleus into a helium-4 nucleus and a free proton:

$$^{2}_{1}H + {}^{3}_{2}He \rightarrow {}^{4}_{2}He + {}^{1}_{1}p + 18.3 \text{ MeV}$$

   The mass of ^{4}He is 4.00260 u.
7. What is the vacuum wavelength of the 0.186 MeV γ-ray emitted by radon-226?

12

RELATIVITY, PARTICLE PHYSICS, AND COSMOLOGY

Science VU/DOE/Visuals Unlimited, Inc.

Figure CO-12 Positron emission tomography (PET) scans of a normal human brain (left) and that of a cocaine abuser (right) showing where cocaine interferes with glucose metabolism. Red shows the highest level of glucose utilization and blue the smallest. A comparison of the two scans reveals that the cocaine-addicted brain does not use glucose as effectively (note the reduction in the red- and yellow-colored areas on the right), a circumstance that can lead to the disruption of brain function.

CHAPTER INTRODUCTION: Antimatter, Available at a Medical Facility Near You

Antimatter. The word conjures up propulsion systems in science fiction tales of deep-space exploration or exotic particles produced by giant atom smashers. For most of us, antimatter remains the stuff of imagination, the purview of scientists who study it in isolated laboratories. Yet antimatter is finding its way into our daily experience because of its potential to save human life through *positron emission tomography* or PET (**Figure CO-12**).

PET is a medical imaging technique that uses positrons— antielectrons—to monitor biochemical processes in the body. Small amounts of tracer compounds containing positron-emitting isotopes with short half-lives, such as carbon-11, oxygen-15, and fluorine-18, are ingested or in-haled by the patient. As the radioactive nuclei decay, they

release positrons that quickly annihilate with electrons in the surrounding tissue to produce two high-energy gamma rays. As they leave the body, the gamma rays are observed with specialized detectors, and an image is cre-ated that reveals the location and concentration of the radioisotope in the body part scanned.

PET has revolutionized medical research and treatment in several areas, especially neurology, where, for example, it has dramatically enhanced our understanding of brain chemistry and function. Because they expose patients to very low doses of radiation for short spans of time, PET scans can be performed rapidly and repeatedly and offer the potential for monitoring the efficacy of new drug treatments. Just as important, all of the treatments take place while the patient remains comfortable and alert.

For patients, PET provides a means for diagnosing and treating their diseases. For students, PET offers a chance to study the application of particle physics in everyday life and to learn what distinguishes matter from antimatter. It also affords the chance to investigate the role matter–antimatter interactions may have played in the early his-tory of the universe.

Our goal in this chapter is to develop an understanding of what particle physicists call the *Standard Model* and to forge links between submicroscopic physics and cosmol-ogy. As we shall see, the Holy Grail of high-energy physics is the unification of all the known forces of Nature into one grand theory of everything. We start with one of the seminal theories in this unification process—the special theory of relativity—developed by Albert Einstein.

12.1 Special Relativity: The Physics of High Velocity

Imagine the following experiment. You are seated in the cargo area of a small pickup truck moving directly away from a stationary companion at a constant velocity of 20 km/h. Your friend tosses a baseball to you with a

Figure 12.1 If you are traveling at 20 km/h, and a friend throws a ball toward you at 50 km/h, you see the ball approaching at 30 km/h.

horizontal velocity of 50 km/h (**Figure 12.1**). From your point of view, what is the ball's velocity? Put another way, from your seat in the truck, how fast does the ball approach you? We have dealt with situations like this before in Section 1.2 when we considered the addition of velocity vectors. And, if your answer to the question posed was 30 km/h, you're right. From your perspective—or as a physicist might say, in your *frame of reference*—the ball is traveling 30 km/h. Of course, from the perspective of the pitcher—in his or her frame of reference—the ball is moving with a velocity of 50 km/h. The difference between what you and your friend measure the ball's velocity to be is clearly due to the fact that your reference frame in the truck is moving relative to the frame of the stationary observer at a velocity of 20 km/h. What the two ballplayers "see" thus depends upon the *relative motion* of their frames of reference. In this instance, the results are squarely in alignment with the physics of Galileo and Newton that we studied in Chapters 1 and 2. We might go so far as to call this Galilean (or Newtonian) "relativity" theory.

Now consider a second hypothetical experiment. You enter a spacecraft and leave Earth, traveling uniformly at a velocity of 200,000 km/s. After a time, a friend sends out a light ray, which moves at the speed of light—300,000 km/s—in your direction (**Figure 12.2**). When the light ray reaches you, what would you measure its speed to be? Adopting the same physics that worked so well in the last example—that is, Galilean relativity—you might be inclined to offer as your answer the "obvious" choice: 100,000 km/s. However, if you did, you'd be wrong. Strange as it may seem, you would find the speed of the light to be 300,000 km/s, just what it is for your friend back on Earth. Evidently, *the speed of light does not depend upon the motion of the observer.* Light (and all other forms of electromagnetic radiation), behaving as Maxwell predicted, does *not* act as we would expect based on the physics of Newton and Galileo. The theory of Newtonian mechanics and the theory of electromagnetism appear to be in conflict.

12.1a Postulates of Special Relativity

Albert Einstein recognized the contradiction between the predictions of classical mechanics and those of electromagnetism as regards the propagation of light. He set about reconciling the two by adopting these two postulates.

1. The speed of light, $c = 300,000$ km/s, is the same for all observers, regardless of their relative motion.

Figure 12.2 Light approaches you at 300,000 km/s, *even if you are moving away* from the source at 200,000 km/s. (Drawing not to scale.)

300,000 km/s   200,000 km/s

Like the gravitational constant G in Newton's law of universal gravitation or Planck's constant, h, in the equation quantizing the energy of atomic oscillators, c is a fundamental constant of Nature. The fact that the speed of light is constant was just starting to be accepted when Einstein began his studies in the early 1900s. Over the past 100 years or more, precise experiments have demonstrated beyond doubt that the speed of light is invariant under all circumstances. For example, measurements of the speed of photons emitted in the decay of subatomic particles called *pions* (π mesons, see Section 12.3) traveling at $0.9998c$, instead of producing $\sim 2c$ as expected from Newtonian kinematics, yield c to within 0.02 percent, in excellent agreement with predictions based on Einstein's first postulate.

2. The laws of physics are the same for all observers moving uniformly—that is, at a constant velocity. This is the **principle of relativity**.

    This means that if two observers traveling toward one another at a constant speed perform identical experiments, they will get identical results. Moreover, *no* experiment can be performed by either observer that will indicate whether they are moving or what their speed is. Two people playing air hockey in the lounge of a 747 jet plane cannot tell from the motion of the puck on the table whether they are aloft and traveling uniformly at 800 km/h or sitting at rest on the airport tarmac. You can play billiards or shuffleboard on a cruise ship and not be able to tell the ship is in motion so long as its velocity is constant. In each case, the results of the "experiments" (playing air hockey or shuffleboard) cannot be used to demonstrate uniform motion relative to Earth because the laws of mechanics (and of physics more generally) are the same for you and an Earth-bound observer. But, you say, you can look out the window of the jet or the ship and see that you are in motion. True, but how can you *prove* that you are not really at rest and that, by some magic, the clouds, trees, mountains, and so forth are not moving uniformly past you in the opposite direction? In point of fact, you can't! The principle of relativity has been confirmed numerous times in particle-collision experiments where energy and momentum measurements show excellent agreement with predictions based on this postulate.

Using these two postulates, Einstein developed his **special theory of relativity**, which was published in 1905. It describes how two observers, in uniform relative motion, perceive space and time differently. One of the interesting aspects of this theory is that once you have accepted the experimentally verified postulates on which it is based, the fundamental predictions can be understood with only elementary algebra. The equations of special relativity are only a little more difficult than Newton's laws of motion or his law of universal gravitation.

Before examining the predictions of special relativity, it is important to emphasize just how revolutionary Einstein's theory was and how it set the direction for much of theoretical and experimental physics for the next 100 years. With his bold adoption of only two simple postulates, both of which were just beginning to achieve acceptance within the physics community in the first decade of the 20th century, Einstein opened up whole new areas of physics research focused on the verification and exploitation of the predictions of special relativity—predictions that at the time seemed nothing short of unbelievable, if not flat-out wrong and crazy, yet would be proven correct in experiments whose clever design and deft execution rivaled the elegance and precision of special relativity theory itself. As you reflect upon the material presented in the next few subsections, try to shake off what we now accept as common knowledge and to approach these predictions with the same astonishment and, perhaps, skepticism that students in, say, 1910 must have felt when first confronted with this new paradigm for the physical universe.

Keep in mind as well the comments made in the Prologue in connection with the scientific method (see Section P.4): "Sometimes 'thought experiments' (i.e., theories) were required because the technology of the times didn't allow for 'real' experiments. Newton predicted artificial satellites, and Einstein unlocked relativity in this way." The ultimate confirmation of the correctness of any theory rests on observation and experimentation—the "how we know what we know." The experimental validation of the predictions of Einstein's theory of special relativity (and later his theory of general relativity) in different circumstances using multiple techniques is in its own way as great an achievement as the development of the predictions themselves. This is especially true because the relativistic effects of high velocity (or those of strong gravity, as we'll see in the next section) are not easily produced in the laboratory or easily measured in astronomical settings. To do so successfully requires exceptional ingenuity, perseverance, and physical intuition, but when achieved can yield considerable reward, as a casual glance at the list of Nobel Prizes in physics (see Appendix A) awarded for technical breakthroughs will reveal.

12.1b Predictions of Special Relativity

Let's consider one of the most important yet puzzling predictions of special relativity, something called **time dilation**. Imagine that we construct two identical "clocks" consisting of a flashbulb, a mirror, and a light-sensitive detector (**Figure 12.3**). The flashbulb emits flashes of light at some preset rate. Each light pulse is reflected by the mirror into the detector. Each time the detector receives a pulse, an audible click is emitted like that of a standard wall clock. Let's now synchronize our two clocks and give one to some friends who are to travel in a spaceship with velocity v relative to you on Earth. The question is, will the two clocks keep the same time? In other words, will they continue to tick at the same rate? The answer seems obvious: yes! This is the answer Newton would have provided. Unfortunately, given the postulates of special relativity, it is the wrong answer. Let's see why.

Consider the clock in the spaceship. When your friends took it on board, everyone agreed that it was a properly working "standard" clock. Consequently, they note nothing peculiar in its performance as they travel along. Indeed, they *cannot* identify anything different about the clock, because if they did, they could know they were moving. Such a circumstance would violate the principle

Figure 12.3 (a) A "light clock" at rest in a laboratory on Earth. (b) The light clock on board a spacecraft traveling uniformly with velocity v, as it would be seen by an observer on Earth.

(a)

(b)

of relativity, which says that physics is the same for all uniformly moving observers. The clock on the spacecraft, *as seen by your friends aboard the craft,* ticks along at the same rate it did when they first received it.

But what about the clock on board the spaceship as seen by you, an external observer? If you track the motion of the clock, say, with a powerful telescope, you see that the light, in going from flashbulb to detector, follows a zigzag path, because the pulse (moving with the spaceship) has a sideways component to its velocity in addition to a vertical component (Figure 1.12). Evidently, the path that the light travels in the moving clock is longer than the path it follows in your laboratory clock. Consequently, because the speed of light is the same in both cases, you conclude that the time it takes the light to reflect back to the detector is longer for the moving clock than for your clock. In other words, the moving clock is running slow: the rate at which it ticks is slower than that for your clock. Or, put yet another way, the time *intervals* between ticks of the moving clock have been dilated or expanded.

Of course, if your friends read your clock from their spaceship, it is *your* clock that appears to be running slow, because from their vantage point, your laboratory appears to be moving with uniform velocity *v* in the *opposite direction.* The symmetry between the observations made on Earth and in the spaceship is guaranteed by the principle of relativity. But who is *really* right, you ask? Whose clock is *really* running slow? Both observers are right, and each clock is really running slow when compared to the other. The observers perceive the rate of flow of time differently because of their relative motion. But because no experiment can determine which observer is really in uniform motion, each observer's perception is as good or as true as the other's. *Time, then, is not an absolute:* it depends on the observer and their state of *relative* motion.

By how much will the interval between successive ticks differ for the two clocks discussed above? Not very much, it turns out, unless *v* is very close to the speed of light. If we let Δt be the time between ticks on the clock at rest with an observer and let $\Delta t'$ be the observed time between ticks on the clock moving with velocity *v* relative to the observer, then Δt and $\Delta t'$ are related by the following equation:

$$\Delta t' = \frac{\Delta t}{\sqrt{1 - v^2/c^2}}$$

Figure 12.4 Graph of the time between clicks ($\Delta t'$) on a clock moving relative to an observer versus the clock's velocity *v*. At speeds even up to 0.5*c*, the time $\Delta t'$ is nearly the same as Δt, the time between ticks when the clock is at rest with respect to the observer.

Figure 12.4 is the graph of $\Delta t'$ for different velocities *v*. Even for velocities as high as one-half the speed of light, 150,000 km/s, $\Delta t'$ is not much larger than Δt; the clocks tick at very nearly the same rate. Effects of time dilation are virtually unknown in our daily lives, because we do not experience extremely high speeds. However, these effects are as real as any other phenomena in physics, and they have been observed in a very interesting manner.

A subatomic particle called the *muon* decays spontaneously into an electron plus some other particles in an average time of 0.000002 seconds. These muons are produced in large numbers by collisions between high-energy particles from space (*cosmic rays*) and atmospheric molecules some 10 or more kilometers above Earth. Given their short lifetimes, we should find very few muons reaching the ground, even though they cover the distance between their place of production and Earth's surface at nearly the speed of light. To the contrary, however, we detect great numbers of muons at ground level. How can this be?

A resolution to this paradox comes by applying special relativity theory. Because the muons are traveling so rapidly, their internal clocks, which regulate their rate of decay, appear to us to be running some 10 times too slow. Consequently, from our perspective, there is ample time for them to reach the ground and to be detected—which is what happens. Of course, from the point of view of the muons, they do decay in 0.000002 seconds according to their own clocks, and, again from their perspective, it is our clocks that are running 10 times too slow.

EXAMPLE 12.1 What is the mean lifetime of a muon as measured in the laboratory if it is traveling at $0.90c$ with respect to the laboratory? The mean lifetime of a muon at rest is 2.2×10^{-6} seconds.

SOLUTION If an observer were moving along with the muon, then the muon would appear to be at rest to such an observer. The muon would decay in an average time $\Delta t = 2.2 \times 10^{-6}$ seconds as seen by this observer. For an observer in the laboratory, the muon lives longer because of time dilation. Applying our equation, we find that the average muon lifetime, $\Delta t'$, as determined in the laboratory is

$$\Delta t' = \frac{\Delta t}{\sqrt{1 - v^2/c^2}} = \frac{2.2 \times 10^{-6}\,\text{s}}{\sqrt{1 - (0.90c)^2/c^2}} = \frac{2.2 \times 10^{-6}\,\text{s}}{\sqrt{0.19}} = 5.0 \times 10^{-6}\,\text{s}$$

This is about 2.3 times the average lifetime of a muon at rest. To the laboratory observer, the muon's clock appears to be running more than two times too slow. How fast would the muon have to be traveling for its clock to appear to be running 10 times too slow, as mentioned in the text?

Time dilation is one verified prediction of special relativity. (See Challenge 7 at the end of the chapter for another astonishing prediction.) There are two other important ones. The first is **length contraction**, in which moving rulers are shortened in the direction of motion. A convenient way to measure a distance is to time how long it takes light to traverse it. But if moving clocks run slow, so that the elapsed light-travel time is smaller, then moving rulers must be too short in the direction of motion. (Remember, distance = velocity × time.) The length of a meter stick moving relative to an observer is decreased by the same factor as the time between ticks in the two light clocks. Thus, moving observers disagree on issues involving *both* length *and* time.

Like time dilation, length contraction isn't ordinarily observed because the speeds at which we normally travel are so very small compared to the speed of light. But it is a real effect. When high-speed electrons move through the Stanford Linear Accelerator in California, their electric field lines are compressed in the direction of motion by length contraction. As the relativistic electrons pass through coils of wire arrayed along the accelerator, they produce a brief signal that is demonstrably different from that of more slowly moving electrons. The observed difference is precisely accounted for in terms of the special relativistic predictions of length contraction. Another similar example involves collision experiments in the Relativistic Heavy Ion Collider at Brookhaven National Laboratory, where particle speeds of up to 99.99% of c are achieved. Results of these investigations can only be explained if the shapes of the heavy ions are flattened from spheres into pancake-like shapes in the direction of motion by length contraction. In extreme cases, the ion radius parallel to the motion may be reduced by as much as a factor of 100 (**Figure 12.5**).

Another consequence of special relativity, and the most important one for particle physics, is the equivalence of energy and mass that was introduced in Section 11.5. If moving observers disagree on matters involving length and

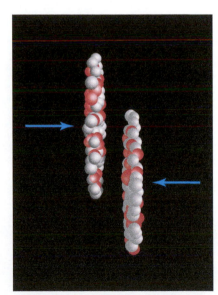

Figure 12.5 Ultra-relativistic quantum molecular dynamics simulation of a gold-gold ion collision. Protons are in red; neutrons are in white. Gold is a nearly spherical ion with a radius of ~7 fm. At relative speeds of $0.99c$, the gold ions are length-contracted in their direction of motion by a factor of 100, yielding a front-to-back length of only 0.14 fm.

time, then they will also disagree on the velocities of material particles. For example, if a collision between two electrons occurs in one laboratory setting, the initial and final velocities of the particles as determined by an observer in that laboratory will not agree in general with those determined by another observer moving uniformly relative to the first. Yet both observers *must* agree that the physics of the collision is the same. (This is the principle of relativity again.) In particular, both must agree that momentum and energy are conserved during the collision.

It turns out that for the laws of conservation of momentum and energy to be preserved in such cases, the observers must each include in their calculations the **rest energy** E_0 of the particles, given by Einstein's famous equation

$$E_0 = mc^2$$

where m is the ordinary mass of each particle (sometimes called the *rest mass*) and c is the speed of light. Thus, as Einstein wrote in 1921, "Mass and energy are therefore essentially alike; they are only different expressions for the same thing."

In special relativity theory, then, the total relativistic energy of a particle as measured by an observer is comprised of two parts: the rest energy, mc^2, of the particle plus whatever additional energy the particle has from its motion—that is, its kinetic energy. The rest energy of a particle is clearly the same for all observers, but the kinetic energy (and hence the total energy) of the particle is not. It depends upon the frame of reference of the observer. Specifically, the energy of a particle of mass m moving with velocity v relative to a particular observer is equal to

$$E_{\text{rel}} = KE_{\text{rel}} + mc^2 = \frac{mc^2}{\sqrt{1 - v^2/c^2}}$$

Solving for the relativistic kinetic energy, we find

$$KE_{\text{rel}} = \frac{mc^2}{\sqrt{1 - v^2/c^2}} - mc^2$$

For very low particle speeds, this equation can be shown to reduce to the familiar Newtonian form $\frac{1}{2}mv^2$ (see Challenge 1 at the end of this chapter). However, for velocities approaching the speed of light, the energy increases without limit. Thus, to accelerate a particle to the speed of light would require an infinite amount of energy. This is the reason why no *material* particle can ever travel at the speed of light. The energy and power demands for doing so simply cannot be met! The speed of light is not only a constant for all observers, it is also an absolute speed barrier that no object can cross.

The fact that the (rest) mass of a particle is the same (that is, *invariant*) for all observers, does *not* mean that the mass cannot change. Quite the contrary! In an inelastic collision, mass frequently changes and is transformed into energy. Conversely, in the course of such encounters, energy may be converted into mass. This state of affairs is the direct result of Einstein's recognition of the equivalence of these two things. Thus, when two balls of clay collide and stick together, some of the initial energy is converted to internal energy. Within the framework of special relativity, the energy that has gone into internal energy (and any other forms of internal excitation present in the final system) is measured exactly by the increase in the rest mass of the final system over that of the initial. Now in most practical situations in everyday life, the changes in mass that accompany interactions of this type are far too small to be detected. However, we have discussed some examples in which this type of conversion does lead to very dramatic effects in connection with nuclear reactions in Section 11.5. As we shall see, similar conversions taking place in inelastic collisions involving high-speed particles are the bread and butter of experimental high-energy physics and lead to the creation of exotic new species seldom seen in Nature.

EXAMPLE 12.2 In an x-ray tube (Figure 8.34), an electron with mass $m = 9.1 \times 10^{-31}$ kilograms is accelerated to a speed of 1.8×10^8 m/s. How much energy does the electron possess? Give the answer in joules and in MeVs (million electronvolts).

SOLUTION The total relativistic energy of the electron, E_{rel}, is its relativistic kinetic energy plus its rest energy. From our equation, we see that

$$E_{rel} = KE_{rel} + mc^2 = \frac{mc^2}{\sqrt{1 - v^2/c^2}}$$

Let's first determine at what fraction of the speed of light the electron is moving.

$$\frac{v}{c} = \frac{1.8 \times 10^8 \text{ m/s}}{3.0 \times 10^8 \text{ m/s}} = 0.60$$

$$v = 0.60c$$

The electron travels at 60 percent of the speed of light.

Evaluating the square root in the equation for E_{rel} gives

$$\sqrt{1 - v^2/c^2} = \sqrt{1 - (0.60)^2} = 0.80$$

The energy of the electron is then given by

$$E_{rel} = \frac{(9.1 \times 10^{-31} \text{ kg})(3.0 \times 10^8 \text{ m/s})^2}{0.80} = 1.02 \times 10^{-13} \text{ J}$$

But 1 joule $= 6.25 \times 10^{18}$ electron volts (see the *Table of Conversion Factors*), so

$$E_{rel} = (1.02 \times 10^{-13})(1 \text{ J}) = (1.02 \times 10^{-13})(6.25 \times 10^{18} \text{ eV})$$

or

$$E_{rel} = 637{,}500 \text{ eV} = 0.638 \text{ MeV}$$

The energy of the electron is approximately 0.638 MeV.

Notice, with E in MeV, the *equivalent* mass of the electron could be given as 0.638 MeV/c^2. This is a frequently used and very convenient way of representing subatomic particle masses because it eliminates the small numbers that necessitate the cumbersome exponential notation.

Let's compare the relativistic kinetic energy of the electron to that given by classical physics. The rest energy of the electron, mc^2, may be easily shown to be 0.511 MeV following the model above. Then

$$KE_{rel} = E_{rel} - mc^2 = 0.638 \text{ MeV} - 0.511 \text{ MeV} = 0.127 \text{ MeV}$$

According to Newtonian mechanics,

$$KE_{classical} = \frac{1}{2} mv^2 = \frac{1}{2}(9.1 \times 10^{-31} \text{ kg})(1.8 \times 10^8 \text{ m/s})^2$$

$$= 1.47 \times 10^{-14} \text{ J} = 92{,}100 \text{ eV} = 0.092 \text{ MeV}$$

The classical result *underestimates* the electron's kinetic energy by almost 30 percent.

If we reflect on the special theory of relativity, we see that it accomplishes a profound unification in physics: it reconciles the physics of low speeds with that of high speeds. It is a better, more comprehensive system of mechanics than that formulated by Newton because it works for all particles, regardless of their relative velocities. In the limit of small velocities, we recover the laws of classical mechanics as we specified them in Chapters 1–3; for high velocities, we find that Einstein's theory predicts new effects not contained in Newton's physics that are confirmed experimentally. We will return to this theme of unification

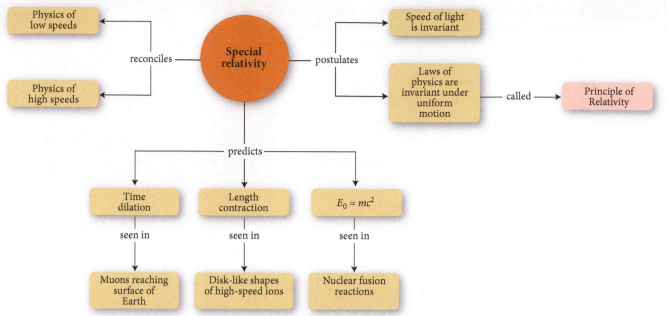

in Section 12.5 after we consider elementary particles and the forces they mediate, because it has been, and remains today, one of the overarching goals of physical science.

Concept Map 12.1 summarizes the basic postulates and implications of Einstein's special theory of relativity.

Learning Check

1. According to Einstein's special theory of relativity, when compared to an identical clock at rest, a clock moving with constant velocity will run
   (a) slow.
   (b) fast.
   (c) at the same rate.
   (d) alternately fast and slow.

2. The _____ states that the laws of physics are the same for all observers moving uniformly.

3. An observer traveling with speed $0.5c$ moves directly toward a beacon emitting photons with

speed c in her direction. The observer measures the speed of the approaching photons to be
   (a) $0.25c$
   (b) $0.5c$
   (c) c
   (d) $1.5c$

4. (True or False.) To someone moving horizontally at a speed of $0.99c$ past a vertical meter stick, the stick will appear to be 100 cm long.

5. A horizontal stretched spring weighs _____ (more than, less than, the same as) an identical spring that is unstretched.

ANSWERS: 1. (a) **2.** principle of relativity **3.** (c) **4.** True **5.** more than

12.2 General Relativity

12.2a Introduction

Albert Einstein called it "the greatest blunder of my life." But was it? The *cosmological constant*, Λ, has a long and checkered history in physics. Introduced by Einstein in 1917 as part of his **general theory of relativity**, it provided a repulsive force between galaxies, keeping them from collapsing together under their mutual gravitational attraction and thereby stabilizing the universe on the largest possible scale. Ten years later, when Edwin Hubble discovered that the galaxies were all rushing apart and that the universe was expanding (see Section 6.2), Einstein discarded his cosmological constant as being inconsistent with observation. It was reintroduced in 1948 by Fred Hoyle (who coined the term "Big Bang" to describe the cataclysmic event that produced our universe) and his collaborators, only to be rejected again. Now, this cosmological parameter has once more taken center stage as a result of work by two independent research groups seeking to understand the nature of the expansion of the universe by studying supernova explosions.

The groups, the High-Z Supernova Search Team and the Supernova Cosmology Project (SCP), used complementary but independent analysis techniques to determine the distances and velocities to particular types of supernovae. Because the galaxies that host these *type Ia supernovae* (**Figure 12.6**) are all receding from us, the frequencies of their spectral lines are all reduced from what they would be if the galaxies were stationary (cf. Section 6.2 and the Doppler effect). The wavelengths of the lines are thus increased and shifted to the red end of the EM spectrum. The amount of the redshift, z, is given by

Courtesy of High-Z Supernova Search Team, HST, NASA

Figure 12.6 Mosaic of four type Ia supernovae in distant galaxies discovered with the Hubble Space Telescope (HST).

$$z = \frac{\lambda_{observed} - \lambda_{emitted}}{\lambda_{emitted}}$$

A redshift of $z = 1$ implies that the observed wavelength is twice as large as the emitted wavelength. For speeds less than about 150,000 km/s, the recessional velocity, v, of the galaxies is related to the redshift as

$$v = cz$$

where c is the speed of light. For higher speeds the relationship between v and z is more complicated and model dependent. Nevertheless, the quantity cz can still be used as an appropriate surrogate for the expansion speeds of even the most distant galaxies.

Based on the work of these scientists, it is now possible to extend the Hubble plot (Figure 6.28) out to distances and look-back times (remember, because the speed of light is constant, the farther out in space we look, the further back in time we see) where departures from a straight-line relationship between speed and distance would be expected. The question is, how do we interpret such departures?

If gravity were the only influence on the expansion of the universe, then we'd expect to see the galaxies gradually slowing down as time goes on because of the attractive forces between them. The greater the mass–energy density in the universe, the greater the attraction and the more rapid the slowing. Because looking farther out in space is equivalent to looking further back in time to earlier epochs in the universe's history, we should see the most remote galaxies moving faster—that is, having higher redshifts—than the nearby ones. A Hubble plot of velocity (*y*-axis) versus distance (*x*-axis) should begin to curve upward. The greater the mass–energy density, the greater the upward curvature.

But look at **Figure 12.7**! It shows a Hubble plot that includes data for type Ia supernovae investigated by the High-Z and SCP teams, and it extends out to redshifts beyond $z = 1.75$. (For reference, $z = 0.5$ corresponds to a look-back time extending to about one-third the present age of the universe, or ~4.5 billion years.) Clearly, the data appear to be turning unmistakably *downward*, not upward, suggesting that the cosmic expansion has been *accelerating*, not decelerating, at least in the epoch since $z \leq 1$. Even for unreasonably small values of the universe's mass–energy density, the data seem to demand that a *repulsive*

Figure 12.7 Hubble plot of velocity versus distance including data for high-redshift galaxies containing type Ia supernovae. The solid line is an extension of the linear Hubble relation $v = H_0 d$ for nearby (very low redshift) galaxies. The dashed line follows the trend in the recent data, revealing that galaxies moved slower in the distant past than they do now. The expansion of the universe has accelerated. (For reference, 1 Mly = 10^6 ly = 9.46×10^{18} km.)

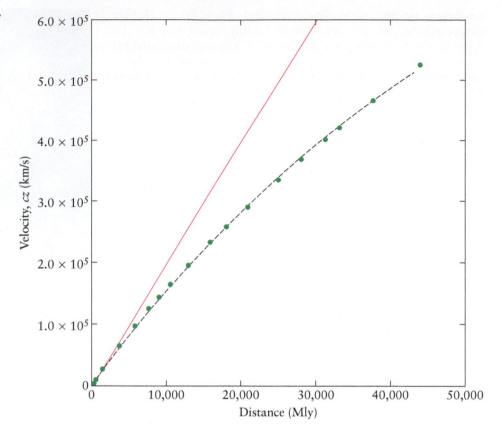

force be included in our cosmological models. This is precisely the kind of term Einstein envisioned when he introduced the cosmological constant. For their discovery of the acceleration of the expansion of the universe using type Ia supernovae, Saul Perlmutter, leader of the Supernova Cosmology Project at Lawrence Berkeley National Laboratory, and Brian Schmidt and Adam Reiss, both of the High-Z Supernova Search Team at the Center for Astrophysics at Harvard University, shared the 2011 Nobel Prize in physics.

Although the negative cosmic pressure seemingly required by the type Ia supernova observations may not be constant over all space and time like Einstein's Λ, the fact of its existence, now confirmed by supernova observations out to $z > 1.7$, will continue to fuel what Washington University physicist Clifford Will has called a "renaissance" in general relativity. Among the most interesting questions spawned by this work is, what is the source of this cosmic repulsion? And, while there are even now many speculative answers to this question, at this juncture in our inquiry into physics, the bigger question is what exactly is this general theory of relativity that forms the basis for any discussion of cosmology?

In what follows, we will consider several important features of general relativity, as well as some implications of this theory for the structure and evolution of the universe. In so doing, we will discover how the various branches of physics (mechanics, thermodynamics, electromagnetism, quantum mechanics, etc.) must be drawn together in order to even begin to address the most fundamental and interesting questions about the cosmos.

EXAMPLE 12.3 On average, the Coma cluster of galaxies is 325 Mly distant. (a) According to the Hubble law, how fast is this cluster receding from us? (b) What is the mean redshift of the cluster? (c) The spectrum of once ionized calcium contains a strong line with wavelength 393.3 nm in terrestrial laboratory sources. If such a feature were to be observed in the spectrum of one of the galaxies in the Coma cluster, what would its wavelength measure?

SOLUTION (a) The Hubble law (cf. Section 6.2) connects the distance, d, to a galaxy with its recessional velocity, v, as follows:

$$v = H_0 d$$

where H_0 is the Hubble number: $H_0 = 20.8$ km/s/Mly. The relative speed with which the Coma cluster is moving away is then:

$$v = (20.8 \text{ km/s/Mly}) (325 \text{ Mly}) = 6,760 \text{ km/s}$$

(b) The redshift associated with the cluster is given by

$$z = v/c$$

where $c = 3 \times 10^8$ m/s. Plugging in:

$$z = (6760 \text{ km/s})/(3 \times 10^8 \text{ m/s}) = (6.76 \times 10^6 \text{ m/s})/(3 \times 10^8 \text{ m/s})$$

$$z = 0.0225$$

(c) An alternative representation for the redshift in terms of wavelength is

$$z = \frac{\lambda_{observed} - \lambda_{emitted}}{\lambda_{emitted}}$$

The emitted, laboratory wavelength is 393.3 nm and, from part (b), the redshift is 0.0225. Inserting the numbers gives:

$$z = 0.0225 = \frac{\lambda_{observed} - 393.3 \text{ nm}}{393.3 \text{ nm}}$$

Solving for $\lambda_{observed}$ yields:

$$\lambda_{observed} = (393.3 \text{ nm})(0.0225) + 393.3 \text{ nm} = 8.86 \text{ nm} + 393.3 \text{ nm}$$

$$\lambda_{observed} = 402.16 \text{ nm}$$

As expected for a positive value of z and recessional motion, the observed wavelength is longer than the emitted.

12.2b Einstein's Theory of General Relativity

The **general theory of relativity**, developed by Einstein beginning in 1915, is basically a theory of gravity. It incorporates special relativity theory (see Section 12.1) and permits us to understand the motion of material particles and photons traveling in strong gravitational fields where Newton's law of universal gravitation (compare Section 2.7) gives only approximately correct answers. There exist several well-documented examples wherein the superiority of Einstein's theory of gravity to that of Newton has been clearly demonstrated. We will consider some of these in due course. Let's first consider some ways in which Einstein's conception of gravity differs fundamentally from Newton's.

A basic postulate of general relativity is the **principle of equivalence**. It asserts that in a uniform gravitational field, all objects, regardless of their size, shape, or composition, accelerate at the same rate (refer to Galileo's experiments with falling bodies in Chapter 1). Einstein recognized that this condition would make it impossible to physically distinguish the motion of a freely falling object near Earth's surface from the motion of the same object released in a laboratory that was accelerating upward at a rate of 9.8 m/s² (1g) *in the absence of a local gravitational field* (as in empty space). As far as the laws of free fall are concerned, an accelerated laboratory is equivalent to one that is unaccelerated but possesses gravity. Einstein even went one step further and argued that not only are the laws of freely falling bodies the same for two such laboratories but also *all* the laws of physics are the same in such circumstances. The adoption of this so-called *strong* principle of equivalence leads to a very striking consequence: If the path of a light ray is observed to deviate from a straight line in

Figure 12.8 A light beam enters a rocket that is accelerating vertically upward and passes through a sequence of equally spaced frosted glass window panes. In empty space, the light travels across the spaceship in a straight line as seen by an external observer. However, as the rocket accelerates upward, the light strikes the glass panes successively closer and closer to the bottom. In the reference frame of an observer on the spaceship, the path of the light is a deviated from a straight line into a parabola like that of a projectile experiencing gravity on Earth.

an accelerating "laboratory" like the rocket in **Figure 12.8**, then the path of a light ray will be similarly deflected in a gravitational field produced, say, by a massive object like the Sun.

Einstein's interpretation of the cause of the deflection of a light ray in a gravitational field was quite different from that which Newton might give. Newton would likely speak of the attractive force acting between the mass of the gravitating body and the effective mass of the photons making up the light (compare Section 10.2). (Recall that photons have no rest mass. Their mass derives from the energy they possess according to Einstein's equation $E = mc^2$; see Section 12.1.) This force produces a centripetal acceleration of the photon and a subsequent deviation in its motion from the straight line predicted by Newton's first law (see Section 2.2). Indeed, one can calculate, using Newtonian methods, the expected deflection for such a light ray, but the result, sadly, turns out to be equal to only one-half the observed deflection.

The remaining deviation, and the correct explanation for the effect, belong to Einstein, who viewed the situation in a revolutionary way. In particular, he argued that there was no need to describe the deflection in terms of forces but noted instead that the paths followed by the photons are merely their natural or *geodesic* trajectories in a curved space (or, more properly, *spacetime*, because in relativity space and time are inextricably linked) that has been distorted by the presence of mass (and/or energy). Princeton physicist John A. Wheeler has contrasted the views of Einstein and Newton on this subject as follows.

According to Newton: *Force tells mass how to accelerate.*
                          Mass tells gravity how to exert force.

According to Einstein: *Curved spacetime tells mass–energy how to move.*
                          Mass–energy tells spacetime how to curve.

Perhaps a fable will help distinguish these two viewpoints more clearly. Imagine you are covering the final hole of the Master's Golf Championship from a helicopter above the eighteenth green. Sharing the reporting responsibilities with you are none other than Isaac Newton and Albert Einstein. As the three of you watch, Jordan Spieth prepares to putt out for a birdie. After lining up his shot, he holes the putt and waves triumphantly to the cheering crowd. The trajectory followed by the ball to the cup, as seen from your aerial vantage point, is shown in **Figure 12.9a**.

Noting that the path deviated from a straight line, Newton remarks that there must have been a force acting on the ball to accelerate it along the curved

(a)

(b)

Figure 12.9 (a) Path of a golf ball as seen from high above the surface of the green. The deviation from straight-line motion appears to require the action of some (unknown) force. (b) Close-up view of the surface of the green on which the ball moved. In rolling into the cup, the ball followed the natural contours of the green, including a small mound that deflected it toward the hole.

trajectory it followed to the cup. He speculates that perhaps a strong cross wind diverted the ball; or that Spieth was using a steel-core ball and there was an iron ore deposit beneath the surface of the green that provided a magnetic force on the ball; or . . . At this juncture, we can imagine Einstein interrupting to point out that all this talk of forces is unnecessary because the ball was just following the natural contours of the curved surface of the eighteenth green on which it moved. To prove this, Einstein suggests that the helicopter be landed and the green be inspected close up. The result is depicted in **Figure 12.9b**.

Einstein knew what all good golfers know: The surfaces of most greens are not flat planes but are instead rolling contours designed to challenge the skill of the players at "reading the green" and putting the ball so as to take advantage of the dips and curves. In making his final shot, Jordan Spieth relied on no mysterious forces to move the ball but simply recognized that the ball would be traveling along a warped surface. He accurately assessed the natural path the ball should take in this space to reach the hole. This is how Einstein would have us view the effects of what we call "gravity."

12.2c **Motion in Curved Spacetime**

As in all parables, there are limitations to the fictional story just told, but it serves to emphasize the fundamental differences in the world views of Einstein and Newton. Thus enlightened, we may now return to the issue of the deflection of light and address the source of the curvature: mass–energy. In Einstein's theory of general relativity, matter and energy warp space (and alter time). The greater the density of matter and energy present, the greater the warping—that is, the stronger the curvature. Again, an analogy in two dimensions may be helpful. Consider a large rubber sheet stretched taut between supporting posts. If we place a large marble near the center of the sheet, its presence will deform the surface only very slightly in its immediate vicinity (**Figure 12.10a**). The space surrounding the marble will remain essentially flat, and the path of a small ball

Figure 12.10 The warping of space by massive objects causes deviations in the paths of particles moving nearby. The greater the mass, the larger the deformation of the space surrounding it. (a) The marble (red) produces only a small warp in the two-dimensional surface of a rubber sheet and consequently a negligible alteration in the motion of the ball bearing (green) from a straight line. (b) The bowling ball creates a considerable depression in the surface of the sheet so that the path of the ball bearing deviates significantly from that of a straight line. In each case, the ball bearing is following a *geodesic* path along the surface of the sheet, but in (b) the surface has been warped from a flat plane by the action of a massive object.

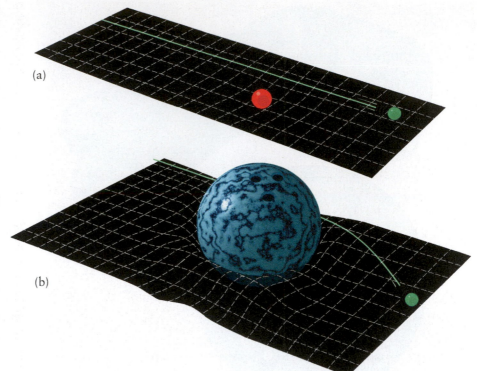

(a)

(b)

bearing rolled past the marble will be deviated by just the tiniest amount from a straight line due to the bending of the surface of the sheet. However, if we replace the marble with a bowling ball, the rubber will be stretched to a much greater degree, and the deformation of the space surrounding the bowling ball from a flat plane will be far larger and extend more widely around the ball. A ball bearing now rolled past the bowling ball will experience a space exhibiting considerable curvature, and its path will deviate significantly from a "straight" line (Figure 12.10b).

A similar deflection effect happens in three dimensions for the light from distant stars that passes near the Sun or the light from remote galaxies that passes near a foreground cluster of galaxies. The first detection of this phenomenon was made in 1919 when the positions of stars close to the Sun on the sky were measured during the relative darkness provided by a solar eclipse. The observed differences in the positions of the stars during and outside the eclipse (when the Sun was far removed from them on the sky) agreed extremely well with the predictions of general relativity but only poorly with those of Newtonian gravitation.

> ### Physics To Go 12.1
>
> You can reproduce the effects of space warping in a manner similar to that shown in Figure 12.10 as follows: Remove the bed linens from a mattress with a smooth surface. (A baby's crib mattress or the mattresses in many hide-a-beds works well.) Place a heavy object with dimensions that are small compared to those of the mattress at the center of the surface. A bowling ball serves the purpose very nicely, but it need not be a spherical object; a large brick or even a heavy book will do, depending on the firmness of the mattress. What is important is that the mattress surface be noticeably depressed near the object. Take a small, lightweight, spherical object, like a marble or a ping-pong ball, and roll it across the mattress surface. Begin by rolling the ball at a moderate speed parallel and close to one edge of the mattress. Describe the path followed by the object as it moves across the surface. Gradually set the starting point for the ball's motion nearer and nearer to the central axis of the mattress. What happens to the trajectory of the object? Describe its motion now. Vary the speed with which the object rolls. How does this affect the object's path along the surface? Can you explain what is happening using the principles of general relativity?

(a)

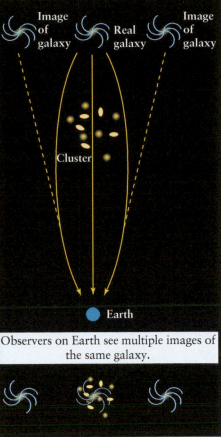

Observers on Earth see multiple images of the same galaxy.

(b)

W.N. Colley and E. Turner (Princeton University), J.A. Tyson (Bell Labs, Lucent Technologies) and NASA

Figure 12.11 (a) HST picture of a cluster of yellow, elliptical galaxies acting like a gravitational lens. The blue ovals surrounding the cluster are distorted images of a more distant galaxy lying almost directly behind the cluster. (b) The large mass of the cluster warps the space in its vicinity and bends the paths of light rays passing nearby from a background galaxy. Light arriving at Earth from different directions produces multiple, distorted images of the more distant galaxy. (Image not to scale.)

More recently, astronomers have turned the deflection of light by massive objects from a test of the validity of general relativity to a tool to assess mass. Figure 12.11a shows an HST image of a yellow cluster of galaxies acting as what is called a *gravitational lens*. The mass of this cluster has distorted the "fabric" of space in its vicinity and, like a lens, bends the light from a more distant blue galaxy to produce multiple images of it arrayed along circular arcs surrounding the cluster (Figure 12.11b). Careful analyses of the images of a lensed galaxy in pictures like this using general relativity permit astronomers to establish the mass of the foreground cluster. In cases where comparisons can be made between this method and others used to determine cluster masses, the agreement has been found to be highly satisfactory. This lends credence to the view that we are correctly interpreting the nature of the observations, and it gives us confidence in the masses established via gravitational lensing alone in circumstances where other methods cannot be employed.

Another prediction of Einstein's general theory is that, because of the warping of space by mass, the orbital path of one spherical object about another will not be a closed ellipse as predicted by the inverse square law of Newton (see Section 2.7). Instead, it will be an open curve for which the point of closest approach between the two bodies slowly advances in space (an effect called *precession*). This is observed in the solar system in the motion of Mercury about the Sun (Figure 12.12): The *perihelion* point of Mercury's orbit precesses at a rate of 43 arcseconds (0.012 degrees) per century due to general relativity. The agreement between the predictions of general relativity and the observations for Mercury's orbit was one of the early successful tests of Einstein's theory.

Figure 12.12 Display of the precession of the perihelion of Mercury. This effect arises because the gravitational attraction of the Sun for Mercury deviates from a strict $1/r^2$ force law due, in part, to general relativistic effects. The actual precession of Mercury is small, accumulating to only about 16° every 10,000 years from all sources (including contributions caused by gravitational perturbations associated with nearby planets).

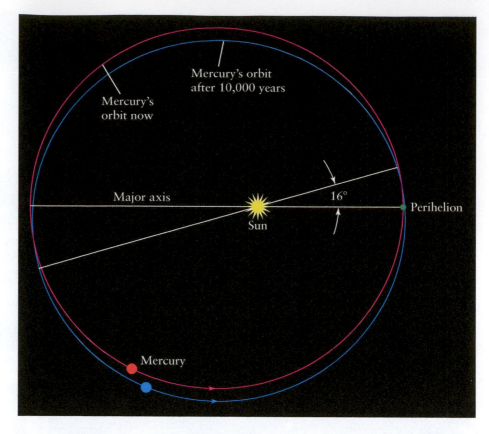

A more dramatic example of this is seen in the motion of two pulsars (see Section 3.8) discovered to be orbiting one another by Russell Hulse and Joseph Taylor in 1974. The objects in this system (catalogued as PSR 1913 + 16) experience gravitational fields 10,000 times stronger than the Sun's field at Mercury, rendering the general relativistic effects far more significant than they are in the solar system. Indeed, the precession rate of the *periastron* point in PSR 1913 + 16 is a bit more than 4° *per year!*

By associating the observed precession in this binary with general relativistic effects, Hulse and Taylor were able to deduce the masses of the two stars and then conduct a very sensitive test of another aspect of general relativity theory: the prediction of a gradual decrease in the orbital period of the system due to energy losses in the form of *gravitational radiation.* This is something that has no analog in Newtonian theory.

Recall from Section 8.5 that when an electric charge is accelerated, it radiates electromagnetic waves, which propagate outward at the speed of light. Similarly, according to general relativity, if a massive object is accelerated, it, too, radiates energy in the form of *gravitational waves* that also travel at the speed of light. Compared to more common electromagnetic waves, gravitational waves are much less intense and very difficult to detect directly.

For a system like the binary pulsar, general relativity predicts that, because of the gravitational radiation emitted from the centripetally accelerated masses, the two stars should gradually spiral inward toward one another, their orbital period getting shorter by a minuscule 75 *microseconds* per year. (Put another way, the size of their mutual orbit is shrinking by a tiny 3.1 mm per orbit.) Astonishingly, by 1983, data on this highly unusual double star had accumulated to yield a measured orbital decay rate of 76 ± 2 microseconds per year in near perfect agreement with general relativity. Thus, although only an indirect test of the general relativistic prediction of gravitational waves, their existence, based on the observations of PSR 1913 + 16, was scarcely in doubt. For these achievements, Taylor and Hulse were selected as the 1993 recipients of the Nobel Prize in physics.

On 14 September 2015 at 5:51 AM Eastern Daylight Time, an extraordinary event occurred that eclipsed the discoveries of Taylor and Hulse and put the icing on the cake, so to speak, in the celebration of the 100th anniversary of the publication of Einstein's 1915 general theory of relativity. For the first time, scientists obtained *direct* evidence of the existence of the gravitational waves thus confirming a major prediction of this theory. The source of the gravitational waves appears to have been a merger of two black holes to form a single, more massive black hole. The detection was made using the twin Laser Interferometer Gravitational-wave Observatory (LIGO) instruments, one in Hanford, Washington, and a second in Livingston, Louisiana. Each observatory consists of 1.2-meter-diameter vacuum pipes arranged in an **L**-shape with 4-kilometer-long arms at the ends of which are suspended test masses fitted with mirrors that reflect ultra-stable laser beams. The beams are used to monitor the distance between the mirrors. According to Einstein's theory, the separation between the mirrors will change by a tiny amount when a gravitational wave passes through the detector. In their present configurations, the LIGO detectors are capable of measuring a change in the path length between the mirrors as small as one ten-thousandth the diameter of a proton, about 10^{-19} meters.

Based on the observed signals (**Figure 12.13**), LIGO investigators estimate that the two black holes involved in this collision were each about 30 times more massive than the Sun and that the merger took place some 1.3 billion years ago. Given the signal strength, scientists estimate that an energy equivalent to three times the mass of the Sun was released according to Einstein's

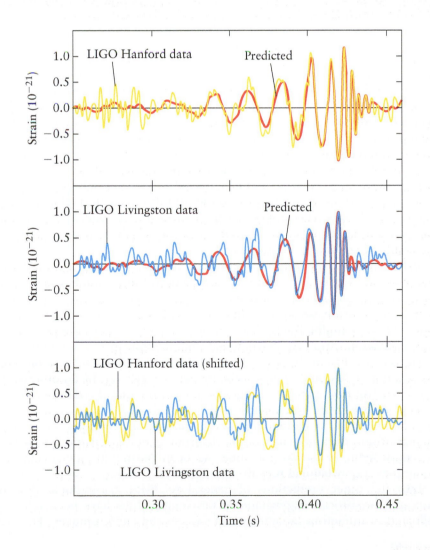

Figure 12.13 Plots showing the gravitational wave signals detected by the twin LIGO observatories. The top two graphs display the data received at Hanford, Washington, and Livingston, Louisiana, respectively, along with the shapes of the waveforms predicted by Einstein's general theory of relativity for the coalescence of two black holes. The vertical axis ("Strain") measures the fractional amount by which the distances between the LIGO mirrors are altered with the passing of the gravitational waves. As each plot shows, the LIGO data closely match the general relativistic predictions. The third graph compares the data from the two LIGO detectors. The Hanford data have been inverted to account for differences in the orientations of the receivers at the two sites. The data have also been shifted to correct for the travel time of the signals between Livingston and Hanford, which arrived first at the Louisiana site and then seven thousandths of a second later at the Washington state site. The correspondence between the two displays is remarkable and leaves little doubt that the two detectors witnessed the same event.

formula $E = mc^2$ as gravitational waves in a fraction of a second. Although a triumphant vindication of Einstein's general theory, it is somewhat ironic that Einstein himself believed that gravitational waves were too weak to ever be detected and that black holes did not exist. As Bruce Allen, managing director of the Max Planck Institute for Gravitational Physics (Albert Einstein Institute), has said, however: "I don't think he [Einstein] would have minded being wrong."

The predictions of general relativity and their observational validations (e.g., gravitational lensing, perihelion precession, and gravitational waves) on which we have focused so far highlight the "space" portion of *spacetime*, insofar as they emphasize the deviations in the paths followed by objects in strong gravitational fields or under the influence of gravitational radiation. But the "time" part of *spacetime* is also affected in regions of strong gravity due to the presence of very massive objects. One important time-related prediction to emerge from the general theory of relativity is that a clock in a strong gravitational field will run slower (that is, the interval between successive "ticks" will be longer) than an identical clock in a region characterized by weak gravity. This time dilation phenomenon is similar to what we saw in Section 12.1 in our discussion of special relativity, only instead of being due to the relative motion of two clocks (a kinematical property), it is due to the difference in the gravitational fields experienced by two clocks. Consequently, this effect is referred to as **gravitational time dilation**.

Like the equation for kinematical time dilation, the relationship between the time interval, Δt_d, between two events measured by an observer at a distance d from a mass M compared to that measured by an observer located very far away from the mass, Δt_f, is an algebraic one:

$$\Delta t_\mathrm{f} = \frac{\Delta t_\mathrm{d}}{[1 - (2GM/dc^2)]^{1/2}} = \frac{\Delta t_\mathrm{d}}{[1 - (d_\mathrm{S}/d)]^{1/2}}$$

Here, G is the gravitational constant that appears in Newton's law of universal gravitation, c is the speed of light, and d_S is the *Schwarzschild radius*: $d_\mathrm{S} = (2GM/c^2)$. As may be seen, for a given mass, as the distance d increases, the term in the denominator approaches unity, and there is no difference in the time intervals measured by two observers each located very far from the gravitating mass. Conversely, for a given distance, as the mass M increases, the term $[1 - (2GM/dc^2)]^{1/2}$ becomes increasingly smaller than one. This means that the time interval "clocked" by the observer at infinity equals that "clocked" by the observer at distance d multiplied by a number *greater than* one. Thus, the distant observer infers that the clock at d is ticking more slowly than his own; the time interval has been increased, or dilated.

As was noted in the *Time Out!* article at the end of Section 1.1, modern scientific clocks employ atoms as their timekeepers, the second being defined as so many oscillations of light with a characteristic wavelength (cf. Section 1.1). In the simplest case, we might define one "tick" of such an atomic clock as the time interval spanning the passage of one wave cycle; that is, the period of the wave oscillation. Applying gravitational time dilation to this situation, we are led to the conclusion that the period of the wave emitted by an atom near a massive object M would appear to be longer as measured by an observer who is very far from the mass M. Longer periods imply that the distance between successive wave crests, which define one cycle of the wave, has increased; that is, the wavelength of the emitted light has increased. Because longer wavelengths (at least within the visible part of the electromagnetic spectrum) are perceived as redder light, we say the radiation has been redshifted. Specifically, it has undergone a **gravitational redshift**.

Like the other predictions of general relativity, gravitational time dilation and the associated gravitational wavelength shifts have been tested and validated in numerous laboratory and space-based circumstances. The classic terrestrial test was carried out in 1960 by R. Pound and G. Rebka who placed a radioactive sample of iron-57 at the top of an evacuated 22.6-m-high tower and measured the shift in the frequency (recall $f = c/\lambda$) of 14.4 keV γ-rays emitted in the decay of the iron atoms. Because the γ-radiation was traveling down the shaft, it was moving from a region of lower gravity to one of higher gravity as it neared Earth's surface. Thus, instead of a redshift in the wavelength of the radiation, a blueshift (toward shorter wavelengths and higher frequencies) was predicted. The expected blueshift was $z = -2.46 \times 10^{-15}$; the Pound-Rebka value was $z = -(2.57 \pm 0.26) \times 10^{-15}$, in exceptionally good agreement with theory.

The Gravity Probe A, launched in 1976, has verified time dilation due to general relativity with a precision of 70 parts per million, and corrections for general relativistic effects in the operation of atomic clocks aboard GPS satellites orbiting Earth are routinely made to maintain proper precision. Additional astronomical corroboration of the reality of gravitational time dilation is found in observations of the star Sirius B, a small, dense endpoint of stellar evolution called a *white dwarf*. The gravitational redshift suffered by a photon leaving the surface of such a dense object is predicted by general relativity to be 2.8×10^{-4}. This is in excellent agreement with the measured gravitational redshift in absorption lines in the spectrum of Sirius B, namely $(3.0 \pm 0.5) \times 10^{-4}$. Together, these and other tests of the predictions of the "time" aspects of general relativity leave no doubt as to its correctness as a fundamental physical theory.

EXAMPLE 12.4 The Sun has a mass of 1.99×10^{30} kg and a radius of 6.96×10^{8} m. If an observer near the solar "surface" observes a flare event that lasts 30 minutes, how long does the flare last as witnessed by an observer on Earth?

SOLUTION Assuming the Earth-bound observer at a distance of 1.50×10^{11} m from the Sun is "far" away, we may use the gravitational time dilation equation to answer the question. We first calculate the Schwarzschild radius of the Sun:

$$d_S = \frac{2GM}{c^2} = \frac{2(6.67 \times 10^{-11} \text{ N-m}^2/\text{kg}^2)(1.99 \times 10^{30} \text{ kg})}{(3 \times 10^8 \text{ m/s})^2}$$

$$d_S = 2{,}950 \text{ m} = 2.95 \text{ km}$$

Substituting into the gravitational time dilation formula with $\Delta t_d = 30$ minutes:

$$\Delta t_f = \frac{\Delta t_d}{[1 - (d_s/d)]^{1/2}} = \frac{30 \text{ min}}{[1 - (2{,}950 \text{ m})/(6.96 \times 10^8 \text{ m})]^{1/2}}$$

$$\Delta t_f = \frac{30 \text{ min}}{[1 - (4.24 \times 10^{-6} \text{ m})]^{1/2}} = \frac{30 \text{ min}}{[0.99999576]^{1/2}}$$

$$\Delta t_f = (1800 \text{ s})/(0.99999788) = 1800.0038 \text{ s}$$

The observer on Earth sees the flare to have lasted 38 ten-thousands of a second longer than the observer at the Sun. Although the effect of gravitational time dilation is seemingly small, such time differences are easily measured and must be compensated for as is done for atomic clocks on GPS satellites orbiting Earth.

Once we acknowledge the validity of general relativity and that mass and energy deform space(time), it becomes possible to inquire about the effect that *all* the matter and energy contained in the universe have on its global structure, evolution, and geometry. After all, the largest possible concentration of matter

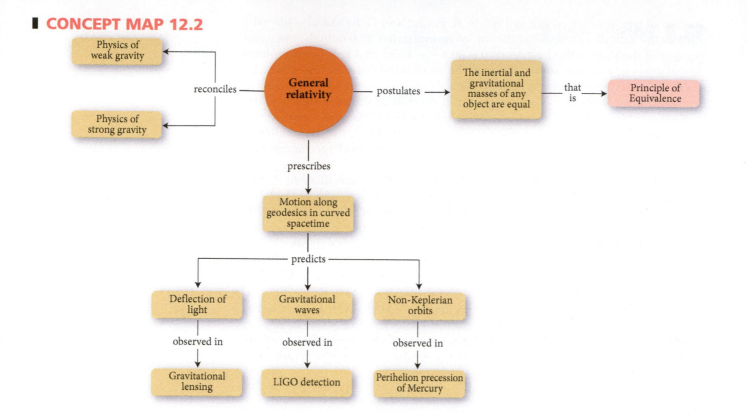

and energy is that present in the *entire* universe! Is the total amount of "stuff" in the universe large enough to significantly alter the "shape" or geometry of the universe from the flat space of Euclid, much as the presence of a heavy bowling ball deforms a flat rubber sheet? Questions like this one bring us into the realm of cosmology: the search for an understanding of what our universe is like on the grandest of scales, how it evolved to this state from its beginnings in the Big Bang (see Section 8.6), and what the future holds for us some 5 or 10 or 100 billion years from now. We shall return to these questions in Section 12.6 after first considering some aspects of particle physics which may have important influences on their answers.

Concept Map 12.2 summarizes the principal elements of the theory of general relativity and some examples of its experimental or observational validation.

■ Learning Check

1. The _____ asserts that in a uniform gravitational field, all objects accelerate at the same rate.

2. (True or False) Accelerating masses radiate energy in the form of gravitational waves.

3. Recent observations of distant supernovae indicate that
   (a) the universe is expanding with increasing speed.
   (b) the universe is expanding with decreasing speed.
   (c) the universe is expanding at a constant speed.
   (d) the universe is contracting at a constant speed.
   (e) the universe is in equilibrium, neither expanding nor contracting.

4. (True or False) Evidence for the superiority of Einstein's theory of gravity (general relativity) over that of Newton is provided by its correct prediction of the deflection of light passing near a massive object.

5. According to (Newton, Einstein [select one]), the planets' orbits are shaped by the curved spacetime in the vicinity of the Sun, while according to (Newton, Einstein), the planets' orbits are produced by the gravitational force exerted on them by the Sun.

ANSWERS: 1. principle of equivalence **2.** True **3.** (a) **4.** True **5.** Einstein; Newton

12.3 Forces and Particles

12.3a The Four Forces: Natural Interactions among Particles

At the end of Section 2.7, we introduced the four fundamental forces of Nature. Table 12.1 lists these forces and includes some properties of each. It is important to acknowledge that *all* the interactions that occur in our environment result from these forces. They produce the beauty, variety, and change that we witness daily in the world around us.

In Chapter 2, we defined a force as a push or pull acting on a body that usually causes a distortion or a change in velocity (or both). This is a perfectly good description of what we mean by a force in classical physics, but to investigate the realm of particle physics, we must broaden our definition to include every change, reaction, creation, annihilation, disintegration, and so on, that particles can undergo. Thus, when a radioactive nucleus spontaneously decays (see Section 11.2), we will describe this decay in terms of a force that acts between the parent nucleus and its decay products. Similarly, when two particles collide and undergo a nuclear reaction to create new particles (see Section 11.4), we say that there is a force responsible for this transformation.

Because the roles played by forces in particle physics are somewhat different from those traditionally assigned to them in classical physics, it is often the case that they are referred to as the *four basic interactions* of Nature instead of the four basic forces. In this context, we use the word "interaction" to mean the mutual action or influence of one or more particles on another. With this in mind, we return to Table 12.1 and discuss each of the four fundamental interactions briefly, beginning with the most familiar, the gravitational interaction.

Gravity, a very important force in our everyday lives, has been investigated at some length in Chapter 2. Several aspects of this interaction as it pertains to particle physics should be reviewed. First, although gravity affects *all* particles, its importance in particle physics is entirely negligible because its strength is so feeble when compared to the other interactions that can occur. To get a feel for just how inconsequential gravity is on a subatomic level, we can compare the strength of the gravitational attraction between two protons bound in a nucleus with the electrical repulsion between these charged particles. A simple calculation (see Challenge 2 at the end of this chapter) shows the electrical interaction to be more than 10^{36} times stronger than the gravitational interaction. Comparisons between the strength of gravity and the other interactions of Nature are given in Table 12.1. In each case, the effects of gravity are too small to be considered seriously.

Table 12.1 The Four Fundamental Forces or Interactions

| Name (example) | Relative Strength | Range (m) | Carrier | | |
			Particle	Mass (MeV/c^2)	Spin
Strong (binds the nucleus)	1	$\approx 10^{-15}$	Meson[a]	$>10^2$	1
Electromagnetic (binds atoms and molecules)	10^{-2}	Infinite	Photon	0	1
Weak[b] (produces beta decay)	10^{-7}	$\leq 10^{-18}$	Z^0, $W^\pm$	$\leq 10^5$	1
Gravitational (binds planets to the Sun)	10^{-38}	Infinite	Graviton	0	2

[a]At the level of the nucleons, we may regard the messengers of the strong force to be the mesons, although, as discussed in Section 12.5, the true carriers of this interaction are the massless, chargeless *gluons*.

[b]All the messengers of the basic forces are uncharged, except for two of those of the weak force. The W^+ particle carries one unit of positive charge, whereas the W^- holds one unit of negative charge. The other carrier of the weak force, the Z^0, is electrically neutral.

Before moving to a discussion of the other forces, it is worth remarking upon two aspects of gravity that do have important consequences for *large-scale* interactions. First, gravitational interactions may dominate in circumstances where charge neutrality prevails. If many particles interact together at once and the number of positive charges balances the number of negative ones, electrical forces may cancel out, leaving gravity the dominant interaction. Unlike the electrical interaction, gravity cannot be shielded out or eliminated because there is only one kind of mass. And second, gravity is a long-range interaction. The gravitational force varies inversely as the distance squared, and although it grows ever weaker with separation, it never completely disappears. Thus, gravitational effects may reach over vast regions of space, accumulating in such a fashion as to affect the structure and evolution of the entire universe (see Section 12.6).

Aside from gravity, the next most familiar force or interaction is the **electromagnetic interaction**, discussed in Chapter 8. Unlike the gravitational force that is always attractive, the electromagnetic force can be either attractive or repulsive, depending on the signs of the interacting charges. But, like gravity, the electromagnetic force is a long-range force, becoming smaller as the distance separating the charges increases. The electromagnetic force also manifests itself in the magnetic forces associated with moving charges. This interaction is responsible for all the various kinds of electromagnetic radiation, from gamma rays to radio waves, investigated in Chapter 8.

It is important to note that although the electric and magnetic forces act only between charged particles, the electromagnetic interaction can have an influence on uncharged particles as well. For example, a photon is not a charged particle, but the absorption or emission of a photon by an atom is an electromagnetic process.

Next among Nature's interactions is the **weak nuclear force**, which is responsible for beta decay (the conversion of a neutron to a proton within the nucleus; see Section 11.2). The term *weak* may be interpreted in a variety of ways. For example, this interaction is weak in the sense that it is effective only over very short distances: at least 100 times smaller than the range of the strong nuclear force and infinitesimal compared to the ranges of gravity and electromagnetism. The weak force is also "weak" because the probability is quite small that interactions involving this force occur. Indeed, particle interactions involving the weak force generally happen only as a last resort when all other interaction mechanisms are prohibited.

Although it is not apparent from what we have said so far, a very close relationship exists between the electromagnetic and weak interactions. The similarity between these two was first noted in the late 1950s, and further studies of the connections between the weak force and electromagnetism have led to a unification of these two interactions into one, the **electroweak interaction**. This is much like the unification between electricity and magnetism in Maxwell's theory. It is now recognized that the source of the electromagnetic and weak forces is the same but that their practical manifestations differ considerably, leading to a separate classification for each. More will be said about the issue of unification of forces in Section 12.5.

The last of the forces of Nature is the **strong nuclear force**. The strong force is responsible for holding the nuclei of atoms together and is involved in nuclear fusion reactions (see Section 11.7). It is a short-range, attractive interaction that does not depend upon electric charge. (The strong nuclear force is sometimes said to be "charge blind.") Considering the probability that two colliding particles will interact by the strong force as opposed to any one of the other three basic interactions, the strong force is indeed quite "strong," some 100 times as effective in bringing about a reaction as the electromagnetic force. Comparisons like these between the relative probability that a reaction will occur via a particular interaction have been used to establish the measures of the relative strengths of the four forces given in Table 12.1.

Having compared the properties of the four forces, let's now consider *how* we have acquired our knowledge about their characteristics. Einstein's theory of special relativity includes the postulate that the speed of light is finite; 300,000 km/s is the maximum speed attainable by particles in the universe. This is also the maximum speed at which information may be propagated through the universe. The fact that a star 150 million kilometers away suddenly explodes *cannot* be known to us until at least 500 seconds (about 8 minutes) later because the particles ejected from the event require that long to make their way to Earth. Information about this explosion thus comes to us through the intermediary of particles that race out from the interaction site carrying data about the nature of the event to our location.

In the same way that we come to understand the details of a stellar explosion by the particles emitted during the event, particle physicists come to know the characteristics of the four forces of Nature by the particles produced in these interactions. In fact, current theories associate with each force a **carrier**, or mediator, of the interaction. These carriers are exchanged between the particles experiencing the forces, and they communicate the interaction between the reactants and the products. For example, if we wiggle an electron, the change in its electric field will propagate outward as a wave traveling at the speed of light. The disturbance in the field produces forces on other charged particles in the neighborhood of the electron and communicates to them information about the electron's motion. The role of the wave is that of a messenger, and this has led physicists to conceive of the influence of the electric field as being conveyed or carried by a messenger particle—the photon (recall Section 10.5 and the discussion of wave-particle duality). In this view, all the effects of an electromagnetic field may be explained by the exchange of photons. (The photons that mediate the electromagnetic interaction are not *real* photons like those produced in the emission of light by excited atoms, but what are called *virtual photons*. These particles are undetectable in the traditional sense, but are responsible for the transmission of forces between other ordinary, observable particles. Thus, even though the virtual photons themselves cannot be directly detected, the *effects* of the virtual photons as carriers of the electromagnetic interaction can be directly seen.)

Table 12.1 includes the carriers of the four basic interactions as well as their masses (measured in equivalent energy units; see Example 12.2). In the next section, we will explore the characteristics of these and other elementary particles in greater detail. Before doing so, we show **Figure 12.14**, which depicts how a particle physicist might represent the weak interaction that converts a neutron inside a nucleus into a proton (a beta decay) and how modern physics views the repulsion between two electrons.

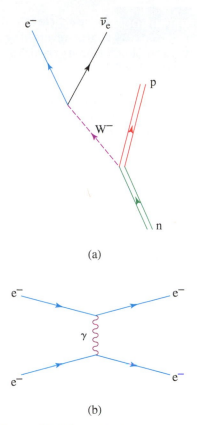

Figure 12.14 (a) Modern representation of the beta decay of a neutron. The neutron transmutes into a proton after emitting a W⁻ particle, and the W⁻ subsequently decays into an electron (e⁻) and an antineutrino ($\bar{\nu}_e$). (b) In particle physics, the electromagnetic repulsion between two electrons is viewed as being caused by the exchange of (virtual) photons (γ). This process is shown in what is called a "Feynman diagram," after Nobel Prize-winning theoretical physicist Richard P. Feynman (1912–1988) who pioneered its use.

12.3b Classification Schemes for Particles

In Chapter 4, we classified matter into solid, liquid, gas, and plasma phases. We also classified matter according to the number and kinds of atoms that are present: elements, compounds, mixtures, and so forth. We were further able to categorize the properties of matter on the basis of the forces that acted between the constituents: large forces in solids, smaller forces in liquids, and so on.

Just as we could classify bulk matter in several different ways, it is possible to classify elementary particles using different schemes. In this section, we will consider three ways of doing so: on the basis of spin, on the basis of interaction, and on the basis of mass. Before going any further in this discussion, however, it is worth defining what we mean by an **elementary particle** and by an **antiparticle**.

> **DEFINITION** **Elementary Particles** The basic, indivisible building blocks of the universe. The fundamental constituents from which all matter, antimatter, and their interactions derive. They are believed to be true "point" particles, devoid of substructure or measurable size.

DEFINITION **Antiparticle** A charge-reversed version of an ordinary particle. A particle of the same mass (and spin) as its cogener but of opposite electric charge (and certain other quantum mechanical "charges").

Every known particle has a corresponding antiparticle. There are antielectrons (more commonly called **positrons** [see Chapter Introduction]), antiprotons, antineutrons, and so on. Collections of antiparticles form antimatter, just as collections of ordinary particles form (ordinary) matter. The first antiparticle, the positron, was discovered in 1932 by Carl Anderson in cosmic rays. The antiproton was discovered in 1955, and the antineutron the year following. Particle physicists now routinely create and store small quantities of antimatter with high-energy accelerators.

When matter and antimatter meet, mutual annihilation results, accompanied by a burst of energy in the form of gamma rays (Figure 12.31). The positrons used in PET scans are produced by the decay of radioactive isotopes, such as fluorine-18, which are first generated by bombarding other nuclei with light particles. For example, fluorine-18 may be created by colliding oxygen-18 nuclei with high-speed protons. In 1995, using similar high-energy physics techniques, scientists at the Conseil Europeén pour la Recherche Nucléaire (CERN) were successful in producing antihydrogen, an atom of antimatter composed of an antiproton and a positron.

In what follows, we shall have several occasions to examine the creation and annihilation of particles and antiparticles. In those reactions, we will distinguish antiparticles using the same symbol as that for the corresponding particle but with a "bar" over it. Thus, for example, if n designates a neutron, then $\bar{n}$ (pronounced "en-bar") represents an antineutron. There are, however, several noteworthy exceptions to this rule. For example, an antielectron (or positron) is symbolized e^+, not $\bar{e}^-$. Likewise, an antimuon is denoted μ^+, not $\bar{\mu}^-$. See Table 12.3 for some other cases in which the "bar" rule is not adhered to strictly.

The kinds of particles that have been termed "elementary" have gradually changed with time. Before 1890 and the discovery of the electron, atoms were regarded as the smallest units of matter, and they were believed to possess no internal structure of their own. In the 1930s, it was believed that the basic building blocks of Nature consisted of the proton, the neutron, the electron, the positron, the photon, and the neutrino (extremely low-mass particles, typically produced in beta decay reactions, that have no charge and only very weakly interact with ordinary matter). Circa 1934, these were the "atoms" (indivisible particles) sought by the ancient Greeks. In the last 85 years, particle physicists have discovered that not only are there more than six "elementary" particles in Nature but also that some of the original six are not really elementary at all! They themselves are composed of still more basic and elusive particles. At the time of this writing, there exist well over 100 different subatomic particles (**Figure 12.15**), and it is well accepted that the proton and the neutron (among others) are not the ultimate constituents of matter. A good fraction of the remainder of this chapter will be spent developing the story of our changing perspective on what constitutes a truly "elementary" particle. But, first, a bit more about the properties of subatomic particles.

12.3c Spin

Spin, like mass and charge, is an intrinsic property of all elementary particles and measures the angular momentum carried by the particle. If we treat an elementary particle as a simple hard sphere, then we can picture spin as resulting from the rotation of the particle about an axis through itself. Unlike, say, a basketball whirling at the end of one's finger, which can have any amount of spin, the spin of elementary particles is quantized (see Section 10.5) in units of $h/2\pi$, where h is Planck's constant. The spins of all known particles are either integral or half-integral multiples of this basic unit. In other words, spin can take on values of 0, $\frac{1}{2}$, 1, $\frac{3}{2}$, and so on, in units of $h/2\pi$. Experiments have shown, for example, that the spin of the electron and the proton is $\frac{1}{2}$, but that of the photon is 1.

"PARTICLES, PARTICLES, PARTICLES."

Sidney Harris

Figure 12.15 Tidying up at CERN?

Particles possessing half-integral spins are called **fermions**, after Italian-American scientist Enrico Fermi (see Profiles in Physics in Chapter 11), who carefully investigated the behavior of collections of such particles. Particles with integer spins are called **bosons**, after Indian physicist Satyendra Bose, who, with Einstein, developed the laws describing their collective behavior. The main difference between these two types of particles is that the former obey the Pauli exclusion principle (Section 10.6) whereas the latter do not. This law, for which Wolfgang Pauli won the Nobel Prize in 1945, states that no two interacting fermions of the same type can be in exactly the same quantum state. They must be distinguishable in some manner. Thus, in a normal helium atom, when the two electrons are in the lowest atomic energy state (the ground state), the exclusion principle demands that these spin $\frac{1}{2}$ particles differ in some way. How can this be achieved? Isn't one electron just like any other? Same mass, same charge, same spin? Yes. But let's return to our rotating sphere model for a moment. Relative to the axis of rotation, a marble may spin either clockwise or counterclockwise. For a given total angular momentum, two distinct spin states associated with the directions of rotation exist (**Figure 12.16**). In the same way, one can associate with the electron two different spin configurations, called *spin-up* and *spin-down*, each with the same total amount of spin, $h/4\pi$. With this addition, it is now possible to satisfy the Pauli principle for helium by requiring one of the electrons to be in a spin-up state while the other is in a spin-down state. The state of every electron in every atom can be accounted for by this principle.

Bosons, by contrast, do not obey the Pauli principle. They are truly indistinguishable. An unlimited number of bosons can occupy a given energy state and so be concentrated in any given volume of space without violating any physical laws. This accounts for the fact that there is no restriction on the number of photons (spin 1 particles) that can be packed into a beam of light and hence no (theoretical) limit to the intensity of the beam. There is also no limit to the number of force-carrying particles that can be exchanged in a given reaction, because *all* the mediators of the fundamental interactions have integral spins (Table 12.1). It is worth mentioning that the phenomena of superfluidity (see Section 4.5) and superconductivity (see Section 7.3) both result from the collective behavior of large numbers of bosons occupying the same state in what is called a *Bose-Einstein condensate.*

12.3d Elementary Particle Lexicon

Separating particles according to spin divides them into two groups. Establishing additional selection criteria further subdivides these two groups. A very useful way of doing so is to identify those particles that participate in strong force interactions and those that do not. We thus distinguish the four groups of particles given in **Table 12.2**: the baryons, the leptons, the mesons, and the **intermediate** (or *gauge*) **bosons**. Examples of several particles of each type are included in the table. An electron is a lepton with spin $\frac{1}{2}$ that

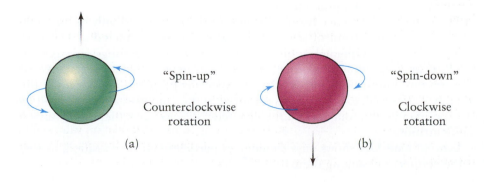

"Spin-up"

Counterclockwise rotation

(a)

"Spin-down"

Clockwise rotation

(b)

Figure 12.16 Spin-up, (a), and spin-down, (b), configurations for a rotating marble.

Table 12.2 Classification of Particles

Spin Group	Particles Interacting Via the Strong Force (Hadrons)	Particles Unaffected by the Strong Force
Fermions (Half-integer spin)	*Baryons* (Protons, neutrons, lambdas, sigmas, . . .)	*Leptons* (Electrons, muons, neutrinos, . . .)
Bosons (Integer spin)	*Mesons* (Pions, kaons, etas, . . .)	*Intermediate (or gauge) bosons* (Photons, Z^0, $W^\pm$, gravitons)

is unaffected by the strong force; a proton is a baryon with spin $\frac{1}{2}$ that does interact via the strong force. Photons are bosons that do not participate in strong interactions.

The names for these groups derive largely from earlier classification schemes based on experimentally determined masses for these particles. The word **baryon** comes from the Greek *barys* meaning heavy, whereas the word **lepton** means "light one" in Greek. **Mesons** refers to the "middle ones" with intermediate mass. At the time these groups were named, the heaviest particles known were found among the baryons and the lightest were included among the leptons. Recent discoveries, however, have revealed mesons and at least one lepton with masses larger than those of protons and neutrons. Thus, it is no longer possible to specify the correct class of an elementary particle by its mass alone, although for most species this is still a useful guide.

The force-carrying particles, the intermediate bosons, exhibit a wide variety of mass. Because they are bosons, there is no limit to the number that can be exchanged in any interaction, but there is a close correlation between the range of the force they mediate and their mass. If the carrier particles have a high mass, it will generally be difficult to produce them and to exchange them over long distances. Thus, a force carried by massive particles will have only a short range. This is the case for the W and Z particles that mediate the weak interaction. However, if the carrier particles have no mass of their own (like the photon and the graviton), the forces associated with them (in these cases, the electromagnetic and gravitational forces, respectively) will be of long range.

One additional bit of nomenclature that is used commonly in connection with elementary particles should be introduced here as well, the word *hadron*. This word is also of Greek origin, coming from *hadros*, meaning "thick" or "strong." Baryons and mesons are collectively referred to as hadrons (Table 12.3) because they interact by the strong force, a distinction not shared by the leptons or the gauge particles.

Among the leptons, the most well-known member is the ubiquitous electron, whose antiparticle, the positron, we have met before (see the Chapter Introduction). Less familiar is the muon and its anti (cf. Example 12.1) and the tau and the anti-tau. In addition, each of these particles has its own distinctive neutrino (and anti-neutrino): ν_e, ν_μ, and ν_τ. All of these entities are spin one-half fermions and can be organized into "families": the electron family; the muon family, and the tau family. Table 12.4 presents data on some of the properties of the members of each lepton family.

The first direct evidence for the tau neutrino was reported only in July 2000 by an international team of scientists in experiments conducted at Fermilab outside Chicago. High-energy protons were driven into a tungsten target to produce (among numerous other particles) tau neutrinos that were detected through rare interactions with nuclei in a specially prepared emulsion. Of the roughly 200 candidate interactions captured in the emulsion after five months of experimentation, four had all the characteristics to mark them as involving tau neutrinos.

Concept Map 12.3 shows ways elementary particles may be classified, as well as the connections among them.

Table 12.3 Properties of Some Long-Lived Hadrons

Class	Particle Name	Symbol (Quark Content)[a]	Antiparticle	Mass (MeV/c^2)	S^c	Lifetime (s)
Baryons ⎰ Spin = $\frac{1}{2}$	Proton	$p(uud)$	$\bar{p}$	938.3	0	Stable(?)
	Neutron	$n(udd)$	$\bar{n}$	939.6	0	886
	Lambda	$\Lambda^0(uds)^b$	$\bar{\Lambda}^0$	1,115.7	−1	2.6×10^{-10}
	Sigma	$\Sigma^+(uus)$	$\bar{\Sigma}^-$	1,189.4	−1	0.8×10^{-10}
		$\Sigma^0(uds)^b$	$\bar{\Sigma}^0$	1,192.6	−1	7.4×10^{-20}
		$\Sigma^-(dds)$	$\bar{\Sigma}^+$	1,197.5	−1	1.5×10^{-10}
	Xi	$\Xi^0(uss)$	$\bar{\Xi}^0$	1,315	−2	2.9×10^{-10}
		$\Xi^-(dss)$	$\bar{\Xi}^+$	1,321	−2	1.6×10^{-10}
Spin = $\frac{3}{2}$	Omega	$\Omega^-(sss)$	$\bar{\Omega}^+$	1,672	−3	0.8×10^{-10}
Mesons ⎰ Spin = 0	Pion	$\pi^+(u\bar{d})$	π^-	139.6	0	2.6×10^{-8}
		$\pi^0(u\bar{u},d\bar{d})^f$	Self[e]	135.0	0	8.4×10^{-17}
	Kaon[d]	$K^+(u\bar{s})$	K^-	493.7	+1	1.2×10^{-8}
		$K^0(d\bar{s},\bar{d}s)^f$	$\bar{K}^0$	497.7	+1	0.9×10^{-10}
						5.2×10^{-8}
	Eta	$\eta^0(u\bar{u},d\bar{d},s\bar{s})^f$	Self	547.8	0	5.5×10^{-19}
	"Dee-plus"	$D^+(\bar{d}c)$	D^-	1,869	0	1.0×10^{-12}
	"Bee-plus"	$B^+(\bar{b}u)$	B^-	5,279	0	1.7×10^{-12}
Spin = 1	Psi	$\psi^0(c\bar{c})$	Self	3,097	0	7.6×10^{-21}
	Upsilon	$Y^0(b\bar{b})$	Self	9,460	0	1.2×10^{-20}

[a]Superscripts to the right of the particle symbols indicate the charge carried by the particle in units of the proton charge; see Sections 12.4 and 12.5 for discussions of the quark composition of baryons and mesons.
[b]While having the same quark content, these two baryons differ in terms of an internal quantum number called *isospin* and so represent different states of the same quark system.
[c]Strangeness (see Section 12.4).
[d]K^0 mesons decay by the weak force with two characteristic times, one 575 times shorter than the other, due to what are called *neutral particle oscillations* (cf.a similar phenomenon for neutrinos in Section 12.5b).
[e]Some neutral particles are their own antiparticles. Thus, when two π^0 meet, they annihilate one another to form γ-rays.
[f]A few mesons are composed of admixtures of two or three quark–antiquark states.

Table 12.4 Lepton Families[a]

Family	Particle Name	Symbol[b]	Antiparticle	Mass (MeV/c^2)
Electron	Electron	e^-	e^+ (positron)	0.511
	Neutrino	ν_e	$\bar{\nu}_e$	$\sim 0(\leq 2\times10^{-6})$
Muon	Muon	μ^-	μ^+	105.7
	Neutrino	ν_μ	$\bar{\nu}_\mu$	$\sim 0(<0.17)$
Tau	Tau	τ^-	τ^+	1,777
	Neutrino	ν_τ	$\bar{\nu}_\tau$	$\sim 0(<15.5)$

[a]The spins of all the leptons, regardless of family, are $\frac{1}{2}$. For this reason, we have not separately listed this property for each particle.
[b]Superscripts to the right of the particle symbols indicate the charge carried by the particle in units of the proton charge. Thus, the electron possesses one unit of negative charge, and the antimuon carries one unit of positive charge. The neutrinos carry no charge.

CONCEPT MAP 12.3

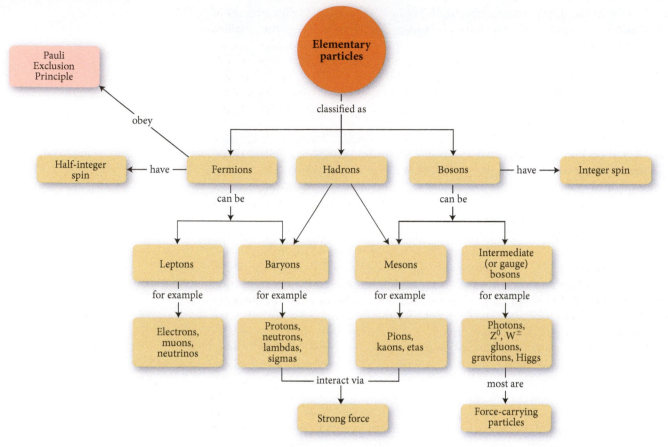

Learning Check

1. Which of the following is/are *not* fundamental forces of Nature?
   (a) friction
   (b) gravity
   (c) tension (as in a spring)
   (d) strong nuclear
   (e) electromagnetism

2. Particles with integer spins are called _____, whereas those with half-integer spins are called _____.

3. Match each item in column A with its description from column B. Each entry in column A has only one correct match from column B.

A	B
(i) baryons	(a) strongly interacting, spin 0 or 1 particles
(ii) mesons	(b) carrier particles for the fundamental forces

   (iii) leptons
   (iv) intermediate bosons
   (c) strongly interacting, spin $\frac{1}{2}$ or $\frac{3}{2}$ particles
   (d) spin $\frac{1}{2}$ particles that do not interact by the strong force

4. (True or False.) A positron has the same mass as an electron, but its charge is $+e$ instead of $-e$.

5. Interactions that have a range of only 10^{-18} m or less are most likely ones that involve
   (a) the gravitational force.
   (b) the electromagnetic force.
   (c) the weak force.
   (d) the strong force.
   (e) None of the above.

ANSWERS: 1. (a), (c) 2. bosons, fermions 3. (a) (ii), (b) (iv), (c) (i), (d) (iii) 4. True 5. (c)

12.4 Quarks: Order Out of Chaos

Beginning in the 1950s and continuing into the 1970s, there occurred a rapid proliferation of newly discovered subatomic particles. Among the earliest found were the kaons (K), the lambdas (Λ), and the sigmas (Σ), all of which exhibited some very strange and previously unobserved characteristics. For example, these *strange particles*, as they came to be called, were always observed to be created in pairs, yet once formed they persisted for unusually long times (by particle physics standards) before decaying to other hadrons in times of the order of 10^{-10} to 10^{-8} seconds.

In 1953, Murray Gell-Mann and Kazuhiko Nishijima independently proposed that certain particles possess another quantum number or another type of "charge" termed **strangeness** (S). Table 12.3 lists the strange charges for some of the better-known strange particles. Their anti's possess strangeness in equal magnitude but of opposite sign. Nonstrange particles, such as protons and neutrons, have zero strangeness.

With the introduction of strangeness and the conditions under which the new "charge" could be changed in particle reactions involving the strong, weak, and electromagnetic interactions, the impact of the Gell-Mann – Nishijima theory was immediate and revolutionary. It completely solved all the puzzles presented by the strange particles, and, as we shall see, it also opened the door for the development of new theories of particle interactions that posit the existence of still other "charges" or intrinsic quantum numbers for subatomic particles to account for the outcomes of high-energy collision reactions involving them.

As the number and complexity of the subatomic species grew, many physicists to wonder if "elementary particles" were really so "elementary" after all. Maybe they themselves were composites of still smaller entities, the "*really elementary*" particles. In 1961, Gell-Mann and Yuval Ne'eman independently gave a new way to classify hadrons. They introduced an additional property of subatomic particles called *unitary spin*. This quantum characteristic has eight components, each of which is a combination of the quantum numbers we have seen before (plus a few others that are too esoteric for us to consider at this level). Because of the eight-part structure of the basic "charge" in this theory and the theory's potential for leading to a deeper understanding of Nature, Gell-Mann christened this the "eightfold way," in analogy with "the noble eightfold way" of Buddhism that leads to nirvana.

Many of the particles listed in Table 12.3 had not yet been discovered at the time the Gell-Mann–Ne'eman theory was initially proposed, but in 1963, this model received its first major experimental validation when the spins of the Σ^0 and $\Sigma^\pm$ were finally measured and discovered to be $\frac{1}{2}$, in agreement with the eightfold way. Additional support for the Gell-Mann–Ne'eman model came a little later, with the discovery of the omega minus, Ω^-. Based on the eightfold way, the existence of a single, massive particle with charge -1, strangeness -3, and spin $\frac{3}{2}$, which had never been seen before, was predicted. Particle physicists began to search for this species in new, higher-energy collision experiments, and in February 1964, the successful detection of a particle having all the predicted characteristics was announced by scientists at the Brookhaven National Laboratory. Figures 12.17 and 12.18 show the original photograph of the particle tracks in the discovery experiment and the analyses that led to its interpretation in terms of the Ω^-.

12.4a Quarks

Further investigation of the implications of the eightfold way led to a refinement of the theory in 1964. At that time Gell-Mann and George Zwieg postulated that all hadrons were formed from three

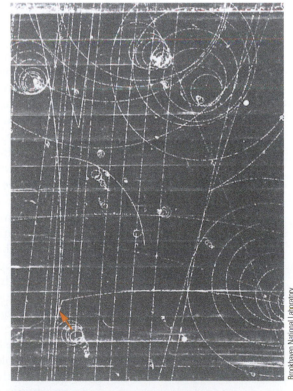

Figure 12.17 The first photograph of the Ω^- particle, taken at the Brookhaven National Laboratory in 1964. The path of the Ω^- is marked with an arrow at the lower left of the picture.

Brookhaven National Laboratory

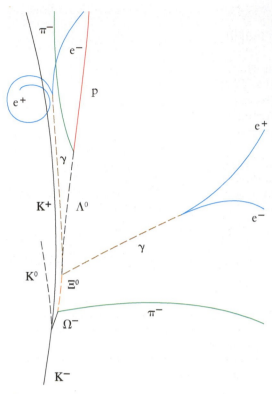

Figure 12.18 A schematic reconstruction of some of the particle tracks shown in Figure 12.17. Solid lines indicate the trajectories of charged particles, and dashed lines give the paths of neutral particles (which do not show on the original photograph). The formation and decay of the Ω^- involve the following reactions: (1) $K^- + p \rightarrow \Omega^- + K^+ + K^0$; (2) $\Omega^- \rightarrow \Xi^0 + \pi^-$.

Table 12.5 Some Properties of the Originally Proposed Quarks[a]

Quark	Electric Charge (Q)[b]	Strangeness (S)
u	$+\frac{2}{3}$	0
d	$-\frac{1}{3}$	0
s	$-\frac{1}{3}$	-1
$\bar{u}$	$-\frac{2}{3}$	0
$\bar{d}$	$+\frac{1}{3}$	0
$\bar{s}$	$+\frac{1}{3}$	$+1$

[a]The quarks are all spin $\frac{1}{2}$ particles.
[b]These values are in units of the proton charge. Thus the charge of an up quark in SI units would be $(\frac{2}{3})(1.6 \times 10^{-19} \text{ C}) = 1.07 \times 10^{-19}$ C.

fundamental particles, which Gell-Mann called **quarks**, and their anti's. The quarks were designated u (for "up"), d (for "down"), and s (for "strange"). **Table 12.5** gives the properties of these three particles and their anti's. Notice that all the quarks have spin $\frac{1}{2}$ and charge $\pm\frac{1}{3}$ or $\pm\frac{2}{3}$ (in units of the proton charge). These are fractionally charged particles, unlike anything we have dealt with before!

Surprising as this may appear, the introduction of these noninteger charged particles enabled physicists to describe perfectly all the hadrons discovered prior to about 1970 and, with some extensions, all the heavy particles found since. The two rules governing the formation of hadrons from quarks are simple:

1. Mesons are composed of quark-antiquark pairs, like $u\bar{d}$ and $d\bar{s}$.

2. Baryons are constructed out of three-quark combinations; antibaryons are made up of three antiquarks.

For example, a π^+ meson is equivalent to a $u\bar{d}$ pair with their spins oppositely directed. This combination gives a particle of spin 0, and charge $+1$ ($= \frac{2}{3} + \frac{1}{3}$; see Table 12.5) as required. A K^0 meson may be shown to consist of a combination of a d quark and an $\bar{s}$ quark.

Similarly, a proton is a collection of uud quarks, whereas the neutron is a udd quark combination. (You should take a few minutes to assure yourself that the addition of quarks as indicated gives the usual properties of the proton and the neutron that you are familiar with. The data in Tables 12.3 and 12.5 will be helpful in this regard.) Other three-quark arrangements give other baryons: $\Sigma^+ = $ uus, $\Lambda^0 = $ uds, and $\Omega^- = $ sss.

It is important to emphasize at this point that the quark model *applies only* to hadrons. Leptons, like the electron, the muon, and the neutrinos, are *not* made of quarks. Indeed, the status of the leptons and the quarks is much the same within the realm of particle physics: both are groups consisting of fundamental, irreducible spin $\frac{1}{2}$ fermions that together make up the matter in the universe.

The quark model has several distinct advantages over any competing theories of subatomic particles. First, it explains why mesons all have integral spins—that is, why mesons are bosons. Because mesons are two-quark combinations, the mesons, as a class, can only have total spin 0 (when the spins of the two constituent quarks point in opposite directions) or 1 (when the two spins are aligned). Second, the model also accounts for the fact that baryons all are fermions, half-integral spin particles: Any arrangement of three quarks will always yield either a spin $\frac{1}{2}$ or a spin $\frac{3}{2}$ particle, depending on whether the spins of two of the three are paired (spin $\frac{1}{2}$) or whether all three spins are parallel (spin $\frac{3}{2}$). By the same token, the quark model permits a new interpretation of strangeness and its conservation. Strangeness is just the difference between the number of strange antiquarks and the number of strange quarks making up the particle.

Given the abundant success of the quark model, it was not too long before particle physicists began looking for evidence of the existence of quarks. Their challenge has been a formidable one because, according to prevailing theories, quarks are believed to be inescapably bound within hadrons by what is called the *color force*. Within each hadron, the color force is mediated by exchange particles called *gluons*, and the "strings" that confine the individual quarks comprising the hadrons are called *gluon tubes*. (See Section 12.5 for more on gluons and the color force.) Remarkably, the force that binds the quarks actually gets *stronger* as the separation between the quarks in the hadrons *increases*! This counterintuitive property of the color force, called "asymptotic freedom," was discovered by David Gross, H. David Politzer, and Frank Wilczek, who shared the 2004 Nobel Prize in physics for their work. Thus, the strong force that exists between hadrons is now revealed to be just a shadow of the "*really* strong force"—the color force—that permanently conceals the quarks inside their host hadrons. This is so because the work required to overcome the binding force and to free the quarks would, in principle, be infinite as the color force gets greater and greater as the distance between a quark and its parent hadron increases. But, if all this is true and *free* quarks are not predicted, we are still left with our original question: What evidence is there for the existence of quarks even within hadrons?

The experimental evidence for quarks comes from two sources: first, the scattering of high-energy electrons off protons and, second, the observation of jets of hadrons coming from collisions between electrons and positrons. In the first instance, a case for protons being composed of smaller point particles has been made on much the same basis as that used by Rutherford to argue for the existence of the nucleus (see Profiles in Physics in Chapter 11). Specifically, the number of electrons deflected through large angles after suffering a head-on collision with point-like quarks inside protons is in good agreement with predictions of the quark model. The second type of evidence for quarks involves the products of high-energy collisions of beams of electrons and positrons. In these reactions, the energy derived from the e^- - e^+ annihilation goes into the creation of q-$\bar{q}$ pairs. As these particles move off in opposite directions (as required to conserve linear momentum), they produce a stream of new q-$\bar{q}$ pairs. The quarks in this bidirectional jet quickly combine to give various hadrons, so that at the level of the experiment, what is seen is two trains of heavy particles, oriented at 180°, traveling away from one another (**Figure 12.19**). These observations also provide persuasive evidence for the overall correctness of the quark model and the existence of these elusive particles.

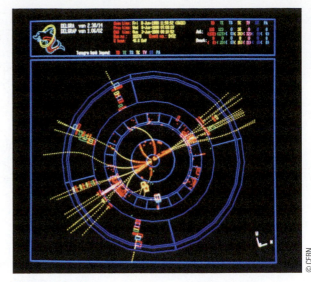

© CERN

Figure 12.19 Computer reconstruction of data collected from the DELPHI detector on the Large Electron-Positron (LEP) collider at CERN. Here a Z^0 particle is produced in a collision between an electron and a positron, which then decays into a quark–antiquark pair. The quark pair yield two opposing streams of new q-$\bar{q}$ pairs that ultimately combine to form a pair of hadron jets.

> ## Physics To Go 12.2

Probing the hidden structure of a system by bombarding it with small but highly energetic particles is nothing new. After all, this is how we learn about the condition of teeth and bones from a medical or dental "x-ray" (see Section 8.5). Rutherford used this technique to explore the structure of the atom and to estimate the size of the nucleus. Particle physicists have used similar means to discern the existence of pointlike quarks within hardrons. Let's try a scattering experiment of our own to illustrate the principles employed in such investigations and to "discover" the size of two-dimensional circular "atoms" in a rectangular target.

Make an enlargement at least twice the size of **Figure 12.20**, including the rectangular boundary enclosing the circles (the atoms). Attach a piece of carbon paper, carbon side down, to the pattern so that it is completely covered, and place the two sheets on a smooth, flat, hard surface. A tabletop or a tiled floor will do. Drop a marble (or other small hard sphere) onto the paper stack from a height of 30 to 60 cm, *making sure to catch the marble on the rebound so that it only strikes the sheets once*. Repeat this process at least 100 times, covering the entire area of the pattern as completely as possible.

Remove the carbon paper from the target sheet and count the total number of marble "hits" (indicated by black carbon dots) that lie within the rectangular boundary defining the target. Next, count only the "hits" that lie wholly within the circles in the target. Using a millimeter ruler, measure the length and width of the target and compute the total target area. Finally, count the number of circles present in the target.

If the circles/atoms are all uniform and the "hits" randomly distributed, then we expect the ratio of the area of all the circles to the total area of the target to be equal to the ratio of the number of hits within the circles to the total number of hits on the target. Compute the ratio of hits from your data. Find the area of all the circles by multiplying this result by the total target area you determined from your millimeter measurements. Dividing by the number of circles yields the area of a single circle/atom in the target. Because the area of a circle is π times the radius of the circle squared, we are now in a position to compute the radius, r, of one of the circles/atoms in the target.

What do your measurements give for this value? How does this indirectly determined value of r compare to what you get from a direct measurement of the radius of one of the circles using a ruler?

Figure 12.20 Target sheet for the scattering experiment described in Physics to Go 12.2.

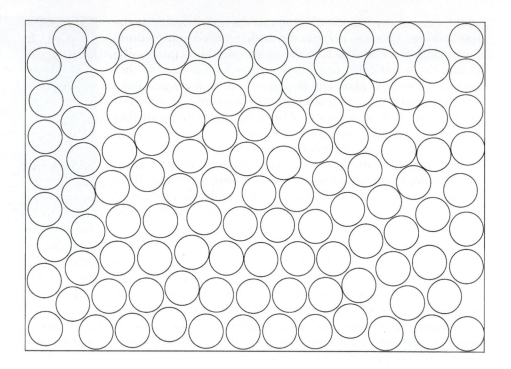

Learning Check

1. Three quarks combine to form
   (a) a baryon.
   (b) an electron.
   (c) a meson.
   (d) a photon.
2. (True or False.) Quarks are elementary particles possessing electric charge equal to $\pm\frac{1}{3}$ or $\pm\frac{2}{3}$ the magnitude of the charge on a proton.

3. Mesons are
   (a) always composed of either two u quarks or two d quarks.
   (b) always composed of two different types of quarks.
   (c) always composed of a quark and an antiquark.
   (d) always composed of a quark and a lepton.
   (e) None of the above.
4. _____ are the carriers of the color force that binds the quarks inside hadrons.

ANSWERS: 1. (a) **2.** True **3.** (c) **4.** Gluons

12.5 The Standard Model and GUTs

12.5a Quark Colors and Flavors

Quarks are fermions. They have half-integral spins, and they should therefore obey the Pauli exclusion principle. But if this is true, how do we explain the existence of the Ω^-? This particle is composed of three strange quarks, all of which share the same mass, spin, charge, and so on—that is, the same quantum numbers. It would appear that the Ω^- is made up of three identical, interacting s quarks. Doesn't this violate the exclusion principle? How can we reconcile these circumstances?

One way would be to argue that quarks are somehow different from other fermions and, consequently, don't have to conform to the Pauli principle. When particles such as the Ω^- were discovered, some theoretical physicists did suggest this explanation to resolve the dilemma. However, other scientists were reluctant to make exceptions to the exclusion principle and instead proposed that quarks carry an additional property that makes them distinguishable within hadrons such as the Ω^-. For this scheme to work, this new characteristic had to come in three varieties to permit, for example, the three s quarks in the Ω^- to be different from each other. The name that particle theorists gave to this new quantum property or number was *color*, although it has absolutely nothing to do with what we commonly refer to as color—that is, our subjective perception of certain frequencies in the electromagnetic spectrum. The three quark colors were labeled after the three primary colors of the artist's palette: red, blue, and green. Antiquarks are colored antired, antiblue, and antigreen (**Figure 12.21**).

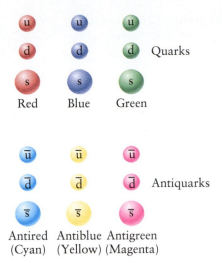

Figure 12.21 Each quark (and antiquark) comes in three colors (or anticolors).

The fact that color is not an observed property of hadrons indicates that these particles are "colorless." If they are composed of colored quarks, then the way the colors come together within the hadron must be such as to produce something with no net color, something that is color neutral. For this to be true, the three quarks that make up baryons must each possess a color different from their companions, one red, one blue, and one green. The addition of the primary colors produces the result we call "white," so baryons containing three different colored quarks are considered to be white or neutral as concerns the color charge. In an analogous manner, for mesons to be color-neutral requires them to be made up of a quark of one color and an antiquark possessing the corresponding anticolor. Returning to our previous example of the π^+ meson, we see in the light of this new color physics that if the u quark is red, the $\bar{d}$ quark must be antired.

You have probably noticed that the language of particle physics is rather whimsical in comparison with the other areas of physics that we have studied. To carry this whimsy one step further, we note that often the different types of quarks, u, d, s (and others that we will shortly introduce), are designated quark *flavors*. Quarks come in six flavors (not counting the antiflavors), and each flavor comes in three colors.

Color is another example of an internal quantum number, like unitary spin. It is not directly observed in hadrons, and it cannot be used to classify them. Moreover, it does not influence the interactions among hadrons. So what is the significance of the color charge? What possible role does it play in particle physics?

Current theories now suggest that the color charge is the source of the inter-quark force, much as ordinary electric charge is the source of the electrical force. And just as the electrical force between charged particles is carried by the photon, the color force between the bound quarks is carried by the *gluons* (see the discussion near the end of Section 12.4). Gluons are electrically neutral, zero mass, spin 1 particles that are exchanged between the bound quarks in hadrons. There are eight gluons in all, and the emission or absorption of a gluon by a quark can lead to a change in the color of the quark. Such changes occur very rapidly and continuously among the quarks forming a given hadron, but always in such a way that the overall color of the hadron remains neutral. The color force is sometimes referred to as the *chromodynamic force*, from the Greek word *chroma*, meaning "color."

12.5b Charm, Truth, and Beauty

In the early 1960s, three quarks were enough to account for the known hadrons, but at the time, four different leptons were known: the electron and its neutrino and the muon and its neutrino. (The tau was not discovered until 1975.) Because of this asymmetry in the numbers of quarks and leptons, it was suggested that there should exist a fourth quark to balance out the leptons. Other, "harder" evidence for the existence of a fourth quark came in the form of certain rare, weak interactions involving hadrons in which their charges remained constant but their strangeness changed. Together, these bits of data convinced particle physicists that another quark, the c quark, was present in

Nature and that it carried a new quantum "charge" called *charm*. Theory indicated that the charmed quark should have electric charge $+\frac{2}{3}$, strangeness 0, and a mass greater than any of the previously identified quarks.

The search for particles containing the c quark—so-called charmed particles—began in earnest in 1974 at Brookhaven and at the Stanford Linear Accelerator Center (SLAC) in California. In November 1974, Samuel Ting, leader of the Brookhaven group, and Burton Richter, head of the SLAC team, jointly reported the discovery of a short-lived particle produced in e^- - e^+ annihilation events that seemed to "fit the bill." The particle, called J by Ting's group and ψ (psi) by Richter's, is now believed to be a $c\bar{c}$ combination—one state of several belonging to what has been referred to as "charmonium"—having a mass of 3.1 GeV/c^2. In 1976, Ting and Richter shared the Nobel Prize in physics for their work.

But even more surprises were in store for particle physicists. In 1977, in experiments conducted at Fermilab outside Chicago, a very heavy meson, called the Υ (upsilon), was discovered. It could not be accounted for in terms of the four previously postulated quarks. So to explain the properties of this hadron, researchers proposed that a fifth quark, the b or "bottom" quark (early on referred to as *beauty*), existed. The Υ was taken to be a $b\bar{b}$ pair.

In 1975, the announcement of the discovery of the tau and its (presumed) neutrino (recall the latter particle was not observed in collision reactions until 2000) brought the number of leptons to six. With the identification of the b quark, the number of quarks once again lagged that for the leptons by one. To maintain the symmetry between the two groups, the need to have a sixth quark was pointed out. The presence of this newest quark, the t or "top" quark (sometimes called *truth*), permits the quarks to be paired off into three families much as the leptons are: (u,d); (s,c); and (b,t).

In March 1995, based on data from proton–antiproton collision experiments carried out using the Fermilab Tevatron accelerator, representatives of a consortium of over 400 scientists and engineers announced the discovery of more than 100 events out of several trillion associated with the production of the top quark with a probable error of less than 1 in 500,000. **Figure 12.22** shows a schematic representation of a typical "top event" resulting from a p-$\bar{\text{p}}$ collision and the subsequent decay of the top and antitop quarks into other particles. Beginning in 2007, "single" top quarks in $t\bar{b}$ meson combinations have been reported in experiments at Fermilab, offering unequivocal proof of the reality of the top quark. Based on the analyses of these events, the current value of the top quark mass is 174.3 GeV/c^2, a value greater than the mass of the nucleus of an atom of gold!

In March 2010, the nearly $4 billion Large Hadron Collider (LHC) successfully carried out collisions between counter-rotating beams of protons each having an energy of 3.5 *tera*-electronvolts (TeV; tera = 10^{12}). Within a year, the proton beam experiments succeeded in raising the beam energies to 7 TeV each, yielding relative energies near 14 TeV for the opposing particle streams. At such enormous energies, proton collisions produce as many as two t quarks a second, or over 2 million such species during typical large experimental runs. Thus, the prospects for using the top quark to answer many questions that still remain about matter and the forces that govern the physical universe, including the existence of the Higgs particle (see Section 12.5c), have never been more promising. **Table 12.6** summarizes the properties of the six quarks.

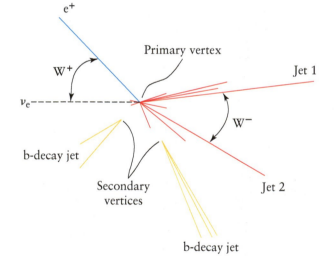

Figure 12.22 Candidate event for top-antitop production. Each top quark decays at the p$\bar{\text{p}}$ primary collision vertex into a W boson plus a bottom quark. The W^+ decays into a positron e^+ plus an invisible neutrino (ν_e), and the W^- decays into a quark and an antiquark, which appear as two jets of hadrons (jets 1 and 2). Each bottom (b) quark becomes a neutral B meson that travels a few millimeters from the production vertex before its decay creates a hadron jet. From "Top-ology" by Chris Quigg, *Physics Today* 50, no. 5 (p. 24). Copyright © 1997 by the American Institute of Physics. Used by permission.

Table 12.6 Summary of Basic Characteristics of u, d, s, c, b, and t Quarks

		Quantum Number or Charge				
Quark[a]	Electric Charge	Strangeness	Charm	Bottomness	Topness	Mass[b] (MeV/c^2)
u	$+\frac{2}{3}$	0	0	0	0	2.5
d	$-\frac{1}{3}$	0	0	0	0	4.8
c	$+\frac{2}{3}$	0	1	0	0	1,275
s	$-\frac{1}{3}$	-1	0	0	0	95
t	$+\frac{2}{3}$	0	0	0	1	174,300
b	$-\frac{1}{3}$	0	0	-1	0	4,180

[a]All the quarks have spin $\frac{1}{2}$.
[b]Values given are approximate upper limits to quark mass ranges.

Are there more than six quarks in Nature? More than six leptons? Should we expect to see an expansion in the numbers of families of these particles beyond three in the future? And if so, does this provide evidence that these "elementary" particles may themselves be composites of still more fundamental species? The answers to all of these questions cannot be given at this time, but the results of experiments done within the last few years seem to indicate that there cannot be more than three distinct families of quarks and leptons. With the confirmation of the existence of the top quark, it appears that we have found *all* of the fundamental building blocks of Nature. **Figure 12.23** summarizes the properties of the three quark and lepton families, as well as those of their anti's, in what is now referred to as the *Standard Model* of elementary particle physics.

Figure 12.23 Standard Model of elementary particles.

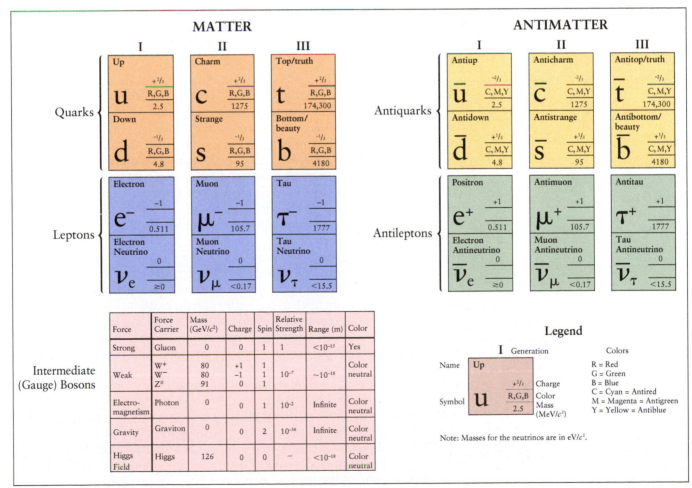

Although highly successful at accounting for much of what is known about elementary particles, the Standard Model has limitations. For example, it predicts that the masses of the neutrinos should all be precisely zero. Laboratory experiments have placed upper limits on these some masses (see Table 12.4), but the most compelling evidence that neutrinos must have some mass comes from recent observations at the Sudbury Neutrino Observatory (SNO). Unlike other neutrino detectors, SNO has the capacity to distinguish the type of neutrino (electron, muon, or tau) being measured, and recent results indicate that only about one-third of the total number of neutrinos reaching Earth are of the electron variety. This discovery confirms results reported by Raymond Davis Jr. and collaborators nearly 50 years ago showing that the number of electron neutrinos detected from the Sun is smaller than what is expected on the basis of our best models for the solar interior. For his pioneering work in what was called the "solar neutrino problem," Davis shared the 2002 Nobel Prize in physics.

Because almost *all* the neutrinos produced by the Sun are electron neutrinos, two-thirds of them must have changed flavor during their approximately eight-minute flight to Earth. Because massless neutrinos cannot spontaneously change type, the implication of this discovery is that neutrinos must have nonzero mass. Theoretical models to explain these *neutrino oscillations*, as they are called, do not provide explicit values for the individual neutrino masses, only their mass differences. Current values for the mass differences, although highly model dependent, are generally small, ranging from about 10^{-1} to as little as 10^{-4} eV/c^2. If the scale of the neutrino masses is set by the electron neutrino (~ 1 eV/c^2), then all the neutrino masses are very small. But the fact that these elusive particles have any mass at all remains something that the Standard Model cannot explain. At present, much of the effort in theoretical particle physics is devoted to revising the Standard Model to remedy this and other deficiencies. The 2015 Nobel Prize in physics was awarded to Arthur B. McDonald and Takaaki Kajita for their discovery of neutrino oscillations showing that neutrinos possess mass.

12.5c The Electroweak Interaction and GUTs

In Section 12.3, we mentioned that similarities between the electromagnetic and the weak nuclear forces first noted in the 1950s have led scientists to the conclusion that these two interactions are really just two different manifestations of a more basic interaction called the **electroweak interaction**. The theory describing this interaction was developed independently by Steven Weinberg (1967) and Abdus Salam (1968), and is often referred to as the Weinberg-Salam theory. Weinberg and Salam, together with Sheldon Glashow, who had also made contributions to the unification of these two forces, shared the Nobel Prize in physics for 1979.

According to the Weinberg-Salam model, at moderate to high energies (of the order of 100 billion electronvolts—100 GeV), the electroweak interaction is carried by four massless bosons and is described by equations that are symmetric (i.e., they remain unchanged under certain mathematical operations on some specific characteristics of these particles). As the energy of the system is lowered, however, the symmetry is broken spontaneously, with the result that the original family of four massless bosons separates into two subfamilies, the first consisting of the massless photon, which mediates what we call the electromagnetic interaction, and the second including the very massive intermediate (or gauge) bosons ($W^{\pm}$, Z^0), which carry the weak interaction.

The reasons why some of the originally massless carriers of the electroweak interaction acquire mass as the symmetry of the system is broken involves what are called *gauge fields* and *Higgs mechanisms*, but before exploring these concepts in any detail, an example involving broken symmetry in another context might help remove some of the mystery associated with this process.

This example has been given by Stephen Hawking and connects the idea of symmetry within a system to its energy. Consider a roulette ball on a roulette wheel (**Figure 12.24**). At high energies when the wheel is spun rapidly, the ball behaves in basically one way: it rolls around and around in one direction in the groove of the wheel. We might say that at high energies there is only one state in which the ball can be found, and all rapidly spinning roulette wheels display the same state: They're all symmetrical with respect to one another. However, as the wheel slows and the ball's energy decreases, it eventually drops into one of the 38 slots molded into the wheel; at low energies, it appears that there are 38 different states in which the ball can exist. If we surveyed roulette wheels solely at low energies, we might be led to the conclusion that these 38 possibilities were the only ones allowed to the ball. We would miss the fact that at high energies these different states merge into one. This is analogous to what happens to the carrier family for the electroweak interaction. At high energy, these carriers are all massless and behave like one another. But at low energy when the symmetry is broken, this group appears as two families of completely different particles.

At the time Weinberg and Salam first published their theory, particle accelerators did not exist that could produce the energies required to test their predictions, particularly those concerning the existence of the massive $W^{\pm}$ and Z^0 particles that carried the weak nuclear force. By the early 1980s, however, machines were beginning to come on line that could achieve the needed energies, and, in January 1983, Carlo Rubbia of CERN reported the detection of the W particles in nine events out of over a million recorded p-$\bar{\text{p}}$ collisions. In July 1984, the CERN group also announced the discovery of the Z^0 particle together with additional evidence to support their earlier detection of the W particles. The 1984 Nobel Prize in physics was awarded to Rubbia and his collaborator Simon van der Meer for their search and discovery of the carriers of the weak interaction. These results completely vindicated the Weinberg-Salam model and left little doubt about its correctness.

One of the long-standing challenges to the Standard Model has been the elucidation of the explicit means by which the symmetric electroweak interaction was cleaved and some of the originally massless carriers took on mass. In the simplest version of the theory, the initial symmetry of the system is broken by the production of a new particle, the *Higgs boson*, a massive, neutral, spin 0 excitation of an all-pervasive Higgs field that interacts with the intermediate gauge bosons, as well as the leptons and the quarks. The latter particles acquire mass in the outcome, the values being determined by how strongly each "couples" to or interacts with the Higgs boson. The Higgs particle is named after Peter Higgs, one of six physicists who independently proposed this mass-acquisition mechanism in 1964. For the last 50+ years, proving the existence of the Higgs boson (as a surrogate for the Higgs field) has been among the highest priorities for high energy particle physicists as it represented one of the critical, unverified aspects of the Standard Model. However, because of the expected high mass of the Higgs, it has only been with advent of the LHC that chances of observing this elusive particle have become possible.

On 4 July 2012, on the basis of two independent experiments involving high-energy collisions between protons and lead ions, researchers at CERN reported the detection of a new particle with a mass in the range of 125 to 127 GeV/c^2 believed to be the Higgs boson. By March 2013, the interaction and decay characteristics of the particle were found to conform to the predictions of the Standard Model for the Higgs in terms of symmetry, spin, and color. Refinements in the measurements yielded an improved mass of 125.09 GeV/c^2 with an uncertainty of less than 0.5%, and a lifetime of 1.56×10^{-22} s. Although the estimated frequency of production of the Higgs boson is only 1 per 10 billion collisions, sufficient data were collected by the CERN experimenters to make the probability of obtaining the observed signal by chance less than 1 in 3 million. With the detection of the Higgs particle confirmed and the original theory proposing its existence validated, the Nobel Prize in physics was awarded

Figure 12.24 High (top) and low (bottom) energy states at the roulette table. When the wheel spins rapidly, there exists only one state of the system—with the ball whirling around in the groove. When the wheel slows down, there appear to be 38 possible states for the system corresponding to the 38 slots into which the ball can lodge. The lower figure shows the ball in one such state, corresponding to slot 23.

Figure 12.25 François Englert (left) and Peter Higgs at CERN on 4 July 2012, on the occasion of the announcement of the discovery of a Higgs boson. Englert and Higgs shared the Nobel Prize in physics in 2015 for their work in the theoretical development of a mechanism for understanding the origin of mass of subatomic particles. (Image: Maximilien Brice/CERN)

to Peter Higgs and François Englert on 10 December 2013 (**Figure 12.25**).

The success of the unification of the electromagnetic and the weak nuclear forces caused many physicists to wonder whether it might not also be possible to unite the strong nuclear force with the electroweak force to produce what has come to be called a **grand unified theory**, or **GUT**. The basic idea of GUTs is the following: just as the electromagnetic and weak forces represent different aspects of the same basic force that coalesce at high energies, the electroweak and strong nuclear forces are thought to be different manifestations of yet another even more fundamental force that emerges at still higher energies. The energy at which all three of these forces become fused into a single force is called the *grand unification energy*. Its value is not well known but would probably have to be at least 10^{15} billion electron-volts, far above the energies ever likely to be attainable with terrestrial laboratory particle accelerators.

Because of the impossibility of reaching the energies needed to test GUTs directly in the laboratory, scientists have begun searching for indirect, low-energy consequences of these theories. One search pertains to the possible decay of the proton. Prior to the development of the quark theory, the proton was believed to be absolutely stable because there exist no lighter baryons to which it can decay. Similar arguments apply to the electron: It is stable because there are no lighter leptons to which it can decay. However, unlike electrons, protons have substructure and are composed of quarks. At grand unification energies, the quarks that make up the proton are indistinguishable from leptons, so that the spontaneous decay of a proton (through quark conversion) to lighter particles such as positrons and mesons becomes possible. The chances of this happening, however, are very small. If you could watch a single proton, hoping to catch it in the act of decaying as described, you'd have to wait and watch for something like a million million million million million (or 10^{30}) years! Clearly, this is not practical. After all, the entire universe has been in existence for only about 14 billion years or so!

To increase your chances of witnessing the decay of the proton and, hence, of validating one version of GUTs, instead of watching only one proton, you could watch lots of them. You could monitor a large quantity of matter containing an enormous number of protons—maybe 10^{31} of them. If this were done for a year, you would expect, on average, to see as many as 10 proton decays. Increasing the number of protons in the sample obviously improves your chances of seeing a decay.

This is precisely the approach taken by researchers running the Super-Kamiokande experiment in Japan. Using an underground tank filled with 50 kilotons of water (each water molecule contributes 10 protons) and equipped with thousands of light-sensitive devices, experimenters look for the flashes of radiation emitted by the high-speed positrons and low mass mesons believed to accompany the decay of the protons. The results of this experiment (and its more recent elaborations) have been discouraging. In 2014, after 17 years of data collection dating back to 1996, investigators affiliated with the Super-Kamiokande experiment (**Figure 12.26**) reported no statistically significant signature of the decay of protons.

Figure 12.26 The Super-Kamiokande neutrino detector consists of a stainless-steel tank, 39.3 m in diameter and 41.4 m tall, filled with 50,000 tons of ultra pure water. About 13,000 photo-sensitive devices are installed on the tank wall. The detector is located 1,000 m underground in the Kamioka Mine in Japan. This image shows technicians checking some of the photon receptors prior to the tank being completely filled with water.

Based on these data, the lower limit to the lifetime of a proton, to 90% confidence, is 5.9×10^{33} years. This value is larger than that predicted by the simplest GUTs, but is close to the value of 10^{34} years given by so-called *supersymmetric models*.

One thing is clear. If the proton is found to be unstable, even on such long time frames, then eventually all the matter in the universe will evaporate. This is so because the positrons, as products of the proton decay, will annihilate with electrons to produce gamma rays. Assuming the universe survives for the requisite amount of time, it will ultimately be stripped of all matter and reduced to a cold, dark space filled with photons (which steadily lose energy) and a few isolated electrons and neutrinos (which escaped destruction)—a grim fate indeed, but one that may well await our now-glorious universe. More so than most, this example demonstrates how the physics of the subatomic domain can influence that of the largest scale known. But the reverse is also true.

We have seen that it is not feasible to expect that terrestrial laboratory experiments will ever achieve the energies required to verify directly the predictions of GUTs. Indeed, one might well ask, "Have such energies ever been available in the entire history of the universe?" And the answer is, "Yes, in the very first few moments after the birth of the universe in the Big Bang!"

Thus we begin to see that the ultimate tests of GUTs may rest in the field of cosmology, the study of the structure and evolution of the universe on the largest scale possible. The justification of theories involving the smallest of physical entities, quarks and electrons, may depend on our understanding of the largest of physical structures, galaxies and clusters of galaxies.

Learning Check

1. A quark can change color by
   (a) absorbing a W^+ boson.
   (b) emitting a photon.
   (c) absorbing a gluon.
   (d) emitting a neutrino.
   (e) None of the above.
2. The recent discovery of the _____, believed to play a critical role in the mechanism by which particles acquire mass, validated a key prediction of the Standard Model of particle physics.
3. Grand unified theories (GUTs) have as their goal the merging of which of the following fundamental forces of Nature:
   (a) the electric and magnetic forces.
   (b) the strong and weak nuclear forces.
   (c) the strong, weak, and electromagnetic forces.
   (d) the strong, weak, electromagnetic, and gravitational forces.
4. (True or False.) Recent experiments suggest that the lifetime of the proton is greater than 10^{33} years, a factor of more than 10^{20} longer than the current age of the universe.
5. The ultimate tests of grand unified theories may come in the field of _____, the study of the structure and evolution of the universe as a whole.

ANSWERS: 1. (c)   2. Higgs boson   3. (c)   4. True   5. cosmology

12.6 Cosmology

12.6a Geometry, Dark Matter and Dark Energy

Let us now return to the issues introduced at the end of Section 12.2, issues pertaining to the origin, structure, and evolution of the universe as a whole, issues that are at the heart of the subject of **cosmology**. Before the voyages of exploration made by Columbus and others more than 500 years ago, many otherwise educated and urbane people believed that the world was flat, like a table top. They shunned traveling too far from home for fear that it would lead them to fall off the edge of the world. The view of the global geometry of the world that they adhered to was strictly a bounded Euclidean one. You may recall from your high school math courses that the geometry of Euclid is based on five postulates, one of which states that through a point not on a line, one and only one line may be drawn that is parallel to the original line. In two-dimensions, Euclid's postulates lead to what is called "plane geometry"—the geometry of flat planes. Real space is not, of course, two-dimensional, but a three-space that satisfies Euclid's postulates is often referred to as a "flat" space nonetheless.

Although our appreciation for the vastness and complexity of the cosmos has increased over the last 500 years, it is remarkable that humankind is still struggling to understand the geometry of the universe and that the early savants may have been correct after all: On the largest of scales, the universe may well be Euclidean, although it is clearly without boundaries or edges.

According to general relativity (Section 12.2), the curvature of the universe is dependent on its total mass–energy density, symbolized here as Ω_T, including the energy contained in the field associated with the cosmic repulsion, now commonly called *dark energy*. If, in the dimensionless units employed in general relativity, Ω_T equals unity, space is flat on a cosmological scale. Any deviation from this *critical value* $\Omega_T = 1$ means that the geometry of the universe will not be that of a three-dimensional Euclidean space.

So what other possible spaces exist? Besides the geometry of Euclid, two other general types of geometry exist that could describe the global structure of the universe. They differ from Euclidean geometry only in terms of the parallel-line postulate. One is called *Riemannian geometry,* after mathematician Georg F. B. Riemann, and it adopts the postulate that through any point not on a line, *no* parallels may be drawn. This leads to a *positively curved* space with geometrical characteristics in three dimensions analogous to those in two dimensions on the surface of a sphere. The other type of geometry demands that through any point not on a line, an *infinite* number of parallel (that is, nonintersecting) lines may be constructed. The characteristics of this *negatively curved* geometry were developed by Nikolai Lobachevski and, in three dimensions, are similar to the properties of two-dimensional surfaces shaped like saddles or the bells of trombones. **Figure 12.27** displays two-dimensional analogs of each of the three possible universal geometries.

A variety of observations in the last few years suggests that the contribution to Ω_T from matter (particles), Ω_M, is about 30% of the critical value—that is, $\Omega_M \sim 0.3$. Of this, most (perhaps 99%) of the matter is "dark" (that is, nonluminous), and a significant fraction (more than about 85%) is likely to be nonbaryonic (compare with the discussion in Section 12.3). Particle physics offers three prime candidates for such material: (1) *axions*, theoretical particles whose masses range from ~ 1 eV/c^2 to as low as 10^{-6} eV/c^2; (2) *neutralinos*, hypothetical so-called *weakly interacting massive particles* (WIMPs), with masses between about 10 and 500 GeV/c^2; and (3) ordinary neutrinos of mass ≤ 1 eV/c^2. Current estimates place the neutrino contribution to the mass–energy density of the universe at between 0.1 and 5%, far too small to account for the bulk of the nonbaryonic contribution to Ω_T. Neutralinos are electrically neutral particles that interact with normal matter only through gravitation. Such

Figure 12.27 Two-dimensional visualizations of the possible global geometries of space. As a consequence of their differing parallel line postulates, triangles formed from geodesics ("straight lines") in these spaces exhibit different characteristics. For example, in a Euclidean geometry, the sum of the interior angles of a triangle equals 180°, while in a positively curved space this sum is greater than 180°; in a negatively curved space, the sum is less than 180°.

Positively curved space

Negatively curved space

Flat (Euclidean) space

particles, if they exist, would likely attach themselves to galaxies and form halos around them. All searches for WIMPs, including neutralinos, have yet failed to produce any conclusive results, although a small number of events possibly attributable to WIMPs with masses at the lower end of the theoretical limits have been reported. Similarly, searches for axions with the CERN Axion Solar Telescope through 2012 have also yielded no definitive detections. The nature of the dark matter thus remains elusive.

Matter thus provides only about a third of the critical mass–energy density required for a flat universe. What about the dark-energy contribution, Ω_Λ? Based on observations of the type Ia supernovae, (Section 12.2) the value for Ω_Λ is about 70%. Remarkably, the sum $(\Omega_M + \Omega_\Lambda)$ is very close to unity (**Figure 12.28**). Consequently, the universe is very close to being Euclidean in structure. If the present mass–energy density of the universe is even within a factor of 2 of the critical density after ~14 billion years of expansion, then, extrapolating back to the time of the Big Bang, the original mass–energy density must have agreed with the critical value to better than a few parts in 10^{61}! Thus, it would seem that the universe appears flat now because it was flat at the beginning.

The precise nature of the dark-energy field that appears to be responsible for the observed acceleration in the expansion of the universe is not yet known. Clues to the nature of the dark-energy source will likely be found in the fine details of the history of the cosmos, because different dark-energy models provide different expansion rates. The differences, however, are extremely small and will require measurements of supernovae brightnesses that are more accurate by up to a factor of 10, as well as the extension of the observations to larger redshifts. This is a big challenge but one that may well be met in the next decade with the advent of new studies involving the Dark Energy Camera (DECam) on the 4-meter Blanco telescope in Chile, as well as space missions with emphasis on dark energy investigations like the 6.5-meter James Webb Space Telescope, NASA's replacement for the HST now scheduled for launch in 2018, and the 2.4-meter Wide-Field Infrared Survey Telescope (WFIRST) with a planned launch date of 2024.

☐ Dark energy   ☐ Baryonic matter

☐ Nonbaryonic dark matter

Figure 12.28 Pie chart showing the composition of the universe. Most of the mass–energy density of the cosmos is tied up in dark energy, the source of the negative pressure responsible for the acceleration of the universe. The remainder is mostly composed of non-baryonic dark matter. Only about 4% of the mass-energy of the universe is made up of ordinary matter, and most of it also is "dark," consisting of gas distributed around and between galaxies. The luminous material in the universe (stars, nebulae, and galaxies) constitutes only about 1% of its total mass–energy density.

12.6b The Inflationary Universe

But why should the universe have started out with just exactly the right amount of matter and energy to render it flat instead of some other value that would have caused it to be curved in a different manner? How are we to interpret the fact that the universe seems to have begun with this carefully selected initial condition? The original Big Bang scenarios (see the Astronomical Application at the end of this section) had no good answers to these questions. And physicists are always suspicious of specially arranged circumstances that do not themselves emerge naturally from the physics of the situation but have to be imposed from outside the theory. This is the essence of what has been called the *flatness problem*.

The currently proposed solution to this problem avoids the necessity of prescribing special initial conditions by introducing some new physics. It provides the missing link that led to the deficiency in the explanatory power of the initial Big Bang models of the universe. Elementary particle physicists have come to the rescue of cosmologists and have developed a new approach that has solved the flatness problem (and several other difficulties inherent in the original Big Bang scenarios). The model for the universe that has emerged from the marriage of the ultrasmall and the ultralarge is known as the **inflationary universe**.

As discussed in Section 12.5, grand unified theories (GUTs) seek to amalgamate the fundamental interactions of Nature into a unified whole. Such a unification, in which the four forces become indistinguishable, can occur only at very high energies, about 10^{16} to 10^{19} GeV. Such energies are unattainable in terrestrial laboratories but were amply present at the creation of the universe in the Big Bang. The era in which these enormous energies prevailed, some 10^{-43} s after the Big Bang, is known as the *TOE* (or Theory of Everything) *epoch*. Initial attempts to develop a TOE focused on *superstring theory*, in which the fundamental particles correspond to different vibrational frequencies of 10-dimensional strings. More recent efforts have succeeded in merging the competing string theories into a grander, 11-dimensional theory called *M-Theory* (variously for membrane, master, magic, or mystery). According to this model, our universe is "glued" to a three-dimensional "brane" (short for membrane) embedded in a higher-dimensional hyperspace. Many other parallel universes may exist on other branes, but they remain invisible to us because photons, in this conception, cannot cross brane boundaries. Only gravitons (cf. Table 12.1) can do so, indicating that we may be able to "feel" the influence of these other universes (and the matter in them) through the gravitational force. This has been a suggested solution for the dark-matter problem.

Fantastic as these predictions are, M-Theory does have one distinguishing characteristic lacking in the earlier string theories: Some of its predictions appear to be testable. For example, if strong sources of gravitational radiation are detected that cannot be easily identified with other, more prosaic sources (like colliding black holes), then these sources could be candidates for concentrations of gravitons transmitted to our brane from normal matter on another adjacent brane. As further refinements in both theory and observation are made, we can expect to learn more about the degree to which M-Theory fulfills its promise to unify the realm of microphysics (quantum mechanics) with that of macrophysics (general relativity).

At the very beginning of time, then, we believe there was complete symmetry among the forces of Nature. As the universe expanded, the energy density declined, and a spontaneous break in the symmetry occurred, as one after the other of the forces separated out and assumed its own identity. These episodes of spontaneously broken symmetry corresponded to phase changes in the universe. Several things happened to the structure and evolution of the early universe, assuming that such symmetries existed but then were broken. One of them was that the universe suffered a period of rapid and enormous inflation in size that effectively rendered it flat, regardless of its initial curvature. Let's see how this developed.

As the universe expanded and cooled, the first phase transition that occurred was the one that split off the gravitational force. The next one, and the one that is important for our discussion, was that which pared the strong nuclear force from the remaining electroweak force. This phase change may be compared to the solidification of water (refer to Sections 4.1 and 5.6). If a vessel of water is slowly cooled and the container is not disturbed during the process, it is possible to *supercool* the water below 0°C without it freezing. The system gets stuck in a metastable liquid phase. Eventually, some perturbation will occur, and nucleation will begin (**Figure 12.29**). Very quickly, ice crystals will form and spread through the volume, converting the symmetric liquid phase (in which the molecules show no preferred orientation with respect to any particular direction) to a distinctly asymmetric solid phase (which exhibits characteristic anisotropies along different crystalline directions). The same kind of thing is believed to have happened in the universe when the strong nuclear force was "frozen out."

Like the water in the preceding example, cosmologists think the universe got stuck in a metastable state called a *false vacuum state* as it expanded and cooled. If this phase was prolonged, the energy density (there was no mass in the universe at this very early period) consisted principally of that associated with the vacuum state. Only a small part of the energy density was tied up in radiation. The nature of a vacuum energy density may be likened to that of the latent heat of fusion in a liquid (see Section 5.6). The latent heat of fusion is

Source: misterseng/youtube.com

Figure 12.29 The supercooled liquid state is not very stable. A small perturbation is all it takes to initiate a spontaneous phase transition from the liquid to the solid state. The figure shows a container of supercooled liquid water that has been tapped on the counter below it. This jolt has caused the water molecules to align in such a way that ice crystals have begun to form and to condense out of the liquid phase, giving the fluid its cloudy appearance. They rapidly fill the entire bottle in just a few seconds.

the amount of energy that must be present at the melting point to maintain the molecules of a substance in the liquid phase. In other words, it is the amount of internal energy required to prevent the molecules from binding together to form a solid. Similarly, the vacuum energy density was the energy needed to maintain the universe in a symmetric phase wherein the strong and electroweak forces were joined. It was this vacuum energy that ultimately drove the inflation.

When the universe eventually underwent the phase transition that severed the strong interaction from its electroweak sibling, enormous amounts of energy were released, exactly as happens when water freezes and energy is transferred to the surroundings. The energy thus provided produced a rapid and extraordinary expansion of the scale of the universe. In fact, the equations of general relativity show that the inflation occurred in an exponential fashion (refer to Section 3.6) causing the universe to grow by a factor of at least 10^{25} (and perhaps as much as 10^{50}) in the period when the universe was about 10^{-32} seconds old (Figure 12.32). The era of inflation was very short (only about 10^{-34} seconds or so long) and ended when the transition to the asymmetric state was fully achieved and the vacuum energy was completely depleted. After this, the universe resumed its normal expansion characteristics, but the effects of the inflationary period were profound.

In particular, if regions of spacetime in the very early universe were highly curved before the phase change, then the inflation reduced their curvature, rendering them flat. A vivid two-dimensional analogue that helps explain how this could happen involves the surface of a balloon that is being inflated (**Figure 12.30**). At the beginning of the process, the balloon's surface is quite small and highly curved. As air gradually enters the balloon, the surface expands and, as perceived by a local observer who samples only a small portion of the space, becomes less highly curved. Imagine now blowing into the balloon very hard so that it inflates rapidly to extremely large dimensions. After the period of inflation is over, the surface of the balloon is so distended that *all* local observers *anywhere* in the two-dimensional space of the surface of the fabric see a flat space. Regardless of what the curvature of the balloon's surface was at the start, once the inflation is complete, it looks flat. A similar circumstance is believed to have occurred in the early universe, and it is for this reason that in the present epoch we see the mass–energy density so close to the critical density for a flat, Euclidean world. A nice aspect about this explanation for the flatness problem is that it requires no special initial conditions for its success. *Whatever* the initial state of the universe, after the inflation, it always arrives at a final state that is flat in terms of its spacetime curvature. Put another way, Ω_T naturally approaches unity due to the inflation with no need for *ad hoc* assumptions about what conditions in the Big Bang were like prior to an age of 10^{-32} seconds.

The inflationary cosmology, developed in the 1980s by Alan Guth, Paul Steinhardt, and Andre Linde, has also provided solutions to several other

Figure 12.30 By inflating a balloon quickly to very large dimensions, the curvature of the surface can be reduced to a point where it is difficult, by local measurements, to distinguish it from a flat, Euclidean space. The same effect may have occurred in the early universe if the inflationary hypothesis is correct. If so, then the flatness problem that plagued early versions of the Big Bang can be solved.

nagging problems associated with the original Big Bang picture (such as the relative smoothness in the distribution of cosmic microwave background radiation; see Section 8.6). The success of this model derives partly from its reliance on physical theories designed to explain the fundamental structure of matter on the tiniest of dimensions. The physics of the smallest scale has thus led to the elucidation of the physics of the largest scale. Elementary particle physics and cosmology have been inextricably linked, and advances in one area have fostered advances in the other. This type of synergy is at the heart of modern scientific exploration, whether in physics, chemistry, biology, medicine, environmental science, or any other field.

In this section, we have focused on certain elements of relativity theory that have opened up new avenues for exploring and understanding the cosmos on both local (laboratory) and global (cosmological) scales. If you carefully review the discussion, particularly noting the many references to earlier sections and topics in the text, you can see just how much "other" physics is required to fully appreciate the perspectives on the universe that relativity offers us. It is not easy to acquire such an appreciation by considering the various physical concepts introduced in this book in isolation from one another. To succeed in this endeavor demands an integration and a synthesis of principles and concepts from the entire body of theory and experiment that constitutes what we call "physics." This, then, is the summary statement with which we wish to leave you. We hope we have piqued your curiosity about the natural world so that, confident in the knowledge you have gained from studying this material, you will continue your own personal inquiry into physics.

Learning Check

1. The composition of the universe is dominated by
   (a) ordinary (baryonic) matter.
   (b) exotic (nonbaryonic) matter.
   (c) dark energy.
   (d) the cosmic background radiation.
   (e) None of the above.
2. (True or False) Current observations suggest that the geometry of the universe follows that developed by Georg Riemann and exhibits positive curvature.

3. The _____ model for the universe suggests that within about 10^{-32} seconds of its creation the universe underwent a short-lived, but enormous expansion, increasing its size by a factor of 10^{25} or more.

4. _____ is the study of the creation, structure and evolution of the entire universe on the largest distance scales.

ANSWERS: 1. (c)   2. False   3. inflationary   4. Cosmology

ASTRONOMICAL APPLICATION   Big Bang Nucleosynthesis

In Section 11.7, we discussed how the elements of the periodic table are built up in the cores of stars from hydrogen by fusion reactions, a process called *nucleosynthesis*. But where, one might ask, did the hydrogen come from that serves as the fuel for the nuclear fires in stars? The same question may be asked about the origins of other light elements such as deuterium (hydrogen-2), helium-3, and helium-4, commonly involved in fusion reactions studied terrestrially. This is an important issue because deuterium and helium-3 are consumed, not produced, in stars during the usual sequence of nuclear reactions that occur. Even helium-4, the end result of hydrogen fusion, is largely used up in the creation of heavier elements by additional

nuclear reactions. And yet we *do* find substantial amounts of these light nuclear species in the cosmos. Where did they come from if not from stars?

Because nuclear processes usually demand high temperatures and high densities for their operation, we seek another place, besides stars, where such conditions prevail *or once prevailed*. A natural source of such extreme temperatures and densities is the Big Bang (see Section 8.6). Indeed, the answer to our question now seems to be that these isotopes were manufactured during the first 3 minutes after the creation of the universe. The source of these nuclei was the enormous energy present in the Big Bang. As we noted earlier in

Section 12.1 energy and mass are alike, and one may be converted into the other. Fission and fusion reactions convert mass into energy, but during the first few seconds after the birth of the universe, energy (in the form of very short wavelength gamma rays) was converted into matter. This matter was generally in the form of particles with equal masses but opposite electric charges, so-called *particle-antiparticle pairs* (**Figure 12.31a**; for more on antiparticles, see Section 12.3). Thus, for example, when the universe was only about a thousandth of a second old, with a temperature greater than 10^{12} K, it consisted of a hurly-burly of high-energy photons and fast-moving particles and their anti's, specifically protons and antiprotons, neutrons and antineutrons, and electrons and antielectrons (positrons).

As the universe expanded and cooled, a point was reached when the temperature fell below 10^{12} K and the photon energies were no longer sufficient to permit the creation of proton–antiproton and neutron–antineutron partners. The existing, slower-moving pairs then began to collide with and annihilate one another, transforming matter back into energy and producing more lower-frequency photons (**Figure 12.31b**). If the number of particles had exactly matched the number of antiparticles, all the mass originally created would have been destroyed, leaving the universe devoid of matter. As it happened, there existed a slight excess of particles over antiparticles. The difference was small but significant: about $10^9 + 1$ protons for every 10^9 antiprotons, for example. This asymmetry left a residue of protons and neutrons after the era of annihilation with about one neutron for every seven protons.

When the universe reached the age of a few seconds and its temperature fell below 10^{10} K, the average photon energy was no longer sufficient to create even the lighter electrons and positrons, and these particles began to annihilate one another like the protons and neutrons and their anti's before them. Again, because of the slight asymmetry of matter over antimatter, the net outcome of this era of annihilation was a residue of electrons. Some 10^5–10^6 years later when the temperature of the universe was between 3,000 and 4,000 K, these electrons combined with the protons to produce neutral hydrogen atoms.

However, when the universe was only a few seconds old, the positrons were effectively gone as were many of the original electrons. As a result, beta decay of the remaining free neutrons into protons and electrons began to dominate over the compensating production of neutrons from the merging of free protons and electrons (cf. beta decay in Section 11.2). During the same period, many of the free neutrons combined with protons to produce deuterium according to the following reaction:

$$n + p \rightarrow {}^2H + \gamma$$

The newly formed deuterium persisted in reasonable numbers only after the temperature in the early universe dropped to less than 10^9 K; that is, after the universe was about 100 seconds old.

Before this time, deuterium quickly underwent a series of nuclear reactions that converted much, but not all, of it into helium-4. A small fraction was transformed into helium-3, and some of it remained unreacted. The fusion reactions opening Section 11.7 show the types of interactions that consumed deuterium to yield helium-3 and helium-4. Thus, after about 200 seconds, all the free neutrons had disappeared, and what remained were protons, electrons, deuterium nuclei, and helium nuclei, the raw material from which future stars and galaxies would be constructed. **Figure 12.32** summarizes the various stages of element production in the early universe.

The currently observed abundances of deuterium, helium-3, and helium-4 are all relatively small compared to the abundance of hydrogen. For example, there is only about 1 helium nucleus for every 10 hydrogen nuclei in the cosmos and a paltry 1 deuterium nucleus for every 50,000 hydrogen nuclei. The helium-3 to hydrogen number ratio is even smaller: about 1 per 500,000. The sequence of events that has been described here represents our best understanding of how the light elements were introduced into the universe. To check whether this scenario makes sense, comparisons between the observed abundances and those calculated from models based on this picture have been made. Although the computed results for the rate of production of the light isotopes are quite sensitive

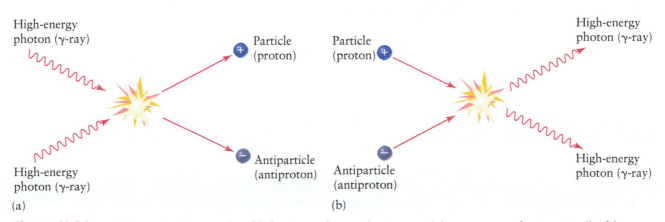

(a)   (b)

Figure 12.31 (a) Pair production reaction. High-energy photons (gamma rays) interact to produce matter (in this case a proton) and antimatter (an antiproton). (b) Pair annihilation. A particle and an antiparticle collide and annihilate one another to produce photons (gamma rays). For each reaction to occur, the total energy of the gamma rays must at least be equal to the mass–energy of the proton and antiproton. Any additional energy manifests itself as kinetic energy of the proton–antiproton pair in (a). Similarly in (b), the annihilation of matter in motion yields photons whose energies sum to more than the rest mass energies of the interacting particle–antiparticle pair.

Figure 12.32 The evolution of the universe from the Big Bang until the present, with particular emphasis on the production of the lightest chemical elements.

to the total density of nucleons at the time of their formation, for reasonable ranges of this parameter, good agreement between the model predictions and the observations can be achieved. Such results constitute the strongest evidence for the correctness of the Big Bang origin of the universe beyond that provided by the existence of the cosmic microwave background radiation (cf. the discussion in Section 8.6).

■ **QUESTIONS**

1. Describe the process of pair production. What is the nature of the basic transformation taking place in pair production reactions?
2. Suppose that in 1 cubic meter of interstellar space there are 50 primordial hydrogen atoms. How many He-4 atoms would be expected to occupy this volume according to Big Bang nucleosynthesis calculations? How many He-3 atoms?

Profiles in Physics Murray Gell-Mann, Quark-atech

Physics, like almost any other discipline, is full of colorful individuals who, by their powerful personalities and intellects, have exerted far-reaching influence on the progress of the science. In elementary particle physics, few people have done more to shape the development of the field or the language in which it is expressed than Murray Gell-Mann (**Figure 12.33**).

Murray Gell-Mann was born in 1929 in New York City and entered Yale University at the age of 15. After graduating in physics,

he attended the Massachusetts Institute of Technology (MIT), where he received his doctorate in theoretical physics when he was only 21. After several short postdoctoral and faculty appointments, Gell-Mann moved to Caltech in 1957, where he remained until his retirement in 1993. He is currently a Distinguished Fellow at the Santa Fe Institute, a not-for-profit research facility that he cofounded in 1984 and that is devoted to the study of complexity theory. Gell-Mann received the Albert Einstein Medal in 2005, and was awarded the

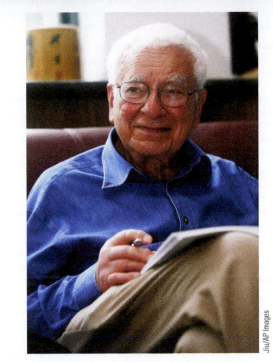

Figure 12.33 Theoretical physicist Murray Gell-Mann, one of the architects of the quark model.

Helmholtz Medal from the Berlin Academy of Sciences and Humanities in 2014.

A man of great intellectual prowess and wide-ranging interests, Gell-Mann is well acquainted with Eastern and Western religions and literature, knows several languages, including Swahili, and, of course, is highly skilled in applying abstract mathematical theory to problems in physics. Logical progression and mathematical rigor are the hallmarks of his work.

A brief summary of the major contributions to theoretical particle physics made by Murray Gell-Mann is appropriate here: 1953—introduced the idea of "strangeness" to explain the unusual properties of kaons and the lambda; 1961—developed the eightfold way to organize the proliferating "zoo" of elementary particles into a few families having simple, regular symmetries; 1964—introduced quarks as a more fundamental way of explaining the symmetries exhibited in the eightfold way; 1971—introduced "colored" quarks to differentiate three possible varieties of each basic quark type; 1973—recognized "color" as the source of the strong force and coined the name *quantum chromodynamics* to describe the theory of the strong interaction. For his accomplishments up to that time, Gell-Mann received the 1969 Nobel Prize in physics, and he justly deserves to be called "one of the heroes of our story," as John Polkinghorne refers to him in his chronicle of the evolution of particle physics from the Second World War to 1980 entitled *Rochester Roundabout*.

As an illustration of the deeply perceptive yet whimsical way in which Gell-Mann approaches physics, the tale of the quarks, as told by Michael Riordan in his book *The Hunting of the Quark*, is perhaps the best. In March 1963, Gell-Mann visited Columbia University to give a series of three talks. At a luncheon during his stay in New York, Gell-Mann was asked why a fundamental group of three particles—a triplet—did not appear in Nature, whereas larger groupings, like those containing eight particles (an octet), did. Gell-Mann replied "That would be a funny quirk," and he proceeded to show why: such a triplet, if it existed, would have to be composed of particles with fractional charges equal to $\frac{2}{3}$, $-\frac{1}{3}$, and $-\frac{1}{3}$ the magnitude of the electron charge, respectively. No such fractionally charged entities had ever been observed.

Later that day and into the following morning, Gell-Mann reflected on this situation and recognized that such fractionally charged species were not completely out of the question *if* they remained permanently trapped inside hadrons and never appeared as free particles in Nature. In his second lecture at Columbia, Gell-Mann referred to the members of this triplet as "quarks," a nonsense word he had used before and by which he meant "those funny little things." Although relatively little time was devoted to "quarks" in the talk, they were very much the subject of the postlecture coffee hour.

In the fall of that year, Gell-Mann worked out the details of how to construct baryons and mesons out of this fractionally charged triplet of particles and wrote a short, two-page article that was published in a new European journal, *Physics Letters*. By this time, Gell-Mann had discovered a passage from James Joyce's novel *Finnegans Wake*, which lent legitimacy to his use of the term *quarks* to designate the fictional triplet:

> Three quarks for Muster Mark!
> Sure he hasn't got much of a bark,
> And sure any he has it's all beside the mark.
> But O, Wreneagle Almighty, wouldn't un be a sky of a lark
> To see that old buzzard whooping about for uns shirt in the dark
> And he hunting round for uns speckled trousers around by Palmerston Park?

The poem relates the dream of one of Joyce's protagonists as he lies passed out on the floor of a Dublin tavern. The last three lines had, no doubt, a particularly appealing irony in them for Gell-Mann in view of the parallel circumstances of the hapless Muster Mark vainly searching for his clothes in the dark and of the frantic experimentalists equally vainly searching for free quarks. Indeed, Gell-Mann was one of the last to believe that quarks were anything but mathematical entities. At a conference in Berkeley in 1966, he spoke of "those hypothetical and probably fictitious spin $\frac{1}{2}$ quarks."

Murray Gell-Mann's achievements have been compared to those of Dmitri Mendeleev (refer to Section 4.1). Both men brought order out of confusion, Mendeleev by organizing the elements into the periodic table based on their chemical properties and Gell-Mann by establishing a kind of periodic table of mesons and baryons based on their group symmetry properties. In each case, gaps in the patterns led to predictions of missing members that were subsequently discovered experimentally. The direction and impetus provided to chemists and atomic physicists by Mendeleev's work is strongly paralleled by that given to elementary particle physicists by Gell-Mann's.

■ QUESTION

1. List three major contributions made by Murray Gell-Mann to the theory of elementary particles.

SUMMARY

» The **special theory of relativity** extends classical mechanics to systems moving with speeds near that of light. It postulates that the speed of light is a constant for all observers and that the laws of physics are the same for all observers moving uniformly.

» **Time dilation**, in which moving clocks run slow, and **length contraction**, in which moving rulers appear shortened in the direction of motion, are experimentally confirmed predictions of this theory.

» Special relativity also establishes the equivalence of mass and energy, according to which the **rest energy** of a stationary object of mass m is given by $E_0 = mc^2$.

» The **general theory of relativity**, developed by Einstein in 1915, incorporates the principles of special relativity and accounts for the behavior of particles and photons in strong gravitational fields. Founded on the **principle of equivalence**, which asserts that in a uniform gravitational field all objects accelerate at the same rate, general relativity holds that particles move along the natural contours of a spacetime that is shaped by the presence of mass-energy. The warping of spacetime and the subsequent deviations of particle trajectories from those of Newtonian physics that are predicted by general relativity have been confirmed by astronomers in observations of the planet Mercury (perihelion precession) and of remote galaxies (gravitational lensing). Other predictions involving the existence of gravitational time dilation and gravitational waves have also been validated in the laboratory and in studies of binary stars.

» There are **four fundamental forces** or interactions in Nature: **gravity**, **electromagnetism**, the **weak nuclear force**, and the **strong nuclear force**. Only the last three of these are important in elementary particle physics. These forces are mediated through the exchange of **carrier particles** such as photons and gravitons.

» All subatomic particles may be classified as either **bosons** or **fermions**, depending on whether their spins are integral or half-integral multiples of the basic unit of intrinsic angular momentum, $h/2\pi$. Particles may also be segregated by mass and by the types of fundamental interactions in which they participate. For every particle, there exists an **antiparticle**, having the same mass, but opposite electric charge. For example, the **positron** is an anti-electron.

» **Quarks** are fractionally charged particles that come in six **flavors:** up, down, strange, charmed, bottom, and top. **Baryons** are composed of three quarks and **mesons** are made up of a quark-antiquark pair. The six quarks and their anti's, together with the six **leptons** and their anti's, comprise all the matter and antimatter in the universe according to the **Standard Model** of particle physics. A major confirmation of the Standard Model and the mechanism by which particles acquire mass has been provided by the discovery of the **Higgs boson** in 2012.

» An important goal in modern physics is the unification of the fundamental forces of Nature. In the 1960s, the electromagnetic force was successfully joined with the weak force to produce the **electroweak interaction**. Scientists are now seeking ways to unify the strong force with the electroweak force in what are called **grand unified theories (GUTs)**. Although far from complete, current versions of GUTs make predictions that may ultimately be tested only by cosmological observations of the universe that connect to its origin in the Big Bang.

» **Cosmology** is the study of the origin, structure, and evolution of the universe as a whole. General relativity posits that the geometry of the universe, specifically its curvature, is determined by the total mass-energy content. Surveys by astronomers have revealed that the composition of the universe is dominated by **dark energy** ($\sim 73\%$) and by nonbaryonic **dark matter** ($\sim 23\%$), the precise natures of which have yet to be established. Current measurements of the mass-energy density of the universe support the view that the geometry of spacetime conforms to that developed by Euclid centuries ago, viz. that spacetime is flat. The fact that the universe has zero curvature on the largest length scales may be explained by the **inflationary model** in which the universe underwent a short, but intense period of expansion shortly after the Big Bang that effectively removed any initial curvature that may have been present, much as the curvature of a balloon is reduced by rapidly inflating it to a large radius.

IMPORTANT EQUATIONS

Equation	Comments
$\Delta t' = \dfrac{\Delta t}{\sqrt{1 - v^2/c^2}}$	Time dilation
$E_0 = mc^2$	Definition of rest energy; equivalence of mass and energy
$E_{\text{rel}} = \dfrac{mc^2}{\sqrt{1 - v^2/c^2}}$	Relativistic total energy
$KE_{\text{rel}} = \dfrac{mc^2}{\sqrt{1 - v^2/c^2}} - mc^2$	Relativistic kinetic energy
$p + \bar{p} \leftrightarrow \gamma + \gamma$	Matter-antimatter annihilation
$z = \dfrac{\lambda_{\text{obs}} - \lambda_{\text{emit}}}{\lambda_{\text{emit}}} = \dfrac{v}{c}$	Definition of redshift parameter
$v = H_0 d$	Hubble relation (see Section 6.2)
$\Delta t_{\text{f}} = \dfrac{\Delta t_{\text{d}}}{[1 - (2GM/dc^2)]^{1/2}}$	Gravitational time dilation

1. The development of the *Standard Model* of elementary particles is one of the most significant achievements in physics in the 20th century. While refinements of this description of the subatomic world continue to be made, an understanding of the basic components and structure of this model serves as a foundation for appreciating the new discoveries in particle physics that frequently occur.

   One way to organize the elements of the standard model is in a display like that shown in Figure 12.23. Another way is to use a concept map. Examine Figure 12.23 closely and, if necessary, reread the material leading up to it. Then, using the principles of concept mapping that we have employed so often in past chapters, try to construct a map that captures the essential features and relationships embodied in the standard model of particle physics. When you have completed your map, exchange it for one that a classmate has prepared. Critically examine your friend's map. Does it include the most important concepts contained in the Standard Model? Does it properly exhibit/explain the relationships between the various elements of this model in a manner consistent with Figure 12.23 and the material introduced in the chapter? What are the strengths and weaknesses of your colleague's map? Discuss these issues with the person with whom you exchanged maps and together try to come to create a better CM display and, in the process, a stronger understanding of the Standard Model.

2. A review of Chapter 12 provides the following list of key words or items relating to the field of cosmology:

redshift	cosmological constant
dark energy	dark matter
WIMPs	inflationary model
Big Bang	flatness problem
false vacuum state	expanding universe
Hubble law	critical density
Type Ia	supernova

   In the spirit of previous *Mapping It Out!* exercises, organize these concepts into a meaningful hierarchy according to their relative importance, significance or scope, linking them as appropriate by words or short phrases that reflect the relationship between/among the various items, to form a concept map for this portion of Chapter 12. Does the structure of your map manifest a particular "flow," for example, large → small; early → late; simple → complex; etc. If not, do you see any opportunities to re-organize your map to produce another model that might have such thematic trends, thereby making it easier to understand or master the information?

(■ **Indicates a review question, which means it requires only a basic understanding of the material to answer. Questions without this designation typically require integrating or extending the concepts presented thus far.)**

1. ■ What does the acronym PET stand for? Why is PET a good example with which to begin a discussion of elementary particle physics?

2. ■ Describe the two fundamental postulates underlying Einstein's special theory of relativity.

3. Suppose you were traveling toward the Sun at a constant velocity of $0.25c$. With what speed does the light streaming out from the Sun go past you? Explain your reasoning.

4. Light travels in water at a speed of 2.25×10^8 m/s. Is it possible for particles to travel through water at a speed $v > 2.25 \times 10^8$ m/s? Why or why not? Explain.

5. ■ In your own words, define what is meant by *time dilation* in special relativity theory. Provide a similar definition for *length contraction*. Give an example in which the effects of time dilation are actually observed.

6. Galileo used his pulse like a clock to measure time intervals by counting the number of heartbeats. If Galileo were traveling in a spaceship, moving uniformly at a speed near that of light, would he notice any change in his heart rate, assuming the circumstances of his travel produced no significant physiological stress on him? If someone on Earth were observing Galileo with a powerful telescope, would he or she detect any change in Galileo's heart rate relative to the resting rate on Earth? Explain your answers.

7. Newton wrote: "Absolute, true, and mathematical time, of itself, and from its own Nature, flows equally without relation to anything external." Comment on the significance of this statement for two timekeepers in relative motion. In the light of special relativity, is Newton's statement valid? Explain.

8. Why don't we generally notice the effects of special relativity in our daily lives? Be specific.

9. Does $E_0 = mc^2$ apply only to objects traveling at the speed of light? Why or why not?

10. If a horseshoe is heated in a blacksmith's furnace until it glows red hot, does the mass of the horseshoe change? If a spring is stretched to twice its equilibrium length, has its mass been altered in the process? If so, explain how and why in each case.

11. ■ Given the Newtonian view of gravity, why is it reasonable to expect that the rate at which the universe is expanding should be decreasing with time? Do observations of the motions of remote galaxies support this position? If not, what do such observations suggest about how the rate of expansion is changing?

12. ■ The principle of equivalence underlies the general theory of relativity. What does this principle assert about the motion of objects in a uniform gravitational field?

13. After landing on the planet Mars, two astronauts awaken from a long induced hibernation inside their windowless spacecraft. Before emerging, is there any way they can determine whether their individual body weights are the result of gravitation or accelerated motion?

14. ■ In what way(s) is Einstein's general theory of relativity superior to Newton's theory of universal gravitation? Give an example of a case where Einstein's theory provides a more accurate description of physical phenomena than does Newton's.

15. In Chapter 9, we discussed the deviations in the path of a beam of light after passing through transparent media (refraction) and the role such deviations play in common natural phenomena like rainbows and halos. Why aren't we as familiar with the deviations in the path of light caused by gravity?

16. Figure 12.34 shows the trajectory of a comet passing near the Sun. Describe how Newton would explain the deviation in

the comet's path from a straight line. Repeat the explanation as it might be given by Einstein.

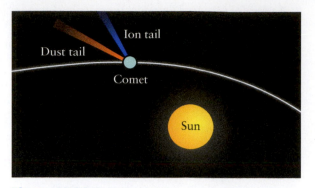

Figure 12.34 Question 16.

17. ■ List three astronomical examples in which the validity of the predictions of general relativity has been demonstrated.
18. ■ What are gravitational waves? How are they produced? What evidence is there to substantiate the existence of such radiation?
19. Do pulsating variable stars, that is, stars that rhythmically expand and contract as a result of thermal instabilities in their atmospheres, generate gravitational waves in this process? Why or why not? Explain.
20. ■ Describe the phenomenon of gravitational time dilation. What experimental or observational evidence exists to support the reality of this prediction of the general theory of relativity?
21. In Jules Verne's classical science fiction tale *Journey to the Center of the Earth*, a group of scientists and adventurers descend deep into the interior of Earth. Among the equipment they carry with them is a rugged and carefully calibrated clock. Imagine a 21st-century update of this story in which colleagues on the surface of Earth possess an identical clock and remain in contact with the explorers throughout their descent via ground penetrating EM waves. As the journey unfolds, would the underground travelers notice any change in the rate at which their clock ticked from what they experienced at the start of their trip? What about those monitoring the expedition from Earth's surface? Would they note any deviation in the rate of ticking of the subterranean clock relative to their own? If so, would the remote clock be seen to be running too slow? Too fast? Explain.
22. A friend alleges that Buddhist monks residing in monasteries high in the Tibetan Himalayans age more slowly than lobstermen fishing off the coast of Maine. Do you accept her statement as true? Why or why not?
23. ■ List the four fundamental interactions of Nature, and discuss their relative strengths and effective ranges.
24. ■ What common feature of the electromagnetic and gravitational interactions requires that their carrier (or exchange) particles be massless?
25. ■ What is an antiparticle? What may happen when a particle and its anti collide?
26. Some neutral particles, such as the π^0, are their own antiparticles, but not the neutron. In what ways are n and $\bar{n}$ the same? Speculate on how they might be different.
27. According to Table 12.4, the rest mass of an electron is 0.511 MeV/c^2. What is the rest mass of a positron?
28. ■ Distinguish between fermions and bosons in as many different ways as you can.

29. ■ Give some ways by which physicists classify elementary particles.
30. ■ In which of the four basic interactions does an electron participate? A neutrino? A proton? A photon?
31. ■ What is a quark? How many different types of quarks are now known? What are some of the basic properties that distinguish these quarks?
32. ■ Describe the kinds of evidence that have led scientists to conclude that quarks exist.
33. ■ How many quarks form a baryon? A meson? What is the relationship (if any) between a quark and a lepton (e.g., an electron)?
34. In the quark model, is it possible to have a baryon with strangeness -1 and electric charge $+2$? Explain.
35. What kind of a particle (baryon, meson, or lepton) corresponds to a $t\bar{t}$—that is, to a top-antitop quark combination? Describe some of the properties such a particle would have.
36. ■ Quarks are said to possess "color." What does this mean? Are physicists really suggesting that quarks look red like ripe strawberries or blue like the cloudless daytime sky? Explain.
37. ■ Describe the *Standard Model* of elementary particle physics.
38. A bumper "snicker" on a car belonging to the chairperson of a physics department reads: "Particle physicists have GUTs!" Explain in your own words the meaning of this little joke or "play on words."
39. ■ Unification of its basic laws and theories has long been a goal in physics. Describe some ways in which physicists have been successful in unifying certain forces and theories. In what area(s) of physics is (are) the process(es) of unification still ongoing?
40. If a proton can decay, then its lifetime is of the order of 10^{34} years, far longer than the current age of the universe. Does this necessarily imply that a proton decay has not yet occurred in the entire history of the universe? Explain.
41. ■ What is dark matter and how much of the total mass-energy budget of the universe consists of dark matter? Give two dark matter candidates that have been proposed by particle physicists.
42. ■ What is dark energy and what role does it play in our understanding of the structure and evolution of the universe?
43. ■ Describe the role of inflation in cosmology. How does it help to explain why the geometry of the universe is flat? What is the source of the energy that drove the rapid expansion of the universe during the brief inflationary era in its history?
44. **Figure 12.35** shows the appearance of three spherical space pods as seen by an Earth-bound observer. The pods are traveling along the same direction, but have different speeds relative to the observer. Rank the speeds of the pods from highest to lowest and explain the rationale for your choices.

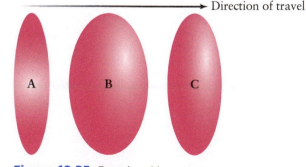

Figure 12.35 Question 44.

PROBLEMS

[For many of the exercises, referring to Table 12.3 and Figure 12.23 will be very helpful.]

1. The lifetime of a certain type of elementary particle is 2.6×10^{-8} s. If this particle were traveling at 95 percent the speed of light relative to a laboratory observer, what would this observer measure the particle's lifetime to be?

2. How fast would a muon have to be traveling relative to an observer for its lifetime as measured by this observer to be 10 times longer than its lifetime when at rest relative to the observer?

3. The lifetime of a free neutron is 886 s. If a neutron moves with a speed of 2.9×10^8 m/s relative to an observer in the lab, what does the observer measure the neutron's lifetime to be?

4. A computer in a laboratory requires 2.50 μs to make a certain calculation, as measured by a scientist in the lab. To someone moving past the lab at a relative speed of $0.995c$, how long will the same calculation take?

5. The formula for length contraction gives the length of an interval on a ruler moving with velocity v relative to an observer as $\sqrt{1 - v^2/c^2}$ times the length of the same interval on a ruler at rest with respect to the observer. By what fraction is the length of a meter stick reduced if its velocity relative to you is measured to be 95 percent the speed of light?

6. If an electron is speeding down the two-mile-long Stanford Linear Accelerator at 99.98 percent the speed of light, how many meters long is the trip as seen from the perspective of the electron? (See Problem 5.)

7. Calculate the rest energy of a proton in joules and MeVs. What is the mass of a proton in MeV/c^2?

8. The tau is the heaviest of all the known leptons, having a mass of 1,777 MeV/c^2. Find the rest energy of a tau in MeVs and joules. What is the mass of the tau in kilograms? Compare your result with the mass in kilograms of an electron.

9. If a 1.0-kg mass is completely converted into energy, how much energy, in joules, would be released? Compare this value to the amount of energy released when 1.0 kg of liquid water at 0°C freezes.

10. In a particular beam of protons, each particle moves with an average speed of $0.8c$. Determine the total relativistic energy of each proton in joules and MeVs.

11. A particle of rest energy 140 MeV moves at a sufficiently high speed that its total relativistic energy is 280 MeV. How fast is it traveling?

12. If the relativistic kinetic energy of a particle is 9 times its rest energy, at what fraction of the speed of light must the particle be traveling?

13. If a proton and an antiproton, both at rest, were to completely annihilate each other, how much energy would be liberated?

14. Determine the redshift of a photon whose observed wavelength is 656.9 nm if its emitted wavelength is 656.3 nm.

15. Spectral measurements of a distant galaxy show it to have a redshift $z = 0.2$. If the wavelength of an ionized calcium line emitted by the galaxy is 393.3 nm, what is the wavelength of this line as detected on Earth?

16. How many years would you have to wait to observe a 1° angular shift in the perihelion of Mercury due to general relativistic effects?

17. If a galaxy is found to have a recessional speed of 140,000 km/s, about how far away is it? The current estimate

of the precision of the Hubble parameter is approximately ±5%. Determine the range within which the true distance to the galaxy lies.

18. Sirius B has a mass and radius of 2.1×10^{30} kg and 5.5×10^6 m, respectively. Assuming Sirius B is perfectly spherical in shape, compute the ratio of the $\Delta t_f / \Delta t_d$. Using this result, confirm that the redshift quoted in Section 12.2 for this star if $z = \Delta t_f / \Delta t_d - 1$.

19. A compact neutron star has a mass of 2.8×10^{30} kg (about 1.4 times the mass of the Sun) but a radius of only 10^4 m (approximately 6.2 mi!). If a clock on the surface of this exotic star marks the passage of 1 h of time, how much time is observed to pass on an identical clock located a very large distance from the neutron star?

20. A 0.1-kg ball connected to a fixed point by a taut string whirls around a circular path of radius 0.5 m at a speed of 3 m/s. Find the angular momentum of the orbiting ball. Compare this with the intrinsic angular momentum of an electron. How many times larger is the former than the latter?

21. What is the quark combination corresponding to an antineutron? An antiproton?

22. Give three (3) combinations of quarks (not antiquarks!) that will give baryons with charge: (a) +1; (b) −1; and (c) 0.

23. Give all the quark-antiquark pairs that result in mesons that have no charge, no strangeness, no charm, and no bottomness. How do you think such particles might be distinguished from one another?

24. Distinguish a particle composed of a combination of $\overline{d}\,\overline{u}\,s$ quarks from one composed of dus quarks in as many ways as you can. Do the same for a dss combination and a $\overline{d}\,\overline{s}\,\overline{s}$ combination.

25. The Δ^- (delta minus) is a short-lived baryon with charge −1, strangeness 0, and spin $\frac{3}{2}$. To what quark combination does this subatomic particle correspond?

26. The F$^-$ meson possesses −1 unit of charge, −1 unit of strangeness, and −1 unit of charm. Identify the quark combination making up this rare subatomic particle.

27. The D$^-$ meson is a charmed particle with charge −1, charm −1, and strangeness 0. Work out the quark combination for this species of elementary particle.

28. What is the quark composition of the Δ^{++} baryon? It has no strangeness, no charm, and no topness or bottomness. Its spin is $\frac{3}{2}$.

29. Analyze the following reactions in terms of their constituent quarks:
    (a) $\pi^+ + p \to \Sigma^+ + K^+$ (c) $p + p \to p + p + p + \overline{p}$
    (b) $\gamma + n \to \pi^- + p$ (d) $K^- + p \to K^+ + K^0 + \Omega^-$

30. Analyze the following decays in terms of the quark contents of the particles:
    (c) $\Sigma^- \to n + \pi^-$ (c) $K^{*+} \to K^0 + \pi^+$
    (d) $\Lambda^0 \to p + \pi^-$ (d) $\Omega^- \to \Lambda^0 + K^-$

31. The properties of the top quark have been confirmed with the 2007 discovery of mesons composed of a t quark and a $\overline{b}$ quark. What charge does such a particle carry? What is its spin? How many units of topness and bottomness does this particle have?

32. If the average lifetime of a proton was 10^{33} years, about how many protons would you have to assemble together and observe simultaneously to witness a total of 100 proton decays in one year? Explain the reasoning that led to your conclusion.

1. Show that in the limit of low speeds the expression for the relativistic kinetic energy (see Important Equations) reduces to the familiar one from classical mechanics, $KE = (1/2)mv^2$. [*Hint:* For speeds v much smaller than c, the quantity $(1 - v^2/c^2)^{-1/2}$ is approximately equal to $1 + v^2/2c^2$. You should check this for yourself using typical speeds given in the exercises in Chapter 3.]

2. Compare the strength of the electric force to that of the gravitational force as suggested in Section 12.3. In particular, find the ratio of the Coulomb force of repulsion between two protons separated by 10^{-15} m to their gravitational force of attraction for the same distance. The mass and charge of the proton are given in Table 11.1.

3. When a proton–antiproton pair at rest annihilates, two photons are created. Find the wavelengths of these two light quanta, given that they share equally in the annihilation energy. In what portion of the electromagnetic spectrum do such wavelengths appear?

4. If the average lifetime of the proton were 10^{16} years, estimate the number of protons per kilogram of human body mass that would decay radioactively each year. Assume that a human being is made entirely of water molecules and that each molecule contributes 10 protons. (*Hint:* The molar weight of water is 18 g, and each mole of any substance contains 6.02×10^{23} particles.)

   If each proton decay released 1 MeV of energy, how much energy per kilogram would be produced *in total* by the decaying protons in your body in 1 year? Give the result in joules per kilogram. If the energy released is mostly in the form of kinetic energy of the positrons produced in the decay, then 1 J/kg of this type of ionizing radiation is rated at about 1 Sv (the *sievert* [Sv] is the SI unit of radiation dose). Compare the total exposure in sieverts of a person to the radiation produced by his/her own body to the maximum safe dose of 1 mSv/yr for the general public from all artificial sources.

5. Although doubly charged baryons have been found (baryons with net charge +2), no doubly charged mesons have yet been identified. What effect on the quark model would there be if a meson of charge +2 were discovered? How could such a meson be interpreted within the quark model?

6. In the electroweak theory, symmetry breaking occurs on a length scale of about 10^{-17} m. Compute the *frequency* of particles with de Broglie wavelengths (see Section 10.5) equal to this value, and, using this result, determine their energies by the Planck formula $E = hf$. Show that the mass of particles having such energies is on the order of the mass of the $W^{\pm}$ particles (Figure 12.23).

7. One interesting (unbelievable?) implication of time dilation is contained in what is called the *twin paradox*. Imagine identical twins who decide to perform an experiment to test the accuracy of Einstein's predictions about time dilation. After synchronizing their (identical) clocks, one of the twins enters a spaceship and travels to a distant star at a speed 95 percent that of light. Upon arrival, the traveling twin immediately turns around and heads home at the same speed. According to the twin on Earth, the clock aboard the spaceship runs slower than the one on Earth, so that the traveling twin ages less than the Earth-bound sibling. But according to the space-faring twin, it is the clock on Earth that is running too slowly, making the stay-at-home twin the younger of the two. This is the paradox: each twin argues that the other will be younger at the end of the trip. How can this paradox be resolved?

   Which of the two twins is correct in their analysis, and why? If the trip requires a total time of 5 years to make as measured by the clock on Earth, what will be the time recorded for the trip by the clock on board the spaceship?

8. Suppose you perform a scattering experiment in which you fire BBs into a slab of jello in which a layer of randomly distributed but identical marbles are suspended. Of the BBs entering the jello, 90 percent pass cleanly through without scattering off any marbles. If the total target area presented to the incoming projectiles is 200 cm² and there are 25 marbles hidden in the jello, what is the cross-sectional area for scattering for *each* marble? What is the radius of a typical marble? Describe how such scattering experiments could be used to probe the structure of subatomic particles. (See Physics to Go 12.2.)

9. The concept of a *black hole*, that is, an object whose gravity is so strong that not even light, traveling at a speed of 3×10^8 m/s, can escape from it, dates back to at least the 18th century and the work of Pierre-Simon Laplace, who referred to them as "dark stars." From Newtonian mechanics, the escape velocity (see Section 3.5) from a spherical mass is given by the equation

$$v_e = (2GM/R)^{1/2}$$

where M and R are the mass and radius of the body, respectively. G is the gravitational constant, 6.67×10^{-11} N-m²/kg².

   (a) Compute the escape velocity from the surface of the Earth, and confirm the result quoted at the end of Section 3.5, viz., $v_e \sim 11{,}200$ m/s.

   (b) Ignoring for the moment that the physics of strong gravity in the vicinity of a black hole requires general relativity and not Newtonian mechanics for its proper understanding, use this relationship to show that the radius of a spherical object of mass M whose escape velocity is the speed of light must be equal to

$$R = 2GM/c^2$$

   Notice that this is precisely the same equation as that for *Schwarzschild radius*, d_s, introduced in Section 12.2 in connection with the time dilation formula. Indeed, the Schwarzschild radius is the distance from a nonrotating spherical mass at which the escape velocity reaches the speed of light and thus may be taken to be the boundary of Laplace's dark star and our black hole. This is one example where the predictions of Newtonian physics happen to match those of general relativity.

   (c) Find the Schwarzschild radius of Earth and then calculate how large the density Earth would have to be for it to become a black hole. Compare your result to the current density of Earth, approximately 5514 kg/m³.

10. Show that time stands still at the Schwarzschild radius of a black hole. Specifically, demonstrate that $\Delta t_f \rightarrow \infty$ when $d = d_s$ in the time dilation equation for any value of Δt_d. The implications of this result are clear: As one nears the Schwarzschild radius of a black hole, gravitational time dilation stretches the interval between successive ticks on the clock as seen by a distant observer until they become infinitely long, effectively freezing the motion just as the Schwarzschild boundary is crossed. This essentially prevents the distant observer from ever obtaining any information about conditions *inside* the black hole.

Winners of the Nobel Prize in Physics

Since 1901, the Nobel Prizes have been awarded nearly annually (some prizes were not presented during the two world wars or when suitable candidates were not identified) in the fields of physics, chemistry, medicine and physiology, and literature, as well as for peace, to individuals whose work has been of major benefit to humankind. (A Nobel Prize in economics was added in 1969.) The prizes commemorate the life and work of Swedish chemist and industrialist Alfred B. Nobel (1833–1896). Nobel, who had little more than an elementary school education, was a highly imaginative and restless person. At the time of his death, he had never held permanent residency in any country, had become a noted linguist (reading, writing, and speaking five languages) who wrote poetry in English, and had accumulated more than 355 patents in different countries around the globe. His best-known inventions were the *Nobel lighter,* a device developed in 1862 that made it possible to safely and reliably harness the explosive power of nitroglycerine, and later a product he called *dynamite,* or nitroglycerine in the form of a paste wrapped in a paper cylinder. However, he also invented other types of high explosives (including blasting gelatin and *ballistite,* a smokeless powder and precursor to today's cordite), designed prototypes for rockets and aerial torpedoes, and worked on producing artificial rubber and leather, as well as artificial silk (which was later perfected by others as *rayon*). During his life, Nobel amassed great wealth from the royalties on his many inventions, and, in his will dated 27 November 1895, he stipulated that the income from his estate should be used each year to support the awarding of what have come to be the most prestigious civic honors in the world.

Through 2015, 573 Nobel Prizes have been awarded to 870 different individuals representing more than 80 countries; in addition, 23 different organizations have been recognized as Nobel laureates. (Nobel expressly stated that in awarding the various prizes, "no consideration whatever be paid to the nationality of the candidate, so that the most worthy receives it.") Each prize winner receives a specially designed diploma, a gold medal (Figure A.1), and a cash award. In physics and chemistry, the selection of Nobel Prize recipients is made by the Royal Swedish Academy of Science based on nominations presented by former prize winners and members of recognized institutions of higher learning and by official national academies of science of any country. The following list gives the Nobel laureates in physics, the year of their award, and the discovery or achievement that led to their selection.

Science & Society Picture Library/Getty Images

Figure A.1 Image of the Nobel Prize medal awarded to J. J. Thomson in 1906.

1901	*Wilhelm Roentgen* Discovery of x-rays.
1902	*Hendrik A. Lorentz and Pieter Zeeman* Research on the relationship between magnetism and radiation.
1903	*Antoine Henri Becquerel* Discovery of radioactivity.
	Pierre Curie and Marie Curie Research in radioactivity.
1904	*John William Strutt (Lord Rayleigh)* Discovery of argon.
1905	*Philipp E. A. von Lenard* Research on cathode rays.
1906	*Joseph J. Thomson* Studies on conduction of electricity in gases.
1907	*Albert A. Michelson* Optical instruments and investigations using them.
1908	*Gabriel Lippmann* Interference method in color photography.
1909	*Guglielmo Marconi and Carl Ferdinand Braun* Contributions to wireless communications.
1910	*Johannes D. van der Waals* Equation of state for gases and liquids.
1911	*Wilhelm Wien* Studies of blackbody radiation.
1912	*Nils G. Dalén* Automatic regulators for use in coastal lighting.
1913	*Heike Kamerlingh Onnes* Research in low-temperature physics.
1914	*Max von Laue* Diffraction of x-rays by crystals.
1915	*William H. Bragg and William L. Bragg* Studies of crystal structure using x-rays.
1916	No prize
1917	*Charles G. Barkla* Discovery of characteristic x-ray spectra of elements.
1918	*Max Planck* Discovery of quantization of energy.
1919	*Johannes Stark* Effect of electric fields on spectral lines.
1920	*Charles-Édouard Guillaume* Discovery of anomalies in nickel-steel alloys.
1921	*Albert Einstein* Explanation of the photoelectric effect.
1922	*Niels Bohr* Quantum model of the atom.
1923	*Robert A. Millikan* Work on the electron's charge and on the photoelectric effect.
1924	*Karl M. G. Siegbahn* Research in x-ray spectroscopy.
1925	*James Franck and Gustav Hertz* Studies of collisions between electrons and atoms.
1926	*Jean B. Perrin* Studies on the discontinuous structure of matter.
1927	*Arthur H. Compton* Scattering of electrons and photons.
	Charles T. R. Wilson Invention of the cloud chamber to study the paths of charged particles.
1928	*Owen W. Richardson* Work on the emission of electrons from hot objects.
1929	*Louis-Victor de Broglie* Discovery of the wave nature of electrons.
1930	*Chandrasekhara Venkata Raman* Studies of the scattering of light by atoms and molecules.
1931	No prize
1932	*Werner Heisenberg* Creation of quantum mechanics.
1933	*Erwin Schroedinger and Paul A. M. Dirac* Contributions to quantum mechanics.
1934	No prize
1935	*James Chadwick* Discovery of the neutron.
1936	*Victor F. Hess* Discovery of cosmic rays.
	Carl D. Anderson Discovery of the positron.
1937	*Clinton J. Davisson and George P. Thomson* Discovery of diffraction of electrons by crystals.

1938   *Enrico Fermi*   New radioisotopes produced by neutron irradiation.

1939   *Ernest O. Lawrence*   Invention of the cyclotron.

1940–
1942   No prize

1943   *Otto Stern*   Discovery of the magnetic moment of the proton.

1944   *Isador I. Rabi*   Nuclear magnetic resonance.

1945   *Wolfgang Pauli*   Discovery of the exclusion principle.

1946   *Percy W. Bridgman*   Contributions to high-pressure physics.

1947   *Edward V. Appleton*   Reflection of radio waves by the ionosphere.

1948   *Patrick M. S. Blackett*   Discoveries in nuclear physics.

1949   *Hideki Yukawa*   Prediction of mesons.

1950   *Cecil F. Powell*   Photographic method of studying nuclear processes.

1951   *John D. Cockroft and Ernest T. S. Walton*   Transmutation of atomic nuclei by accelerated particles.

1952   *Felix Bloch and Edward M. Purcell*   Research on the magnetic fields of nuclei.

1953   *Frits Zernike*   Invention of the phase-contrast microscope.

1954   *Max Born*   Research in quantum mechanics.
        Walther Bothe   Development of the coincidence method.

1955   *Willis E. Lamb*   Discoveries concerning the spectrum of hydrogen.
        Polykarp Kusch   Determination of the magnetic moment of the electron.

1956   *William Shockley, John Bardeen, and Walter H. Brattain*   Discovery of the transistor.

1957   *C. N. Yang and T. D. Lee*   Prediction of the violation of conservation of parity.

1958   *Pavel A. Čerenkov, Ilya M. Frank, and Igor Tamm*   Discovery and interpretation of the Čerenkov effect.

1959   *Emilio G. Segrè and Owen Chamberlain*   Discovery of the antiproton.

1960   *Donald A. Glaser*   Invention of the bubble chamber.

1961   *Robert Hofstadter*   Studies of electron scattering by nuclei.
        Rudolf L. Mössbauer   Studies of emission and absorption of gamma rays in crystals.

1962   *Lev D. Landau*   Theoretical work on condensed matter, especially liquid helium.

1963   *Eugene P. Wigner*   Laws governing interactions between protons and neutrons in the nucleus.
        Maria Goeppert-Mayer and J. Hans D. Jensen   Discoveries related to nuclear shell structure.

1964   *C. H. Townes, Nikolai G. Basov, and Alexander M. Prochorov*   Discovery of the principle behind lasers and masers.

1965   *Sin-Itiro Tomonaga, Julian Schwinger, and Richard P. Feynman*   Development of quantum electrodynamics.

1966   *Alfred Kastler*   Optical methods for studying atomic structure.

1967   *Hans A. Bethe*   Theory of nuclear reactions in stars.

1968   *Luis W. Alvarez*   Contributions to elementary particle physics.

1969   *Murray Gell-Mann*   Discoveries concerning elementary particles.

1970   *Hannes Alfvén*   Work in magnetohydrodynamics.
        Louis Néel   Discoveries concerning ferrimagnetism and antiferromagnetism.

1971   *Dennis Gabor*   Holography.

1972   *John Bardeen, Leon N. Cooper, and J. Robert Schrieffer*   Explanation of superconductivity.

1973   *Leo Esaki*   Discovery of tunneling in semiconductors.
        Ivar Giaever   Discovery of tunneling in superconductors.

Brian D. Josephson   Prediction of properties of supercurrent through a tunnel-barrier.

1974   *Antony Hewish*   Discovery of pulsars.
        Martin Ryle   Pioneering work in radio astronomy.

1975   *Aage Bohr, Ben Mottelson, and James Rainwater*   Discovery of the relationship between collective and particle motion in nuclei.

1976   *Burton Richter and Samuel C. C. Ting*   Discovery of the J/ψ particle.

1977   *John N. Van Vleck, Nevill F. Mott, and Philip W. Anderson*   Studies of electrons in magnetic solids.

1978   *Arno A. Penzias and Robert W. Wilson*   Discovery of the cosmic microwave background radiation.
        Peter L. Kapitza   Work in low-temperature physics.

1979   *Sheldon L. Glashow, Abdus Salam, and Steven Weinberg*   Work on the unification of the weak and electromagnetic interactions.

1980   *Val L. Fitch and James W. Cronin*   Experimental proof of charge-parity (CP) violation, which has implications for cosmology.

1981   *Nicolaas Bloembergen and Arthur L. Schawlow*   Contributions to the development of laser spectroscopy.
        Kai M. Siegbahn   Contributions to the development of high-resolution electron spectroscopy.

1982   *Kenneth G. Wilson*   Theory of critical phenomena in connection with phase transitions.

1983   *Subrahmanyan Chandrasekhar and William A. Fowler*   Work relating to the evolution of stars.

1984   *Carlo Rubbia and Simon van der Meer*   Discovery of the W and Z particles.

1985   *Klaus von Klitzing*   Discovery of the quantized Hall effect.

1986   *Ernst Ruska*   Design of the first electron microscope.
        Gird Binning and Heinrich Rohrer   Design of the scanning tunneling microscope.

1987   *J. George Bednorz and K. Alex Müller*   Discovery of superconductivity in a new class of materials.

1988   *Leon Lederman, Melvin Schwartz, and Jack Steinberger*   Experiment that established the existence of a second kind of neutrino and that employed the first beam of neutrinos produced in a laboratory.

1989   *Norman Ramsey, Hans Dehmelt, and Wolfgang Paul*   Contributions to the development of precision atomic spectroscopy, which laid the basis for cesium atomic clocks and other devices.

1990   *Jerome Friedman, Henry Kendall, and Richard Taylor*   Experimental verification of the existence of quarks.

1991   *Pierre-Gilles de Gennes*   Development of methods for studying complex systems at the atomic level.

1992   *Georges Charpak*   Invention and development of particle detectors.

1993   *Russell Hulse and Joseph Taylor*   Discovery of a binary pulsar.

1994   *Bertram N. Brockhouse and Clifford Shull*   Development of neutron-scattering techniques.

1995   *Frederick Reines and Martin L. Perl*   Pioneering experimental contributions to lepton physics.

1996   *David M. Lee, Douglas D. Osheroff, and Robert C. Richardson*   Discovery of superfluidity in helium-3.

1997   *Steven Chu, Claude Cohen-Tannoudji, and William Phillips*   Development of methods to cool and trap atoms with laser light.

1998   *Robert Laughlin, Horst Störmer, and Daniel Tsui*   Discovery and explanation of the fractional Hall effect.

1999   *Gerardus 't Hooft and Martinus J. G. Veltman*   Contributions to mathematical aspects of the standard model of particle physics.

2000 *Zhores I. Alferov, Herbert Kroemer, and Jack S. Kilby* Invention and development of devices crucial to information technology.

2001 *Eric A. Cornell, Wolfgang Ketterle, and Carl E. Wieman* Creation of Bose-Einstein condensates.

2002 *Raymond Davis Jr., Masatoshi Koshiba, and Riccardo Giacconi* Contributions to astrophysics.

2003 *Alexei A. Abrikosov, Vitaly L. Ginzburg, and Anthony J. Leggett* Contributions to superconductivity and superfluidity.

2004 *David J. Gross, H. David Politzer, and Frank Wilczek* Theoretical discoveries about the strong nuclear force.

2005 *Roy J. Glauber, John L. Hall, and Theodor W. Hänsch* Advancements in optics.

2006 *John C. Mather and George F. Smoot* Characterization of the cosmic background radiation.

2007 *Albert Fert and Peter Grünberg* Discovery of giant magnetoresistance effects that are the basis for data retrieval techniques from hard drives.

2008 *Yoichiro Nambu, Makoto Kobayashi, and Toshihinde Maskawa* Research on broken symmetry in subatomic physics and the existence of three families of quarks.

2009 *Charles K. Kao, Willard S. Boyle, and George E. Smith* Achievements in communications using optical fibers and the development of charge-coupled devices.

2010 *Andre Geim and Konstantin Novoselov* Experiments characterizing a new class of two-dimensional graphitic materials.

2011 *Saul Perlmutter, Brian P. Schmidt, and Adam G. Riess* For the discovery of the accelerating expansion of the universe through observations of distant supernovae.

2012 *Serge Haroche and David J. Wineland* For ground-breaking experimental methods that enable the measuring and manipulation of individual quantum systems.

2013 *François Englert and Peter W. Higgs* For the theoretical discovery of a mechanism that provides an understanding of the origin of mass of subatomic particles.

2014 *Isamu Akasaki, Hiroshi Amano, and Shuji Nakamura* For the invention of efficient blue light-emitting diodes which enabled the development of bright, energy-saving white light sources.

2015 *Takaaki Kajita and Arthur B. McDonald* For the discovery of neutrino oscillations, which shows that neutrinos have mass.

APPENDIX B

Math Review

The following is a review of the math employed in this book that goes beyond the simple arithmetic you use nearly every day. Although most of it should be familiar to you, you may have forgotten some of it if you have not needed it in some time. Keep in mind that the mathematics appearing in this book is used primarily to concisely express and apply relationships between physical quantities that must first be understood conceptually.

Basic Algebra

Algebra is arithmetic with letters used to represent numbers. Let's say, for example, that you form an equal partnership with three other people. Your share of any income or expense is then one-fourth of each amount. When an item is purchased for $60, your share of its cost is $15. This situation can be represented by an algebraic relationship—an equation. If we let the letter S stand for your share and the letter A stand for the amount of the income or expense, then S is equal to one-fourth of A, or 0.25 multiplied by A. The equation is

$$S = 0.25 \times A$$

or

$$S = 0.25A$$

The multiplication sign, $\times$, is usually omitted in algebra. Whenever a number and a letter, or two letters, appear side by side, they are multiplied together.

S and A are called *variables* because they can take on different values (that is, they can be different numbers). An algebraic equation expresses the mathematical relationship that exists between variables.

To find your share, S, of a $200 payment, you replace A with $200 and perform the multiplication:

$$S = 0.25A$$
$$S = 0.25 \times \$200$$
$$S = \$50$$

The use of algebra in this situation really isn't necessary. All you have to do is to remember to multiply the total payment by 0.25 to compute your share. The algebraic equation is just a shorthand way of expressing the same thing.

This approach can be broadened for use with larger or smaller partnerships. We can let F represent the fraction of ownership for the partnership. In the above example, F is 0.25. For a partnership of two people, F is one-half, or 0.5. For 10 people, it is 1/10th or 0.1. Then the share, S, of an amount, A, for each partner is given by

$$S = FA$$

For a partnership of 10 people, the share of a $200 income is

$$S = 0.1 \times \$200$$
$$S = \$20$$

You might be able to calculate this mentally, but the equation is a formal statement of the rule that you use to compute the answer.

This type of equation, in which one quantity equals the product of two others, is the one that appears most often in this book. A large number of important relationships in physics take this form.

The above equation can be used in additional ways. For example, if the share S is known, the equation can be used to find the original amount A. This can be done because any equation can be manipulated in certain ways without altering the precise relationship it expresses. The basic rule is:

A mathematical operation can be performed on an equation without altering its validity as long as *the exact same process is carried out on both the left- and right-hand sides of the equals sign.*

The following examples illustrate the kinds of operations that can be useful. Numerical values are also used in the examples to show that the relationship expressed by each equation is still correct.

Addition

A number or a variable can be added to both sides of an equation. Let's say we have the equation

$$A = B - 6$$

This means that when B is known to be 8, A is 2:

$$2 = 8 - 6$$

If B is 11, A is 5; if $B = 83$, $A = 77$; and so on.

We can add any quantity to both sides of the equation without affecting its validity. For example, we can add 4 to both sides of the equation $A = B - 6$:

$$A + 4 = B - 6 + 4$$
$$A + 4 = B - 2$$

The relationship has not been changed. This means that any pair of values for A and B that satisfied the original equation will also satisfy the "new" equation. The values $B = 8$ and $A = 2$, which worked in the first equation, also work in the second:

$$2 + 4 = 8 - 2$$
$$6 = 6$$

Why would we want to add something to both sides of an equation? We can do this to change the focus of the equation. The original equation is used to find A when B is the known, or given, quantity. When B is known (or chosen) to be 8, the equation indicates that A must then be 2. If, instead, the value of A is known and we wish to find B, we can change the equation to yield B in terms of A. In this case, we simply add 6 to both sides:

$$A = B - 6$$
$$A + 6 = B - 6 + 6$$
$$A + 6 = B$$

or

$$B = A + 6$$

The original relationship is preserved because we get 8 for B when A is known to be 2:

$$B = 2 + 6$$
$$B = 8$$

What we have done is *solve the equation for* B, or, equivalently, *express B in terms of* A. This sort of thing is generally the goal when equations are manipulated.

Here's another example, except in this case a variable is added:

$$C = D - E$$

For example, if $D = 13$ and $E = 7$, then C is 6:

$$C = 13 - 7$$
$$C = 6$$

If, instead, we know C and E, then we might want to find the value of D. To do so, we add E to both sides:

$$C + E = D - E + E$$
$$C + E = D$$

or

$$D = C + E$$

With C equal to 6 and E equal to 7, we find that D is 13, as before:

$$D = 6 + 7$$
$$D = 13$$

Subtraction

The same variable or number can be subtracted from both sides of an equation.

Subtraction is very similar to addition. We can use it, for example, to reverse the first addition example.

$$B = A + 6$$

We can subtract 6 from both sides.

$$B - 6 = A + 6 - 6$$
$$B - 6 = A$$

or

$$A = B - 6$$

Multiplication

Both sides of an equation can be multiplied by the same variable or number. For example,

$$G = \frac{H}{5}$$

The right side is the standard way to represent division in algebra. In this case, it means "H divided by 5."

If H is known to be 45, then

$$G = \frac{45}{5}$$
$$G = 9$$

We can multiply both sides of the original equation by 5 if we wish to solve it for H:

$$G \times 5 = \frac{H}{5} \times 5$$

$$5G = \frac{H5}{5}$$

The 5's to the right of the equals sign can be canceled, leaving H.

$$5G = \frac{H5}{5} = \frac{H\cancel{5}}{\cancel{5}}$$

$$5G = H$$

or

$$H = 5G$$

For $G = 9$, H has to be 45, as before.

$$H = 5 \times 9$$
$$H = 45$$

Choosing to multiply both sides by 5 allows us to solve the equation for H.

This process also works for multiplication by a variable.

$$I = \frac{J}{B}$$

This equation gives a value for I when the values for J and B are known. If $J = 42$ and $B = 7$, then the equation tells us that I must be 6. If we want to express J in terms of I and B, we multiply both sides by B:

$$I \times B = \frac{J}{B} \times B$$

As before, the Bs can be canceled.

$$IB = \frac{JB}{B} = \frac{J\cancel{B}}{\cancel{B}}$$

$$IB = J$$

$$J = IB$$

If $I = 6$ and $B = 7$, then J will be 42, as required.

Division

We can divide both sides of an equation by the same variable or number.

Let's go back to the partnership. Suppose one of your partners tells you that your share of a bill is \$75. If you want to know what the amount of the bill is, you can solve the "share equation" for A by using division. (Remember that A = amount, F = fraction, and S = your share.)

$$S = FA$$

We can divide both sides by F.

$$\frac{S}{F} = \frac{FA}{F}$$

$$\frac{S}{F} = \frac{\cancel{F}A}{\cancel{F}} = A$$

$$A = \frac{S}{F}$$

So, the original amount A when your share is \$75 is

$$A = \frac{\$75}{0.25}$$

$$A = \$300$$

Sometimes it is necessary to use different operations in sequence. For example,

$$T = \frac{U}{C}$$

This tells us that T must be 50 if $U = 400$ and $C = 8$. But what if we are given the values of T and U so that C is unknown? We solve the equation for C by first multiplying both sides by C and then dividing both sides by T:

$$C \times T = \frac{U}{C} \times C$$

$$CT = U$$

$$\frac{CT}{T} = \frac{U}{T}$$

$$C = \frac{U}{T}$$

If we knew that $U = 400$ and $T = 50$, we could find the correct value for C:

$$C = \frac{U}{T}$$

$$C = \frac{400}{50}$$

$$C = 8$$

There is a nice shorthand way of looking at what we've done. The two forms of the equation (before and after) are

$$T = \frac{U}{C} \quad \text{(before)}$$

$$C = \frac{U}{T} \quad \text{(after)}$$

Note that in the first equation, C is below on the right and T is above on the left. In the second equation, the C and T have been switched: C is above on the left, and T is below on the right. This is a general result for this kind of equation: to move a variable (or number) to the other side of an equation, you must change it from above to below, or from below to above.

Here is another example of using operations in sequence. We will solve the following equation for M. To do this, we use the previous rules to move everything to one side of the equation except M.

$$P = 3M + 7$$

First, we subtract 7 from both sides:

$$P - 7 = 3M + 7 - 7$$
$$P - 7 = 3M$$

Then we divide both sides by 3:

$$\frac{P - 7}{3} = \frac{3M}{3} = M$$

$$M = \frac{P - 7}{3}$$

Powers

Both sides of an equation can be raised to the same power—that is, we can square or cube both sides, we can take the square root of both sides, and so on.

If we have

$$V = WD$$

Then

$$V^2 = (WD)^2$$

Using numbers, if $W = 3$ and $D = 2$,

$$V = 3 \times 2$$
$$V = 6$$

After the equation is squared:

$$V^2 = (3 \times 2)^2$$
$$V^2 = 6^2$$
$$V^2 = 36$$

So $V = 6$ also fits this equation.

If we start with the equation

$$X^2 = YZ$$

we can take the square root of both sides to find X.

$$\sqrt{X^2} = \sqrt{YZ}$$
$$X = \sqrt{YZ}$$

Or X could be

$$X = -\sqrt{YZ}$$

In most situations in physics, the positive square root is used.

Exercises

Solve the following equations for X.

1. $A = X - 7$
2. $B = X - 25.2$
3. $C = 12 + X$
4. $C = X + 87$
5. $E = X - F$
6. $G = X + 2H$
7. $I = \dfrac{X}{12}$
8. $J = -6X$
9. $K = \dfrac{X}{L}$
10. $M = NX$
11. $O = \dfrac{4}{X}$
12. $P = \dfrac{20}{X}$
13. $R = 3X - 5$
14. $S = 2X + T$
15. $U = \sqrt{X}$
16. $V = 4X^2$

Geometry

Angles

When straight lines intersect, they create angles (**Figure B.1**). The angles A and B are called *vertical angles*, and all vertical angles are equal: angle A = angle B. Angles C and D are also vertical angles, as can be seen by rotating Figure B.1 clockwise to place angle C at the top. Thus, angle C = angle D, as well.

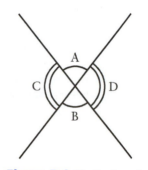

Figure B.1 Vertical angles.

A *triangle* is a closed geometrical figure formed by three intersecting lines (**Figure B.2**). The sum of the interior angles of any triangle is 180°: A + B + C = 180°.

Figure B.2 A triangle.

A *right triangle* has one angle that equals 90° (**Figure B.3**). The sum of the other two angles must then also be 90°.

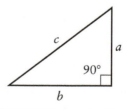

Figure B.3 A right triangle.

The side of a right triangle opposite the right angle is called the *hypotenuse* of the triangle. For any right triangle, the *Pythagorean theorem* establishes the following relationship between the sides of the triangle (refer to Figure B.3):

$$a^2 + b^2 = c^2$$

In words, for any right triangle, the square of the length of the hypotenuse is equal to the sum of the squares of the other two sides.

Measures of Common Shapes

1. A triangle with base b and height h (**Figure B.4**) has an area A where

$$A = \frac{1}{2}bh$$

2. A circle of radius r has a circumference C where

$$C = 2\pi r$$

and an area A such that

$$A = \pi r^2$$

3. A sphere of radius r has a surface area S where

$$S = 4\pi r^2$$

and a volume V such that

$$V = \frac{4}{3}\pi r^3$$

4. A right circular cylinder with radius r and height h (**Figure B.5**) has a volume V where

$$V = \pi r^2 h$$

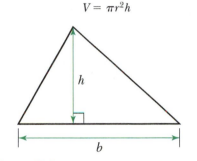

Figure B.4 Finding the area of a triangle.

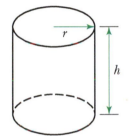

Figure B.5 Right circular cylinder.

Proportionalities

Much of physics involves studying the relationships between different variables such as speed and distance. It is very useful to know, for example, how the value of one quantity is affected if the value of another is changed. To illustrate this, imagine that you go into a candy store to buy several chocolate bars of the same kind. Because we can choose from different-sized bars with different prices, the number of bars that you buy will depend on the price of each as well as on the amount that you spend. The following equation can be used to tell you the number N of bars costing C dollars each that you will get if you spend D dollars:

$$N = \frac{D}{C}$$

If you spend $2 on chocolate bars that cost $0.50 (50 cents) each, the number that you buy is

$$N = \frac{\$2}{\$0.50}$$

$$N = 4$$

Let's look at this relationship in terms of how N changes when one of the other quantities takes on different values. For example, how is N affected when you increase the amount you spend on a particular type of chocolate? Clearly, the more you spend, the more you get. At 50 cents each, $2 will let you buy four bars, $3 will let you buy six bars, $4 will let you buy eight bars, and so on. This relationship is called a *proportionality:* we say that "N is proportional to D," (or "N is directly proportional to D"). If D *is doubled*—as from $2 to $4—then N *is also doubled*, from 4 to 8 in this case. If D is tripled, N becomes three times as large as well.

In a similar way, if the amount you spend, D, is decreased, the number N that you get is also decreased. One dollar will let you buy only two bars at 50 cents each. If D is halved, then N is halved.

$D(\$)$	N
1	2
2	4
3	6
4	8

Now let's say you decide to spend only $2. How does the cost of the bar you decide to buy affect the number of bars that you get? We know that you get four bars when they cost 50 cents each. If you buy bars that cost $1 each instead, the number you get is two.

$$N = \frac{\$2}{\$1}$$

$$N = 2$$

If you choose bars that cost $2 each, you can get only one bar. Thus, when C is increased, N is decreased. In particular, *if* C *is doubled,* N *becomes one-half as large.* This is called an *inverse proportionality.* We say that "N is inversely proportional to C." If C is decreased, N is increased. At 25 cents each, you can buy eight chocolate bars for $2.

$C(\$)$	N
0.25	8
0.50	4
1	2
2	1

Scientific Notation

Scientific notation is a convenient shorthand way to represent very large and very small numbers. It is based on using powers of 10. Because 100 equals 10×10 and 1,000 equals $10 \times 10 \times 10$, we can abbreviate them as follows:

$$100 = 10^2 \qquad (= 10 \times 10)$$
$$1{,}000 = 10^3 \qquad (= 10 \times 10 \times 10)$$

The numbers 100, 1,000, and so on, are powers of 10 because they are equal to 10 multiplied by itself a certain number of times. The number 2 in 10^2 is called the *exponent.* To represent any such power of 10 in this fashion, the exponent equals the number of zeros. So:

$$10 = 10^1$$
$$10{,}000 = 10^4$$
$$1{,}000{,}000 = 10^6$$

In this same fashion, since 1 has no zeros it is represented as

$$1 = 10^0$$

Decimals like 0.1 and 0.001 can also be represented this way by using negative exponents: 10 with a negative exponent equals 1 divided by 10 to the corresponding positive number.

$$0.01 = \frac{1}{100} = \frac{1}{10^2} = 10^{-2}$$

Similarly,

$$0.001 = \frac{1}{1,000} = \frac{1}{10^3} = 10^{-3}$$

To represent decimals by 10 to a negative exponent, the number in the exponent equals the number of places that the decimal point is to the left of the 1. For 0.01, the decimal point is two places to the left of 1, so $0.01 = 10^{-2}$. Other examples are as follows:

$$0.1 = 10^{-1}$$
$$0.0001 = 10^{-4}$$
$$0.000000001 = 10^{-9}$$

To review the process, 10 with a positive exponent equals 1 followed by that number of zeros. Ten with a negative exponent equals a decimal in which the number of zeros between the decimal point and the 1 is equal to the number in the exponent *minus one*. So, 10^5 equals 1 followed by five zeros—100,000. The number 10^{-5} equals the decimal with four zeros $(5 - 1)$ between the decimal point and 1—0.00001.

$10^4 = 10,000$	$10^{-1} = 0.1$
$10^3 = 1,000$	$10^{-2} = 0.01$
$10^2 = 100$	$10^{-3} = 0.001$
$10^1 = 10$	$10^{-4} = 0.0001$
$10^0 = 1$	

When two powers of 10 are multiplied, their exponents are added:

$$100 \times 1,000 = 100,000$$
$$10^2 \times 10^3 = 10^5$$

Similarly,

$$10^4 \times 10^8 = 10^{12}$$
$$10^{-3} \times 10^7 = 10^4$$
$$10^{-2} \times 10^{-5} = 10^{-7}$$

When a power of 10 is divided by another power of 10, the second exponent is subtracted from the first:

$$10^6 \div 10^2 = \frac{10^6}{10^2} = 10^{6-2} = 10^4$$
$$10^4 \div 10^9 = 10^{4-9} = 10^{-5}$$
$$10^3 \div 10^{-5} = 10^{3-(-5)} = 10^8$$
$$10^{-7} \div 10^{-1} = 10^{-7-(-1)} = 10^{-6}$$

Scientific notation utilizes powers of 10 to represent any number. A number expressed in scientific notation consists of a number (usually between 1 and 10) multiplied by 10 raised to some exponent. For example,

$$23,400 = 2.34 \times 10,000$$
$$23,400 = 2.34 \times 10^4$$

When expressing a number in scientific notation, the exponent equals the number of times the decimal point is moved. For numbers like 23,400, the decimal point is understood to be to the right of the last digit—as in 23,400.—so it was moved four places to the left to get it between the 2 and the 3. *If the decimal point is moved to the left, the exponent is positive*, as in the above example. Similarly,

$$641 = 6.41 \times 10^2$$
$$497,500,000 = 4.975 \times 10^8$$

If the decimal point is moved to the right, the exponent is negative.

$$0.0053 = 5.3 \times 0.001 = 5.3 \times 10^{-3}$$
$$0.07134 = 7.134 \times 10^{-2}$$
$$0.000000964 = 9.64 \times 10^{-7}$$

When two numbers expressed in scientific notation are multiplied or divided, the "regular" parts are placed together, and the powers of 10 are placed together.

For multiplication,

$$(4.2 \times 10^3) \times (2.3 \times 10^5) = (4.2 \times 2.3) \times (10^3 \times 10^5)$$
$$= 9.66 \times 10^8$$
$$(1.36 \times 10^4) \times (2 \times 10^{-6}) = (1.36 \times 2) \times (10^4 \times 10^{-6})$$
$$= 2.72 \times 10^{-2}$$

For division,

$$(6.8 \times 10^7) \div (3.4 \times 10^2) = (6.8 \div 3.4) \times (10^7 \div 10^2)$$
$$= 2.0 \times 10^5$$
$$(7.3 \times 10^{-5}) \div (2 \times 10^{-8}) = (7.3 \div 2) \times (10^{-5} \div 10^{-8})$$
$$= 3.65 \times 10^3$$

Exercises

Express the following numbers in scientific notation.

17. 253,000
18. 93,000,000
19. 5632.5
20. 0.0072
21. 0.00000000338

Use scientific notation to find answers to the following operations:

22. $(4 \times 10^2) \times (2.1 \times 10^7)$
23. $(6.1 \times 10^{11}) \times (1.4 \times 10^{-5})$
24. $(3.2 \times 10^4) \times (5 \times 10^{-6})$
25. $(7 \times 10^{-3}) \times (8.3 \times 10^{-8})$
26. $(6 \times 10^9) \div (2 \times 10^2)$
27. $(9.3 \times 10^7) \div (4.1 \times 10^{-4})$
28. $(5.23 \times 10^{-8}) \div (3 \times 10^{-6})$
29. $(4.5 \times 10^{-3}) \div (2 \times 10^{-5})$
30. $(2.4 \times 10^4) \div (6 \times 10^9)$

Significant Figures

The numbers that are used in physics and other sciences are typically the results of measurements. A measured distance of, say, 96.4 miles is really some value between 96.35 miles and 96.45 miles. This imprecision must be kept in mind when such numbers are multiplied, divided, and so on. For example, if a car travels 96.4 miles in 3 hours, the speed is found by dividing the distance by the time:

$$\text{speed} = \frac{\text{distance}}{\text{time}} = \frac{96.4 \text{ miles}}{3 \text{ hours}}$$

The strictly mathematical answer is

$$\text{Speed} = 32.1333333333 \text{ mph}$$

But the uncertainty in the distance measurement means that the exact speed can't be known precisely. Most likely it is some value between 32.1 mph and 32.2 mph. The value for the time is also imprecise. For this reason, the answer should not indicate more precision than the input numbers warrant.

In this case, the first three digits (3, 2, and 1) are said to be *significant figures*, while the remaining digits (all 3s) are *insignificant*. The significant figures represent the limit to the accuracy or precision in the answer. Our proper answer is one that shows only the significant figures—the number 32.1333333333 rounded off to 32.1:

$$\text{Speed} = 32.1 \text{ mph}$$

Mathematical rules can be used to determine how many significant figures there are in an answer, based on the accuracy of the input data. But for the sake of simplicity, you may assume there are three or four significant figures in the answers to the Problems and Challenges in this book. This means that you should round off your answers to three or four digits. Some examples:

$$6.4768 \rightarrow 6.48$$
$$25,934 \rightarrow 25,900$$
$$0.4575 \rightarrow 0.458$$

Answers to Odd-Numbered Exercises

1. $X = A + 7$
3. $X = C - 12$
5. $X = E + F$
7. $X = 12I$
9. $X = KL$
11. $X = \dfrac{4}{O}$
13. $X = \dfrac{R + 5}{3}$
15. $X = U^2$
17. 2.53×10^5
19. 5.6325×10^3
21. 3.38×10^{-9}
23. 8.54×10^6
25. 58.1×10^{-11} or 5.81×10^{-10}
27. 2.27×10^{11}
29. 2.25×10^2

Answers

This appendix contains answers to the Physics to Go activities, the odd-numbered Problems, and the odd-numbered Challenges that are problems. Worked-out solutions are given for the odd-numbered Problems, except those that are nearly identical to Examples or other Problems.

Answers to Odd-Numbered Problems and Challenges

CHAPTER 1 Problems

1. $d = 20$ m
   From conversion table: 1 m = 3.28 ft
   $d = 20$ m $= 20 \times 1$ m $= 20 \times 3.28$ ft
   $d = 65.6$ ft

3. The prefix *milli-* means $^{1}/_{1000}$ or 0.001. So 1 second = 1000 milliseconds
   0.0452 s $= 0.0452 \times 1$ s $= 0.0452 \times 1000$ ms
   0.0452 s $= 45.2$ ms

5. $T = 0.8$ s (period)
   $$f = \frac{1}{T} = \frac{1}{0.8 \text{ s}}$$
   $f = 1.25$ Hz

7. average speed $= \dfrac{\text{distance}}{\text{time}}$
   $$v = \frac{d}{t} = \frac{1{,}200 \text{ mi}}{2.5 \text{ h}}$$
   $v = 480$ mph

9. average speed $= \dfrac{d_{\text{final}} - d_{\text{initial}}}{t_{\text{final}} - t_{\text{initial}}}$
   $$v = \frac{20 \text{ mi} - 5 \text{ mi}}{2 \text{ h}} = \frac{15 \text{ mi}}{2 \text{ h}}$$
   $v = 7.5$ mph

11. In Figure 1.13c, the length of the resultant velocity arrow is 0.66 times the length of the arrow representing 8 m/s (v_1). Therefore, the resultant velocity is 0.66×8 m/s = 5.3 m/s.
    In Figure 1.13d, the resultant velocity is 12.5 m/s.

13. $d = vt$
    $d = 25$ m/s $\times 5$ s
    $d = 125$ m
    For $v = 250$ m/s: $d = vt$
    $\quad\quad\quad\quad\quad\quad\quad d = 250$ m/s $\times 5$ s
    $\quad\quad\quad\quad\quad\quad\quad d = 1{,}250$ m

15.

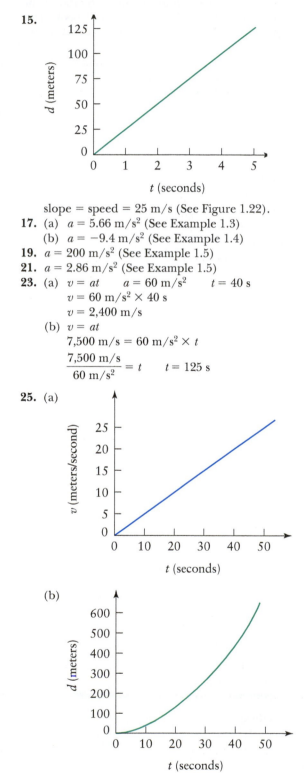

slope = speed = 25 m/s (See Figure 1.22).

17. (a) $a = 5.66$ m/s^2 (See Example 1.3)
    (b) $a = -9.4$ m/s^2 (See Example 1.4)

19. $a = 200$ m/s^2 (See Example 1.5)

21. $a = 2.86$ m/s^2 (See Example 1.5)

23. (a) $v = at \quad\quad a = 60$ m/s$^2 \quad\quad t = 40$ s
    $\quad\quad v = 60$ m/s$^2 \times 40$ s
    $\quad\quad v = 2{,}400$ m/s
    (b) $v = at$
    $\quad\quad 7{,}500$ m/s $= 60$ m/s$^2 \times t$
    $\quad\quad \dfrac{7{,}500 \text{ m/s}}{60 \text{ m/s}^2} = t \quad\quad t = 125$ s

25. (a)

(b)

27. (a) $v = at$ $a = g$ $t = 3$ s
$$v = gt = 9.8 \text{ m/s}^2 \times 3 \text{ s}$$
$$v = 29.4 \text{ m/s}$$

(b) $d = \frac{1}{2}at^2$ $d = \frac{1}{2}gt^2$

$$d = \frac{1}{2}(9.8 \text{ m/s}^2) \times (3 \text{ s})^2$$

$$d = 4.9 \text{ m/s}^2 \times 9 \text{ s}^2$$
$$d = 44.1 \text{ m}$$

29. $a = 4.9 \text{ m/s}^2$

(a) $v = at$ $t = 3$ s
$$v = 4.9 \text{ m/s}^2 \times 3 \text{ s}$$
$$v = 14.7 \text{ m/s}$$

(b) $d = \frac{1}{2}at^2$

$$d = \frac{1}{2}(4.9 \text{ m/s}^2) \times (3 \text{ s})^2$$

$$d = 2.45 \text{ m/s}^2 \times 9 \text{ s}^2$$
$$d = 22.05 \text{ m}$$

31. At "a": $a = \text{slope} = \dfrac{500 \text{ m/s}}{0.001 \text{ s}}$
$$a = 500,000 \text{ m/s}^2$$
At "b": $a = 0 \text{ m/s}^2$
At "c": $a = -2,500,000 \text{ m/s}^2$

33. $a = \dfrac{\Delta v}{\Delta t} = \dfrac{300 \text{ mph}}{5 \text{ s}} = 60 \text{ mph/s} = 2.7 \text{ g}$

CHAPTER 1   Challenges

3. (a) For the car: $d = \frac{1}{2}at^2$ with $a = 4.0 \text{ m/s}^2$.
For the truck: $d = vt$ with $v = 12$ ms.

t (s)	d_{car} (m)	d_{truck} (m)
1.0	2.0	12
3.0	18	36
5.0	50	60
7.0	98	84

(b)

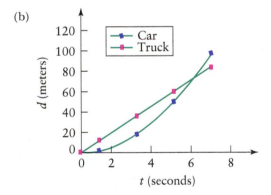

The car overtakes the truck at a time just after they both have traveled the same distance. This occurs at approximately 6 s when their common distance is about 72 m.

This can be confirmed by setting $d_{car} = d_{truck}$ and solving the resulting equation for t: $\frac{1}{2}at^2 = vt$

$$t = \frac{2v}{a} = \frac{2 \times 12 \text{ m/s}}{4.0 \text{ m/s}^2}$$

$$t = \frac{24 \text{ m/s}}{4.0 \text{ m/s}^2} = 6.0 \text{ s}$$

5. $d = \frac{1}{2}at^2$ $d = 2.0$ m $t = 0.71$ s

$$a = \frac{2d}{t^2} = \frac{2 \times 2.0 \text{ m}}{(0.71 \text{ s})^2}$$

$$a = \frac{4.0 \text{ m}}{0.504 \text{ s}^2} = 7.9 \text{ m/s}^2$$

7. $v = \sqrt{2ad}$

9. (a) $v_{\text{final}} = 0 \text{ m/s}$ $v_{\text{initial}} = 282.4 \text{ m/s}$ $\Delta t = 1.4$ s

$$a = \frac{(v_{\text{final}} - v_{\text{initial}})}{\Delta t} = \frac{(0 \text{ m/s} - 282.4 \text{ m/s})}{1.4 \text{ s}}$$

$$a = \frac{(-282.4 \text{ m/s})}{1.4 \text{ s}} = -201.7 \text{ m/s}^2 = -20.6 \text{ g}$$

(b) $d = v_{\text{initial}} t + \frac{1}{2}at^2$

$$d = (282.4 \text{ m/s} \times 1.4 \text{ s}) + \frac{1}{2}(-201.7 \text{ m/s}^2) \times (1.4 \text{ s})^2$$

$$d = 395.4 \text{ m} - 197.7 \text{ m} = 197.7 \text{ m}$$

CHAPTER 2   Problems

1. One example: $W = 150$ lb 1 lb $= 4.45$ N
$$W = 150 \times 1 \text{ lb} = 150 \times 4.45 \text{ N}$$
$$W = 667.5 \text{ N}$$
$$W = mg$$
$$667.5 \text{ N} = m \times 9.8 \text{ m/s}^2$$
$$\frac{667.5 \text{ N}}{9.8 \text{ m/s}^2} = m$$
$$m = 68.1 \text{ kg}$$

3. (a) $W = mg = 30 \text{ kg} \times 9.8 \text{ m/s}^2$
$$W = 294 \text{ N}$$

(b) $W = 294$ N 1 N $= 0.225$ lb
$$W = 294 \times 1 \text{ N} = 294 \times 0.225 \text{ lb}$$
$$W = 66.15 \text{ lb}$$

5. $F = 36,000 \text{ N}$ (See Example 2.1)

7. $F = ma$
$$10 \text{ N} = 2 \text{ kg} \times a$$
$$\frac{10 \text{ N}}{2 \text{ kg}} = a$$
$$a = 5 \text{ m/s}^2$$

9. $F = ma$
$$20,000,000 \text{ N} = m \times 0.1 \text{ m/s}^2$$
$$\frac{20,000,000 \text{ N}}{0.1 \text{ m/s}^2} = m$$
$$m = 200,000,000 \text{ kg}$$

11. (a) $W = mg$ $W = 800 \text{ N}$ $g = 9.8 \text{ m/s}^2$
$$800 \text{ N} = m \times 9.8 \text{ m/s}^2$$
$$\frac{800 \text{ N}}{9.8 \text{ m/s}^2} = m$$
$$m = 81.6 \text{ kg}$$

(b) The scale measures the total force required to support the person's weight plus that needed to accelerate the person upward at 2 m/s^2.
$$F = W + ma = 800 \text{ N} + 81.6 \text{ kg} \times 2 \text{ m/s}^2$$
$$F = 800 \text{ N} + 163 \text{ N} = 963 \text{ N}$$

(c) When moving at a constant speed of 5 m/s, $a = 0$. Therefore, the force measured by the scale again just equals the person's weight, 800 N.

13. $F = ma = 0.7 \text{ kg} \times 3,500 \text{ m/s}^2$
$$F = 2,450 \text{ N} = 551 \text{ lb}$$

15. (a) $a = -6,000 \text{ m/s}^2$ (See Example 1.3)
$$a = -612 g$$

(b) $F = -870 \text{ N}$ (See Example 2.1)

17. $a = 2g$ $g = 9.8$ m/s^2 $m = 2,000$ kg
$a = 2 \times 9.8$ m/s$^2 = 19.6$ m/s^2
$F = 39,200$ N   (See Example 2.1)

19. $a = 10g$ $g = 9.8$ m/s^2 $m = 50$ kg
$a = 10 \times 9.8$ m/s$^2 = 98$ m/s^2
(a) $F = 4,900$ N   (See Example 2.1)
(b) $W = mg = 50$ kg $\times 9.8$ m/s^2
$W = 490$ N   1 N $= 0.225$ lb
$W = 490$ N $\times 0.225$ lb/N
$W = 110$ lb
$F = 10W = 1,100$ lb

21. $v = 10$ m/s $r = 8$ m $m = 120$ kg

(a) $F = \dfrac{mv^2}{r} = \dfrac{120 \text{ kg} \times (10 \text{ m/s})^2}{8 \text{ m}}$

$F = \dfrac{12,000 \text{ kg-m}^2/\text{s}^2}{8 \text{ m}}$

$F = 1,500$ N

(b) $F = ma$

$a = \dfrac{F}{m}$

$a = \dfrac{1,500 \text{ N}}{120 \text{ kg}} = 12.5$ m/s^2

23. $r = 50$ m $F = 8,000$ N $m = 1,000$ kg

$F = \dfrac{mv^2}{r}$

$8,000$ N $= \dfrac{1,000 \text{ kg} \times v^2}{50 \text{ m}} = 20v^2$

$400 = v^2$
$v = 20$ m/s   (maximum speed)

25. $m = 2,000$ kg $F = -400$ N $\Delta v = -1000$ m/s
$F = ma$
-400 N $= 2,000$ kg $\times a$

$a = \dfrac{-400 \text{ N}}{2,000 \text{ kg}} = -0.20$ m/s^2

$a = \dfrac{\Delta v}{\Delta t}$

-0.20 m/s$^2 = \dfrac{-1000 \text{ m/s}}{\Delta t}$

$\Delta t = \dfrac{1000 \text{ m/s}}{0.20 \text{ m/s}^2} = 5,000$ s $= 83.3$ min

27. $F_1 = 89.0$ N $\Delta x_1 = 0.0191$ m $\Delta x_2 = 0.0508$ m
$F = -k\Delta x$

$\dfrac{F_1}{F_2} = \dfrac{\Delta x_1}{\Delta x_2}$

$\dfrac{89.0 \text{ N}}{F_2} = \dfrac{0.0191 \text{ m}}{0.0508 \text{ m}}$

$F_2 = \dfrac{(89.0 \text{ N} \times 0.0508 \text{ m})}{0.0191 \text{ m}} = \dfrac{4.52 \text{ N-m}}{0.0191 \text{ m}}$

$F_2 = 237$ N

$k = \dfrac{F_1}{\Delta x_1} = \dfrac{89.0 \text{ N}}{0.0191 \text{ m}} = 4660$ N/m

CHAPTER 3 Problems

1. $mv = 650$ kg-m/s   (See Section 3.2)

3. $F = \dfrac{\Delta mv}{\Delta t} = \dfrac{(mv)_{\text{final}} - (mv)_{\text{initial}}}{\Delta t}$

$mv_{\text{initial}} = 1,000$ kg $\times 0$ m/s
$mv_{\text{initial}} = 0$ kg-m/s
$mv_{\text{final}} = 1,000$ kg $\times 27$ m/s

$mv_{\text{final}} = 27,000$ kg-m/s
$\Delta t = 10$ s

$F = \dfrac{(mv)_{\text{final}} - (mv)_{\text{initial}}}{\Delta t} = \dfrac{27,000 \text{ kg m/s} - 0 \text{ kg m/s}}{10 \text{ s}}$

$F = \dfrac{27,000 \text{ kg-m/s}}{10 \text{ s}}$

$F = 2,700$ N

5. $m = 0.75$ kg $v_{\text{before}} = -13$ m/s $\Delta t = 25$ ms
Impulse $= F\Delta t = \Delta(mv)$ m is constant
Impulse $= m\Delta v = m \times (v_{\text{after}} - v_{\text{before}})$
Impulse $= 0.75$ kg $\times (0$ m/s $- [-13$ m/s$])$
Impulse $= 0.75$ kg $\times 13$ m/s $= 9.75$ kg-m/s
Impulse $= F\Delta t$
9.75 kg-m/s $= F \times 25$ ms

$F = \dfrac{9.75 \text{ kg-m/s}}{25 \times 10^{-3} \text{ s}} = 390$ N

7. $v = 39$ m/s   (See Example 3.2)

9. $mv_{\text{before}} = 50$ kg $\times 5$ m/s
$mv_{\text{before}} = 250$ kg-m/s
$mv_{\text{after}} = (40$ kg $+ 50$ kg$) \times v$
$mv_{\text{after}} = 90$ kg $\times v$
$mv_{\text{before}} = mv_{\text{after}}$
250 kg-m/s $= 90$ kg $\times v$

$\dfrac{250 \text{ kg-m/s}}{90 \text{ kg}} = v$

$v = 2.78$ m/s

11. $mv_{\text{before}} = 0$ kg-m/s
$mv_{\text{after}} = (mv)_{\text{gun}} + (mv)_{\text{bullet}}$
$mv_{\text{after}} = 1.2$ kg $\times v + 0.02$ kg $\times 300$ m/s
$mv_{\text{after}} = 1.2$ kg $\times v + 6$ kg-m/s
$mv_{\text{before}} = mv_{\text{after}}$
0 kg-m/s $= 1.2$ kg $\times v + 6$ kg-m/s
-6 kg-m/s $= 1.2$ kg $\times v$

$\dfrac{-6 \text{ kg-m/s}}{1.2 \text{ kg}} = v$

$v = -5$ m/s   (opposite direction of bullet)

or use: $\dfrac{v_{\text{gun}}}{v_{\text{bullet}}} = \dfrac{-m_{\text{bullet}}}{m_{\text{gun}}}$

13. Work $= 30,000$ J   (See Example 3.3)

15. $PE = 2,156$ J   (See Example 3.7)

17. $KE = 45,000$ J   (See Example 3.6)

19. $KE = \dfrac{1}{2} mv^2$

$60,000$ J $= \dfrac{1}{2} \times 300$ kg $\times v^2$

$\dfrac{60,000 \text{ J}}{150 \text{ kg}} = v^2$

400 J/kg $= v^2$
$v = 20$ m/s

21. $F = 180$ N $m = 0.021$ kg $d = 0.50$ m

(a) $W_{\text{avg}} = \dfrac{1}{2}[(Fd)_{\text{final}} - (Fd)_{\text{initial}}] = \dfrac{1}{2}(180$ N $\times 0.50$ m$)$

$W_{\text{avg}} = 90$ N $\times 0.50$ m $= 45$ N-m
(b) $KE =$ Work done

$\dfrac{1}{2} mv^2 = 45$ N-m

$v^2 = \dfrac{(2 \times 45 \text{ N-m})}{0.021 \text{ kg}} = 4286$ m^2/s^2

$v = 65.5$ m/s

(c) $v^2 = 2ah$ $a = g$
$4286 \text{ m}^2/\text{s}^2 = 2 \times 9.8 \text{ m/s}^2 \times h$
$h = \dfrac{4286 \text{ m}^2/\text{s}^2}{19.6 \text{ m/s}^2} = 218.7 \text{ m} = 717.25 \text{ ft}$

23. (a) $m = 25 \text{ kg}$ $k = 8750 \text{ N/m}$
$F = W = kd$
$W = mg = 25 \text{ kg} \times 9.8 \text{ m/s}^2$
$W = 245 \text{ N}$
$245 \text{ N} = 8750 \text{ N/m} \times d$
$d = \dfrac{245 \text{ N}}{8750 \text{ N/m}} = 0.028 \text{ m} = 2.8 \text{ cm}$

(b) $PE_{\text{spring}} = \frac{1}{2}kd^2 = \frac{1}{2} \times 8750 \text{ N/m} \times (0.028 \text{ m})^2$
$PE_{\text{spring}} = 3.43 \text{ J}$

25. $v^2 = 2gd$
$(7.7 \text{ m/s})^2 = 2 \times 9.8 \text{ m/s}^2 \times d$
$59.3 \text{ m}^2/\text{s}^2 = 19.6 \text{ m/s}^2 \times d$
$\dfrac{59.3 \text{ m}^2/\text{s}^2}{19.6 \text{ m/s}^2} = d$
$d = 3 \text{ m}$ (or use Table 3.2)

27. (a) $PE = 58,210 \text{ J}$ (See Example 3.7)
(b) $v = 50.9 \text{ m/s}$ (114 mph; see Example 3.9)

29. (a) $KE = 4,000 \text{ J}$ (See Example 3.6)
(b)

KE at bottom $= PE$ when stopped on hill
$4,000 \text{ J} = mgd = 80 \text{ kg} \times 9.8 \text{ m/s}^2 \times d$
$\dfrac{4,000 \text{ J}}{784 \text{ kg-m/s}^2} = d$
$d = 5.1 \text{ m}$ or use $d = \dfrac{v^2}{2g}$

31. It would have to be thrown vertically, with just enough speed that it would slow to a stop just as it reached the ceiling: $v = \sqrt{2gd}$
$= \sqrt{2 \times 9.8 \text{ m/s}^2 \times 20 \text{ m}} = 19.8 \text{ m/s}$ (=44 mph)

33. $KE_{\text{before}} = \dfrac{1}{2}mv^2 = \dfrac{1}{2} \times 50 \text{ kg} \times (5 \text{ m/s})^2$
$KE_{\text{before}} = 25 \text{ kg} \times 25 \text{ m}^2/\text{s}^2$
$KE_{\text{before}} = 625 \text{ J}$
$KE_{\text{after}} = \dfrac{1}{2}mv^2 = \dfrac{1}{2} \times 90 \text{ kg} \times (2.78 \text{ m/s})^2$
$KE_{\text{after}} = 45 \text{ kg} \times 7.73 \text{ m}^2/\text{s}^2$
$KE_{\text{after}} = 348 \text{ J}$
$KE_{\text{lost}} = 625 \text{ J} - 348 \text{ J}$
$KE_{\text{lost}} = 277 \text{ J}$

35. $P = \dfrac{\text{Work}}{t}$
$200 \text{ W} = \dfrac{10,000 \text{ J}}{t}$
$t = \dfrac{10,000 \text{ J}}{200 \text{ W}}$
$t = 50 \text{ s}$

37. $m = 1,040 \text{ kg}$ $PE = 11,780 \text{ J}$
$PE = mgd$
$11,780 \text{ J} = 1,040 \text{ kg} \times 9.8 \text{ m/s}^2 \times d$
$11,780 \text{ J} = 10,192 \text{ kg-m/s}^2 \times d$
$d = \dfrac{11,780 \text{ J}}{10,192 \text{ kg-m/s}^2} = 1.16 \text{ m}$
$P = \dfrac{PE}{t} = \dfrac{11,780 \text{ J}}{0.62 \text{ s}} = 19,000 \text{ W}$ $1 \text{ W} = 0.00134 \text{ hp}$
$P = 19,000 \text{ W} \times 0.00134 \text{ hp/W} = 25.5 \text{ hp}$

39. $m = 65 \text{ kg}$ $d = 1,050 \text{ ft} = 320.3 \text{ m}$ $t = 9 \text{ min } 33 \text{ s} = 573 \text{ s}$
(a) Work $= Fd = mgd$
Work $= 65 \text{ kg} \times 9.8 \text{ m/s}^2 \times 320.3 \text{ m}$
Work $= 204,000 \text{ J}$
(b) $P = \dfrac{\text{Work}}{t} = \dfrac{204,000 \text{ J}}{573 \text{ s}}$
$P = 356 \text{ W}$ $1 \text{ W} = 0.00134 \text{ hp}$
$P = 356 \text{ W} \times 0.00134 \text{ hp/W} = 0.48 \text{ hp}$
$P = 0.48 \text{ hp} \times 550 \text{ ft-lb/s/hp} = 262.4 \text{ ft-lb/s}$

41. $m_1 = m_2 = 0.25 \text{ kg}$
$r_1 = r_2 = \frac{1}{2} \times 0.5 \text{ m} = 0.25 \text{ m}$
$v_1 = v_2 = 0.75 \text{ m/s}$
Total angular momentum $= m_1 v_1 r_1 + m_2 v_2 r_2$
Total angular momentum $= 2 \times 0.25 \text{ kg} \times 0.25 \text{ m} \times 0.75 \text{ m/s}$
Total angular momentum $= 0.094 \text{ kg-m}^2/\text{s}$

CHAPTER 3   Challenges

3. (a) $m_{\text{truck}} = 4,000 \text{ kg}$ $v_{\text{truck}} = 10 \text{ m/s}$ east
$m_{\text{car}} = 1,000 \text{ kg}$ $v_{\text{car}} = 20 \text{ m/s}$ west
Let east represent the positive direction.
$(mv)_{\text{before}} = m_{\text{truck}} \times v_{\text{truck}} + m_{\text{car}} \times v_{\text{car}}$
$= 4,000 \text{ kg} \times 10 \text{ m/s} + 1,000 \text{ kg} \times (-20 \text{ m/s})$
$= 40,000 \text{ kg-m/s} - 20,000 \text{ kg-m/s}$
$= 20,000 \text{ kg-m/s}$
$(mv)_{\text{after}} = (m_{\text{truck}} + m_{\text{car}}) \times v$
$(mv)_{\text{after}} = (4,000 \text{ kg} + 1,000 \text{ kg}) \times v$
$= 5,000 \text{ kg} \times v$
$(mv)_{\text{before}} = (mv)_{\text{after}}$
$20,000 \text{ kg-m/s} = 5,000 \text{ kg} \times v$
$v = 4 \text{ m/s}$
The wreckage moves with a speed of 4 m/s in an eastward (positive) direction.
(b) $(KE)_{\text{initial}} = \frac{1}{2}m_{\text{truck}}v^2_{\text{truck}} + \frac{1}{2}m_{\text{car}}v^2_{\text{car}}$, respectively.
$(KE)_{\text{initial}} = \frac{1}{2} \times 4,000 \text{ kg} \times (10 \text{ m/s})^2 + \frac{1}{2} \times 1,000 \text{ kg} \times (-20 \text{ m/s})^2$
$= 2,000 \text{ kg} \times 100 \text{ m}^2/\text{s}^2 + 500 \text{ kg} \times 400 \text{ m}^2/\text{s}^2$
$= 200,000 \text{ kg-m}^2/\text{s}^2 + 200,000 \text{ kg-m}^2/\text{s}^2$
$= 400,000 \text{ kg-m}^2/\text{s}^2$
$(KE)_{\text{final}} = \frac{1}{2} \times (m_{\text{truck}} + m_{\text{car}}) \times v^2$
$(KE)_{\text{final}} = \frac{1}{2} \times 5,000 \text{ kg} \times (4 \text{ m/s})^2$
$= 2,500 \text{ kg} \times 16 \text{ m}^2/\text{s}^2$
$= 40,000 \text{ J}$
$\Delta KE = (KE)_{\text{final}} - (KE)_{\text{initial}}$
$= 40,000 \text{ J} - 400,000 \text{ J} = -360,000 \text{ J}$
360,000 J of kinetic energy are lost during the collision.

7. Apply the law of conservation of angular momentum: angular momentum at closest point = angular momentum at farthest point
$(mvr)_{\text{closest}} = (mvr)_{\text{farthest}}$ m is constant
$(vr)_{\text{closest}} = (vr)_{\text{farthest}}$
$54,500 \text{ m/s} \times d = v \times 60d$
$54,500 \text{ m/s} = 60 \times v$
$v = \dfrac{54,500 \text{ m/s}}{60} = 908 \text{ m/s}$

9. The system is shown in the figure.

$m = 0.1$ kg $T = 0.3$ s

(a) distance = circumference = $2\pi r$

$C_1 = 2\pi r_1 = 2 \times 3.14 \times 0.1$ m = 0.628 m

$C_2 = 2\pi r_2 = 2 \times 3.14 \times 0.2$ m = 1.25 m

$C_3 = 1.88$ m $C_4 = 2.51$ m $C_5 = 3.14$ m

(d) distance/time = speed

$$\frac{C}{T} = v$$

$$v_1 = \frac{C_1}{T} = \frac{0.628 \text{ m}}{0.3 \text{ s}} = 2.08 \text{ m/s}$$

$$v_2 = \frac{C_2}{T} = \frac{1.25 \text{ m}}{0.3 \text{ s}} = 4.17 \text{ m/s}$$

$v_3 = 6.28$ m/s $v_4 = 8.37$ m/s $v_5 = 10.47$ m/s

(c) angular momentum = mvr for each mass

$mv_1 r_1 = 0.1$ kg $\times 2.08$ m/s $\times 0.1$ m = 0.0209 kg-m^2/s

$mv_2 r_2 = 0.1$ kg $\times 4.17$ m/s $\times 0.2$ m = 0.0834 kg-m^2/s

$mv_3 r_3 = 0.188$ kg-m^2/s $mv_4 r_4 = 0.335$ kg-m^2/s

$mv_5 r_5 = 0.524$ kg-m^2/s

Total angular momentum = $mv_1 r_1 + mv_2 r_2 + mv_3 r_3 + mv_4 r_4 + mv_5 r_5 = 1.151$ kg-m^2/s

CHAPTER 4   Problems

1. $p = \dfrac{F}{A} = \dfrac{2,000,000 \text{ lb}}{400 \text{ ft}^2}$

$p = 5,000 \dfrac{\text{lb}}{\text{ft}^2}$

$1 \text{ ft}^2 = 144 \text{ in.}^2$ $p = 5,000 \dfrac{\text{lb}}{\text{ft}^2} = 5,000 \dfrac{\text{lb}}{144 \text{ in.}^2}$

$$p = \frac{5,000 \text{ lb}}{144 \text{ in.}^2}$$

$$p = 34.7 \text{ psi}$$

3. $F = pA$

$F = 80$ psi $\times 1,200$ in.^{2}

$F = 96,000$ lb

5. $F = 1,176$ lb   (See Example 4.2)

7. (a) $D = \dfrac{m}{V} = \dfrac{393 \text{ kg}}{0.05 \text{ m}^3}$

$D = 7,860$ kg/m^3

$D_W = D \times g = 7,860$ kg/m^3 $\times 9.8$ m/s^2

$D_W = 77,028$ N/m^3

(b) Iron   (From Table 4.4)

9. (a) $D_W = \dfrac{W}{V}$

water: $D_W = 62.4$ lb/ft^3 (See Table 4.4)

$62.4 \text{ lb/ft}^3 = \dfrac{40,000 \text{ lb}}{V}$

$V = \dfrac{40,000 \text{ lb}}{62.4 \text{ lb/ft}^3}$

$V = 641$ ft^3

(b) gasoline: $D_W = 42$ lb/ft^3

$D_W = 42 \text{ lb/ft}^3 = \dfrac{40,000 \text{ lb}}{V}$

$$V = \frac{40,000 \text{ lb}}{42 \text{ lb/ft}^3}$$

$$V = 952 \text{ ft}^3$$

11. $m = 162$ kg   (See Example 4.4 and Table 4.4)

13. (a) $m = 2,190$ kg   (See Example 4.4 and Table 4.4)

(b) $W = mg = 2,190$ kg $\times 9.8$ m/s^2

$W = 21,500$ N = 4,830 lb

15. $p = 0.433$ $d = 0.433 \times 12$ ft

$p = 5.196$ psi

17. $p = 3,030,000$ Pa   (See Example 4.6)

19. About 7 psi

21. $F_b = W_{\text{water displaced}} = (D_W)_{\text{water}} \times V$

$F_b = 62.4$ lb/ft^3 $\times 12$ ft^3

$F_b = 749$ lb

23. (a) $W_{\text{He}} = D_W \times V = 0.011$ lb/ft^3 $\times 200,000$ ft^3 = 2,200 lb   (D_W from Table 4.4)

(b) $F_b = W_{\text{air}} = D_W \times V = 0.08$ lb/ft^3 $\times 200,000$ ft^3 = 16,000 lb

(c) $F_{\text{net}} = F_b - W_{\text{He}} = 16,000$ lb $-$ 2,200 lb = 13,800 lb

25. (a) $W = 468$ lb   (See Example 4.5 and Table 4.4)

(b) $F_b = W_{\text{water}} = (D_W)_{\text{water}} \times V = 62.4$ lb/ft^3 $\times 3$ ft^3

$F_b = 187.2$ lb

(c)

$F_{\text{net}} = 468$ lb $-$ 187.2 lb

$F_{\text{net}} = 280.8$ lb   (downward)

27. (a) $W_{\text{ice}} = 5,720,000$ lb   (See Example 4.5 and Table 4.4)

(b) $W_{\text{seawater}} = F_b = W_{\text{ice}}$ (since it floats)

$W_{\text{seawater}} = D_w V$ D_w(seawater) = 64.3 lb/ft^3

5,720,000 lb = 64.3 lb/ft^3 $\times V$

$\dfrac{5,720,000 \text{ lb}}{64.3 \text{ lb/ft}^3} = V$

$V_{\text{seawater}} = 88,960$ ft^3

(c) $V_{\text{out}} = V_{\text{total}} - V_{\text{underwater}}$
$= 100,000$ ft^3 $-$ 88,960 ft^3

$V_{\text{out}} = 11,040$ ft^3 $= \dfrac{1}{9} V_{\text{total}}$

29. Scale reading = $W_{\text{Al}} - F_b = 100$ N $- F_b$

$F_b = W_{\text{water}} = D_{\text{water}} \times g \times V_{\text{Al}}$

$W_{\text{Al}} = D_{\text{Al}} \times g \times V_{\text{Al}}$ $V_{\text{Al}} = 0.00378$ m^3

$F_b = 37.0$ N

Scale reading = 63.0 N   (See Table 4.4 for densities)

31. $F_{\text{piston}} = 45$ N $F_{\text{plunger}} = 1900$ N

$\dfrac{F_{\text{piston}}}{A_{\text{piston}}} = p = \dfrac{F_{\text{plunger}}}{A_{\text{plunger}}}$

$\dfrac{A_{\text{plunger}}}{A_{\text{piston}}} = \dfrac{F_{\text{plunger}}}{F_{\text{piston}}} = \dfrac{1900 \text{ N}}{45 \text{ N}} = 42.2$

33. $F_{\text{lift}} = 75,000$ N $A_{\text{wing}} = 12$ m^2

$F_{\text{lift}} = F_{\text{bottom}} - F_{\text{top}}$

$\dfrac{F_{\text{lift}}}{A_{\text{wing}}} = \dfrac{F_{\text{bottom}}}{A_{\text{wing}}} - \dfrac{F_{\text{top}}}{A_{\text{wing}}} = \Delta p$

$\Delta p = \dfrac{75,000 \text{ N}}{12 \text{ m}^2} = 6250$ Pa

35. $A_1v_1 = A_2v_2 \qquad v_1 = 5.0 \text{ ft/s}$
$A_1 = 8 \text{ in.} \times 8 \text{ in.} = 64 \text{ in.}^2 = 0.44 \text{ ft}^2$
$A_2 = 1.0 \text{ ft}^2$
$$v_2 = \frac{A_1v_1}{A_2} = \frac{0.44 \text{ ft}^2 \times 5.0 \text{ ft/s}}{1.0 \text{ ft}^2}$$
$$v_2 = \frac{2.22 \text{ ft}^3/\text{s}}{1.0 \text{ ft}^2} = 2.22 \text{ ft/s}$$

CHAPTER 4 Challenges

3. $D_{Al} = 2,700 \text{ kg/m}^3$
specific gravity = 2.7
$(D_W)_{Moon} = 27.6 \text{ lb/ft}^3$

5. About $0.014 \text{ lb} = 0.062 \text{ N}$ lighter.

15. Fluid element at surface of container with mass m has
$PE_i = mgh$ relative to the bottom of the container ($h = 0$);
at the surface, fluid is at rest: $KE_i = 0$.
At location of tap, fluid element has $h = 0$: $PE_f = 0$
At tap opening, fluid is in motion with speed v as it leaves
container:
$$KE_f = \frac{1}{2}mv^2.$$
Apply conservation of energy: $E_i = E_f$
$E_i = KE_i + PE_i = 0 + mgh = mgh$
$E_f = KE_f + PE_f = \frac{1}{2}mv^2 + 0 = \frac{1}{2}mv^2$
$mgh = \frac{1}{2}mv^2$
$gh = \frac{1}{2}v^2$
$v^2 = 2gh \qquad v = \sqrt{2gh} \qquad$ (See Section 3.5a)

CHAPTER 5 Problems

1. $30°C = 86°F \qquad$ (no jacket needed)

3. $\Delta l = -0.336 \text{ ft} \qquad$ (it is shorter) $\qquad$ (See Example 5.1 and Table 5.2)
$l = 699.664 \text{ ft}$

5. Need to reduce the diameter from 5 mm to 4.997 mm
$\Delta l = -0.003 \text{ mm}$
$\Delta l = \alpha l \Delta T \qquad \alpha = 12 \times 10^{-6}/°C$
$\quad - 0.003 \text{ mm} = 12 \times 10^{-6}/°C \times 5 \text{ mm} \times \Delta T$
$$\frac{-0.003 \text{ mm}}{60 \times 10^{-6} \text{ mm}/°C} = \Delta T$$
$$\frac{-0.003}{60} \times 10^6 \,°C = \Delta T$$
$-0.00005 \times 10^6 \,°C = \Delta T$
$\qquad\qquad \Delta T = -50°C$

7. Apply the ideal gas law:
$$\frac{(p_1 V_1)}{T_1} = \text{constant} = \frac{(p_2 V_2)}{T_2}$$
$p_1 = 1.75 \times 10^5 \text{ Pa} \qquad V_1 = 2.75 \text{ m}^3 \qquad V_2 = 4.20 \text{ m}^3$
$T_1 = 16°C = 16 + 273.15 = 289.15 \text{ K}$
$T_2 = 26.4°C = 299.55 \text{ K}$
$$\frac{(1.75 \times 10^5 \text{ Pa}) \times (2.75 \text{ m}^3)}{289.15 \text{ K}} = \frac{p_2 \times (4.20 \text{ m}^3)}{299.55 \text{ K}}$$
$1664 = p_2 \times 0.0140 \qquad$ (SI units)
$$p_2 = \frac{1664}{0.0140} \text{ Pa} = 1.19 \times 10^5 \text{ Pa}$$

9. $\Delta U = \text{work} + Q \qquad Q = -2 \text{ J} \qquad$ (heat lost)
$\qquad\qquad \text{work} = Fd = 50 \text{ N} \times 0.1 \text{ m}$
$\qquad\qquad \text{work} = 5 \text{ J} \qquad$ (Work done *on* gas)
$\qquad\qquad \Delta U = \text{work} + Q = 5 \text{ J} + (-2 \text{ J})$
$\qquad\qquad \Delta U = 3 \text{ J}$

11. $Q = 1,081,000 \text{ J} \qquad$ (See Example 5.4 and Table 5.3)

13. (a) $Q = Cm\Delta T$
$\qquad C = 4,180 \text{ J/kg-}°C \qquad$ for water
$\qquad \Delta T = 100°C$
$\qquad Q = 4,180 \text{ J/kg-}°C \times 1 \text{ kg} \times 100°C$
$\qquad Q = 418,000 \text{ J}$
(b) $Q = 2,260,000 \text{ J}$ (see Section 5.6)

15. (a) $KE = 375,000 \text{ J} \qquad$ (See Example 3.6)
(b) $Q = Cm\Delta T \qquad Q = KE$
$\qquad KE = Cm\Delta T \qquad C = 460 \text{ J/kg-}°C \qquad$ for iron
$\qquad 375,000 \text{ J} = 460 \text{ J/kg-}°C \times 20 \text{ kg} \times \Delta T$
$$\qquad \frac{375,000 \text{ J}}{9,200 \text{ J}/°C} = \Delta T$$
$\qquad\qquad \Delta T = 40.8°C$

17. $\Delta T = 47.4°C \qquad$ (See Example 5.3)

19. (a) Rel. Hum. = 62.5% $\qquad$ (See Example 5.8 and Table 5.5)
(b) Rel. Hum. = 5.78% $\qquad$ (See Example 5.8 and Table 5.5)

21. The amount of water vapor in the air is given by the water vapor density, which is the humidity.
$$\text{Rel. Hum.} = \frac{\text{Humidity}}{\text{Sat. Den.}} \times 100\%$$
At 20°C, Sat. Den. = 0.0173 kg/m^3
$$40\% = \frac{\text{Humidity}}{0.0173 \text{ kg/m}^3} \times 100\%$$
$$\frac{40\% \times 0.0173 \text{ kg/m}^3}{100\%} = \text{Humidity}$$
Humidity = 0.00692 kg/m^3
There is 0.00692 kg of water vapor in each m^3 of air.

23. $m = DV \qquad V = 150 \text{ m}^3$
$$\text{Rel. Hum.} = \frac{\text{Humidity}}{\text{Sat. Den.}} \times 100\%$$
$$60\% = \frac{D}{0.0228} \times 100\% \qquad D = 0.0137 \text{ kg/m}^3$$
$m = 0.0137 \text{ kg/m}^3 \times 150 \text{ m}^3$
$m = 2.1 \text{ kg}$

25. Humidity = 0.0094 kg/m^3 $\qquad$ (See Problem 21.)
From Table 5.5, air with this humidity will be saturated when cooled to 10°C. Therefore the air must be cooled 20°C, which means it must rise 2,000 m.

27. Carnot eff. = 34.9% $\qquad$ (See Example 5.9)

29. (a) Eff. $= \dfrac{\text{work output}}{\text{energy input}} \times 100\%$
$\qquad$ Eff. $= \dfrac{3,000 \text{ J}}{15,000 \text{ J}} \times 100\%$
$\qquad$ Eff. = 20%
(b) Carnot Eff. = 88% $\qquad$ (See Example 5.9)

31. $\Delta E = \Delta S \times T_R = (1.67 \text{ J/K}) \times (220 \text{ K})$
$\Delta E = 367.4 \text{ J}$

CHAPTER 5 Challenges

3. (a) $Q = -(2.42 \times 10^6 \text{ J/kg}) \times m$
$\qquad Q = -(2.42 \times 10^6 \text{ J/kg}) \times 0.175 \text{ kg} = -4.24 \times 10^5 \text{ J}$
Heat is transferred *out* of the athlete.
(b) $\Delta U = \text{work} + Q$
$\qquad \Delta U = -1.25 \times 10^5 \text{ J} - 4.24 \times 10^5 \text{ J} = -5.49 \times 10^5 \text{ J}$
Work is done *by* the athlete.
(c) One (food) Calorie = 1000 calories = 4184 J

To replace 5.49×10^5 J, the athlete must consume

$$\frac{5.49 \times 10^5 \, J}{4184 \, J/Cal} = 131 \, Cal$$

5. $m = DV$ $V = 60 \, m^3$ $T_i = 35°C$

$$Rel. \, Hum. = \frac{Humidity}{Sat. \, Den.} \times 100\%$$

$$77\% = \frac{D}{0.0396 \, kg/m^3} \times 100\%$$

$D = 0.0305 \, kg/m^3$
$m = 0.0305 \, kg/m^3 \times 60 \, m^3 = 1.83 \, kg$
Cooling to $T_f = 25°C$ and assuming the air is fully
saturated (relative humidity = 100%), $D = 0.0228 \, kg/m^3$
$m' = 0.0228 \, kg/m^3 \times 60 \, m^3 = 1.37 \, kg$ of water in the
room
Amount of water condensed out is:
$\Delta m = m - m' = 1.83 \, kg - 1.37 \, kg = 0.46 \, kg$

CHAPTER 6   Problems

1. (a) $\rho = 0.167 \, kg/m$ (See Example 6.1)
   (b) $v = 15.5 \, m/s$ (See Example 6.1)
3. $v = 388 \, m/s$ (See Example 6.2)
5. $v = f\lambda$
   $v = 4 \, Hz \times 0.5 \, m$
   $v = 2 \, m/s$
7. $v = f\lambda$
   $80 \, m/s = f \times 3.2 \, m$
   $\dfrac{80 \, m/s}{3.2 \, m} = f$
   $f = 25 \, Hz$
9. For $f = 20 \, Hz$:
   $\lambda = 17.2 \, m = 56.4 \, ft$ (See Example 6.3)
   For $f = 20,000 \, Hz$
   $\lambda = 0.0172 \, m = 0.677 \, in.$ (See Example 6.3)
11. (a) $\lambda = 1.315 \, m$ (See Example 6.3)
   (b) $v = f\lambda$ $v = 1,440 \, m/s$ (water, from Table 6.1)
   $1,440 \, m/s = 261.6 \, Hz \times \lambda$
   $\dfrac{1,440 \, m/s}{261.6 \, Hz} = \lambda$
   $\lambda = 5.505 \, m$
13. $v = 347 \, m/s$ (See Problem 5)
   $v = 20.1 \times \sqrt{T} = 347 \, m/s$ (See Example 6.1.)
   $(20.1)^2 \times T = (347 \, m/s)^2$
   $T = 298 \, K = 25°C$
15. This is destructive interference, so the difference in the
   two distances equals one half the wavelength of the sound.
   $7.2 \, m - 7 \, m = 0.2 \, m = \dfrac{1}{2}\lambda$
   $\lambda = 0.4 \, m$
   $v = 344 \, m/s$ (sound speed at room temperature)
   $f = 860 \, Hz$
17. (a) Total distance sound travels: $d = vt = 320 \, m/s \times$
      $0.03 \, s = 9.6 \, m$ (v from Table 6.1)
      Distance to snow $= \dfrac{1}{2} d = 4.8 \, m$
   (b) Depth of snow $= 5 \, m - 4.8 \, m = 0.2 \, m$
19. $d = vt$ $d = 150 \, m$
   For sound, $v = 344 \, m/s$
      $150 \, m = 344 \, m/s \times t$
      $\dfrac{150 \, m}{344 \, m/s} = t$
         $t = 0.436 \, s$

For light, $v = 3 \times 10^8 \, m/s$, and
         $t = 5 \times 10^{-7} \, s$
Compared to sound, the light travel time is essentially
instantaneous.
21. $t = 1.74 \, s$ (See Problem 19)
23. For each 10 dB, sound is 2 times louder.
   Sound level difference = 95 dB − 65 dB = 30 dB
   30 dB = 3 × 10 dB
   Loudness is $2 \times 2 \times 2 = 8$ times greater
25. Harmonics have frequencies that are 2, 3, 4, . . . times the
   frequency of the note itself.
   The first four harmonics of middle C are:
      $2 \times 261.6 \, Hz = 532.2 \, Hz$
      $3 \times 261.6 \, Hz = 784.8 \, Hz$
      $4 \times 261.6 \, Hz = 1,046 \, Hz$
      $5 \times 261.6 \, Hz = 1,308 \, Hz$

CHAPTER 6   Challenges

1. Amplitude is decreased.
3. In air: $t = 0.581 \, s$ (See Problem 19 and Table 6.1)
   In steel: $t = 0.038 \, s$
   Jack hears the sound in the rail 0.543 seconds before he
   hears the sound in the air.
7. 1,000,000 times
9. C# above middle C: $n = 1$ half-step higher
   $f = \left(\sqrt[12]{2}\right)^1 \times 261.6 \, Hz = (1.05946) \times 261.6 \, Hz$
   $f = 277.2 \, Hz$
   F above middle C: $n = 5$ half-steps higher
   $f = \left(\sqrt[12]{2}\right)^5 \times 261.6 \, Hz = (1.3348) \times 261.6 \, Hz$
   $f = 349.2 \, Hz$
   A above middle C: $n = 9$ half-steps higher
   $f = \left(\sqrt[12]{2}\right)^9 \times 261.6 \, Hz = (1.6817) \times 261.6 \, Hz$
   $f = 439.9 \, Hz$ Compare with Figure 6.45 ($f = 440 \, Hz$)

CHAPTER 7   Problems

1. $F \propto \dfrac{q_1 q_2}{d^2}$ $F = 16 \, N$

   If $d \to \frac{1}{2}d$, then $d^2 \to \frac{1}{4}d^2$
   F increases by a factor of 4.
   $F = 4 \times 16 \, N = 64 \, N$
   To yield a force of 64 N for the original distance,
   either q_1 or q_2 would have to be 4 times larger.
3. Electric field strength $= \dfrac{F}{q}$ $q = 2.0 \times 10^{-6} \, C$ $F = 8 \, N$
   upward
   Electric field strength $= \dfrac{8 \, N}{2.0 \times 10^{-6} \, C} = 4.0 \times 10^6 \, N/C$

   For a positive charge, the direction of F is along the
   direction of the electric field. Thus, the field is upward
   in this case. For a negative charge, the magnitude of the
   electric field would remain the same (4 million N/C),
   but its direction would be downward.

5. $I = \dfrac{q}{t}$ $q = 100 \, C$ $t = 0.05 \, s$
   $I = \dfrac{100 \, C}{0.05 \, s} = 2,000 \, A$

7. $I = \dfrac{q}{t}$ $I = 0.0001 \, A$ $t = 5 \, min = 300 \, s$
   $0.0001 \, A = \dfrac{q}{300 \, s}$
      $q = 0.0001 \, A \times 300 \, s$
         $= 0.03 \, C$

9. $V = IR$ $V = 120$ V $I = 20$ A

120 V $= 20$ A $\times R$

$R = \dfrac{120\ \text{V}}{20\ \text{A}} = 6\ \Omega$ (minimum resistance)

11. $V = IR$ $V = 120$ V $R = 80\ \Omega$

120 V $= I \times 80\ \Omega$

$I = \dfrac{120\ \text{V}}{80\ \Omega} = 1.5$ A

13. $V = IR$ $R = 150\ \Omega$

  (a) $I = 0.3$ A

     $V = 0.3$ A $\times 150\ \Omega$

       $= 45$ V

  (b) $V = 18$ V

    18 V $= I \times 150\ \Omega$

    $I = \dfrac{18\ \text{V}}{150\ \Omega} = 0.12$ A

15. $P = IV$ $I = 0.5$ A $V = 400$ V

$P = 0.5$ A $\times 400$ V

$= 200$ W

17. $P = IV$ $V = 120$ V

  (a) $I_{max} = 20$ A

      $P = 120$ V $\times 20$ A

        $= 2,400$ W   (maximum power)

  (b) $P_{max} = 2,400$ W

      $= $ (number of appliances) $\times 1,200$ W each

Two appliances can be plugged in without blowing the fuse.

19. (a) $P = 40$ W $I = 3.33$ A   (See Example 7.5)

      $P = 50$ W $I = 4.17$ A   (See Example 7.5)

  (b) $P = 40$ W $R = 3.6\ \Omega$ (See Problem 9.)

      $P = 50$ W $R = 2.88\ \Omega$ (See Problem 9.)

21. $E = 9,600,000$ J   (See Example 7.6)

or: $P = 4$ kW $t = 40$ min $= \dfrac{2}{3}$ h

$E = Pt = 4$ kW $\times \dfrac{2}{3}$ h

$E = 2.67$ kWh

23. Hair dryer: $E = 360,000$ J   (See Example 7.6)

     Lamp: $E = 2,160,000$ J   (See Example 7.6)

25. (a) $P = 1,080$ W   (See Example 7.4)

  (b) $E = 64,800$ J   (See Example 7.6)

     or: $P = 1.08$ kW $t = 1$ min $= \dfrac{1}{60}$ h

     $E = Pt = 1.08$ kW $\times \dfrac{1}{60}$ h

     $E = 0.018$ kWh

  (c) $R = 13.3\ \Omega$ (See Problem 9)

27. (a) $P = IV$

     $1,000,000,000$ W $= I \times 24,000$ V

     $\dfrac{1,000,000,000\ \text{W}}{24,000\ \text{V}} = I$

          $I = 41,670$ A

  (b) $E = Pt$ $t = 24$ h $= 24 \times 3600$ s

               $t = 86,400$ s

     $E = 1,000,000,000$ W $\times 86,400$ s

     $E = 8.64 \times 10^{13}$ J

     in kWh: $E = 1,000,000$ kW $\times 24$ h

                   $E = 24,000,000$ kWh

  (c) $10¢ = \$0.1$ Revenue $= \$2,400,000$

29. $P = 6.0$ W   (See Problem 15)

$E = Pt$ $E = 40,000$ J

$40,000$ J $= 6.0$ W $\times t$

$\dfrac{40,000\ \text{J}}{6.0\ \text{W}} = t$

$t = 6,667$ s 1 hr $= 3,600$ s

$t = 1.85$ hr

31. $I = 12$ A   (See Problem 11)

$P = IV$

$P = 12$ A $\times 120$ V

$= 1,440$ W

$E = Pt$ $t = 30$ min $= 1,800$ s

$E = 1,440$ W $\times 1,800$ s

$= 2,592,000$ J

CHAPTER 7   Challenges

1. $F = -8.2 \times 10^{-8}$ N

3. Number of electrons $= 1.25 \times 10^{18}$

7. $V = 12$ V: $I = 100$ A

   $V = 30$ V: $I = 40$ A

   $V = 60$ V: $I = 20$ A

   $V = 120$ V: $I = 10$ A

The lower voltage dryers would have to have successively larger wires to handle the larger currents.

9. $P = I^2 R$

4 times as much (40 kWh per day)

CHAPTER 8   Problems

1. $\dfrac{V_o}{V_i} = \dfrac{N_o}{N_i}$

120 V input, 5 V output

$\dfrac{5\ \text{V}}{120\ \text{V}} = \dfrac{N_o}{N_i}$

$N_o = N_i \times \dfrac{5}{120} = 1,000 \times \dfrac{5}{120} = 41.7$ turns

3. Example: $f = 92.5$ MHz $= 92,500,000$ Hz

            $c = f\lambda$

3×10^8 m/s $= 9.25 \times 10^7$ Hz $\times \lambda$

$\dfrac{3 \times 10^8\ \text{m/s}}{9.25 \times 10^7\ \text{Hz}} = \lambda$

          $\lambda = 3.24$ m

5. $c = f\lambda$

3×10^8 m/s $= f \times 0.0254$ m

$\dfrac{3 \times 10^8\ \text{m/s}}{0.0254\ \text{m}} = f$

          $f = 1.18 \times 10^{10}$ Hz $= 11,800$ MHz

7. UV band: $f = 7.5 \times 10^{14}$ Hz to $f = 10^{18}$ Hz

λ's:

          $c = f\lambda$ $c = f\lambda$

3×10^8 m/s $= 7.5 \times 10^{14}$ Hz $\times \lambda$ 3×10^8 m/s $= 10^{18}$ Hz $\times \lambda$

$\dfrac{3 \times 10^8\ \text{m/s}}{7.5 \times 10^{14}\ \text{Hz}} = \lambda$ $\dfrac{3 \times 10^8\ \text{m/s}}{1 \times 10^{18}\ \text{Hz}} = \lambda$

      $\lambda = 4 \times 10^{-7}$ m $\lambda = 3 \times 10^{-10}$ m

9. Energy emitted $\propto T^4$. From 300 K to 3000 K, T increases 10 times.

Energy output increases $10^4 = 10,000$ times

11. Power $= 5.67 \times 10^{-8} \times T^4 \times A$ $T = 450$ K   Power $= 175$ J/s

$175 = 5.67 \times 10^{-8} \times (450)^4 \times A$ (SI units)

$175 = 5.67 \times 10^{-8} \times 4.10 \times 10^{10} \times A$

$175 = 2,325 \times A$

$\dfrac{175}{2,325} = A$ in SI units

     $A = 0.075$ m^2

13. $\lambda_{max} = \dfrac{0.0029\ \text{m-K}}{T}$

$\lambda_{max} = \dfrac{0.0029\ \text{m-K}}{10,000,000\ \text{K}}$

$\lambda_{max} = 2.9 \times 10^{-10}$ m   X-ray (See Figure 8.29)

15. $\lambda_{max} = \dfrac{0.0029}{T}$

$T \times \lambda_{max} = T \times 0.0000012$ m $= 0.0029$

$T = 2{,}420$ K

CHAPTER 8 Challenges

7. $P_o = P_i$ $N_o/N_i = V_o/V_i$

$I_o V_o = I_i V_i$

$\dfrac{V_o}{V_i} = \dfrac{I_i}{I_o}$

$\dfrac{N_o}{N_i} = \dfrac{I_i}{I_o}$

$N_o I_o = N_i I_i$

CHAPTER 9 Problems

1. $S = 1.00$ m $a = 0.200$ mm $\lambda = 633$ nm

$\Delta x = \left(\dfrac{S}{a}\right)\lambda$ $\Delta x = \left(\dfrac{1.00 \text{ m}}{2.00 \times 10^{-4} \text{ m}}\right) \times (633 \times 10^{-9} \text{ m})$

$\Delta x = (5.00 \times 10^3) \times (633 \times 10^{-9} \text{ m}) = 0.0032$ m $= 3.2$ mm

3. By the law of reflection, the angle of reflection equals the angle of incidence, 50°. Therefore the reflected ray makes an angle of 50° with respect to the normal.

For the refracted ray, the light slows upon entering the glass, so the ray is bent toward the normal—the angle of refraction is smaller than the angle of incidence. The actual angle is found using Figure 9.34. The refracted ray makes an angle of about 32° with respect to the normal.

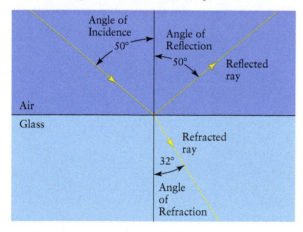

5.

Angle of Incidence (°)	Angle of Refraction (°)*
5	3
10	7
20	14

*From Figure 9.34

Doubling the angle of incidence approximately doubles the angle of refraction. Doubling the angle of incidence from 20° to 40° should result in an angle of refraction of about $2 \times 14°$ or 28°. Figure 9.34 gives an angle of refraction of about 27° for this result, which is in good agreement with this estimate.

7. $p = \dfrac{s \times f}{s - f}$ $f = 30$ cm $s = 200$ cm

$p = \dfrac{(200 \text{ cm} \times 30 \text{ cm})}{(200 \text{ cm} - 30 \text{ cm})} = \dfrac{6000 \text{ cm}^2}{170 \text{ cm}}$

$p = 35.3$ cm

9. $M = \dfrac{\text{image size}}{\text{object size}} = \dfrac{6 \text{ cm}}{2 \text{ cm}}$

$M = +3$ Image is upright and 3 times larger.

11. $M = -\dfrac{p}{s}$ $s = 200$ cm $p = 35.3$ cm

$M = -\dfrac{35.3 \text{ cm}}{200 \text{ cm}} = -0.17$ Image is inverted and reduced in size.

13. $p = \dfrac{s \times f}{s - f}$ $s = 8.0$ cm $f = 10$ cm

$p = -40$ cm   Image is virtual.   (See Problem 7.)

$M = -\dfrac{p}{s} = -\dfrac{(-40 \text{ cm})}{8.0 \text{ cm}} = +5$ Image is upright and enlarged five times.

15. $p = -14.3$ cm $M = -0.29$

Image is virtual, inverted, and reduced in size.   (See Problems 7 and 11.)

17. $p = \dfrac{s \times f}{s - f}$ $s = 16$ cm $f = 8$ cm

$p = 16$ cm   Image is located as far in front of the mirror as the object.

$M = -\dfrac{p}{s} = -\dfrac{(16 \text{ cm})}{(16 \text{ cm})} = -1$ Image is inverted and the same size as the object.

19. Scattered light ratio $\propto \lambda^{-4}$

If $\lambda \to 2\lambda$, the ratio changes by $2^{-4} = 0.0625$.

The amount of scattered light at the new (longer) wavelength is smaller than that at the old (shorter) wavelength.

CHAPTER 9 Challenges

1. The point in an interference pattern where constructive interference occurs depends on the wavelength of the light. Because different colors have different wavelengths, interference with white light results in each color having its interference maxima occur at slightly different locations on the observing screen. The colors no longer overlap completely at all points (as in white light), but are separated on the observing screen: Dispersion has taken place. The shorter wavelengths yield constructive interference at points closer to the middle of the pattern, so the spectrum of colors starts with violet closest to the middle of the pattern and red farthest away.

7. $f = 10$ cm

s (cm)	p (cm)	s (cm)	p (cm)
2	−2.50	25	16.67
4	−6.67	50	12.50
8	−40.00	100	11.11
9.9999	−999990.00	200	10.53
12	60.00	500	10.20
14	35.00	1000	10.10
18	22.50	10000	10.01
20	20.00		

Object-image plot

When p is negative, the image is virtual and on the same side of the lens as the object. The image distance, p, changes most rapidly for values of the object distance, s, close to the focal length of the lens. As s gets large and approaches infinity, p gets closer and closer to 10 cm, the focal length of the lens. When the object is far from the lens, the rays from any point on the object are parallel to one another and are all brought to a focus at a distance equal to the focal length of the lens: $p = f$. See Figure 9.42.

9. $f = 200$ mm $\qquad s = 3500$ mm

$$\frac{1}{p} = \frac{1}{f} - \frac{1}{s} = \frac{1}{200 \text{ mm}} - \frac{1}{3500 \text{ mm}}$$

$$\frac{1}{p} = (5.00 \times 10^{-3} \text{ mm}^{-1}) - (2.86 \times 10^{-4} \text{ mm}^{-1}) = 4.71 \times 10^{-3} \text{ mm}^{-1}$$

$p = 212.1$ mm

$$M = -\frac{p}{s} = -\frac{212.1 \text{ mm}}{3500 \text{ mm}} = -0.061$$

Image height = 100 mm × 0.061 = 6.1 mm

The kitten's image is inverted and reduced to about 6% of its original size.

CHAPTER 10 Problems

1. $E = 41.4$ eV $\qquad$ (See Example 10.1)

3. First find the frequency; then use Figure 10.7.

$E = hf \qquad h = 6.63 \times 10^{-34}$ J/Hz

9.5×10^{-25} J $= 6.63 \times 10^{-34}$ J/Hz $\times f$

$$\frac{9.5 \times 10^{-25} \text{ J}}{6.63 \times 10^{-34} \text{ J/Hz}} = f$$

$$\frac{9.5}{6.63} \times \frac{10^{-25}}{10^{-34}} \text{ Hz} = f$$

$f = 1.43 \times 10^9$ Hz $\qquad$ low-frequency microwave

1 eV = 1.6×10^{-19} J $\qquad 1 \text{ J} = \dfrac{1 \text{ eV}}{1.6 \times 10^{-19}}$

$E = 9.5 \times 10^{-25}$ J

$$E = 9.5 \times 10^{-25} \times \left(\frac{1 \text{ eV}}{1.6 \times 10^{-19}}\right)$$

$$E = \frac{9.5 \times 10^{-25}}{1.6 \times 10^{-19}} = \frac{9.5}{1.6} \times \frac{10^{-25}}{10^{-19}}$$

$E = 5.94 \times 10^{-6}$ eV

5. $\lambda = 9.1 \times 10^{-12}$ m $\qquad$ (See Example 10.2)

$\lambda = 0.0091$ nm

7. $\lambda = \dfrac{h}{mv}$

$$v = \frac{h}{m\lambda} = \frac{6.63 \times 10^{-34} \text{ J/Hz}}{(9.11 \times 10^{-31} \text{ kg}) \times (6.7 \times 10^{-7}\text{m})}$$

$v = 0.109 \times 10^{(-34 + 31 + 7)}$ m/s

$v = 1{,}090$ m/s

9. (a) $r_4 = 0.847$ nm $= 8.47 \times 10^{-10}$ m $\qquad n = 4$

$n\lambda = 2\pi r_4$

$4\lambda = 2\pi \times 8.47 \times 10^{-10}$ m

$\lambda = (\pi/2) \times 8.47 \times 10^{-10}$ m

$\qquad = \dfrac{3.14 \times 8.47}{2} \times 10^{-10}$ m

$\qquad = 13.3 \times 10^{-10}$ m $= 1.33 \times 10^{-9}$ m

$\qquad = 1.33$ nm

(b) $mv = 5.0 \times 10^{-25}$ kg-m/s $\qquad$ (See Problem 7.)

11. (a) To be ionized, the atom's energy must be $E_\infty = 0$ because the electron goes to the $n = \infty$ level. If it starts in the $n = 2$ level, the energy it must gain, ΔE, is:

$\Delta E = E_\infty - E_2$

From Figure 10.30:

$E_2 = -3.40$ eV

$\Delta E = 0 - (-3.40 \text{ eV})$

$\Delta E = 3.40$ eV

(b) The energy of a photon that would ionize the atom equals ΔE.

$E_{\text{photon}} = \Delta E = 3.40$ eV

To find the frequency:

$$E_{\text{photon}} = hf$$

$3.40 \text{ eV} = 4.136 \times 10^{-15}$ eV/Hz $\times f$

$$\frac{3.40 \text{ eV}}{4.136 \times 10^{-15} \text{ eV/Hz}} = f$$

$$f = 8.22 \times 10^{14} \text{ Hz}$$

13. Using Figure 10.32 as a model, the possible energy-level transitions are shown in the figure.

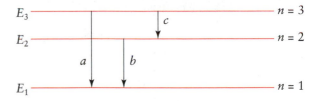

Only *downward* transitions lead to *emission* of photons. The frequency of each photon is proportional to the change in energy of the electron during the transition. On the energy-level diagram, this energy is proportional to the length of the arrow showing the transition. Since a's arrow is longest and c's is shortest,

$$E_a > E_b > E_c$$
$$f_a > f_b > f_c$$

The photon emitted in the $n = 3$ to $n = 1$ level transition has the highest frequency, followed by the $n = 2$ to $n = 1$ photon, and the $n = 3$ to $n = 2$ photon.

15. Number of electrons = $2n^2$

With $n = 1$, 2 electrons may be accommodated.

With $n = 2$, 8 more electrons may be added.

With $n = 3$, 18 additional electrons may be included.

With $n = 3$, a total of $(2 + 8 + 18)$ or 28 electrons can be accommodated in the ground state. To account for 30 electrons in zinc, the minimum value of n must be 4.

17. $K_\alpha = 0.154$ nm and $K_\beta = 0.139$ nm for Cu.

$\Delta E = hf = hc/\lambda$

$$\frac{\Delta E_\alpha}{\Delta E_\beta} = \frac{hc/\lambda_\alpha}{hc/\lambda_\beta} = \frac{hc \times \lambda_\beta}{hc \times \lambda_\alpha}$$

$$\frac{\Delta E_\alpha}{\Delta E_\beta} = \frac{\lambda_\beta}{\lambda_\alpha} = \frac{0.139 \text{ nm}}{0.154 \text{ nm}} = 0.90$$

19. $E = hf = hc/\lambda$ $\lambda = 0.072$ nm $= 0.072 \times 10^{-9}$ m

$$E = \frac{6.63 \times 10^{-34} \text{ J/Hz} \times 3 \times 10^8 \text{ m/s}}{0.072 \times 10^{-9} \text{ m}}$$

$= 2.76 \times 10^{-15}$ J 1 J $= 6.25 \times 10^{18}$ eV

$= 17{,}266$ eV

21. $\Delta E = 0.117$ eV

(a) $\Delta E = hf = 4.136 \times 10^{-15}$ eV/Hz $\times f$

0.117 eV $= 4.136 \times 10^{-15}$ eV/Hz $\times f$

$$f = \frac{0.117 \text{ eV}}{4.136 \times 10^{-15} \text{ eV/Hz}}$$

$= 2.83 \times 10^{13}$ Hz

(b) This radiation is in the infrared region of the spectrum.   (See Figure 10.7.)

23. $f_{\text{K-shell}} \propto Z^2$ The element with the largest value of Z will have the largest (highest) frequency; that element with the smallest value of Z will have the smallest (lowest) frequency.

(a) Highest frequency: Iridium ($Z = 77$)

(b) Lowest frequency: Calcium ($Z = 20$)

CHAPTER 10   Challenges

5. *Circ.* $= 3{,}140$ nm $= 3.14 \times 10^{-6}$ m

$f = 6.5 \times 10^9$ Hz, $T = 1.5 \times 10^{-10}$ s

$v = 21{,}000$ m/s   (much smaller than c)

CHAPTER 11   Problems

1. (a) carbon-14

$A = 14$ $Z = 6$ $N = 8$

6 protons and 8 neutrons

(b) calcium-45

$A = 45$ $Z = 20$ $N = 25$

20 protons and 25 neutrons

(c) silver-108

$A = 108$ $Z = 47$ $N = 61$

47 protons and 61 neutrons

(d) radon-225

$A = 225$ $Z = 86$ $N = 139$

86 protons and 139 neutrons

(e) plutonium-242

$A = 242$ $Z = 94$ $N = 148$

94 protons and 148 neutrons

3. $^{110}_{47}\text{Ag} \to {}^{110}_{48}\text{Cd} + {}^{0}_{-1}\text{e}$

$\uparrow$
daughter

5. $^{210}_{84}\text{Po} \to {}^{206}_{82}\text{Pb} + {}^{4}_{2}\text{He}$

$\uparrow$
daughter

7. $^{107}_{47}\text{Ag*} \to {}^{107}_{47}\text{Ag} + \gamma$

$\uparrow$
daughter

9. ${}^{1}_{0}\text{n} + {}^{235}_{92}\text{U} \to {}^{236}_{92}\text{U*} \to {}^{143}_{57}\text{La} + {}^{A}_{Z}\text{[?]} + 3{}^{1}_{0}\text{n}$

A: $1 + 235 = 236 = 143 + A + 3 = 146 + A$

$236 - 146 = A$

$A = 90$

Z: $0 + 92 = 92 = 57 + Z + 0$

$92 - 57 = Z$

$Z = 35$

$^{90}_{35}\text{Br}$ bromine

11. $^{15}_{7}\text{N} + {}^{1}_{1}\text{H} \to {}^{12}_{6}\text{C} + {}^{4}_{2}\text{He}$

13. 4,000 cts/min to 1,000 cts/min in 12 min

$\frac{1}{4}$ as large   2 half lives   12 min $= 2t_{1/2}$

$t_{1/2} = 6$ min.

15. After 1 half-life, count rate is reduced to $\frac{1}{2}$.

After 2 half-lives, count rate is reduced to $\frac{1}{4}$.

After 3 half-lives, count rate is reduced to $\frac{1}{8}$.

Three half-lives must elapse for count rate to drop to a safe level.

$t_{1/2} = 3$ days

Thus, 9 days must elapse for safety.

17. $E = mc^2$ $m = 1.0$ gm $= 10^{-3}$ kg $c = 3 \times 10^8$ m/s

$E = 10^{-3}$ kg $\times (3 \times 10^8$ m/s$)^2$

$= 9 \times 10^{13}$ J

19. $^{48}_{20}\text{Ca} + {}^{244}_{94}\text{Pu} \to {}^{289}_{114}\text{Fl} + 3{}^{1}_{0}\text{n}$

$A = 48 + 244 - 3 = 48 + 241 = 289$

21. Mass defect $= 3p + 4n - m_{\text{Li-7}}$

In amu: Mass defect $= (3 \times 1.00730$ u$) + (4 \times 1.00869$ u$)$
$- 7.01600$ u

Mass defect $= 3.0219$ u $+ 4.0348$ u $- 7.0160$ u

Mass defect $= 7.0567$ u $- 7.0160$ u $= 0.0407$ u

In kg: 1 u $= 1.66 \times 10^{-27}$ kg

0.0407 u $= 0.0407 \times (1.66 \times 10^{-27}$ kg$)$

Mass defect $= 6.7496 \times 10^{-29}$ kg

CHAPTER 11   Challenges

3. (a) uranium-238: 8 alpha decays
uranium-235: 7 alpha decays

(b) 1:1

(c) approximately 1:80
(80 ${}^{207}_{82}$Pb nuclei for each ${}^{235}_{92}$U nucleus)

(d) approximately 1:1.7
(138 ${}^{207}_{82}$Pb nuclei for every 80 ${}^{206}_{82}$Pb nuclei)

5. $m(^2\text{H}) = 2.01410$ u

$^{2}_{1}\text{H}$: $A = 2$ $Z = 1$ $N = 1$

$\Delta m = 0.00244$ u $= 4.05 \times 10^{-30}$ kg   (See Problem 21.)

$E_B = 3.65 \times 10^{-13}$ J $= 2.28$ MeV

Binding energy per nucleon $= 1.14$ MeV/nucleon (See Figure 11.16)

7. $E_\gamma = 0.186$ MeV $= 186{,}000$ eV 1 J $= 6.25 \times 10^{18}$ eV

$$E_\gamma = \frac{186{,}000 \text{ eV}}{6.25 \times 10^{18} \text{ eV/J}} = 2.98 \times 10^{-14} \text{ J}$$

$E_\gamma = \dfrac{hc}{\lambda}$ $\lambda = \dfrac{hc}{E_\gamma}$

$$\lambda = \frac{(6.63 \times 10^{-34} \text{ J/Hz}) \times (3 \times 10^8 \text{ m/s})}{2.98 \times 10^{-14} \text{ J}}$$

$$= \frac{1.99 \times 10^{-25} \text{ J-m}}{2.98 \times 10^{-14} \text{ J}}$$

$\lambda = 6.68 \times 10^{-12}$ m $= 6.68$ pm (1 pm $= 10^{-12}$ m)

CHAPTER 12   Problems

1. $\Delta t' = 8.3 \times 10^{-8}$ s   (See Example 12.1)

3. $\Delta t' = 3{,}460$ s $= 57.7$ min.   (See Example 12.1)

5. $L' = L\sqrt{(1 - v^2/c^2)} = 1.0$ m $\sqrt{(1 - (0.95c)^2/c^2)}$

$= 1.0$ m $\sqrt{(1 - 0.9025)} = 0.312$ m

The meter stick's length is reduced to 31.2 cm, or shortened by a factor of about 3.2.

7. $E_0 = mc^2 = (1.673 \times 10^{-27} \text{ kg}) \times (3 \times 10^8 \text{ m/s})^2$
$$= 1.506 \times 10^{-10} \text{ J}$$
$E_0 = 1.506 \times 10^{-10} \text{ J} \times (6.25 \times 10^{18} \text{ eV/J}) = 9.41 \times 10^8 \text{ eV}$
$$= 9.41 \times 10^8 \text{ eV} \times 10^{-6} \text{ MeV/eV}$$
$$= 9.41 \times 10^2 \text{ MeV} = 941 \text{ MeV}$$
$m = 941 \text{ MeV}/c^2$ (See Table 12.3)

9. $E_0 = mc^2 = 1.0 \text{ kg} \times (3 \times 10^8 \text{ m/s})^2$
$$= 9.0 \times 10^{16} \text{ J}$$
$E =$ (mass of water) $\times$ (latent heat of fusion)
$$= 1.0 \text{ kg} \times (334{,}000 \text{ J/kg}) = 3.34 \times 10^5 \text{ J}$$
$$\frac{E_0}{E} = \frac{9.0 \times 10^{16} \text{ J}}{3.34 \times 10^5 \text{ J}} = 2.69 \times 10^{11} \text{ (269 billion times larger)}$$

11. $E_{\text{rel}} = \dfrac{mc^2}{\sqrt{(1 - v^2/c^2)}} = \dfrac{140 \text{ MeV}}{\sqrt{(1 - v^2/c^2)}} = 280 \text{ MeV}$
$\sqrt{(1 - v^2/c^2)} = 0.5$
$1 - v^2/c^2 = 0.25$
$v^2/c^2 = 0.75$
$v = 0.87\,c$

13. $E_0 = mc^2 = (2m_pc^2) = 2(1.673 \times 10^{-27} \text{ kg}) \times$
$(3 \times 10^8 \text{ m/s})^2 = 3.01 \times 10^{-10} \text{ J}$
$$= 3.01 \times 10^{-10} \text{ J} \times (6.25 \times 10^{18} \text{ eV/J}) = 1.88 \times 10^9 \text{ eV}$$
$$= 1.88 \times 10^9 \text{ eV} \times 10^{-6} \text{ MeV/eV} = 1.88 \times 10^3 \text{ MeV}$$
$$= 1{,}880 \text{ MeV}$$

15. $z = 0.2 \qquad \lambda_{\text{emit}} = 393.3 \text{ nm}$
$z = \dfrac{\lambda_{\text{obs}} - \lambda_{\text{emit}}}{\lambda_{\text{emit}}}$
$0.2 = \dfrac{\lambda_{\text{obs}} - 393.3 \text{ nm}}{393.3 \text{ nm}}$
$0.2 \times 393.3 \text{ nm} = \lambda_{\text{obs}} - 393.3 \text{ nm}$
$78.7 \text{ nm} = \lambda_{\text{obs}} - 393.3 \text{ nm}$
$\lambda_{\text{obs}} = 78.7 \text{ nm} + 393.3 \text{ nm} = 472.0 \text{ nm}$
$\lambda_{\text{obs}} > \lambda_{\text{emit}} \Rightarrow$ redshift

17. $v = 140{,}000 \text{ km/s} \qquad H_0 = 20.8 \text{ km/s/Mly}$
$v = H_0 d$
$140{,}000 \text{ km/s} = (20.8 \text{ km/s/Mly}) \times d$
$d = \dfrac{140{,}000 \text{ km/s}}{20.8 \text{ km/s/Mly}} = 6{,}730 \text{ Mly}$
$H_0 + 0.05H_0 = 21.8 \text{ km/s/Mly} \Rightarrow d_{\min} = 6{,}420 \text{ Mly}$
$H_0 - 0.05H_0 = 19.8 \text{ km/s/Mly} \Rightarrow d_{\max} = 7{,}070 \text{ Mly}$
$6{,}420 \text{ Mly} \le d \le 7{,}070 \text{ Mly}$

19. $\Delta t_d = 1 \text{ h} = 3600 \text{ s} \qquad d = 10^4 \text{ m} \qquad M_{\text{WD}} = 2.8 \times 10^{30} \text{ kg}$
$\Delta t_f = \dfrac{\Delta t_d}{\left\{1 - \left[\dfrac{2GM}{dc^2}\right]\right\}^{1/2}}$
$$= \dfrac{3600}{\left\{1 - \dfrac{(2 \times 6.67 \times 10^{-11} \times 2.8 \times 10^{30})}{(10^4 \times (3 \times 10^8)^2)}\right\}^{1/2}} \text{ (SI units)}$$
$\Delta t_f = \dfrac{3600}{\left\{1 - \dfrac{3.74 \times 10^{20}}{9 \times 10^{20}}\right\}^{1/2}} = \dfrac{3600}{\{1 - 0.416\}^{1/2}} = \dfrac{3600}{\{0.584\}^{1/2}}$
$\Delta t_f = \dfrac{3600}{0.764} = 4710 \text{ s} = 1.31 \text{ h}$

21. $\bar{\text{n}} = \bar{\text{d}}\,\bar{\text{d}}\,\text{u};\ \bar{\text{p}} = \bar{\text{u}}\,\bar{\text{u}}\,\text{d}$

23. $\text{u}\bar{\text{u}}, \text{d}\bar{\text{d}}, \text{t}\bar{\text{t}}, \text{c}\bar{\text{c}}, \text{s}\bar{\text{s}}, \text{b}\bar{\text{b}}, \text{u}\bar{\text{t}}$, and $\bar{\text{u}}\text{t}$

25. d d d; all three quarks must have their spins aligned to yield a net spin of 3/2

27. $\text{d}\bar{\text{c}}$

29. (a) $\pi^+ + \text{p} \rightarrow \Sigma^+ + \text{K}^+$ (Use Table 12.3)
$(\text{u}\bar{\text{d}}) + (\text{uud}) \rightarrow (\text{uus}) + (\text{u}\bar{\text{s}})$
Canceling the common quarks on both sides of the arrow yields:
$(\text{d} + \bar{\text{d}}) \rightarrow (\text{s} + \bar{\text{s}})$
The annihilation of the d and anti-d quarks produces energy that is used to create a new $(\text{s} + \bar{\text{s}})$ combination.
(b) $\gamma + \text{n} \rightarrow \pi^- + \text{p}$
$\gamma + (\text{udd}) \rightarrow (\text{d}\bar{\text{u}}) + (\text{uud})$
$\gamma \rightarrow (\text{u} + \bar{\text{u}})$
The energy of the gamma ray is converted into mass as a $(\text{u} + \bar{\text{u}})$ pair.
(c) $\text{p} + \text{p} \rightarrow \text{p} + \text{p} + \text{p} + \bar{\text{p}}$
$(\text{uud}) + (\text{uud}) \rightarrow (\text{uud}) + (\text{uud}) + (\text{uud}) + (\bar{\text{u}}\,\bar{\text{u}}\,\bar{\text{d}})$
KE (of colliding protons) $\rightarrow (\text{uud}) + (\bar{\text{u}}\,\bar{\text{u}}\,\bar{\text{d}})$
$KE \rightarrow 2(\text{u} + \bar{\text{u}}) + (\text{d} + \bar{\text{d}})$
The original KE is used to create two $(\text{u} + \bar{\text{u}})$ pairs plus a $(\text{d} + \bar{\text{d}})$ pair.
(d) $\text{K}^- + \text{p} \rightarrow \text{K}^+ + \text{K}^0 + \Omega^-$
$(\text{s}\bar{\text{u}}) + (\text{uud}) \rightarrow (\text{u}\bar{\text{s}}) + (\text{d}\bar{\text{s}}) + (\text{sss})$
$(\text{u} + \bar{\text{u}}) \rightarrow 2(\text{s} + \bar{\text{s}})$
The annihilation energy of the $(\text{u} + \bar{\text{u}})$ pair is used to create two $(\text{s} + \bar{\text{s}})$ pairs.

31. The $(\text{t}\bar{\text{b}})$ meson has $Q = (+2/3) + (+1/3) = +1$, topness equal to $+1$, and bottomness equal to $+1$. The spin of the particle will be either 0 if the spins of the quark and antiquark are opposed or $+1$ if the spins are parallel. Either case is consistent with the spins of the mesons having integer values.

CHAPTER 12 Challenges

1. $KE_{\text{rel}} = \dfrac{mc^2}{\sqrt{(1 - v^2/c^2)}} - mc^2$
For $v \ll c$, $(1 - v^2/c^2)^{-1/2} = 1 + \frac{1}{2}v^2/c^2$
Thus, $KE_{\text{rel}} = mc^2 \times (1 + \frac{1}{2}v^2/c^2) - mc^2$
$$= mc^2 + \tfrac{1}{2}mv^2 - mc^2 = \tfrac{1}{2}mv^2$$

3. $E_p + E_{\bar{p}} = 2E_\gamma = 2hf$
$m_pc^2 + m_{\bar{p}}c^2 = 2hf$
Since $m_{\bar{p}} = m_p$, $2m_pc^2 = 2hf \qquad$ Or,
$m_pc^2 = hf$
$1.67 \times 10^{-27} \text{ kg} \times (3 \times 10^8 \text{ m/s})^2 = 6.63 \times 10^{-34} \text{ J/Hz} \times f$
$\dfrac{1.67 \times 10^{-27} \text{ kg} \times (3 \times 10^8 \text{ m/s})^2}{6.63 \times 10^{-34} \text{ J/Hz}} = f$
$2.27 \times 10^{23} \text{ Hz} = f$
Since $\lambda f = c$, $\lambda = c/f = \dfrac{3 \times 10^8 \text{ m/s}}{2.27 \times 10^{23} \text{ Hz}}$
$\lambda = 1.32 \times 10^{-15} \text{ m}$
These photons have wavelengths in the gamma-ray region of the spectrum.

5. Mesons are now believed to be composed of a quark and an antiquark. The largest possible magnitude of total charge (in units of the basic charge $e = 1.60 \times 10^{-19}$ C) for such a combination is $(2/3 + 1/3) = 1$. If mesons with charge $+2$ were discovered, the composition of this class of subatomic particles within the quark model would demand 4 quarks, combined in $q\bar{q}$ pairs in ways that preserved the properties of the already known mesons having charge 0 and ± 1. Such a discovery would also permit the existence of mesons with spin = 2, formed by the 4 quarks, all with spin $\frac{1}{2}$, aligned together in spin-up or spin-down configurations.

7. If the clock at rest on Earth measures an interval of time equal to 5 years, the shipboard clock of the traveling (younger) twin will record a time of only 5 years $\times \sqrt{1 - (0.95)^2}$ or 1.56 years.

9. $M_E = 5.98 \times 10^{24}$ kg $R_E = 6.38 \times 10^6$ m

(a) $v_e = \left(\dfrac{2GM}{R}\right)^{1/2}$

$v_e = \left[\dfrac{(2 \times 6.67 \times 10^{-11} \times 5.98 \times 10^{24})}{6.38 \times 10^6}\right]^{1/2}$ (SI units)

$v_e = \left[\dfrac{(7.98 \times 10^{14})}{6.38 \times 10^6}\right]^{1/2} = [1.25 \times 10^8]^{1/2}$

$v_e = 1.12 \times 10^4$ m/s $= 11{,}200$ m/s

(b) $v_e = c = \left(\dfrac{2GM}{R}\right)^{1/2}$

$c^2 = \dfrac{2GM}{R} \Rightarrow R = \dfrac{2GM}{c^2} \equiv d_S$

(c) $d_{SE} = \dfrac{2GM_E}{c^2}$

$d_{SE} = \dfrac{(2 \times 6.67 \times 10^{-11} \times 5.98 \times 10^{24})}{(3 \times 10^8)^2}$ (SI units)

$d_{SE} = 8.86 \times 10^{-3}$ m $= 8.86$ mm

density $=$ mass/volume $V = \left(\tfrac{4}{3}\right)\pi r^3$ $r = d_{SE}$

density $= \dfrac{5.98 \times 10^{24}}{\left(\tfrac{4}{3}\right)\pi (8.86 \times 10^{-3})^3}$ (SI units)

density $= \dfrac{5.98 \times 10^{24}}{2.91 \times 10^{-6}} = 2.05 \times 10^{30}$ kg/m^3

ratio $= \dfrac{2.05 \times 10^{30} \text{ kg/m}^3}{5514 \text{ kg/m}^3} = 3.72 \times 10^{26}$ times greater

Answers to Physics to Go Activities (Some activities do not have explicit answers.)

Chapter 1

Physics to Go 1.1

1. The number is the period of the motion.
2. The number is the frequency of the motion.
3. The period $= \dfrac{1}{\text{frequency}}$, and the frequency $= \dfrac{1}{\text{period}}$
4. A longer string makes the period larger and the frequency smaller.

Physics to Go 1.2

Runners can keep up their fastest pace for only about 20 seconds; then fatigue sets in. Friction makes swimming speeds much lower than running speeds. (More on these matters in Chapters 2 and 3.)

Physics to Go 1.3

Light travels almost a million times faster than sound. The lightning is approximately 1 mile away for every 5 seconds it takes the sound of the thunder to reach your ears. (Or 1 kilometer away for every 3 seconds.)

Physics to Go 1.4

The acceleration in g's $= 1 \div$ number of seconds it takes the speed to change by 22 mph.

Chapter 2

Physics to Go 2.1

1. No. Things with rougher surfaces (more friction) require a steeper book.
2. The book angle is usually much less steep. Force of rolling friction is usually much smaller than that of static friction.

Physics to Go 2.2

To the left or right, depending on which direction you whirled the sock. (See Figure 2.12)

Physics to Go 2.3

Higher speeds require larger forces on the string. Centripetal force increases if speed increases.

Physics to Go 2.4

45°, if the nozzle is near the ground.

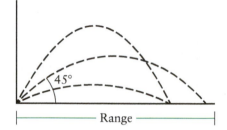

Physics to Go 2.5

1. No. The wadded filter hits the floor first. No. Size/area also matters.
2. No. A larger object can fall faster than a smaller one if it is much heavier; it requires a greater distance to reach its terminal speed.

Physics to Go 2.7

2. Place the pins right next to each other.

Chapter 3

Physics to Go 3.1

As you are moving horizontally, your feet exert horizontal forces on the floor, but little work is done because the floor is not appreciably displaced or distorted as a result of these applied forces; as a consequence, you should not feel especially tired. Your applied force is perpendicular to the gravitational force (your weight), so your legs do no work against gravity. As you are going up stairs, your legs do work lifting your body weight vertically, so you tire at a rate that depends on your physical condition and how high you climb.

Physics to Go 3.2

1. The farther up the ramp the battery is when released, the faster it is going when it reaches the bottom. It starts with more potential energy, so it gains more kinetic energy.
2. It is a tie. The lower battery ends up going one-half as fast as the upper one because its vertical drop is one-fourth as much ($v = \sqrt{2gh}$). But the total distance it has to travel is considerably less. The total time it takes each battery to get to the backstop is the same (ignoring kinetic friction). The lower battery wins (barely) if it is started farther up the ramp, and the upper one wins if the lower one is started below one-fourth of the way up.

Physics to Go 3.3

The power output depends on how much you weigh and how fast you go up the stairs. It is typically between 250 W (one-third hp) and 1,500 W (2 hp).

Physics to Go 3.4

You speed up—spin faster—when you pull your arms in. Each part of your arms has a certain amount of angular momentum as it moves in a circle with some radius. When you pull your arms in, the radius decreases, which makes the angular momentum decrease. Consequently, the rest of your body and the chair spin faster so that the total angular momentum remains constant. The reverse happens when you move your arms out: you slow down.

Chapter 4

Physics to Go 4.1

Depending on how hard you squeeze, the point will cause a stronger sensation or even pain. The area of the point is much smaller than that of the other end, so the pressure is much larger. Nerves in your fingers respond to this higher pressure.

Physics to Go 4.2

The can is crushed by the atmosphere. The boiling causes steam in the can to drive out much of the air. When the can is inverted into the water, much of the steam quickly condenses back to liquid water, causing the pressure inside the can to drop. The air pressure on the outside of the can is no longer balanced by pressure on the inside, so the can is crushed.

Physics to Go 4.3

One simple way is to choose packages with the same weight (or mass), like one pound of flour and one pound of cereal. Then the material in the package with the smaller volume has the higher density. Glass containers should be avoided because glass adds a lot more to the mass and weight of the package than does cardboard or plastic. Substances with nearly the same density would require measurements (weight, volume) to determine which has higher density.

Physics to Go 4.4

The water shoots out farthest, and therefore is going fastest, from the bottom hole. It is going slowest from the top hole. This is because the pressure is highest at the lowest hole (the greatest depth).

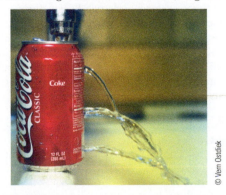

© Vern Ostdiek

Physics to Go 4.5

The weight of the water causes the cardboard or paper to bulge downward slightly. This increase in volume lowers the pressure inside the glass, so the (higher) atmospheric pressure outside causes a large enough upward force on the cardboard or paper to support the water and keep it from spilling out.

Physics to Go 4.6

The rubber band is stretched less with the cup in the water, because the water causes an upward buoyant force on the cup. Therefore, the rubber band has to supply a smaller upward force to support the weight of the cup.

Physics to Go 4.7

The ice cube floats at the boundary of the gasoline and the water. Since ice's density is greater than that of gasoline but less than that of water (see Table 4.4), it sinks in gasoline and floats in water.

Physics to Go 4.8

The moving air above the paper is at a lower pressure, so the paper is pushed up by the higher pressure below it.

Chapter 5

Physics to Go 5.1

Unless some of them are heated and others not (for example, some in sunlight, or some near a heating or cooling vent), they should show about the same temperature. Typically, the readings will vary by a degree or two because the thermometers are not designed to be precision instruments.

Physics to Go 5.2

See Figure 5.8. The two pieces of roadway pull away from each other a bit as their temperature drops.

Physics to Go 5.3

The water level is lower after the balloon has been removed from the hot water because the air in the balloon occupies a larger volume (so it displaces more water) when its temperature is higher. The air expands as it is warmed.

Physics to Go 5.4

Parts of the pump should be warmer after you pump up the tire. The work you do in compressing the air heats it. The heated air, in turn, warms the metal parts of the pump.

Physics to Go 5.5

Your hand should warm up faster in #2. In #1, your hand warms up because of heat transferred by radiation. In #2, both radiation and convection of the heated air warm your hand.

Physics to Go 5.6

The time in #4 should be roughly seven times that in #3. To boil a given amount of water to steam, it must gain much more internal energy than it takes to raise its temperature from room temperature to the boiling point. Note: While the water is boiling, the can and water stay at about 100°C, so the paint doesn't get hot enough to smoke. Once the water is boiled away, the can's temperature rises rapidly.

Chapter 6

Physics to Go 6.1

1. They travel with the same speed.
2. Depending on how much friction there is between the Slinky and the surface, the pulse can reflect and return—perhaps only partway—to your end. The pulse is "inverted" by the reflection: It switches from right to left or from left to right (look ahead to Figure 6.13).

Physics to Go 6.2

The tone or pitch of the bottle with CO_2 in it is lower, because sound travels slower in CO_2 than it does in air.

Physics to Go 6.3

The echoes arrive sooner after the clap as you get closer to the building. (When 15 m away the echo arrives about 0.09 s after the clap; at 10 m, the time is about 0.06 s.)

The ear can distinguish two successive sound pulses only if the second comes about 0.02 s or more after the first. That corresponds to a distance of about 3.5 m from the building.

Physics to Go 6.4

1. The sound is loudest when you are directly in front of the opening.
2. The diffracted sound—what you hear when you are off to one side—has less treble: the higher frequencies are not diffracted as much.

Physics to Go 6.5

2. Your voice should sound muffled, with much less treble in it than in #1.
3. Your voice should have much more treble in it than in #1.
   Some of the sound that you hear from your voice travels through the air to your ears, and some travels through the bones and other tissues in your head. In the latter case, much of the higher-frequency sound does not make it to your inner ear. When you cover your ears, you block the sound reaching your ears through the air.

Physics to Go 6.6

2. Each vowel sound requires a unique positioning of the jaw, tongue, and lips. When you try to make the sound of one vowel using the positioning for another, the result usually does not sound like what you intended.

Physics to Go 6.7

2. As you move farther apart, most of the sound each of you hears is reverberant sound. Each successive spoken syllable is drowned out by the reverberating sound from the preceding one.
3. When you speak slowly, the syllables don't "overlap" as much and it is easier to be understood.
4. The reverberation time is shorter. You can understand each other when you are farther apart.

Physics to Go 6.8

1. The tone quality should be different when your ears are blocked because the higher frequencies present in the complex tone are attenuated.
2. Since a steady whistle is a pure tone consisting of just one frequency, the tone quality should be the same when your ears are blocked.

Chapter 7

Physics to Go 7.1

The pieces of paper and thread should be attracted to the cup, perhaps strongly enough to be pulled up to the cup and stick to it. If the latter happens, the electric force was stronger than the force of gravity.

Physics to Go 7.2

The foil wads should repeatedly jump up to the cup, and then back to the metal. This should last a few seconds. The speed depends on how heavy the wads are and how well charged the cup is. The wads are attracted to the cup, but as soon as they touch it some of the extra electrons on it move onto the wads. This

makes the wads and the cup both negatively charged, so they repel each other and the wads go back down (see Section 7.2). When they touch the metal, the extra electrons flow into them so the wads are again attracted to the cup. The fact that electrons can readily move in metals is the key to this exercise.

Physics to Go 7.3

The cups repel each other because they are both negatively charged. The hanging cup should swing away from the other cup a bit. The closer you place the cups, the larger the force and the farther the string will be deflected from the vertical.

Physics to Go 7.4

The stream of water should be deflected toward the cup.

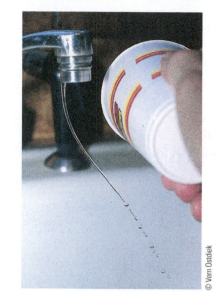

© Vern Ostdiek

Physics to Go 7.5

The rising smoke should be attracted to the Styrofoam cup. This is the principle behind electrostatic precipitators.

Physics to Go 7.6

Fluorescent lights emit light only when a current is flowing through them. When AC is used, the current flows for a short time, is nearly zero momentarily, then flows in the other direction. Because of this, a fluorescent light actually emits "bursts" of light 120 times each second (two bursts during each cycle, 60 cycles each second—in the U.S.). This causes the "stroboscope" effect you see. Incandescent light bulbs do not show this effect nearly as well because they rely on a glowing filament to emit the light. In the short time that the alternating current is nearly zero, the filament does not cool off enough to reduce its light output appreciably.

Chapter 8

Physics to Go 8.1

Depending on how good your compass is, it should agree within 5 degrees or so.

Physics to Go 8.2

The needle should turn quickly to a direction perpendicular to the cable. As an automobile is being started, a huge current flows through the battery and starter motor—typically around 100 amperes. This produces strong magnetic fields around the cables that carry the current, which will deflect a compass needle.

Physics to Go 8.4

The paper should appear yellowish when the lights are very dim. The dimmer reduces the current to the lights, which makes the filaments cooler. This shifts the peak wavelength of their equivalent blackbody radiation to larger values, so there is relatively more red light present.

Physics to Go 8.5

If you examine Aldebaran and Sirius, you should find that the former appears orange in color, while the latter looks blue-white. Aldebaran is about 1000 K hotter than Betelgeuse, and Sirius has about the same temperature as Rigel. The majority of the brightest-appearing stars in the sky will look blue-white in color. Although relatively rare, because of their high surface temperatures, these stars are very luminous and can be seen over great distances. By contrast, the more numerous cool stars are not intrinsically very bright and so cannot be seen very far out in space. Our visual perception of the dominant type of stars in the sky is biased by how strongly the power emitted by a star depends on its temperature (P is proportional to T^4).

Chapter 9

Physics to Go 9.1

Align the images with your head on one side of the normal (perpendicular) direction to the mirror. Use a straightedge to draw a line on the cardboard base along your line of sight. Move your head to the opposite side of the normal and realign the images. Again, use a straightedge to draw a line on the cardboard along your new sight line. Extend the two lines until they meet at the mirror's surface. Now draw in the normal to the mirror at the point of intersection. Use a protractor to measure the angle of incidence and the angle of reflection. You should find they are equal to within small errors of measurement.

Physics to Go 9.2

The green and red beams combine to give the color yellow (additive mixing). Looking through a red filter at a brightly lit scene produces a red image of the scene. If a green filter is now placed behind the red one, the image will go dark (black): No light gets through the second filter. The red filter passes only light with wavelengths near 650 nm. The green filter passes only light with wavelengths near 520 nm. The light emerging from the red filter does not have the proper wavelengths to be transmitted by the green filter, and little or no light passes through it.

Physics to Go 9.3

The intensity of the overhead lights should not change as you rotate the Polaroid lens because the light emitted by fluorescent tubes or incandescent bulbs is unpolarized. The intensity of the reflected light ("glare") should vary as you turn the Polaroid lens, with the relative amount of change depending on the angle of incidence of the light on the reflecting surface. The greatest extinction of the reflected light should occur when the sunglasses are oriented in their "normal" fashion for wearing. The extinction should be the least when the glasses are at 90° to this orientation. Using two pairs of polarizing lenses, you should see a similar effect. This suggests that the reflected light is polarized, although generally not to the same extent as the light leaving the first lens of the crossed sunglasses.

Physics to Go 9.4

The light intensity from the LCD should vary much like that of the reflected light in Physics to Go 9.3 indicating that the LCD emission is polarized too. Performing the same experiment using sky light as the source, you should once again see the same variation in intensity as the Polaroid lens is rotated. This shows that sky light is also partially polarized.

Physics to Go 9.5

The image of your right hand in a mirror is that of a left hand.

Physics to Go 9.6

With an angle of 45°, you should see 8 images, 7 reflected ones and the real one of the original object. At 60°, you should see 6 images, 5 reflections and the original. For a mirror angle of 30°, a total of 12 images should be seen. The total number of images (including the original) equals 360° divided by the angle between the mirrors. The accompanying figure shows a simple design for a kaleidoscope using two mirrors at 60° cemented in a cylindrical tube. For convenience of viewing, one end of the tube is closed off, save for a small peephole that permits visual access to the region between the mirrored surfaces.

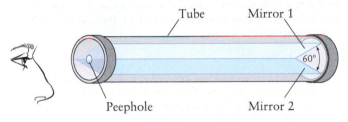

Physics to Go 9.7

If your finger is close enough to the spoon's concave surface, you should see an upright, magnified image. With the spoon at a comfortable distance from your face, the image of your face should be inverted. Using the reverse (convex) side of the spoon, you will see an upright image of your face (and surroundings) that is reduced in size because the image is further behind the spoon's surface than your face is in front of it.

Physics to Go 9.8

You should see a magnified image of your hand or the lamp (see the accompanying figure). As you move your hand/lamp up and down, the size of the image (its magnification) will change accordingly. At higher revolution speeds, the curvature of the oil's surface is greater, yielding a smaller focal length for the concave mirror. For a given object distance, the image distance will then be smaller and so will the resultant magnification.

© Vern Ostdiek

Physics to Go 9.9

Looking obliquely through the side of the beaker at the light emerging from the water's surface, you will not be able to see the entire circle; a portion of it along the perimeter closest to you will be unobservable. As you move your head up and down, the fraction of the circle that "goes missing" decreases and increases, respectively. Light from the unobservable part of the circle strikes the water's surface at an angle greater than the critical angle for total internal reflection at the air-water interface. This light is reflected back into the water and does not cross the boundary between the two media. Thus, no rays from this portion of the circle reach your eyes.

Physics to Go 9.10

The image of the source should be inverted and reduced in size.

Physics to Go 9.11

Employing the technique shown in Figure 9.47, the following table summarizes what you will see as the distance between the source and the magnifying (converging) lens is changed.

Object	Image			
Location	Type	Location	Orientation	Relative Size
$\infty > s > 2f$	Real	$f < p < 2f$	Inverted	Reduced
$s = 2f$	Real	$p = 2f$	Inverted	Same size
$f < s < 2f$	Real	$\infty > p > 2f$	Inverted	Magnified
$s = f$	-	∞	-	-
$s < f$	Virtual	$-p > s$	Upright	Magnified

When the lens is closer than one focal length to the source, the image is virtual and not able to be projected on the screen. Looking back through the lens at the source yields an upright, magnified image.

Physics to Go 9.12

With the screen inside the position of best focus, the outside edges of the blurred image should appear reddish-orange. Moving the screen beyond the point of best focus results in a blurred image whose edges are bluish in color. Compare with Figure 9.53.

Physics to Go 9.13

To create your own personal rainbow, the Sun must be behind you and the water spray generally in front of you (in the direction of your shadow). The rainbow is most easily seen using a mist of small droplets that tend to remain suspended in air and produce a more stable display. The outer portion of the rainbow arc will be colored red, while the inside will have a blue-violet hue. The angular radius of the bow should be about 40° or about two spread-hands in size (see Figure 9.65) providing that the drops are more than a few arm-lengths distant.

Chapter 10

Physics to Go 10.1

You should see a continuous spectrum of all the colors (as in Figure 9.63), from blue (closest to the image of the bulb) through red.

Physics to Go 10.2

You should see separate images of the compact fluorescent lamp (with some overlap), each a different color: blue, two shades of green, yellow, and red. The light from compact fluorescent lamps mainly consists of several narrow bands of color, rather than just sharp lines (as in an emission-line spectrum) or all colors (as in a continuous spectrum).

Physics to Go 10.4

The greater the number of energy levels present, the richer the spectrum—that is, the greater the number of lines of either type (absorption or emission) in the spectrum. As with hydrogen, spectral lines may be aggregated in *series*, groups of transitions all of which end on the same lower level. For a given series, as the upper energy levels crowd together, the wavelength (or frequency) separation between the spectral lines becomes smaller and the lines also get closer together; eventually they merge into what is called the *series limit*, the shortest wavelength (highest frequency) line in the series. Thus, the general trends observed in the spectrum of hydrogen are played out in other more complicated atoms and ions, but the number of allowed series is typically much larger because so many more energy states exist to accommodate the greater numbers of electrons in more massive species.

Chapter 11

Physics to Go 11.1

5. Still 50%. It's the same for each throw, regardless of how many times in a row a coin has turned up "heads."

Chapter 12

Physics to Go 12.1

When rolling the ball along the edge of the mattress starting from a point well removed from the central object and the depressed surface of the mattress, the path should remain very close to a straight line parallel to the boundary of the mattress; the speed of the ball should be approximately constant save for any slight slowing due to frictional effects between the ball's surface and that of the mattress. As the release point moves inward toward the central axis of the mattress, the path of the ball should begin to veer toward the central massive object as it begins to respond to the curvature of the mattress's surface in the vicinity of the large mass. If the perpendicular distance of release point from the symmetry axis is close to the "radius" of the *central attractor*, the ball may actually roll into the depressed space near this mass and be captured by it. As the ball moves toward the attractor, you should notice that its speed increases: it accelerates. As the release point moves inward, in order to avoid the ball being captured, you should find that the initial speed (and kinetic energy) of the ball must be increased. All these effects attend the motion of celestial objects in space—motion that Newton would ascribe to the action of forces (gravity), but that Einstein would assign to the curvature of spacetime.

Physics to Go 12.2

The value for r that emerges from this exercise depends, of course, on the degree to which Figure 12.20 is enlarged. Higher magnifications (within reason) favor more accurate results. With careful execution, it is common for the experimental value of r to agree to better than 10% with the value directly measured from the target.

GLOSSARY

This glossary includes definitions and laws that have been highlighted in the text, along with selected words or phrases that are important and appear in several places in the text. Terms that are narrow in scope or are used in only one section or chapter (such as fiber optics and reverberation) are not included here.

Acceleration (a) Rate of change of velocity. The change in velocity divided by the time elapsed. A vector quantity.

Alternating current (AC) An electric current consisting of charges flowing back and forth.

Amplitude Maximum displacement of points on a wave, measured from the equilibrium position.

Angular momentum The mass of an object times its speed, times the radius of its path.

Antiparticle A charge-reversed version of an ordinary particle. A particle of the same mass (and spin) but of opposite electric charge. The positron (anti-electron) is an example.

Archimedes' principle The buoyant force acting on a substance in a fluid at rest is equal to the weight of the fluid displaced by the substance.

Atom The smallest unit or building block of a chemical element.

Atomic spectrum A display of the discrete set of wavelengths (or frequencies) contained in the dispersed light from a hot gas of atoms. Unique, characteristic set of emission (or absorption) features contained in luminous atomic sources. May be used to determine the composition of gaseous mixtures.

Baryon A strongly interacting particle that is composed of three quarks. Protons and neutrons are examples. Baryons are fermions.

Bernoulli's principle When the speed of a moving fluid *increases,* the pressure *decreases,* and vice versa.

Big Bang The explosive event at the beginning of the universe nearly 14 billion years ago. The expansion produced by this event continues today and is reflected in the movement of distant galaxies away from one another.

Binding energy Amount of work required to disassemble a composite system and separate its constituent parts to infinity. Amount of energy released when a system of particles is assembled under their mutual forces of attraction. Energy released during a nuclear fission or fusion reaction.

Blackbody A hypothetical object that absorbs all light and other electromagnetic waves that strike it. The radiation it emits is called *blackbody radiation* (BBR).

Boson A particle with integral spin. Bosons do not obey the Pauli exclusion principle.

Buoyant force The upward force exerted by a fluid on a substance partly or completely immersed in it.

Centripetal acceleration Acceleration of an object moving along a circular path. It is directed toward the center of the circle.

Centripetal force Name applied to the force acting to keep an object moving along a circular path. It is directed toward the center of the circle.

Compound A pure substance consisting of two or more elements combined chemically.

Conduction The transfer of heat by direct contact between atoms and molecules.

Convection The transfer of heat by buoyant mixing in a fluid.

Cosmology The study of the structure and evolution of the universe as a whole.

Coulomb's law The force acting on each of two charged objects is directly proportional to the net charges on the objects and inversely proportional to the square of the distance between them.

Current (I) Rate of flow of electric charge. The amount of charge that flows past a given point per second. The unit of measure is the ampere.

Diffraction The bending of a wave as it passes around the edge of a barrier. Diffraction causes a wave passing through a gap or a slit to spread out into the shadow regions.

Direct current (DC) An electric current that is a steady flow of charge in one direction.

Dispersion The variation in the speed of propagation of a periodic wave with the frequency of the wave. Dispersion is responsible for the separation of white light into a spectrum of colors by a prism.

Doppler effect The apparent change in frequency of a wave due to motion of the source of the wave, the receiver, or both.

Echolocation Process of using the reflection of a wave to locate objects.

Elastic collision Collision in which the total kinetic energy of the colliding objects is the same before and after the collision.

Electric charge (q) An inherent physical property of certain subatomic particles that is responsible for electrical and magnetic phenomena. The unit of measure is the coulomb.

Electric field The agent of the electric force. It exists in the space around charged objects and causes a force to act on any other charged object. A vector quantity.

Electromagnetic force (or interaction) The force that acts between charged particles or objects. It can be either attractive or repulsive, and it is responsible for most of the common forces in our everyday life like contact forces, friction, and elastic forces.

Electromagnetic wave A transverse wave consisting of a traveling combination of oscillating electric and magnetic fields.

Electron Negatively charged particle, usually found orbiting the nucleus of an atom.

Element One of the 118 different fundamental substances that are the simplest and purest forms of common matter.

Elementary particles The basic, indivisible building blocks of the universe. The fundamental constituents from which all matter, antimatter, and their interactions derive. They are believed to be true "point" particles devoid of internal structure or measurable size.

Energy (E) The measure of a system's capacity to do work. That which is transferred when work is done.

Fermion A particle with half-integral spin. Fermions obey the Pauli exclusion principle.

First law of thermodynamics The change in internal energy of a substance equals the work done on it plus the heat transferred to it.

Force (F) A push or a pull acting on a body. Usually causes some distortion of the body, a change in its velocity, or both.

Frequency (f) The number of cycles of a periodic process that occurs per unit time. The unit of measure is the hertz (Hz).

Frequency (f) (of a wave) The number of cycles of a wave passing a point per unit time. It equals the number of oscillations per second of the wave.

Friction A force of resistance to relative motion between two bodies or substances.

Gamma rays The highest-frequency electromagnetic waves. One type of nuclear radiation.

Gravitational force The attractive force that acts between all pairs of objects.

Half-life The time it takes for one-half of the nuclei of a radioisotope to decay. The time interval during which each nucleus has a 50% probability of decaying.

Heat (Q) The form of energy that is transferred between two substances at different temperatures.

Heat engine A device that converts heat into mechanical energy or work. It absorbs heat from a hot source such as burning fuel, converts some of this energy into usable mechanical energy or work, and outputs the remaining energy as heat at some lower temperature.

Hubble relation (or law) A mathematical expression showing that the farther a galaxy is from us, the faster it is moving away. One implication of this relation is that the universe is expanding.

Humidity The mass of water vapor in the air per unit volume. The density of water vapor in the air.

Inelastic collision Collision in which the total kinetic energy of the colliding bodies after the collision is not equal to the total kinetic energy before.

Infrared (IR) Electromagnetic waves with frequencies just below those of visible light. Infrared is the main component of heat radiation.

Interference The consequence of two waves arriving at the same place and combining. Constructive interference occurs wherever the two waves meet in phase (peak matches peak); the waves add together. Destructive interference occurs wherever the two waves meet out of phase (peak matches valley); the waves cancel each other.

Intermediate (or gauge) boson A force-carrying elementary particle that mediates the fundamental interactions of Nature. The photon is an example.

Internal energy (U) The sum of the kinetic energies and potential energies of all the atoms and molecules in a substance.

Isotopes Isotopes of a given element have the same number of protons in the nucleus but different numbers of neutrons.

Kinetic energy (KE) Energy due to motion. Energy that an object has when it is moving.

Kinetic friction Friction between two substances that are in contact and moving relative to each other.

Laser   A beam of coherent monochromatic light. Acronym for "light amplification by stimulated emission of radiation."

Law of conservation of angular momentum   The total angular momentum of an isolated system is constant.

Law of conservation of energy   Energy cannot be created or destroyed, only converted from one form to another. The total energy in an isolated system is constant.

Law of conservation of linear momentum   The total linear momentum of an isolated system is constant.

Law of fluid pressure   The (gauge) pressure at any depth in a fluid at rest equals the weight of the fluid in a column extending from that depth to the "top" of the fluid divided by the cross sectional area of the column.

Law of reflection   The angle of incidence equals the angle of reflection.

Law of refraction   A light ray is bent toward the normal when it enters a transparent medium in which light travels more slowly. It is bent away from the normal when it enters a medium in which light travels faster.

Lepton   An elementary particle that does not interact via the strong nuclear force. Electrons and neutrinos are examples. Leptons are fermions.

Linear momentum   The mass of an object times its velocity.

Longitudinal wave   Wave in which the oscillations are along the direction the wave travels.

Magnetic field   The agent of the magnetic force. It exists in the space around magnets and causes a force to act on each pole of any other magnet. A vector quantity.

Mass (m)   A measure of an object's resistance to acceleration. A measure of the quantity of matter in an object.

Mass density (D)   The mass per unit volume of a substance. The mass of a quantity of a substance divided by the volume it occupies.

Mass number (A)   The total number of protons and neutrons in a nucleus.

Meson   A particle that is composed of a quark and an antiquark. Mesons are bosons.

Microwaves   Electromagnetic waves with frequencies between those of radio waves and infrared light.

Molecule   Two or more atoms combined to form the smallest unit of a chemical compound.

Neutrino   An elementary particle with no charge and little mass. Many important processes in astronomy release neutrinos, including supernova explosions and fusion reactions in stars. Neutrinos are fermions.

Neutron   Electrically neutral particle residing in the nucleus of an atom.

Neutron number (N)   The number of neutrons contained in a nucleus.

Newton's first law of motion   An object will remain at rest or in motion with constant velocity unless acted on by a net external force.

Newton's law of universal gravitation   Every object exerts a gravitational pull on every other object. The force is proportional to the masses of both objects and inversely proportional to the square of the distance between them.

Newton's second law of motion   An object is accelerated whenever a net external force acts on it. The net force equals the object's mass times its acceleration.

Newton's third law of motion   Forces always come in pairs: when one object exerts a force on a second object, the second exerts an equal and opposite force on the first.

Nuclear fission   The splitting of a large nucleus into two smaller nuclei. Free neutrons and energy are also released.

Nuclear fusion   The combining of two small nuclei to form a larger nucleus, accompanied by the release of energy.

Ohm's law   The current in a conductor is equal to the voltage applied to it divided by its resistance.

Orbit   The path of a body as it moves under the influence of a second body. An example is the path of a planet or a comet as it moves around the Sun.

Pascal's principle   Pressure applied to an enclosed fluid is transmitted undiminished to all parts of the fluid and to the walls of the container.

Pauli exclusion principle   Two interacting fermions of the same type, such as electrons in an atom, cannot be in exactly the same state.

Period (T)   The time for one complete cycle of a periodic process.

Photon   The quantum of electromagnetic radiation. The energy of a photon equals Planck's constant times the frequency of the radiation.

Plasma   An ionized gas, often referred to as the fourth state of matter. It consists of positive ions and electrons.

Polarization   A term used to describe the orientation of the plane of vibration of a transverse wave. Light can be polarized by passing it through a Polaroid filter that absorbs all waves except those vibrating along a direction parallel to its transmission axis.

Potential energy (PE)   Energy due to an object's position. Energy that a system has because of its configuration.

Power (P)   The rate of doing work. The rate at which energy is transferred or transformed. Work done divided by the time. Energy transferred divided by the time.

Pressure (p)   The force per unit area for a force acting perpendicular to a surface. The force acting on a surface divided by its area.

Proton   Positively charged particle residing in the nucleus of an atom.

Quantum mechanics   A mathematical model that describes the behavior of particles on an atomic and subatomic scale. This theory demonstrates that matter and energy are quantized (come in small discrete bundles) on smallest scales imaginable.

Quark   A fractionally charged elementary particle found only in combination with other quarks in mesons and baryons. Quarks are fermions.

Radioactivity (or radioactive decay)   A process wherein an unstable nucleus spontaneously emits particles or electromagnetic radiation (or both). Results in a change in the nuclear atomic number or the release of energy, or both. Common examples include alpha (α), beta (β), and gamma (γ) decay.

Refraction   The bending of a wave as it crosses the boundary between two media with different densities.

Relative humidity   Humidity expressed as a percentage of the saturation density.

Relativity   Theory developed by Einstein connecting measurements of space, time, and motion made by one observer to those made by another observer in a different environment. The special theory of relativity relates experimental results for two observers moving at high speed with respect to one another; the general theory of relativity relates observations made in strong gravitational fields to those found in weak fields.

Resistance (R)   A measure of the opposition that a substance has to current flow. The unit of measure is the ohm.

Second law of thermodynamics   No device can be built that will repeatedly extract heat from a heat source and deliver mechanical work or energy without ejecting some heat to a lower-temperature reservoir.

SI   An internally consistent system of units within the metric system.

Speed (v)   Rate of movement. Rate of change of distance from a reference point. The distance that something travels divided by the time elapsed.

Spin   Intrinsic angular momentum possessed by subatomic particles. It is quantized in integral or half-integral multiples of $h/2\pi$.

Strong nuclear force   One of the four fundamental forces in Nature. It is the force that holds neutrons and protons together in the nucleus.

Superconductor   A substance that has zero electrical resistance. Electrical current can flow in it with no loss of energy to heat.

Temperature (T)   The measure of hotness or coldness. The temperature of matter depends on the average kinetic energy of atoms and molecules.

Total internal reflection   Process whereby a light ray is refracted back into the incident medium at the boundary separating the incident medium from a denser transmitted medium. Only occurs for incident angles exceeding a critical angle for which the angle of refraction is 90°.

Transverse wave   Wave in which the oscillations are perpendicular to the direction the wave travels.

Ultraviolet radiation (UV)   Electromagnetic waves with frequencies just above those of visible light.

Velocity   Speed with direction. A vector quantity.

Voltage (V)   The work that a charged particle can do divided by the size of the charge. The energy per unit charge given to charged particles by a power supply. The unit of measure is the volt.

Wave   A traveling disturbance consisting of coordinated vibrations that carry energy with no net movement of matter.

Wavelength (λ)   The distance between two successive "like" points on a wave. An example is the distance between two adjacent peaks or two adjacent valleys.

Weak nuclear force   One of the four fundamental forces in Nature. It is responsible for beta decay.

Weight (W)   The force of gravity acting on a substance.

Weight density (D_W)   The weight per unit volume of a substance. The weight of a quantity of a substance divided by the volume it occupies.

Work   The force that acts times the distance moved in the direction of the force.

X-rays   Electromagnetic waves with frequencies in the range from higher-frequency ultraviolet to low-frequency gamma rays.

Note: Tables are indicated by *t* following the page number.

Tone color, 242
Tone quality, 241–42, 245–46
Torque, 113
Torricelli, Evangelista, 161
Torsion balance, 68
Total internal reflection, 348–51
Transducers, 298
Transformers, 277, 297
Transistors, 264, 271
Transmission axis, 333
Transmutation, 435
Transverse waves, 214–15, 302, 333
Tribology, 48
Tritium, 421, 447–49
Triton, 172
Truth, quarks and, 489–92
Tsunamis, 444
Two-slit interference, 331–32, 395
Tycho's supernova, 446
Type Ia supernovae, 465, 497

U

Ultrahigh frequency (UHF), 304
Ultrasonic lithotripsy, 234
Ultrasound, 233–34
Ultraviolet light, 307, 320, 379, 382
Ultraviolet photons, 401
Ultraviolet radiation, 304, 307, 317
Uncertainty principle, 398
Uniform circular motion, 92
Unit of measure, 8
Unitary spin, 485
Universal gravitation
  development of, 75
  Earth and, 67–69
  energy converters, 99t, 99–100
  gravitational field, 71–72, 171
  law of, 46, 67–72
  orbits, 70–71
  surface gravity in our solar system, 70t
  tides and, 73
Unstable chain reaction, 440
Uranium, 420, 426, 433, 435, 439–42, 450, 452
Usable energy, 204–5

V

van der Meer, Simon, 493
Variometer, 147
Vector, 22–24
Vector addition, 22–24
Vega, 313
Velocity. *See also* Speed
  acceleration and, 25
  air and, 22–24
  arrows to represent, 27
  constant velocity, 30–31
  definition of, 21
  displacement and, 21–22
  distance and, 22
  escape velocity, 105, 171
  resultant, 22–24
  speed versus, 21–22
  summary of, 29

time and, 32
  vectors and, 23–24
  zero velocity, 30
Venus, 108, 319
Vertical airspeed indicator, 147
Very high frequency (VHF), 304
Very Large Telescope (VLT) array, 344
VHF (Very high frequency), 304
Vibration, waves and, 213
Virtual images, 355
Virtual photons, 479
Visible light, 303, 306, 320, 327, 382
Visible photons, 401
Vitruvius, 247
VLT (Very Large Telescope array), 344
Voice coil, 298
Volta, Alessandro, 279
Voltage, 267t, 267–69, 271–72, 279, 296–97
Voltmeter, 267
Volume, 13, 137–38
Volume density, 142
Von Guericke, Otto, 161
Voyager 2 spacecraft, 108

W

Waggle dances, 336
Warm cloud lightning, 262
Warm objects, detection of, 313
Warping of space, 470–71
Water
  as coolant, 192
  evaporation of, 196–98
  phase transitions, 193–99
  properties of, 127
  specific gravity, 142
  thermal expansion and, 175
  waves in, 221
Water vapor, 196–98, 197t
Watt (W), 110, 273
Watt, James, 207
Wave(s), 213–20. *See also* Light waves; Sound
  amplitude, 217
  bow waves, 225
  compression of, 218
  crests and troughs, 217
  cycles of, 218
  definition of, 214
  diffraction, 225–26
  Doppler effect, 223–25
  expansion of, 218
  frequency, 217–18
  gravitational, 472–73
  interference, 226–28
  medium of, 214
  peaks and valleys, 217
  properties of, 217–20
  reflection, 222–23
  shock waves, 225
  types of, 213–17
  in water, 221
  wave propagation, aspects of, 221–30
  wavelength, 217–18, 306, 306t, 380
Wave functions, 398
Wave propagation, 221–30. *See also* Wave(s)

bow waves, 225
  diffraction, 225–26
  Doppler effect, 223–25
  interference, 226–28
  ray model, 221–22
  reflection, 222–23
  shock waves, 225
  wavefront model, 221
Wave pulse, 214, 221
Wave trains, 408
Waveform, 220, 232, 245–46
Wavefront model, 221
Wavelength, 217–18, 306, 306t, 380
Weak nuclear force, 421, 478
Weakly interacting massive particles (WIMPs), 496–97
Weight, 15, 45–46, 52, 54
Weight density, 141
Weinberg, Steven, 492
Weinberg-Salam model, 492
WFIRST (Wide-Field Infrared Survey Telescope), 497
Wheeler, John A., 468
Whispering chamber, 223
White dwarfs, 188, 475
White noise, 246
Wide-Field Infrared Survey Telescope (WFIRST), 497
WiFi signals, 304
Wilczek, Frank, 486
Wilkinson Microwave Anisotropy Probe (WMAP), 316
Wilson, Robert, 315
WIMPs (Weakly interacting massive particles), 496–97
Wind shear, 234
Wireless remote control units, 306
WMAP (Wilkinson Microwave Anisotropy Probe), 316
Wood, Robert L., 309
Work
  distance and, 90
  energy and, 89–93
  force, 92–93
  force times distance, 90
  gravity and, 93
  joule, 91
  power and, 109–11
  uniform circular motion, 92

X

X-ray spectra, 406–7
X-rays, 303, 307–8, 382, 450

Y

Young, Thomas, 331, 372, 395
Ytterby, Sweden, 132

Z

"Z" machine, 448
Zeppelins, 124, 153
Zero velocity, 30
Zirconium, 132
Zwieg, George, 485

TABLE OF CONVERSION FACTORS AND OTHER INFORMATION

DISTANCE:

1 m = 3.28 ft	1 ft = 0.305 m
1 cm = 0.394 in.	1 in. = 2.54 cm
1 km = 0.621 mi	1 mi = 1.61 km
1 light-year = 5.9×10^{12} mi	1 light-year = 9.5×10^{15} m

AREA:

1 m^2 = 10.8 ft^2	1 ft^2 = 0.093 m^2
1 cm^2 = 0.155 in.^{2}	1 in.^{2} = 6.45 cm^2
1 km^2 = 0.386 mi^2	1 mi^2 = 2.59 km^2

VOLUME:

1 m^3 = 35.3 ft^3	1 ft^3 = 0.028 m^3
	= 7.48 gallons
1 cm^3 = 0.061 in.^{3}	1 in.^{3} = 16.4 cm^3
1 m^3 = 1,000 liters	
1 liter = 1.06 qt	1 qt = 0.95 liter

TIME:

1 min = 60 s
1 h = 60 min = 3,600 s
1 day = 24 h = 86,400 s
1 year = 3.16×10^7 s

SPEED:

1 m/s = 2.24 mph	1 mph = 0.447 m/s
	= 1.47 ft/s
1 m/s = 3.28 ft/s	1 ft/s = 0.305 m/s
= 3.60 km/h	= 0.682 mph
1 km/h = 0.621 mph	1 mph = 1.61 km/h

ACCELERATION:

1 m/s^2 = 3.28 ft/s^2	1 ft/s^2 = 0.305 m/s^2
1 m/s^2 = 2.24 mph/s	1 mph/s = 0.447 m/s^2
1 g = 9.8 m/s^2	1 g = 32 ft/s^2 = 22 mph/s
1 m/s^2 = 0.102 g	

MASS:

1 kg = 0.069 slug	1 slug = 14.6 kg
1 kg = 6.02×10^{26} u	1 u = 1.66×10^{-27} kg

FORCE:

1 N = 0.225 lb	1 lb = 4.45 N
	1 ton = 8,896 N
	= 2,000 lb

ENERGY, WORK, AND HEAT:

1 J = 0.738 ft-lb	1 ft-lb = 1.36 J
= 0.239 cal	= 0.0013 Btu
1 cal = 4.184 J	1 Btu = 778 ft-lb
1 cal = 0.004 Btu	1 Btu = 252 cal
1 J = 0.00095 Btu	1 Btu = 1,055 J
1 kWh = 3,600,000 J	
1 kWh = 3,413 Btu	1 Btu = 0.00029 kWh
1 food calorie (Cal) = 1,000 cal	
1 eV = 1.6×10^{-19} J	1 quad (Q) = 10^{15} Btu
1 J = 6.25×10^{18} eV	

POWER:

1 W = 0.738 ft-lb/s	1 ft-lb/s = 1.36 W
1 W = 0.00134 hp	1 hp = 550 ft-lb/s
	= 746 W
1 W = 3.41 Btu/h	1 Btu/h = 0.293 W

PRESSURE:

1 Pa = 1.45×10^{-4} psi	1 psi = 6,900 Pa
1 Pa = 9.9×10^{-6} atm	1 psi = 0.068 atm
1 atm = 1.01×10^5 Pa	1 atm = 14.7 psi
= 76 cm Hg	= 29.92 in. Hg
= 10.3 m H$_2$O	= 33.9 ft H$_2$O

MASS DENSITY:

1 kg/m^3 = 0.0019 slug/ft^3	1 slug/ft^3 = 515 kg/m^3

WEIGHT DENSITY:

1 N/m^3 = 0.0064 lb/ft^3	1 lb/ft^3 = 157 N/m^3

METRIC PREFIXES:

Power of 10	Prefix	Abbreviation
10^{15}	peta	P
10^{12}	tera	T
10^9	giga	G
10^6	mega	M
10^3	kilo	k
10^{-2}	centi	c
10^{-3}	milli	m
10^{-6}	micro	μ
10^{-9}	nano	n
10^{-12}	pico	p
10^{-15}	femto	f

IMPORTANT CONSTANTS AND OTHER DATA

Symbol	Description	Value
c	Speed of light in vacuum	3×10^8 m/s
g	Acceleration of gravity (Earth)	9.8 m/s^2
G	Gravitational constant	6.67×10^{-11} N-m^2/kg^2
h	Planck's constant	6.63×10^{-34} J-s
q_e	Charge on electron	-1.60×10^{-19} C
m_e	Mass of electron	9.11×10^{-31} kg
m_p	Mass of proton	1.673×10^{-27} kg
m_n	Mass of neutron	1.675×10^{-27} kg

SOLAR SYSTEM DATA

Sun:	mass	1.99×10^{30} kg
	radius (average)	6.96×10^8 m
Earth:	mass	5.979×10^{24} kg
	radius (average)	6.376×10^6 m
Moon:	mass	7.35×10^{22} kg
	radius (average)	1.74×10^6 m
Earth-Sun distance		1.496×10^{11} m
Earth-Moon distance		3.84×10^8 m
Earth orbital period		3.16×10^7 s
Moon orbital period (average)		2.36×10^6 s